ESTUARINE COMPARISONS

Produced by

The Estuarine Research Federation

Sponsored by

U.S. Fish and Wildlife Service
Bureau of Land Management
National Oceanic and Atmospheric Administration
National Marine Fisheries Service
Pacific Northwest River Basins Commission

Proceedings of the Sixth Biennial International
Estuarine Research Conference,
Gleneden Beach, Oregon, November 1–6, 1981

ESTUARINE COMPARISONS

Edited by

VICTOR S. KENNEDY

University of Maryland
Center for Environmental and Estuarine Studies
Horn Point Environmental Laboratories
Cambridge, Maryland

ACADEMIC PRESS 1982
A Subsidiary of Harcourt Brace Jovanovich, Publishers
New York London
Paris San Diego San Francisco São Paulo Tokyo Toronto

ACADEMIC PRESS, INC.
111 Fifth Avenue, New York, New York 10003

United Kingdom Edition published by
ACADEMIC PRESS, INC. (LONDON) LTD.
24/28 Oval Road, London NW1 7DX

Library of Congress Cataloging in Publication Data

International Estuarine Research Conference (6th : 1981 : Gleneden Beach, Or.)
Estuarine comparisons.

Includes index.
1. Estuaries--Congresses. I. Kennedy, Victor S. II. Estuarine Research Federation. III. U.S. Fish and Wildlife Service. IV. Title.
GC96.5.157 1981 574.5'26365 82-16332
ISBN 0-12-404070-5

PRINTED IN THE UNITED STATES OF AMERICA

82 83 84 85 9 8 7 6 5 4 3 2 1

CONTENTS

LIST OF CONTRIBUTORS

Ariathurai, R., Coastal and Oceanographic Engineering Department, University of Florida, Gainesville, FL 32611

Armstrong, N. E., Department of Civil Engineering and Center for Research in Water Resources, University of Texas, Austin, TX 78712

Bandy, A. R., Department of Chemistry, Drexel University, Philadelphia, PA 19104

Bohlen, W. F., Marine Sciences Department, University of Connecticut, Avery Point, Groton, CT 06340

Boicourt, W. C., Chesapeake Bay Institute, 4800 Atwell Road, Shadyside, MD 20764

Bonsang, B., Department of Chemistry, Drexel University, Philadelphia, PA 19104

Boynton, W. R., Chesapeake Biological Laboratory, Box 38, Solomons, MD 20688

Brown, C. A., Department of Chemistry, Drexel University, Philadelphia, PA 19104

Burton, R. S., Department of Biology, University of Pennsylvania, Philadelphia, PA 19104

Chalmers, A. G., University of Georgia Marine Institute, Sapelo Island, GA 31327

Chen Ji-Yu, Institute of Estuarine and Coastal Research, Physics Department, East China Normal University, Shanghai, China

Christy, J. H., Belle W. Baruch Institute for Marine Biology and Coastal Research, University of South Carolina, Columbia, SC 29208

Conner, W. H., Coastal Ecology Laboratory, Center for Wetland Resources, Louisiana State University, Baton Rouge, LA 70803

Cronin, T. W., Department of Biology, Yale University, Box 6666, New Haven, CT 06511

Day, J. W., Jr., Coastal Ecology Laboratory, Center for Wetland Resources, Louisiana State University, Baton Rouge, LA 70803

DeLaune, R. D., Laboratory for Wetland Soils and Sediments, Center for Wetland Resources, Louisiana State University, Baton Rouge, LA 70803

D'Elia, C. F., Chesapeake Biological Laboratory, Box 38, Solomons, MD 20688

Dittel, A. I., Escuela de Biologia, Universidad de Costa Rica, San Pedro de Montes de Oca, Costa Rica

Dixit, J. G., Coastal and Oceanographic Engineering Department, University of Florida, Gainesville, FL 32611

Dunn, E. L., School of Biology, Georgia Institute of Technology, Atlanta, GA 30332

Durkin, J. T., National Marine Fisheries Service, Hammond Biological Field Station, Box 155, Hammond, OR 97121

Dyer, K. R., Institute of Oceanographic Sciences, Crossway, Taunton, Somerset, United Kingdom TA1 2DW

Epifanio, C. E., College of Marine Studies, University of Delaware, Lewes, DE 19958

Feldman, M. W., Department of Biological Sciences, Stanford University, Stanford, CA 94305

Fisher, F. W., California Department of Fish and Game, 4001 North Wilson Way, Stockton, CA 95205

Forward, R. J., Jr., Duke University Marine Laboratory, Beaufort, NC 28516

Frasco, B. R., Environmental Consulting and Testing Services, Cherry Hill, NJ 08034

Fresh, K. L., Washington Department of Fisheries, Research and Development Division, Olympia, WA 98504

Gleason, M. L., Centre College of Kentucky, Danville, KY 40422

Good, N. F., 745 Redman Avenue, Haddonfield, NJ 08033

Good, R. E., Biology Department, Rutgers University, Camden, NJ 08102

Gu Peiyu, Nanjing Hydraulic Research Institute, Ministry of Water Conservancy and Ministry of Communications, 223 Guangzhou Road, Nanjing, China

Hatton, R. S., Laboratory for Wetland Soils and Sediment, Center for Wetland Resources, Louisiana State University, Baton Rouge, LA 70803

Healey, M. C., Department of Fisheries and Oceans, Pacific Biological Station, Nanaimo, British Columbia, Canada V9R 5K6

Hedgecock, D., Bodega Marine Laboratory, University of California, Box 247, Bodega Bay, CA 94923

He Songlin, Institute of Estuarine and Coastal Research, East China Normal University, Shanghai, China

Hirschberg, D. J., Marine Sciences Research Center, State University of New York, Stony Brook, NY 11794

Hobbie, J. E., The Ecosystems Center, Marine Biological Laboratory, Woods Hole, MA 02543

Hoffman, J. F., Oceanography Department, U.S. Naval Academy, Annapolis, MD 21402

Hopkinson, C. S., University of Georgia Marine Institute, Sapelo Island, GA 31327

Horton, H. F., Department of Fisheries and Wildlife, Oregon State University, Corvallis, OR 97331

Howarth, R. W., The Ecosystems Center, Marine Biological Laboratory, Woods Hole, MA 02543

Huish, M. T., North Carolina Cooperative Fishery Research Unit, Box 5577, North Carolina State University, Raleigh, NC 27650

Jeffries, H. P., Graduate School of Oceanography, University of Rhode Island, Kingston, RI 02881

Keefe, C. W., Chesapeake Biological Laboratory, Box 38, Solomons, MD 20688

Kemp, W. M., Horn Point Environmental Laboratories, Box 775, Cambridge, MD 21613

Kirby, R., Institute of Oceanographic Sciences, Crossway, Taunton, Somerset, United Kingdom TA1 2DW

Kjelson, M. A., U.S. Fish and Wildlife Service, 4001 North Wilson Way, Stockton, CA 95205

Kjerfve, B., Belle W. Baruch Institute for Marine Biology and Coastal Research, University of South Carolina, Columbia, SC 29208

Lambert, R. M., Graduate School of Oceanography, University of Rhode Island, Kingston, RI 02881

Lewis, M. R., Department of Biology, Dalhousie University, Halifax, Nova Scotia, Canada B3H 4JI

Lü Quan-Rong, Institute of Estuarine and Coastal Research, East China Normal University, Shanghai, China

Luo Zhaosen, Nanjing Hydraulic Research Institute, Ministry of Water Conservancy and Ministry of Communications, 223 Guangzhou Road, Nanjing, China

Mao Zhi-Chang, Institute of Estuarine and Coastal Research, East China Normal University, Shanghai, China

Maroulis, P. J., Department of Chemistry, Drexel University, Philadelphia, PA 19104

Marozas, M., Belle W. Baruch Institute for Marine Biology and Coastal Research, University of South Carolina, Columbia, SC 29208

McAnally, W. H., Jr., USAE Waterways Experiment Station, Vicksburg, MS 39180

Mehta, A. J., Coastal and Oceanographic Engineering Department, University of Florida, Gainesville, FL 32611

Mohr, A. W., Corps of Engineers, 30 Pryor Street, Atlanta, GA 30303

Montague, C. L., Department of Environmental Engineering Sciences, A. P. Black Hall, University of Florida, Gainesville, FL 32611

Myers, K. W., Fisheries Research Institute, 260 Fisheries Center, University of Washington, Seattle, WA 98195

Parchure, T. M., Coastal and Oceanographic Engineering Department, University of Florida, Gainesville, FL 32611

Parker, W. R., Institute of Oceanographic Sciences, Crossway, Taunton, Somerset, United Kingdom TA1 2DW

Patrick, W. H., Jr., Laboratory for Wetland Soils and Sediment, Center for Wetland Resources, Louisiana State University, Baton Rouge, LA 70803

Platt, T., Marine Ecology Laboratory, Bedford Institute of Oceanography, Dartmouth, Nova Scotia, Canada B2Y 4A2

Polgar, T. T., Martin Marietta Corporation, Environmental Center, 1450 South Rolling Road, Baltimore, MD 21227

Proehl, J. A., Department of Oceanography, University of Washington, Seattle, WA 98195

Raquel, P. F., U.S. Fish and Wildlife Service, 4001 North Wilson Way, Stockton, CA 95205

Rublee, P. A., Department of Biology, Whitman College, Walla Walla, WA 99362

Rulifson, R. A., Center for Environmental Sciences, Unity College, Unity, ME 04988

Salo, E. O., Fisheries Research Institute, University of Washington, Seattle, WA 98195

Schubel, J. R., Marine Sciences Research Center, State University of New York, Stony Brook, NY 11794

Schwing, F. B., Skidaway Institute of Oceanography, Box 13687, Savannah, GA 31406

Seim, H. E., Belle W. Baruch Institute for Marine Biology and Coastal Research, University of South Carolina, Columbia, SC 29208

Shen Huan-ting, Institute of Estuarine and Coastal Research, East China Normal University, Shanghai, China

Sikora, J. P., Coastal Ecology Laboratory, Center for Wetland Resources, Louisiana State University, Baton Rouge, LA 70803

Sikora, W. B., Coastal Ecology Laboratory, Center for Wetland Resources, Louisiana State University, Baton Rouge, LA 70803

Simenstad, C. A., Fisheries Research Institute, University of Washington, Seattle, WA 98195

Stancyk, S. E., Belle W. Baruch Institute for Marine Biology and Coastal Research, University of South Carolina, Columbia, SC 29208

Stevenson, J. C., Horn Point Environmental Laboratories, Box 775, Cambridge, MD 21613

Strathmann, R. R., Friday Harbor Laboratories, University of Washington, Friday Harbor, WA 98250

Sulkin, S. D., Horn Point Environmental Laboratories, Box 775, Cambridge, MD 21613

Thoesen, R. W., Fishery Resources, U.S. Fish and Wildlife Service, Region 4, Atlanta, GA 30303

Van Heukelem, W., Horn Point Environmental Laboratories, Box 775, Cambridge, MD 21613

Wan Jia-ruo, Institute of Estuarine and Coastal Research, Physics Department, East China Normal University, Shanghai, China

Welsh, B. L., Marine Sciences Department, University of Connecticut, Avery Point, Groton, CT 06340

Wetzel, R. L., Virginia Institute of Marine Science, Gloucester Point, VA 23062

Whitlatch, R. B., Marine Sciences Department, University of Connecticut, Avery Point, Groton, CT 06340

Xu Haigen, Institute of Estuarine and Coastal Research, Physics Department, East China Normal University, Shanghai, China

Yun Cai-xing, Institute of Estuarine and Coastal Research, Physics Department, East China Normal University, Shanghai, China

Zhou Yueqin, Institute of Estuarine and Coastal Research, East China Normal University, Shanghai, China

Zhu Hui-fang, Institute of Estuarine and Coastal Research, East China Normal University, Shanghai, China

Zhu Shu-Zheng, Institute of Estuarine and Coastal Research, East China Normal University, Shanghai, China

FOREWORD

How would you be able to cope with life without comparisons? Everyday, you make comparisons which become an integral part of the learning process. *Comparison* is defined as the bringing together of two or more items to ascertain or establish their similarities and dissimilarities. A comparison implies an impartial search for resemblances as well as differences. During the development of the history of science, many comparisons have been made: between plants and animals, between physical and chemical factors, and between many features separated over time and space. These have all contributed to a more comprehensive understanding of our universe. So you might ask, what do comparisons have to do with estuaries?

Another chapter in the learning process, another segment of the estuarine learning curve, was initiated early in November 1981, when the Estuarine Research Federation held its sixth biennial international meeting, celebrated its first decade of existence, held its first meeting on the Pacific coast, and devoted its entire program to comparisons of estuaries. The theme of the conference, and this volume, *Estuarine Comparisons*, not only compared Pacific, Gulf, and Atlantic coast estuaries, but made comparisons of physical, chemical, and biological parameters in estuaries throughout the world. Principal attention was focused on comparing features of North Sea, east and west Atlantic, Gulf, and Pacific estuaries, as well as of pioneering work in the Chang Jiang estuary of China, one of the largest in the world. Comparisons of anadromous fisheries, estuarine microbiology, and many other interactive features over a wide variety of latitudinal and longitudinal variation, set the conference, and this volume, apart from its predecessors.

Estuarine Comparisons represents a first in the efforts of the Estuarine Research Federation to compare the knowledge we have gained about many of the world's estuaries. Over the first decade of the Federation's existence, many valuable conferences have been held, and many publications have been generated which define estuaries, discuss the interactions within estuaries, and document recent advances in a variety of studies connected with specific estuaries. *Estuarine Comparisons*, however, is the first real effort to compare different estuaries. The resultant interaction between estuarine scientists was evident throughout the conference. The diversity of invited presentations summarized in this volume also conveys the comprehensive effort of ERF Vice-President and Program Chairman, Austin Williams, and the Program Committee, in truly comparing, in a comprehensive way over time and space, the many detailed features of estuaries.

Such estuarine comparisons are vital to the future of estuaries. No longer can these sensitive systems be studied in a vacuum, i.e., only on a local scale. All future research, as well as new management concepts, must consider the diversity of estuaries in the world, the various management options available, and the wealth of information gained by comparing estuaries, in order to make environmentally effective and economically workable management decisions. With changes in the availability of research funds, it will no longer be possible to repeat the same study in different parts of the world's estuaries without clear documentation of the similarities and differences which would preclude interpolation or extrapolation of existing study results. With the ever-increasing pressure from people wishing to live on the land surrounding our estuaries, and to locate factories and ports on or near our world's estuaries, it is timely to make vital comparisons so that legally defensible management decisions can be made.

Estuarine Comparisons represents the beginning of a trend that I believe will continue. More comparisons must be made of existing information, and more comparisons of ongoing studies must be made in order to understand and manage this vital renewable natural resource so important to all the people of the world. It is the hope of all the officers of the Estuarine Research Federation, and its Governing Board, representing the affiliates (New England Estuarine Research Society, Atlantic Estuarine Research Society, Southeastern Estuarine Research Society, Gulf Estuarine Research Society, and the Pacific Estuarine Research Society) that this volume, *Estuarine Comparisons*, makes a significant contribution to the world's body of knowledge about estuaries.

Robert J. Reimold
President
Estuarine Research Federation
1979–1981

PREFACE

温故而知新
可以為師矣

The one who looks back to the old
And knows the new
Is worthy to be a scholar

These proceedings of the Sixth Biennial Conference of the Estuarine Research Federation continue the practice of publishing selected papers presented at Invited Sessions during the meetings. The Federation has consistently set a goal of high quality in publishing this series. To that end, over 90 referees provided peer review of the 47 papers submitted at the 1981 Gleneden Beach meeting; in nearly all instances, the successful authors were able to benefit from the comments of these reviewers in the revision of their manuscripts. As has been the case in the past for the series, publication has occurred within a year of the meetings, keeping the papers fresh. In terms of themes, the convenors of the Invited Sessions made strong efforts to organize their sessions around clearly focused topics. This has helped greatly in providing coherency to the different sections of this volume. I believe that the results will be useful over time to researchers in the fields represented here.

In the spirit of the proverb quoted above, the book includes reviews of published literature and presentations of new information. In the first section, a number of comparisons of different kinds are made among estuaries after attention has been drawn to problems associated with the making of measurements on which such comparisons are based. The second section includes review papers on aspects of the diverse interactions that occur on or in marsh soils, coupled with new and complementary information derived from recent studies. A section on anadromous fishes draws together data on salmon from British Columbia to California, and provides insight into factors influencing East and Gulf Coast striped bass, sturgeon, and clupeids.

The matter of retention of invertebrate larvae in seaward-draining estuaries is examined in a fourth section; insights from physical oceanography, behavioral experiments, field sampling, and genetic studies are brought to bear on the problem. There follows consideration of properties of and problems caused by that ubiquitous estuarine material – sediment – that ultimately snuffs out the life of estuaries, and, in the process, hinders commercial and naval ship traffic. The final section draws attention to the Chang Jiang estuary in China, juxtaposing historical data spanning two millenia with recent knowledge derived from use of space satellites.

These latter papers (and one in the sediment section) by Chinese scientists are welcome evidence of the growing international discourse on estuarine science. They join with the papers from Europe and North and Central America to continue the tradition of the Estuarine Research Federation meetings – the coming together of scientists from around the world to exchange ideas and information.

In the editing of these proceedings, in addition to being indebted to the referees and convenors, I owe my thanks to the staff of Economy Printing in Easton, Maryland, for their cooperation in production of camera-ready copy; to Denise Davidson for secretarial assistance; to Jennifer Kennedy for help with the Index; and to Deb, Jen and Chris for pleasant distraction from the pressures of the editorial job.

Victor S. Kennedy, Editor

ECODYNAMIC COMPARISONS AMONG ESTUARIES

Convenor: *B. L. Welsh*
University of Connecticut
Groton, Connecticut

SCALES OF VARIABILITY IN ESTUARINE ECOSYSTEMS

Marlon R. Lewis
Department of Biology
Dalhousie University
Halifax, Nova Scotia, Canada

and

Trevor Platt
Marine Ecology Laboratory
Bedford Institute of Oceanography
Dartmouth, Nova Scotia, Canada

Abstract: Ecological dynamics in estuaries can be characterized in terms of spatial and temporal scales of variability. Data from Bedford Basin and other inlets on the Atlantic Coast of Nova Scotia are used to examine the scales of physical forcing and the scales of biological response. We establish temporal scales for which these inlets appear to be autonomous with respect to the adjacent continental shelf (∼<3 days), and develop theoretical criteria for autonomy based on the distribution of biological variability in the frequency domain. The spatial distribution of phytoplankton biomass is analyzed in the wavenumber domain. A "biological window" in the wavenumber spectrum is suggested, where growth processes are the principal determinants of the variance spectrum. On either side of this window, physical forces override biological processes. It is concluded that in estuaries, where physical forcing is often strong in amplitude and complex in frequency structure, it is particularly important to emphasize the scales of variability for both physical and biological processes and to recognize that time and space scales are intimately related.

Introduction

In the past decade, a conceptual framework has emerged in the biological oceanographic community (Steele 1978) in which the importance of understanding the variability of the marine ecosystem is accorded at least as much emphasis as quantification of the instantaneous or mean values of the system variables. It has also become recognized that ecosystem variance can occur simultaneously on more than one time-scale. Hand-in-hand with the evolution of these ideas has progressed the development of suitable instrumentation (Herman and Platt 1980) and statistical tools (Platt and Denman 1975) to describe the distribution or decomposition of variance in terms of temporal or spatial scale.

Biological processes in estuarine environments are affected by physical forcing at a variety of scales. For a number of years (Platt and Conover 1971;

ISBN 0-12-404070-5

Platt 1975; Platt and Jassby 1976; MacCaull and Platt 1977; Harrison and Platt 1980; Côte and Platt in prep.) we have been involved in a study of the variability of the parameters describing the relationship between irradiance and photosynthesis by phytoplankton in Bedford Basin, Nova Scotia. To illustrate the magnitude of the variability at different scales, we have plotted the rate of light-saturated photosynthesis (P_m^B) normalized to chlorophyll biomass for time segments ranging from hours to a year (Fig. 1). It is clear that, although the amplitude of the variance is somewhat higher on an annual basis, variance on scales of hours to days is decidedly nontrival.

The royal way to treat these data would be an analysis of variance in the frequency domain (spectral analysis). This would permit comparison between the scales of biological response and the potential scales of forcing (coherence analysis). Further, a spectral approach would allow objective comparisons between different estuarine environments at various scales of interest. Unfor-

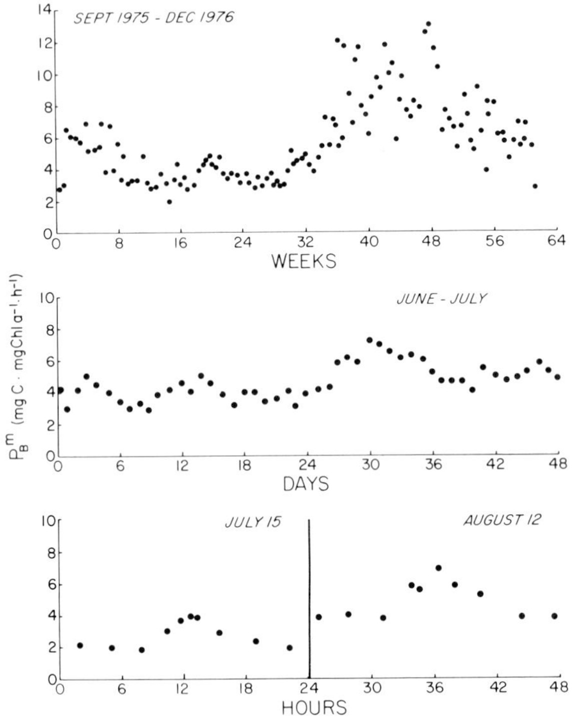

Figure 1. Variability in light saturated specific photosynthesis at scales ranging from hours to a year. Values were obtained from photosynthesis versus irradiance curves as described in Platt and Jassby (1976). Data in the uppermost graph are from Harrison and Platt (1980), the middle from Côte and Platt (in prep.), and the lowermost from MacCaull and Platt (1977).

tunately, the difficulties involved in collecting sufficiently closely spaced measurements over a long enough time, and also the strongly nonstationary nature of the statistical moments of the biological state variables render formal application of time-series analysis problematical. Nevertheless, a number of conclusions and ideas have emerged from these studies. In this paper we discuss these generalizations as they pertain to the ecology of estuaries.

The Functional Autonomy of Estuaries

The degree to which a given areal domain can be said to be autonomous with respect to its ecological dynamics is dependent on the size of the domain and the averaging time over which autonomy is tested. An autonomous system is an ecological unit which can be specified by knowledge of the internal dynamics and the boundary conditions alone; it is independent of the dynamics of events outside the domain. Even the smallest of domains can be considered autonomous over very small averaging periods, while larger domains tend to retain their functional integrity over longer averaging times. For a domain of any size there will exist some characteristic time-scale at which functional autonomy breaks down. Averaging the ecological dynamics over longer periods then requires that we treat the domain as part of a larger system, instead of a self-contained whole. Of course, the elucidation of the characteristic time scale for a domain of a given size is not a simple problem. The best that we can do is to estimate it according to several independent criteria and see how closely these estimates agree.

Time Scales of Estuarine Coupling to Continental Shelves

A particularly relevant question here is the degree to which estuaries are coupled biologically to the adjacent continental shelf. Platt et al. (1972) described the annual cycle of phytoplankton nutrients in two inlets 35 km apart on the Atlantic coast of Nova Scotia. The inlets are both about 20 km long but differ in shape. St. Margaret's Bay has a bigger surface area and is more exposed than Bedford Basin (Fig. 2). At the mouth of Bedford Basin is a 20 m sill and a long, narrow channel; on the other hand, St. Margaret's Bay has only a narrow 35 m sill.

With such different physiography, there is every reason to expect that the evolution of the nutrient dynamics would proceed independently in the two inlets. Indeed, the difference in response to the strong seasonal pulse establishes the autonomy of the inlets: there is one month lag in the timing of the spring bloom in St. Margaret's Bay that is reflected very clearly in the onset of nutrient depletion (Fig. 3).

But there are also many instances where the two inlets apparently respond to the same larger scale forcing with no phase lag at all. It is clear that forces are operating on the adjacent continental shelf that affect both inlets similarly and simultaneously, and that most of the time these forces are strong enough to dominate changes due to the intrinsic biological dynamics unique

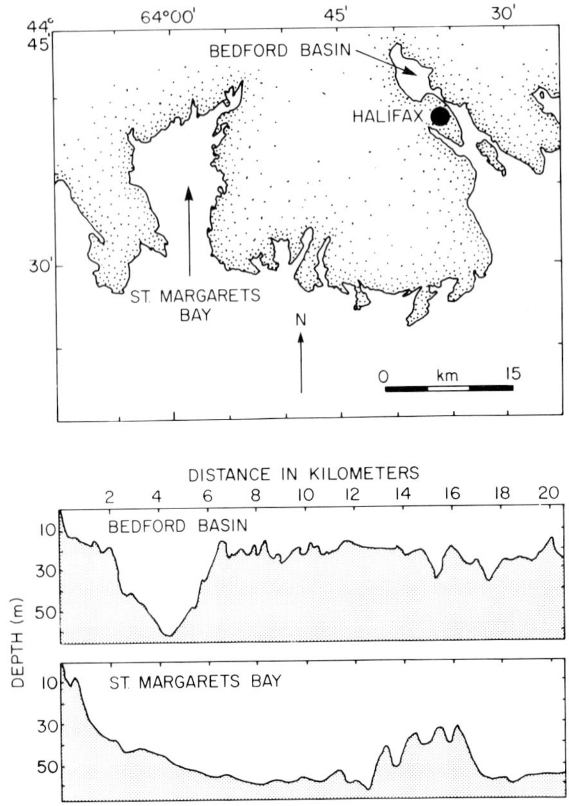

Figure 2. Section of the Nova Scotia coastline depicting the location and bathymetry of the two study sites.

to each inlet. There is little doubt that the responsible mechanism is direct water replacement forced by atmospheric pressure fluctuations (Platt et al. 1972; Heath 1973). These replacements can act over large horizontal scales (Fig. 4).

Flushing of estuaries by this mechanism has been underemphasized in the past (cf Heath 1973; Kjerve et al. 1978; Wang and Elliot 1978; Wang 1979). The typical forcing period by atmospheric fluctuations on the Nova Scotia Coast is about four days (Heath 1973). The resultant motion is an out-of-phase, two-layer flow: shear between the layers may enhance the mixing compared to that set up by the tidal flow. The resultant flushing time is one measure of time-scale of autonomy for the inlet concerned.

In order to evaluate better the magnitude and scale of this process in Bedford Basin, we have analyzed some measurements of current velocities determined from a set of current meters moored at the mouth of the Basin.

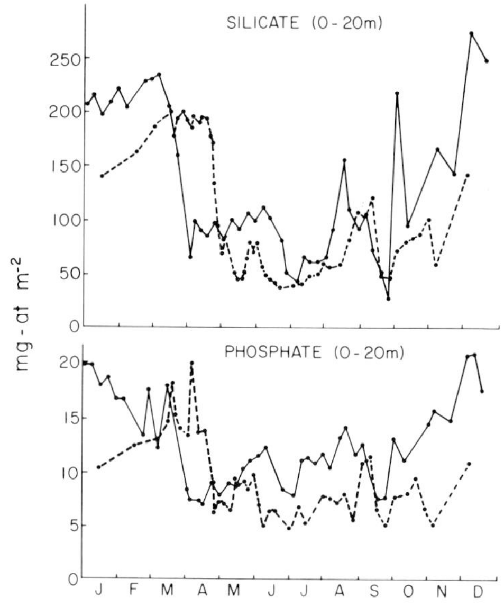

Figure 3. Nutrient concentrations in Bedford Basin (solid line) and St. Margaret's Bay (dotted line) integrated over the upper 20 m in 1967.

Two moorings were deployed each with current meters at 3, 8 and 15 meters. The original data were edited, filtered and reduced to a series of six-hour averages. A harmonic analysis was used to remove principal tidal constituents (M_2, S_2, K_1, O_1,MS_4, M_4) followed by a Cartwright filter with a sharp cutoff at periods slightly more than one day. The resultant residual time-series for the surface and bottom meters in summer (July-August) and winter (November-January) are shown in Fig. 5. A well-developed flow field is evident at these lower frequencies with the surface and bottom records being roughly out of phase. Power spectra of the surface and bottom stations in summer and winter (Fig. 6) reveal that most energy resides in periods of approximately two to eight days, corresponding to the dominant period of transient weather systems. Such variability is coherent along the coast of Nova Scotia as evidenced by sea level records (Sandstrom 1980) and a coherent response of temperature profiles taken along the coast (Fig. 4) to a large scale event. The significance of current variability at this scale relative to that associated with tidal frequencies is not unique to the Maritime coastline: Weisberg (1976) found 48% of the total velocity variance to reside in frequencies lower than tidal in Narrangansett Bay, RI.

We can make an order of magnitude estimate of the flushing associated with the various scales from the variance or power spectrum. Assuming that a two-layer system exists and that the current record at 3 m is representative of the average flow in the upper layer, we first calculate the root mean square

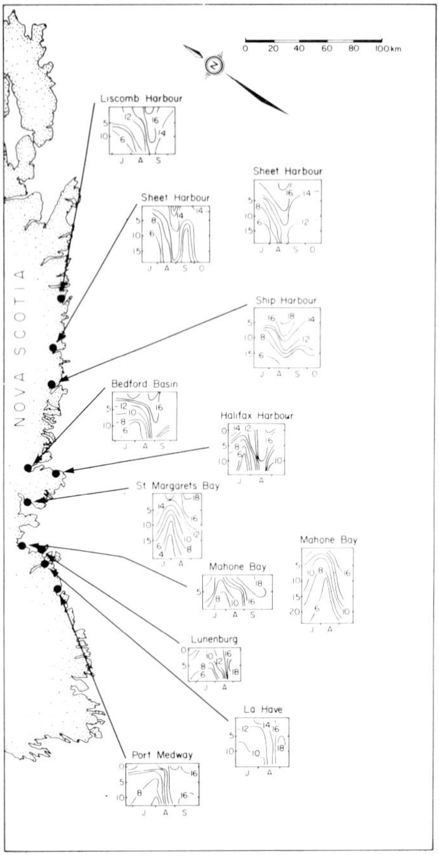

Figure 4. *Temperature variations with depth and time in various inlets along the Nova Scotia coast in late summer 1924. Data from Canadian Oceanographic Data Centre.*

velocity in each frequency band as the square root of the variance spectrum integrated across that particular frequency band:

$$\bar{U}(f) \approx \left(\int_{f - \frac{1}{2}\Delta f}^{f + \frac{1}{2}\Delta f} E_v(f)\, df \right)^{1/2} \qquad (1)$$

where $E_v(f)$ is the current variance spectrum and Δf is the bandwidth. Multiplying by the cross sectional area A of the upper layer yields the volumetric flow, $F(f)$:

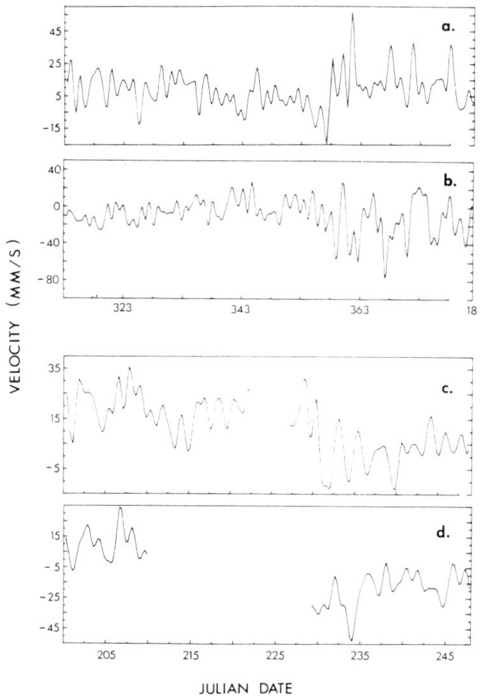

Figure 5. Time series of residual axial current velocities in the mouth of Bedford Basin in the summer and winter, 1970. a) winter 3m; b) winter 18 m; c) summer 3 m; d) summer 18 m; Negative values indicate flow into the Basin.

$$F(f) \approx \bar{U}(f) \bullet A \tag{2}$$

and dividing by the volume gives the flushing rate $R(f)$ of the layer,

$$R(f) \approx F(f)/V \tag{3}$$

assuming uniform mixing.

For Bedford Basin, we can take the thickness of the upper layer to be ~10 m in depth, the cross-sectional area at the Narrows ~10^4 m^2 and the volume of the upper layer of order 10^8 m^3. If we assume that the scales are independent, we can integrate across the low frequency band (0.125 - 1.0 d^{-1}) and, using the variance estimates for the summer record, we calculate an approximate flushing time of 5 days.

This estimate does not include flushing associated with river inputs or that associated with tidal frequencies and it might be of interest to compare the low frequency flushing to these other mechanisms. River input into Bed-

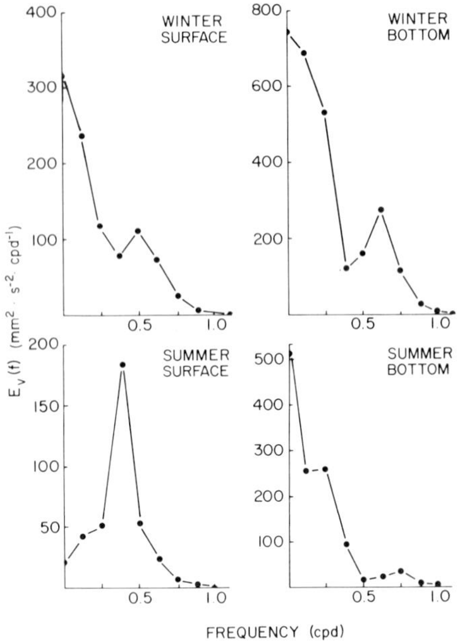

Figure 6. Current variance spectra ($E_v(f)$) computed from the residual current velocities in Fig. 5.

ford Basin is generally low relative to the volume and ranges from 10^3 to 10^7 m^3 d^{-1}. Thus we would expect an order of magnitude or more lower flushing rates associated with the fresh water input in the summer on average. The tidal range in the Basin is ~1 m; using the simplified tidal prism approach (Dyer 1973) and assuming only the layer above the sill depth (~20 m) is involved, we find an approximate flushing time of 5 days. The assumption of complete and rapid mixing is essential to all the above calculations. While this is not generally a good assumption for the barotropic tidal flow, it may be somewhat more realistic for the meteorologically forced oscillations, since the mixing efficiency is enhanced by shear set up by the out-of-phase flows.

Heath (1973) has attempted similar calculations for nearby St. Margaret's Bay and obtained e-folding flushing time-scales associated with the low-frequency component of 10 days. Petrie and Drinkwater (1978) also computed similar values for St. Georges Bay on north-eastern Nova Scotia. Both estuaries are larger than Bedford Basin and consequently somewhat longer flushing times would be expected.

These data lead to the conclusion that for most of the year, nutrient concentrations in the inlets along the Atlantic coast of Nova Scotia are determined more by physical processes offshore than by internal biological processes. The same would apply to the species composition and biomass of phytoplankton. From this perspective, we could not consider the inlets to function as autonomous ecological units over time scales longer than 2-3 days, except at particular times of the year (such as during the spring bloom of phytoplankton) when the biological signal may be strong enough to rise above the background of the larger scale physical forcing. In these inlets the low frequency fluctuations appear to exert as much influence as the tidal flows and warrant further attention.

On the 24 h time scale, an estuary like the Bedford Basin may be treated as an autonomous unit within the limits of measurement error and given the appropriate boundary condition at the mouth. Thus, Platt and Conover (1971) were able to balance the chlorophyll budget for the Basin over a complete tidal cycle. The difference between the initial and final chlorophyll surveys could be explained by a combination of measured phytoplankton production, zooplankton grazing, and net export. Some 58% of the phytoplankton production was exported from the Basin, compared to 34% lost by grazing.

With an hourly sampling schedule, considerable tidal variations in chlorophyll content could be seen (Fig. 7). These are of critical importance in the measurement of export. Calculations based on mean chlorophyll concen-

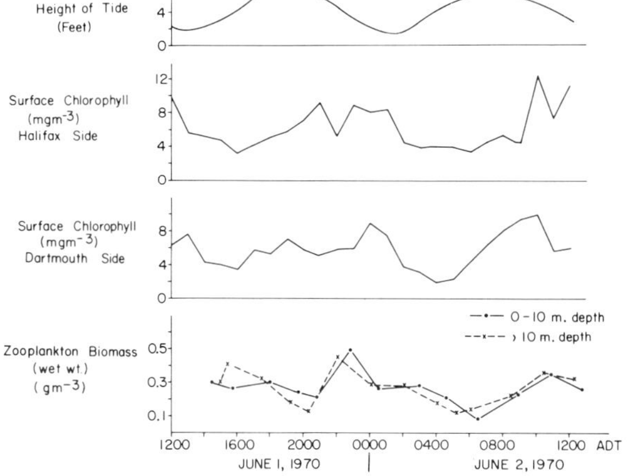

Figure 7. Variability of chlorophyll exchange through the mouth of Bedford Basin, tidal height, surface chlorophyll concentrations, and zooplankton biomass over a 25 h period.

trations and mean flows would account for only 35% of chlorophyll export: 65% was exported in the fluctuating or turbulent mode, an amount that agrees well with the flushing calculations above.

Scales of Autonomy from Biological Variability

The difference between the amounts exported in the mean flow and turbulent modes increases with the correlation between the forcing (transport) and the state variable of interest (in this case, chlorophyll). Correlations between state variables themselves can also be used as criteria of holism or functional autonomy. For example, suppose that the state variable "phytoplankton biomass", B, is replaced by three separate variables, microplankton, M; nanoplankton, N; picoplankton, P (Sieburth et al. 1978),

$$B = M + N + P \tag{4}$$

Over any given averaging time, we can calculate the statistical moments and write

$$\sigma_B^2 = \sigma_M^2 + \sigma_N^2 + \sigma_P^2 + 2[Cov\ (M,N) + Cov\ (N,P) + Cov\ (M,P)] \tag{5}$$

If the separation of the state variables has been done with biological insight, the covariances (*Cov*) may not be zero. This is the case if a positive fluctuation in one variable is compensated by a negative fluctuation in another: that is, if the elements of the phytoplankton biomass appear to exert some internal regulation on each other. Then the covariances will be negative and the ratio of the sum of the component variances (first three terms on the rhs) to the total variance will exceed one:

$$\zeta = \frac{\Sigma\sigma_i^2}{\sigma_B^2} \tag{6}$$

where the σ_i^2 indicate the component variances. Such a ratio has been used by Weiss (1969) to erect an operational definition of an autonomous system; the ecological context has been discussed by Mann (1975). Lewis (1980) attempted to apply this definition in a shallow water microcosm where total community net productivity was partitioned experimentally into that due to phytoplankton, to submerged macrophytes, and to benthic microflora. Over a 17 week period, variance in net community production was exceeded by the sum of the component variances (Fig. 8).

We now return to our point of departure that the relative autonomy of a system is a function of scale and observe that the total variance of B (a function of time) may be decomposed into contributions from processes of different characteristic time-scales (a spectral analysis).

Using primed variables to denote spectral functions of frequency f, and

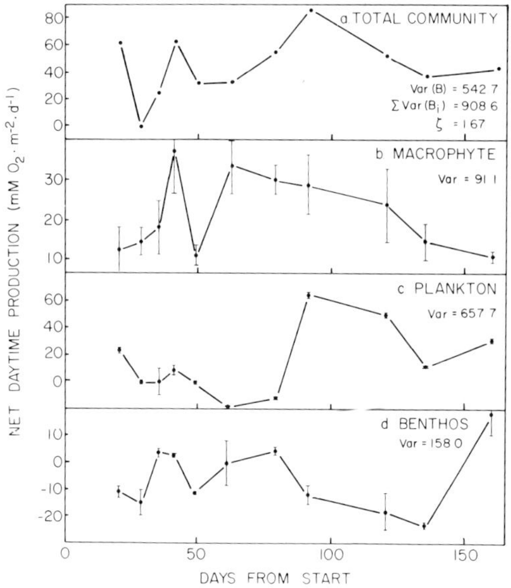

Figure 8. Variability in total net community production, and that associated with the macrophytic vegetation, plankton, and the benthic microflora in a laboratory microcosm.

generalizing to the case where B is fractionated into k variables b_i we can rewrite eqn. (5) as

$$\int_{-\infty}^{\infty} B'(f)df = \int_{-\infty}^{\infty} \left(\sum_{i=1}^{k} b'_{ii}(f) + \sum_{i \neq j}^{k} \sum^{k} b'_{ij}(f) \right) df \qquad (7)$$

where B' is the power spectrum of total biomass; b'_{ii} is the power spectra for fractions; b'_{ij} is the cross-spectrum of b_i with b_j and the integrals are taken over the entire spectrum. Furthermore, eqn (7) is approximately true for each of the individual Fourier modes.

That is,

$$\int_{f_m - \frac{1}{2}\Delta f}^{f_m + \frac{1}{2}\Delta f} B'(f)df = \int_{f_m - \frac{1}{2}\Delta f}^{f_m + \frac{1}{2}\Delta f} \left(\sum_{i=1}^{k} b'_{ii}(f) + \sum_{i \neq j}^{k} \sum^{k} b'_{ij}(f) \right) df \qquad (8)$$

for all Fourier modes f_m. This equation is strictly true for systems where interactions between components are linear; it is a good approximation when the interactions are weakly non-linear, that is where interactions between components at different frequencies are small. If this assumption is satisfied then the variance in any frequency band of the spectrum of total biomass can be found by summing the spectra for the component biomasses and all of their cross-spectra in that frequency band. We can then define a variance ratio that will allow us to judge the degree of internal compensation (system autonomy) *as a function of time-scale*,

$$\zeta(f_m) = \frac{\int_{f_m - \frac{1}{2}\Delta f}^{f_m + \frac{1}{2}\Delta f} (\sum_{i=1}^{k} b'_{ii}(f))df}{\int_{f_m - \frac{1}{2}\Delta f}^{f_m + \frac{1}{2}\Delta f} B'(f)df} \tag{9}$$

The statistic $\zeta(f_m)$ will equal unity in the case that all of the covariances for the partial biomasses vanish; it will be greater than one if there is substantial internal compensation between the partial biomasses; and it will be smaller than one if the partial biomasses fluctuate in phase, for example under large-scale physical forcing. Data of Colebrook (1977) illustrate this last possibility. He found that fluctuations of 17 taxonomic categories of zooplankton in the period 1955–1959 were coherent over large stretches of the California Current. Bernal (1980), examining the total zooplankton volumes for the same region for the period 1949–1969, found that the fluctuations were coherent over all the subareas of the region considered (~1600 × 400 km). The characteristic response frequency, however, varied from subarea to subarea, generally increasing in a N-S direction. These results are consistent with the interpretation that the zooplankton fluctuations are being forced by variation in penetration of subarctic water onto the coast of Southern California.

Conversely, an example of internally compensated fluctuations, apparently independent of large-scale environmental changes, has been reported by Sherman et al. (1981). Here, fluctuations in pelagic fish stocks show similar features on both sides of the North Atlantic. A general decline in stocks of herring and mackerel has been accompanied by an increase in sprat, sand eel and Norway pout for the eastern North Atlantic and in sand eel in the western North Atlantic. According to Sherman et al. (1981), the coincidence of these changes over such a large horizontal scale excludes the possibility of physical forcing.

Interactions between Physical and Biological Scales

Support is increasing for the concept that the dominant time scales of physical forcing are a partial determinant of the species dominance in a particular system through resonance between forcing frequencies and the characteristic time-scales of the organisms. Early work involved analysis of horizontal variation in plankton abundance. In the simplest case (Skellam 1951; Kierstead and Slobodkin 1952), local changes in phytoplankton abundance arise from two opposing mechanisms. The specific rate of growth of the organisms tends to maintain and expand fluctuations in biomass whereas diffusive processes tend to smooth local variances in abundance. In one dimension then,

$$\frac{\partial B}{\partial t} = K_H \frac{\partial^2 B}{\partial x^2} + rB \tag{10}$$

where B is the local concentration of phytoplankton (M), K_H is the horizontal component of diffusivity (L^2T^{-1}), x is length (L), and r is the growth rate (T^{-1}). We can non-dimensionalize eqn. 10 (see Platt 1981 for details) by introducing a characteristic phytoplankton concentration $\bar{B}$, a characteristic length scale l, and a characteristic time-scale τ:

$$\begin{aligned} B &= BB^* \\ x &= lx^* \\ t &= \tau t^* \end{aligned} \tag{11}$$

where the variables which carry the asterisk are dimensionless. Therefore:

$$\frac{1}{\tau} \frac{\partial B^*}{\partial t^*} = \frac{K_H}{l^2} \frac{\partial^2 B^*}{\partial x^{*2}} + rB^* \tag{12}$$

If we set τ to one or the other of the implicit time scales, $l^2K_H^{-1}$ or r^{-1}, we obtain the non-dimensional number, ($r\, l^2\, K_H^{-1}$), which completely specifies the dynamics. When the time scale of growth balances the scale of diffusion, this number (to within a scaling factor of order unity) will equal one. For any combination of K_H or r then, there will exist a length scale l_{cr} where this will occur:

$$l_{cr} \propto (K_H/r)^{1/2} \tag{13}$$

Exact solution of eqn. 10 yields a value of π for the proportionality constant. Increases in K_H or decreases in growth rate r both require larger critical dimensions of the patch. Although this is a simple example, and has been expanded to include grazing effects (Wroblewski et al. 1975), competition (Dubois 1975), two dimensional structure (Skellam 1951), and variation in K_H with length scale (Okubo 1978), the basic conclusion remains sound: the characteristic length scale is determined by the relative importance of the physical scale to the biological reaction scale.

With respect to the autonomy of estuaries, consider the case where l_{cr} is fixed as some length scale characteristic of a particular estuarine system, and allow r to vary as a function of changing diffusivities. For maintenance of a population, or in order that organisms will grow and reproduce within the bounds of the estuary, increases in K_H would have to be compensated by concomitant increases in r. We can replace the diffusive time scale, $l^2K_H^{-1}$, with a time scale appropriate to estuarine flushing in an analogous fashion. Using our estimate above for Bedford Basin, coupled with measured primary production rates in the Basin, we compute a value ~ 1 for the non-dimensional ratio of the flushing rate to the phytoplankton turnover rate.

On one hand, then, large scale, low frequency physical variability can impose a limit to the scale at which biological interactions operate; on the other hand, at frequencies higher than the characteristic biological scale, distributions are controlled primarily by turbulent fluid motion. For example, we can take the approach of Denman and Platt (1976) in a theoretical analysis of the spectral distribution of phytoplankton abundance. Consider the case where a geostrophic current prevails with a time scale $\sim f^{-1}$ (where f is the Coriolis parameter) and with a length scale $\lambda \approx U/f$, where U is a characteristic geostrophic velocity. At scales much smaller than those dominated by variances associated with geostrophic flow, the wavenumber spectrum of velocity variance is dependent only on wavenumber and the rate of energy dissipation as viscous heat. On dimensional grounds then (Kolmogoroff 1941),

$$E_v(k) = A\varepsilon^{2/3}k^{-5/3} \tag{14}$$

where $E_v(k)$ is the velocity variance spectrum (L^3T^{-2}), ε is the rate (per unit mass) of energy dissipation (L^2T^{-3}), k is the wavenumber, and A is a dimensionless constant of order unity. At scales where the geostrophic flow dominates, the rate of turbulent dissipation can be considered unimportant and we can write, again on dimensional grounds,

$$E_v(k) = A'f^2k^{-3}F(\lambda k) \tag{15}$$

Dimensional arguments in this case leave us with an equation that is not completely determinate; F is an unspecified function of the dimensionless group, λk. Airborne observations of sea-surface temperature at scales large relative to λ (Holladay and O'Brien 1975) and theoretical arguments concerned with two-dimensional turbulence (Kraichnan 1967) yield variance spectra with k^{-3} wavenumber dependence thus lending tentative support to the above.

If we now consider the variance spectrum of a passive, but nonconservative, contaminant such as chlorophyll we must specify at least two more parameters. The rate of destruction of biomass variance at small scales, analogous to ε, will be denoted by $\chi_\beta(\beta^2 \bullet T^{-1})$ (where β is local chlorophyll concentration) and the rate of growth by $r(T^{-1})$. At high wavenumbers then, we can write the following which again is not completely determinate since we have one more undetermined index than we do dimensions:

$$E_\beta(k) = A''\chi_\beta r^{-1}k^{-1}F(\varepsilon k^2/r^3) \tag{16}$$

We can specify a critical wavenumber based on the non-dimensional group,

$$k_{cr} \sim (r^3/\varepsilon)^{1/2} \tag{17}$$

This can be interpreted as the scale at which plankton growth and turbulent dissipation assume equal significance. For wavenumbers greater than k_{cr} then, we expect both the growth process and geostrophic influences to be small and that the biomass variance spectrum would be of the same shape as that of the velocity variance spectrum; $E\beta(k)$ would go as the $-5/3$ power of wavenumber. At wavenumbers lower than λ^{-1}, we might expect the variance to be dominated by geostrophic eddies. Only for the region $\lambda^{-1}<k<k_{cr}$, if it exists, would the influence of the biological rate be of importance. In this region, the slope of the spectrum has been predicted to go as the -1 power of k(Denman and Platt 1976). Three regions are predicted in the idealized case: a high wavenumber region where turbulence dominates with a wavenumber dependence of $-5/3$, a low wavenumber region where large geostrophic currents predominate with a spectral slope of ~ -3 and finally, sandwiched in between, a region where biological interactions dominate in which the slope may go as k^{-1}. One would thus expect coherence between measures of biomass and measures of conservative properties such as temperature at scales larger and smaller than that of the biomass turnover time. Such a three phase spectrum has been reported recently (Horwood 1981) for data collected in the North Sea. At high wavenumbers, the Fourier spectrum of fluorescence had a slope (log-log) of -2. A break in the spectrum at length scales of about 10 km was found and related to the scales at which biological events predominate. However, at longer wavelengths, ca. 25 km, the slope of the fluorescence spectrum reverted to that of the small wavelengths. Fluorescence was coherent with temperature at both large and very small scales, but was not coherent in the middle region where growth processes dominate. Expected wavelengths for λ in the North Sea are probably smaller than 25 km and may reflect the inappropriateness of our assumptions for the large scale processes here.

The length of time required for shipboard sampling at these large scales approaches the time for phytoplankton turnover and the time for significant changes in incident irradiance, hence the essence of synopticity is lost. These scales can be better resolved by use of satellite imagery (Gordon et al. 1980; Hovis et al. 1980). Recent results using this technology (Gower et al. 1980) support the idea that at these scales phytoplankton distributions, as indexed by reflectance, behave in a coherent fashion with large scale eddy motion inferred from satellite thermal imagery (Fig. 9). Although studies encompassing such low frequency scales are limited, and the statistical confidence at these bandwidths is small, it appears as if there exists a biological "window" in the spectrum of biomass variance where biological dynamics are strong enough to counter the impositions set by the physical environment. In the North Sea

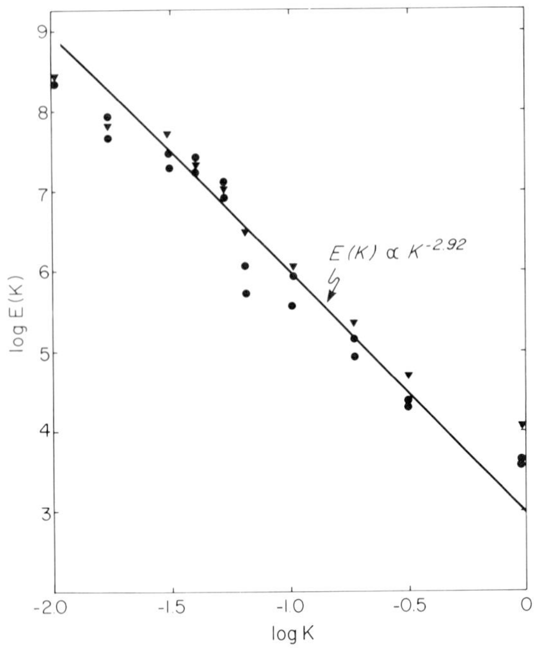

Figure 9. Wavenumber variance spectrum of satellite imaged 'chlorophyll' off Iceland. (From Gower et al. 1980). Reprinted by permission from Nature, Vol. 288:157-159. Copyright ©1980. Macmillan Journals Ltd.

study, this window corresponds to scales of order 10 to 25 km with a corresponding time scale of ~5 to 12 days.

Conclusions

Ecological discussions should always be prefaced by a statement of the time scales that will be admitted in the argument. This is as true for the physical forcing processes as it is for the biological processes that respond to them. In estuarine ecology, where the physical forcing is often relatively strong in amplitude and complex in frequency structure, it is particularly important to retain the idea of time scale. Even an elementary consideration of dynamical equations reveals that time and space scales are entwined in an intimate way, so that whenever the idea of time scale is important, the idea of space scale will be equally so. The extent to which estuaries can be considered an autonomous ecosystem depends on their size. Analysis of ecological data in the frequency or wavenumber domain preserves information that is lost if simple statistical moments are calculated across entire time or space series. In the same way, it is often fruitful to develop theoretical ideas about ecosystems in spectral form, an approach that places due emphasis on the importance of scale.

Acknowledgments

We would like to thank K. H. Mann, B. Conover, W. Silvert, J. Cullen and A. Herman of the Bedford Institute for many illuminating discussions and comments on the manuscript. D. H. Lawrence provided direction in the current analysis. R. A. Myers, K. Denman and R. Ulanowicz also commented on the manuscript and we are grateful for their insight. MRL would like to thank W. M. Kemp and J. C. Stevenson of the Horn Point Environmental Laboratories (U. Maryland) for advice on the microcosm experiment. Portions of this study were supported by the National Research Council (Canada), the U.S. Environmental Protection Agency, and the Horn Point Environmental Laboratories.

References Cited

Bernal, P. A. 1980. Large-scale biological events in the California Current: The low frequency response of the epipelagic ecosystem. Ph.D. Thesis, U. California, San Diego. 184 pp.

Colebrook, J. J. 1977. Annual fluctuations in biomass of taxonomic groups of zooplankton in the California Current 1955-1959. *Fish Bull.* 75:337-368.

Côte, B. and T. Platt. (in prep.) Daily variations in the photosynthetic parameters of coastal marine phytoplankton.

Denman, K. L. and T. Platt. 1976. The variance spectrum of phytoplankton in a turbulent ocean. *J. Mar. Res.* 34:593-601.

Dubois, D. M. 1975. A model of patchiness for prey-predator plankton populations. *Ecol. Modelling* 1:67-80.

Dyer, K. R. 1973. *Estuaries: A Physical Introduction.* John Wiley, London. 140 p.

Gordon, H. R., D. K. Clark, J. L. Mueller and W. A. Hovis. 1980. Phytoplankton pigments from the Nimbus – 7 Coastal Zone Color Scanner: Comparisons with surface measurements. *Science* 210:63-66.

Gower, J. F. R., K. L. Denman and R. J. Holyer. 1980. Phytoplankton patchiness indicates the fluctuation spectrum of mesoscale oceanic structure. *Nature* 288:157-159.

Harrison, W. G. and T. Platt. 1980. Variations in assimilation number of coastal marine phytoplankton: Effects of environmental covariates. *J. Plankton Res.* 2:249-260.

Heath, R. A. 1973. Flushing of coastal embayments by changes in atmospheric conditions. *Limnol. Oceanogr.* 18:849-862.

Herman, A. and T. Platt. 1980. Meso-scale spatial distribution of plankton: co-evolution of concepts and instrumentation, pp. 204-225. *In*: M. Sears and D. Merriman (eds.), *Oceanography: The Past.* Springer-Verlag, New York.

Holladay, C. G. and J. J. O'Brien. 1975. Mesoscale variability of sea surface temperatures. *J. Phys. Oceanogr.* 5:761-772.

Horwood, J. 1981. Variation of fluorescence, particle-size groups and environmental parameters in the southern North Sea. *J. Cons. Int. Explor. Mer.* 39:261-270.

Hovis, W. A., D. K. Clark, F. Anderson, R. W. Austin, W. H. Wilson, E. T. Baker, D. Ball, H. R. Gordon, J. L. Mueller, S. Z. El-Sayed, B. Sturm, R. C. Wrigley and C. S. Yentsch. 1980. Nimbus 7 Coastal Zone Color Scanner: System description and initial imagery. *Science* 210:60-63.

Kierstead, H. and L. B. Slobodkin. 1953. The size of water masses containing plankton blooms. *J. Mar. Res.* 12:141-147.

Kjerve, B. J., J. E. Greer and R. L. Crout. 1978. Low-frequency response of estuarine sea-level to non-local forcing, pp. 497-513. *In*: M. Wiley (ed.), *Estuarine Interactions*. Academic Press, New York.

Kolmogoroff, A. N. 1941. The local structure of turbulence in incompressible viscous fluid for very large Reynolds numbers. *C. R. Acad. Sci. U.R.S.S.* 30:299-303.

Kraichnan, R. H. 1976. Inertial ranges in two-dimensional turbulence. *Phys. Fluids* 10:1417-1423.

Lewis, M. R. 1980. An investigation of some homeostatic properties of model ecosystems in terms of community metabolism and component interactions. M. S. Thesis, U. Maryland, College Park, MD 150 pp.

MacCaull, W. A. and T. Platt. 1977. Diel variations in the photosynthetic parameters of coastal marine phytoplankton. *Limnol. Oceanogr.* 22:723-731.

Mann, K. H. 1975. Relationship between morphometry and biological functioning in three coastal inlets of Nova Scotia, pp. 634-644. *In*: L. E. Cronin (ed.), *Estuarine Research (Vol. I)*. Academic Press, New York.

Okubo, A. 1978. Horizontal dispersion and critical scales for phytoplankton patches, pp. 21-42. *In*: J. H. Steele (ed.), *Spatial Pattern in Plankton Communities*. Plenum Press, New York.

Petrie, B. and K. Drinkwater. 1978. Circulation in an open bay. *J. Fish Res. Bd. Can.* 35:1116-1123.

Platt, T. 1981. Thinking in terms of scale: introduction to dimensional analysis, pp. 112-121. *In*: T. Platt, K. H. Mann and R. E. Ulanowicz (eds.), *Mathematical Models in Ecology*. The UNESCO Press, France. 156 p.

Platt, T. 1975. Analysis of the importance of spatial and temporal heterogeneity in the estimation of annual production by phytoplankton in a small enriched marine basin. *J. Exp. Mar. Biol. Ecol.* 18:99-109.

Platt, T. and R. J. Conover. 1971. Variability and its effect on the 24-hr. chlorophyll budget of a small marine basin. *Mar. Biol.* 10:52-65.

Platt, T. and K. L. Denman. 1975. Spectral analysis in ecology. *Ann. Rev. Ecol. Syst.* 6:189-210.

Platt, T. and A. D. Jassby. 1976. The relationship between photosynthesis and light for natural assemblages of coastal marine phytoplankton. *J. Phycol.* 12:421-430.

Platt, T., A. Prakash and B. Irwin. 1972. Phytoplankton nutrients and flushing of inlets on the coast of Nova Scotia. *Le Naturalist Can.* 99:253-261.

Sandstrom, H. 1980. On the wind-induced sea level changes on the Scotian Shelf. *J. Geophys. Res.* 85:461-469.

Sherman, K., C. Jones, L. Sullivan, W. Smith, P. Berrien and L. Ejsymont. 1981. Congruent shifts in sand eel abundance in western and eastern N. Atlantic ecosystems. *Nature* 291:486-489.

Sieburth, J., V. Smetacek, and J. Lenz. 1978. Pelagic ecosystem structure: Heterotrophic compartments of the plankton and their relationship to plankton size fractions. *Limnol. Oceanogr.* 23:1256-1263.

Skellam, J. G. 1951. Random dispersal in theoretical populations. *Biometrika* 38:196-218.

Steele, J. H. 1978. Some comments on plankton patches, pp. 1-20. *In:* J. H. Steele (ed.), *Spatial Pattern in Plankton Communities*. Plenum Press, New York.

Wang, D. P. 1979. Subtidal sea-level variations in the Chesapeake Bay and relations to atmospheric forcing. *J. Phys. Oceanogr.* 9:413-572.

Wang, D. P. and A. J. Elliot. 1978. Non-tidal variability in the Chesapeake Bay and Potomac River: Evidence for non-local forcing. *J. Phys. Oceanogr.* 8:225-232.

Weisberg, R. H. 1976. The non-tidal flow in the Providence River of Narragansett Bay: A stochastic approach to estuarine circulation. *J. Phys. Oceanogr.* 6:721-734.

Weiss, P. A. 1969. The living system: determinism stratified, pp. 3-55. *In*: A. Koestler and J. R. Smythies (eds.), *Beyond Reductionism*. MacMillan, New York.

Wroblewski, J. S., J. J. O'Brien and T. Platt. 1975. On the physical and biological scales of plankton patchiness in the ocean. *Mem. Soc. Roy. des Sci. Liege 6ᵉ serie*, VII:43-57.

LOCALIZED MIXING OF LOW SALINITY PATCHES IN A PARTIALLY MIXED ESTUARY (SOUTHAMPTON WATER, ENGLAND)

K. R. Dyer
Institute of Oceanographic Sciences
Crossway, Taunton
Somerset, United Kingdom

Abstract: The time scales of many chemical and biological processes are from minutes to hours and, for comparative studies, variations in the physical conditions of the same scale need to be considered. Observations within Southampton Water have shown the presence of intermediate scale patches of low salinity water. These patches, which appear to be caused by river flow variations, mix relatively slowly with the surrounding water masses. The mixing is greatest on the ebb flow at spring tides, and appears to be mainly restricted to a comparatively short reach of the estuary. The influence of these patches on chemical and biological processes could cause contrasts not otherwise apparent in comparative studies based on tidally averaged characteristics.

Introduction

Within estuaries the spatial and temporal variation in salinity and temperature, as well as of many chemical and biological variables, is caused by the interaction of the tide with topographic features of many sizes. Tidal rise and fall of the sea surface forces the water to oscillate, advecting the gradient zone between the salt and fresher water backwards and forwards. The vertical and lateral velocity gradients then cause turbulent mixing which modifies the gradients. The presence of headlands, bends, tributaries and tidal flats also creates the possibility of separation of the flow and formation of intermediate scale patches of water which become entrained and advected by the tidal flow and which gradually undergo turbulent mixing. The timescales are thus interlinked, and range from seconds to minutes and metres in extent for short period turbulence, minutes to hours and hundreds of metres for intermediate scale, and hours to days and kilometres for long period effects.

With many chemical and biological processes, time constants are of the order of minutes to hours. Changes in the dynamics of the estuary of longer time scales than these will simply move the reaction around unaltered. Shorter period changes will modify the situation before the processes get underway. Changes of the same scale, however, could cause drastic alterations in rates of chemical processes and in biological responses. Consequently, for ecodynamic comparisons we need to consider the intermediate timescales. The facts that there is such a large range in topography and that

ISBN 0-12-404070-5

no two estuaries have the same shape make estuarine comparisons difficult, and could account, to a degree, for chemical and biological differences observed between estuaries which have the same tidally averaged physical classification.

Most estuarine comparisons have been carried out on tidally averaged velocity and salinity data taken at a small number of stations over only one or two tidal cycles. The techniques involved in obtaining and averaging the data remove short and intermediate scale effects and channel tortuosity, for instance, becomes a second order effect, the estuary width, depth, and length being dominant. Classification schemes have been constructed using relatively simple parameterizations, such as estuary numbers (Harleman and Abraham 1966), estuarine Richardson numbers (Fischer 1972), or stratification-circulation diagrams (Hansen and Rattray 1966). From these, various aspects of the dynamics can be predicted. Simple models allow calculation of flushing times, for example, and more complicated computer models allow calculation of salinity distributions and currents. Nevertheless, since the normal circulation exists for less than half of the time, the rest being affected by wind events (Elliott 1978), the averaged conditions can only be predicted for part of the time. Also, the greatest fluxes of substances through the estuary occur during occasional floods, and this involves processes for which measurements at more normal conditions may be unrepresentative.

Comparatively little effort has gone into describing and quantifying the intermediate timescale variations in estuaries. Since these variations are likely to be caused by advection within the estuary of water masses of tens to hundreds of metres in extent, a combination of Eulerian and Lagrangian measurements is ideally required. Little of this type of measurement has been carried out. One feature that has been studied, however, is the fronts that are often associated with tributaries in estuaries. The tributaries can often inject a completely different water mass with different mixing characteristics as a plume into the main body of water, and the interface between them can have high static stability. The water masses often have distinctive colour differences and the boundary shows up as a front with a concentration of debris and scum. These plumes have been studied by Garvine (1974) and Garvine and Monk (1974), and modelled by Kao et al. (1977). Spatially and temporally localized mixing can create patches of water of intermediate scale with unique characteristics that are likely to be advected within the estuary, influencing conditions elsewhere. Recent work by Partch and Smith (1978) and Gardner and Smith (1978) have shown that intense mixing occurs on the ebb tide in a 4 km long section of the highly stratified Duwamish Estuary. It appeared that the mixing resulted from an internal hydraulic jump associated with a change in water depth. Another internal hydraulic jump occurred at a constriction created by a bridge. Farmer and Smith (1980) have also observed mixing resulting from an internal jump caused by a sill in a fjord.

This paper reports the results of an investigation into intermediate scale features in a partially mixed estuary. Five days of continuous measurements on salinity and velocity showed the presence of low salinity patches that were gradually mixed with the surrounding water.

Measurements

Southampton Water is a partially mixed estuary. The 10 km portion near the mouth is about 2.5 km wide, with a central 500 m wide channel dredged to about 13 m below mean water level. The research vessel was anchored fore-and-aft at 2 stations on either side of the deep channel (Fig. 1). Station 1 was occupied for 29 h at neap tides (2 April 1978) and for 13 h at near spring tides (6 April 1978), whilst Station 2 was occupied for 48 h (3-5 April 1978).

Four direct-reading Braystoke current meters spaced evenly throughout the water depth measured the current speed averaged over a minute continuously at each level throughout the entire period. Every 20 min a Robertson Research Laboratory conductivity-temperature (CT) probe was profiled through the water column. This failed after about 15 h and thereafter a direct reading temperature-salinity (TS) bridge was traversed every 20 min, with measurements at 1 m intervals from the surface to the seabed.

Additionally a Bissett-Berman 9006 salinity-temperature-depth (STD) sensor was suspended from the vessel, generally at 4 m depth, just below the halocline. These measurements were also continuously recorded throughout the entire period, with a sampling frequency of about 1 Hz.

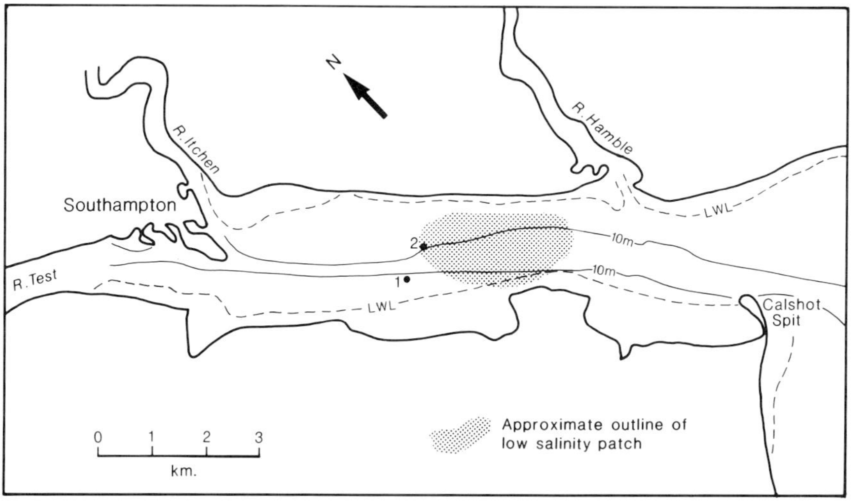

Figure 1. Location of the measurement positions in Southampton Water, showing approximate outline of low salinity patch at low water.

General Features of the Flow

Within Southampton Water the tidal curve is complicated, flood tide lasting about 7 h, but with a 1 h period of almost steady water level in the middle. At high water there is a 2 h period of almost constant water level, and the ebb is restricted to only 3 h of the tidal cycle. Flow velocities on the flood tide are about half those on the ebb, and come in two separate periods. Despite these complications, tidally averaged velocity and salinity profiles (Fig. 2) show features typical of a partially mixed estuary.

The time sequence of salinity and velocity for Station 1 on 2 April and on 6 April are shown in Fig. 3. Plotted velocities are 11 min averages centered on the time of the salinity profiles. Between the two experiments, velocities almost doubled and the first phase of the flood current became better developed towards spring tides. Very similar results were obtained from Station 2.

Patchiness

One distinctive feature of the salinity distributions shown on Fig. 3 was that the minimum surface and seabed salinities occurred during the ebb tide and started rising well before low water. This pattern is what one would expect from a patch, or lens, of less saline water being advected past on the ebb tide. On 2 April the patch appeared at about 0945 hours with an abrupt lowering of salinity. On the sea bed salinity was reduced by about $1^0/_{00}$ and on the surface by about $3^0/_{00}$. This downstream end of the patch had the appearance of a front similar to those described from shelf seas (eg. Pingree et al. 1974). Vertical CT profiles taken at 20 min intervals show that the basic

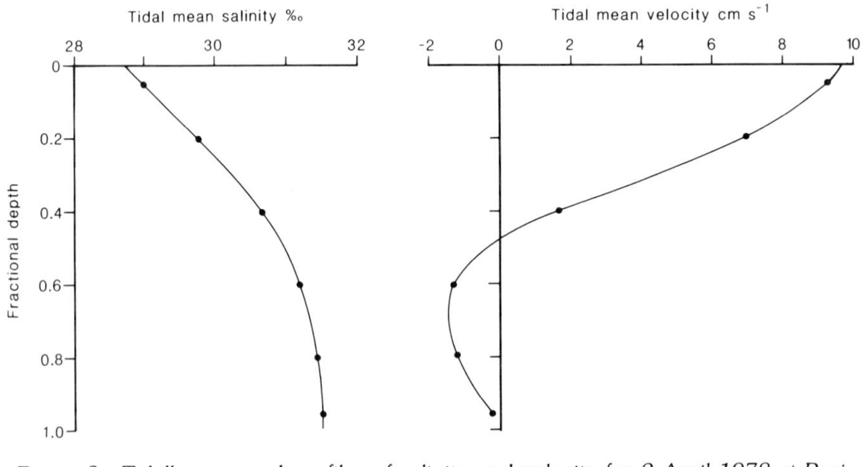

Figure 2. Tidally averaged profiles of salinity and velocity for 2 April 1978 at Position 1.

stratification changed little across the front (Fig. 4) but the profiles were reduced in salinity uniformly by about 2‰. Since current speed at that time was about 25 cm s^{-1}, this front would have been about 450 m wide if it was aligned across the estuary. Also it must have been a fairly steep feature, with a minimum slope of 6/450. The tail of the patch at about 1145 h was less sharply defined than the front. The surface temperature dropped from 9.4°C to 9.0°C at the front of the patch and the seabed temperature decreased from 8.2°C to 7.5°C. However at mid-depth, temperature rose because of a considerable thickening of the warmer surface water.

A patch was also visible on the following ebb tide between about 2230 and 0030 h (Fig. 3) though the surface-to-bottom salinity difference was much larger. On the intervening flood tide a patch was visible between about 1430 and 1745 h. Consequently it is probable that we were seeing the same patch being carried past the measuring station on the tidal oscillation, downstream on ebb and upstream on flood. Assuming that the velocity was uniform along the channel, the patch length and the tidal excursion can be calculated. The patch length appears to have been about 2 km and the tidal excursion was approximately 6 km, the patch travelling 1 km upstream and downstream at Station 1.

At Station 2 on the opposite side of the channel, between 3-5 April, the patch was again apparent. The front of the patch appeared at the same part of the tidal cycle as at Station 1 indicating that the front extended slightly further downstream on the northern side. Patch length was greater, being about 3 km, and the tidal excursion rose from about 8 km to about 10.2 km over the 3 days.

On the last full ebb tide of measurements at Station 2 the patch did not completely pass, whereas subsequently on 6 April at Station 1 (Fig. 3) it passed about 650 m downstream. This indicates that the patch was still longer on the northern side at that stage. However there were quite marked differences between the characteristics of the patch on flood and on ebb, the surface salinities being higher on the flood. This might indicate a considerable change in surface extent of the patch between the two phases of the tide, though the same effect was visible on both sides of the channel.

The detailed form of the patch can be seen by examination of the continuous salinity records from the STD at 4 m depth. There was considerable variability in salinity with periods of less than 20 min visible in the STD records (Fig. 5) and in many instances the salinities corresponding to those obtained by the 20 min profiles were not good representations of the mean values. Much of this variability was due to internal waves. Comparison of the fluctuations in salinity with the vertical profiles shows that these waves have amplitudes of 1-1.5 m and they have been interpreted as due to lateral internal seiching (Dyer 1982). However, there are many broader features that persist from one tide to another, undergoing progressive modification.

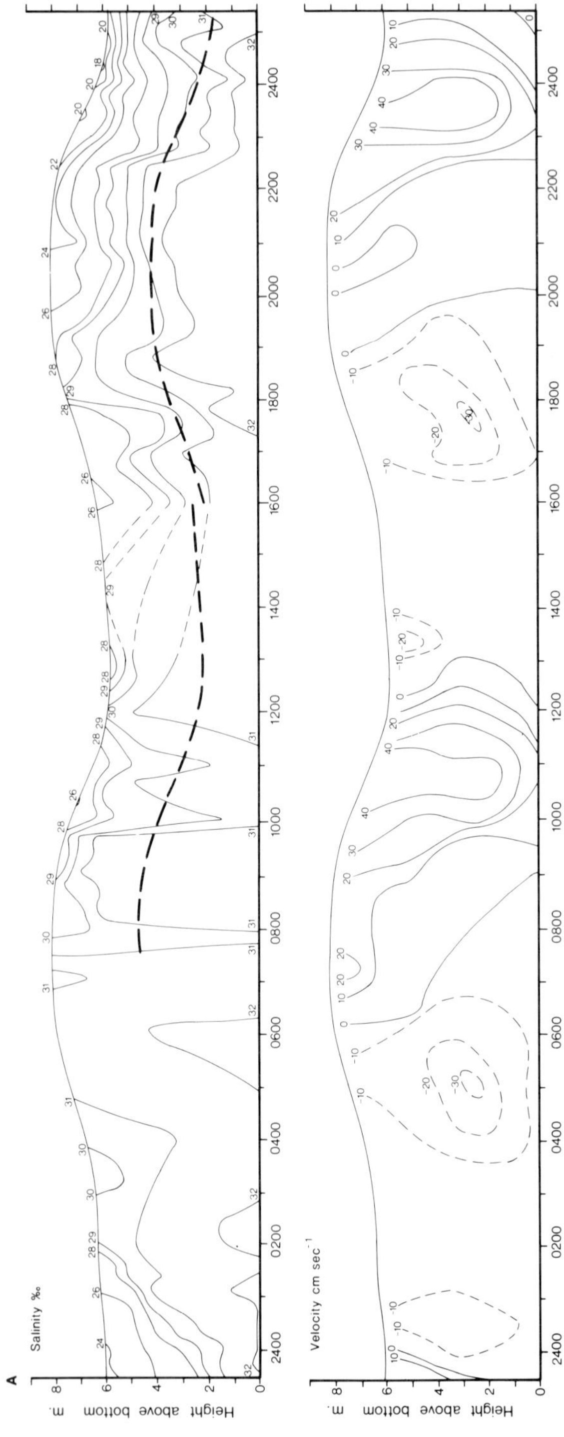

Figure 3(A). Time sequence of salinity and velocity variation at Station 1. Dashed line is level of STD measurements, (2 April 1978).

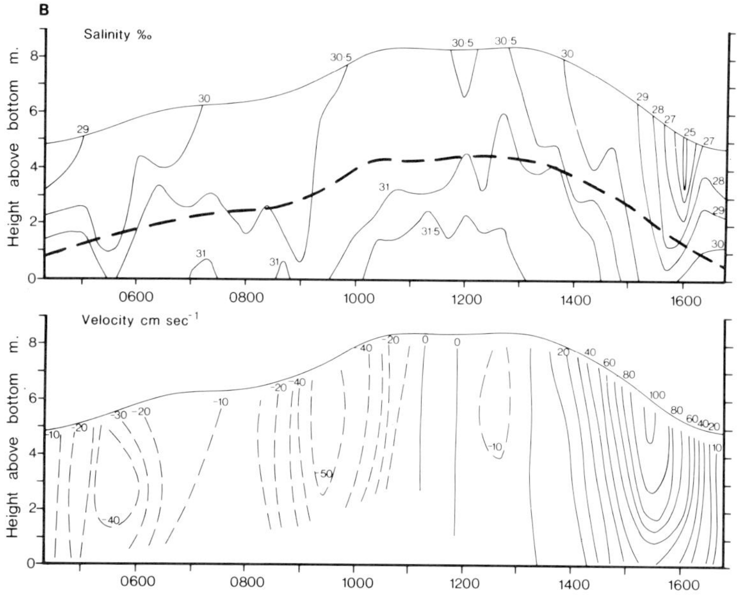

Figure 3(B). As Fig. 3(A), but for 6 April 1978.

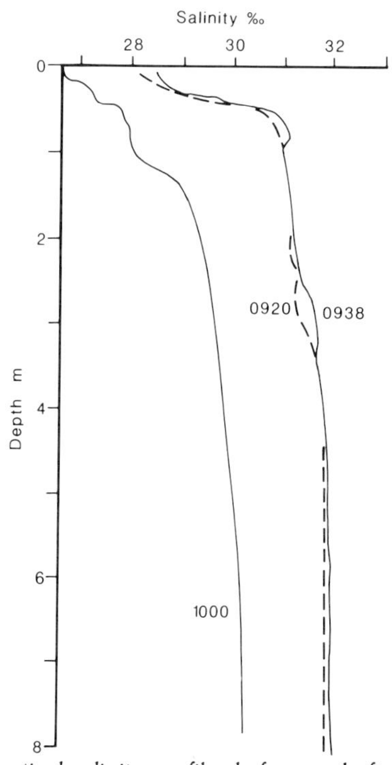

Figure 4. Continuous vertical salinity profiles before and after the appearance of the front. The instants of these measurements compare with those from the STD shown on Fig. 5 (2 April 1978).

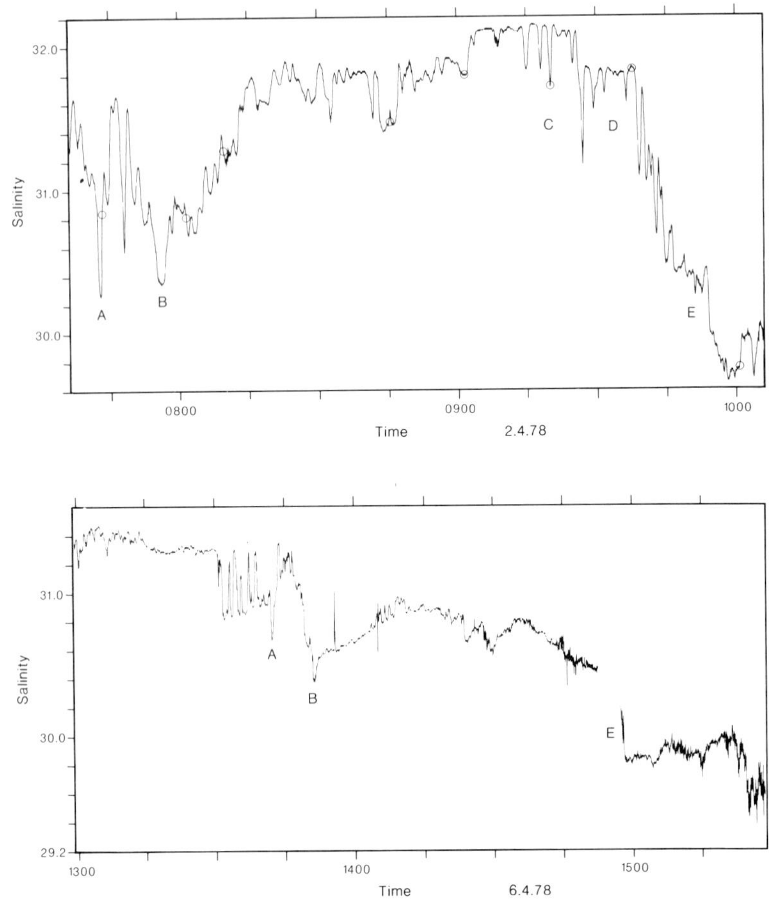

Figure 5. Continuous salinity records from the STD at about mid-depth on 2 April 1978 (upper) and 6 April 1978 (lower), showing corresponding Events A, B and E. Circles are salinity values coincident with those from continuous vertical profiles.

The STD record for the beginning of the first ebb tide on 2 April (Fig. 5) shows that the leading edge of the patch, at a depth of 4 m, was a sharp change in salinity from about $32.1^{0}/_{00}$ to about $29.7^{0}/_{00}$ within half an hour. There was a corresponding change in temperature from 8.25°C to 8.85°C. The first indication of the front was a distinctive step from water mass C, of salinity $32.1^{0}/_{00}$, to water mass D with a salinity of $31.9^{0}/_{00}$. Between the two water masses, at about 0925 h, there was a narrow finger of less saline water. After 1000 h the salinity at 4 m depth gradually rose to a value of $31.0^{0}/_{00}$ at 1145 h. The tail of the patch (the upstream end) was less sharply defined than the front, and was distinguished by a step of about $0.6^{0}/_{00}$.

One tidal cycle later, water mass D appeared to have mixed considerably with C, but a narrow finger of low salinity water had developed between water masses D and E. The front had not transgressed, but its overall steepness had decreased. On the intervening flood tide the salinity at 4 m depth did not fall below 30.9‰, but the returning patch was still visible on the 20 min profiles.

An additional feature on the STD record for 2 April (Fig. 5) is a second low salinity patch about 700 m downstream of the main low salinity patch, centered at about 0800 h, and which contained two distinctive lower salinity features, Events A and B. This patch also had a steep downstream front and a gradual rise to the tail at 0815 h. Events A and B were visible on every ebb tide at both Stations 1 and 2, on either side of the channel. Within the patch the lowest salinities were very similar at both Stations and contained comparable structures. This indicates that there is continuity across the estuary. Event B, for example, must have been only about 100 m wide and yet it extended laterally over the 500 m between Stations. There were large differences, however, between the detailed structure on flood and ebb tides, with Events A and B seldom being distinguishable at flood tides. Nevertheless there was again considerable correspondence between the detailed structure on flood tides between both Stations. Consequently there was likely to have been a vertical movement of the isohalines between flood and ebb tide which was probably associated with a slight lateral displacement of the patch as a whole. This was probably caused by an alteration in the secondary circulation patterns involving the whole estuary width.

A comparison between the STD records for similar parts of the tidal cycle on 2 April and on 6 April (Fig. 5) shows how the features developed over the intervening time at 4 m depth. Events A and B are easily distinguished at both times, and were altered very little during the 3 days. However, the front of the main patch had been considerably degraded and water masses C and D were no longer distinguishable. The broad high salinity period between Events B and E had been considerably eroded. The lowest salinity in the patch after E had not altered; it remained at about 29.5‰. Consequently the front of the main patch was gradually becoming less steep, and even though the distance between Events B and E were not altering, the low salinity periods were lengthening as the two patches merged. One could now effectively define a combined patch as starting at the front preceding Event B at 1350 h instead of at 1457 h.

Consequently it is apparent that patches of low salinity water remained within a fairly restricted section of the estuary for at least 5 days, maintaining their integrity to a very great degree. An approximate outline of the main patch at low water is shown on Fig. 1. At neap tides the tail of this patch at high water would have been at the confluence of the two arms of the estuary, the rivers Test and Itchen. At spring tides following the survey the front of the patch would have just about reached the mouth of the estuary at Calshot Spit at low water.

Origin of the Patch

The salinity deficit of the main patch was approximately $2^0/_{00}$. Taking realistic values for the patch length of 2.5 km, width 700 m and depth 7 m, gives a total freshwater volume of 7.7×10^5 m^3. This is equivalent to 30 h flow of the River Itchen, or 14 h of the River Test. Examination of river flow records (Fig. 6) shows that there was a brief spate in both the Test and Itchen 16 days before the start of the survey. The patch may consequently be the result of one tidal cycle of discharge from the River Test. The travel time of the spate from the gauging station to the head of the estuary was likely to be about 1 day. Consequently the flushing time for this patch within the estuary was at least 20 days. This contrasts with the value of 12 days calculated for this river flow using the salinity distribution and estuarine volumes quoted in Collins (1978). The initial mixing of the patch is thus likely to have taken place at the head of the estuary and the patch then travelled down the estuary as a discrete unit.

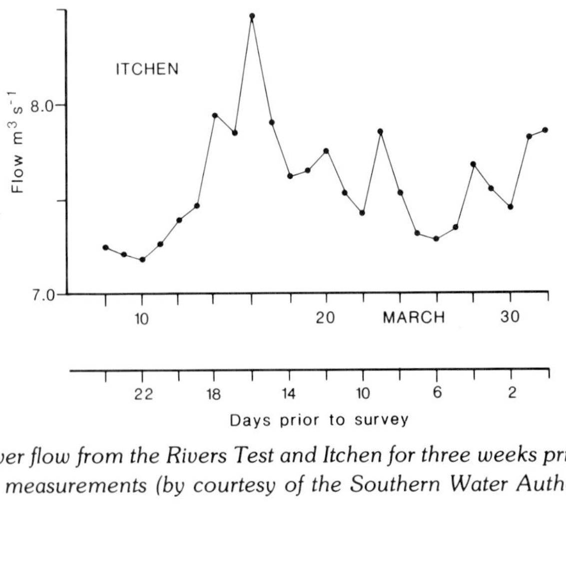

Figure 6. River flow from the Rivers Test and Itchen for three weeks prior to the salinity measurements (by courtesy of the Southern Water Authority).

The low salinity patch following Event B, which has the appearance of a small patch with the same overall shape as the main one, is presumably related to some earlier discharge event, and possibly could be the result of the same spate coming from the somewhat shorter River Itchen.

Mixing

The progressive changes of salinity within the patch, and the comparative stability of such features as Events A and B raise interesting questions about the degree of mixing of a water mass that takes place as it moves around within the estuary. It is obvious that maximum potential for mixing will occur when the tidal currents are at or near their maximum velocity. The amount of mixing that takes place will depend on the density stratification at the time of maximum current shear. If stratification is large then turbulence will be relatively ineffective in modifying it during the short period of the tidal cycle when currents are strong. The balance between the stirring effect of the velocity gradient and the restoring force of the density gradient is expressed by the Richardson number. Based on spectra of salinity fluctuations measured during these experiments, Dyer (1982) has shown that turbulence begins to modify salinity fluctuations when the layer Richardson Number $R_{iL} \cong 20$ where:

$$R_{iL} = \frac{g\dfrac{\Delta\varrho h}{\varrho}}{U^2}$$

h is water depth, $\Delta\varrho$ the surface to bottom density difference, g is gravity and U is the depth mean velocity. This is interpreted as being the onset of effective turbulent mixing.

Figure 7 shows a plot of R_{iL} against time for the whole period of measurements. As expected, the values are dominated by the tidal current velocity which contributes over 90% of the variance of R_{iL}. The lowest values occurred at times of maximum current and with very large values near slack water. Near neap tides the mixing was mainly concentrated on ebb tide, with less on the second stage of flood tide, and very little on the first stage. Towards spring tides, more mixing occurred on the first stage of flood tide. The times of appearance of Event A and the front and tail of the patch are also marked on Fig. 7.

On 2 April mixing over the entire length of the patch would have occurred on ebb tide at 0945-1145 h within the vicinity of Station 1. With uniform currents upestuary, the tail would have mixed for possibly the preceding 2 h (which is equivalent to approximately 2 km), but hardly at all once it passed downestuary of Station 1. With uniform currents downestuary, the front

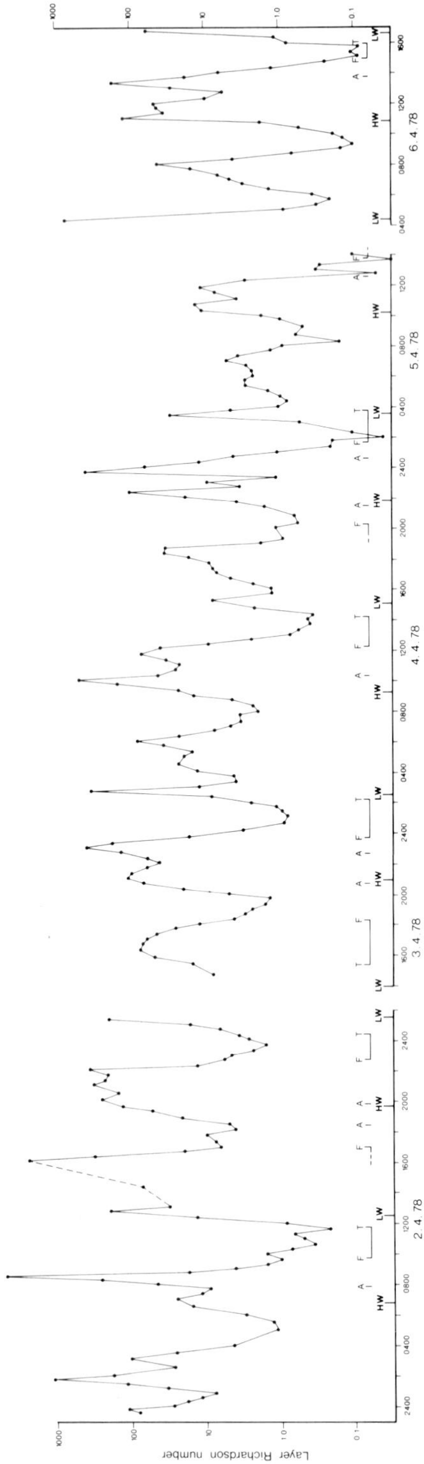

Figure 7. Variation of Layer Richardson Number R_{iL} with time throughout the measurement period. HW and LW, High water and Lower water respectively. A, presence of Event A. F and T, Front and Tail respectively of the main low salinity patch.

would have undergone mixing for the 2 h following its movement past Station 1, and would have also mixed for a short while upestuary of Station 1. Events A and B would only have undergone mixing if uniform conditions existed for in excess of the order of 1 km downstream of Station 1. On the following flood tide some mixing of the water around Events A and B would have been likely at Station 1 and for some 1 h beforehand. However, the lack of significant change in structure around Events A and B at this time indicates the mixing may not be effective more than 1 km downestuary of Station 1. This coincides with the beginning of a natural deep area where the channel widens and deepens.

Comparison between the two ebb tides at this position shows that there is again relatively little change in salinity patterns upstream of the tail of the patch, which suggests that active mixing conditions do not extend beyond about 2 km upestuary of Station 1. Again this coincides with a wider channel area at the confluence of the rivers Test and Itchen. The fact that the steep front on the patch was visible at the beginning of the surveys implies that there was little mixing near the point where the patch was during the previous spring tide.

Towards spring tides, though the gradients of the fronts limiting the patch became reduced, effective mixing around Events A and B was still small. This part of the water mass did not appear to be within the zone of intense mixing at the right part of the ebb tide. The most intense mixing appeared to have taken place between Events B and E, a distance of about 1 km. Consequently the zone of intense mixing is likely to have been about 1 km long and in the immediate vicinity of Station 1. Events A and B must have passed through this zone of mixing on the tidal mean flow during a previous neap tide period when mixing rates were low.

Discussion

This work has shown the presence of two distinct patches of lower salinity water in the estuary which were gradually mixed together towards spring tides, though within the experimental period they did not become indistinguishable. It is possible that the largest patch originated from the River Test whereas the smaller one further downstream came from the Itchen. Consequently it is feasible that concentrations of nutrients, trace elements and other chemical constituents could have differed significantly between the two patches even though they had the same salinity. Also they would have differed from the surrounding water masses. As the two patches gradually merged an average concentration would have been achieved.

Since at any one point in the estuary one will only see events relating to a limited time span – older events having gone past and younger ones still being upstream – along the estuary there may be numerous patches. This has consequences on the results of, for instance, chemical sampling, since each patch may have a different source and mixing history and thus have different characteristics. Presence of several sources could lead to a large noise level in chemical signals. With too slow a sampling rate their presence could go undetected and, particularly in comparative studies or baseline studies, the results could be confusing. Any salinity-or temperature-dependent chemical and biological processes could also proceed at different rates within a comparatively restricted area. Within a water patch, processes could achieve equilibrium since the rate of mixing is low and there would be many days of virtually constant salinity and temperature. For a fixed point on the estuary bed or sides, however, very drastic changes are possible. Salinity changes of $2^0/_{00}$ in 0.5 h followed by over 1.5 h of almost steady salinities occurred as the front of the main patch passed, but $1^0/_{00}$ changes of very brief duration occurred in Event A.

From the motion of the patch and the detailed salinity changes it appears that there is a zone of maximum mixing in the channel of about 1 km in extent within a region of 2 km upstream and 1 km downstream of Station 1. Southampton Water is unique in having such a complicated tidal profile; however, there are comparable features in other estuaries which suggest similar mixing patterns may be common.

In many estuaries, stratification is greatest on ebb tide, reaching a maximum value near low water. Stratification decreases quickly on the flood tide and minimum stratification, and maximum surface salinity, occur close to high water. This effect is particularly noticeable in estuaries for measurements taken in the lower reaches fairly close to the mouth. On the ebb, more highly stratified water is advected down estuary slightly mixed, is mixed further when it has passed the station, and on the flood undergoes mixing before being transported back past the measuring point. Examples are the Vellar estuary described by Dyer and Ramamoorthy (1969) and the Mersey (Hughes 1958; Bowden and el Din 1966). It would appear in both these cases that the measurement position was at the landward end of the intense mixing zone. For the Mersey the measurements were obtained in The Narrows, a constriction where the tidal mixing can be expected to be considerably higher than the wider, though shallower, portion inland. Generally the maximum salinity occurs about 30 min after high water, with the minimum between 1 h before and 1 h after low water. The vertical gradient is greatest near low water, tending to vanish at mid-tide, especially on the flood, and to reach a secondary maximum near high water. One extreme tidal variation in salinity at a neap tide with high river discharge presented by Bowden and el Din (1966, Fig. 7a), shows a maximum stratification about 2 h after low water and a minimum at about high water. This can be explained as a low salinity patch being advected landwards past the measuring position on the flood. Unfortunately there is no indication of its previous or subsequent movement. For the Vellar, the measurement positions were landward of a shallow part of the estuary just within the mouth, where intense mixing is to be expected.

Tributaries and topographic irregularities have important consequences for estuarine dispersion studies, but these results suggest that injection of water masses from tributaries is an intermittent process rather than a steady diffusion, the slow bleeding-back of water into the flow, which Okubo (1973) has modelled. Mixing and the dispersal of pollutants within an estuary are often characterized by use of a dispersion coefficient. They are calculated assuming complete cross-sectional mixing, which takes a period of time after injection, and this can be normalized by the length of the tidal cycle. Fischer (1973) states that: "For small dimensionless times after injection the observed distributions of dispersing clouds are always skewed. The bulk of the contaminant starts out moving as a blob at the uniform velocity, while long contaminant 'tails' stretch out along the boundaries. Only after a sufficient time for approximately complete cross-sectional mixing does the cloud distribution approach normality". The picture of such a blob consequently has an abrupt front and a long more diffuse tail, which is just what was seen in Southampton Water. Even within the Itchen it is unlikely that cross-sectional mixing would have been complete since injection into the main estuary would tend to disperse sharp gradients and ensure a more normal distribution. Since both patches were observed to be skewed it was likely that they were more highly skewed further up the rivers. Thus timescales for cross-sectional mixing become comparable with the total flushing time of the estuary, in which case dispersion coefficients become of limited application.

Conclusions

In estuaries with tributaries, or with topographic irregularities, there may be patches of water with different salinity, temperature, and possibly other characteristics. These patches may maintain their integrity for many days. Additionally, mixing within the estuary is intermittent in space and in time. Mixing appears to be most intense in shallow zones and at constrictions, particularly near the mouth, and occurs mainly at spring tides. This interaction of topography with the flow means that mixing is concentrated in certain areas, and major mixing is of those waters that are in that area at around spring tides. For the rest of the lunar cycle the salinity stratification and distribution may be dominated by advection of these structures to other parts of the estuary. Alteration of the estuarine topography by dredging, etc. in these areas of relatively intense mixing will have a greater effect on the estuarine characteristics, such as stratification, flushing times, and dispersion, than such alterations in less critical areas.

Acknowledgments

I am grateful to many colleagues for assistance with collection of data and for discussion of results, particularly to Dr. A. P. Salkield and Mrs. L. D. M. Marshall for assistance with data analysis. River discharge data were kindly supplied by the Southern Water Authority. I am grateful for the cooperation of the Officers and crew of RV SARSIA and to the Director of the Marine Biological Association Laboratory, Plymouth, UK.

References Cited

Bowden, K. F. and S. H. Sharaf el Din. 1966. Circulation, salinity and river discharge in the Mersey estuary. *Geop. J. Roy. astr. Soc.* 10:383-399.

Collins, K. J. 1978. The fluxes of organic carbon and nutrients in Southampton Water, PhD Dissertation, Oceanography Department, U. Southampton, U. K.

Dyer, K. R. 1982. Mixing caused by lateral internal seiching within a partially mixed estuary. *Estuarine, Coastal and Shelf Science.* (In press).

Dyer, K. R. and K. Ramamoorthy. 1969. Salinity and water circulation in the Vellar Estuary. *Limnol. Oceanog.* 14:4-15.

Elliott, A. J. 1978. Observations of the meteorologically induced circulation of the Potomac Estuary. *Est. Coastal Mar. Sci.* 6:284-299.

Farmer, D. M. and J. D. Smith. 1980. Tidal interaction of stratified flow with a sill in Knight Inlet. *Deep Sea Res.* 27:239-254.

Fischer, H. B. 1972. Mass transport mechanisms in partially stratified estuaries. *J. Fluid Mech.* 53:672-687.

Fischer, H. B. 1973. Longitudinal dispersion and turbulent mixing in open channel flow. *Ann. Rev. Fluid Mech.* 5:59-78.

Gardner, G. B. and J. D. Smith. 1978. Turbulent mixing in a salt wedge estuary, pp.79-106. *In:* J. C. J. Nihoul (ed.), *Hydrodynamics of Estuaries and Fjords.* Elsevier.

Garvine, R. W. 1974. Dynamics of small-scale oceanic fronts. *J. Phys. Ocean.* 4:557-569.

Garvine, R. W. and J. D. Monk. 1974. Frontal structure of a river plume. *J. Geop. Res.* 79:2251-2259.

Hansen, D. V. and M. Rattray, Jr. 1966. New dimensions in estuary classification. *Limnol. Oceanog.* 11:319-325.

Harleman, D. R. F. and G. Abraham. 1966. One-dimensional analysis of salinity intrusion in the Rotterdam Waterway. Delft Hydro. Lab. Pub. 44.

Hughes, P. 1958. Tidal mixing in The Narrows of the Mersey Estuary. *Geop. J. Roy. astr. Soc.* 1:271-283.

Kao, T. W., C. Park and H. P. Pao. 1977. Buoyant surface discharge and small scale oceanic fronts: a numerical study. *J. Geop. Res.* 82:1747-1752.

Okubo, A. 1973. Effect of shoreline irregularities on streamwise dispersion in estuaries and other embayments. *Neth. J. Sea Res.* 6:213-224.

Partch, E. N. and J. D. Smith. 1978. Time dependent mixing in a salt wedge estuary. *Est. Coastal Mar. Sci.* 6:3-19.

Pingree, R. D., G. R. Forster and G. K. Morrison. 1974. Turbulent convergent tidal fronts. *J. Mar. Biol. Assn. UK.* 54:469-479.

TEMPORAL AND SPATIAL CONSIDERATIONS IN MEASURING ESTUARINE WATER FLUXES

Björn Kjerfve
Belle W. Baruch Institute
for Marine Biology and Coastal Research
Department of Geology and Marine Science Program
University of South Carolina
Columbia, South Carolina

and

J. A. Proehl, F. B. Schwing,
H. E. Seim, and M. Marozas
Belle W. Baruch Institute
for Marine Biology and Coastal Research
University of South Carolina
Columbia, South Carolina

Abstract: North Inlet, South Carolina, a tidal estuary with little fresh water input, is the site for a multi-year ecological field study of seasonal water and material fluxes between the coastal ocean and the estuary/salt marsh system. Large tidal currents and small residual fluxes make it difficult to determine budgets with confidence. In addition to large temporal flux variations, estuarine cross-sections exhibit systematic lateral residual current reversals, a surprising find in an estuary classified as well mixed. These tidally-driven circulation features introduce large spatial variations which must be accounted for in field sampling programs. An experiment was undertaken to select an optimum sampling density and rate for discharge measurements in a 320 m wide estuarine cross-section. We found that simultaneous velocity measurements at meter intervals from surface to bottom, at three locations across the section, every 1.5 lunar hours, for at least one tidal cycle, was acceptable. Also, simultaneous discharge measurements, at two cross-sections along the same channel, yielded residual fluxes that could not be separated statistically. This lends more credence to the technique of calculating material fluxes from simultaneous, cross-sectional velocity and concentration measurements.

Introduction

A thorough knowledge of the hydrodynamics of estuaries is necessary to understand processes affecting distributions and fluxes of particulates, pollutants, nutrients, and organisms. It is not sufficient to focus solely on biogeophysical cycles and other traditional biological/ecological aspects. Estuarine circulation, stratification, mixing, flushing, and the temporal scales of these processes need careful attention. When the time scale of fluid transports is rapid relative to biological, geological, and chemical transforma-

ISBN 0-12-404070-5

tions, the hydrodynamics of the system become the dominant consideration (Officer 1980).

It is now widely recognized that estuaries are inherently unsteady marine systems (see Elliot 1976). Variations in distributions and transports (fluxes) vary over turbulent, tidal, weather-induced, seasonal, and longer time scales. In spite of this, it is still common practice to assume that the tidally averaged estuarine processes are steady over finite time periods, i. e. the processes are quasi-steady. This assumption is still rather suspect even in estuaries where the main tidal component (usually M_2) accounts for 75% or more of the total temporal variance. Distributions and fluxes can also vary greatly spatially. Changes in estuarine bathymetry, channel width, location of islands, or presence of fronts due to debouching rivers or tidal creeks can introduce large concentration and flux differences over very short distances.

It is obvious that systems that exhibit great variability, both of a temporal and spatial nature, are difficult to sample. This is almost universally the case with estuaries. The existence of large concentration and velocity variabilities was a major consideration and problem in the measurement of material fluxes into and out of North Inlet, South Carolina (33°N, 79°W). It is the purpose of this paper to focus attention on some of the problems experienced due to large temporal and spatial variabilities in determining net material fluxes between North Inlet and the coastal ocean. In particular, we will attempt to answer whether or not it is possible to measure fluxes with some degree of confidence by making simultaneous measurements of velocities and concentrations in estuarine cross-sections and making these measurements the basis for seasonal budgets.

Physical Description

The North Inlet (Fig. 1) marsh-estuary system in South Carolina covers 32.2 km^2 and consists of *Spartina alterniflora* salt marsh (75.2%), tidal creeks (22.0%), and exposed mudflats (2.5%) at mean tide. The system is located along the Atlantic coast, 70 km northeast of Charleston, SC. It connects with the coastal ocean via a tidal inlet through the barrier island chain that fronts most of the Southeast's Atlantic coast. It is also connected to Winyah Bay, of which Mud Bay is a portion, via three creeks (Jones, No Man's Friend, and Haulover) and, occasionally, water may be exchanged as sheet flow across the marsh. The annual fresh water discharge into the North Inlet system is only 5 m^3/s from ground water input, limited local surface runoff, and direct rainfall (Kjerfve and McKellar 1980). Winyah Bay, on the other hand, receives an annual mean input of approximately 500 m^3/s (US Geological Survey 1981). However, most of the water exchange with the North Inlet system occurs through Town Creek (cross-section A), which on the average is the flow conduit for 79% of the total system tidal prism.

The system is made up of numerous winding creeks with typical channel depths of 3 m. The greatest depth of 7.4 m occurs in cross-section A (Fig.

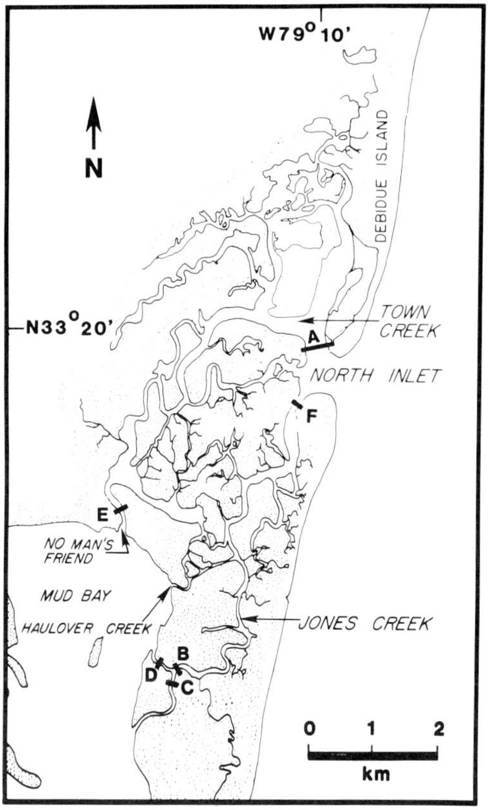

Figure 1. Area map of the North Inlet system. The experimental cross-sections are denoted A (Town Creek), B (South Jones Creek), C (lower South Jones Creek), D (Sign Creek), E (No Man's Friend Creek), and F (North Jones Creek).

1). The tide is primarily semi-diurnal (Palmer et al. 1980) with a mean range of 1.4 m, a typical neap range of 1.1 m, and maximum measured spring tide range of 2.5 m. The tidal prism measures 2.1×10^7 m^3. Tidal currents totally dominate the flow and have been measured to exceed 2.3 m/s. Typical peak tidal flows, however, are 1.3 m/s. Because of the limited fresh water input, the residual (or net) flow is small when integrated over a cross-section. However, the depth-integrated residual flow for a given location is often on the order of 25 cm/s, due to lateral variations of flow within the cross-section. The North Inlet circulation is driven by tidal pumping rather than by gravitational circulation.

The gravitational circulation is close to nonexistent in North Inlet because of the lack of substantial amounts of fresh water input. This is also the reason for high salinities in North Inlet. The annual mean salinity (Fig. 2) in Town Creek (cross-section A) is 34.6 ppt with a monthly standard error of the

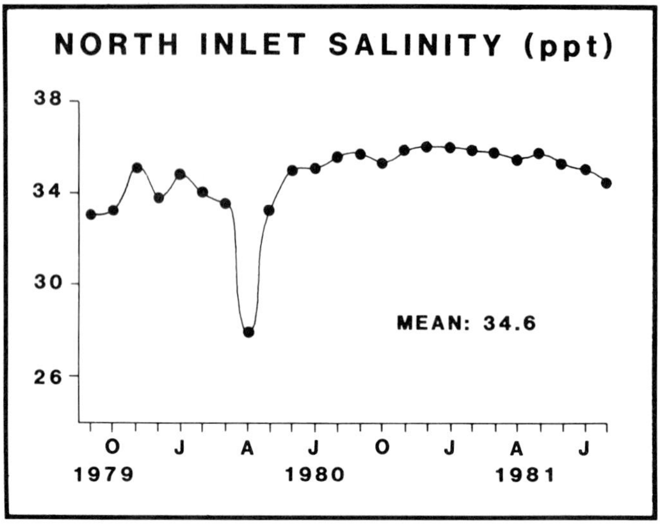

Figure 2. Monthly salinity variations in Town Creek (cross-section A) for a two-year period, based on daily water samples.

mean of 0.4 ppt. Only after prolonged rains can the salinity at this location be expected to dip as low as 28 ppt, which was the case for April 1980.

The combination of low fresh water input and vigorous tidal mixing results in the absence of significant vertical salinity or density stratification. Based on the absence of gravitational circulation (circulation number <1.5) and weak vertical stratification (stratification number $<10^{-2}$), North Inlet fits into category 1a of the classification scheme of Hansen and Rattray (1966). Few other coastal systems described in the estuarine literature are grouped into this category. Still, it is our contention that North Inlet is representative of a series of marsh-estuary systems, located along the Atlantic seaboard from Cape Hatteras, North Carolina, to Cape Canaveral, Florida, and thus warrants further study as a member of an extensive class of similar systems.

Because of mesotidal conditions and shallow depth, North Inlet is a good example of an estuary with rapid hydrodynamic turnover. From simple tidal prism considerations, it is possible to calculate a characteristic residence time, τ, for a water parcel in the system

$$\tau = V/(dV/dt) \tag{1}$$

where V is mean tidal volume and dV/dt is rate of change of water, represented by the combination of tidal prism per half tidal cycle and the fresh water input rate. The assumptions are that the system is well mixed on a time scale which is short relative to the residence time and that the system is in quasi-steady state. The calculation does not take into consideration that a fraction of the ebbing water mass may return to the estuary on the next flood tide. The residence time for North Inlet, according to this representation, is only 10-12 h. Although it may not be possible to put a physical interpretation on this number, a relative comparison can still be made between different estuarine systems which have similarly computed residence times. Such a comparison indicates that North Inlet has an extremely short residence time. It is highly likely that many biogeochemical transformations cannot be completed on an equally short time scale. This implies that the sampling of the velocity becomes more critical for flux determinations than the sampling of concentrations (Kjerfve et al. 1981).

Water Level Variations

Estuarine water level variations and water fluxes can be related analytically through continuity and momentum equations. Whereas fluxes are difficult to measure, water level changes can be measured with ease. Usually, periods of sea level variations and water motions are identical and can be identified by computation of the variance spectrum. The spectrum (Fig. 3) is based on hourly measurements of the North Inlet water level for three consecutive years, September 1978 - August 1981. The time sequence was initially filtered to remove all turbulence and short-term variations with a period less than 4 h. A Fast Fourier Transform algorithm was used for the computations. The resulting spectrum was first frequency-averaged and then segment-averaged (Bendat and Piersol 1971).

The water level oscillations were dominated by the semi-diurnal tide (M_2 and S_2), which accounted for 75% of the overall variance. The diurnal tides (K_1 and O_1) were separated in the spectrum and jointly accounted for 13% of the variance. Several over-tides (particularly M_4 and M_6) due to non-sinusoidal time variation of the tide record were noticeable but accounted for less than a percent of the variance. Much more important were the variances associated with weather-induced (4%) and seasonal fluctuations (7%).

The weather-induced fluctuations had periods from 3 to 12 days but did not show up convincingly in the spectrum because the variations covered a broad frequency band. However, low-pass filtering of the hourly time sequence of water level data from 1979, using a recursive Butterworth filter in both a forward and reverse direction (Roberts and Roberts 1978), produced a spectacular picture (Fig. 4) of non-tidal sea level fluctuations. This 8th order filter produces no phase shifts, has excellent gain characteristics, and effectively suppresses fluctuations with periods less than 48 h.

These 3-12 day oscillations with ranges as great as 90 cm are mainly estuarine manifestations of non-local forcing. This forcing is predominantly in the form of continental shelf waves whose passage in the coastal ocean alternately fill and empty water into and out of estuaries such as North Inlet. Thus, if a system has little fresh water input, discharge measurements during one or a few complete tidal cycles can easily yield net import of water to the system from the coastal ocean. This depends on the phase of the meteorological cy-

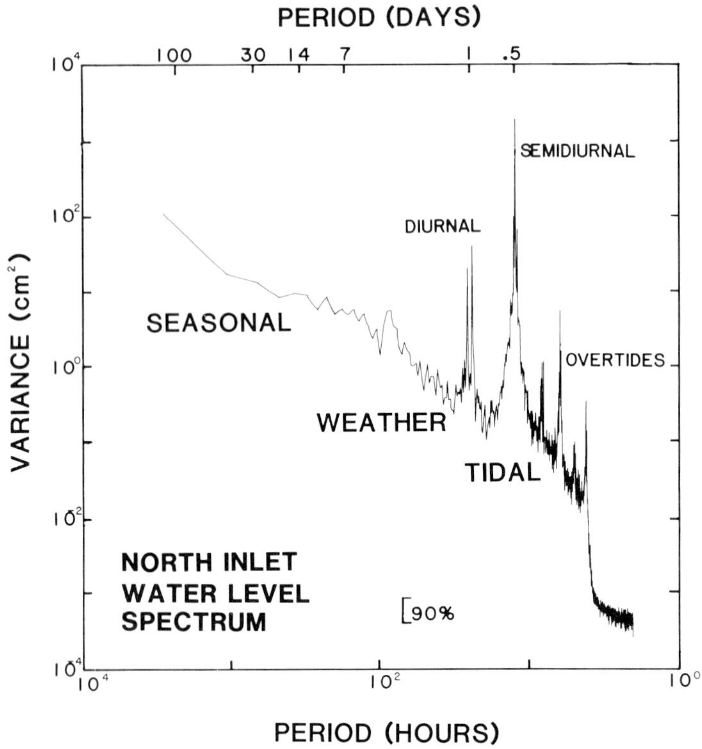

Figure 3. *Variance spectrum of the North Inlet sea level variations based on hourly data from three consecutive years. Spectral frequency and segment averaging was accomplished by accumulating five adjacent spectral estimates for each of the three years, yielding 30 degrees of freedom per estimate.*

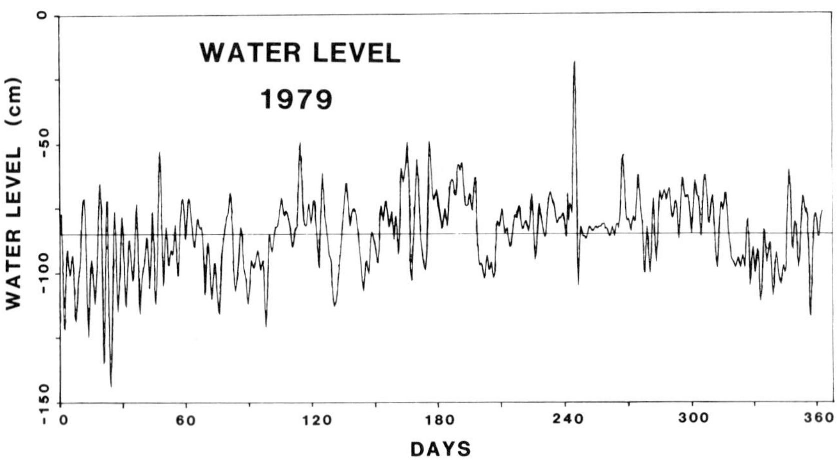

Figure 4. *North Inlet water level for 1979 after low-pass filtering, using an 8th-order Butterworth filter with zero phase shift and a 48 h half-gain period. The spike around Julian Day 250 is associated with passage of Hurricane David.*

cle in which the measurements were made. Thus, estuaries do not necessarily export a net amount of water every tidal cycle. It is reasonable to assume that these weather-induced sea level variations also affect the flux of water-borne materials through tidal inlets. Effects of the seasonal sea level oscillation have previously been discussed by Kjerfve and McKellar (1980).

Effects of Sampling Density and Sampling Rate

The estimation of material fluxes in estuarine cross-sections from simultaneous measurements of velocities and concentrations requires a great deal of effort. First, it is essential to select a suitable *sampling duration*. This is required to determine, with confidence, magnitudes and directions of resultant fluxes in an oscillatory tidal flow, where the resultant flux typically is one to three orders of magnitude smaller than the r. m. s. tidal flux. It is also essential to select a *sampling rate* and a *sampling density* such that any errors are within acceptable limits. This usually requires a compromise between that which is optimal and that which is practical.

Kjerfve et al. (1981) defined and considered the variation of a *relative percentage error*, γ, as sampling density varied. The computed fluxes were water, ammonium-nitrogen, and adenosine 5′-triphosphate (ATP). In each case when a station was occupied, vertical velocity profiles were measured between the water surface and bottom at the same time that water samples were drawn from several depths. The main conclusions of the analysis were:

1. it is necessary to choose a station density of at least one station for every 2×10^6 m^2 of cross-sectional area, computed as the ratio of estuarine tidal prism (P) to mean cross-sectional depth, for γ to be less than 15%;
2. station density as well as precise location of stations in a cross-section must be determined carefully, with at least one station occupying each bathymetric region of the cross-section, e. g. each channel or shallow region;
3. in the case of the 320 m wide Town Creek cross-section (A), three lateral stations, two located in channels and the third in the separating shallow region, were a compromise between the practical and the optimum.

In the case of water flux, the percentage error was defined (Kjerfve et al. 1981) as

$$\gamma = 100 \text{ D/P} \tag{2}$$

where

$$D = \Delta t \left\{ \sum_{k=1}^{n} (Q_{1k} - Q_{mk})^2 \right\}^{1/2} \tag{3}$$

and represents the r. m. s. variation of a particular discharge case. Q_{1k} is the instantaneous discharge for the ideal case at time k; Q_{mk} is the instantaneous discharge at time k for a case, m, utilizing a fewer number of stations than the ideal case; and Δt is the sampling rate. The computed tidal prism for a case m, P_m, was computed from

$$P_m = \Delta t \{ \sum_{k=1}^{n} |Q_{mk}| \} / 2r \tag{4}$$

where n is the total number of sampling times for case m and r is the number of tidal cycles. In the earlier percentage error analysis (Kjerfve et al. 1981), the ideal case consisted of 10 stations in the A cross-section, representing the largest number of stations that could logistically be sampled simultaneously every hour for two consecutive tidal cycles.

The percentage error compares a transport volume error to the tidal prism. The transported volume error is the r. m. s. deviation of the instantaneous flows for a given case, m, relative to the ideal case. The γ-value is an uppermost limit of the size of the error. The actual error can be expected to be an order of magnitude smaller.

Here, we only consider water flux and extend the previous analysis by investigating how γ changes as both sampling density and sampling rate change. The data set for the new ideal case consists of simultaneous velocity profiles made at 10 stations across Town Creek (section A), each sampled from surface to bottom every 0.5 h for three consecutive tidal cycles in November 1977 (see Kjerfve and Proehl 1979).

The net velocity distributions in the A section (after Kjerfve and Proehl 1979) for each of the three tidal cycles, are shown in Fig. 5. These data represent the ideal case in this analysis. The recomputation of the percentage error, γ, as both sampling density and rate vary is shown in Table 1. The sampling rate was varied from 0.5 to 2.5 h in half-hour incremental step increases. The selected cases are the ones yielding the smallest γ-value for a given number of stations.

This analysis indicates that γ is sensitive to both sampling rate and sampling density as can be expected. It appears, however, that improvement in γ as the sampling rate is decreased from 1.5 h to 1.0 h is very small, and if the rate is further decreased to 0.5 h, the gain is still small. When Δt increases beyond 1.5 h, especially with three or less stations in the cross-section, the sampling rate is too long as evidenced by a significant increase in γ. The effect of changing sampling density is particularly obvious as the number of stations decreases from six to five and from three to two. An optimum choice for continued flux sampling in the A section would have been case #9 with six stations sampled once every 1.5 h. The corresponding γ-value is only 3.7% with stations A2, A3, A6, A8, A9, and A10 (see Fig. 5) making up case #9. Unfortunately, six stations in the A cross-section represents too great a seasonal effort in establishing water and material budgets. Because of logistic constraints, three stations was the maximum feasible limit. The gain in sampling more frequently than once every 1.5 h is marginal. Thus, we selected case

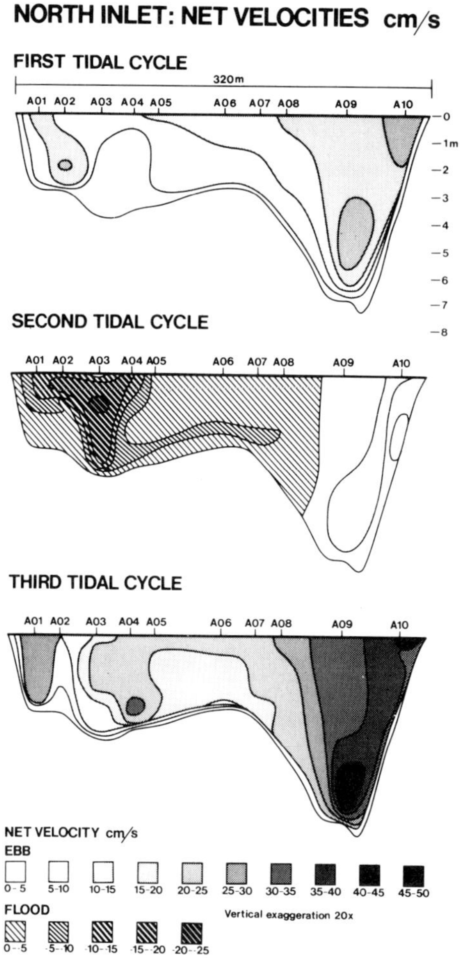

Figure 5. *Isotachs of time-averaged (net) currents for each of three consecutive tidal cycles in the A cross-section (after Kjerfve and Proehl 1979) in November 1977. Velocity profiles were measured simultaneously at 10 stations every 0.5 h from surface to bottom. This view shows the A cross-section from the ocean side.*

#26, consisting of stations A2, A6, and A9, for continued seasonal flux measurements in this cross-section. The corresponding γ-value is 19.9%.

Lateral Variations

North Inlet is the kind of well mixed system that most often would be treated analytically as a 1-dimensional estuary. This implies that concentrations and velocities are almost uniform in the cross-section. Boundary shear

Table 1. Percentage error, γ, for different sampling rates and number of stations in Town Creek cross-section (A), based on measurements during three consecutive tidal cycles in November 1977. The case numbers correspond to those listed in Kjerfve et al. (1981).

	Sampling Rate (h)				
	0.5	*1.0*	*1.5*	*2.0*	*2.5*
9 stations (Case #4)	2.9	4.3	5.0	5.7	5.4
8 stations (Case #6)	2.4	3.2	2.6	5.2	5.2
7 stations (Case #7)	3.1	4.2	3.8	6.9	5.3
6 stations (Case #9)	3.1	4.1	3.7	6.8	5.9
5 stations (Case #12)	8.6	12.0	13.6	15.4	19.0
4 stations (Case #19)	10.2	14.0	16.8	18.5	24.0
3 stations (Case #26)	12.9	17.7	19.9	25.4	27.6
2 stations (Case #28)	24.6	33.8	39.8	47.7	55.4
1 station (Case #29)	21.1	31.0	35.0	41.3	50.1

effects and other temporal-spatial variations are usually lumped into effective cross-sectional diffusivities in solving for material fluxes. However, Fig. 5 indicates that although North Inlet is classified as well mixed, there exist large variations in the cross-sectional velocity distribution. The lateral changes are particularly spectacular. The second tidal cycle in the A section (Fig. 5) appears to correspond most closely with the long-term tidally averaged flow distribution. Net ebb currents occur in the main channel and net flood currents in the secondary channel. This laterally reversing net flow has been discussed by Kjerfve (1978) and Kjerfve and Proehl (1979) and is called tidal pumping. It is caused by non-linear interactions between flow and estuarine bathymetry/geometry via field acceleration terms in the momentum balance equation (cf. Tee 1976).

Isotach plots of the tidally averaged currents in the South Jones cross-sections (B, C, and D; Fig. 6) and the North Jones section (F; Fig. 7) show somewhat different features. Whereas Town Creek on the average exports water, North Jones Creek typically imports water. The effect of this is that deep channel flow on the average is flood directed and the secondary region

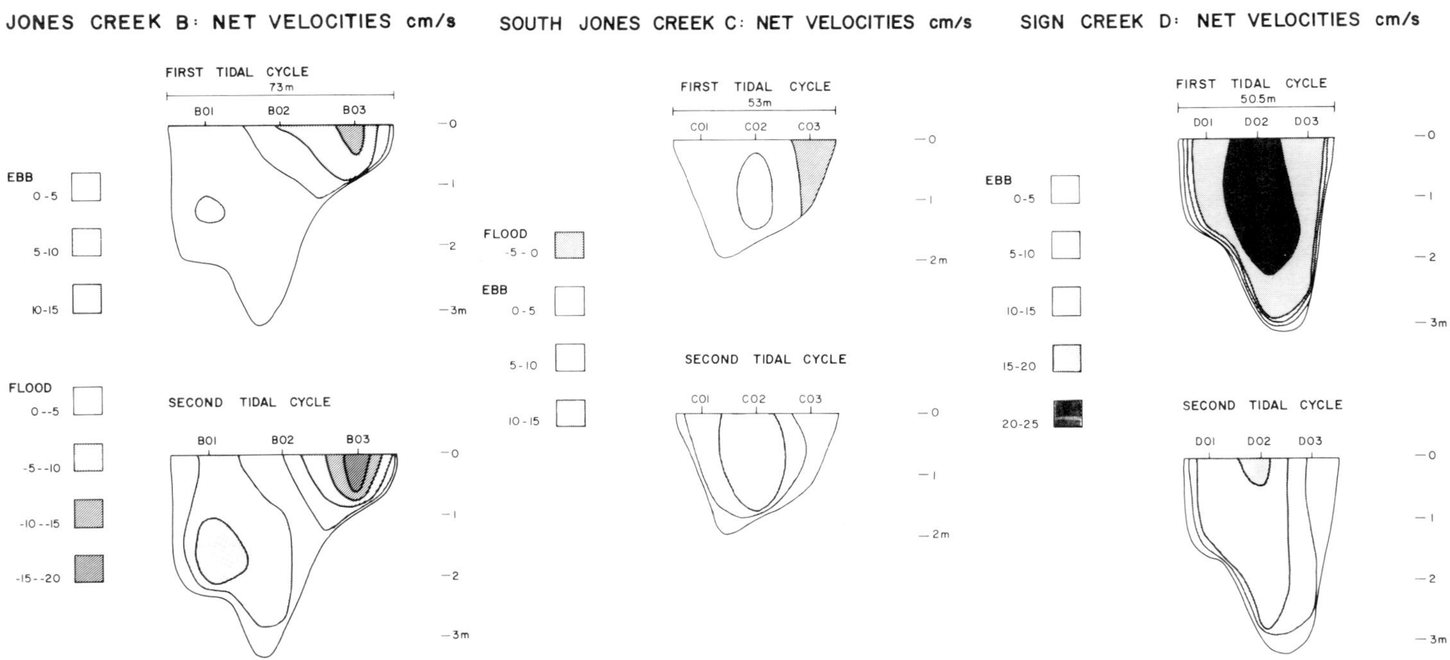

Figure 6. *Isotachs of net currents in the three South Jones cross-sections (B, C, and D) for the same two tidal cycles in March 1978. Each graph is based on velocity profiles measured at three stations in each cross-section every 0.5 h from surface to bottom. The cross-sections are viewed from Winyah Bay.*

experiences ebb directed net currents. This is directly opposite to the situation in Town Creek, although both cross-sections have flows that reverse directions laterally.

The three cross-sections in South Jones (Fig. 6) lack the complex lateral bathymetry and have no tidally averaged current reversals. Gravitational circulation is obviously absent and a 1-dimensional model treatment may be appropriate here. Maximum tidal currents in these sections are significantly less than maximum currents at A and F. It may be that the existence of lateral reversals in net currents can be predicted either based on the lateral bathymetry (Kjerfve 1978) or the flow characteristics.

The Reynolds number is a non-dimensional measure of dynamic flow stability. As the flow becomes more turbulent, the Reynolds number increases, and it appears that the lateral flow coherence breaks down. We define a lateral Reynolds number as

$$Re_{\ell} = P/\tau \nu h_o \tag{5}$$

where P is the tidal prism from eq. 4; τ is the duration of half a tidal cycle; ν is

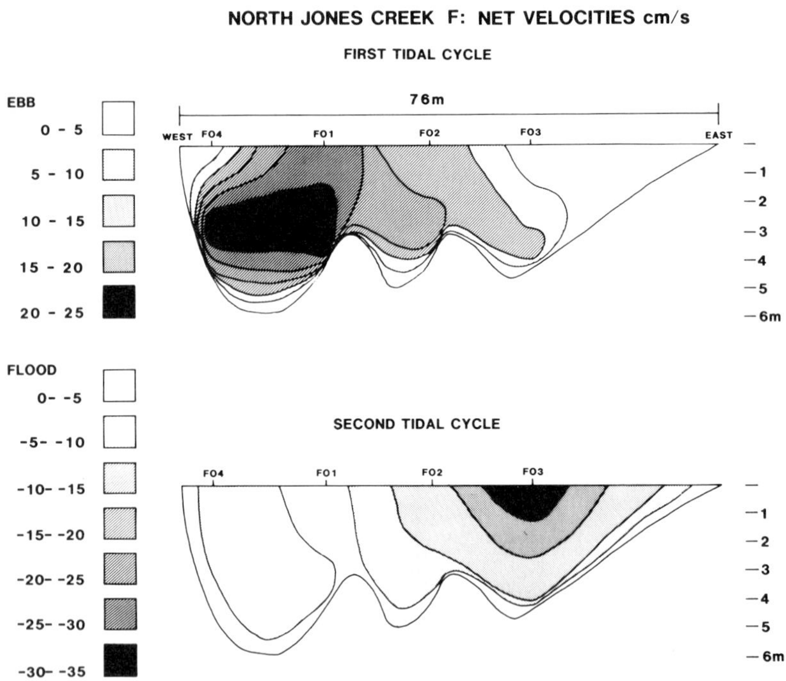

Figure 7. Isotachs of net currents in the North Jones cross-section (F) for two tidal cycles in June 1978 based on three and four lateral stations, respectively. In each case, velocity profiles were measured from surface to bottom simultaneously every 0.5 h. The cross-section is viewed from the North.

the kinematic viscosity, approximately constant at 10^{-6} m^2/s; and h_o is the cross-sectional and tidal averaged water depth. This definition of a Reynolds number implies that the cross sectional width is the critical length scale and that a representative tidal velocity, P/τ, is critical to the existence of single or multiple net current cores in a cross-section.

Computation of Re_ℓ yields 213×10^6 for cross-section A; 51×10^6 for F; 17×10^6 for B; and values 5 to 10×10^6 for C and D. Both cross-sections B and C were located close to a channel bend, and the development of the oppositely directed velocity cores in these sections (Fig. 6) may be due to the curvature. However, in the case of sections A and F, we suggest that the reason for the existence of oppositely directed velocity cores in the sections (Figs. 5 and 7) is due to a breakdown of the lateral flow coherence in the time-average. This may be due to the shallowness of the estuary, the pre-existence of large variations in lateral topography, or the dynamics of the system. In either case, the lateral Reynolds number is large. In fact, whenever Re_ℓ exceeds 15×10^6, a cross-section seems to develop more than one tide-averaged velocity core. When Re_ℓ is small only a single tide-averaged velocity core exists. This is the case in section D and in data from a Louisiana estuary (Kjerfve 1976) where Re_ℓ is 2×10^6. Thus, as Re_ℓ exceeds some critical number, maybe 15×10^6, the lateral coherence worsens and the net flow becomes confined into laterally positioned high velocity cores which may flow in either the same or opposite directions. This may serve as a useful guideline in determining necessary sampling density for flux measurements in an estuarine cross-section.

Flow Continuity

Because estuaries typically exhibit a great amount of both spatial and temporal variance, can water fluxes be measured directly with accuracy and precision? In particular, is it possible to determine tidally averaged fluxes with some degree of confidence? Tidally averaged fluxes are usually one or two orders of magnitude smaller than maximum instantaneous fluxes. This question is of special interest to the Outwelling Project in North Inlet, where a multidisciplinary investigation has been carried out since 1977 in an attempt to establish seasonal budgets for water and some 30 dissolved constituents. The approach has been to make simultaneous measurements of current velocities and material concentrations at an optimum or logistically practical number of locations in critical cross-sections in North Inlet (A, B, and F). The measurements were collected at a fixed sampling rate of 1.5 h for eight tidal cycles during four seasons. The present study was a preliminary calibration study to aid in the design of Outwelling.

The field experiment in March 1978 in South Jones Creek was conducted to test whether the sampling design and analysis techniques would yield flow continuity. By measuring velocities and calculating cross-sectional water flux every 0.5 h for two tidal cycles at sections B, C, and D, we hoped that the data would indicate that equal amounts of water moved through

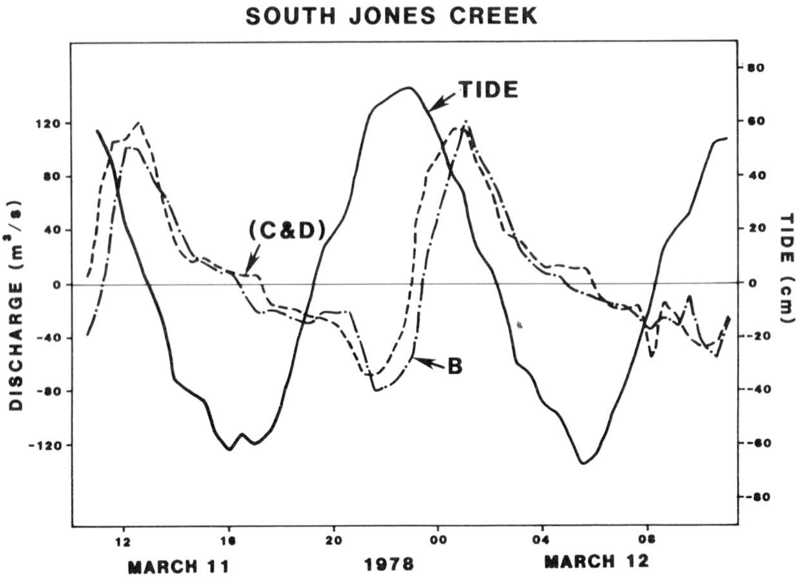

Figure 8. *Water flux and tide time series in South Jones Creek in March 1978. The discharge through cross-section B is plotted along with the combined discharge through sections C and D.*

cross-section B as compared to the sum of water fluxes through sections C and D. The flux and tidal water elevation variations are shown in Fig. 8. The covariation of the water fluxes is remarkably good.

The real test, however, concerns the time-means of the two flux series. We will test the null hypothesis that the flux means are equal, H_o: $\overline{Q}_B = \overline{Q}_{C+D}$. Initially, it is tempting to use a paired t-test. This, however, is not the correct test, as the water fluxes at a given time are not measures of the same water. B is separated from the other sections by approximately 0.7 km, the maximum current seldom exceeds 0.4 m/s, and flux lags of up to an hour between the sections should be expected. Figure 8 indicates that this indeed is the case. Thus, the appropriate test is the unpaired t-test, each series consisting of 50 data points. The means and standard deviations are $\overline{Q}_B = 3.44 \pm 49.24$ m^3/s and $\overline{Q}_{C+D} = 13.82 \pm 52.07$ m^3/s. An F-test indicates that the variances are equal, implying that the appropriate test statistic, t, is $t = (3.44\text{-}13.82)/50.67(2/50)^{1/2} = -1.02$ with 98 degrees of freedom. The tabulated critical t-value is $\pm$ 1.66 with 98 d. f. for $\alpha = 0.10$. We fail to reject the null hypothesis and conclude that the tidally averaged water fluxes are identical.

The simple field experiment is not the final answer to the feasibility of measuring small net estuarine fluxes with confidence. The calculations have only been carried out for water fluxes. However, concentrations generally vary little in cross-sections by comparison and the results are encouraging. The technique does yield flow continuity and thus should be applicable for seasonal budget measurements as long as a sufficient number of measurements are made in space and time.

Acknowledgments

This study was supported by National Science Foundation Grant No. DEP76-83010 and DEB80-04275 and is Contribution No. 433 from the Belle W. Baruch Institute for Marine Biology and Coastal Research. Sampling would not have been possible without the enthusiastic and competent field assistance by several hundred University of South Carolina undergraduate and graduate students, high school students from Spring Valley High School in Columbia, SC, undergraduates from Queen's College, Charlotte, NC, and other volunteers.

References Cited

Bendat, J. S. and A. G. Piersol. 1971. *Random Data: Analysis and Measurement Procedures*. John Wiley & Sons, New York. 407 pp.

Elliot, A. J. 1976. A study of the effect of meteorological forcing on the circulation in the Potomac estuary. *Special Report 56. Chesapeake Bay Institute*. The Johns Hopkins University, Baltimore, MD 32 pp.

Hansen, D. V. and M. Rattray, Jr. 1966. New dimensions in estuary classification. *Limnol. Oceanogr.* 1:319-326.

Kjerfve, B. 1976. Circulation and salinity distribution in coastal Louisiana estuaries. *Contr. Mar. Sci.* 20:1-10.

Kjerfve, B. 1978. Bathymetry as an indicator of net circulation in well-mixed estuaries. *Limnol. Oceanogr.* 23:816-821.

Kjerfve, B. and H. N. McKellar, Jr. 1980. Time series measurements of estuarine material fluxes, pp. 341-357. *In:* V. S. Kennedy (ed.), *Estuarine Perspectives*. Academic Press, New York.

Kjerfve, B. and J. A. Proehl. 1979. Velocity variability in a cross-section of a well-mixed estuary. *J. Mar. Res.* 37:409-418.

Kjerfve, B., L. H. Stevenson, J. A. Proehl, T. H. Chrzanowski and W. M. Kitchens. 1981. Estimation of material fluxes in an estuarine cross-section: a critical analysis of spatial measurement density and errors. *Limnol. Oceanogr.* 26:325-335.

Officer, C. B. 1980. Box models revisited, pp. 65-114. *In:* P. Hamilton and K. B. McDonald (eds.), *Estuarine and Wetland Processes with Emphasis on Modelling*. Plenum Press, New York.

Palmer, M. A., B. Kjerfve, and F. B. Schwing. 1980. Tidal analysis and prediction in a South Carolina estuary. *Contr. Mar. Sci.* 23:17-23.

Roberts, J. and T. D. Roberts. 1978. Use of the Butterworth low-pass filter for oceanographic data. *J. Geophys. Res.* 83:5510-5514.

Tee, K. T. 1976. Tide-induced residual current, a 2-D nonlinear numerical tidal model. *J. Mar. Res.* 34:603-628.

US Geological Survey. 1981. Water resources data for South Carolina, water year 1980. Water-Data-Report SC-80-1. Columbia, SC 322 pp.

RELATIONSHIP BETWEEN PHYSICAL CHARACTERISTICS AND ORGANIC CARBON SOURCES AS A BASIS FOR COMPARING ESTUARIES IN SOUTHERN NEW ENGLAND

B. L. Welsh, R. B. Whitlatch and W. F. Bohlen

Marine Sciences Department
University of Connecticut
Avery Point, Groton, Connecticut

Abstract: In an attempt to compare estuaries holistically, we hypothesized that contributions of various carbon sources to total production were strongly correlated to estuarine physical attributes (eg., size, area: volume characteristics, or tidal characteristics). Based on a compilation of data from nine systems in southern New England, we found that only the area:volume ratio correlated with any of the production components, but that those relationships were very strong with respect to benthic production ($R^2 = 95\%$), total production ($R^2 = 92\%$), and the benthic:pelagic production ratio ($R^2 = 82\%$). We conclude that the relationships answer the need for an easily-obtained underlying physical parameter for comparing estuaries in terms of their autochthonous production components, and that these patterns are so strong that the area:volume ratio would have to be taken into consideration when seeking any additional relationships with other parameters.

Introduction

This paper addresses a quest for underlying physical relationships for comparing whole estuaries as unit ecosystems. We use the relative importance of various categories of organic carbon sources as the biological units for comparison, weighing them against physical characteristics which can be obtained from governmental charts and publications if they are not available in the scientific literature. We have confined our considerations to estuaries in southern New England.

The idea of linking organic carbon supply to physical characteristics of estuaries is not new. Ketchum (1954) hypothesized that the success of phytoplankton populations within estuaries would depend on the doubling rates of the particular species and the tidal exchange ratio of the estuary. Mann (1975) compared three estuaries in Nova Scotia and concluded that the relative contributions of phytoplankton and benthic macrophytes to total production depended on estuarine morphometry and the local hydrodynamics governed by river runoff, tides, and winds.

With respect to contributions to production by autochthonous carbon, Nixon (1980) reviewed a broad range of research dealing with the net ex-

ISBN 0-12-404070-5

changes of carbon between tidal marshes, estuaries, and coastal waters, and showed that the results from study to study were highly conflicting. Odum et al. (1979), using a theoretical approach to carbon import, hypothesized that the direction of net exchange depended on estuary configuration: 1) a basin connected to a larger body of water by a narrow, shallow channel would favor import, 2) a v-shaped basin gradually deepening and widening toward the mouth would favor export, and 3) an intermediate basin with a mouth partially constricted by bars and/or shallow sills might import or export depending on the relative strength of tidal and freshwater flows.

At present there appears to be no unifying concept by which to compare estuaries in terms of their total allochthonous and autochthonous carbon budgets. In this paper we explore the hypothesis that differences in abundance and relative dominance of carbon sources are strongly related to simple physical attributes such as size, area:volume characteristics, or tidal characteristics, which are readily available parameters. If basic relationships could be shown to exist, they would provide a valuable tool for describing, predicting, and comparing the ecological structures of estuaries. For these comparisons, we bound the estuaries at the midtide level, which includes intertidal mudflats but excludes marshes. This delineation is appropriate for New England estuaries where embayment areas are usually greater than marsh areas.

Site Selection

There appears to be no estuary in southern New England or elsewhere for which a complete carbon budget is available based on measured values.

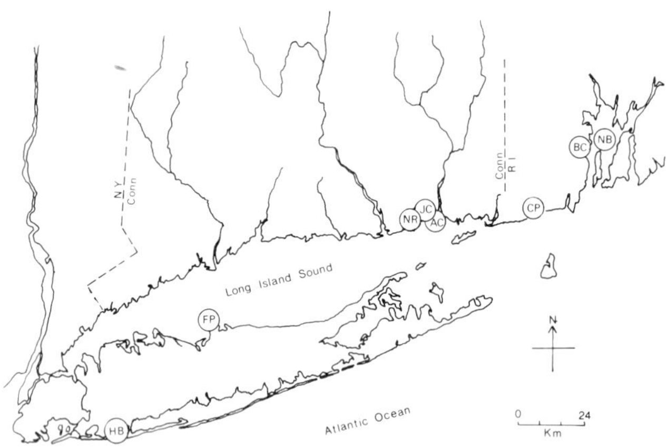

Figure 1. Location of estuaries (including Long Island Sound) compared in this study. HB = Hempstead Bay, FP = Flax Pond, NR = Niantic River, JC = Jordan Cove, AC = Alewife Cove, CP = Charlestown Pond, BC = Bissel Cove, NB = Narragansett Bay. Coastline runs from 74°00′ W, 40°33′ N to 71°10′W, 41°30′N.

We found only nine systems (Fig. 1) for which there appeared to be enough data to begin to construct comparisons. These systems vary widely in size and shape (Fig. 2; Table 1), but all conformed to categories (1) or (3) of the Odum et al. (1979) classification described in the Introduction. Since the original data were generated by different investigators using a variety of techniques, they required a number of manipulative procedures to convert them to a common production unit (g C $m^{-2}yr^{-1}$).

Methods of Budget Construction

Phytoplankton Production (Table 2)

Net phytoplankton production for Bissel Cove (180 g O_2 $m^{-2}yr^{-1}$; Nixon and Oviatt 1973) was converted to carbon using 0.31 g C (g O_2)$^{-1}$ (Oviatt et al. 1981). For Flax Pond, a total system estimate of 5.871×10^6 g C yr^{-1} based on ^{14}C (Moll 1974) was prorated over the embayment area of 10 ha (Brinkhuis 1975). Since Moll prorated over embayment plus marsh (50 ha), his area-weighted estimate of 12 g C $m^{-2}yr^{-1}$ was substantially lower than ours. Alewife and Jordan Cove chlorophyll concentrations were obtained for three stations representing upper, middle, and lower portions of each estuary (Welsh unpub.) using standard fluorescence procedures (Baillie and Welsh 1980). Alewife Cove data were available for April, May, July, August, October, and December, 1981, and January 1982, and were supplemented

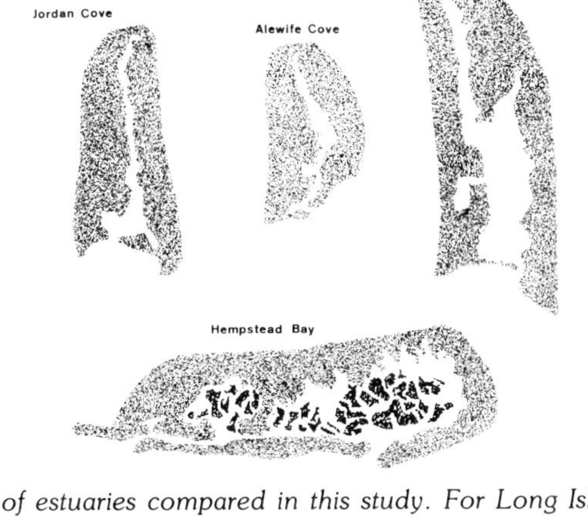

Figure 2. Shapes of estuaries compared in this study. For Long Island Sound and Narragansett Bay, refer to Fig. 1.

Table 1. Physical characteristics of selected estuarine embayments in southern New England. Charlestown Pond includes Ninigret and Greenhill Ponds. Exchange ratios are given as ranges of spring and neap conditions if known, otherwise as mean tidal exchange. References, by system, are: (1) Nixon and Oviatt 1973; (2) Woodwell et al. 1977; (3) (4) Welsh and Whitlatch 1980, Welsh 1980; (5) Kollmeyer 1972; (6) Nixon and Lee 1980; Thorne-Miller and Harlin 1982; (7) Udell et al. 1969; (8) Hicks 1959; (9) Riley 1952.

Estuary	*Size (ha)*[a]	*Area:Volume*[b]	*Exchange Ratio*[c]	*Marsh:Embayment*
(1) Bissel Cove	0.66	4.0	0-0.59	2.6
(2) Flax Pond	10	1.6	0.92	5.3
(3) Alewife Cove	18	1.2	0.20-0.40	0.55
(4) Jordan Cove	48	0.83	0.41-0.50	0.18
(5) Niantic River	335	0.62	0.34-0.45	0.025
(6) Charlestown Pond	800	0.90	0.088	0.14
(7) Hempstead Bay	4654	0.50	0.25	0.56
(8) Narragansett Bay	32,413	0.11	0.12-0.15	0.020
(9) Long Island Sound	318,540	0.05	0.08	0.022

[a] area bounded by mean tide level; [b] mean low water volume; [c] tidal volume/(tidal volume + mean low water volume)

with March data from a 14 station survey in 1977. Intervening months were estimated by interpolation. Jordan Cove data were available for April, May, July, and August, 1981. Annual production was assumed to be in the same proportion to those estimates as it was for Alewife Cove, which lies 2 km to the east and was sampled synoptically. The Niantic River estimates were based on chlorophyll concentrations sampled biweekly at five stations between April and December, 1970-1972 (D. A. McGill, U. S. Coast Guard Academy, unpub.) and analyzed by fluorometry (Strickland and Parsons 1972). They were supplemented with data from Marshall and Wheeler (1965) for the January-March period.

Production was estimated from the above chlorophyll data by applying the Eppley (1972) relationships for maximum net production per unit chlorophyll based on temperature and the cellular C:Chl ratio. Field temperatures and a cellular C:Chl ratio of 30:1 to 60:1 were used. This maximum rate was assumed to correspond to the 4 h midday rates described in Oviatt et al. (1981), and their factor of 1.75 was used to correct the 4 h rate to net daytime production. Nighttime respiration was assumed to be 10% of net daytime production (Peterson 1980).

Production for Charlestown Pond was taken as the mean of two estimates, 51 g C $m^{-2}yr^{-1}$ (Smayda 1962) and 32 g C $m^{-2}yr^{-1}$ (Nixon and Lee 1980), both based on chlorophyll and *in situ* ^{14}C incubations. Estimates for Narragansett Bay (Oviatt et al. 1981) were also based on chlorophyll and *in situ* ^{14}C incubations. Those for Hempstead Bay (Udell et al. 1969) and Long Island Sound (Riley 1972) were based on chlorophyll and *in situ* oxygen metabolism.

Macrophyte Production (Table 2)

Benthic macrophyte production for Bissel Cove (4814 kcal $m^{-2}yr^{-1}$;

Table 2. Autochthonous production budgets for southern New England estuaries. B:P is the total benthic production:phytoplankton ratio. All estimates are area-weighted. Units are g C m^{-2} yr^{-1}.

Estuary	*Phytoplankton*	*Benthic Macrophytes*	*Benthic Microalgae*	*Total Production*	*B:P*
Bissel Cove	56	712	52	820	13.6
Flax Pond	60	307	52	419	6.0
Alewife Cove	162	141	45	348	1.1
Jordan Cove	66	218	41	325	3.9
Niantic River	72	180	32	284	2.9
Charlestown River	42	233	41	316	6.5
Hempstead Bay	177	42	62	281	0.59
Narragansett Bay	242	106	–	348	0.44
Long Island Sound	205	–	–	–	–

Nixon and Oviatt 1973) was converted to carbon using 0.80 organic:total weight as determined in our laboratory by combustion and 0.53 organic carbon:organic dry weight (Zieman and Wetzel 1980). The Flax Pond values (Brinkhuis 1975) were based on growth measurements of fucoids and *Ulva lactuca*, which are estimated to comprise about 90-95% of total macroalgal production (Brinkhuis, pers. comm.).

Alewife and Jordan Cove production was estimated from areal distributions of the major plant communities, *Zostera marina*, *Ulva lactuca* and an *Enteromorpha* spp./ *Cladophora* spp. complex (Welsh 1980). *Zostera* biomass in channel areas with dense beds was estimated at 570 gdw m^{-2}, the mean of a range of 80-1120 gdw m^{-2} reported for Charlestown Pond (Thorne-Miller et al. 1982). This estimate is low compared with Long Island systems (760 gdw m^{-2}; Burkholder and Doheny 1968). For basin areas with thinner beds we used 116 gdw m^{-2} based on field measurements of above-ground biomass in Alewife Cove (Welsh unpub.). Associated macroalgae were 446 g m^{-2} in the latter measurements, but we used 200 g m^{-2} based on field measurements from *Zostera marina* beds of similar density in Charlestown Pond (Thorne-Miller et al. 1982) because the large import of drifting macroalgae to Alewife Cove probably inflated our measurement. *Zostera marina* production was calculated as 2.09 times its biomass based on the mean of reported P/B ratios (1.1 - 2.8; Thorne-Miller and Harlin 1982). The basin estimate was corrected for below-ground production using a factor of 1.5 (Thorne-Miller and Harlin 1982). For the associated algae, mostly *Gracilaria* spp, we used a P/B ratio for that genus of 1.1 (Thorne-Miller and Harlin 1982). Dry weight production was converted to carbon using an organic:total weight ratio of 0.81 and a C:organic weight ratio of 0.46 (Zieman and Wetzel 1980). Biomass of *Ulva lactuca* was estimated at 600 gdw m^{-2} based on our measured value in Alewife Cove of 598 gdw m^{-2}*Ulva lactuca* with 28 gdw m^{-2} of associated macroalgae. Biomass for the *Enteromorpha/Cladophora* complex was estimated at 250 gdw m^{-2} based on early-season measurements in Alewife Cove of 100 gdw m^{-2} (Welsh and Whitlatch 1980) and the lower end of the density range in Charlestown Pond (412-1240 gdw m^{-2}; Thorne-Miller et al. 1982). A production:biomass ratio of 3 (Thorne-Miller and Harlin 1982) was applied and the dry weight units were converted to carbon as for Bissel Cove. Niantic River estimates were based on areal distributions of *Zostera marina* determined from aerial photographs. They were converted to carbon production by the above procedures.

Charlestown Pond estimates (Thorne-Miller and Harlin 1982) were based on growth, including underground portions, and on areal surveys. They were converted to carbon by the procedure stated above. Estimates for Hempstead Bay were derived from standing crops (Udell et al. 1969) using the P/B ratio of 3 (Thorne-Miller and Harlin 1982). For Narragansett Bay, estimates for *Laminaria* spp. were taken from Brady-Campbell (1982) based on growth. For other macroalgae, standing crop estimates were obtained from M. Harlin (Botany Dept., U. Rhode Island, unpub.), and conversions to carbon production were made as outlined above. There are no known published estimates for benthic production in Long Island Sound.

Benthic Microalgal Production (Table 2)

Winter and summer estimates were calculated as the means of November-April (17.2 ± 9.9 mg C $m^{-2}hr^{-1}$; n = 12) and May-October (58 ± 39.6 mg C $m^{-2}hr^{-1}$; n = 11) based on ^{14}C measurements by Marshall (1970). These measurements are the only data available for southern New England. The values were converted to daily net production following the rationale developed for phytoplankton, i.e. the values were applied as a 4 h mid-day rate equivalent to 57% of net daytime production (Oviatt et al. 1981), then a correction of 10% was made for nighttime respiration (Peterson 1980). The results were carried over 182 days per season yielding a total winter production of 20 g C m^{-2} and a summer production of 65 g C m^{-2}. The estimate is close to Marshall's annual estimate of 90 g C m^{-2}, which did not correct for nighttime respiration. It is low with respect to our estimate for epipelic algae on intertidal mudflats (129 g C $m^{-2}yr^{-1}$; Baillie and Welsh 1980). The winter and summer rates allowed us to apply these values on a seasonal basis to areas free of dense macrophyte growth and within the 1.5 m depth contour, which was taken as an arbitrary limiting depth.

Allochthonous Contributions (Table 3)

Estimates for Bissel Cove were derived from Nixon and Oviatt (1973) using their conversion factors of 3.3 kcal g^{-1} and 0.5 g C g^{-1} for *Spartina alterniflora*. Flax Pond estimates were obtained from Woodwell et al. (1977). For

Table 3. Allochthonous contributions (g C m^{-2} yr^{-1}) to southern New England estuaries. T = terrestrial debris; M = marsh grass; S = seaweed; POC = unidentified particulate organic carbon; DOC = dissolved organic carbon. All estimates are area-weighted to whole estuary or to inner basins only if labelled (I).

	Source of Materials			
Estuary	*Freshwater*	*Lateral*	*Marine (Tidal)*	*Net Total*
Bissel Cove	3.8 (DOC)	36 (M)	–	40
Flax Pond	–	780-1521 (M)	61-349 (POC)	841-1870 (POC)
			– (8-48) (DOC)	793-1822
Alewife Cove	0.34 (T)	8 (T)	5.2 (S)	14
		16 (T,I)		
Jordan Cove		10 (T)		10
		22 (T,I)		
Niantic River		5 (T)		5
		10 (T,I)		
Charlestown Pond		15 (M)	– 19 (POC)	– 4
Hempstead Bay		82 (M)	–	82

marine imports, the lower estimate represents a minimum value because it assumes even deposition over the embayment plus marsh. The higher estimate assigns all import to the embayment. For contributions from the marsh, it was assumed that only the tall *Spartina alterniflora* was imported to the embayment (Nixon and Oviatt 1973), and its biomass was calculated as 75% of the total for the species (Woodwell et al. 1977). That percentage was based on area, so it probably underestimates the tall *Spartina alterniflora* biomass. The lower estimate corrects for assumed export to Long Island Sound, although that estimate was not a direct measurement but one derived by subtracting estimates of deposition and decomposition within the Pond (Woodwell et al. 1977).

For Charlestown Pond and Hempstead Bay, estimates were based on biomass (Oviatt et al. 1977; Udell et al. 1969) assuming that tall *Spartina alterniflora* constituted 49% of total marsh production. This percentage, representing the mean of 23% for Bissel Cove (Nixon and Oviatt 1973) and 65% for Flax Pond (Woodwell et al. 1977), was considered a good compromise for systems which contained both upstream pocket marshes and marshes near their outlets.

Alewife Cove imports were based on collections of benthic detritus (Welsh and Whitlatch 1980). Only terrestrial materials and brown seaweeds which were known to originate outside the estuary were included. Dry weights were converted to organic weight using organic:total weight ratios of 0.62 for terrestrial litter and 0.80 for macroalgae (Whitlatch unpub.). Carbon was calculated from organic weights using a carbon:organic weight ratio of 0.5.

Terrestrial imports for Jordan Cove and the Niantic River were calculated by prorating lateral imports for Alewife Cove (96% of total additions; Welsh and Whitlatch 1980) over the linear measurement of wooded shoreline, then applying that import rate to the wooded shorelines in the other systems and prorating over their basin areas.

Results and Discussion

Autochthonous Comparisons

Despite the wide variety of sources, the data, when reduced to common units, showed a remarkable consistency (Table 2). Total production varied by a factor of less than 3, which is consistent with the uniformity reported in a comparison of world-wide lagoon systems (Nixon 1982) and Texas bay systems (Odum 1967). Within this range, however, there were some strong patterns of correlation (Table 4). Total production was strongly coupled to benthic macrophyte production (R^2 = 90%), which, on the average, constituted 62% of the budget. Of the three physical parameters tested, only the area:volume ratio was significantly related to production components ($p<.05$). Area:volume ratio produced strong correlations with benthic production, total production and the benthic:pelagic ratio (Fig. 3). This result is not surprising, since the area:volume ratio is the reciprocal of

Table 4. Regressions between components of production and physical factors reaching a significance level of 5%. R^2 is the coefficient of determination. Production units are g C m^{-2} prorated over the entire estuary.

Variables (Y vs X)	*Equation*	R^2
1. Benthic Production vs Area-Volume Ratio	$Y = 74 + 171X$	95%
2. Total Production vs Area-Volume Ratio	$Y = 220 + 141X$	92%
3. Total Production vs Benthic Macrophytes	$Y = 194 + 0.82X$	90%
4. Benthic:Pelagic Ratio vs Area-Volume Ratio	$Y = 0.346 + 3.3X$	82%
5. Total Production vs Benthic:Pelagic Ratio	$Y = 238 + 35X$	75%

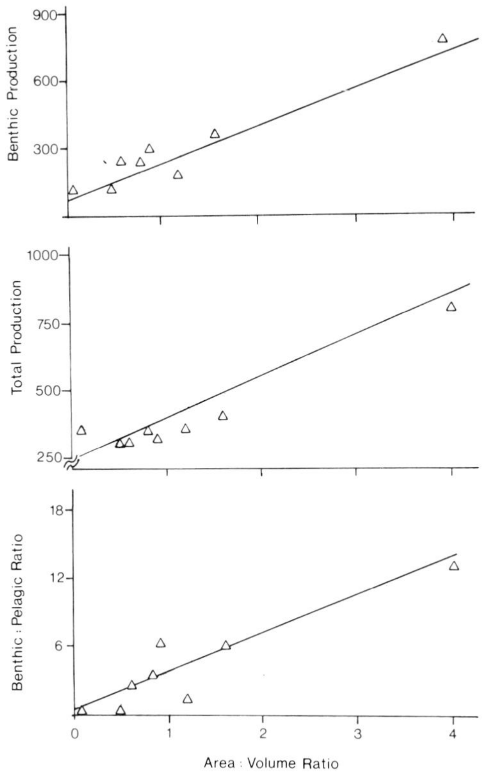

Figure 3. Correlations of benthic production, total production, and benthic:phytoplankton production ratios with area:volume ratio. Production units are g C $m^{-2}yr^{-1}$. Equations and R^2 are given in Table 4. To identify systems, refer to the area:volume ratios in Table 1 (Long Island Sound not included).

mean depth, a parameter already used to characterize lake systems (Reid 1961).

It might be argued that these results are artifacts of the calculations, since the original investigators obtained their estimates from a combination of growth measurements and areas of habitat. For all the estuaries except Alewife and Jordan Coves and the Niantic River, however, either production or standing crop was obtained independently. In all the systems the areal distribution was obtained from direct observations of plant type, not from the bathymetric analysis used for computing the area:volume ratio. Only the benthic microalgal estimates were based on bathymetry, and while they did constitute a supplement to benthic macrophyte performance in each estuary, their contributions averaged only 16% of the benthic term and their correlation with the area:volume ratio was not significant. As a final argument, phytoplankton production, with a calculation tied directly to volume and area terms, did not correlate significantly with the area:volume ratio.

The relationships between production components and the area: volume ratio appear to satisfy the quest for a predictor variable which is easily obtained. For components which deviate strongly from the relationship (Alewife Cove phytoplankton production, for example), they urge reexamination of the adequacy of the data base or a search for unusual conditions within the system. Where no data exist, such as Long Island Sound benthic production, they supply predictions of 83 g C $m^{-2}yr^{-1}$ based on the benthic plant equation and 106 g C $m^{-2}yr^{-1}$ based on the benthic:pelagic ratio equation (Table 4). Assuming a 5 m photic zone (Riley and Schurr 1959), which is 46.2% of total Sound area (Riley 1952), the prediction would require intensive productivities averaging 180-229 g C $m^{-2}yr^{-1}$ over the lighted areas. These are entirely reasonable based on the benthic plant contributions in Table 2.

Moreover, these relationships, though purely empirical, have strong implications for production in these estuaries. The influence of shallow water appears to hold, regardless of the shape of the system and despite the low freshwater input. This finding is further evidence that the driving force for production is regeneration of nutrients from the sediments, as in the accounts of Nixon et al. (1976) and Zeitzschel (1980).

The strength of these relationships is remarkable given the wide variation in morphology and size of the systems. We are not denying potential influences of other parameters, such as freshwater inflow, tidal height, or morphology (Mann 1975) on production or its distribution within the estuary. By confining our comparisons to the southern New England region, and by dealing with whole systems, we have normalized these factors to a certain extent. We would expect the coefficients of the equations of Table 4 to vary from region to region. The strength of the correlations, however, show that the components of estuarine production are under strong physical control so that influences of weaker factors would have to be tested with respect to the area:volume influence.

Allochthonous Contributions

In contrast to the strong relationships found within a relatively weak data set for autochthonous production, the allochthonous budget provides few patterns. Quantitative estimates of imported organic carbon to the selected estuaries are fragmentary (Table 3). Contributions by freshwater inflow appear to be negligible, if the two available measurements are at all representative. Lateral imports from marshes or upland terrestrial sources appear to be far greater. Marsh imports from Rhode Island systems (Bissel Cove and Charlestown Pond) were only 2-4% of autochthonous production, whereas those to Long Island systems (Hempstead Bay and Flax Pond) were 30-36%. Flax Pond had both the highest marsh:embayment ratio (Table 1) and the highest import values, but our sample size is too small to demonstrate any consistency between marsh imports and marsh:embayment ratios.

Lateral imports of terrestrial litter are known only for the Connecticut estuaries, and all quantification was done in Alewife Cove (Welsh and Whitlatch 1980). For Jordan Cove, we have qualitative observations from a previous study which provide some additional support (Welsh 1978). These imports were only 2-3% of autochthonous production, but they may exert a disproportionately strong influence on the inner basins of these estuaries where they become trapped and decompose slowly. Whitlatch (unpub.) has determined that oak, which comprises the bulk of the imports (Welsh and Whitlatch 1980) requires 3-5 years to decompose, which is about 33% longer than in adjacent freshwater areas. About 50% of the bottom debris was material from previous years. There does not appear to be a high BOD associated with the leaf debris (P. Herring, U. Connecticut, unpub.), probably because of its slow decomposition rate.

Marine imports appear to make a substantial contribution only to Flax Pond, where they reach 14-83% of autochthonous production. As noted in the Introduction, the quantification of tidal fluxes is a controversial issue (Nixon 1980). There are indications, however, that tidal import of particulates could be substantial for southern New England estuaries. Current velocity profiles for estuaries in this region characteristically exhibit strong time and velocity asymmetries between flood and ebb tides and a strong attenuation of energy between mouth and inner basins as shown in Fig. 4 (Welsh and Whitlatch 1980). These characteristics occur to different degrees in Bissel Cove (Nixon and Oviatt 1973), coastal ponds of Rhode Island (Nixon 1981), Flax Pond (Woodwell et al. 1977), Jordan Cove (Welsh 1978), and the Niantic River (Kollmeyer 1972). The asymmetries have been modelled as a phenomenon called tidal choking, which is characteristic of basins with small constricted inlets (Glenne et al. 1971).

The differences in velocities alone suggest that there is a category of larger-sized particles which can be transported into the estuary but not back out again. In addition, during the period of slackening velocities at the end of flood tide, an additional size-group of particles will be deposited. If data from Alewife Cove are representative, then throughout most of the inner basin areas, under normal conditions, sheer velocities on the ebb never reach the minimum required for resuspension of even the most readily resuspended inorganic or organic materials (Inman 1949; Sternberg 1971; Fisher et al.

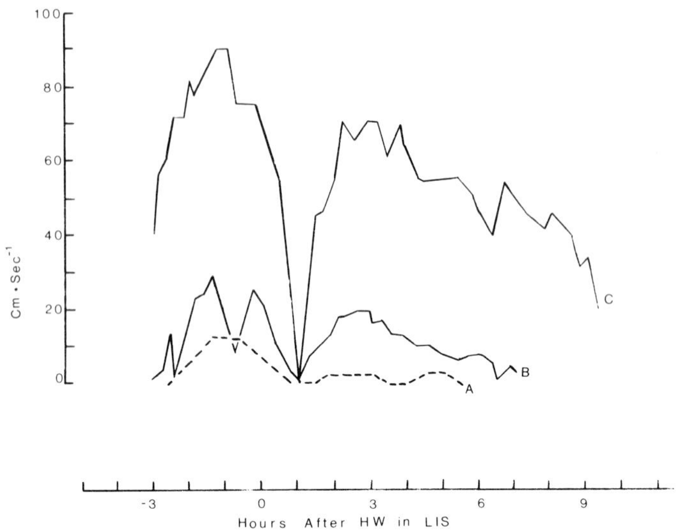

Figure 4. *Current velocity profiles in Alewife Cove during a spring tide. C = mouth of Cove, B = entrance to middle basin area of Cove, A = entrance to inner basin area of the Cove; HW = high water; LIS = Long Island Sound.*

1979; Dyer 1980). Again, using Alewife Cove profiles, the size-classes of particles expected to be deposited through this mechanism include many biogenic categories such as fecal pellets and vascular plant debris (Haven and Morales-Alamo 1968; Fisher et al. 1979). Deposits of allochthonous macroalgae are found in Alewife Cove (Table 3) and are reported for Flax Pond (Woodwell et al. 1977), providing additional evidence that the estuary serves as a settling basin for tidal waters as well as freshwater runoff.

Conclusions

Based on these comparisons we conclude that the relationship between simple physical characteristics and autochthonous carbon sources is strong, and that the area:volume ratio provides that strong single predictor. The case for the relationships between allochthonous contributions and physical predictors is weak despite the basic notion that there ought to be a relation; that they ought to relate to increasing shoreline:area ratios in smaller systems, or to marsh:estuary ratios, or to asymmetrical tidal regimes.

This latter weakness appears to stem from the fragmentary and inconsistent data for making these comparisons, despite the attention that particulate flux studies have received in recent years. The prospects for securing estimates with any degree of precision appear to be dim, given the methodological problems (Kjerfve and Proehl 1979; Nixon 1980). It appears that our characterization of southern New England systems as settling basins for runoff waters, lateral input, and tidal waters is qualitatively correct, but the phenomenon may be a regional one related to our small, relatively young estuaries, moderate tidal range, low freshwater inflow, and rising sealevel. Southern New England is not unique, however, because similar conditions have been observed in other northern temperate areas such as the estuaries bordering the North Sea (Postma 1961).

Acknowledgments

We are grateful to M. Harlin, V. Lee, N. Marshall, S. Nixon, C. Oviatt, and B. Thorne-Miller for their advice and assistance in constructing budgets for the Rhode Island systems; to B. H. Brinkhuis and P. H. Rich for assistance and aerial photos of Flax Pond; and L. Stewart for providing aerial photos and D. A. McGill for sharing unpublished chlorophyll data for the Niantic River system. The authors take sole responsibility, however, for the final budgets and any errors therein. We also thank C. Hall and V. Kennedy for their helpful review of the manuscript, and D. A. Johnson for computer analyses and drafting.

References Cited

Baillie, P. W. and B. L. Welsh. 1980. The effect of tidal resuspension on the distribution of intertidal epipelic algae in an estuary. *Est. Coastal Mar. Sci.* 10:165-180.

Brady-Campbell, M. K. 1982. An assessment of the productivity of two species of *Laminaria* in Narragansett Bay and Rhode Island Sound. MS Thesis, U. Rhode Island, Kingston, RI

Brinkhuis, B. H. 1975. Seasonal variations in macroalgae productivity in a temperate salt marsh. Ph.D. Dissertation, State U. New York, Stony Brook, NY.

Burkholder, P. R. and T. E. Doheny. 1968. The biology of eelgrass. Publ. Dept. Conservation and Waterways. Town of Hempstead, Long Island, NY. 120 pp.

Dyer, K. R. 1980. Velocity profiles over a rippled bed and the threshold of movement of sand. *Est. Coastal Mar. Sci.* 10:181-201.

Eppley, R. W. 1972. Temperature and phytoplankton growth in the sea. *Fish. Bull.* 70:1063-1085.

Fisher, J. S., J. Pickral and W. E. Odum. 1979. Organic detritus particles: initiation of motion criteria. *Limnol. Oceanogr.* 24:529-531.

Glenne, B., C. R. Goodwin and C. F. Glanzman. 1971. Tidal choking. *J. Hydraul. Res.* 9:321-333.

Haven, D. S. and R. Morales-Alamo. 1968. Occurrence and transport of fecal pellets in suspension in a tidal estuary. *Sed. Geol.* 2:141-151.

Hicks, S. D. 1959. The physical oceanography of Narragansett Bay. *Limnol. Oceanogr.* 4:316-327.

Inman, D. 1949. Sorting of sediments in the light of fluid mechanics. *J. Sed. Petrol.* 19:51-70.

Ketchum, B. H. 1954. Relation between circulation and planktonic populations in estuaries. *Ecology* 35:191-200.

Kjerfve, B. and J. A. Proehl. 1979. Velocity variability in a cross-section of a well-mixed estuary. *J. Mar. Res.* 37:409-418.

Kollmeyer, R. C. 1972. A Study of the Niantic River Estuary, Niantic, Connecticut. Phases I and II. Physical and Chemical Aspects of the Niantic River Estuary. Tech. Report RDCGA 18, U.S.C.G. Headquarters, Washington, DC 78 pp.

Mann, K. H. 1975. Relationship between morphometry and biological functioning in three coastal inlets of Nova Scotia, pp. 634-644. *In:* L. E. Cronin (ed.), *Estuarine Research, Vol. 1.* Academic Press, New York.

Marshall, N. 1970. Food transfer through the lower trophic levels of the benthic environment, pp. 52-66. *In:* J. H. Steele (ed.), *Marine Food Chains*, U. California Press, Berkeley, CA.

Marshall, N. and B. M. Wheeler. 1965. The role of coastal and upper estuarine waters contributing phytoplankton to the shoals of the Niantic estuary. *Ecology* 46:665-673.

Moll, R. A. 1974. The phytoplankton community of a temperate zone salt marsh. Ph. D. Dissertation, State U. New York, Stony Brook, NY 122 pp.

Nixon, S. W. 1980. Between coastal marshes and coastal waters. – A review of twenty years of speculation and research on the role of salt marshes in estuarine productivity and water chemistry, pp. 437-525. *In:* P. Hamilton and K. B. MacDonald (eds.), *Estuarine and Wetland Processes*. Plenum Press, New York.

Nixon, S. W. 1982. Nutrient dynamics, primary production and fisheries yields of lagoons. Proceedings, International Symposium on Coastal Lagoons, Bordeaux, France, Sept. 1981. (In press).

Nixon, S. W. and V. Lee. 1980. The flux of carbon, nitrogen and phosphorus between coastal lagoons and offshore waters. *In:* Coastal Lagoons: Present and Future Research, Part II. Unesco Tech. Papers Mar. Sci. (in press).

Nixon, S. W. and C. A. Oviatt. 1973. Ecology of a New England salt marsh. *Ecol. Monogr.* 43:463-498.

Nixon, S. W., C. A. Oviatt and S. S. Hale. 1976. Nitrogen regeneration and the metabolism of coastal marine bottom communities, pp. 269-284. *In:* J. M. Anderson and A. Macfadyen (eds.), *The Role of Terrestrial and Aquatic Organisms in Decomposition Processes*. Blackwell Scientific Publ., Oxford.

Odum, H. T. 1967. Biological circuits and the marine systems of Texas, pp. 99-157. *In:* T. A. Olson and F. J. Burgess (eds.), *Pollution and Marine Ecology*. Interscience Publishers, New York.

Odum, W. E., J. S. Fisher and J. C. Pickral. 1979. Factors controlling the flux of particulate organic carbon from estuarine wetlands, pp. 69-80. *In:* R. J. Livingston (ed.), *Ecological Processes in Coastal and Marine Systems*. Plenum Press, New York.

Oviatt, C. A., B. Buckley and S. Nixon. 1981. Annual phytoplankton metabolism in Narragansett Bay calculated from survey field measurements and microcosm observations. *Estuaries* 4:167-175.

Oviatt, C. A., S. W. Nixon and J. Garber. 1977. Variation and evaluation of coastal salt marshes. *Environmental Management* 1:201-211.

Peterson, B. J. 1980. Aquatic primary production and the ^{14}C-CO_2 method: a history of the productivity problem. *Ann. Rev. Ecol. Syst.* 11:359-385.

Postma, H. 1961. Transport and accumulation of suspended matter in the Dutch Wadden Sea. *Neth. J. Sea Res.* 1:148-190.

Reid, G. K. 1961. *Ecology of Inland Waters and Estuaries*. Van Nostrand Co., New York. 375 pp.

Riley, G. A. 1952. Hydrography of the Long Island and Block Island Sounds. *Bull. Bingham Oceanogr. Coll.* 13:5-39.

Riley, G. A. 1972. Patterns of production in marine ecosystems, pp. 91-112. *In:* J. A. Weins (ed.), *Ecosystem Structure and Function*. Oregon State U. Press, Corvallis, OR

Riley, G. A. and H. M. Schurr. 1959. Transparency of Long Island Sound waters. *Bull. Bingham Oceanogr. Coll.* 17:66-85.

Smayda, T. J. 1962. Some quantitative aspects of primary production in a Rhode Island coastal salt pond, pp. 123-125. *In:* D. S. Gorsline (ed.). Proceedings 1st National Coastal and Shallow Water Research Conference, NSF and ONR, Tallahassee, FL

Sternberg, R. W. 1971. Incipient motion of sediment particles in the environment. *Mar. Geol.* 10:113-119.

Strickland, J. D. H. and T. R. Parsons. 1972. *A Practical Handbook of Seawater Analysis*. Fish Res. Bd. of Canada, Bull. 167.

Thorne-Miller, B. and M. M. Harlin. 1982. Macrophyte production and biomass fluctuations in shallow coastal lagoons in Rhode Island, USA. *Aquat. Bot.* (in review).

Thorne-Miller, B., M. M. Harlin, G. B. Thursby, M. M. Brady-Campbell and B. A. Dworetzky. 1982. Macrophyte distribution and standing crop biomass in shallow coastal lagoons in Rhode Island, USA. *Aquat. Bot.* (in review).

Udell, H. F., J. Zarudsky, T. E. Doheny and P. R. Burkholder. 1969. Productivity and nutrient values of plants growing in the salt marshes of the town of Hempstead, Long Island. *Bull. Torrey Bot. Club* 96:42-51.

Welsh, B. L. 1978. The effect of reduced wetlands and storage basins on the size, stability and productivity of the watershed mixing zone. Tech. Report, Inst. of Water Resources. U. Connecticut, Storrs, CT 43 pp.

Welsh, B. L. 1980. Jordan Cove: its hydraulic character, sediment and macrophyte distributions as related to shoaling. Tech. Report, Marine Sciences Inst., U. Connecticut, Storrs, CT 85 pp.

Welsh, B. L. and R. B. Whitlatch. 1980. A study of Alewife Cove: its siltation, eutrophication and hydraulic character. Tech. Report, Marine Sciences Inst., U. Connecticut, Storrs, CT 102 pp.

Woodwell, G. M., R. A. Houghton, C. A. S. Hall, D. E. Whitney, R. A. Moll and D. W. Juers. 1977. The Flax Pond ecosystem study: the annual metabolism and nutrient budgets of a salt marsh, pp. 491-511. *In:* R. L. Jeffries and A. J. Davy (eds.), *Ecological Processes in Coastal Environments*. Blackwell Scientific Publ., Oxford.

Zeitzschel, B. 1980. Sediment-water interactions in nutrient dynamics, pp. 195-218. *In:* K. R. Tenore and B. C. Coull (eds.), *Marine Benthic Dynamics*. U. South Carolina Press, Columbia, SC

Zieman, J. C. and R. G. Wetzel. 1980. Productivity in seagrasses: methods and rates, pp. 87-116. *In:* R. C. Phillips, and C. P. McRoy (eds.), *Handbook of Seagrass Biology: An Ecosystem Perspective*. Garland STPM Press, New York.

A COMPARATIVE ANALYSIS OF NUTRIENTS AND OTHER FACTORS INFLUENCING ESTUARINE PHYTOPLANKTON PRODUCTION

W. R. Boynton
Chesapeake Biological Laboratory
University of Maryland
Solomons, Maryland

W. M. Kemp
Horn Point Environmental Laboratories
University of Maryland
Cambridge, Maryland

and

C. W. Keefe
Chesapeake Biological Laboratory
University of Maryland
Solomons, Maryland

Abstract: We reviewed data concerning phytoplankton production, chlorophyll *a*, and associated physical and chemical variables from 63 different estuarine systems. Data were analyzed statistically to test hypotheses regarding algal productivity and factors regulating temporal patterns. Prior to statistical analysis, estuarine systems were classified into four groups based on criteria of physical circulation and geomorphology. Analysis of grouped data indicated that algal production and biomass were consistently high in warm periods of the year in a broad spectrum of estuaries and that ratios of available nitrogen to phosphorus were low during periods of high production, except in highly eutrophic systems. In general, phytoplankton production and biomass exhibited weak correlations with a variety of physical and chemical state variables, perhaps indicating the significance of rate processes as opposed to standing stocks in regulating these important features of estuarine systems. A six-year time series of measurements of algal production and chlorophyll *a* at stations in middle Chesapeake Bay exhibited considerable year-to-year variability, with a three-fold range in peak values. Summertime maxima were strongly related to annual loadings of both nitrogen (N) and phosphorus (P) but annual production appeared to be sustained primarily on recycled nitrogen and phosphorus. To generalize from these findings, N and P loading rates were estimated for 14 different estuarine systems, and a significant positive relationship was obtained between phytoplankton production and nitrogen (but not phosphorus) inputs.

Introduction

Information concerning the magnitude and seasonal patterns of phytoplankton production in estuarine ecosystems has been accumulating at an

ISBN 0-12-404070-5

ever-increasing rate over the last 50 years, particularly with the development of various automated analytical techniques. Phytoplankton production is of central importance in estuarine environments because of its role in supporting many food webs (Haines 1976; Thayer et al. 1978; Kemp and Boynton 1981). Strickland (1960) emphasized this in his suggestion that the effects of a lack of interest in algal production would "not be unlike that which would prevail if livestock raisers lacked knowledge of, or interest in, pasture conditions." Thus, developing a clear understanding of factors regulating phytoplankton production in various estuaries is an important goal.

Despite the extensive number of productivity and associated physical and chemical measurements currently available, the degree to which the factors regulating productivity are understood varies considerably among different types of aquatic systems. It appears that more success has been attained in developing manageable paradigms concerning regulation of algal production in lakes. For example, in a variety of lakes, strong correlations have been obtained between annual plankton production and phosphorus and nitrogen loading (Schindler 1978; Smith 1979), while Fee (1979) reported a good regression of annual production and an index of benthic nutrient recycling potential. Brylinsky and Mann (1973) were reasonably successful in predicting production as a function of insolation for a global sampling of lakes.

In contrast to the synthetic efforts published on lakes, little comparable work has been produced for estuarine areas. Williams (1972) concluded that phytoplankton production in some shallow estuaries along the southeastern coast of the U.S. tended to follow the seasonal cycle in water temperature and that available nitrogen (rather than phosphorus) commonly limited production. Riley (1972) compared the magnitude of productivity in selected estuarine, coastal, and marine ecosystems and concluded that the rate of supply of nutrients to surface waters was of overwhelming importance. Recently, Nixon (1981) reviewed productivity patterns in a group of estuarine systems and inferred that high rates of production common to estuarine systems were the result of relatively complete and rapid heterotrophic nutrient recycling. While the conclusions reached in these studies are useful in developing concepts of estuarine productivity, the number of estuaries reviewed was small in all cases, and hence, the generality of the findings remains uncertain.

The purpose of this paper is to examine a comparative data base compiled from 63 estuaries and to consider, by statistical inference, the relative importance of various physical and chemical factors influencing phytoplankton production and biomass in these estuarine systems. Specifically, we have (1) analyzed this data base for relationships between production and various physical and chemical factors, (2) summarized temporal patterns of productivity as related to algal biomass and nutrient status, (3) examined year-to-year variability in phytoplankton production at one location in a particular estuary, and (4) explored influences of nutrient additions from the watershed on the magnitude and temporal patterns of algal production.

Analytical Approach and Data Sources

In developing the data base, we used as a guide the conceptual model of factors influencing phytoplankton production shown in Fig. 1. While somewhat more specific, our approach was not unlike that adopted by Brylinsky and Mann (1973) in their review of factors influencing productivity in lakes. In brief, inputs of energy and materials or physical characteristics (morphology) common to all estuarine systems are shown in circles. The rectangles in the center of the diagram represent the mechanisms through which the inputs affect primary production. For example, turbidity is shown as influencing primary production and, in turn, is caused by algal biomass and sediments from both external (riverine and others) and internal (resuspension due to wind and tides) sources. Ultimately, these variables are shown to influence the rate of primary production and the generation of algal biomass.

The specific variables for which we obtained quantitative values are listed in the caption of Fig. 1. These variables were selected because data

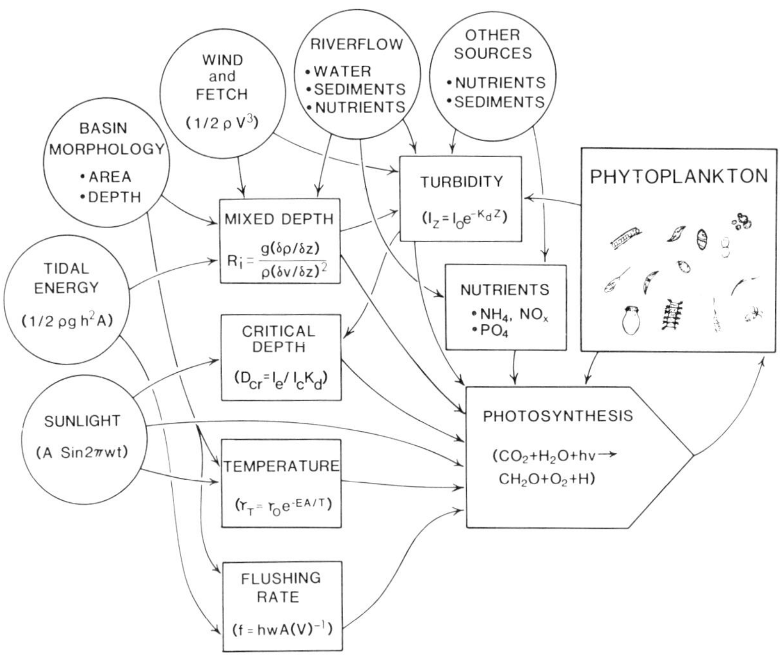

Figure 1. A conceptual model indicating sources of materials and mechanisms influencing phytoplankton production rates in estuarine ecosystems. Specific variables for which data were collected included latitude, insolation, temperature, extinction coefficient, mean depth, stratification depth, mixed depth, critical depth, surface area, drainage area, freshwater input, tidal range, salinity, nutrient concentrations, nutrient loading rate, chlorophyll a *concentration, and phytoplankton production rate.*

were available for many estuarine systems and were consistent with the idea that both internal (state variables) and external (inputs) factors potentially play important regulatory roles. Moreover, if an approach requiring more detailed information were adopted, then a comparative study would not have been possible simply due to a lack of data. This analysis assumes at the outset that data commonly collected in phytoplankton production studies have relevance in understanding the emergent patterns and magnitudes of algal production.

Details concerning the data base are described elsewhere (Keefe et al. 1981) but a few explanatory comments are appropriate here. Data concerning the variables listed in Fig. 1 were obtained from a wide variety of sources including journal publications, research reports, unpublished data, and personal communications. In many cases, multiple sources were available for a particular estuary and these were carefully cross-checked for consistency. In cases where there were conflicts, authors were contacted so that a more direct resolution was possible. After reviewing this material we concluded that we could maximize the number of estuarine systems included, keep the analysis tractable, and still adequately address issues concerning major temporal patterns if we dealt with seasonal mean values rather than more intensive samplings. Because of this, the range in a variable reported here is equal to or less than the maximum range of a variable reported in the original publication. In any case, our data base is composed of four values for each variable (where available).

Variables were also differentiated according to depth and, where appropriate, values from surface and bottom portions of the water column were entered in the data base. Methodologies concerning conversions of oxygen-based productivity data to carbon equivalents, expansion of volumetric productivity rates to areal rates, and formulas used in calculating flushing time are given in Keefe et al. (1981). The estuaries considered in this analysis, along with literature citations, are given in the appendix table.

Classification of Estuaries

Qualitative inspection of the data base and results of preliminary correlation analyses led us to conduct further analyses on groups of estuaries with similar characteristics rather than on the entire data base. The considerable similarity of primary production rates among estuaries with widely varying physical and chemical attributes suggested that there were either more general features of estuarine systems sustaining high production rates, as Nixon (1981) has suggested, or that there were multiple mechanisms which are active in different estuaries.

To avoid obscuring important relationships present in some (but not all) estuaries, we assigned each site to one of four categories (fjord, lagoon, embayment, or river dominated). *Fjords* were defined as having a shallow sill and deep basin waters with a slow exchange with adjacent sea waters (Pritchard 1955). *Lagoons* were taken as those systems which are shallow, well-mixed, slowly flushed, and only slightly influenced by riverine inputs. *Embayments* were considered to be deeper than lagoons, often stratified, only slightly influenced by freshwater inputs, and having good exchange with the ocean. The category, *river dominated*, contains a more diverse group of systems, but all in this group exhibit seasonally depressed salinities due to riverine inputs and variable degrees of stratification. The quantitative criteria used for depth, salinity, stratification, flushing, tidal amplitude, and river flow in this classification system are given in Keefe et al. (1981).

While it was tempting to use traditional classification systems based on physiography or stratification and circulation (i.e. Pritchard 1955; Hansen and Rattray 1966), we consider this grouping more appropriate because it includes (directly or indirectly) the primary mechanisms known to influence phytoplankton production, including flushing rate, nutrient inputs and light regime. Ultimately, we were more interested in a scheme which considers the interaction of biological processes with a host of physical and chemical factors, rather than a classification which considers physical and physiographic features only.

The statistical strength of these groupings was examined with the aid of a step-wise discriminate analysis (Fig. 2), a multivariate technique in which weighted, linear combinations of variables are used to force groups to be as distinct as possible. The distribution of groupings in discriminate space agrees

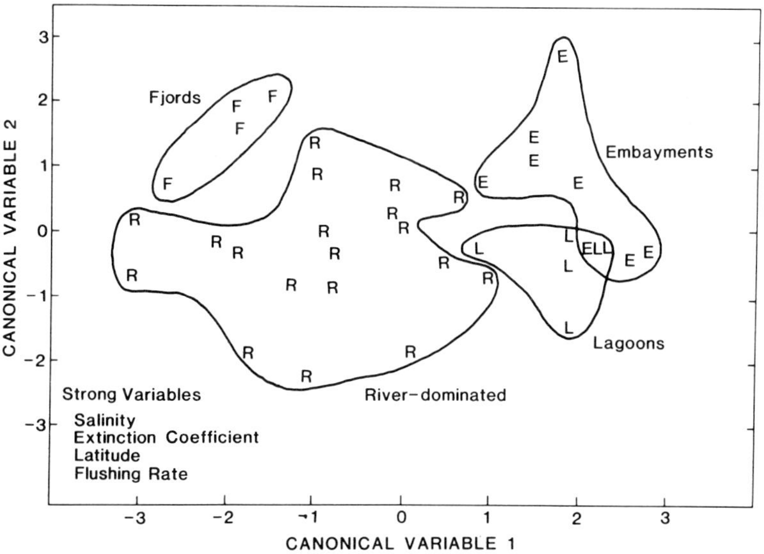

Figure 2. Results of a typical discriminate analysis where four types of estuarine systems are depicted in discriminate space. Solid lines enclosing each group are provided as a visual aid and do not represent statistically significant differences between groups.

reasonably well with the classification criteria. Fjords are clearly quite different from either lagoons or embayments, but river dominated estuaries are loosely clustered, which is to be expected in view of the widely ranging criterion of river flow within this single classification. We allowed the discriminate analysis program to survey the data base and select variables which contributed most strongly to separating the groups. Important variables entering the analyses were salinity, extinction coefficient, latitude, and flushing rate, suggesting the importance of some physical attributes in determining characteristics of various estuarine types. Neither phytoplankton production, chlorophyll *a*, nor nutrient concentrations entered the analysis, which is consistent with our earlier observation that in widely disparate estuaries similar rates of production and algal biomass are often observed.

Production, Biomass, and N:P Ratios

Mean daily rates of primary production were available for 45 of the 63 estuarine systems (Fig. 3). To our knowledge, this represents the most comprehensive review currently available. Average seasonal rates ranged from near zero to 2.5 g C m^{-2} d^{-1}. The mean of all annual average rates was 0.52 g C m^{-2} d^{-1} (190 g C $m^{-2}yr^{-1}$), a value intermediate between the 100 g C m^{-2} yr^{-1} reported by Ryther (1963) for coastal areas and the 300 g C m^{-2} yr^{-1} estimated for upwelling areas. Despite the large range in freshwater input, physical morphology, insolation and other factors, maximum production rates always occurred during the warm periods of the year (June-Sept.) while minimum rates generally occurred during the winter. This pattern has been interpreted to indicate that temperature-regulated metabolism strongly influences nutrient recycling processes and planktonic growth rates which are, in turn, important factors maintaining high photosynthetic rates (Williams 1966; Flemer 1970; Eppley 1972; Nixon 1981). Recent calculations concerning the coupling of nutrients and production in Chesapeake Bay indicate that recycling is also dependent on a source of labile organic matter which comes mainly from phytoplankton production, again indicating the tight coupling of heterotrophic and autotrophic processes (Kemp et al. 1982).

While there are considerable differences in the physical structure of the systems shown in Fig. 3, the range of production values observed within one group overlaps those in other groups. Annual means for river dominated estuaries, embayments, lagoons, and fjords were 0.58±0.37, 0.36±0.23, 0.49±0.23 and 0.62±0.53 g C m^{-2} d^{-1}, respectively. In short, estuaries with different physical characteristics commonly have comparable rates of primary production, suggesting that there are system-specific biotic and physical mechanisms operative in different estuaries.

Annual average concentrations of chlorophyll *a* were available for 39 estuarine systems and these data are shown in Fig. 4. Seasonal ranges are shown by the horizontal bars and the chlorophyll *a* concentration at the time of maximum productivity is indicated by a solid triangle. While this sampling is

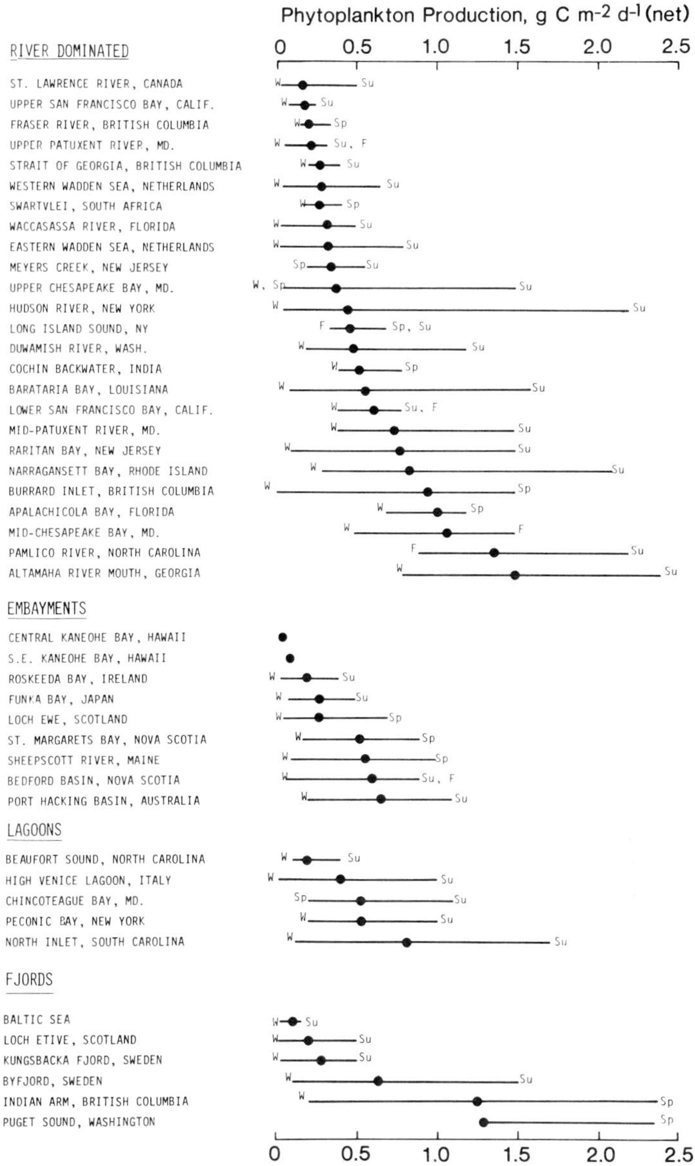

Figure 3. *Summary of average daily phytoplankton production rates (solid dot) in 45 estuarine systems. Horizontal bars indicate annual ranges. Season in which maximum and minimum rates occurred is also indicated (W, winter; Sp, spring; Su, summer; F, fall).*

not as extensive as that for production, several interesting patterns emerge. In most cases, chlorophyll *a* concentrations were highest during the warm seasons of the year (May-Oct.) and, in all but one case, chlorophyll *a* concen-

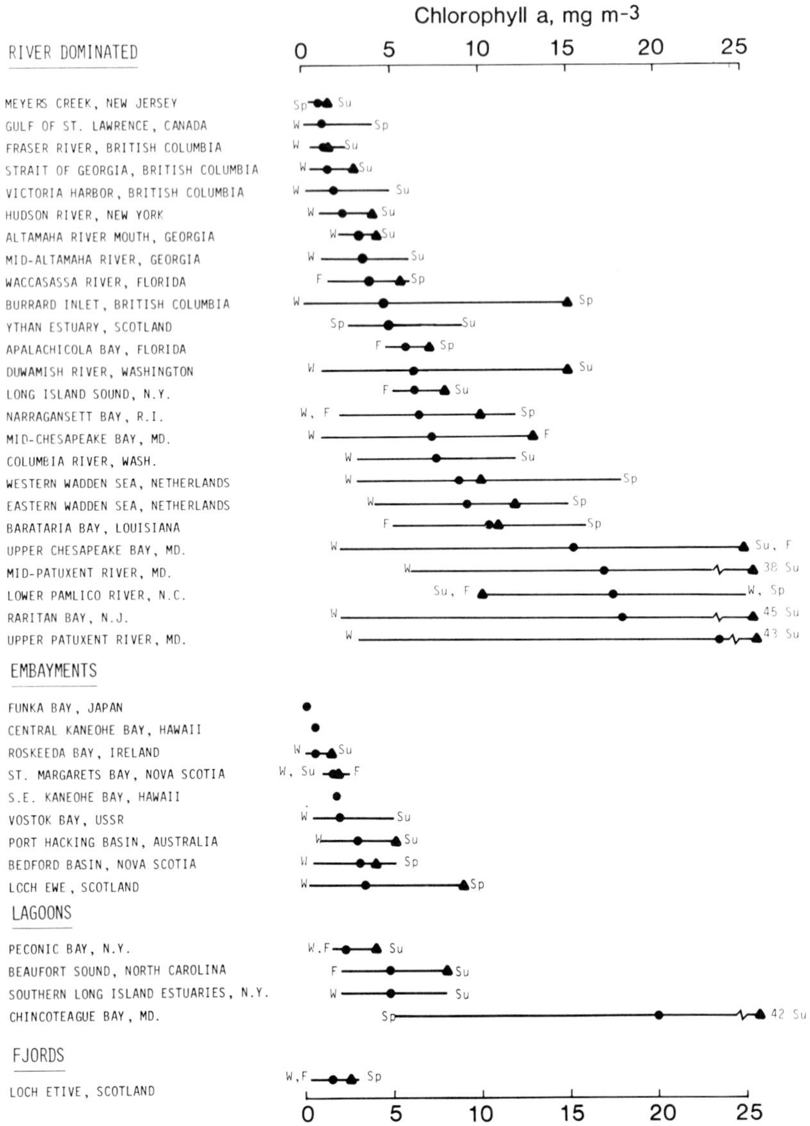

Figure 4. *Summary of chlorophyll* a *concentrations in 39 estuarine systems. Annual ranges and seasons in which maximum and minimum concentrations occurred are indicated as in Fig. 3. Solid triangle indicates chlorophyll* a *concentration at time of maximum production.*

trations were well above the annual mean at the time when peak productivity occurred. Thus, there is a reasonably strong pattern wherein chlorophyll *a* concentrations parallel those of phytoplankton production. Finally, there were some substantial differences in mean chlorophyll *a* levels within some of the groups. Chlorophyll *a* concentrations were clearly higher in some river dominated systems suggesting that those estuaries with strong couplings to riverine nutrients develop the highest standing stocks.

Ratios of dissolved inorganic nitrogen (DIN = $NO_3^- + NO_2^- + NH_4^+$) to dissolved inorganic phosphorus (DIP) are shown for 28 estuarine systems in Fig. 5. The horizontal line represents annual ranges while the solid triangle in-

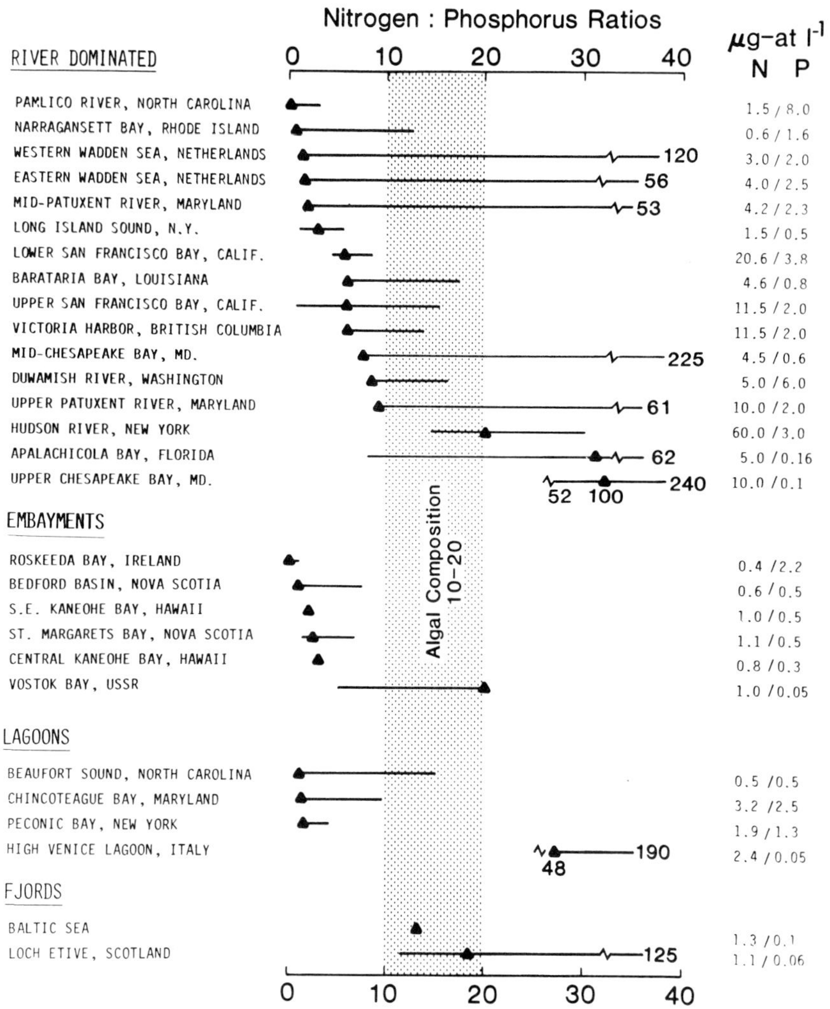

Figure 5. *Summary of DIN:DIP ratios (seasonal mean ratios) from 28 estuarine ecosystems. Horizontal bars indicate the annual ranges in DIN:DIP ratios; solid triangle represents ratio at time of maximum productivity. Absolute concentrations (μg-at ℓ⁻¹) at time of peak production are on the right. Vertical band represents the typical range of algal composition ratios.*

dicates the ratio at the time of peak production. The significance of this ratio lies in the fact that algal production is constrained by (among other things) the need for nitrogen and phosphorus in proportions (atomic) of 16:1, respectively (Redfield 1934). Water column nutrient concentrations of DIN:DIP less than the "Redfield Ratio" indicate that nitrogen is less abundant than phosphorus in terms of phytoplankton demand while values in excess of the ratio indicate that phosphorus is less abundant. Thus, if phytoplankton demand was sufficient to utilize available nutrients, either N or P would be expended first, depending on the ambient ratio. Studies by Parsons et al. (1961) and Rhee (1978) indicated that there was some variability in algal composition ratios and hence it may be more appropriate to view composition ratios as ranging from 10:1 to 20:1.

Over the past decade, many studies (e.g. Ryther and Dunstan 1971) have concluded that nitrogen rather than phosphorus is more likely to limit production in coastal and estuarine systems. The data in Fig. 5 support this notion in that nitrogen is consistently less abundant than phosphorus during periods of peak productivity in a wide variety of estuarine ecosystems. In most cases, those that do not follow this pattern are heavily enriched by point and diffuse nutrient sources throughout the year (e.g. High Venice Lagoon, Hudson River). On the right side of Fig. 5, actual concentrations of DIN and DIP at the time of peak production are shown. Clearly, actual concentrations vary considerably between various estuaries and it is an open question whether these concentrations are limiting. However, nitrogen enrichment in estuarine areas often stimulates algal growth, indicating that despite relatively high ambient concentrations, nitrogen limitation of phytoplanktonic production can occur (Ryther and Dunstan 1971; Williams 1972; Goldman et al. 1973; Thayer 1974).

Also shown in Fig. 5 are the substantial annual excursions observed in N:P ratios, particularly in the river dominated group. Commonly, N:P ratios were quite high in such river dominated systems as Chesapeake Bay during periods of high river flow (>60:1) and the high ratios were due to very high concentrations of nitrogen (>50 μg-at ℓ^{-1}) rather than low concentrations of phosphorus. However, by late summer (period of peak production), ratios were dramatically reduced in most cases, as were concentrations in the water column. The above suggests that there are active mechanisms which rapidly remove DIN from or add DIP to the water column. In terms of nitrogen, several authors have found denitrification to be a significant loss term in estuarine nitrogen budgets (Seitzinger et al. 1980; Kemp et al. 1982). In addition, we have found that a substantial fraction (20–30%) of the nitrogen input to some estuaries may be buried in the sediments as particulate organic nitrogen (Boynton et al. 1981) and thus be unavailable to the water column. The seasonal decline in N:P ratios may also be caused by summer inputs of phosphorus from anoxic sediments as reported in several systems (Taft and Taylor 1976; Boynton et al. 1980).

Relationships between Phytoplankton Production, Biomass, and Environmental Variables

Investigators have routinely measured several environmental variables in addition to phytoplankton production, generally with the intention of conducting statistical tests on these data in order to infer which variables may influence production rates. We performed a set of statistical tests designed to explore correlative relationships between phytoplankton production and chlorophyll *a*, and several physical, chemical, and biological variables in a broad spectrum of estuaries to determine if reasonably good and generalized predictors emerged from this set of variables (Table 1).

Few of these commonly measured variables were significant predictors of phytoplankton production and biomass. Of all correlations attempted, 73% were non-significant, 18% were significant but accounted for less than 50% of the variability in the dependent variable (production or biomass), and only 9% accounted for more than 50% of the variability (but never more than 75%). Single factor correlations were more successful when applied to embayments, lagoons, and fjords than in river dominated systems, but it is not clear whether this is a reflection of the larger and more diverse group of river dominated estuaries analyzed or if phytoplankton productivity and biomass regulation is dominated by fewer variables in the former set of systems. In any case, when viewed as a group (i.e. lagoons, embayments, fjords, and river dominated systems) very few of these environmental data were good predictors of phytoplankton production or biomass. There were even fewer significant correlations when all estuarine systems were included as a single group in the analysis.

Perhaps these results are not surprising. As Walsh (1976) has said, "simple consideration of single control factors of . . . primary production such as light, nutrients, mixing . . . can lead to rather naive and perhaps inaccurate conceptualizations of the dynamics of non-linear aquatic ecosystems." Thus, we are reminded once again of the complexity of estuarine ecosystems. Nevertheless, this correlative analysis provides some sense of appropriate directions for further analysis. It appears that we should either utilize a multivariate approach with these state variables or consider rate variables such as nutrient recycling or nutrient inputs.

Variability in Annual Estuarine Production

To this point we have emphasized inter-estuarine comparisons. We now turn to the issue of temporal variability at a single estuarine location. How much year-to-year variability might be expected? Do temporal patterns of production change between years? If there is significant year-to-year variability, what factors might be responsible?

In this regard, there is a considerable literature concerning phytoplankton responses to changing environmental conditions which are physiological in nature and occur over time periods of from minutes to days (e.g. McCarthy

Table 1. Results of correlation analysis between phytoplankton production and biomass, and several physical, nutrient, and biological variables. (+) and (−) signs represent significant positive and negative correlations, respectively. Large and small signs represent r²'s greater and lesser than 50%, respectively. Surface-bottom gradients were calculated as the difference between surface and bottom nutrient concentrations divided by the difference in surface and bottom salinity.

LOCATION	PHYSICAL FACTORS								NUTRIENTS												
									SURFACE CONCENTRATIONS mg-at m^{-3}				TOTAL WATER COLUMN mg-at m^{-2}				SURFACE-BOTTOM GRADIENTS See legend				
	DEPTH, M	TEMPERATURE, C	SUNLIGHT, LANGLIES	EXTINCTION, M^{-1}	MIXED/MEAN DEPTH	CRITICAL/MIXED DEPTH	RIVER FLOW/AREA, M DAY^{-1}	FLUSHING TIME, DAYS	NO_3	NH_4	DIN	PO_4	NO_3	NH_4	DIN	PO_4	NO_3	NH_4	DIN	PO_4	CHLOROPHYLL, MG M^{-3}
RIVER DOMINATED																					
PRODUCTION (SEASONAL)	+	−	−	+		−	+		+		+		+								−
PRODUCTION (ANNUAL)		−	−	+					+		+										
CHLOROPHYLL (SEASONAL)	+	−	−	−	−		+	+					+	+	+						
CHLOROPHYLL (PEAK)				−	−	+	+											−			
EMBAYMENTS																					
PRODUCTION (SEASONAL)			−	−	+	−								−	−						−
PRODUCTION (ANNUAL)					+									−	−						
CHLOROPHYLL (SEASONAL)			−									+	+	−							
CHLOROPHYLL (PEAK)							−		−		−			−							
LAGOONS																					
PRODUCTION (SEASONAL)		−	−	−	+					−		−			−						−
PRODUCTION (ANNUAL)																					
CHLOROPHYLL (SEASONAL)				−				−		−		−		−	−						
CHLOROPHYLL (PEAK)																					
FJORDS																					
PRODUCTION (SEASONAL)			−		+		−		−			−				−	−				−
PRODUCTION (ANNUAL)																					
CHLOROPHYLL (SEASONAL)	−			+	−			−				−				−				−	
CHLOROPHYLL (PEAK)																					

and Goldman 1979; Falkowski 1980; Li 1980). While results of the above studies are interesting, the time-scales are too short for our purposes. It appears that there are very few studies in which phytoplankton productivity was measured weekly or monthly for periods approaching a decade.

Between January 1972 and December 1977, Mihursky et al. (1977) measured phytoplankton production at six stations in central Chesapeake Bay. Measurements were made routinely at monthly intervals using the ^{14}C technique. Several interesting points are immediately apparent (Fig. 6). First, over this 6-year period there was considerable variation in estimates of integrated annual production, where annual production in 1973 was some 2.3 times that in the least productive year (1977). Furthermore, peak productivity varied by a factor of 3.6 over the six-year period. Other investigators have also measured phytoplankton production for single-year periods in this section of the Chesapeake and the range in values reported (361-858 g C m^{-2} yr^{-1}) was quite similar to those shown in Fig. 6 (Flemer 1970; Taft et al. 1980). Thus, at least in this river dominated estuary, substantial yearly variability exists.

While amplitude of peak production is different in each year, the timing, shape, and frequency of events are nearly identical among the six annual sequences. For example, in all years production rates were in excess of 1 g C m^{-2} d^{-1} for 4-8 mo and maximum rates were always observed in the period of July-September. In addition, a moderate spring peak was observed in the March-May period in 5 of the 6 years studied.

It is, indeed, unfortunate that such data bases are so rare. It remains an open question as to whether substantial year-to-year variability exists in most estuaries or is particularly distinctive in those, such as the Chesapeake, where inputs of nutrients from watershed sources vary markedly with meteorological events and resulting river discharge.

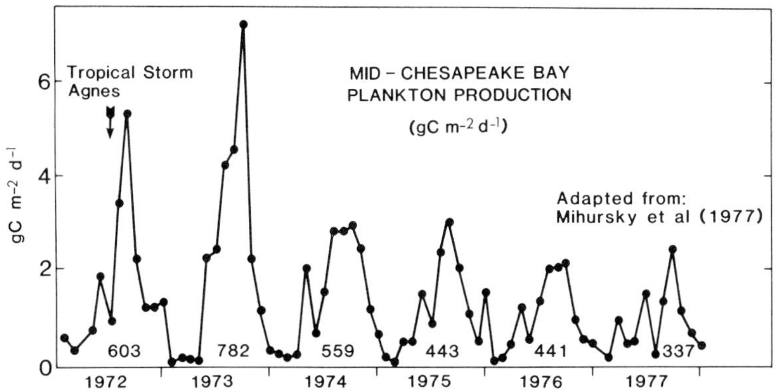

Figure 6. Mean monthly phytoplankton production rates at 6 stations in central Chesapeake Bay, 1972-1977. Values below peaks represent estimates of annual phytoplankton production (g C m^{-2} yr^{-1}).

Responses to Nutrient Inputs

Patterns evident in Fig. 6 suggest the possibility that changes in annual river-borne nutrient loads may explain some of this yearly variation in phytoplankton production. The data presented in Fig. 6 represent a natural experiment resulting from the major storm (Agnes) which passed through the watershed region in early June 1972, near the beginning of the summer period of high productivity. Inputs of nitrogen and phosphorus were 2 to 3 times higher in that year than in preceding or subsequent years and phytoplankton production was also high, 603 g C m^{-2} yr^{-1}. However, maximum annual production did not occur until 1973 (782 g C m^{-2} yr^{-1}), a year in which N and P loadings were more typical of average conditions. Taft et al. (1980) have suggested a possible explanation in that much of the organic matter which is decomposed in the Bay each summer appears to have been derived from inputs and phytoplankton production of the previous year. Similarly, some of the nitrogen and phosphorus associated with inputs in 1972 may have not become available until the summer of 1973 and, coupled with the 1973 inputs, supported very high rates of phytoplankton production. Cory (1978) observed a similar response pattern in the Rhode River, a small tributary of Chesapeake Bay. By 1975, productivity had decreased to annual rates which were comparable to those reported by Flemer (1970) prior to this major event (300-400 g C m^{-2} yr^{-1}).

While most of the annual nutrient load to Chesapeake Bay normally occurs during the late winter and early spring (Guide and Villa 1972), maximum productivity rates are not apparent until much later in the year after high nutrient concentrations associated with riverine sources have decreased to low levels. We suggest that high summer rates of production appear to be supported primarily on recycled nutrients (most importantly nitrogen), some fraction of which is introduced to the estuary during the spring runoff period. The spring pulse in primary production which is frequently observed may be the result of direct utilization of "new nitrogen" (NO_3^-) in the sense of Dugdale and Goering (1967), but most of the annual production is probably supported by recycled nitrogen. This observation is consistent with the calculations of Carpenter et al. (1969) who reported that the nitrogen associated with the spring freshet in Upper Chesapeake Bay was sufficient to support phytoplankton demand for several weeks only. We do not have empirical data which conclusively indicate the mechanisms by which nutrient inputs are trapped in the estuary and thus become available for recycling, but rapid uptake of dissolved nutrients by phytoplankton and subsequent sedimentation of particulates are undoubtedly involved.

We next attempted to address this issue of nutrient input effects on phytoplankton production more directly. During this period of record (1972-1977) there were no major additions of nutrient point-sources in this region of Chesapeake Bay. Hence, we focused on diffuse, watershed sources. Using the data as presented in Fig. 6, and calculated loading rates

(Guide and Villa 1972), we found that annual production and annual loadings of N were indeed related but that less than half of the variability in production could be explained by N loading ($r^2 = 0.38$). We then looked for potential relationships between annual loadings of N and P and other features of phytoplankton dynamics such as peak values of production and chlorophyll *a*. Results of this analysis (Fig. 7) suggest that there is a strong relationship, especially if 1973 data are excluded (this was done because we were not able to account adequately for 1973 inputs plus the residual inputs associated with the tropical storm of 1972). Several authors have previously suggested that peak standing crops were determined by initial nutrient concentrations (e.g. Pratt 1965), but to our knowledge strong relationships have not been reported for peak production rates. Since algal biomass is constrained (within limits) by nutrient composition ratios (Parsons et al. 1961; Rhee 1978) it is not surprising that biomass and loadings are related. We have already argued that high production levels were supported mainly on recycled nitrogen and so it is possible that peak production rates are related to the amount of nitrogen available to be recycled. Riley (1972) and Nixon (1981) have suggested that autotrophic rates are inherently more rapid than heterotrophic

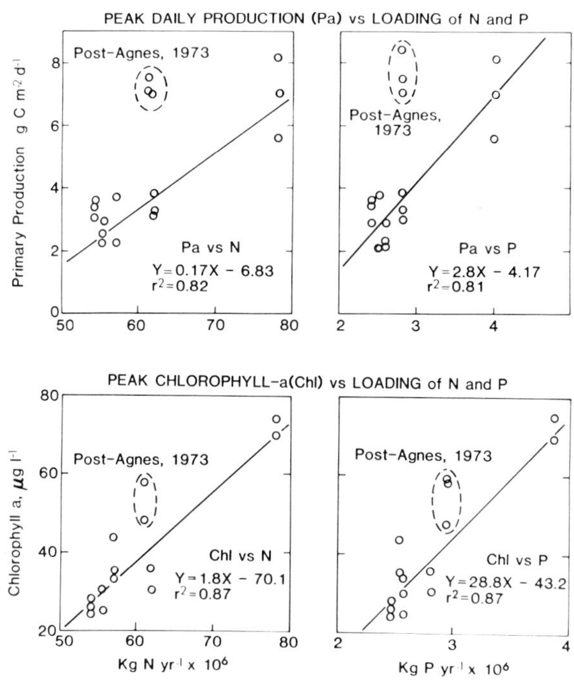

Figure 7. *Regression plots of peak daily production and chlorophyll* a *concentrations vs. annual loadings of nitrogen and phosphorus. Data used in the analysis are from 1972-1977, excluding 1973. In most cases, data from three stations were available for each year.*

processes (remineralization) and hence may be regulated by the slower process. Conversely, it would seem that nutrient loading *per se* has little effect on the timing of phytoplankton productivity.

While results such as these are provocative, the question of the generality of these findings naturally arises. We addressed this issue by obtaining estimates of annual N and P inputs and production for as many estuaries as possible, and were successful in developing such budgets for 14 estuarine areas. Both loadings and production were expressed as unit areal values (m^{-2}); loadings were apportioned over the area in which there was riverine influence by using salinity as a tracer. The results of this exercise (Fig. 8) indicated that in this broad sampling of estuarine environments a reasonable relationship existed between loading and production for nitrogen ($r^2 = 0.60$) but not for phosphorus ($r^2 = 0.08$). This result is consistent with our earlier findings in that there were few significant correlations observed between production and phosphorus concentrations (Table 1) and that nitrogen to

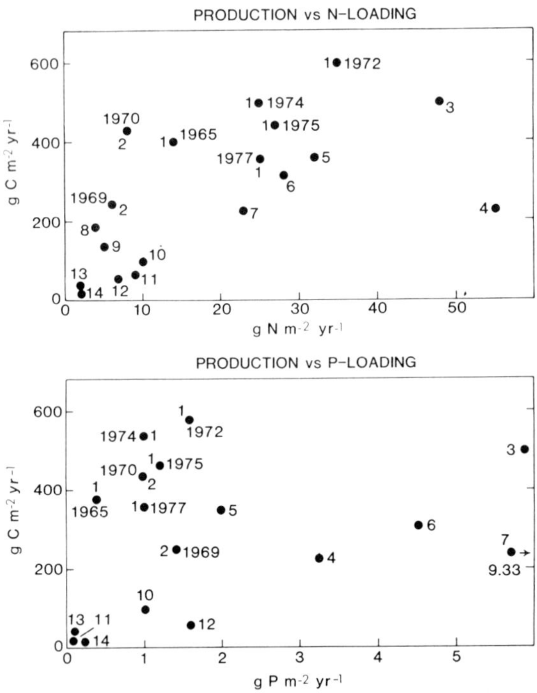

Figure 8. *Regression plots relating nitrogen and phosphorus loadings to annual phytoplankton production in a variety of estuarine ecosystems: (1) Chesapeake Bay, (2) Patuxent River, (3) Pamlico River, (4) Byfjord, (5) Apalachicola Bay, (6) Narragansett Bay, (7) San Francisco Bay, (8) St. Margarets Bay, (9) Long Island Sound, (10) Kungsbacka Fjord, (11) Loch Etive, (12) St. Lawrence River, (13) Baltic Sea, and (14) Kaneohe Bay.*

phosphorus ratios were low in most estuaries during periods of peak production. All of this suggests that nitrogen dynamics are more important in regulating estuarine phytoplankton production than are those of phosphorus. In addition, the residual variability evident in Fig. 8 indicates that factors other than nitrogen loading/recycling are also influential in regulating estuarine primary production and these certainly deserve careful consideration.

Conclusions

Obviously, our conclusions rely heavily on inference rather than experimental testing and hence results must be viewed with some caution. Despite this limitation, comparative approaches such as this have the great advantage of being comprehensive, providing the opportunity to see if hypothesized cause-effect relationships observed in one estuary extend to broader groupings of estuarine ecosystems.

In brief, we found that (1) phytoplankton production and chlorophyll *a* concentrations tended to be highest during the warmer seasons of the year (May-Oct.) in a broad spectrum of estuarine systems; (2) there were few, if any, environmental state variables which alone satisfactorily predicted either phytoplankton production or chlorophyll *a* concentrations in large groups of estuarine systems; (3) temporal patterns of primary production in one large estuary did not vary substantially from year to year although there were considerable differences in the magnitude of annual production; and (4) in 14 estuarine systems, where N and P loading-rate data were available, there was a reasonably good relationship between N loading (but not P loading) and annual production. Based on these findings, we are led to believe that nitrogen rather than phosphorus dynamics are of more central importance in regulating phytoplankton production in estuarine systems and that rate processes, such as nutrient loading rates and recycling rates, seem to be more useful than standing stock values for predicting productivity.

One obvious question that arises concerns appropriate directions for future research and analysis. In light of our current findings we believe that additional analysis of the nutrient loading-phytoplankton production relationship is warranted. It may be that some of the residual variability in this relationship could be accounted for by inclusion of such variables as insolation, euphotic depth, flushing rate, and nutrient recycling rates. In addition, it would be helpful if the estuarine area influenced by watershed-derived nutrients could be better defined. Thus, better prediction of annual production in estuarine systems may be possible using multivariate techniques emphasizing rate rather than state variables. Finally, the question of year-to-year variability in the magnitude and temporal patterns of phytoplankton production remains somewhat unresolved due to the scarcity of long-term data bases. We would encourage the development of carefully designed, decade-long, monitoring programs as a tool for addressing such questions as year-to-year variability, responses to catastrophic events (e.g. hurricanes), and eutrophication tendencies of the phytoplankton component of estuarine ecosystems.

Acknowledgments

Discussions with S. Nixon, J. Day, D. Flemer, C. D'Elia, R. Wetzel and T. Fisher were particularly helpful in this review. We gratefully acknowledge the time and guidance we received from the many researchers contacted during this study. We sincerely thank J. Mihursky for allowing us the use of a substanial productivity data base from Chesapeake Bay and K. Wood for assistance in many aspects of this work. Statistical analyses were conducted by W. Caplins with support from the University of Maryland Computer Science Center. The manuscript was typed by B. Ashby and the figures drafted by F. Younger. Contribution No. 1245, Center for Environmental and Estuarine Studies, University of Maryland.

References Cited

Boynton, W. R., W. M. Kemp and C. G. Osborne. 1980. Nutrient fluxes across the sediment-water interface in the turbid zone of a coastal plain estuary, pp. 93-109. *In:* V. S. Kennedy (ed.), *Estuarine Perspectives.* Academic Press, New York.

Boynton, W. R., W. M. Kemp, K. V. Wood, C. W. Keefe and C. F. D'Elia. 1981. The significance of nutrient and carbon flux across the sediment-water interface along major estuarine gradients. Unpublished report. UMCEES Ref. No. 81-169 CBL. Chesapeake Biological Laboratory, Solomons, MD 20688.

Brylinsky, M. and K. H. Mann. 1973. An analysis of factors governing production in lakes and reservoirs. *Limnol. Oceanogr.* 18:1-14.

Carpenter, J. H., D. W. Pritchard and R. C. Whaley. 1969. Observations of eutrophication and nutrient cycling in some coastal plain estuaries, pp. 210-221. *In: Eutrophication: causes, consequences, correctives.* Nat. Acad. Sci., Washington, DC.

Cory, R. L. 1978. Open-water metabolism in the Rhode River Estuary, MD. 1971-1975, pp. 567-580. *In:* Proc. Symposium on Technical, Environmental, Socioeconomic and Regulatory Aspects of Coastal Zone Management. San Francisco, CA.

Dugdale, R. C. and J. J. Goering. 1967. Uptake of new and regenerated forms of nitrogen in primary productivity. *Limnol. Oceanogr.* 12:196-206.

Eppley, R. W. 1972. Temperature and phytoplankton growth in the sea. *Fish. Bull.* 70:1063-1085.

Falkowski, P. G. 1980. Light-shade adaptation in marine phytoplankton, pp. 99-120. *In:* P. G. Falkowski (ed.), *Primary Productivity in the Sea.* Plenum Press, New York.

Fee, E. J. 1979. A relation between lake morphometry and primary productivity and its use in interpreting whole-lake eutrophication experiments. *Limnol. Oceanogr.* 24:401-416.

Flemer, D. A. 1970. Primary production in the Chesapeake Bay. *Chesapeake Sci.* 11:117-129.

Goldman, J. C., K. R. Tenore and H. I. Stanley. 1973. Inorganic nitrogen removal from waste water: effect on phytoplankton growth in coastal marine waters. *Science* 180:955-956.

Guide, V. and O. Villa, Jr. 1972. Chesapeake Bay Nutrient Input Study. Annapolis Field Office, United States Environmental Protection Agency Technical Report 47. 80 pp.

Haines, E. B. 1976. Relation between the stable carbon isotope composition of fiddler crabs, plants and soils in a salt marsh. *Limnol. Oceanogr.* 21:880-883.

Hansen, D. V. and M. Rattray, Jr. 1966. New dimensions in estuary classification. *Limnol. Oceanogr.* 11:319-326.

Keefe, C. W., W. R. Boynton and W. M. Kemp. 1981. A review of phytoplankton processes in estuarine environments. Unpublished report. UMCEES Ref. No. 81-193 CBL. Chesapeake Biological Laboratory, Solomons, MD 20688.

Kemp, W. M. and W. R. Boynton. 1981. External and internal factors regulating metabolic roles of an estuarine benthic community. *Oecologia* 51:19-27.

Kemp, W. M., R. L. Wetzel, W. R. Boynton, C. F. D'Elia and J. C. Stevenson. 1982. Nitrogen cycling and estuarine interfaces: Some current concepts and research directions, pp. 209-230. *In:* V. S. Kennedy (ed.), *Estuarine Comparisons.* Academic Press, New York.

Li, W. K. W. 1980. Temperature adaptation in phytoplankton: cellular and photosynthetic characteristics, pp. 259-280. *In:* P. G. Falkowski (ed.), *Primary Productivity in the Sea.* Plenum Press, New York.

McCarthy, J. J. and J. C. Goldman. 1979. Nitrogenous nutrition of marine phytoplankton in nutrient depleted waters. *Science* 203:670-672.

Mihursky, J. A., D. R. Heinle and W. R. Boynton. 1977. Ecological effects of nuclear steam electric station operations on estuarine systems. Unpublished report. UMCEES Reference No. 77-28-CBL. Chesapeake Biological Laboratory, Solomons, MD 20688.

Nixon, S. W. 1980. Between coastal marshes and coastal waters – A review of twenty years of speculation and research on the role of salt marshes in estuarine productivity and water chemistry, pp. 437-525. *In:* P. Hamilton and K. B. MacDonald (eds.), *Estuarine and Wetland Processes.* Plenum Press, New York.

Nixon, S. W. 1981. Freshwater inputs and estuarine productivity, pp. 31-57. *In:* R. Cross and D. Williams (eds.), *Proceedings of the National Symposium on Freshwater Inflow to Estuaries.* U.S. Fish and Wildlife Service, Office of Biological Services. FWS/OBS-81/04.

Parsons, T. R., K. Stevens and J. D. H. Strickland. 1961. On the chemical composition of eleven species of marine phytoplankton. *J. Fish. Res. Bd. Canada* 18:1001-1016.

Pratt, D. M. 1965. The winter-spring diatom flowering in Narragansett Bay. *Limnol. Oceanogr.* 10:173-184.

Pritchard, D. W. 1955. Estuarine circulation patterns. *Proc. Amer. Soc. Civ. Eng. 81,* 717/1-717/11.

Redfield, A. C. 1934. On the proportions of organic derivatives in seawater and their relation to the composition of plankton. *James Johnstone Memorial Volume,* Liverpool, pp. 177-192.

Rhee, G.-Y. 1978. Effects of N:P atomic ratios and nitrate limitation on algal growth, cell composition, and nitrate uptake. *Limnol. Oceanogr.* 23:10-25.

Riley, G. A. 1972. Patterns of production in marine ecosystems, pp. 91-112. *In:* J. A. Wiens (ed.), *Ecosystem Structure and Function.* U. Oregon Press, Corvallis, OR.

Ryther, J. H. 1963. Geographical variations in productivity, pp. 347-380. *In:* M. N. Hill (ed.), *The Sea, Vol II.* Wiley Interscience, New York.

Ryther, J. H. and W. H. Dunstan. 1971. Nitrogen, phosphorus, and eutrophication in the coastal marine environment. *Science* 171:1008-1013.

Schindler, D. W. 1978. Factors regulating phytoplankton production and standing crop in the world's freshwaters. *Limnol. Oceanogr.* 23:478-486.

Seitzinger, S., S. W. Nixon, M. E. Q. Pilson and S. Burke. 1980. Denitrification and N_2O production in near-shore marine sediments. *Geochimica et Cosmochimica Acta* 44:1853-1860.

Smith, V. H. 1979. Nutrient dependence of primary productivity in lakes. *Limnol. Oceanogr.* 24:1051-1064.

Strickland, J. D. H. 1960. Measuring the production of marine phytoplankton. *Bull. No. 122. Fisheries Res. Bd. Canada.* 172 pp.

Taft, J. L., E. O. Hartwig and R. Loftus. 1980. Seasonal oxygen depletion in Chesapeake Bay. *Estuaries* 3:242-247.

Taft, J. L. and W. R. Taylor. 1976. Phosphorus dynamics in some coastal plain estuaries, pp. 79-89. *In:* M. L. Wiley (ed.), *Estuarine Processes, Vol. I,* Academic Press, New York.

Thayer, G. W. 1974. Identity and regulation of nutrients limiting phytoplankton production in the shallow estuaries near Beaufort, N.C. *Oecologia* 14:75-92.

Thayer, G. W., P. L. Parker, M. W. LaCroix and B. Fry. 1978. The stable carbon isotope ratio of some components of an eelgrass, *Zostera marina*, bed. *Oecologia* 35:1-12.

Walsh, J. J. 1976. Herbivory as a factor in patterns of nutrient utilization in the sea. *Limnol. Oceanogr.* 21:1-13.

Williams, R. B. 1966. Annual phytoplankton production in a system of shallow temperate estuaries, pp. 699-717. *In:* H. Barnes (ed.), *Some Contemporary Studies in Marine Sciences.* George Allen and Unwin Ltd., London.

Williams, R. B. 1972. Nutrient levels and phytoplankton productivity in the estuary, pp. 59-89. *In:* R. A. Chadwick (ed.), *Proc. Coastal Marsh and Estuary Management Symposium.* La. State Univ. Div. of Continuing Education, Baton Rouge, LA.

Appendix Table. A listing of estuarine systems and data sources utilized in this study.

Altamaha River, Georgia: L.R. Pomeroy, et al., Limnol Oceanogr Spec. Symp. 1, 274(1972): J.P. Thomas, M.S. Thesis U. of Ga. 88 pp. (1966).

Amur Bay, USSR: I.F. Rassashko, Oceanology 14, 557(1973).

Apalachicola Bay, Florida: R.H. Estabrook, M.S. Thesis Florida State U., 166 pp. (1973).

Auke Bay, Alaska: R.L. Iverson, et al., Limnol Oceanogr. 19, 271(1974); D.M. Schell, Limnol Oceanogr. 19, 260(1974); Iverson (pers. commun.).

Baltic Sea: R. Sen Gupta, Marine Biol. 17, 82(1972); R. Sen Gupta, Mar. Chem. 1, 267(1973); McKellar (pers. commun.).

Barataria Bay, Louisiana: B.B. Barrett, et al., Louisiana Dept. of Wildlife & Fisheries Tech. Bull. 27, 197 pp. (1978); J.W. Day, et al., Louisiana State U. Publ. No. LSU-SG-72-04, 79 pp. (1973).

Beaufort Sound, North Carolina: G.W. Thayer, Chesapeake Sci. 12, 240(1971); G.W. Thayer, Oecologia 14, 75(1974); R.B. Williams & M.B. Murdoch, Limnol Oceanog. 11, 73(1966).

Bedford Basin, Nova Scotia: S.A.M. Conover, Marine Biol. 32, 147(1975); T. Platt, J. exp. mar. Biol. Ecol. 18, 99(1975); T. Platt, et al., Naturaliste Can. 99, 253(1972); S. Taguchi, et al., Fish. Res. Bd. Canada Tech. Rept. No. 587, 63 pp. (1975).

Burrard Inlet, British Columbia: J.G. Stockner & D.D. Cliff, J. Fish. Res. Bd. Canada 36, 1(979).

Byfjord, Sweden: R. Rosenberg, et al., Marine Biol. 42, 99(1977).

Caloosahatchee River, Florida: A. Dragovich, et al., U.S. Fish and Wildlife Service, Fish. Bull. 66, 463(1968).

Chesapeake Bay, Maryland: D.A. Flemer, Chesapeake Sci. 11, 117(1970); V. Guide & O. Villa, EPA Region III Tech. Rept. 47, 80 pp. (1972); J.J. McCarthy et al., Limnol. Oceanogr. 22, 996(1977); J.A. Mihursky et al., Chesa. Biol. Lab. Ref. No. UMCEES 77-28-CBL (1977); J.L. Taft et al., in Estuarine Interactions, M. Wiley, Ed. (Academic Press, 1978) p. 115; J.L. Taft et al., Estuaries 3, 242(1980); J.L. Taft and W.R. Taylor, in Estuarine Processes, M. Wiley, Ed. (Academic Press, 1976) p. 79; R.C. Whaley et al., Ches. Bay Inst. Johns Hopkins U. Spec. Rept. 12, 88 pp.(1966); USGS Streamflow data; Macknis (pers. commun.).

Chincoteague Bay, Maryland: W.R. Boynton, M.S. Thesis U. of North Carolina, 142 pp. (1974).

Cochin Backwater, India: S.Z. Qasim & C.V.G. Reddy, Bull. Mar. Sci. 17, 95(1967); V.N. Sankaranarayanan & S.Z. Qasim, Marine Biol. 2, 236(1969); T. Wyatt & S.Z. Qasim, Limnol. Oceanogr. 18, 301(1973).

Columbia River, Washington: L. Haertel, et al., Ecology 50, 962(1969); V.T. Neal, in The Columbia River and Adjacent Ocean Waters, A.T. Pruter & D.L. Alverson, Ed. (U. of Washington Press, 1972) p. 19.

Duplin River, Georgia: E.B. Haines, Oikos 29, 254(1977); R.A. Ragotzkie, Publ. of the Inst. of Mar. Sci. 6, 146(1959).

Duwamish River, Washington: E.B. Welch, J. Water Poll. Contr. Fed. 40, 1711(1968); E.B. Welch, et al., Limnol. Oceanogr. 17, 731(1972).

Ems Estuary, Netherlands: A.J. van Bennekom, et al., Neth J. Sea Res. 8, 174(1974); G.C. Cadee & J. Hegeman, Neth. J. Sea Res. 8, 240(1974); G.C. Cadee & J. Hegeman, Neth. J. Sea Res. 8, 260(1974); W. Helder, Neth. J. Sea Res. 8, 154(1974); V.N. de Jonge & H. Postma, Neth. J. Sea Res. 8, 139(1974).

Flax Pond, New York: R.A. Moll, Ph.D. Thesis State U. of N.Y. at Stony Brook, 122 pp. (1974); R.A. Moll, Marine Biol. 42, 109 (1977).

Fraser River, British Columbia: T.R. Parsons et al., J. exp. Mar. Biol. Ecol. 3, 27(1969); M. Takahashi et al., Marine Biol. 19, 102(1973); M. Waldichuk, J. Fish Res. Bd. Canada 14, 321 (1957).

Funka Bay, Japan: Y. Maita & M. Yanada, Marine Biol. 44, 329(1978); M. Yanada & Y. Maita, Estuarine & Coastal Marine Sci. 6, 523(1978).

Gulf of St. Lawrence, Canada: D.M. Steven, McGill U. Marine Sciences Centre M.S. Rept. 26, 116 pp. (1974).

High Venice Lagoon, Italy: A. Vatova, J. Cons. Perm. Intern. Explor. Mer 26, 148(1961).

Hudson River, New York: M.J. Bowman & L.D. Wunderlich. Limnol. Oceanogr. Spec. Symp. 2, 58(1976); C. Garside, et al., Estuarine & Coastal Marine Sci. 4, 281(1976); T.C. Marlone, Limnol. Oceanogr. Spec. Symp. 2, 260(1976); T.C. Malone, Estuarine & Coastal Marine Sci. 5, 157(1977); H.J. Simpson, et al., in Estuaries, Geophysics and the

Appendix Table (continued).

Environment (National Research Council, 1977) p. 94; D.L. Sirois & S.W. Fredrick, Estuarine & Coastal Marine Sci. 7, 413(1978).
Indian Arm Inlet, British Columbia: M. Gilmartin, J. Fish Res. Bd. Canada 19, 921(1962); M. Gilmartin, J. Fish. Res. Bd. Canada 21, 505(1964).
Kaneohe Bay, Hawaii: W.A. Harvey & J. Caperon, Pac. Sci. 30, 329(1976); S.V. Smith, et al. Hawaii Institute of Marine Biology, Univ. of Hawaii at Manoa Rept. 166 p. (1978).
Kungsbacka Fjord, Sweden: I. Olsson & E. Olundh, Marine Biol. 24, 17(1974).
Little Manatee River, Florida: A. Dragovich, et al., U.S. Fish & Wildlife Service, Fish. Bull. 66, 463(1968).
Loch Creran, Scotland: P. Tett & A. Wallis, J. Ecol. 66, 227(1978).
Loch Etive, Scotland: L. Solorzano & B. Ehrlich, J. exp. Mar. Biol. Ecol. 29, 45(1977); B.J.B. Wood, et al., J. Ecol. 61, 569(1973).
Loch Ewe, Scotland: J.M. Davies, Marine Biol. 31, 353(1975); J.H. Steele & I.E. Baird, Mem. Ist. Ital. Idrobiol. 29 (Suppl.), 73(1972); Baird (pers. commun.).
Long Island Estuaries, New York: J.P. Barlow, et al., Limnol. Oceanogr. 8, 251(1963); E.F. Mandelli, et al., Marine Biol. 7, 153(1970).
Long Island Sound, New York: M.J. Bowman, Estuarine & Coastal Marine Sci. 5, 531(1977); E. Harris, Bull. Bingham oceanogr. Coll. 17, 31(1959); G.A. Riley, Bull. Bingham oceanogr. Coll. 15, 324(1956); G.A. Riley & S.A.M. Conover, Bull. Bingham oceanogr. Coll. 15, 47(1956); G.A. Riley & H.M. Schurr, Bull. Bingham Oceangr. Coll. 17, 66(1959).
Manatee River, Florida; A. Dragovich, et al., U.S. Fish & Wildlife Service, Fish. Bull. 66, 463(1968).
Meyers Creek, New Jersey: J.R. Durand, et al., Rutgers, The State U. of New Jersey Pub. No. NJ/RU-DEP-11-9-79-1, 194 pp. (1979); T. Sugihara, et al., Rutgers, The State U. of New Jersey Pub. No. NJ/RU-DEP-11-9-79-3(1979).
Mississippi River Plume, Louisiana: W.H. Thomas & E.G. Simmons, in Recent Sediments, Northwest Gulf of Mexico, 1951-1958 (Am. Assoc. Petrol., Geol., Tulsa, Okla., 1960) p. 103.
Myakka River, Florida: A. Dragovich, et al., U.S. Fish & Wildlife Service, Fish. Bull. 66, 463(1968).
Narragansett Bay, Rhode Island: M.J. Furnas, et al., in Estuarine Processes I., M. Wiley, Ed. (Academic Press, 1976) p. 118; J.N. Kremer & S.W. Nixon, A Coastal Marine Ecosystem. Simulation & Analysis (Springer-Verlag, 1978) 217 pp.; C. Oviatt, et al., Estuaries 4. 167(1981); D.M. Pratt, Limnol. Oceanogr. 10, 173(1965).
North Inlet, South Carolina: K.G. Sellner, et al., Botanica Marina 19, 119(1976); Zingmark (pers. commun.).
Pamlico River, North Carolina: J.E. Hobbie,et al., in Estuarine Research I, L.E. Cronin, Ed. (Academic Press, 1975) p. 287; E.J. Kuenzler, et al., Water Resources Research Institute of North Carolina State U. Rept. 139, 163 pp. (1979).
Patuxent River, Maryland: D.A. Flemer et al., Chesa. Biol. Lab. Ref. No. 71-6, 217 pp. (1970); V. Guide & O. Villa, EPA Region III Tech. Rept. 47, 80 pp. (1972); USGS Streamflow data; Macknis (pers. commun.).
Peace River, Florida: A. Dragovich, et al., U.S. Fish & Wildlife Service, Fish. Bull. 66, 463(1968).
Peconic Bay, New York: S.F. Bruno, et al., Estuarine & Coastal Marine Sci. 10, 247(1980).
Piscataqua River, New Hampshire: J.R. O'Shea, M.S. Thesis U. of New Hampshire, 64 pp. (1978).
Port Hacking Basin, Australia: B.D. Scott, Aust. J. Mar. Freshwater Res. 29, 31(1978); B.D. Scott, Aust. J. Mar. Freshwater REs. 29, 803(1978); B.D. Scott, Aust. J. Mar. Freshwater Res. 30, 449(1979).
Puget Sound, Washington: D.F. Winter, et al., Marine Biol. 29, 139(1975).
Raritan Bay, New Jersey: A.L. Bleecker, Ph.D. Thesis, Rutgers, The State U. of New Jersey, 472 pp. (1971); H.P. Jeffries, Limnol. Oceanogr. 7, 21(1962); B. Patten, Limnol. Oceanogr. 6, 369(1961).
Roskeeda Bay, Ireland: R.C.T. Raine & J.W. Patching, J. exp. Mar. Biol. Ecol. 47, 127(1980); Raine, pers. commun.
St. Lawrence Estuary, Canada: A.R. Coote & P.A. Yeats, J. Fish Res. Bd. Canada 36, 122(1979); J.-M. Sevigny et al., J. Fish Res. Bd. Canada 36, 187(1979); M. Sinclair, J. Fish Res. Bd. Canada 35, 1171(1978); M. Sinclair, et al., J. Fish Res. Bd. Canada 33, 1271(1976); D.M. Steven, McGill U. Marine Sciences Centre M.S. Rept. 26, 116 pp. (1974); Sinclair (pers. commun.).
St. Margarets Bay, Nova Scotia: T. Platt, J. Cons. perm. int. Explor. Mer 33, 324(1971); T. Platt & B. Irwin, Fish. Res. Bd. Canada Tech. Rept. 77, 123 pp. (1968); W.H. Sutcliffe, J. Fish. Res. Bd. Canada 29, 357(1972).
San Francisco Bay, California: B.E. Cole & R.E. Herndon, USGS Menlo Park, Calif. Open File Rept. 79-983, 120 pp. (1979); T.J. Conomos & D.H. Peterson, M'emoires de l'Institut de Geologie du Bassin d'Aquitaine 7, 45(1974); T.J. Conomos & D.H. Peterson, in Proceedings of the Workshop on Algal-Nutrient Relationships in the San Francisco Bay & Delta, R.L. Brown, Ed. (San Francisco Bay & Estuarine Assoc., 1975) p. 103; T.J. Conomos & D.H. Peterson, in Estuarine Processes II, M. Wiley, Ed. (Academic Press, 1977) pp. 82; D.H. Peterson, et al., Estuarine & Coastal Marine Sci. 3, 1(1975); E.P. Scrivani, in Proceedings of the Workshop on Algal-Nutrient Relationships in the San Francisco Bay & Delta, R.L. Brown, Ed. (San Francisco Bay & Estuarine Assoc., 1975) p. 127.
Sheepscott River, Maine: C. Garside, et al., in Estuarine Interactions, M. Wiley, Ed. (Academic Press, 1978) p. 565.
Strait of Georgia, British Columbia: M. Takahashi, et al., Marine Biol. 19, 102(1973); M. Waldichuk, J. Fish. Res. Bd. Canada 14, 321(1957).
Swartvlei, South Africa: R.D. Robarts, J. exp. Mar. Biol. Ecol. 24, 93(1976).
Sydney Harbor, Australia: N. Revelante & M. Gilmartin, Aust. J. Mar. Freshwater Res. 29, 9(1978).

Appendix Table (continued).

Vancouver Island Inlet, British Columbia: M. Waldichuk, J. Water Poll. Contr. Fed. 41, 745 (1969).
Vellar River, India: A. Purushothaman & G.P. Bhatnagar, Arch. Hydrobiol. 77, 37(1976); R.C. Subbaraju & K. Krishnamurthy, Marine Biol. 14, 25(1972).
Vostok Bay, USSR: M.V. Propp, Helgolander wiss. Meeresunters. 30, 598(1977).
Waccasassa River, Florida: H.D. Putnam, in Advances in Water Pollution Res., Proc. Int. Water Pollution Conf. III, Munich (Pergamon Press, 1966) p. 121.
Wadden Sea, Netherlands: A.J. van Bennekom, et al., Neth. J. Sea Res. 8, 174(1974); G.C. Cadee & J. Hegeman, Neth. J. Sea Res. 8, 240(1974); G.C. Cadee & J. Hegeman, Neth. J. Sea Res. 8, 260(1974); G.C. Cadee & J. Hegeman, Neth. J. Sea REs. 13, 224(1979); W. Helder, Neth. J. Sea Res. 8, 154(1974); V.N. de Jonge & H. Postma, Neth. J. Sea Res. 8, 139(1974); H. Postma & J.W. Rommets, Neth. J. Sea Res. 4, 470(1970).
Wassaw Sound, Georgia: R.E. Turner, et al., Estuarine & Coastal Marine Sci. 9, 603(1979).
Ythan Estuary, England: J.H. Leach, J. Mar. Biol. Assn. U.K. 51, 137(1971).
Zaire River, Zaire: A.J. van Bennekom, et al. Neth. J. Sea Res. 12, 296(1978); G.C. Cadee, Neth. J. Sea Res. 12, 368(1978).

BIOCHEMICAL CORRELATES OF STRUCTURE AND STABILITY IN DIVERGENT PLANKTON COMMUNITIES

H. Perry Jeffries and Rose M. Lambert
Graduate School of Oceanography
University of Rhode Island
Kingston, Rhode Island

Abstract: Fatty acid composition of macrozooplankton (>240 μm) revealed responses specific to habitat type along a transect extending from Rhode Island Sound to the inner reaches of Narragansett Bay. A medium chain saturate (stearic acid) characterized riverine zooplankton, a direct response to foods rich in natural detritus and raw sewage. A long chain polyunsaturate (docosahexaenoic acid) denoted affinities in tissue composition more genetically determined than environmentally based. Zooplankton biomass and chlorophyll *a* concentrations over two summers varied in relation to the same two acids, indicating an association between structure at the chemical level and function within the ecosystem. A discriminant function based on these two acids clearly separated estuarine from offshore macrozooplankton. Compared with published data, these results suggest that zooplankton contains more stearic acid and less docosahexaenoic acid in European waters (North Sea, Rias of Northwest Spain, Mediterranean Sea) than in comparable areas of Rhode Island and Newfoundland. Within the Rhode Island area, tendency to change consistently (stability) increased in the order of physical environment, planktonic community, and biochemical composition. Thus, not only are fatty acids ubiquitous in nature and especially important energetically in zooplankton, but dynamically these monomeric compounds are also most conservative. Consequently, they serve well in comparative study at various levels of natural organization.

Introduction

Structural comparisons between natural communities are made difficult by the observations that ecologists make – the species present and their abundances. Any attempted ordination runs afoul of divergences in numbers and kinds of species. This problem becomes acute when we compare populations distributed across the continental shelf. Recently, however, close correspondences between organic composition and ecological properties of plankton and benthos have been demonstrated (Jeffries 1979; Jeffries and Lambert 1980). These similarities allow us to compare chemical species in addition to taxonomic species. The approach is based upon statistical properties that fatty acid patterns reveal within entire communities.

Fatty acids comprise a lipid class that is especially important in plankton. Temporal variation among the major fatty acids produces a message of sorts, which although noisy, reveals processes vital to the survival of a community as a definable entity in nature. During seasonal succession, for exam-

ISBN 0-12-404070-5

ple, fatty acid pattern in plankton changes quickly, responding first to environmental influences and then to genetic change as species representation shifts among biogeographic groups.

In this paper, biochemical correlates are compared in planktonic communities distributed along an inshore-offshore transect. We show that correspondence between ecological characterization (Jeffries 1967) and biochemical pattern is sufficiently close to permit a broad-scale system of comparison based upon monomers common throughout nature.

Methods

Zooplankton was sampled biweekly for two summers (1979-1980) along a six-station transect extending from Rhode Island Sound, through Narragansett Bay, and into its polluted tributary, the Providence River (Fig. 1). A No. 6 mesh, 0.5-m net (240 μm aperture), with a General Oceanics flow meter mounted off-center in its mouth was towed obliquely for 5 to 10 min periods. A surface water sample was filtered at the same time for fluorimetric determination of chlorophyll *a*. Filters and plankton samples were

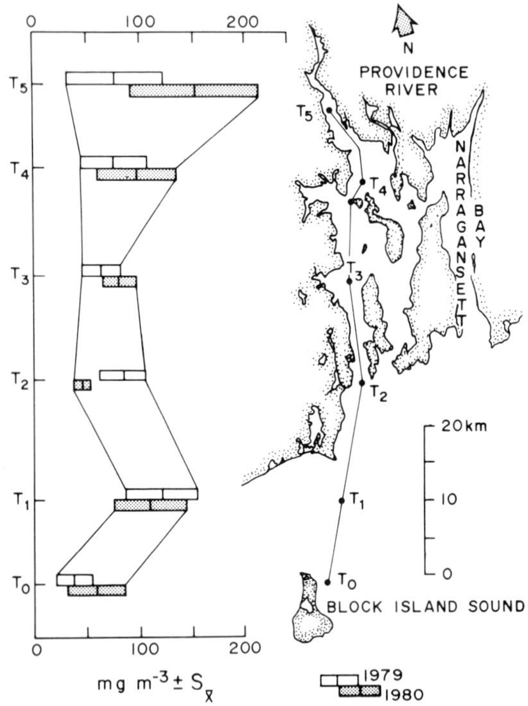

Figure 1. Station locations in the Rhode Island Sound-Narragansett Bay area and zooplankton standing crop (mg dry wt m^{-3} ± standard error of the mean) in summer 1979 and 1980.

frozen immediately in liquid nitrogen and later held at −20 C in a vacuum desiccator.

Fatty acids were extracted directly from freeze-dried zooplankton; they were analyzed as methyl esters (Schultz and Quinn 1972) by gas chromatography on packed columns (2 m × 3.1 mm, SP-2330 on Chromosorb W-AW, 100/120). Only the major fatty acids (>2% of the total) were considered. Analyses were believed to be precise within ± 2% of stated values, based upon repeated analyses of duplicate extracts. For statistical analysis, percentages of total fatty acid methyl esters were transformed to their arc sin equivalents.

Results

Environmental properties at the mid-sound and river stations are shown in Table 1. Chlorophyll *a* was much higher in the river than offshore, a response to organic pollutants (Oviatt 1981). Zooplankton also responded to this enrichment, with standing crops ranging up to 372.6 mg dry wt m^{-3}, but the mean value was statistically the same in the river as offshore (Fig. 1), presumably due to heavy predation by the ctenophore *Mnemiopsis leidyii* during mid-summer (Kremer 1976).

Average fatty acid composition of zooplankton at either end of the transect is given in Table 2. Note that the component designated as 18:0 (number of n-carbon atoms:number of double bonds, stearic acid) was more common at the river station than offshore. The same was true for 16:0 and 17:0, additional homologues in the saturated series. Both offshore and inshore, however, the characteristically marine, long-chain polyunsaturates (20:5 and 22:6) accounted for more than 25% of the total fatty acid composition.

Table 1. Environmental properties of Rhode Island Sound and Providence River, summer 1979, 1980.

		Station	
Property		*Sound*	*River*
Surface temperature (°C)			
	$\overline{X}$	18.0	23.2
	range	14.4-22.0	17.6-27.8
Surface salinity ($^0/_{00}$)			
	$\overline{X}$	31.8	23.3
	range	31.0-32.4	20.9-27.7
Chlorophyll *a* ($\mu g\ l^{-1}$)			
	$\overline{X}$	0.8	26.6
	range	0.2-1.8	3.0-61.4
Zooplankton (mg dry wt m^{-3})			
	$\overline{X}$	116.8	110.7
	range	7.1-232.3	0.2-372.6

Table 2. Fatty acid composition of summer zooplankton in Rhode Island Sound and Providence River as mean percentages of total fatty acid methyl esters (standard errors in parentheses).

	Station	
Fatty acid	*Sound*	*River*
14:0[a]	5.8 (1.2)	5.6 (0.6)
16:0	13.8 (1.4)	21.4 (1.2)
16:1	6.6 (1.0)	4.9 (0.3)
17:0	1.0 (0.1)	1.7 (0.2)
18:0	2.6 (0.4)	6.0 (0.6)
18:1	8.0 (0.9)	7.6 (1.0)
18.3	2.7 (0.3)	2.9 (0.3)
18:4	4.5 (0.7)	4.0 (0.7)
20:5	13.7 (1.3)	14.4 (1.2)
22:6	15.6 (1.6)	14.3 (1.4)
Other[b]	25.7	17.2

[a] Short hand notation for number of n-carbon atoms:number of double bonds;
[b] Individual values <2%.

Principal components analysis (Dixon and Brown 1979) on the entire fatty acid data set (11 samples, 10 fatty acids at each of six stations) produced the eigenvectors in Table 3. Grouping fatty acids by their patterns of ranking first and second (sign disregarded) within the eigenvector matrix yielded the following five groups:

Group	*Component fatty acids*
I	14:0, 16:1, 17:0, 18:0
II	16:0, 18:3
III	18:1
IV	18:4, 20:5
V	22:6

The tendency for these variables to cluster, though weakly defined, may be due to chain length (shorter chain length in group I, longer chains in groups IV and V) as well as unsaturation (more double bonds in groups IV and V than in groups I-III). This grouping is consistent with previous results that show unsaturation and chain length increasing in the food web at the herbivore level and decreasing with the formation of detritus (Jeffries 1972).

Linear additive transforms (Green 1979) calculated from the first and second eigenvectors are given in Fig. 2. Although principal components analysis is not really intended for classification (Maxwell 1977), grouping resulted for stations in the sound, bay and river. Thus, macrozooplankton produced distinctive chemical patterns in response to specific features of each

Table 3. Eigenvectors corresponding to the first four eigenvalues (principal components). For each fatty acid the two largest eigenvectors are underlined (sign disregarded). Arc sin transforms expressed in radians.

	Eigenvalue			
Fatty acid	*1*	*2*	*3*	*4*
14:0	0.36	0.41	0.13	0.12
16:0	0.34	0.00	0.22	−0.40
16:1	0.35	0.53	−0.03	0.14
17:0	0.35	−0.43	−0.17	−0.17
18:0	0.42	−0.30	−0.10	−0.25
18:1	0.00	0.15	0.02	−0.70
18:3	−0.39	0.07	0.00	−0.31
18:4	−0.28	0.15	0.57	−0.25
20:5	0.30	0.07	0.48	0.08
22:6	0.01	−0.47	0.57	0.23

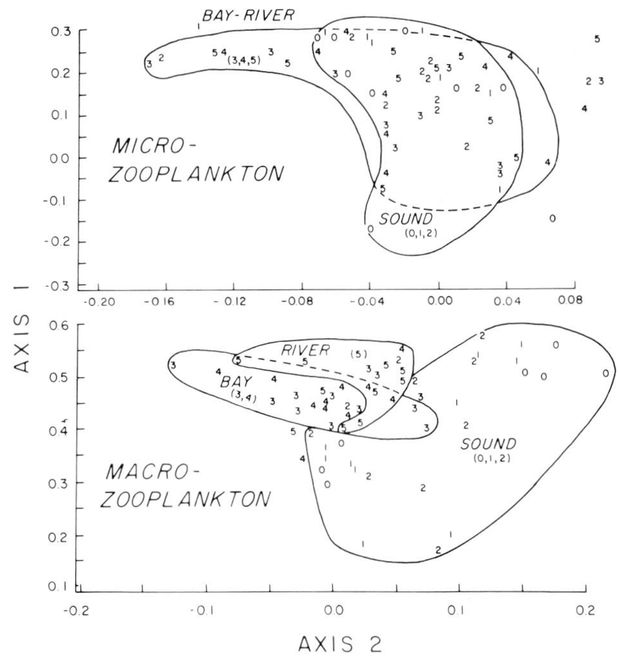

Figure 2. Station groupings of microzooplankton (73–153 μm) and macrozooplankton (> 240 μm) on the first two principal components, as sound ($T_0 - T_2$), bay ($T_3 - T_4$) and river (T_5); based on distributions of 10 major fatty acids. Collections made over summer 1979, 1980 in the Rhode Island Sound-Narragansett Bay area.

major habitat. Corresponding patterns in the microzooplankton (73-153μm) were examined in samples taken with two concentrically mounted nets (Nos. 10 and 20) towed at the surface. The chief components were copepod nauplii, benthic larvae (polychaetes, lamellibranchs) and diatoms. In this smaller-sized fraction of mixed ecological representation, chemical separation between habitats was not evident: patterns were broadly superimposed upon one another.

Even when fatty acid matrices are reduced to their major components, they are still complex. We can simplify the situation to include only 18:0 (stearic acid), which represents the saturated, shorter-chain length homologues, and 22:6 (docosahexaenoic acid), the longest and most unsaturated component (group V). Spatially, only the mid-sound (T_1) and river (T_5) stations need to be considered. (Station T_0, although farthest offshore, was so close to Block Island that T_1 was a better representative of conditions in the sound). In this form, comparison was made by discriminant analysis of a pooled covariance matrix.

In Davis' (1973) notation, the discriminant index, R_o, is the midpoint between two centers of distribution, R_s (sound) and R_r (river) mean values. For arc sin transformed data expressed in degrees:

$$
\begin{aligned}
R_s &= \lambda_1\bar{A}_1 + \lambda_2\bar{A}_2 \\
&= -0.846\ (8.95)\ -0.005\ (22.36) \\
&= -7.68
\end{aligned}
$$

λ_1 and λ_2 are the coefficients for 18:0 and 22:6
$\bar{A}_1$ and $\bar{A}_2$ are mean values of 18:0 and 22:6 at sta T_1.

In the river:

$$
\begin{aligned}
R_r &= \lambda_1\bar{B}_1 + \lambda_2\bar{B}_2 \\
&= -0.846\ (14.06)\ -0.005\ (21.96) \\
&= -12.01
\end{aligned}
$$

$\bar{B}_1$ and $\bar{B}_2$ are mean values of 18:0 and 22:6 at sta T_5.

The midpoint between the two stations is:

$$
\begin{aligned}
R_o &= \lambda_1\psi_1 + \lambda_2\psi_2 \\
&= -0.846\ (11.505)\ -0.005\ (22.16) \\
&= -9.84
\end{aligned}
$$

ψ_1 and ψ_2 are the averages of mean 18:0 and of mean 22:6 at T_1 and T_5.

Position of each habitat in relation to the midpoint R_o is shown in Fig. 3. High values to the right (the estuarine tendency) resulted chiefly from high stearic acid concentration, which in turn reflected abundances of this particular fatty acid in a diet containing a high proportion of non-living organic matter.

To broaden the comparison, mean summer locations of macrozoo-

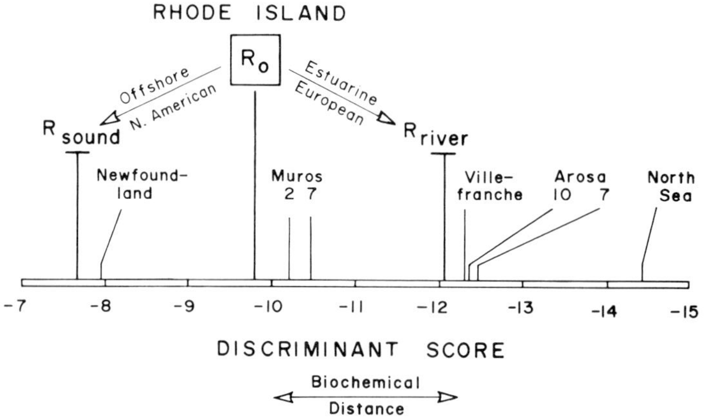

Figure 3. Ordination of North Atlantic zooplankton plotted on a biochemical scale derived from 18:0 and 22:6 concentrations (percentages of total fatty acid methyl esters, arc sin transformation in degrees) for Rhode Island Sound and the Providence River. Positions for the Rias de Arosa and Muros et Noya in northern Spain were calculated from concurrently obtained samples (> 240 μm). Stations Arosa 10 and Muros 7 were approximately at mid-length in each Ria; stations Arosa 7 and Muros 2 were at the mouths of the respective Rias.

plankton in two Spanish estuaries – Ria de Arosa and Ria de Muros et Noya, adjacent bays on the Atlantic Coast of northern Spain (Tenore and Gonzalez 1975; Tenore et al. 1982) – were positioned on the Rhode Island discriminant function. Although the Spanish samples had greater species diversity than zooplankton in estuaries bordering the Western North Atlantic, they did fall on the estuarine side of R_o.

Also indicated on the discriminant function diagrammed in Fig. 3 are locations of samples from: Placentia Bay, Newfoundland, chiefly *Temora longicornis* (Ackman and Hooper 1970); Villefranche Bay in the Mediterranean, average of three samples (Goutx and Saliot 1980); and the North Sea, outside the Wadden Sea, average of six copepod species (Kattner et al. 1981). Again the European zooplankton came out on the estuarine side of R_o, whereas the Newfoundland sample resembled more closely the offshore, Rhode Island condition.

Thus, in general, Fig. 3 suggests two tendencies: offshore/North American and estuarine/European. Location is set on this particular function chiefly by the 18:0 concentration; 22:6 is included as a representative of the long chain polyunsaturated component. In future studies, a step-wise, multiple descriminant analysis should be attempted.

We expressed stability (consistency) of an ecological property by its pattern of increase and decrease over time. For each variable in a system there is a matrix of transition probabilities; stability is simply the determinant of the matrix (Patten 1961). The determinants at each level that one wishes to consider in a system are averaged, which gives estimates of consistency ranging from −1.0 (least stable) to 1.0 (most stable).

Stability at three levels was calculated: temperature and salinity at the environmental level; zooplankton biomass and chlorophyll *a* at the community level; and fatty acids 18:0 and 22:6 at the chemical level. Environmental stability was lower offshore than inshore, but samples were taken here at the surface early in the morning. Integrated temperature and salinity values over the water column would merit more refined calculation (Patten 1962); however, the order of increasing stability among levels probably would remain the same.

Stability was lowest at the environmental level, greater at the planktonic (community) level, and greatest at the chemical level (Table 4). It follows that comparisons drawn biochemically are most conservative, yet they reflect metabolic processes dynamically tuned to a changing environment.

Discussion

Although much maligned, species diversity indices remain as one of the few quantitative measures available for comparisons of community structure. In their application to zooplankton, however, we run across a problem in trying to decide what to call an ecologically meaningful species. Adult copepods, for example, may feed on particles over a wide size range, whereas immature copepodites are limited to a comparatively narrow range (Paffenhofer and Knowles 1978). Feeding becomes increasingly complex among nauplii (Marshall and Orr 1956; Fernandez 1979). Indeed, the nauplii of several estuarine copepod species may not graze selectively (Allen et al. 1977). Consequently, environmental requirements differ more within what we identify and enumerate as a species population than between similarly defined, coexisting populations. Then there are taxonomic problems in our attempts to include myriad developmental stages, both holo- and meroplanktonic.

The above problems with traditional methods make continuous ordination along an environmental gradient so difficult that it is infrequently attempted across broad sections of the ocean. Biochemical correlates of community structure provide a logically sound alternative. Not only are fatty acids ubiquitous in nature, but as shown here, their ecological behavior is conservative compared to the environmental and organismic levels of ecosystem

Table 4. Comparative stability estimates; summer conditions.

	Station	
Ecosystem level	*Sound*	*River*
Environmental	−0.27	−0.19
Organismic	−0.05	−0.03
Chemical	0.07	0.05

organization. Thus, we can separate and classify widely ranging communities on a chemical scale, in a manner analogous to the comparative anatomist's arrangements of increasing structural complexity and functional dependencies on an evolutionary scale.

The ordination in Fig. 3 was derived from biochemical distances separating estuarine and coastal zooplankton. It was calibrated with respect to mean distances during summer conditions in Rhode Island. Superimposed on this scale were positions for the North Sea and the Mediterranean, tending strongly toward the estuarine side of the scale. The European data from which these R values were calculated (Kattner et al. 1981, Table 1; Goutx and Saliot 1980, Table IV) indicate higher 18:0 and lower 22:6 levels than the corresponding Rhode Island values given in Table 2. Also at the high or estuarine end of the R scale were the two Spanish rias. Thus in European zooplankton, this 18:0 effect extended throughout estuarine reaches and beyond headlands. But in the Western North Atlantic, a Newfoundland copepod population (Ackman and Hooper 1970) was clearly in the offshore sector of the ordination, even though its 22:6 concentration was 10% lower than that of comparable Rhode Island zooplankton, perhaps because this particular polyunsaturate decreases in cooler waters (Jeffries 1970).

The ordination must be confirmed, but for now, we speculate that the underlying cause for high 18:0 in zooplankton resulted from detritus in the diet. Compared to living phytoplankton, detritus is rich in 18:0 and low in long-chain polyunsaturates (Jeffries 1972; Schultz and Quinn 1977). Thus, as the proportion of detritus to total particulate matter increases inshore, so does the nutritional intake of 18:0 by zooplankton. An additional 18:0 input is sewage emanating from sources that are usually well within estuarine reaches, e.g. downstream from a malfunctioning sewage treatment plant, as was the case in the Providence River. Because fatty acid distributions in the diet are passed on partially intact to consumers, we expect that tissue 18:0 concentrations in primary consumers would be highest where the detrital load is the heaviest. In a trophic sense then, European zooplankton appears closer to Providence River zooplankton than it does to the condition characterizing Rhode Island Sound.

The polyunsaturate 22:6 may be synthesized internally from C 18-20 unsaturates in the diet (Gunstone 1967). A canonical correlation (Cooley and Lohnes 1971) agreed with this route. Macrozooplankton biomass and chlorophyll *a* on one side were significantly associated with macrozooplankton 18:0 and 22:6 concentrations on the other. More generally, as a result of food chain enrichment, the Providence River's organic wastes showed up in a luxurient zooplankton crop.

To cross levels of natural organization, stability was estimated at both ends of the transect. Stabilities at the organismic and biochemical levels did not differ appreciably. This similarity was important because the river received up to 60 mgd (227,100 m^3 day^{-1}) of raw sewage during both summers investigated. Neither the macrozooplankton, microzooplankton nor chlorophyll *a* concentrations gave indication of adverse effect. Rather, organic enrichment apparently provided additional food and stimulated plankton growth, until predation by the ctenophore *Mnemiopsis leidyii* drastically reduced zooplankton crops in late summer (Kremer 1976).

Once a community has been cast on a biochemical ordination, a single number replaces all the complexity in a data matrix that could include the entire suite of major fatty acid variables numbering 10 or more. Perhaps this is a partial answer to Pielou's (1972) question: "...can any single number – a descriptive statistic, coefficient, or index – be envisaged that would measure the amount of structure, or the degree of 'structuredness', of an ecological community?" In the future, biochemical scales should be calibrated with respect to areas or regions covered by the particular application. Even now, however, the ordination raises a speculative question about comparative aspects of food web dynamics on either side of the North Atlantic.

If further sampling verifies the relative positions shown in Fig. 3, a possible explanation would be that in European near-shore waters the food of macrozooplankton contains a higher proportion of detritus, rich in stearic acid (18:0), than does the food available to grazers in comparable, North American habitats.

Acknowledgments

This investigation was supported by DOE Grant No. DE-AC02-78EV04941, and a cooperative grant between the United States and Spain to K. R. Tenore, Skidaway Institute of Oceanography. For the rias samples we are indebted to Jerónimo Corral and Jose Cabanas-Lopez, Instituto Español de Oceanografia.

References Cited

Ackman, R. G. and S. N. Hooper. 1970. Analyses of fatty acids from Newfoundland copepods and sea water with remarks on the occurrence of arachidic acid. *Lipids* 5:417-421.

Allen, J. D., S. Richman, D. R. Heinle and R. Huff. 1977. Grazing in juvenile stages of some estuarine copepods. *Mar. Biol.* 43:317-331.

Conover, R. J. 1956. Oceanography of Long Island Sound, 1952-1954 VI. Biology of *Acartia clausi* and *A. tonsa*. *Bull. Bingham. Oceanogr. Coll.* 15:156-233.

Cooley, W. W. and P. R. Lohnes. 1971. *Multivariate Data Analysis*. Wiley, New York. 364 p.

Davis, J. C. 1973. *Statistics and Data Analyses in Geology*. Wiley, New York. 550 p.

Dixon, W. J. and M. B. Brown. 1979. BMDP-79, *Biomedical Computer Programs*. P-Series. U. California Press, Berkeley, CA 880 p.

Fernandez, F. 1979. Nutrition studies in the nauplius larva of *Calanus pacificus* (Copepods: Calanoida). *Mar. Biol.* 53:131-147.

Goutx, M. and A. Saliot. 1980. Relationship between dissolved and particulate fatty acids and hydrocarbons, chlorophyll *a* and zooplankton biomass in Villefranche Bay, Mediterranean Sea. *Mar. Chem.* 8:299-318.

Green, R. H. 1979. *Sampling Design and Statistical Methods for Environmental Biologists*. Wiley, New York. 257 p.

Gunstone, F. D. 1967. *An Introduction to the Chemistry and Biochemistry of Fatty Acids and their Glycerides*, pp. 179-183. Chapman and Hall, London. 209 p.

Jeffries, H. P. 1967. Saturation of estuarine zooplankton by congeneric associates, pp. 500-508. *In:* G. Lauff (ed.). *Estuaries.* Am. Assoc. Adv. Sci., Washington, DC

Jeffries, H. P. 1970. Seasonal composition of temperate plankton: fatty acids. *Limnol. Oceanogr.* 15:419-426.

Jeffries, H. P. 1972. Fatty acid ecology of a tidal marsh. *Limnol. Oceanogr.* 17:433-440.

Jeffries, H. P. 1979. Biochemical correlates of seasonal change in marine communities. *Am. Nat.* 113:643-658.

Jeffries, H. P. and R. Lambert. 1980. Biochemical correlates of complexity and change in plankton communities. *Int. Council Expl. Sea.* C.M. 1980/L:68, 10 p. mimeo.

Kattner, G., M. Krause and J. Trahans. 1981. Lipid composition of some typical North Sea copepods. *Mar. Ecol. Progr. Ser.* 4:69-74.

Kremer, P. 1976. Population dynamics of a pulsed zooplankton predator, *Mnemiopsis leidyii*, pp. 197-215. *In:* M. L. Wiley (ed.) *Estuarine Processes, Vol. I.* Academic Press, New York.

Marshall, S. M. and A. P. Orr. 1956. On the biology of *Calanus finmarchicus* IX. Feeding and digestion in the young stages. *J. Mar. Biol. Assoc. U.K.* 35:587-603.

Maxwell, A. E. 1977. *Multivariable Statistics in Behavioral Research.* Chapman and Hill, London. 164 p.

Oviatt, C. A. 1981. Some aspects of water quality in and pollution sources to the Providence River. Unpublished report., Marine Ecosystems Research Laboratory, U. Rhode Island, Kingston, 236 p. mimeo.

Paffenhofer, G.-A. and S. C. Knowles. 1978. Feeding of marine planktonic copepods on mixed phytoplankton. *Mar. Biol.* 48:143-152.

Patten, B. C. 1961. Preliminary method for estimating stability in plankton. *Science* 134:1010-1011.

Patten, B. C. 1962. Improved method for estimating stability in plankton. *Limnol. Oceanogr.* 7:266-268.

Pielou, E. C. 1972. Measurement of structure in animal communities, pp. 113-135. *In:* J. A. Weins (ed.), *Ecosystems Structure and Function.* Oregon State U. Press, Corvallis, OR

Schultz, D. M. and J. G. Quinn. 1972. Fatty acids in surface particulate matter from the North Atlantic. *J. Fish. Res. Board Can.* 29:1482-1486.

Schultz, D. M. and J. G. Quinn. 1977. Suspended material in Narragansett Bay: fatty acid and hydrocarbon composition. *Organic Geochem.* 1:27-36.

Tenore, K. R. and N. Gonzalez. 1975. Food chain patterns in the Ria de Arosa, Spain: an area of intense mussel aquaculture. *10th European Symp. Mar. Biol.* 2:601-619.

Tenore, K. R. and 17 authors. 1982. Coastal upwelling in the Rias Bajas, northwest Spain: contrasting the benthic regimes of the Rias de Arosa and de Muros. *J. Mar. Res.* (in press).

RESPONSES OF TEXAS ESTUARIES TO FRESHWATER INFLOWS

Neal E. Armstrong
Department of Civil Engineering
and
Center for Research in Water Resources
The University of Texas at Austin
Austin, Texas

Abstract: Six of the seven major estuarine systems on the Texas coast are compared using geomorphic, hydrologic, hydrographic, nutrient loading, and commercial fishery harvest indices as common bases. While the geomorphic characteristics of these systems are relatively similar, major differences are found in annual freshwater inflows, precipitation and evaporation patterns, and hydraulic residence times, as well as tidal exchange influenced residence times. Biotic composition is relatively similar except where salinity tolerance limitations reduce populations of finfish and shellfish. Nutrient budgets show that nutrient inputs are dominated by freshwater inflows, with exchange from saltwater and delta marshes and precipitation providing a small fraction of the inflow. Finfish and shellfish harvests are shown to be functions of average salinity of the estuaries as well as of areal loading rates of nutrients.

Introduction

In ecological system theory, some of the most important concepts relate to controlling environmental variables such as temperature, moisture, depth, and altitude. Historically, concern has focused on habitats found along gradients of these variables and on the structure and function of the ecosystems found within these habitats.

Aquatic ecologists have examined changes in lake, estuarine and ocean systems with temperature, with depth (particularly in the ocean), and with salinity (in estuaries). Considerable research has been devoted to the roles of temperature and especially salinity in governing estuarine communities. The Remane (1934) diagram of species composition change with salinity change is classic. On the Gulf Coast one of the early workers to give attention to the importance of salinity, temperature, and organism distribution was Gunter (1950, 1967) and Gunter et al. (1964) who examined the role of salinity and freshwater inflows in the distribution and production of penaeid shrimp. Many other workers, notably Copeland (1966) and Copeland et al. (1972), have also studied the role of freshwater inflows as contributors of nutrients to Texas estuaries and their importance to primary and secondary production. Others like Parker (1955) and Kutkuhn (1966) have been interested in the paleon-

ISBN 0-12-404070-5

tological significance of salinity as an organism distribution determinant and the importance of salinity for estuary management, respectively.

Here, Texas estuaries are considered as examples of widely variant systems in order to delineate roles of freshwater inflows as governors of salinity and carriers of nutrients and to examine how secondary production responds to those inputs.

Texas Estuaries

Bay Descriptions

Texas has about 600 km of open-ocean or Gulf shoreline and 2290 km of bay shoreline (Diener 1975). Seven major estuarine systems and three minor estuaries lie along this shoreline. These estuarine systems have a total open-water surface area of more than 607,000 ha with more than 445,000 ha of adjacent marshlands and tidal flats (Fig. 1).

The largest and most urbanized of these estuaries is the Trinity-San Jacinto estuary, more commonly called Galveston Bay. Houston, with over 3 million inhabitants, lies to the north and west. Lavaca-Tres Palacios estuary, or Matagorda Bay, lies approximately at the midpoint of the Texas Coast and is the second largest estuary and perhaps the least urbanized of the major systems. The area around this Bay is developing, however. Sabine-Neches estuary (Sabine Lake) lies on the border between Texas and Louisiana and is heavily developed industrially with the cities of Port Arthur, Beaumont, and Orange being in close proximity. Guadalupe estuary (San Antonio Bay) and

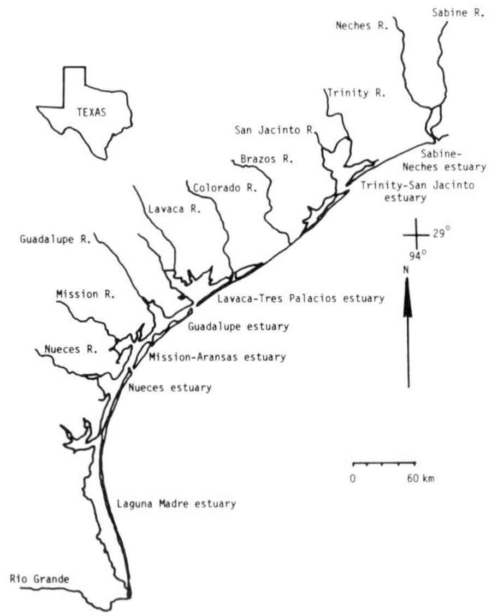

Figure 1. Location of estuaries on Texas coast.

Mission-Aransas estuary (Copano Bay and Aransas Bay) are smaller systems with tidal access to the Gulf through Aransas Pass. The Nueces estuary (Corpus Christi Bay) is the southern-most major estuary considered here. Corpus Christi is located on its western shores.

South of Corpus Christi Bay is Laguna Madre (and Baffin Bay). This system is shallow and hypersaline, lying behind the barrier island with very limited tidal access to the Gulf of Mexico. The minor estuaries on the Texas Coast include Brazos River, Colorado River, and Rio Grande River estuaries. Only the six estuaries from Corpus Christi Bay northward are considered in this paper.

Geomorphic Features

The geomorphic features of these six estuarine systems are given in Table 1. Sabine-Neches estuary is the smallest with a surface area of 17,798 ha while Trinity-San Jacinto estuary is the largest at 143,153 ha. These same estuaries are the smallest and largest respectively in terms of volume. Texas estuaries are quite shallow with average depths ranging from 1.4 m in Guadalupe estuary to 2.4 m in Nueces estuary.

Hydrologic Features

Hydrologic features of Texas estuaries are of particular importance in determining their salinity regimes and nutrient inflows. Precipitation on Sabine-Neches estuary averages 151.7 cm yr^{-1} while that on Nueces estuary is just less than half that amount, or 74.2 cm yr^{-1} (Table 1). This change in precipitation from southeast Texas to the lower coast is representative of the general pattern of precipitation in Texas. Highest rainfalls occur in the eastern portion of the state diminishing to near zero in the western portion. Annual rainfall amounts diminish 10 mm for ever 9.5 km as one moves east to west. Evaporation rates follow a reverse trend, amounting to 112.4 cm yr^{-1} in Sabine-Neches estuary and increasing to 150.7 cm yr^{-1} in Nueces estuary (Table 1).

Freshwater inflows to Texas estuaries follow the pattern of precipitation and evaporation. Sabine-Neches estuary receives the highest combined inflow (includes overland runoff, return flows, and diversion) of 16.05 km^3 yr^{-1} whereas Mission-Aransas estuary receives the lowest at 0.48 km^3 yr^{-1} (Table 1). Taking into account evaporation from water surfaces in the estuaries, net inflows (combined inflows plus direct precipitation minus evaporation) range from 16.14 km^3 yr^{-1} in Sabine-Neches estuary to 0.19 km^3 yr^{-1} in Mission-Aransas area. The surplus of precipitation over evaporation in the Sabine-Neches and Trinity-San Jacinto estuaries produces a higher net inflow than combined inflow while the reverse is true in the other four estuaries (Table 1).

Water Quality

Water quality has been monitored through a number of studies conducted primarily by the Texas Department of Water Resources and to a lesser

Table 1. Geomorphic, hydrographic, and hydrologic features of Texas estuarine systems.

Estuarine System	*Surface Area (ha)*	*Volume*[2] *(km^3)*	*Average Depth*[3] *(m)*	*Precipitation*[1,5] *(cm yr^{-1})*	*Evaporation*[1,5] *(cm yr^{-1})*	*Combined Inflow*[1,6] *(km^3 yr^{-1})*	*Net Inflow*[1,7] *(km^3 yr^{-1})*
Sabine-Neches	17,798	0.326	1.8[4]	151.7	112.4	16.05	16.14
Trinity-San Jacinto	143,153	2.911	2.1	134.8	118.8	12.05	12.28
Lavaca-Tres Palacios	101,368	2.134	2.3	105.6	143.0	3.62	3.24
Guadalupe	56,162	0.754	1.4	97.9	142.4	2.80	2.55
Mission-Aransas	46,279	0.925	2.0	88.6	151.3	0.48	0.19
Nueces	44,451	1.147	2.4	74.2	150.7	0.84	0.51

[1]From Texas Department of Water Resources (1980a, b; 1981a, b, c).
[2]From Currington et al. (1966) except for Sabine Estuary
[3]Calculated as volume/surface area except for Sabine Estuary
[4]From Ward (1973)
[5]Calculated as precipitation or evaporation volume per year divided by surface area
[6]Freshwater inflows from overland runoff, return flows, and diversions
[7]Combined inflows plus precipitation minus evaporation

extent by other government entities in the state. These studies have been summarized recently by the Department in a series of reports (Texas Department of Water Resources 1980a, b; 1981a, b, c) dealing with the importance of freshwater inflows, and in other ongoing studies (Ward and Armstrong 1979). Based on the data summarized in those reports, the average concentrations of salinity, total organic carbon (TOC), total nitrogen (principally inorganic nitrogen, N), and total phosphorus (P) are summarized in Table 2.

Average salinities range from a low of 2 ppt in the Sabine-Neches estuary to 27 ppt in Nueces estuary. Average TOC concentrations were less than 10 mg l^{-1} in the three estuaries for which data were available. Average total inorganic nitrogen concentrations ranged from below 0.04 mg l^{-1} in the Lavaca-Tres Palacios estuary to a high of 0.3 mg l^{-1} in the Sabine-Neches estuary. If one takes into account the organic nitrogen fraction, Sabine-Neches estuary nitrogen concentrations would approach 1 mg l^{-1} as would the values for the Trinity-San Jacinto estuary. Average phosphorus concentrations mirrored to some extent the nitrogen concentrations with concentrations reaching 0.27 mg l^{-1} and falling to 0.06 mg l^{-1}. The lowest values, however, are still high enough to support substantial populations of algae in these systems. It has been determined through many previous studies (e.g. Davis 1973) that Texas estuaries are largely nitrogen limited, and thus Sabine-Neches and Trinity-San Jacinto estuaries and possibly Guadalupe estuary are the only ones that have nitrogen concentrations high enough to support substantial algae populations.

Biota

Recent information on phytoplankton, zooplankton, and benthic animal populations in Texas estuaries is limited primarily to those studies supported by the Texas Department of Water Resources (1980a, b; 1981a, b, c). These studies were conducted over different time periods and bay areas (e.g. only on the upper reaches of the Guadalupe and Lavaca-Tres Palacios estuaries), so comparisons presented below are imperfect. Substantial

Table 2. Annual average water quality in Texas estuarine systems. – = no data.

	Nutrients			
Estuarine System	*TOC* (*mg l^{-1}*)	*Total N* (*mg l^{-1}*)	*Total P* (*mg l^{-1}*)	*Salinity* (*g l^{-1}*)
Sabine-Neches	9	0.3	0.07	2.3
Trinity-San Jacinto	8	0.12	0.27	17.6
Lavaca-Tres Palacios	5	0.04	0.05	23
Guadalupe	–	0.20	0.18	11
Mission-Aransas	–	0.06	0.07	13
Nueces	–	0.06	0.06	27

amounts of information pertaining to shellfish and finfish populations are available from Texas Parks and Wildlife studies and commercial fisheries landings tabulated by the National Marine Fisheries Service. While catch data are often considered suspect for estimating yields of finfish and shellfish from estuaries, the data for Texas estuaries have been summarized to a great extent and evaluated for their reliability in the report series prepared by the Texas Department of Water Resources, and that information will be used here.

Phytoplankton populations consisted primarily of diatoms and green algae in the Sabine-Neches, Trinity-San Jacinto, Guadalupe, and Nueces estuaries. Green algae were the dominant forms in Guadalupe estuary while phytoflagellates dominated Lavaca-Tres Palacios estuary. Only in the Mission-Aransas estuary did blue-green algae dominate, and these were primarily filamentous forms. Average phytoplankton population densities are given in Table 3 and are shown to be largest in the upper reaches of Guadalupe and upper Lavaca-Tres Palacios estuaries.

Zooplankton populations were dominated by arthropods, primarily *Acartia tonsa*. The only estuaries in which this was not the case were Lavaca-Tres Palacios and Guadalupe estuaries, but freshwater forms influenced populations significantly. Again, average population densities are given in Table 3 and appear to be relatively constant from estuary to estuary. In Sabine estuary there was a definite correlation between population density and river inflow (Texas Department of Water Resources 1981c) while in other estuaries this correlation was evident during major flood events only (Texas Department of Water Resources 1980a, b; 1981a, b).

Benthic animal populations consisted primarily of polychaetes and bivalve and gastropod molluscs. Average population densities were 640 organisms m^{-2} or less in all estuaries except for Trinity Bay in Trinity-San Jacinto estuary where population densities averaged 9400 organisms m^{-2}. Considerable density variations were found spatially and temporally.

Table 3. Average concentrations of phytoplankton, zooplankton, and benthic animals in Texas estuaries.

Estuarine System	*Date of Study*	*Phytoplankton (10^6 cells l^{-1})*	*Zooplankton (10^3 ind. m^{-3})*	*Benthic Organisms (10^3 ind. m^{-2})*
Sabine-Neches	1974-75	0.11	7.2	0.31
Trinity-San Jacinto	1975-76	0.17	10	9.94
Lavaca-Tres Palacios	1973-75	3.7	12.5	0.18
Guadalupe	1973-74	8.88	12	0.17
Mission-Aransas	1972-75	0.10	10	0.14
Nueces	1972-75	0.28	10	0.64

The major commerical shellfish organisms in Texas estuaries are shrimp, blue crab (*Callinectes sapidus* Rathbun), and the bay oyster (*Crassostrea virginica* Gmelin). Of the shrimp, the white shrimp (*Penaeus setiferus* Linnaeus) is taken in the largest numbers, with brown (*P. aztecus* Ives) and pink shrimp (*P. duorarum* Burkenroad) in lower numbers (Table 4). Trinity-San Jacinto estuary has yielded an average of 3.9 million kg of shellfish per year from 1942 to 1976. Next in magnitude are Lavaca-Tres Palacios, Guadalupe, Mission-Aransas, and Nueces estuaries. On an areal basis, Trinity-San Jacinto estuary had the highest yield of 27.2 kg ha^{-1} yr^{-1} followed by Sabine-Neches, Guadalupe, Mission-Aransas, and Lavaca-Tres Palacios estuaries. The "small" yield of the Lavaca-Tres Palacios estuary has been attributed to underfishing (Copeland in Currington et al. 1966). Of the shellfish harvest, white shrimp and blue crab dominated except in Trinity-San Jacinto estuary where oysters made up about one-third of the total catch (Table 4).

Finfish harvest was substantially less than shellfish harvest (Table 5). The highest average annual bay yield was in Mission-Aransas estuary with 0.27 million kg yr^{-1} while the lowest yield was from Sabine-Neches estuary with <0.01 million kg yr^{-1}. On an areal basis, Mission-Aransas estuary pro-

Table 4. Average commercial shellfish harvest in Texas estuarine systems from 1942-1976[1]. A = 10^6 kg yr^{-1}; B = kg ha^{-1} yr^{-1}; C = % of Total.

Estuarine System	*Units*	*White Shrimp*	*Brown & Pink Shrimp*	*Blue Crab*	*Bay Oyster*	*Total*
Sabine-Neches	A	0.0811	0.0063	0.333	–	0.420
	B	4.56	0.354	18.71	–	23.6
	C	19.3	1.5	79.3	–	97.9[2]
Trinity-San Jacinto	A	1.405	0.482	0.729	1.282	3.898
	B	9.82	3.37	5.09	8.96	27.23
	C	36.0	12.4	18.7	32.9	96.1
Lavaca-Tres Palacios	A	0.722	0.186	0.354	0.114	1.376
	B	7.12	1.83	3.49	1.12	13.57
	C	52.5	13.5	25.8	8.3	91.0
Guadalupe	A	0.381	0.118	0.389	0.100	0.981
	B	6.78	2.10	6.92	1.78	17.46
	C	38.9	12.1	39.6	10.2	90.1
Mission-Aransas	A	0.307	0.126	0.313	0.0097	0.756
	B	6.63	2.72	6.76	0.210	16.34
	C	40.6	16.7	41.4	1.3	73.4
Nueces	A	0.158	0.073	0.036	–	0.267
	B	3.56	1.64	0.81	–	6.01
	C	59.2	27.4	13.5	–	68.9

[1]Taken from Texas Department of Water Resources (1980a, b; 1981a, b, c).
[2]Percent of total commercial fishery harvest (including finfish and shellfish) – see Table 5

Table 5. Average commercial finfish harvest in Texas estuarine systems from 1942-1976[1]. A = 10^6 kg yr^{-1}; B = kg ha^{-1} yr^{-1}; C = % of Total.

Estuarine System	*Units*	*Spotted Seatrout*	*Red Drum*	*Black Drum*	*Other*	*Total*
Sabine-Neches	A	0.00442	0.00254	0.00039	0.00172	0.00907
	B	0.248	0.143	0.0221	0.0966	0.510
	C	48.7	28.0	4.3	19.0	2.1[2]
Trinity-San Jacinto	A	0.0752	0.0165	0.0215	0.0440	0.1572
	B	0.525	0.115	0.150	0.307	1.098
	C	47.8	10.5	13.7	28.0	3.9
Lavaca-Tres Palacios	A	0.0610	0.0331	–	0.0420	0.136
	B	0.602	0.326	–	0.414	1.342
	C	44.8	24.3	–	30.9	9.0
Guadalupe	A	0.0291	0.0368	0.0270	0.0146	0.108
	B	0.518	0.655	0.481	0.260	1.922
	C	27.0	34.2	25.1	13.5	9.9
Mission-Aransas	A	0.0925	0.0798	0.056	0.0450	0.274
	B	2.000	1.724	1.21	0.972	5.921
	C	33.8	22.2	20.6	16.4	26.6
Nueces	A	0.030	0.029	0.038	0.024	0.120
	B	0.675	0.652	0.855	0.54	2.70
	C	24.5	24.1	31.6	20.0	31.1

[1]Taken from Texas Department of Water Resources (1980a, b; 1981a, b, c)
[2]Percent of total commercial fishery harvest (including finfish and shellfish) – see Table 4

vided the highest yield of 5.92 kg ha^{-1} yr^{-1}, followed by Nueces, Guadalupe, Lavaca-Tres Palacios, Trinity-San Jacinto, and Sabine-Neches estuaries. Major components of the finfish harvest were spotted seatrout (*Cynoscion nebulosus* Cuvier), red drum or redfish (*Sciaenops ocellata* Linnaeus), and black drum (*Pogonis cromis* Linnaeus), while the balance of the catch was croaker (*Micropogon undulatus* Linnaeus), flounder (*Paralichthys* spp.), sea catfish (*Arius felis* Linnaeus), and sheepshead (*Archosargus probatocephelus* Walbaum). Spotted seatrout dominated the catch in Sabine-Neches, Trinity-San Jacinto and Lavaca-Tres Palacios estuaries while the catch was relatively evenly divided among spotted seatrout and drums in the other bays.

Shellfish dominated the commercial fishery harvest (Tables 4, 5). Almost 98% of the total commercial catch by weight in the Sabine-Neches estuary was shellfish; less than 3% was finfish. The lowest percentage for shellfish was 69% in Nueces estuary.

These changes in catch composition among the Texas estuaries reflect the importance of salinity as a controlling variable for distributions of organisms. As shown in Fig. 2, the percent of catch that is shellfish increases with increased freshwater content of the estuary while finfish harvest decreases. Influencing the shellfish trend most are the white shrimp at higher

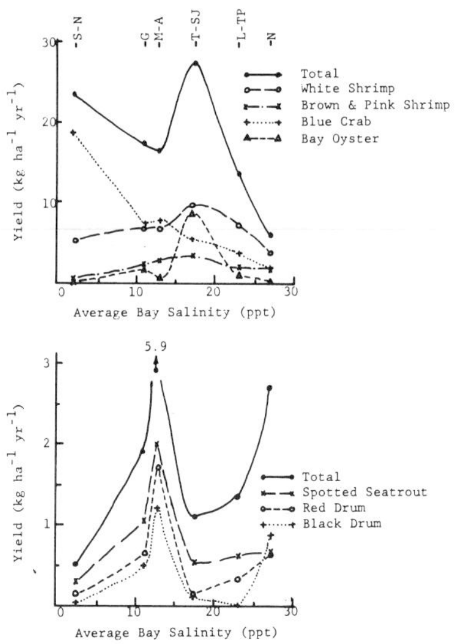

Figure 2. Average annual yield of shellfish and finfish in Texas estuaries as a function of average bay salinity.

salinities and blue crab at lower salinities. The bay oyster is taken in large quantities in Trinity-San Jacinto estuary and obscures the pattern there. Finfish catch is strongly influenced by spotted seatrout, black drum, and red drum. Childress et al. (1975) have addressed some of these relationships in the Guadalupe estuary.

Estuarine Comparisons

Framework

Comparison of the Texas estuaries will be made in several ways. First, geomorphic and hydrologic characteristics of each will be considered using hydraulic residence times with and without the influence of tidal exchange. Second, a comparison will be made of the nutrient budget for each estaury on the assumption that primary and secondary productivity of the systems are, in part at least, dependent on nutrients brought into each estuary by freshwater inflows, precipitation, and exchange with fresh, brackish, and saltwater marsh systems. Third, comparisons will be made based on yields of commercial fisheries, specifically shellfish and finfish, as a function of the salinity regime of the estuary and of nutrient loading.

Hydrographic Comparisons

The hydrographic regime of an estuary is primarily a function of freshwater inflow and tidal exchange, and Pritchard (1967) has shown that estuaries may be classified using these factors. Indeed, one classification scheme uses the ratio of freshwater inflow during a tidal cycle to the tidal prism, and states that estuaries with a ratio of inflow to tidal prism greater than one are the saltwedge type, while those with lower ratios are partially mixed or completely mixed systems.

The first step in comparing Texas estuaries hydrographically might be to compare these hydraulic residence times. If one calculates residence time simply as the volume of the estuary, V, divided by the combined freshwater inflow, Q (except for precipitation), then one obtains the average residence time, t_o, of water flowing through this estuary as if it were a simple, well-mixed reactor with the only mechanism of flushing being freshwater inflow. Under such an assumption (admittedly a large assumption because geomorphic characteristics are greatly simplified), residence times in Texas estuaries range from 0.02 yr, or just over 7 days, in the Sabine-Neches estuary up to 1.93 yr in the Mission-Aransas estuary (Table 6). If one accounts for the inflow of freshwater as well as precipitation and evaporation, or net inflow (Q′), residence times (t_o') range from 0.02 yr in Sabine-Neches estuary to 4.87 yr in the Mission-Aransas estuary (Table 6). Note that residence times have shortened for Sabine-Neches and Trinity-San Jacinto estuaries since precipitation exceeds evaporation in those two systems, but they have increased in the other four estuaries because evaporation exceeds precipitation.

While these figures may be of interest, it is unrealistic to describe residence time in estuaries based on this simple hydraulic calculation; one must take into account additional dilution due to tidal exchange. To do so one may use the single layer model of Bowden (1963) with which one calculates the amount of freshwater in the estuary based on the salinity of the system. One then computes the amount of freshwater it would take to flush that freshwater from the system. The equation to be used is

$$t_f = f \frac{V}{Q'}$$

where t_f = residence time (yr); V = bay volume (km^3); Q′ = net freshwater inflow ($km^3\ yr^{-1}$); f = average freshwater content = $(S-S_o)/S_o$; S =

Table 6. Residence times in Texas estuaries. $t_o = V/Q$; $t'_o = V/Q'$; $f = (S - S_o)/S_o$; $t_f = f\,V/Q'$

	Hydraulic Residence Times		*Single Layer Model*		
Estuarine System	t_o *(yr)*	t'_o *(yr)*	*f*	*Salinity* *(g l^{-1})*	t_f *(yr)*
Sabine-Neches	0.02	0.02	0.94	2.3	0.02
Trinity-San Jacinto	0.24	0.23	0.48	17.6	0.11
Lavaca-Tres Palacios	0.59	0.65	0.32	23	0.21
Guadalupe	0.27	0.28	0.68	11	0.19
Mission-Aransas	1.93	4.87	0.62	13	3.02
Nueces	1.36	2.25	0.20	27	0.46

average bay salinity (g l^{-1}); and S_o = seawater salinity (34 g l^{-1} for Gulf of Mexico coastal water). Using this procedure, residence times range from 0.02 yr in the Sabine-Neches estuary to 3.0 yr in the Mission-Aransas estuary. Residence time in Sabine-Neches estuary has changed little from the above values due almost entirely to the fact that tidal exchange is very small in that estuary compared to river flow. On the other hand, residence time in the Mission-Aransas estuary has been reduced considerably (Table 6).

The very short residence time in the Sabine-Neches system implies that allochthonous materials brought into this system are flushed to the Gulf with little reaction occurring except for the organic or inorganic nutrient forms readily available for biological uptake. In Lavaca-Tres Palacios estuary with a residence time of 0.19 yr, or about 2.5 months, and Guadalupe and Nueces estuary with larger residence times, allochthonous materials certainly experience reactions before being removed from the system. With the long residence time in Mission-Aransas estuary, it is likely that internal nutrient cycling mechanisms dominate. Mission-Aransas estuary is somewhat unique in that it interacts strongly with Guadalupe and Nueces estuaries, and in fact much of the water which passes through the lower portion of Mission-Aransas estuary may have come from the Guadalupe estuary.

Nutrient Loading

The nutrient budget compiled for each of the estuarine systems is given in Table 7. These nutrient budgets were compiled for carbon, nitrogen, and phosphorus derived from river inflows, precipitation, and marshes using a number of data sources, especially the Texas Department of Water Resources reports (1980a, b; 1981a, b, c) and reports being prepared for Matagorda Bay (Ward and Armstrong 1979; Armstrong and Gordon 1981) as part of a U.S. Fish and Wildlife Service supported, three-year study of that system. To calculate nutrient inflows from rivers, data for nutrient concentrations in rivers and streams feeding each estuary were determined from available literature, combined with the respective annual average freshwater inflows from 1942 to 1976 (totals given in Table 1), and flux values calculated. These flux values are approximations only since freshwater inflows not only vary considerably seasonally and historically but nutrient concentrations are highly variable from month to month.

Nutrients from marsh systems were considered to be derived by two primary mechanisms. The first was through tidal (solunar and wind) inundation of saltwater and brackish water marshes and the concomitant transfer of nutrients from the marsh to adjacent estuarine waters. Exchange rates used have been determined through laboratory microcosm experiments (Dawson and Armstrong 1975; Armstrong and Brown 1977; Armstrong and Gordon 1978a, b; Armstrong et al. 1978), nutrient flux measurements in the field (Armstrong et al. 1975; Espey Huston and Assoc. 1977a, b) and summaries of work elsewhere (Armstrong and Hinson 1978). Exchange rates measured

Table 7. Carbon, nitrogen and phosphorus budgets for Texas estuaries. All values are 10^6 kg yr^{-1}, except for Areal Loading (g m^{-2} yr^{-1}). – = no data available.

Nutrient	Estuarine System	Freshwater Inflows	Marshes		Precipitation	Total	Areal Loading
			Tidal	Flood			
Carbon	Sabine-Neches	115.70	2.50	1.445	–	119.64	672.2
	Trinity-San Jacinto	103.44	4.02	0.143	–	107.60	75.2
	Lavaca-Tres Palacios	75.75	5.35	0.137	–	81.62	80.5
	Guadalupe	17.95	0.882	0.300	–	19.13	34.1
	Mission-Aransas	5.98	–	–	–	5.98	12.9
	Nueces	8.21	0.091	0.419	–	17.74	39.9
Nitrogen	Sabine-Neches	9.32	0.0251	0.361	0.043	9.749	54.8
	Trinity-San Jacinto	11.58	0.116	0.036	0.344	12.08	8.4
	Lavaca-Tres Palacios	3.584	0.0535	0.0341	0.510	4.330	4.3
	Guadalupe	5.802	0.0088	0.075	0.135	6.021	10.7
	Mission-Aransas	0.443	–	–	0.111	0.554	2.0
	Nueces	0.546	0.018	0.005	0.107	0.676	1.52
Phosphorus	Sabine-Neches	0.745	0.50	0.413	0.005	1.213	6.82
	Trinity-San Jacinto	3.632	0.091	0.041	0.043	3.807	2.66
	Lavaca-Tres Palacios	1.309	0.107	0.039	0.037	1.542	1.52
	Guadalupe	1.002	0.0176	0.086	0.017	1.123	2.00
	Mission-Aransas	0.0706	–	–	0.014	0.081	0.18
	Nueces	0.219	0.182	0.120	0.013	0.534	1.20

in all studies were quite variable, and several of the studies indicated that the marshes were acting as sinks for nutrients rather than as sources. It was assumed for this study that tidal inundation nutrient exchange rates were 0.05 kg TOC ha^{-1} d^{-1}, 0.005 kg N ha^{-1} d^{-1}, and 0.01 kg P ha^{-1} d^{-1} (positive values imply export). These rates were applied to all estuarine systems except Mission-Aransas which has little inundatable marsh.

The other mechanism for transfer of nutrients from marsh to adjacent waters was by flooding. At the head of most of the Texas estuaries lie extensive river deltas in which brackish and freshwater marsh may be found. During the year, especially in spring, high rainfalls cause floods which inundate varying amounts of the delta marshes for several days. It is expected that nutrients in the particulate form are transported to the estuary during the inundation and the subsequent dewatering of the delta marshes. The Texas Department of Water Resources through its various studies of the estuaries developed inundation models to predict the amount of delta marsh inundated for a given flood flow. Using the exchange rates derived from laboratory microcosm studies by Dawson and Armstrong (1975), Armstrong and Gordon (1981) estimated that, over a three-day inundation period, 42 kg TOC ha^{-1}, 10.5 kg N ha^{-1}, and 12 kg P ha^{-1} would be derived. Through a statistical analysis of historical floods and using the median number of times these floods occurred, the marsh area inundated for a given flood, and the nutrients exchanged per unit area, it was possible to estimate the mass of nutrients transported from these delta marshes each year due to marsh inundation and dewatering.

Nutrient flux to the estuary water surface via precipitation was determined through rainfall sampling at several coastal points (Armstrong and Gordon 1981; H. M. Liljestrand, University of Texas at Austin, pers. comm.). Flux rate for total nitrogen was 0.425 kg N ha^{-1} yr^{-1} and for phosphorus was 0.031 kg P ha^{-1} yr^{-1}. These flux rates were multiplied by the water surface area to obtain the mass of nutrients derived from that source.

Nutrients reaching the estuaries from these three sources were summed for each estuary and that total divided by the estuary surface area to determine the areal loading in units of g nutrients m^{-2} yr^{-1}. Nutrients from other sources such as seagrasses, tidal exchange, and waste discharges are small compared to the above three sources (Armstrong 1979; Armstrong and Gordon 1981) and have not been included here.

Several interesting comparisons may be made among the Texas estuaries based on nutrient loading (Table 7). First, because of its large freshwater inflows, Sabine-Neches estuary receives the highest areal loading of any of the six estuaries. In fact, total organic carbon loading to this system is 672 g m^2 yr^{-1} compared to the next highest rates of 80.5 g m^{-2} yr^{-1} in Lavaca-Tres Palacios and 75.2 g m^{-2} yr^{-1} for Trinity-San Jacinto estuary. Similar patterns exist for nitrogen and phosphorus. The decrease in areal loading as one moves from Sabine-Neches estuary to Nueces estuary reflects primarily the reduction of freshwater inflows.

Second, the nutrients derived from freshwater inflows dominate the nutrient budget of these estuaries. In all cases, freshwater inflows account for over 80% of the nutrients reaching each system. Marshes contribute only a small fraction of the nutrients reaching the estuaries either through tidal exchange or through flood inundation and dewatering. Precipitation is also very low but in some cases contributes as much or more nitrogen and phosphorus than do marshes.

Third, when one compares the nutrients derived by flooding and dewatering of delta marshes with nutrients derived from tidal exchange, it appears that the former is a much smaller source of nutrients. Only in the case of Sabine-Neches estuary for nitrogen and phosphorus and in Guadalupe estuary for phosphorus does the flood-dewatering mechanism produce more nutrients than tidal inundation. To be sure, floods occur during a portion of the year only (spring) and during that short portion could provide a larger pulse of nutrients, perhaps in a chemical form more amenable to use by the estuary, than would tidal inundation. However, in both cases it is the particulate form of nutrients which are exported from the marsh rather than the inorganic forms.

What the nutrient budgets show is that freshwater inflows do indeed contribute the majority of nutrients reaching Texas estuarine waters and that marshes, while contributing substantial amounts of nutrients, do not make up a significant portion of this budget. Their role as suppliers of nutrients has been brought into question recently by several studies on the East and Gulf coast as discussed by de la Cruz (1980). Earlier summaries by Day et al. (1973) and Odum et al. (1973) discussed the importance of marshes as nutrient sources and the latter mentioned freshwater inflows specifically but provided no quantitative comparisons. Further, Haines (1976) has shown through carbon ratio studies that organics being exported from the marshes are predominantly in the form of algae. Cooper (1974) has postulated that most primary production of the higher marsh macrophytes may actually go into peat production rather than being exported. The small overall contribution of nutrients by Texas marshes may be primarily due to the small marsh areas available for tidal inundation and the relative infrequency of delta marsh inundation. The role of marshes as finfish and shellfish habitat is not diminished by this small role in nutrient inflows.

Nutrient Loading Versus Fishery Yields

It was shown earlier that the composition of the shellfish and finfish harvest is greatly dependent on the salinity regime. Some variability is noted in these patterns which might be related to nutrient loading. If, for example, salinity and the areal loading rates of carbon, nitrogen, and phosphorus are correlated with fishery yields on a response surface plot, some patterns may be discerned. One can express the salinity regime as the fraction, f, of

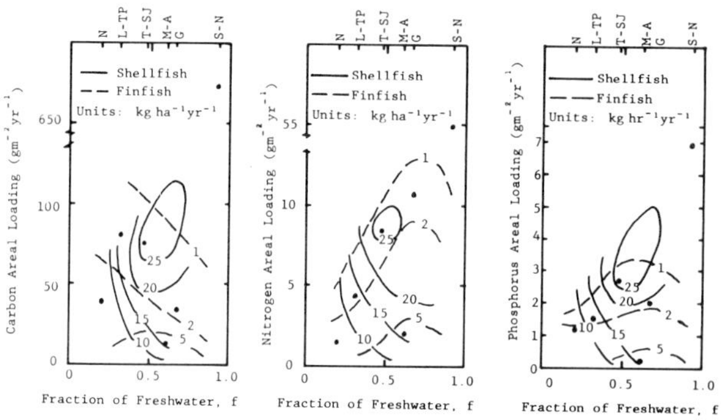

Figure 3. Influence of annual areal loading rates of nutrients and estuarine freshwater content on average annual areal yields of finfish and shellfish.

freshwater in the estuary, i.e., $(S - S_o)/S_o$; the nutrient areal loading rate as grams nutrient reaching the estuary per square meter of water surface per year; and finfish and shellfish yields as kilograms per hectare per year, and develop response surfaces (Fig. 3). The general pattern that results is as might be expected – an increase in shellfish yield with an increase in freshwater content (decrease in salinity) due primarily to the increase in yield of white shrimp, and a concurrent increase in yield with increase in nutrient areal loading. This pattern demonstrates again that salinity is a major environmental controlling variable and that nutrient loading stimulates directly or indirectly the detrital food chain through which the shellfish feed. The correlations with salinity also imply a correlation with freshwater inflow. An inverse pattern is indicated for finfish with lowest yields coming from Sabine-Neches estuary with its very low salinity, and highest yields occurring in the long residence times systems such as Mission-Aransas estuary. Not only do these estuaries contain the higher salinities preferred by the species making up the majority of the finfish harvest, but the systems contain the forage fish and zooplankton populations which provide the bulk of their food.

Conclusions

This brief examination of some Texas estuaries provides the following conclusions:

1. The Texas estuaries are widely variant primarily in the amounts of freshwater that reach them and as result of the precipitation and evaporation patterns over the coastline;

2. Average residence times in the estuaries vary from 0.02 yr to 3.02 yr taking into account the hydraulic flushing of the estuary and tidal exchange, and salinity regimes in these estuaries reflect these resident times;

3. Nutrient loadings to the Texas estuaries come primarily from freshwater inflows with small contributions from marshes and precipitation;

4. Annual average commercial bay shellfish harvests are directly

related to increasing freshwater content of the estuaries and to nutrient loading; and

5. Annual average commercial bay finfish harvests appear to be inversely related to increasing freshwater content of the estuaries and the nutrient loading, but internal nutrient cycling mechanisms may be more important in high salinity systems offsetting lack of external nutrient sources.

References Cited

Armstrong, N. E. 1979. Nutrient exchange studies on the seagrasses of Texas. Center for Research in Water Resources, U. Texas at Austin. CRWR-161. 91 pp.

Armstrong, N. E. and B. A. Brown. 1977. Exchange of carbon, nitrogen, and phosphorus in Lavaca Bay, Texas marshes. Volume 1. The role of sediments and nutrients exchange in the Lavaca Bay brackish marsh system. Center for Research in Water Resources, U. Texas at Austin. CRWR-147. 82 pp.

Armstrong, N. E. and V. N. Gordon. 1978a. Exchange rates for carbon, nitrogen, and phosphorus in the Colorado River delta marshes. Center for Research in Water Resources, U. Texas at Austin. CRWR-153. 37 pp.

Armstrong, N. E. and V. N. Gordon. 1978b. Exchange rates for carbon, nitrogen, and phosphorus in Nueces and San Antonio Bay marshes. Center for Research in Water Resources, U. Texas at Austin. CRWR-152. 144 pp.

Armstrong, N. E. and V. N. Gordon. 1981. Nutrient budget for Matagorda Bay. Center for Research in Water Resources, U. Texas at Austin. CRWR-181. 92 pp.

Armstrong, N. E., S. E. Harris and V. N. Gordon. 1978. Exchange rates for carbon, nitrogen, and phosphorus in the Trinity River delta marshes. Center for Research in Water Resources, U. Texas at Austin. CRWR-154. 109 pp.

Armstrong, N. E. and M. O. Hinson, Jr. 1978. Influence of flooding and tides on nutrient exchange from a Texas marsh, pp. 365-379. *In:* M. L. Wiley (ed.), *Estuarine Interactions*. Academic Press, New York.

Armstrong, N. E., M. O. Hinson, Jr., J. A. Collins and E. G. Fruh. 1975. Biogeochemical cycling of carbon, nitrogen, and phosphorus in saltwater marshes of Lavaca Bay, Texas. Report to the Texas Water Development Board by the Center for Research in Water Resources, U. Texas at Austin, CRWR-122. 235 pp.

Bowden, K. F. 1963. Circulation and diffusion, pp. 15-36. *In:* G. H. Lauff (ed.), *Estuaries*. Amer. Assoc. Adv. Sci., Washington, DC.

Childress, R. E., E. Haggen, E. Bradley and S. Williamson. 1975. The effects of freshwater inflows on hydrological and biological parameters in the San Antonio Bay system, Texas. Coastal Fisheries Branch, Texas Parks and Wildlife Department, Austin, TX 184 pp.

Cooper, A. W. 1974. Salt marshes, pp. 55-98. *In: Coastal Ecological Systems of the United States, Volume 2*. The Conservation Foundation, Washington, DC.

Copeland, B. J. 1966. Effects of decreased river flow on estuarine ecology. *Wat. Poll. Ctrl. Fed.* 38:1831-1839.

Copeland, B. J., H. T. Odum and D. C. Cooper. 1972. Water quantity for preservation of estuarine ecology, pp. 107-126. *In:* E. F. Gloyna and W. L. Butcher (eds.), *Conflicts in Water Resources Planning*, Proceedings of Water Resources Symposium Number 5, Austin, Texas, November 1970, Center for Research in Water Resources, U. Texas at Austin.

Currington, H. W., D. M. Wells, F. D. Masch, B. J. Copeland and E. F. Gloyna. 1966. Return flows – impact on the Texas bay systems. Report prepared for the Texas Water Development Board by Bryant-Currington, Engineers, Austin, TX 419 pp.

Davis, E. M. 1973. An investigation of tertiary estuary productivity and nutrient optimization in

Texas coastal water. Report to the Texas Water Development Board, Austin, TX 355 pp.

Dawson, A. J. and N. E. Armstrong. 1975. Exchange of carbon, nitrogen, and phosphorus in Lavaca Bay, Texas marshes. Volume II. The role of plants in nutrient exchange in Lavaca Bay brackish marsh system. Report to the Texas Department of Water Resources, Center for Research in Water Resources, U. Texas at Austin, CRWR-129. 125 pp.

Day, J. W., Jr., W. G. Smith and C. S. Hopkinson, Jr. 1973. Some trophic relationships of marsh and estuarine areas, pp. 115-130. *In:* R. H. Chabreck, (ed.), *Proceedings of the Coastal Marsh and Estuary Management Symposium*. Division of Continuing Education, Louisiana State U., Baton Rouge, LA.

de la Cruz, A. A. 1980. Recent advances in our understanding of salt marsh ecology, pp. 51-65. *In:* P. L. Fore and R. D. Peterson (eds.), *Proceedings of the Gulf of Mexico Coastal Ecosystems Workshop*, U.S. Fish and Wildlife Service, Coastal Ecosystems Project and Office of Environment, Albuquerque, NM.

Diener, R. A. 1975. Cooperative Gulf of Mexico estuarine inventory and study – Texas: area description. National Marine Fishery Service Circular 393. 129 pp.

Espey, Huston & Assoc. Inc. 1977a. Marsh biota and nutrient exchange in three Texas estuaries. Report to the Texas Water Development Board, Austin, TX 453 pp.

Espey, Huston and Assoc. Inc. 1977b. Marsh plant production and potential detritus in Lavaca, San Antonio and Nueces Bays. Report to the Texas Water Development Board, Austin, TX 53 pp.

Gunter, G. 1950. Seasonal population changes and distributions as related to salinity, of certain invertebrates of the Texas coast, including the commercial shrimp. *Publ. Inst. Mar. Sci.* 1:7-51.

Gunter, G. 1967. Some relationships of estuaries to the fisheries of the Gulf of Mexico, pp. 621-638. *In:* G. H. Lauff (ed.), *Estuaries*. Amer. Assoc. Adv. Sci., Washington, DC.

Gunter, G., J. Y. Christmas and R. Killebrew. 1964. Some relations of salinity to population distributions of local estuarine organisms, with special reference to penaeid shrimp. *Ecology* 45:181-185.

Haines, E. B. 1976. Stable carbon isotope ratios in the biota, soils and tidal water of a Georgia salt marsh. *Est. Coastal Mar. Sci.* 4:609-616.

Kutkuhn, J. H. 1966. The role of estuaries in the development and perpetuation of commercial shrimp resources. American Fishery Society, Special Publications 3:16-36.

Odum, W. E., J. C. Zieman and E. J. Heald. 1973. The importance of vascular plant detritus to estuaries, pp. 91-114. *In:* R. H. Chabreck (ed.), *Proceedings of the Coastal Marsh and Estuary Management Symposium*, Div. Continuing Education, Louisiana State U., Baton Rouge, LA.

Parker, R. H. 1955. Changes in the invertebrate fauna, apparently attributable to salinity changes, in the bays of Central Texas. *J. Paleontology* 29:193-211.

Pritchard, D. W. 1967. Observations of circulation in coastal plain estuaries, pp. 37-44. *In:* G. H. Lauff (ed.), *Estuaries*. Amer. Assoc. Adv. Sci., Washington, DC.

Remane, A. 1934. Die Brackwasserfauna. *Verh. dt. Zool. Ges.* 36:34-74.

Texas Department of Water Resources. 1980a. Lavaca-Tres Palacios estuary: a study of the influence of freshwater inflows. LP-106, Austin, TX 349 pp.

Texas Department of Water Resources. 1980b. Guadalupe estuary: a study of the influence of freshwater inflows. LP-107, Austin, TX 344 pp.

Texas Department of Water Resources. 1981a. Nueces and Mission-Aransas estuaries: a study of the influence of freshwater inflows. LP-108, Austin, TX 381 pp.

Texas Department of Water Resources. 1981b. Trinity-San Jacinto estuary: a study of the influence of freshwater inflows. LP-113, Austin, TX 439 pp.

Texas Department of Water Resources. 1981c. Sabine-Neches estuary: a study of the influence of freshwater inflows. LP-116, Austin, TX 327 pp.

Ward, G. W., Jr. 1973. Hydrodynamics and temperature structure of the Neches estuary. Volume 1, Physical hydrography. Prepared for Gulf States Utilities by Tracor, Inc., Austin, TX 182 pp.

Ward, G. W., Jr. and N. E. Armstrong. 1979. Phase I Final Report. For studies on the effects of alternations of freshwater inflows into Matagorda Bay area, Texas. Prepared for U.S. Fish and Wildlife Service, Division of Ecological Services, Albuquerque, NM 391 pp.

AN ANALYSIS OF ENVIRONMENTAL FACTORS REGULATING COMMUNITY METABOLISM AND FISHERIES PRODUCTION IN A LOUISIANA ESTUARY

John W. Day, Jr.
Coastal Ecology Laboratory
Center for Wetland Resources
Louisiana State University
Baton Rouge, Louisiana

Charles S. Hopkinson
University of Georgia
Marine Institute
Sapelo, Georgia

and

William H. Conner
Coastal Ecology Laboratory
Center for Wetland Resources
Louisiana State University
Baton Rouge, Louisiana

Abstract: Using information from the Barataria Basin, Louisiana, we address questions regarding controls on aquatic primary productivity (APP) and the role of wetlands in the organic budget of this estuary and in supporting fisheries in Barataria Basin and the nearshore Gulf of Mexico. Gross APP ranged from 1058 to 3286 g O_2 m^{-2} yr^{-1}. The highest values were in lakes with direct upland runoff, suggesting that these nutrient sources are important regulators of APP. Most of the water bodies of the Basin are heterotrophic and, to satisfy measured community respiration, significant inputs of allochthonous organic carbon are required. The importance of allochthonous inputs is qualitatively supported by the lack of correlation between temporal patterns of organic carbon and chlorophyll concentration and aquatic primary production in the water column. Most marine nekton species of the Basin spend only part of their life cycle in the estuary and they use the estuary in systematic and predictable ways. Shallow-water marsh areas have nekton biomass 7 to 12 times higher than that of open water areas. Young and juvenile nekton seek out these areas actively.

Introduction

Over the past 15 years there has been an extensive series of studies of Barataria Basin in coastal Louisiana. The studies were of a number of different habitats and involved specialists from several disciplines. The primary objectives of the program are to understand the ecological function of the dif-

ISBN 0-12-404070-5

ferent habitats, to learn how the different areas are coupled physically and ecologically, and to understand the impact of human activities on the area.

During the period that these studies took place, there were some remarkable shifts in ecologists' perceptions of estuaries. There are at present a number of strong controversies concerning some basic questions of estuarine ecology. These include the factors that control aquatic primary productivity and the sources of nutrients that support production (Haines 1979b; Turner et al. 1979a; Nixon 1981) and the role of intertidal wetlands in the organic budget of estuaries and coastal waters and in supporting coastal fisheries (Turner et al. 1979b; Nixon 1980; Odum 1980). In this paper we address these issues within the context of the Barataria Basin data base. Our hypothesis is that in the Basin, wetlands and uplands exert a strong influence on community metabolism and on fisheries.

Area Description

The Barataria Basin is a large (>400,000 ha) interdistributary estuarine-wetland system located between the natural levees of the Mississippi River and Bayou Lafourche, an abandoned river channel (Fig. 1). Rainfall, which averages 156 cm yr^{-1}, is the only major freshwater input. The rate of water movement through the Basin is controlled by tide range (approximately 0.3 m at the coast), wind, precipitation (especially upland runoff), and the gradual slope of the land from the swamps to the Gulf (approximately 1.0 cm

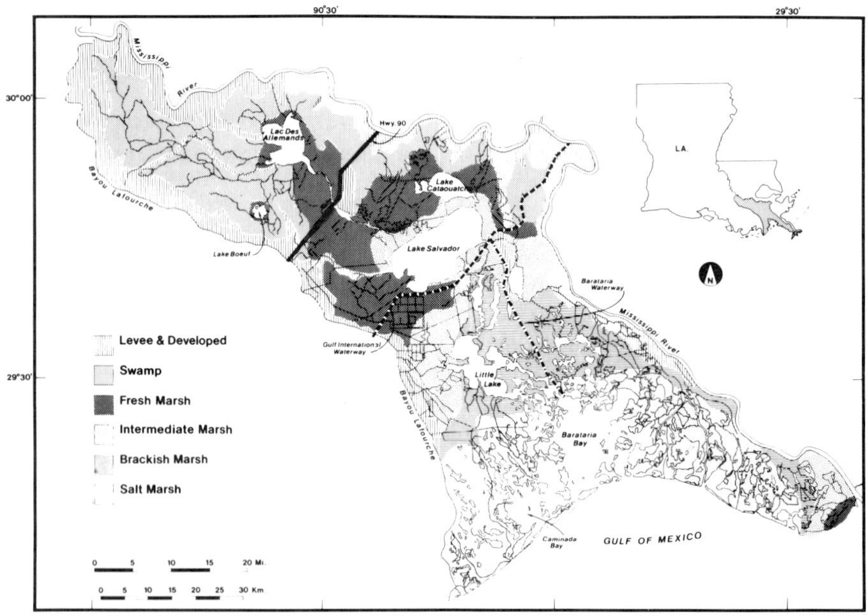

Figure 1. Vegetation zones of Barataria Basin.

per km). These conditions generally create sluggish bayous and promote overland sheetflow through the wetlands. Fresh water from the Mississippi River influences the Basin only indirectly as it affects nearshore Gulf waters (Day et al. 1973). Five broad wetland vegetation zones in the Basin (Fig. 1, Chabreck 1972) are primarily a result of decreasing salinity from the Gulf of Mexico.

There are several changes throughout the Basin that are pertinent to the analysis in this paper (Table 1). Turnover in water bodies is lowest in the mid Basin, intermediate in the upper Basin, and highest near the coast. The area of uplands is large relative to wetland and water in the upper Basin (Table 1, Fig. 1). The uplands are also highly developed (urban areas and agriculture), resulting in rapid runoff of turbid, nutrient-laden water. These factors result in high aerial water loading (and, thus, turnover), high nutrient concentrations, and turbid waters in the upper Basin (Day et al. 1977; Hopkinson and Day 1979, 1980a,b; Seaton 1979). Water turnover is frequent in the lower Basin because of tidal flushing, but upland impacts are low.

Factors Affecting Aquatic Primary Productivity

Studies of aquatic primary production have been conducted in all major water bodies of the Basin and the adjacent Gulf of Mexico. Water bodies in the upper Basin are characterized by high productivity, pronounced seasonality, and net heterotrophy. The more saline areas are less productive, lack consistent seasonality, and are not heterotrophic. In this section, we will briefly describe levels and seasonal patterns of aquatic primary productivity and discuss factors regulating productivity.

Lac des Allemands is a fresh, turbid lake (Table 1) that is influenced by adjacent wetlands and upland drainage. It is extremely productive (Table 2) and has a pronounced seasonal pattern of productivity (Fig. 2). For nine to ten months of the year, there are dense blue-green algae blooms. The lake is

Table 1. Physical characteristics of water bodies in the Barataria Basin.

Ecological Zone	*Example*	*Turn-overs Per Year*	*Average Depth (m)*[4]	*Salinity (‰)*[4]	*Secchi Depth (cm)*[4]	*Tidal Range (cm)*	*Upland to Wetland + Water Ratio*[8]
Upper Basin	Lac des Allemands	4.6[1]	2.0	0	33	3.2[5]	1:2.3
Middle Basin	Lake Cataouatche	1.5[1]	2.0	0-3	30	NA[2]	1:6.7
	Lake Salvador	1.0[1]	2.5	0-6	72	8.5[6]	
Lower Basin	Little Lake	NA[2]	1.5	0-15	72	12[7]	1:33.3
	Lower Barataria Bay	14.6[3]	2.0	10-35	68	30[7]	

[1]Craig and Day 1977; [2]NA = not available; [3]Happ et al. 1977; [4]Witzig and Day 1982; [5]Byrne et al. 1976; [6]Interpolated between Barataria and des Allemands stations; [7]Marmer 1948; [8]Wax et al. 1978.

Table 2. Comparative aquatic productivity (g O_2 m^{-2} yr^{-1}) and mean annual chlorophyll *a* (mg m^{-3}) in Barataria Basin from freshwater bayous to offshore areas. NDP = net daytime photosynthesis; NR = nighttime respiration; GP = gross production; NCP = net community production.

Lake	*Chl* a	*NDP*	*NR*	*GP*	*NCP*	*Reference*
Bayous	25	316	446		−130	Day et al. 1977
Lac des Allemands	65	1418	1868	3286	−450	Day et al. 1977
Lake Cataouatche	50	876	1205	2222	−350	Hopkinson and Day 1979
Lake Salvador	12	402	602	1058	−198	Hopkinson and Day 1979
Little Lake	10	639	753	1307	−117	Hopkinson and Day 1979
Brackish-Saline	10	940	910	1850	0 to + 54	Allen 1975; Day et al. 1973
Offshore	7.6	732				Happ et al. 1977; Sklar and Turner 1981

extremely eutrophic, primarily as a result of high nutrient loading from upland runoff (Day et al. 1977; Seaton and Day 1979). Waters in and around the lake are heterotrophic (Table 2), reflecting the high proportion of wetlands in the upper Basin and the export of organic matter to the water bodies. Bayous entering the lake are much less productive than the lake (Table 2) and there is no measurable phytoplankton production in waters overlying the swamp surface (McNamara 1978). Since nutrients are abundant in these areas (Kemp 1979), phytoplankton is undoubtedly light-limited because of shading by trees, floating vegetation, and high turbidity.

Productivity patterns in the three lakes of the central Basin partially reflect the degree to which upland runoff affects the lakes (Fig. 2). Gross production is highest in Lake Cataouatche (2222 g O_2 $m^{-2}yr^{-1}$), a result of high nutrient loading from upland runoff (Hopkinson and Day 1979). Mean annual chlorophyll in the lake is about 50 mg m^{-3}. Gross production in the other two lakes is significantly lower, though Little Lake is more productive than Lake Salvador (1307 g O_2 as compared with 1058). Mean annual chlorophyll in Salvador and Little Lake is about 12 and 10 mg m^{-2}, respectively. Nutrient levels in the two lakes are about the same (Hopkinson and Day 1979). The higher production in Little Lake is probably because of production by benthic algae. Hopkinson and Day (1979) found that, in Little Lake, Secchi depth was strongly correlated to the magnitude of production. When water transparency increased (February and April, for example), significantly greater production took place (Fig. 2). Lake Salvador, by comparison, has similar water transparency, but is deeper and light never reaches the bottom. All three lakes were heterotrophic, ranging from −350 g O_2 $m^{-2}yr^{-1}$ for Cataouatche to −117 for Little Lake (Table 2). Lake Cataouatche with high nutrient inputs and turbidity has a similar seasonal pattern of productivity to Lac des Allemands. Production is high from May through September. Lake Salvador also shows distinct seasonality but to a lesser extent. There is no

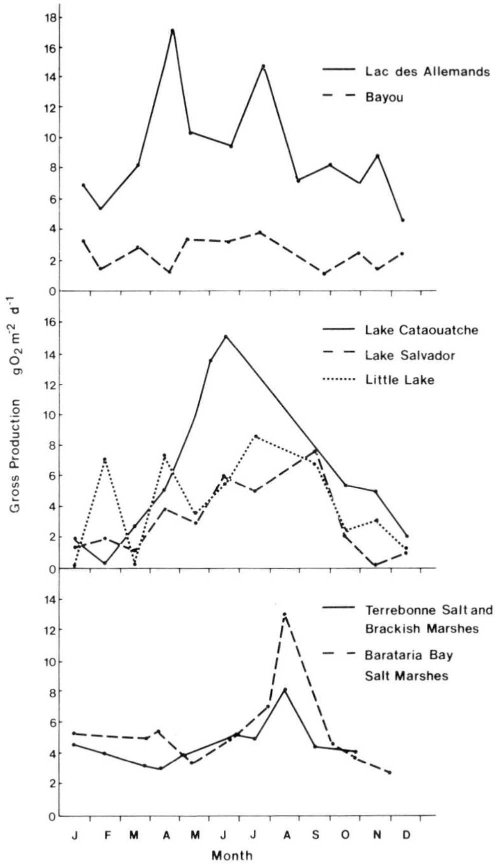

Figure 2. *Gross production of major lakes and bayous in Barataria Basin going from fresh (upper graph) to salt water conditions (lower graph) [after Day et al. 1973; Allen 1975 (used by permission); Day et al. 1977 (copyright 1977, Academic Press, N. Y.); and Hopkinson and Day 1979 (copyright 1979, Plenum Publishing Corporation, New York)].*

consistent seasonal pattern in Little Lake primarily as a result of significant benthic production during times of high water clarity.

Aquatic production in the saline waters of lower Barataria Basin is somewhat higher than in Little Lake (Table 2), but the seasonal pattern is similar in that there was no consistent change (Day et al. 1973). Seasonal patterns and levels of aquatic productivity of saline and brackish waters in the Terrebonne Basin (west of the Barataria Basin, Allen 1975) are similar to those found in Little Lake and the lower Barataria Basin. Both of these studies showed no heterotrophic conditions. In their study, Day et al. (1973) reported that production by benthic algae was about 20% higher than by phytoplankton. Thus, the results from both Little Lake and Airplane Lake indicate that benthic algal production is significant.

The studies of aquatic productivity indicate that, on the basis of temporal patterns, degree of eutrophy, and degree of heterotrophy, the Basin can be divided into two regions. The upper Basin is characterized by clear seasonal patterns which are lacking in the lower Basin. Water bodies in the upper Basin are quite heterotrophic and eutrophic while in the lower basin P/R is close to one. Lac des Allemands and Lake Cataouatche are examples from the upper Basin. Little Lake and Airplane Lake are examples from the lower Basin. Chlorophyll data also show this division of the Basin into two parts (Table 2).

Sklar and Turner (1981) measured phytoplankton production in the nearshore zone of the Gulf of Mexico adjacent to the Basin. Primary productivity, chlorophyll *a* concentrations, and assimilation numbers all peaked in spring, when riverflow reached a maximum, and were correlated with riverflow. Thus, there is a clear influence of the river; however, the impacts of estuarine outwelling on the nearshore phytoplankton are not clear. Sklar and Turner (1981) found that during the winter months when northerly winds decrease the water levels in the Basin, there was evidence that the Barataria marshes exported nutrients, and this increased the primary production of the coastal waters.

These results indicate that the factors controlling productivity change from the upper Basin to the Gulf. In the upper Basin, nutrient loading from upland runoff seems to be important in controlling both seasonal patterns and the magnitude of production. In the lower Basin the water clarity and depth are important. Offshore, riverflow is the dominant factor controlling productivity. We believe that in the absence of direct upland runoff, Basin water bodies would have two patterns of productivity depending on whether light reaches the bottom. Shallow lakes throughout the Basin would have seasonal patterns similar to those observed in Little Lake and Airplane Lake, while deeper lakes would be similar to Lake Salvador. The importance of upland runoff is substantiated by the results of Witzig and Day (1982) who showed a relationship between trophic state index (TSI) and primary production (Fig. 3). Locations with a high TSI had high inputs of nutrients from upland runoff. Gael and Hopkinson (1979) reported that TSI was significantly correlated with canal density, a measure of the intensity of upland runoff.

New Versus Recycled Nutrients

In an effort to estimate the importance of "new" nutrients, we calculated the amounts of both "new" and remineralized nutrients for each of the major water bodies in the Basin (Table 3). In all cases, estimates of remineralized nitrogen were greater than for "new" nitrogen. The "new" inputs were from 16% to 42% of the net demand by phytoplankton. The data also suggest that higher new inputs stimulate remineralization. This is to be expected, since not

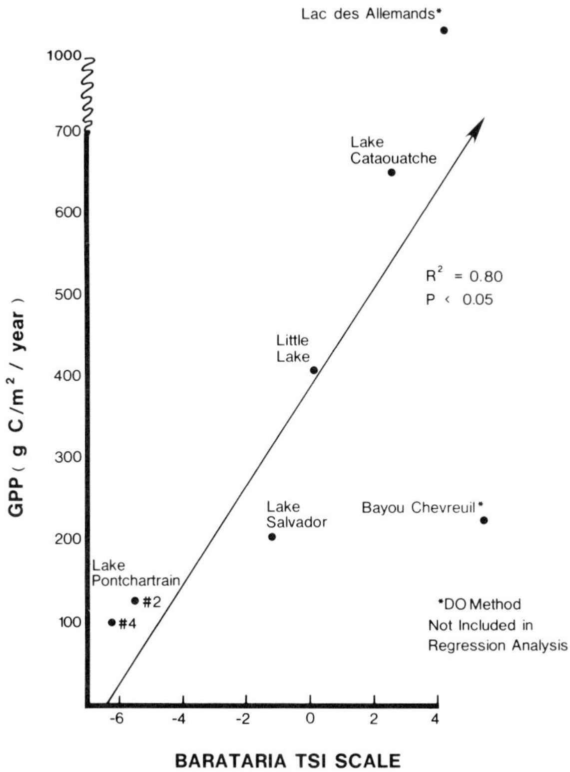

Figure 3. *Relationship between trophic state index (TSI) and aquatic primary production. TSI is a statistically derived scale based on measurements of total phosphorus, total inorganic nitrogen, total organic nitrogen, secchi depth, and chlorophyll* a *(Witzig and Day 1982).*

only will "new" nutrients stimulate *in situ* production, but runoff high in nitrogen is also high in organic matter, which leads to increased metabolism and remineralization. These results suggest that remineralization is the most important immediate source of nutrients. However, allochthonous sources are important because they can play an important role in determining local metabolism and mineralization rates.

We must stress that these estimates were all calculated indirectly, and although useful in the overall analysis, they must be considered tentative. The only validation we were able to make is that estimated benthic respiration agreed closely with direct measurements from lower Barataria Bay (Hopkinson et al. 1978a).

Wetlands and the Estuarine Carbon Budget

With the tremendous expanse of periodically flooded marsh and

Table 3. Estimates of new and remineralized nitrogen as compared to nitrogen demand by phytoplankton in the Barataria Basin (all values in g N m^{-2} yr^{-1}).

Location	*Net Demand by Phytoplankton*[1]	*Remineralized N*		*New N*[4]
		Benthic[2]	*Water Column*[3]	
Lac des Allemands	77	17	28	32
Lake Cataouatche	47	16	20	11
Lake Salvador	22	2	16	4
Barataria Bay	51	12	31	8

[1]Using net photosynthesis from Table 2 and assuming a C:N ratio of 6:1.

[2]Estimated from relationships between total organic input and benthic respiration, and benthic respiration and ammonia regeneration [from Figures 2 and 3, respectively, in Nixon (1981)]. Total organic input calculated as the sum of watershed input (from Table 4) and primary production (Table 2).

[3]Calculated as the difference between net demand by phytoplankton and the sum of benthic remineralization plus new inputs.

[4]Lac des Allemands from Day et al. 1977; net water flow between the different water bodies is as described in Table 4 and was calculated as the annual surplus (R. Baumann, Center for Wetland Resources, per. comm.). Nitrogen flux was calculated as surplus water flux times average total N concentrations [from Seaton (1979) and Hopkinson and Day (1979)]. Nitrogen export from salt marshes from DeLaune and Patrick (1980).

swamp areas in the Barataria Basin, one might expect these wetlands to play a major role in controlling or augmenting the productivity of the estuarine system. We constructed an annual organic carbon budget for Barataria Basin from a combination of direct and indirect measurements of carbon fluxes (Table 4). By dividing the Basin into two headwater systems (Lac des Allemands and Cataouatche) and three sequential downstream systems (fresh, brackish, and saline), we show that all aquatic habitats are strongly dependent on allochthonous organic inputs from adjacent watersheds and that upstream habitats are significant sources of organic matter for downstream habitats.

The portion of wetland primary production exported to adjacent water bodies is lowest in the swamp (~2%) and greatest in the salt marsh (30%, see Table 4). Total and areal loading also increases in a downstream direction. Total loading increases from about 1 to 30 × 10^{12} g C yr^{-1} and areal loading from 269 to 373 g C m^{-2} of water body yr^{-1} (swamp to salt marsh).

In situ production plus allochthonous carbon inputs exceed respiration and sedimentation in all aquatic habitats of the Barataria Basin. Annual precipitation exceeds evapotranspiration by one third, resulting in a water surplus. This surplus is a driving mechanism that transports carbon from upstream to downstream habitats. In Lac des Allemands, a net export of carbon was measured in conjunction with water export (Day et al. 1977); the value is in close approximation to the carbon export calculated by mass balance (Table 4). Downstream exports from other habitats in Barataria Basin could not be measured directly because of tidal current interactions, so they

Table 4. Annual organic carbon budget for Barataria Basin (all carbon flows in g C $yr^{-1} \times 10^{12}$).

Habitat	Area[1] ($m^2 \times 10^{10}$)	Total Net Production[2]	Export[3]	Respiration[4]	Sedimentation[5]
Des Allemands Swamp	10.30	69.94	1.75		
Des Allemands Aquatic	0.65	3.15	0.65	4.04	0.21
Lake Cataouatche Wetlands	2.20	19.75	1.58		
Lake Cataouatche Aquatic	0.37	1.10	1.04	1.52	0.12
Fresh Wetlands	4.20	37.72	3.01		
Fresh Aquatic	1.90	2.59	2.21	3.91	1.17
Brackish Wetlands	5.50	115.22	14.97		
Brackish Aquatic	4.83	10.49	10.41	12.43	4.83
Saline Wetlands	6.34	67.52	20.26		
Saline Aquatic	8.21	26.23	12.38	25.55	18.97
Nearshore Gulf of Mexico	15.00	39.90[6]			

[1]Hopkinson and Day 1980 a, b; Wax et al. 1978. Nearshore Gulf assumes a width of 15 km.
[2]Wetlands = aboveground net plant production. Aquatic = net daytime photosynthesis (from Table 1). Wetland net aboveground production by habitat (g C m^{-2} yr^{-1})

Swamp	769	Day et al. 1977; Conner et al. 1981
Fresh marsh	898	Hopkinson et al. 1978b; Chabreck 1972
Brackish marsh	2095	Hopkinson et al. 1978b; Chabreck 1972
Saline marsh	1065	Kirby and Gosselink 1976

[3]Wetland export measured at 2.5% of net aboveground production in Des Allemands swamp (Day et al. 1977) and 30% of net aboveground production in salt marsh (Hopkinson et al. 1978b). A linear interpolation between these values was used to estimate export from fresh (8%) and brackish (13%) marshes. Aquatic export refers to the transfer of organic carbon to the adjacent downstream aquatic habitat. The downstream order is: Lac des Allemands and Lake Cataouatche systems drain into the fresh zone which drains into the brackish, then saline, then offshore aquatic systems.
[4]Nighttime respiration of aquatic ecosystem using free water diurnal technique (Table 2).
[5]W. H. Patrick, Jr., Center for Wetland Resources, LSU (pers. comm.): 32 g C m^{-2} yr^{-1} in des Allemands and 231 g C m^{-2} yr^{-1} for a saline lake. Areal sedimentation in Cataouatche is assumed similar to that in des Allemands. Sedimentation in other fresh and brackish areas was assumed to be between that measured in saline and swamp habitats and be proportional to the percentage export from wetlands (61.6 g C m^{-2} yr^{-1} fresh; 100.1 g C m^{-2} yr^{-1} brackish).
[6]Net production per m^2 (Sklar 1976).

were estimated indirectly by balancing mass budgets for each aquatic system from the headwater region down.

The magnitude and importance of carbon inputs from upstream to the receiving water body increase downstream. Inputs are less than 2.5×10^{12} g C yr^{-1} in the freshwater region and exceed 12×10^{12} g C yr^{-1} to the Gulf of Mexico. Upstream allochthonous carbon inputs are not as quantitatively important to any aquatic region as *in situ* production and export from adjacent marshes. Upstream inputs provide from 9% to 30% of the total carbon inputs to the different water habitats. Total allochthonous inputs from all sources (wetland export plus upstream inputs) make up > 50% of the total organic inputs for all inshore aquatic habitats except Lac des Allemands. Allochthonous inputs to the nearshore region of the Gulf of Mexico (0-15 km) provide 25% of the total carbon inputs to the system.

The mass balance technique we used to calculate a carbon budget for the Barataria Basin depends somewhat on fluxes calculated by difference and, as such, the budget is not rigorously validated. However, considerable data do exist that support the importance of "outwelling" from Barataria Basin to the Gulf, the importance of downstream carbon transfer between aquatic regions, and export from marshes to adjacent water bodies. We know that the downstream export of carbon must occur because of the net runoff of surplus precipitation through the Basin to the Gulf. The product of net water flow and total organic carbon concentration in the water column sets a conservative lower limit on carbon flux.

Happ et al. (1977) calculated a range of values for annual carbon flux from Barataria Bay (saline aquatic zone) that encompasses our mass balance estimate. They measured a statistically significant gradient of suspended organic carbon levels that decreased from areas deep in the marsh to open bays to the Gulf of Mexico. The organic carbon flux, based on the measured organic carbon gradient and water renewal rates for Barataria Bay, was calculated to be between 4.3 and 28×10^{12} g C yr^{-1}. Our estimate of 12.4×10^{12} g C compares favorably with this range of values.

Much evidence qualitatively supports the concept of organic export from marsh to bay. In a modeling study, Hopkinson and Day (1977) indicated that maximum export was to be expected in late winter and early spring. This prediction was based on relationships among the timing of maximum dead marsh vegetation, rising flushing energy, and rising temperature. The importance and timing of marsh export simulated in the salt marsh model parallel the data of Happ et al. (1977). They showed that for most of the year dissolved organic carbon (DOC), total suspended organic carbon (TOC), particulate carbon (POC), and chlorophyll levels were fairly constant. Peaks occurred in winter, and the timing of these peaks suggested that the marsh and not *in situ* production was the major source of the organic material. Dissolved organic carbon peaked in January, chlorophyll in February, and TOC in January and March. The increase in TOC during January was entirely because of DOC. In March, however, the peak was not DOC and was therefore attributed to particulate organic matter since, by definition, TOC = DOC + POC. The greatest mass of dead marsh vegetation occurs in winter. High DOC values were attributed to leaching and flushing of the labile contents of these dead plants. During fall and early winter, the salt marsh is flooded for the greatest percentage of time in Louisiana and this promotes rapid decomposition. Because the frequency of flooding is low at this time, hydraulic energies are adequate only for moving dissolved detritus from the marsh. In spring, flooding frequency increases and energies are available to move marsh grass detritus that has decayed over winter into adjacent creeks, lakes, and bays.

The organic-rich nature of small marsh creeks and lake edges further suggests an influence of organic export from adjacent marshes. Happ et al. (1977) found a gradient of organic matter within the saline estuary, decreasing from 8.5 mg l^{-1} TOC in water bodies deep in the marsh to 2.8 mg l^{-1} TOC in the nearshore Gulf. Seaton (1979) observed that natural tidal creeks within marsh sites in Barataria Basin were more enriched than open water, presumably by organics exported from fringing marsh. Hopkinson et al. (1978a) associated high benthic oxygen uptake rates along a lake edge with increased food supply, again presumably from marsh inputs.

The Role of the Marsh-Estuary in Fisheries Production

Louisiana has the greatest area of coastal wetlands (Turner and Gosselink 1975) and the largest commercial fishery in the U.S. (see U.S. Fishery statistics). It is a commonly held belief that the coastal wetlands play an important role in supporting the fisheries (Lindall and Saloman 1977). We believe that there is strong evidence which shows coupling between fisheries and the marsh estuarine system, and in this section we review several lines of evidence from the Barataria Basin.

There have been over 20 studies in the Basin of nekton community composition, biomass distribution, and migratory patterns (see Chambers 1980 for a review). Seven to eight species comprise 80% to 95% of the total numbers and biomass (Gunter 1936, 1938a, 1938b; Perret et al. 1971; Wagner 1973; Sabins and Truesdale 1974; Chambers 1980):

Anchoa mitchilli	(bay anchovy)	*Leiostomus xanthuros*	(spot)
Micropogon undulatus	(croaker)	*Brevoortia patronus*	(menhaden)
Arius felis	(sea catfish)	*Menidia beryllina*	(silverside)
Mugil cephalus	(striped mullet)	*Penaeus* sp.	(shrimp)

Anchoa mitchilli is considered to be an estuarine resident that normally completes its entire life cycle within the estuary. The other species spawn offshore and use the estuary as a nursery and feeding ground.

The studies also indicate that there are fairly specific, repeating patterns in the way nekton species use the Barataria Basin. For example, Sabins and Truesdale (1974) identified over 80 species of juvenile fishes that migrated through Caminada Pass. They identified a "warm water fauna" composed mainly of the young of inshore spawners and a "cold water fauna" composed predominantly of immigrating young of offshore spawners. Wagner (1973) classified the majority of fishes in a number of marsh and estuarine habitats in the Caminada Bay area as seasonal migrants using the estuary for spawning, feeding, or as a nursery.

Chambers (1980) presented a diagram of the pattern of use of the Basin by four different nekton groups (Fig. 4): (1) euryhaline larvae, postlarvae, and juveniles of marine nekton which spawn offshore, migrate far up the Basin in the late winter and spring, and then gradually move downbay as they grow, eventually emigrating to the Gulf in later summer and fall; (2) juvenile and adult freshwater species which move southward in the fall into oligohaline areas as they become fresher and replace the emigrating marine species, returning to fresh areas in the late winter; (3) during the warmer months, mesohaline juveniles of certain marine species move up to the mid Basin during periods of high salinity, and later return to the lower bays and Gulf in the late fall and winter as salinities decrease; (4) some euryhaline species spend their entire life cycles in the estuary and often may be found anywhere from the freshwater swamps to the lower bays and barrier islands bordering the Gulf. The data of both Wagner (1973) and Chambers (1980) suggest that euryhaline marine-spawned juveniles preferentially migrate into waters with low salinity and slowly move into waters of higher salinity as they grow. For example, Wagner found that less saline areas were characterized by high densities of postlarvae and juveniles. By contrast, biomass was higher

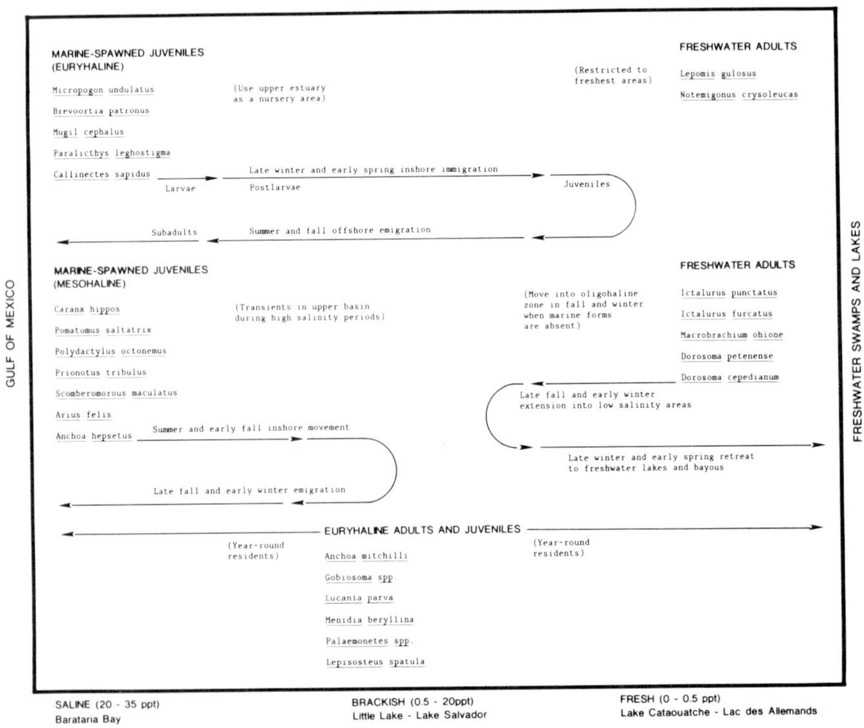

Figure 4. Patterns of estuarine use by nektonic organisms (from Chambers 1980, used by permission).

in the more saline areas because of the presence of large juveniles and adults.

Thus far, results of studies from Barataria Basin show that many nekton species spend only a part of their life cycle in the Basin and that these species use the Basin in very predictable ways. The evidence also suggests that nekton species, especially larval and juvenile forms, preferentially seek out shallow water adjacent to wetlands, such as marsh ponds, tidal creeks, and the marsh edge in general. Data from Barataria Basin and Lake Pontchartrain, LA, show that nekton biomass is 7 to 12 times higher in shallow water marsh areas as compared with open waters (Table 5). It has also been demonstrated on the east coast that shallow tidal creeks and marsh shoals harbor dense populations of juvenile marine species (Shenker and Dean 1979; Weinstein 1979; Bozeman and Dean 1980) and that young fish actively seek creek headwaters.

It may be that it is primarily the shallow nature of these waters that attracts nekton seeking either food or refuge from predators. This is undoubtedly at least partially true, since these areas are practically all <1 m deep. However, most open waters of Barataria Basin are <2 m deep and many are <1 m. The lakes and bays of the Basin have essentially flat, shallow bottoms. Since there is a distinct biomass difference between open waters and near marsh waters, it must be caused by more than the shallow nature of these areas. Shallow-water marsh areas in Barataria Basin seem to satisfy the three major requirements outlined by Joseph (1973) for a nursery area: (1) physiologically suitable in temperature, salinity, and other physiochemical parameters; (2) abundant suitable food with a minimum of competition at critical trophic levels; and (3) a degree of protection from predators.

A number of workers have shown correlations among estuaries and wetlands and fisheries. Turner (1977) correlated shrimp yield (kg/ha) and intertidal wetland areas on a worldwide basis. On a regional basis (the northern

Table 5. Comparison of estimated nekton standing crops in shallow, marsh areas and open-water areas. All fish collected by otter trawl, except * = Antimycin.

Reference	*Area*	*Standing Crop (g wet wt m^{-2})*
Chambers 1980	Upper Barataria Basin, shallow water	3.41
	Upper Barataria Basin, open water	0.50
Wagner 1973	Caminada Bay, shallow marsh ponds	13.8 to 46.1*
	Caminada Bay, open water	1.19
B. A. Thompson, LSU Center for Wetland Resources, pers. comm.	Lake Pontchartrain, shallow water	2.57
	Lake Pontchartrain, open water	0.32

Gulf of Mexico), he found that yields of inshore shrimp are directly related to the area of estuarine vegetation whereas they are not correlated with area, average depth, or volume of estuarine water. Moore et al. (1970) presented data on distribution of demersal fish off Louisiana and Texas. Their data suggest that the greatest fish populations occur offshore from extensive wetlands with a high freshwater input.

The evidence from the Barataria Basin suggests that wetlands enhance fisheries productivity. The picture is certainly not complete and we know that very few nekton species are absolutely dependent on estuaries or wetlands. However, the available data do suggest specific ways that nekton use wetlands. It is probable that many nektonic species have evolved behaviorial patterns that allow them to exploit wetlands as both food sources and habitat.

Acknowledgments

Funding for this analysis was supplied primarily by the Louisiana Sea Grant College Program. Support was also provided by the Sapelo Island Research Foundation. Special thanks to Jane Caffrey for her help in analyzing the new and remineralized nitrogen data. This is a joint publication of the Coastal Ecology Laboratory, Center for Wetland Resources (Contribution No. LSU-CEL-81-36) and the University of Georgia Marine Institute (Contribution No. 456).

References Cited

Allen, R. L. 1975. Aquatic primary productivity in various marsh environments in Louisiana. M.S. thesis, Louisiana State U., Baton Rouge, LA 50 pp.

Bozeman, E. L. and J. M. Dean. 1980. The abundance of estuarine larval and juvenile fish in a South Carolina creek. *Estuaries* 3:89-97.

Byrne, P., M. J. Borengasser, G. Drew, R. A. Muller, B. L. Smith, Jr. and C. L. Wax. 1976. Barataria Basin: hydrologic and climatologic processes. Louisiana State U., Center for Wetland Resources, Baton Rouge, LA. Final Report to Louisiana State Planning Office, Baton Rouge, LA

Chabreck, R. H. 1972. Vegetation, water, and soil characteristics of the Louisiana coastal region. Louisiana State U., Baton Rouge, LA. Agric. Exp. Sta. Bull. No. 664. 72 pp.

Chambers, D. G. 1980. An analysis of nekton communities in the upper Barataria Basin, Louisiana. M.S. thesis, Louisiana State U., Baton Rouge, LA 286 pp.

Conner, W. H., J. G. Gosselink and R. T. Parrondo. 1981. Comparison of the vegetation of three Louisiana swamp sites with different flooding regimes. *Amer. J. Bot.* 68:320-331.

Craig, N. J. and J. W. Day, Jr. 1977. Cumulative impact studies in the Louisiana coastal zone: eutrophication and land loss. Center for Wetland Resources, Louisiana State U., Baton Rouge, LA. Final Report to Louisiana State Planning Office, Baton Rouge, LA 157 pp.

Day, J. W. Jr., W. G. Smith, P. Wagner and W. Stowe. 1973. Community structure and carbon budget in a salt marsh and shallow bay estuarine system in Louisiana. Louisiana State U., Center for Wetland Resources, Baton Rouge, LA. Sea Grant Publ. No. LSU-SG-72-04.

Day, J. W. Jr., T. J. Butler and W. H. Conner. 1977. Productivity and export studies in a cypress swamp and lake system in Louisiana, pp. 255-269. *In:* M. L. Wiley, (ed.) *Estuarine Processes Vol. II.* Academic Press, New York.

DeLaune, R. D. and W. H. Patrick, Jr. 1980. Nitrogen and phosphorus cycling in a Gulf coast salt marsh, pp. 143-151. *In:* V. S. Kennedy, (ed.), *Estuarine Perspectives.* Academic Press, New York.

Gael, B. T. and C. S. Hopkinson. 1979. Drainage density, land-use, and eutrophication in

Barataria Basin, Louisiana, pp. 147-163. *In:* J. W. Day, Jr., D. D. Culley, Jr., R. E. Turner, and A. J. Mumphrey, Jr. (eds.), *Proc. Third Coastal Marsh and Estuary Management Symposium*, Louisiana State U. Press, Baton Rouge, LA

Gunter, G. 1936. Studies of the destruction of marine fish by shrimp trawlers in Louisiana. *La. Cons. Rev.* 5:18-24, 45-56.

Gunter, G. 1938a. The relative numbers of species of marine fish on the Louisiana coast. *Amer. Nat.* 72:77-83.

Gunter, G. 1938b. Seasonal variations in abundance of certain estuarine and marine fishes in Louisiana with particular reference to life histories. *Ecol. Monogr.* 8:314-346.

Haines, E. B. 1979a. Interactions between Georgia salt marshes and coastal waters: a changing paradigm, pp. 35-46. *In:* R. J. Livingston (ed.), *Ecological Processes in Coastal and Marine Systems.* Plenum Press, New York.

Haines, E. B. 1979b. Nitrogen pools in Georgia coastal waters. *Estuaries* 2:34-39.

Happ, G., J. G. Gosselink and J. W. Day, Jr. 1977. The seasonal distribution of organic carbon in a Louisiana estuary. *Est. Coastal Mar. Sci.* 5:695-705.

Hopkinson, C. S. and J. W. Day, Jr. 1977. A model of the Barataria Bay salt marsh ecosystem, pp. 235-265. *In:* C. A. S. Hall and J. W. Day, Jr., (eds.), *Ecosystem Modeling in Theory and Practice.* John Wiley and Sons, New York.

Hopkinson, C. S. and J. W. Day, Jr. 1979. Aquatic productivity and water quality at the upland-estuary interface in Barataria Basin, Louisiana, pp. 291-314. *In:* R. Livingston, (ed.), *Ecological Processes in Coastal and Marine Systems.* Plenum Press, New York.

Hopkinson, C. S. and J. W. Day, Jr. 1980a. Modeling the relationship between development and storm water and nutrient runoff. *Environ. Mgmt.* 4:315-324.

Hopkinson, C. S. and J. W. Day, Jr. 1980b. Modeling hydrology and eutrophication in a Louisiana swamp forest ecosystem. *Environ. Mgmt.* 4:325-335.

Hopkinson, C. S., J. W. Day, Jr. and B. T. Gael. 1978. Respiration studies in a Louisiana salt marsh. *An. Centro Cienc. del Mar y Limnol. Univ. Nal. Auton. México* 5:225-238.

Hopkinson, C. S., J. G. Gosselink and R. T. Parrondo. 1978b. Aboveground production of seven marsh plant species in coastal Louisiana. *Ecology* 59:760-769.

Joseph, E. B. 1973. Analysis of a nursery ground, pp. 118-121. *In:* A. L. Pacheco, (ed.), Proceedings of a workshop on egg, larval, and juvenile stages of fish in Atlantic coast estuaries. Tech. Publ. No. 1, NMFS. Mid. Atl. Coast. Fish. Cent., Highlands, NJ 338 p.

Kemp, G. P. 1978. Agricultural runoff and nutrient dynamics of a swamp forest in Louisiana. M.S. thesis, Louisiana State U., Baton Rouge, LA 58 pp.

Kirby, C. J. and J. G. Gosselink. 1976. Primary production in a Louisiana gulf coast *Spartina alterniflora* marsh. *Ecology* 57:1052-1059.

Lindall, W. N., Jr., and C. H. Saloman. 1977. Alteration and destruction of estuaries affecting fishery resources of the Gulf of Mexico. *Mar. Fish. Rev.* 39(9):1-7.

Marmer, H. A. 1948. The currents in Barataria Bay. Texas A. and M. Research Foundation, College Station, TX Project 9, 30 pp.

McNamara, S. J. 1978. Metabolism measurements of a flooded soil community in a Louisiana swamp forest. M.S. thesis, Louisiana State U., Baton Rouge, LA 66 pp.

Moore, D., H. A. Busher and L. Trent. 1970. Relative abundance, seasonal distribution and species composition of demersal fishes off Louisiana and Texas, 1962-1964. *Contr. Mar. Sci.* 15:45-70.

Nixon, S. W. 1980. Between coastal marshes and coastal waters – a review of twenty years of speculation and research on the role of salt marshes in estuarine productivity and water chemistry, pp. 437-525. *In:* P. Hamilton and K. B. Macdonald (eds.), *Estuarine and Wetland Processes with Emphasis on Modeling.* Plenum Press, New York.

Nixon, S. W. 1981. Remineralization and nutrient cycling in coastal marine ecosystems, pp. 111-138. *In:* B. J. Neilson and L. E. Cronin, (eds.), *Estuaries and Nutrients.* Humana Press, New Jersey.

Odum, E. P. 1980. The status of three ecosystem-level hypotheses regarding salt marsh estuaries: tital subsidy, outwelling, and detritus-based food chains, pp. 485-495. *In:* V. S. Kennedy (ed.), *Estuarine Perspectives.* Academic Press, New York.

Perrett, W. S., W. R. Latapie, J. F. Pollard, W. R. Mack, G. B. Adkins, W. J. Gaidry and C. J. White. 1971. Fishes and invertebrates collected in trawl and seine samples in Louisiana estuaries, pp. 39-105. *In:* Cooperative Gulf of Mexico Estuarine Inventory and Study, Louisiana. Phase IV: Biology. Louisiana Wildlife and Fisheries Commission, New Orleans, LA

Sabins, D. S. and F. M. Truesdale. 1974. Diel and seasonal occurrence of immature fishes in a Louisiana tidal pass. *Proc. Ann. Conf. S. E. Assoc. Game Fish Comm.* 28:161-171.

Seaton, A. M. 1979. Nutrient chemistry in the Barataria Basin – A multivariate approach. M.S. thesis, Louisiana State U., Baton Rouge, LA 78 pp.

Seaton, A. M. and J. W. Day, Jr. 1979. The development of a trophic state index for the quantification of eutrophication in the Barataria Basin. pp. 113-125. *In:* J. W. Day, Jr., D. D. Culley, Jr., R. E. Turner and A. J. Mumphrey, Jr., (eds.), *Proc. Third Coastal Marsh and Estuary Management Symposium,* Louisiana State U. Press, Baton Rouge, LA

Shenker, J. M. and J. M. Dean. 1979. The utilization of an intertidal salt marsh creek by larval and juvenile fishes: abundance, diversity, and temporal variation. *Estuaries* 2:154-163.

Sklar, F. H. 1976. Primary productivity in the Mississippi Delta Bight near a shallow bay estuarine system in Louisiana. M.S. thesis, Louisiana State U., Baton Rouge, LA 120 pp.

Sklar, F. H. and R. E. Turner. 1981. Characteristics of phytoplankton production off Barataria Bay in an area influenced by the Mississippi River. *Contr. Mar. Sci.* 24:93-106.

Turner, R. E. 1977. Intertidal vegetation and commercial yields of Penaeid shrimp. *Trans. Amer. Fish. Soc.* 106:411-416.

Turner, R. E. and J. G. Gosselink. 1975. A note on standing crops of *Spartina alterniflora* in Texas and Louisiana. *Contr. Mar. Sci.* 19:113-118.

Turner, R. E., S. W. Woo and H. R. Jitts. 1979a. Estuarine influences on a continental shelf plankton community. *Science* 206:218-220.

Turner, R. E., S. W. Woo and H. R. Jitts. 1979b. Phytoplankton production in a turbid, temperate salt marsh estuary. *Est. Coastal Mar. Sci.* 9:1-11.

Wagner, P. R. 1973. Seasonal biomass, abundance, and distribution of estuarine dependent fishes in the Caminada Bay system of Louisiana. Ph.D. dissertation. Louisiana State U., Baton Rouge, LA 193 pp.

Wax, C. L., M. J. Borengasser and R. A. Muller. 1978. Barataria Basin: synoptic weather types and environmental responses. Louisiana State U., Center for Wetland Resources, Baton Rouge, LA. Sea Grant Publ. NO. LSU-T-78-001.

Weinstein, M. P. 1979. Shallow marsh habitats as primary nurseries for fish and shellfish, Cape Fear River, North Carolina. *Fish. Bull.* 77:339-357.

Witzig, A. S. and J. W. Day, Jr. 1982. Trophic state index for the Louisiana coastal zone. Louisiana State U., Center for Wetland Resources, Baton Rouge, LA. Final report to Louisiana Water Resources Research Institute, Baton Rouge, LA

DYNAMICS BENEATH THE MARSH SOIL SURFACE

Convenor: *R. R. Christian*
East Carolina University
Greenville, North Carolina

A REVIEW OF PRIMARY PRODUCTION AND DECOMPOSITION DYNAMICS OF THE BELOWGROUND MARSH COMPONENT

Ralph E. Good

Biology Department
Rutgers University
Camden, New Jersey

Norma F. Good

Haddonfield, New Jersey

and

Barry R. Frasco

Environmental Consulting
and Testing Services
Cherry Hill, New Jersey

Abstract: Studies of belowground production and decomposition in marsh systems have lagged behind aboveground studies for technical reasons. Sampling is difficult, especially in species where much of the belowground material consists of large irregularly spaced components such as tubers and rhizomes. Despite these and other problems a growing body of literature on belowground production is emerging. Belowground standing crop and productivity are large, typically exceeding aboveground measurements for the same species. Root/shoot ratios are variable, showing the influence of the species, life history, hydrology/habitat, and climate. The most active portion of the belowground zone appears to be the upper 10 cm (approx.). The pathways and contribution of belowground material to food chains and nutrient cycling are poorly known at present. Opportunities for transport of materials are variable; creekside habitats may contribute relatively larger amounts of nutrients to the estuarine system than those only occasionally flooded. In many marshes a considerable amount of belowground material eventually becomes incorporated into the marsh substrate, thereby maintaining the structure of the marsh and even determining future existence in areas of rising water level or land subsidence.

Introduction

Initial studies on marsh productivity centered almost exclusively on the more accessible aboveground component. As information on the aboveground component became abundant, it was natural for interest in the belowground component to develop. Most of the available belowground literature is less than 10 years old but information is growing rapidly. This paper compares and contrasts data from a variety of estuarine marsh studies as they relate to belowground production; seasonal dynamics; factors affecting pro-

ISBN 0-12-404070-5

duction, decomposition and turnover time; and the fate of belowground material.

Methods of Estimation

The lack of completely reliable sampling techniques and apparatus has contributed to the relatively slower progress in the estimation of belowground biomass as compared to aboveground biomass. Typically, some sort of coring device with or without a plunger is used to obtain samples. Gallagher (1974) described the details of such an instrument. Others have used variations constructed of metal or plastic with core diameters mostly ranging from 6.5-10.8 cm and sample depths of 15.0-75.0 cm (Stroud 1976; Valiela et al. 1976; de la Cruz and Hackney 1977, and others). Samples are returned to the laboratory for the extensive washing needed to remove adhering soil particles and biomass is separated into live and dead components. Separation is usually done on the basis of color and turgidity of rhizome and root material (Stroud 1976; Valiela et al. 1976), usually aided by a staining technique to differentiate live and dead material.

Many investigators have found live-dead separation techniques to be inadequate and report belowground biomass as a combination of live and dead material (de la Cruz and Hackney 1977; Good and Frasco 1979; Smith et al. 1979, and others). Differences in root/rhizome ratios, incomplete washing, the loss of fine root materials during washing, and differences in separation criteria all represent potential sources of error. The change-in-biomass technique, originally used for prairie vegetation (Kucera et al. 1967; Coupland 1975) is often used to estimate productivity. Periodic regression techniques have also been used in conjunction with the change-in-biomass technique in estimating belowground biomass (de la Cruz and Hackney 1977; Smith et al. 1979). Questions of death during the growing season, translocation, and aging of roots and rhizomes have not been addressed seriously so far in relation to their potential to influence productivity estimates. Whigham and Simpson (1978), working with freshwater tidal wetland species, have suggested that regression equations may be useful in estimating belowground biomass from aboveground biomass. The potential of such techniques appears not to have been evaluated for salt marshes so far.

Production Estimates and Seasonal Changes in Biomass

Because belowground sampling is tedious and still not routinely done in marsh productivity studies, data on belowground productivity estimates for many salt marsh species are still fragmentary. The greatest number of estimates are for *Spartina alterniflora* (Table 1). Productivity estimates for this species are quite variable, with estimates for both short and tall forms from Massachusetts (Valiela et al. 1976), New Jersey (Good 1977; Good and Frasco 1979; Smith et al. 1979), and Georgia (Gallagher and Plumley 1979) being in reasonably close agreement with each other (2.0-3.5 kg m^{-2} yr^{-1}) but

Table 1. Belowground productivity estimates for ten saltwater and six freshwater angiosperms for several locations along the western Atlantic coast.

Species	*Productivity* $kg\ m^{-2}\ yr^{-1}$	*Location*	*Source*
Saltwater species			
Borrichia frutescens	0.82	Georgia	Gallagher & Plumley 1979
Distichlis spicata	1.07	Georgia	Gallagher & Plumley 1979
	3.40	Delaware	Gallagher & Plumley 1979
	2.78	New Jersey	Good & Frasco 1979
Juncus gerardi	4.29	Delaware	Gallagher & Plumley 1979
	1.62	Maine	Gallagher & Plumley 1979
Juncus roemerianus	1.36	Mississippi	de la Cruz & Hackney 1977
	4.4-7.6	Alabama	Stout 1978
	3.36	Georgia	Gallagher & Plumley 1979
Phragmites communis	3.65	Delaware	Gallagher & Plumley 1979
(australis)	2.81	New Jersey	Good & Walker 1977
Salicornia virginica	0.43	Georgia	Gallagher & Plumley 1979
	1.43	Delaware	Gallagher & Plumley 1979
Spartina alterniflora	2.1	Georgia	Gallagher & Plumley 1979
(tall form)	0.5	North Carolina	Stroud 1976
	2.9	New Jersey	Good 1977
	3.3	New Jersey	Good & Frasco 1979
	2.4	New Jersey	Good & Walker 1977
	3.5	Massachusetts	Valiela et al. 1976
	0.22	Maine	Gallagher & Plumley 1979
(short form)	2.02	Georgia	Gallagher & Plumley 1979
	0.56	North Carolina	Smith & Odum 1981
	0.46	North Carolina	Stroud 1976
	3.2	New Jersey	Good 1977
	2.4	New Jersey	Good & Frasco 1979
	2.3	New Jersey	Smith et al. 1979
	3.6-6.2	Alabama	Stout 1978
Spartina cynosuroides	2.2	Mississippi	de la Cruz & Hackney 1977
	3.56	Georgia	Gallagher & Plumley 1979
Spartina patens	0.31	Georgia	Gallagher & Plumley 1979
	0.47	Delaware	Gallagher & Plumley 1979
	3.27	New Jersey	Good & Frasco 1979
	2.5	Massachusetts	Valiela et al. 1976
	0.54	Maine	Gallagher & Plumley 1979
Sporobolus virginicus	0.58	Georgia	Gallagher & Plumley 1979
Freshwater species (tidal wetlands only)			
Acorus calamus	0.22	New Jersey	Whigham et al. 1978
Nuphar advena	1.36	New Jersey	Good & Good 1975
Peltandra virginica	2.46	New Jersey	Good & Good 1975
Sagittaria latifolia	0.004	New Jersey	Whigham et al. 1978
Typha angustifolia	1.37	New Jersey	Good & Good 1975
Zizania aquatica	0.61	New Jersey	Good & Good 1975

much higher than estimates (0.46-0.56 kg m^{-2} yr^{-1}) from North Carolina (Stroud 1976; Smith and Odum 1981). The reasons for the lower North Carolina productivity are not clear but the differences appear to be real. Estimates from Maine (Gallagher and Plumley 1979) indicate very low productivity, probably related to climate/latitude. Estimates for Alabama (Stout 1978) are high (4.4-7.6 kg m^{-2} yr^{-1}). Despite the large differences in aboveground productivity for *S. alterniflora* ecophenes, belowground productivity estimates for any given location seem to be very similar and indicate that the total productivity of the short ecophene is not nearly as low as previously judged from consideration of aboveground data only.

Estimates for other salt marsh species are less extensive and hence more difficult to compare (Table 1). Estimates for *Spartina patens* from Georgia, Delaware, and Maine (Gallagher and Plumley 1979) are all low, ranging from 0.31-0.54 kg m^{-2} yr^{-1}, while estimates from Massachusetts (Valiela et al. 1976) and New Jersey (Good and Frasco 1979) are high, ranging from 2.5-3.27 kg m^{-2} yr^{-1}. Estimates for *Distichlis spicata* are limited to Georgia, Delaware, (Gallagher and Plumley 1979) and New Jersey (Good and Frasco 1979) and are rather variable with no apparent trends, although they do indicate that the belowground productivity of this species can be very high under some conditions. Values for *Juncus gerardi* from Maine are less than half those from Delaware (Gallagher and Plumley 1979), again indicating the probable influence of climate/latitude on belowground production. Data on *Juncus roemerianus* for Mississippi (de la Cruz and Hackney 1977), Alabama (Stout 1978), and Georgia (Gallagher and Plumley 1979) are quite different. Data on *Spartina cynosuroides* from Mississippi (Hackney and de la Cruz, pers. comm.) and Georgia (Gallagher and Plumley 1979) also indicate lower belowground production in the Mississippi marshes. Estimates for other salt marsh species are too limited for comparisons but suggest that *Phragmites communis (australis)* has high belowground productivity while species such as *Sporobolus virginicus*, *Salicornia virginica*, and *Borrichia frutescens* have relatively modest belowground production.

The occurrence of peak belowground biomass appears to vary somewhat among species. For *Spartina alterniflora* (all height forms), peak belowground biomass occurs in late spring-summer (May-August) with sites at lower latitudes generally peaking earlier (Stroud 1976; Valiela et al. 1976; White et al. 1978; Good and Frasco 1979; Haines 1979; Smith et al. 1979). A smaller secondary biomass peak at the end of the growing season (October-November) has also been commonly reported (Stroud 1976; White et al. 1978; Haines 1979; Smith et al. 1979) probably reflecting the influence of translocation from aboveground portions. Peak belowground biomass of *Distichlis spicata* has been found to occur in August (Gallagher and Plumley 1979; Good and Frasco 1979). Similar findings have been reported for *Spartina patens* (Good and Frasco 1979). *Spartina cynosuroides* belowground biomass peaked in September while biomass peaks in April and September were found for *Sporobolus virginicus* (Gallagher and Plumley 1979). De la

Cruz and Hackney (1977) reported an April biomass peak with a secondary peak in October for *Juncus roemerianus* while Kruczynski et al. (1978) found a December peak with a secondary peak in April for the same species.

Belowground productivity estimates for freshwater tidal species are much more limited than for salt marsh species although the magnitude and regularity of flooding in some of these wetlands suggests that biomass produced there is probably more available to estuarine waters than that produced on infrequently flooded areas of high salt marsh. Data from New Jersey (Good and Good 1975) indicate that typical perennials in these marshes may have belowground productivity values in the order of 1.36-2.46 kg m^{-2} yr^{-1} (Table 1). The highest value is for *Peltandra virginica* which seems to have a much greater allocation to belowground biomass than other freshwater tidal species (Whigham and Simpson 1978). Similarly, Good and Walker (1977) found *Peltandra virginica* to have very large peak belowground standing crops (11.8 kg m^{-2}) which exceeded all other adjacent salt and freshwater species in a transition zone area on the Mullica River in New Jersey. *Zizania aquatica*, an important dominant in New Jersey freshwater tidal wetlands, had considerably reduced belowground productivity (0.61 kg m^{-2} yr^{-1}; Good and Good 1975) in comparison to perennial species, although *Acorus* and *Sagittaria*, both perennials, may have extremely low belowground productivity (Whigham et al. 1978). Based on these limited data it would appear that the belowground component of freshwater tidal marshes is smaller than that of saltwater marshes in the same geographical area. More data are needed, however, because these marshes are typically floristically richer than salt marshes and many species have not been studied.

Biomass Distribution with Depth

Living belowground biomass is generally quite shallow with the bulk of the material for most species found 30 cm or less in depth. Smith et al. (1979) found only half as much *Spartina alterniflora* short form material in the 20-30 cm depth interval as compared to the 0-10 cm interval, while samples below 40 cm were composed of small and erratic living biomass mixed with material in various states of decomposition. Broome et al. (1975) and Woodhouse et al. (1974) found 75% of *Spartina alterniflora* belowground biomass in the upper 10 cm of the marsh while Valiela et al. (1976) found the majority of *S. alterniflora* and *S. patens* roots and rhizomes in the upper 5 cm of the marsh. De la Cruz and Hackney (1977) found *Juncus roemerianus* to be even more concentrated in the upper soil zones with 94% of the belowground biomass production in the upper 20 cm. The distribution of belowground materials is not uniform, however, and Gallagher and Plumley (1979) distinguished three types of belowground macro-organic matter profiles. In the first type the concentration was relatively uniform with depth; in the second type macro-organic matter was mostly concentrated near the surface and declined rapidly with depth; the third type was characterized by low concentration near the surface, a high concentration about 5-20 cm below the surface, and a low concentration at greater depths. Profile type seemed mostly dependent on species but some habitat effects were also noted. All three types were present in *Spartina alterniflora* assemblages depending on habitat conditions. Type two profiles were most common while type one profiles were present where root-shoot ratios were low.

Root-Shoot Ratios

Root-shoot ratios in marsh species reflect differences in species and habitats. Root-shoot ratios (biomass, Table 2) as high as 50 have been reported for the short *Spartina alterniflora* ecophene (Gallagher 1974) although most estimates for this ecophene are considerably lower: 1.2 and 1.3 (Stroud 1976), 4.7 (Smith et al. 1979), 5.24 (Good and Frasco 1979). Estimates for the tall ecophene are generally lower: 0.4 and 0.3 (Stroud 1976), 1.43 (Gallagher 1974), and 4.53 (Good and Frasco 1979), but Valiela et al. (1976) reported a ratio of 8.25. Other root-shoot ratios for salt marsh species reported in the literature are 4.5 (Good and Frasco 1979) and 7.2 (Gallagher 1974) for *Distichlis spicata;* 5.58 (Good and Frasco 1979) and 4.0 (Valiela et al. 1976) for *Spartina patens;* and 0.80 (de la Cruz and Hackney 1977), 3.26 (Stout 1978), 8.2 (Gallagher 1974) and 3.7 to 8.7 (Kruczynski et al. 1978) for *Juncus roemerianus*. High root-shoot ratios have been considered to be indicative of adaptive mechanisms, with unfavorable

Table 2. Root-shoot ratios (biomass) for four species of saltwater marsh plants for several locations along the western Atlantic coast.

Species	*Ratio*	*Location*	*Source*
Distichlis spicata	7.2	Georgia	Gallagher 1974
	4.5	New Jersey	Good & Frasco 1979
Juncus roemerianus	0.80	Mississippi	de la Cruz & Hackney 1977
	3.26	Alabama	Stout 1978
	3.7-8.7	Florida	Kruczynski et al. 1978
	8.2	Georgia	Gallagher 1974
Spartina alterniflora			
(tall form)	1.43	Georgia	Gallagher 1974
	0.3-0.4	North Carolina	Stroud 1976
	4.53	New Jersey	Good & Frasco 1979
	8.25	Massachusetts	Valiela et al. 1976
(short form)	3.72	Alabama	Stout 1978
	48.9	Georgia	Gallagher 1974
	1.2-1.3	North Carolina	Stroud 1976
	4.7	New Jersey	Smith et al. 1979
	5.24	New Jersey	Good & Frasco 1979
Spartina patens	5.58	New Jersey	Good & Frasco 1979
	4.0	Massachusetts	Valiela et al. 1976

soil conditions requiring greater root surface for each unit of aboveground material (Shaver and Billings 1975). Early studies in Europe (Repp 1939) have demonstrated the common occurrence of high root-shoot ratios in a variety of halophyte species.

Freshwater tidal wetland root-shoot ratios are available for a number of species (Whigham and Simpson 1978). Perennial species typically had a greater allocation to belowground biomass than did annuals with most ratios from 0.55 to 3.64 (7 spp.) except 8.42 for *Peltandra virginica.* Root-shoot ratios for annual species were 0.14 to 0.46 (7 spp.). Annuals in these marshes are typically characterized by rapid initial root development in spring-early summer followed by strong dominance of aboveground allocation for the remainder of the growing season.

Data on seasonal changes in chemical constituents would complement data on biomass, giving qualitative and quantitative information on the nature and phenology of important translocation events. Some data on seasonal biochemical change are available (Stroud 1976; de la Cruz and Hackney 1977; Gallagher and Plumley 1979; Mendelssohn 1979; Smith et al. 1979) but comparisons are difficult because the constituents studied and the methods used are quite variable. De la Cruz and Hackney (1977) found that energy, carbon, hydrogen, nitrogen, and phosphorus did not vary significantly throughout the year for a *Juncus roemerianus* salt marsh in Mississippi but some depth differences were noted. The 0-10 cm layer, consisting primarily of rhizomes, had lower ash-free, energy, and elemental values (except for phosphorus) than 10-20 cm sections which were primarily composed of root materials. Gallagher and Plumley (1979), determined mineral composition for *Distichlis spicata* belowground macro-organic matter in Delaware. Nitrogen, phosphorus, potassium, and zinc decreased with depth while calcium, magnesium, manganese, and copper content did not vary with depth. Smith et al. (1979) found energy content to increase slightly with depth while mineral content was quite stable.

Stroud (1976), Smith et al. (1979), and Lytle and Hull (1980) have attempted more comprehensive studies of belowground constituents in relation to aboveground phenology and biomass accumulation, season, and depth. All studies dealt with *Spartina alterniflora;* Stroud (1976) studied short and tall ecophenes in North Carolina; Smith et al. (1979) studied the short ecophene in New Jersey; and Lytle and Hull (1980) studied both ecophenes in Rhode Island.

Smith et al. (1979) used proximate analysis (Hinds 1975) which differentiates between ash (minerals), crude fat (compounds soluble in ether), crude protein (nitrogen bearing compounds), N-free extract (remaining soluble compounds, mostly carbohydrates), and crude fiber (insoluble compounds, mostly structural carbohydrate). Crude fat was a minor constituent (about 1%) without seasonal trends while protein (about 7%) changes corresponded mostly to biomass changes. N-free extract (about 50%) and crude fiber (about 43%) included the major carbon pathway compounds and their dynamics were inversely related. N-free extract peaks were noted in May, June, July, November, and December with a low in February. Crude fiber was most characteristic of maturation stages with a rapid increase in the latter part of the growing season and low levels early in the growing season.

Stroud (1976) considered the carbohydrate components, especially those which were lumped together as N-free extract by Smith et al. (1979). Non-reducing sugars were the largest carbohydrate component and their dynamics determined the overall patterns of total soluble sugars. The short ecophene showed little annual variation in non-reducing sugars while in the tall form, non-reducing sugar content in roots was lowest at the beginning of the growing season with increases until January-February. Rhizomes had a higher and more variable non-reducing sugar content in both ecophenes. Increases were noted throughout the growing season with translocation to rhizomes evident during periods of high photosynthesis. Reducing sugars were highest in rhizomes, with lowest concentrations at the onset of aboveground growth, slight increases throughout the growing season, and a peak coinciding with aboveground dieback. Starch was essentially absent except for a low content in rhizomes. Carbohydrate patterns were dependent on marsh location, phenophase events, and ecophenal differences.

Lytle and Hull (1980) found sucrose to be the major carbohydrate extracted from culms and rhizomes throughout the year. Peak rhizome starch content was only 4%. Ecophenal differences were pronounced with a rapid fall increase in tall ecophene rhizome carbohydrates coincident with a decline in shoot carbohydrate content in culms. In contrast, the short ecophene retained a relatively high simple sugar content through October with gradual increases in rhizome carbohydrates during autumn and early winter. Lipid content in tall *S. alterniflora* showed a well defined seasonal fluctuation with a spring maximum and summer minimum while lipid content in short *S. alterniflora* was relatively constant throughout the year.

These studies indicate that the dynamics of belowground compounds, particularly carbohydrates, are important in understanding carbon balances in marshes but that much more information on a greater variety of species and locations will be needed before any generalizations can be made.

Mendelssohn (1979) studied several aspects of nitrogen metabolism in the height forms of *Spartina alterniflora* in North Carolina. Concentrations of root nitrogen constituents (total N, soluble TKN, NO_3, NH_4) were generally highest in the tall form, intermediate in the intermediate form, and lowest in the short form. In all height forms, levels of TKN, NO_3, and NH_4 were highest in winter, intermediate during spring-summer, and lowest in fall. Total root nitrogen showed no apparent seasonal dependency. Low NO_3 and nitrate reductase levels in all height forms indicated that NO_3 was not a major nitrogen source for *S. alterniflora*.

Factors Affecting Belowground Productivity

Data available on belowground production indicate that this component responds to environmental conditions quite differently than aboveground production. Thus the more extensive aboveground literature is of only limited value in predicting belowground responses. Seedling response to photoperiod and thermoperiod by *Spartina alterniflora* (Seneca and Broome 1972) and *S. patens* (Seneca 1974a) from North Carolina were found to be quite similar. Seedlings were exposed to five thermoperiods under both short and long-day photoperiods. Belowground biomass increased under both photoperiods as temperatures increased, but was greater in each thermoperiod under long-day conditions. Seneca (1974b) studied seedling photoperiod-thermoperiod responses of 12 populations of *S. alterniflora* from Rhode Island to Texas. While thermoperiods and photoperiods affected rhizome weight and root-shoot ratios of all populations, there were no apparent north-south population response trends.

Belowground production data of salt marsh species do not follow the generally increasing trend from north to south seen for aboveground production (Keefe 1972; Turner 1976). The highest values for both *Spartina alterniflora* and *S. patens* so far reported are from Massachusetts (Valiela et al. 1976) and New Jersey (Good and Frasco 1979). Data available from Maine (Gallagher and Plumley 1979) do, however, indicate that extremely northern salt marshes have generally low belowground productivity. Data from Georgia (Gallagher and Plumley 1979) and Mississippi (de la Cruz and Hackney 1977) indicate substantial but not exceptionally high productivity. It seems possible that the extended growing season in southern marshes and the mild winter conditions may minimize translocation to belowground storage organs while warm summers coupled with cold winters may increase belowground storage. It is also possible that the physical and chemical properties of the rooting zone are much more important than climate in determining belowground production.

Influence of Nutrients

Many studies have shown nitrogen fertilization to increase belowground production of *Spartina alterniflora* (Woodhouse et al. 1974; Broome et al. 1975; Haines and Dunn 1976; Haines 1979; Buresh et al. 1980; Smart and Barko 1980; Linthurst and Seneca 1981). A similar response was noted for *Distichlis spicata* (Smart and Barko 1980). Belowground response, however, was not as great as that observed in aerial portions as reflected by a decrease in the root-shoot ratio under fertilized conditions (Woodhouse et al. 1974; Broome et al. 1975; Gallagher 1975; Haines and Dunn 1976; Buresh et al. 1980; Smart and Barko 1980; Linthurst and Seneca 1981). Valiela et al. (1976) reported no nitrogen effect on rhizome production in either high or low marsh dominated by *Spartina patens* and *S. alterniflora*, respectively, and negative effects on root production, especially in the high marsh.

Fewer studies have looked at the effect of phosphorus on belowground production. Smart and Barko (1980) found that phosphorus was not limiting total production of *S. alterniflora* and *Distichlis spicata* in 15 marsh sediments. Woodhouse et al. (1974) found that phosphorus limited belowground production of *Spartina alterniflora* only when marsh plots were fertilized with nitrogen. The low N and P content of the sandy marsh substrate in this study was the primary reason for P limiting conditions after N fertilization.

Influence of Salinity, Oxygen, and Drainage

In most cases, belowground growth of salt and brackish marsh species is depressed with increasing salinity. Under laboratory conditions, Linthurst and Seneca (1981) found a 61% decrease in *Spartina alterniflora* biomass when soil water salinity was increased from $15^{0}/_{00}$ to $45^{0}/_{00}$. Salinity effects were more pronounced between $30^{0}/_{00}$ and $45^{0}/_{00}$. Similar results were reported by Parrondo et al. (1978). Culture solution salinities of $0\text{-}16^{0}/_{00}$ had no significant effect on *S. alterniflora* root growth but a salinity of $32^{0}/_{00}$ significantly reduced root biomass. Other studies have also shown reduced *S. alterniflora* root growth with increased salinity (Gosselink 1970; Smart and Barko 1978). Haines and Dunn (1976) studied the effect of ammonium nitrogen and salinity on *S. alterniflora* under controlled laboratory conditions. Salinity had a varying effect on root-rhizome growth depending upon ammonium level. At low and intermediate N levels, there was no change in root weight and an increase in rhizome weight with increasing salinity. At the high N level, root weight decreased with increasing salinity while rhizome weight was greatest at intermediate salinity.

Belowground biomass of *Spartina patens*, *Spartina foliosa*, *Spartina cynosuroides*, and *Sporobolus virginicus* also decreases with increased substrate or culture solution salinity (Gosselink 1970; Parrondo et al. 1978; Smart and Barko 1978; Gallagher 1979). Parrondo et al. (1978) found that salinities of $0\text{-}32^{0}/_{00}$ had no significant effect on *Distichlis spicata* root growth. Conversely, Smart and Barko (1978) noted a significant decrease in *D. spicata* belowground biomass when sediment salinity increased from $18^{0}/_{00}$ to $32^{0}/_{00}$.

In most cases, increased salinity reduced aerial production more than belowground production as reflected in an increased root-shoot ratio at high salinities (Haines and Dunn 1976; Parrondo et al. 1978; Smart and Barko 1978). Gosselink (1970) working with *S. alterniflora* and *S. patens* and Gallagher (1979) working with *Sporobolus virginicus* found increased salinity to depress root growth more than aerial growth.

Aeration was found to have a positive effect on the growth of *S. alterniflora* (Linthurst 1980; Linthurst and Seneca 1981). The response of aboveground production to aeration was found to be greater than belowground production response. Linthurst and Seneca (1981) noted that belowground production response to aeration was not equal for three *S. alterniflora* height forms, with the tall form showing the greatest increase in production under aerated conditions.

Several studies have monitored the effect of drainage conditions on growth of salt marsh species, especially *S. alterniflora*. Field studies have shown that reduced drainage conditions result in a decrease in total biomass of *S. alterniflora* (Linthurst and Seneca 1980; Mendelssohn and Seneca 1980). As noted with other environmental factors, aerial production is affected more than belowground production. Decreased drainage reduced root-rhizome growth of three *S. alterniflora* height forms although the reduction was not significant for the short form (Mendelssohn and Seneca 1980). Linthurst and Seneca (1980) found no significant change in belowground biomass of short form *S. alterniflora* grown under reduced drainage.

Laboratory studies, however, have indicated that reduced drainage conditions generally result in increased *S. alterniflora* belowground biomass (Parrondo et al. 1978; Mendelssohn and Seneca 1980). Plants grown under stagnant conditions (standing water over the soil surface), however, had significantly reduced root biomass (Mendelssohn and Seneca 1980). The conflicting results of laboratory and field studies can be resolved somewhat as the stagnant condition of laboratory studies are more like the reduced drainage conditions of field studies, and the reduced drainage conditions in laboratory studies are more like those of normal short form *S. alterniflora* communities in the field (Mendelssohn and Seneca 1980).

Decomposition and Turnover Times

Decomposition of belowground materials in waterlogged soils is usually very slow due to anaerobic conditions (Reddy and Patrick 1975). Detectable decomposition measured by a modified litterbag technique was limited to the upper 20 cm of substrate in a Mississippi study (Hackney and de la Cruz 1980) with decomposition rates approximately twice as fast at 5 cm than at 15 cm. Approximately 20% of the mass was lost in the top 10 cm during the first year. This study also found specific differences, with *Spartina cynosuroides* decomposing faster than *Juncus roemerianus* in the first few months, and differences in tissue types, with *Juncus* rhizomes decomposing faster than *Juncus* roots. Decomposition was relatively constant throughout the year in this climate. No trends in hydrogen, carbon, phosphorus, nitrogen, or caloric content were apparent.

Turnover times derived from dividing maximum biomass by the biomass increment give another measure of relative decomposition rates although they do not provide information on the details of the process. A number of such calculations are available for Georgia, Delaware, and Maine (Gallagher and Plumley 1979) and for New Jersey (Good and Walker 1977; Good and Frasco 1979; Smith et al. 1979). Turnover times calculated by Gallagher and Plumley (1979) for creekside (tall form) *Spartina alterniflora* from Georgia and Maine marshes were 18 and 119 months, respectively, while data from New Jersey (Good and Walker 1977; Good and Frasco 1979) for this form indicated a turnover of 22-31 months, suggesting the probable influence of temperature on decomposition rates. High marsh (short form) *S. alterniflora* turnover time was calculated to be 57 months in Georgia and 62-66 months in New Jersey, indicating the importance of location within the marsh on decomposition rates for this species and the reduced influence of climate under these conditions. Turnover times for *Juncus roemerianus* (44 months) and *Spartina cynosuroides* (28 months) from Georgia support the Hackney and de la Cruz (1980) finding that *S. cynosuroides* decomposes more readily than *J. roemerianus* under comparable conditions. *Phragmites communis (australis)* turnover time was quite similar for Delaware (27 months – Gallagher and Plumley 1979) and New Jersey (23-38 months – Good and Walker 1977). *Juncus gerardi, Borrichia frutescens*, and *Salicornia virginica* all had rather rapid (18-26 month) turnover times in Delaware and Georgia (Gallagher and Plumley 1979) while *Spartina patens* was found to be a more resistant species with a 68 month turnover time for Georgia and a 92 month turnover time for both Delaware and Maine (Gallagher and Plumley 1979). Turnover time for *S. patens* in New Jersey marshes was found to be quite variable, ranging from 30-72 months (Good and Walker 1977; Good and Frasco 1979). Turnover time for *Distichlis spicata* was 40 months in Georgia and Delaware (Gallagher and Plumley 1979) and 54 months in New Jersey (Good and Frasco 1979).

Fate of Belowground Production

Because available data indicate that the belowground component is at least equal to and often greater than the aboveground component, the potential significance of belowground productivity to food chains/webs, nutrient cycling, and maintenance of the structural integrity of the marsh is enormous.

Herbivory

In situ utilization of roots and rhizomes can be substantial as documented by Smith and Odum (1981). Snow goose (*Anser caerulescens* ssp. *atlantica*) utilization of three North Carolina salt marshes dominated by *Spartina alterniflora, S. patens-Distichlis spicata*, and *Scirpus robustus* resulted in removal of approximately 58% of the belowground biomass in the upper 20 cm or so. Some of this material was consumed directly and some was probably left on the marsh surface and made available for transport to estuarine waters. Belowground production was reduced on two of the three sites the following growing season. Lynch et al. (1947) reported that goose "eat-outs" could cause considerable damage to salt and brackish marshes in Louisiana and Texas. Geese were found to prefer roots and rhizomes over aerial portions and could disturb marsh substrate to depths of 20 cm. It was also noted that while geese were prodigious grazers, they consumed only $\simeq$ 10% of the roots and rhizomes they disturbed. While others including Cottam (1935) and Saylor (1941) have documented similar feeding of winter flocks of geese, the overall importance of goose "eat-outs" is not well quantified.

The importance of herbivory of other known belowground utilizers, such as muskrats, is even more poorly known. Lynch et al. (1947) found that muskrat "eat-outs" in Gulf Coast brackish marshes caused considerable damage in localized areas. Although direct *in situ* feeding on the belowground component deserves further study, in most cases it would appear that the total amount of biomass utilized in this fashion on an ecosystem basis would be relatively modest. While phenomena such as "eat-outs" may provide locally important mechanisms for root and rhizome material to enter the detritus food chain, they appear to be spatially and temporally limited so that the overall contribution of belowground materials to the detritus food chain in the form of recognizable material must also be limited.

Marsh invertebrate utilization of belowground materials has also received limited attention. Several studies have shown that larger invertebrate species such as fiddler crabs (*Uca* spp.) can play a role in introducing belowground material to the detrital food chain. Everest and Davis (1979), using laboratory salt marsh microecosystems, found fiddler crabs played an important role in P cycling. The burrowing activity of the crabs brought material buried in the sediment to the marsh surface, thus introducing the material into the food chain. Katz (1980) estimated that fiddler crab burrowing activity in a creekside *Spartina alterniflora* community resulted in an 18% turnover of the upper 15 cm of sediment. Belowground materials have been demonstrated in tidal creek benthic detritus although they constitute only a small portion of the detritus standing stock (Pickral and Odum 1977).

Decomposition processes

The bulk of the belowground material must decompose in place in an undisturbed marsh and the products of this decomposition as microbial biomass or utilizable compounds may enter the food chain. It has become clear that conventional measurement methods which are appropriate for estimating energy flow of aboveground production do not adequately measure energy flow of the belowground component. The belowground habitat includes both aerobic environments, the surface layer of the soil and the rhizosphere, and the deeper anaerobic environments (Pomeroy et al. 1977). Heterotrophic aerobic microbial metabolism mainly involves conversion of organic matter to carbon dioxide and biomass while anaerobic microbial metabolism is more complex. Microbial anaerobic groups include fermenting and dissimilatory nitrogenous oxide reducing bacteria, sulfate reducing bacteria, and methanogenic bacteria (Pomeroy et al. 1977). Early studies of sediment oxygen uptake (Teal 1962) did not adequately monitor input via these pathways and studies monitoring carbon exchange such as Woodwell et al. (1977) have concluded that salt marshes are not large exporters of carbon and annual flows may actually be inward, making the marsh a sink rather than a source. Recently Nixon (1980) has presented a critical review summarizing earlier and current views on coastal marsh-estuarine interactions. Unfortunately, even this current work does not seriously consider the belowground component although this component appears to represent at least one half of marsh productivity by even conservative estimates. Measurements reported by Kaplan et al. (1979) for denitrification in Great Sippewissett Marsh, Massachusetts, accounted for only about 12 g C m^{-2} yr^{-1}. Over 60% of the denitrification occurred in muddy creek bottoms rather than in the *Spartina* vegetated areas of the marsh.

Recently Howarth and Teal (1979, 1980) have focused attention on the sulfate reducing bacteria in Great Sippewissett Marsh after preliminary studies indicated that methanogenic bacteria have modest inputs. Lipschultz (1981) also reports relatively modest rates of methane release from a low salinity mudflat although the rates were higher than those of marsh peat. Sulfate reduction in the marsh peat is estimated to account for the consumption of 1800 g organic C m^{-2} yr^{-1}. These high rates are believed to be related to the high organic input of *Spartina alterniflora* into the substrate, the resupply of sulfate to the peat by tidal waters, and maintenance of sulfide below toxic levels. Iron pyrite is a major end product of sulfate reduction. It is hypothesized that the energy available from reduced sulfur compounds could support substantial chemoautotroph production in nearby creeks and estuaries, thus providing a potential mechanism for energy transfer from the large belowground element to coastal food webs. In most marshes, export of these reduced sulfur compounds via tidal action and groundwater movement is probably limited to creekside sites as highmarsh sites exhibit reduced total flooding as well as a hydraulically stagnant water table (Gardner 1973; Nestler 1977). Mechanisms for energy transfer and measurements of the actual amount of energy transferred verses incorporation into the marsh sediment await further study.

Fluxes

Fluxes of materials between marshes and associated estuarine waters are extremely difficult to quantify and typical tidal cycles may carry only a very small fraction of material compared to major storms. Woodwell et al. (1977) reported a 45-fold difference between average transport and extraordinary high tide transport. Although their occurrence is irregular, it is probable that truly major storm tides may remove very large amounts of materials, both particulate and dissolved, from the marsh.

Marsh Maintenance

A portion of the belowground biomass may contribute to marsh accretion which is essential for marsh maintenance in areas where subsidence or rising sea level dominate (De Laune and Patrick 1980). Much of the accretion can result from entrapment and stabilization of suspended sediments. Water management projects reducing sediment load significantly reduce marsh accretion (De Laune and Patrick 1980). Organic matter was generally less than 20% of soil marsh solids in Louisiana although it was important in providing structural support (De Laune and Patrick 1980). A New Jersey pedological study (Jaworski 1980) found that organic matter generally decreased with depth but tended to accumulate in less frequently flooded areas of the marsh. Organic matter content in this study was generally less than 20% so that the soils were not generally considered to be organic. Sediment supply appears to be more important for accretion in low marsh (intertidal) areas (Redfield 1972; Richard 1978) and areas where significant compaction occurs, such as the Mississippi Delta (De Laune and Patrick 1980). Long-term storage of organic materials may be somewhat limited under such conditions.

The contribution of belowground material to marsh accretion appears to be more important under high marsh conditions (Armentano and Woodwell 1975). Many studies have demonstrated the existence of thick layers (up to 5-10 m) of high marsh peat (Davis 1911; Daddario 1961; Redfield and Rubin 1962, and others). The organic matter content of these peats is high (usually over 20% on a dry weight basis) and is more or less constant with depth (Redfield 1965; Armentano and Woodwell 1975; Frasco 1980). These data coupled with low marsh accretion rates for high marshes of 1 to 2 mm yr^{-1} long term (Redfield and Rubin 1962; Stuvier and Daddario 1963) and 2 to 7 mm yr^{-1} short term (Armentano and Woodwell 1975; Harrison and Bloom 1977; Richard 1978) indicate that a considerable portion of marsh

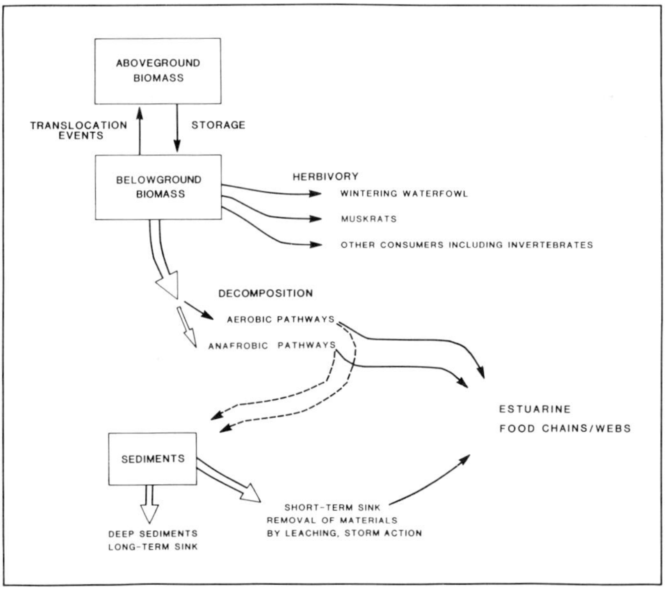

Figure 1. Conceptual Model of Belowground Biomass Pathways.

belowground biomass can form a long-term storage component of organic carbon under appropriate conditions.

The relative contributions of belowground biomass to the various possible pathways (herbivory, aerobic and anerobic decomposition, sedimentation) are unknown at the present time. Although field and laboratory studies in the last ten years or so have provided useful information on the general magnitude of the belowground component and some of the several possible pathways of its utilization or deposition, the data are not extensive enough to construct meaningful quantitative models. We do present a simple qualitative model of energy flow and material flux for belowground environments (Fig. 1). Nixon's (1980) review should dampen the desire to extrapolate and interpolate from a limited data base. The limitations of laboratory data and presently available methods of measurement of such important processes as microbial metabolism should be taken into account in any synthesis of available data. The concept of a "role" exists primarily with the researcher and not with the marsh. It appears likely that a continuum of "roles" exist dependent on such variables as developmental stage of the marsh, tidal amplitude, climate, and other environmental variables.

Acknowledgments

We wish to thank the many ecologists whose laundry procedures have provided the data in this paper. Specific thanks are given to John L. Gallagher, Courtney T. Hackney, Robert W. Howarth and Judy P. Stout for reviewing the manuscript in draft form.

References Cited

Armentano, T. V. and G. M. Woodwell. 1975. Sedimentation rates in a Long Island marsh determined by ^{210}Pb dating. *Limnol. Oceanogr.* 20:452-456.

Broome, S. W., W. W. Woodhouse, Jr. and E. D. Seneca. 1975. The relationship of mineral nutrients to growth of *Spartina alterniflora* in North Carolina: II. The effects of N, P, and Fe fertilizers. *Soil Science Soc. Am. Proc.* 39:301-307.

Buresh, R. J., R. D. DeLaune and W. H. Patrick, Jr. 1980. Nitrogen and phosphorus distribution and utilization by *Spartina alterniflora* in a Louisiana Gulf coast marsh. *Estuaries* 3:111-121.

Cottam, C. 1935. Blue and Snow Geese in the eastern United States in the winter of 1934-5; with notes on their food habits. *Auk* 52:432-441.

Coupland, R. T. 1975. Productivity of grassland ecosystems, pp. 44-49. *In: Productivity of World Ecosystems.* National Academy of Sciences, Washington DC.

Daddario, J. J. 1961. A lagoon deposit profile near Atlantic City, New Jersey. *Bull. N. J. Acad. Sci.* 6:7-14.

Davis, C. A. 1911. Salt marshes, a study in correlation. *Ann. Assoc. Amer. Geogr.* 1:139-143.

de la Cruz, A. A. and C. T. Hackney. 1977. Energy value, elemental composition, and productivity of belowground biomass of a *Juncus* tidal marsh. *Ecology* 58:1165-1170.

DeLaune, R. D. and W. H. Patrick, Jr. 1980. Rate of sedimentation and its role in nutrient cycling in a Louisiana salt marsh, pp. 401-412. *In:* P. Hamilton and K. B. MacDonald (eds.), *Estuarine and Wetland Processes.* Plenum Press, New York.

Everest, J. W. and D. E. Davis. 1979. Studies of phosphorus movement using salt marsh microecosystems. *J. Envir. Quality* 8:465-468.

Frasco, B. R. 1980. Plant ecology of the upland-salt marsh transition zone surrounding several forest islands in southern New Jersey. Ph.D. Dissertation, Rutgers U., New Brunswick, NJ 223 pp.

Gallagher, J. L. 1974. Sampling macro-organic matter profiles in salt marsh plant root zones. *Soil Sci. Soc. Amer. Proc.* 38:154-155.

Gallagher, J. L. 1975. Effect of an ammonium nitrate pulse on the growth and elemental composition of natural stands of *Spartina alterniflora* and *Juncus roemerianus*. *Amer. J. Bot.* 62:644-648.

Gallagher, J. L. 1979. Growth and element compositional responses of *Sporobolus virginicus* (L.) Kunth. to substrate salinity and nitrogen. *Amer. Mid. Nat.* 102:68-75.

Gallagher, J. L. and F. G. Plumley. 1979. Underground biomass profiles and productivity in Atlantic coastal marshes. *Amer. J. Bot.* 66:156-161.

Gardner, L. R. 1973. The effect of hydrologic factors on the pore water chemistry of intertidal marsh sediments. *Southeastern Geol.* 15:17-28.

Good, R. E. 1977. An environmental assessment of the proposed reconstruction of State Route 152 (Somers Point-Longport) Atlantic County, New Jersey. Report to E. Lionel Pavlo Engineering, New York, NY 41 pp.

Good, R. E. and B. R. Frasco. 1979. Estuarine evaluation study; a four year report on production and decomposition dynamics of salt marsh communities: Manahawkin marshes, Ocean County, New Jersey. Report to New Jersey Department Environmental Protection, Div. Fish, Game and Shellfisheries, Trenton, NJ 105 pp.

Good, R. E. and N. F. Good. 1975. Vegetation and production of the Woodbury Creek-Hessian Run freshwater tidal marshes. *Bartonia* 43:38-45.

Good, R. E. and R. Walker. 1977. Relative contribution of saltwater and freshwater tidal marsh communities to estuarine productivity. Final Report (071171) submitted to Rutgers U., Center for Coastal and Environmental Studies, New Brunswick, NJ. Unpublished manuscript.

Gosselink, J. G. 1970. Growth of *Spartina patens* and *S. alterniflora* as influenced by salinity and source of nitrogen. *Coastal Stud. Bull.* 5:97-110.

Hackney, C. T. and A. A. de la Cruz. 1980. *In situ* decomposition of roots and rhizomes of two tidal marsh plants. *Ecology* 61:226-231.

Haines, B. L. and E. L. Dunn. 1976. Growth and resource allocation response of *Spartina alterniflora* Loisel. to three levels of NH_4-N, Fe, and NaCl in solution culture. *Bot. Gaz.* 137:224-230.

Haines, E. B. 1979. Growth dynamics of cordgrass, *Spartina alterniflora* Loisel., on control and sewage sludge fertilized plots in a Georgia salt marsh. *Estuaries* 2:50-53.

Harrison, E. Z. and A. L. Bloom. 1977. Sedimentation rates on tidal salt marshes in Connecticut. *J. Sed. Petrology* 47:1484-1490.

Hinds, W. T. 1975. Energy and carbon balance in cheatgrass: an essay in autecology. *Ecol. Monogr.* 45:367-388.

Howarth, R. W. and J. M. Teal. 1979. Sulfate reduction in a New England salt marsh. *Limnol. Oceanogr.* 24:999-1013.

Howarth, R. W. and J. M. Teal. 1980. Energy flow in a salt marsh ecosystem: the role of reduced inorganic sulfur compounds. *Amer. Nat.* 116:862-872.

Jaworski, A. Z. 1980. A pedologic study of some tidal marsh soils in New Jersey. Ph.D. Dissertation, Rutgers U., New Brunswick, NJ 154 pp.

Kaplan, W., I. Valiela and J. M. Teal. 1979. Denitrification in a salt marsh ecosystem. *Limnol. Oceanogr.* 24:726-734.

Katz, L. C. 1980. Effects of burrowing of the fiddler crab, *Uca pugnax* (Smith). *Est. Coastal Mar. Sci.* 11:233-237.

Keefe, C. 1972. Marsh production: a summary of the literature. *Contr. Mar. Sci.* 16:163-181.

Kruczynski, W. L., C. B. Subrahmanyam and S. H. Drake. 1978. Studies on the plant community of a north Florida salt marsh. Part I. Primary production. *Bull. Mar. Sci.* 28:316-334.

Kucera, C. L., R. C. Dahlman and M. R. Koelling. 1967. Total net productivity and turnover on an energy basis for tallgrass prairie. *Ecology* 48:536-541.

Linthurst, R. A. 1979. The effect of aeration on the growth of *Spartina alterniflora* Loisel. *Amer. J. Bot.* 66:685-691.

Linthurst, R. A. 1980. A growth comparison of *Spartina alterniflora* Loisel. ecophenes under aerobic and anaerobic conditions. *Amer. J. Bot.* 67:883-887.

Linthurst, R. A. and E. D. Seneca. 1980. The effects of standing water and drainage potential on the *Spartina alterniflora*-substrate complex in a North Carolina salt marsh. *Est. Coastal Mar. Sci.* 11:41-52.

Linthurst, R. A. and E. D. Seneca. 1981. Aeration, nitrogen and salinity as determinants of *Spartina alterniflora* Loisel. growth response. *Estuaries* 4:53-63.

Lipschultz, F. 1981. Methane release from a brackish intertidal salt marsh embayment of Chesapeake Bay, Maryland. *Estuaries* 4:143-145.

Lynch, J. J., T. O'Neil and D. W. Lay. 1947. Management significance of damage by geese and muskrats to Gulf Coast marshes. *J. Wildl. Man.* 11:50-76.

Lytle, R. W. Jr. and R. J. Hull. 1980. Annual carbohydrate variation in culms and rhizomes of smooth cordgrass (*Spartina alterniflora* Loisel). *Agron. J.* 72:942-946.

Mendelssohn, I. A. 1979. Nitrogen metabolism in the height forms of *Spartina alterniflora* in North Carolina. *Ecology* 60:574-584.

Mendelssohn, I. A. and E. D. Seneca. 1980. The influence of soil drainage on the growth of salt marsh cordgrass *Spartina alterniflora* in North Carolina. *Est. Coastal Mar. Sci.* 11:27-40.

Nestler, J. 1977. A preliminary study of the sediment hydrology of a Georgia salt marsh using Rhodamine WT as a tracer. *Southeastern Geol.* 18:265-271.

Nixon, S. W. 1980. Between coastal marshes and coastal waters – a review of twenty years of speculation and research on the role of salt marshes in estuarine productivity and water chemistry, pp. 437-525. *In:* P. Hamilton and K. B. MacDonald (eds.), *Estuarine and Wetland Processes.* Plenum Press, New York.

Parrondo, R. T., J. G. Gosselink and C. S. Hopkinson. 1978. Effects of salinity and drainage on the growth of three salt marsh grasses. *Bot. Gaz.* 139:102-107.

Pickral, J. C. and W. E. Odum. 1977. Benthic detritus in a saltmarsh tidal creek, pp. 280-292. *In:* M. L. Wiley (ed.), *Estuarine Processes, Voume II.* Academic Press, New York.

Pomeroy, L. R., K. Bancroft, J. Breed, R. R. Christian, D. Frankenberg, J. R. Hall, L. G. Maurer, W. J. Wiebe, R. G. Wiegert and R. L. Wetzel. 1977. Flux of organic matter through a salt marsh, pp. 270-279. *In:* M. L. Wiley (ed.), *Estuarine Processes, Volume II.* Academic Press, New York.

Reddy, K. R. and W. H. Patrick, Jr. 1975. Effect of alternate aerobic and anaerobic conditions on redox potential, organic matter decomposition and nitrogen loss in a flooded soil. *Soil Biol. Biochem.* 7:87-94.

Redfield, A. C. 1965. The thermal regime in salt marsh peat at Barnstable, Massachusetts. *Tellus* 17:246-259.

Redfield, A. C. 1972. Development of a New England salt marsh. *Ecol. Monogr.* 42:201-237.

Redfield, A. C. and M. Rubin. 1962. The age of salt marsh peat and its relation to recent changes in sea level at Barnstable, Massachusetts. *Proc. Nat. Acad. Sci. (U.S.)* 48:1728-1735.

Repp, G. 1939. Ökologische Untersuchungen im Halophyten-gebiet am Neusiedlersee. *Jahrb. Wiss. Bot.* 88:554-632.

Richard, G. A. 1978. Seasonal and environmental variations in sediment accretion in a Long Island salt marsh. *Estuaries* 1:29-35.

Saylor, W. L. 1941. Winter food of Snow and Blue Geese in Delaware. *Auk* 58:92.

Seneca, E. D. 1974a. Seedling response to photoperiod and thermoperiod by salt meadow cordgrass, *Spartina patens*, from Ocracoke Island, North Carolina. *Chesapeake Sci.* 15:230-232.

Seneca, E. D. 1974b. Germination and seedling response of Atlantic and Gulf Coast populations of *Spartina alterniflora*. *Amer. J. Bot.* 61:947-956.

Seneca, E. D. and S. W. Broome. 1972. Seedling response to photoperiod and temperature by smooth cordgrass, *Spartina alterniflora*, from Oregon Inlet, North Carolina. *Chesapeake Sci.* 13:212-215.

Shaver, G. R. and W. D. Billings. 1975. Root production and root turnover in a wet tundra ecosystem, Barrow, Alaska. *Ecology* 56:401-409.

Smart, R. M. and J. W. Barko. 1978. Influence of sediment salinity and nutrients on the physiological ecology of selected salt marsh plants. *Est. Coastal Mar. Sci.* 7:487-495.

Smart, R. M. and J. W. Barko. 1980. Nitrogen nutrition and salinity tolerance of *Distichlis spicata* and *Spartina alterniflora*. *Ecology* 61:630-638.

Smith, K. K., R. E. Good and N. F. Good. 1979. Production dynamics for above and belowground components of a New Jersey *Spartina alterniflora* tidal marsh. *Est. Coastal Mar. Sci.* 9:189-201.

Smith, T. J., III and W. E. Odum. 1981. The effects of grazing by Snow Geese on coastal salt marshes. *Ecology* 62:98-106.

Stout, J. P. 1978. An analysis of annual growth and productivity of *Juncus roemerianus* Scheele and *Spartina alterniflora* Loisel. in coastal Alabama. Ph.D. Dissertation, U. Alabama, University, AL 95 pp.

Stroud, L. M. 1976. Net primary production of belowground material and carbohydrate patterns of two height forms of *Spartina alterniflora* in two North Carolina marshes. Ph.D. Dissertation, North Carolina State U., Raleigh, NC 140 pp.

Stuiver, M. and J. J. Daddario. 1963. Submergence of the New Jersey coast. *Science* 142:951.

Teal, J. M. 1962. Energy flow in the salt marsh ecosystem of Georgia. *Ecology* 43:614-624.

Turner, R. E. 1976. Geographic variations in salt marsh macrophyte production: a review. *Contr. Mar. Sci.* 20:47-68.

Valiela, I., J. M. Teal and N. Y. Persson. 1976. Production and dynamics of experimentally enriched salt marsh vegetation: belowground biomass. *Limnol. Oceanogr.* 21:245-252.

Whigham, D. F., J. McCormick, R. E. Good and R. L. Simpson. 1978. Biomass and primary production in freshwater tidal wetlands of the Middle Atlantic Coast, pp. 3-20. *In:* R. E. Good, D. F. Whigham, and R. L. Simpson (eds.), *Freshwater Wetlands: Ecological Processes and Management Potential.* Academic Press, New York.

Whigham, D. F. and R. L. Simpson. 1978. The relationship between aboveground and belowground biomass of freshwater tidal wetland macrophytes. *Aquat. Bot.* 5:355-364.

White, D. A., T. E. Weiss, J. M. Trapani and L. B. Thien. 1978. Productivity and decomposition of the dominant salt marsh plants in Louisiana. *Ecology* 59:751-759.

Woodhouse, W. W., Jr., E. D. Seneca and S. W. Broome. 1974. Propagation of *Spartina alterniflora* for substrate stabilization and salt marsh development. Tech. Memo. 46. Coastal Eng. Res. Center, U. S. Army Corps of Eng., Fort Belvoir, VA 155 pp.

Woodwell, G. M., D. E. Whitney, C. A. S. Hall and R. A. Houghton. 1977. The Flax Pond ecosystem study: Exchanges of carbon in water between a salt marsh and Long Island Sound. *Limnol. Oceanogr.* 22:833-838.

BACTERIA AND MICROBIAL DISTRIBUTION IN ESTUARINE SEDIMENTS

Parke A. Rublee
Department of Biology
Whitman College
Walla Walla, Washington

Abstract: Acridine orange direct counts (AODC) are a useful tool for assessing bacterial numbers in fine-grained marsh and estuarine sediments. The number of bacteria in marshes ranges from 1-20 × 10^9 cells/cm^3 in surface sediments and decreases to 1-3 × 10^9 cells/cm^3 at a depth of 20 cm. Most cells are associated with sediment particles and there is a strong correlation of bacterial number with sediment organic carbon content. The average cell volume of sediment bacteria is about 0.2 μm^3 (range <0.05 to>2.0 μm^3). Estimated bacterial biomass in the upper 20 cm of marsh sediments is in the range of 10-20 g carbon/m^2. Total microbial biomass as estimated from adenosine triphosphate (ATP) measurement is in the range of 13-32 g carbon/m^2 to a depth of 20 cm. Benthic algal biomass (estimated from chlorophyll *a*) is extremely variable in marsh sediments, but constitutes the major microbial biomass component in surface sediments (0-1 cm). Fungi, protozoans, and meiofauna are also important in surface sediments. Microbial biomass and diversity decrease with depth and, below the surface few centimeters, bacteria comprise the dominant proportion of the total microbial community.

Introduction

During the past several decades the study of the role of microorganisms in coastal ecosystem processes has achieved a new degree of prominence. Classic studies (e.g. Teal 1962) implicated bacteria and microbial processes as being extremely important in the marsh economy. Although some concepts derived from pioneering studies have been criticized in recent years (e.g., Haines 1979; Nixon 1980), the importance of microbial pathways is not in doubt but simplistic earlier approaches must be replaced by a more complete understanding of the highly complex trophodynamic characteristics of the microbial communities of salt marshes.

The purpose of this paper is to review and compare current information on distribution and biomass of the benthic estuarine microbial community. Emphasis is placed on direct count determinations of bacteria and their relationship to other microbial taxa (benthic algae, fungi, protozoans). I discuss: (1) methodologies for determination of microbial abundance; (2) distribution of bacteria in estuarine sediments; (3) bacteria particle interactions in sediments and bacterial transport; (4) size of sediment bacteria; (5) deter-

ISBN 0-12-404070-5

mination of bacterial biomass; and (6) determination of biomass of total microbial community and its components. A large microbial standing crop would imply, but would not demonstrate, an active role in estuarine systems. In fact, although information on standing crop values is a necessary prerequisite to a complete understanding, it is only when these are combined with rate measures that we can assess definitively the importance of these microbial groups in the overall economy of the system.

Methodology for Determination of Microbial Abundance

Several recent volumes have addressed methods for the determination of characteristics of microbial communities (Costerton and Colwell 1979; Jones 1979; Litchfield and Seyfried 1979). One approach is direct observation of the organisms in question; alternatively, an indirect approach measures chemical components of microorganisms. A complete discussion of these approaches is beyond the scope of this paper, however a brief review of some of the more common methods should yield insight into advantages and disadvantages of various techniques.

Direct Methods

Direct methods include culturing or microscopic observation of the sample. Culture methods rely on questionable assumptions and are generally considered inadequate for community assessment, although they are important in enumeration of specific populations (Costerton and Colwell 1979). Microscopic methods for large protozoans generally involve extraction of organisms from sediment (Fenchel 1969; Dornseif 1976). For smaller microorganisms, samples are viewed directly as a sediment smear or diluted and agitated with particle-free water prior to observation. Stains are often used to enhance visibility. For fungi, additional steps may be necessary to promote optimal viewing (Newell and Hicks 1982).

For bacteria, the direct method most often used is the acridine orange direct count method (AODC) which utilizes a fluorescent dye and epiillumination (Hobbie et al. 1977). A recent review by Daley (1979) discusses this method in detail. Primary advantages are the relatively rapid, inexpensive analysis of samples; up to 20 samples may be counted in a day with a coefficient of variation of 10-30%. Disadvantages include tediousness and subjectivity of counting. This method yields information on total numbers and sizes of cells: no data are generated on the metabolic state of the bacteria. Further, it is difficult to determine cell volumes accurately because many of the naturally occurring microbial cells are near the limits of resolution of the light microscope (0.5 μm in major dimension).

The use of AODC in sediments is difficult since many cells are attached to the substrate. For reproducible counts, cells are first removed from sediment particles, generally by dilution and homogenization in a commercial blender. The dilution is necessary so that particles do not cover the majority of the microscopic field to be counted, but it also must not be so great that it results in too few cells for a statistically significant count. The homogenization does not completely remove cells. Rublee and Dornseif (1978) estimated a 74% efficiency of removal of bacterial cells from particles by quantitatively assessing the number of bacteria remaining on particles after homogenizing. In Maryland, I have found 60–70% removal of cells from particles after a 1 min homogenization with good reproducibility of counts from replicate samples; longer blending did not improve results (Fig. 1). Moriarty (1980) found poor efficiency of removal in sandy sediments which makes counting extremely difficult, but he favors the AODC method in soft clays (D.J.W. Moriarty, CSIRO, Queensland, Australia, pers. comm.). A balance must be struck between the acceptable error, the ease of counting, and processing of samples.

Daley (1979) has pointed out that proof of the accuracy of the AODC method is "(necessarily) indirect but nonetheless compelling" for planktonic samples. Similarly, rigorous testing of the accuracy of AODC in sediments has not been conducted. Nevertheless, at least within fine-grained sediments, the AODC method provides consistent and reproducible results which are not incompatible with information derived by other approaches. Thus, within

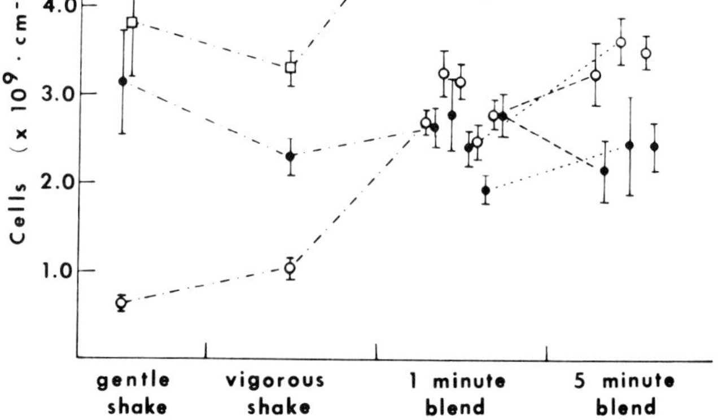

Figure 1. Effect of treatment on proportion of free and attached bacteria and total number of bacteria in replicate marsh sediment samples. O – free cells; ● – cells attached or adjacent to particles; □ – total cell count. Lines connect samples which were taken through successive treatments.

marshes and estuaries where fine sediments predominate the method is appropriate.

Use of fluorochrome stains other than acridine orange may aid in the direct counting of benthic bacteria. Recently, DAPI (Porter and Feig 1980), Hoechst 33258 (Paul 1982), and other nucleic acid specific flourochromes (ethidium bromide, propidium iodide) have been used. Each of these dyes is more specific than acridine orange and therefore sediment particles are not dyed as they are with acridine orange. However, the reduced ability to view any of the background (filter or particles) upon which the cells lie may also reduce the ability to determine if the dilution is appropriate, as well as the opportunity of determining incidental observations such as the nature of the particulate matter (organic vs. inorganic, etc.). Comparisons of DAPI, Hoechst 33258, and AO indicate similar cell counts (Porter and Fieg, 1980; Paul 1982; J. E. Hobbie, MBL, Woods Hole, MA, pers. comm.). Refinement of fluorescent counting techniques will be a continuing process that should improve our abilities to quantify components of microbial communities.

Indirect Methods

Perhaps the most common indirect method for analysis of total living microbial biomass is the use of firefly lantern extract in the determination of ATP and other cellular nucleotide triphosphates. This method is rapid, sensitive, relatively inexpensive, and fraught with problems of both analysis and interpretation, particularly when used in sediments. Briefly, a known quantity of sediment is extracted by use of a boiling buffer, organic solvent, or cold acid. Samples may then be stored until analysis against reagent standards. The ATP content may be converted to a biomass measure by a conversion factor. Several cautions are warranted: First, the ability of various extractant menstruums to recover ATP is variable and intimately related to the characteristics of the samples (Karl 1980). A second problem relates to the ratio of carbon to ATP. Karl (1980) has pointed out that carbon to ATP ratios vary from 43 to 9500, although he suggests that, at least in planktonic systems, the ratio of 250:1 is reasonable. Recently Davis and White (1980) have proposed a chemical method which appears to surmount the extraction and analysis problems, and allows the determination of energy charge as well as total adenylates which appears to be a better indicator of microbial biomass than ATP alone.

Other cell constituents have been used to estimate biomass of specific microbial groups. For example, determination of various pigments, especially chlorophyll *a*, has been used to estimate algal biomass. Conversion of values of chlorophyll *a* to carbon is difficult, however, especially in sediments where algal cells may be buried and surviving heterotrophically. Muramic acid, a constituent of prokaryotic cell walls, has been utilized as an estimator of bacterial biomass (King and White 1977; Moriarty 1980). This method has disadvantages in that gram positive and gram negative bacteria contain different amounts of muramic acid on a unit volume basis, thus reducing the accuracy of a muramic acid to biomass conversion factor. Moriarty (1980) has pointed out, however, that in sediments with large particles (e.g., sandy sediments) the method is preferable to direct count methods.

Recently, White and co-workers, in a series of papers, have suggested a suite of biochemical techniques for determination of microbial biomass and community structure (e.g., White et al. 1979). These methods quantify lipid and fatty acid components that are specific to various microbial taxa. Currently, these methods are proving useful in the characterization of microbial communities and factors which affect them (e.g., White et al. 1980; Bobbie et al. 1981). Some of these methods are cost, time, and labor intensive and thus beyond the scope of some laboratories. Nevertheless, such methods offer great hope as the specific relationships of the biochemical constitutents to microorganisms are more rigorously defined and as they become more widely used in estuarine sediments.

Distribution of Bacteria in Estuarine Sediments

Direct counts of total bacteria in intertidal and subtidal sediments range from about 4-17 $\times$ 10^9 cells/cm^3 in surface sediments and decrease with increasing depth of sediment (Table 1). Seasonal distribution of bacteria numbers has been reported in a study of a North Carolina marsh (Rublee 1982) where variability in surface sediments (0 to 1 cm) was apparent with a high of 14 $\times$ 10^9 cells/cm^3 in late fall and lowest numbers (around 5 $\times$ 10^9 cells/cm^3) in late summer. At 5 cm depth this pattern was repeated but with lower maximum and minimum numbers (7 and 3 $\times$ 10^9 cells/cm^3); at sediment depths of 10 and 20 cm the pattern did not appear to reflect seasonality. Rublee and Dornseif (1978) observed that, in four stations spanning this same marsh from the high tide line to a shallow subtidal station, the number of bacteria was not significantly different among any of the stations to a depth of 20 cm. In the Rhode River estuary of Maryland, similar results were found among six stations including four vegetated marsh sites, a tidal creek site, and a subtidal mud flat. Physical and biological characteristics of the sites differed, but the number of bacteria among stations was not significantly different at the 95% confidence level (F = 2.86; 6, 15 d.f.). However, a trend of lower numbers in the upper layers of creek and subtidal sediments as compared to intertidal sediments was evident (Fig. 2).

Bacterial distribution in intertidal and subtidal sediments may be controlled by chemical and physical sediment characteristics. For example, the trend toward lower numbers of bacteria in subtidal, as compared to intertidal, sediments (Table 1, Fig. 2) parallels a decrease in organic carbon content of those sediments (Dale 1974; Rublee and Dornseif 1976). In contrast, the highest numbers of bacteria are found in intertidal sediments of high (10-50%) organic carbon content. Dale (1974) noted this relationship and a

Table 1. Direct counts of bacteria in some subtidal and intertidal sediments. Values are given as mean (and range) of observations. sfc = surface.

Location/site description	Number of samples	Depth (cm)	Cell numbers × 10^9 per g dry weight	Cell numbers × 10^9 per cm^3	Reference
Marsh Sediments					
Newport River Estuary, NC.	13	0-1	13.8 (8.5-22.0)	8.5 (5.3-13.7)	Rublee (1982)
Spartina alterniflora marsh,	13	5-6	5.7 (3.2- 7.7)	5.4 (3.3- 8.1)	
yearly mean.	13	10-11	2.9 (1.8- 4.4)	3.0 (1.7- 4.2)	
	13	20-21	1.8 (1.1- 3.0)	1.9 (1.0- 2.9)	
Newport River Estuary, NC	4	0-1	19.7 (12.0-34.1)	9.4 (8.4-10.9)	Rublee and
Transect across *Spartina* marsh.	4	5-6	7.2 (4.6- 9.8)	6.1 (5.1- 7.1)	Dornseif (1978)
	4	10-11	3.2 (2.3- 5.0)	3.5 (2.4- 4.4)	
	4	20-21	2.1 (1.6- 2.3)	2.3 (2.2- 2.6)	
Rhode River Estuary, MD.	3	0-1	66.3 (57.2-84.6)	16.0 (13.8-20.4)	this report
Typha angustifolia/Scirpus spp.	3	5-6	35.6 (24.2-48.0)	9.7 (6.6-13.1)	
low marsh.	3	10-11	29.9 (27.9-31.9)	3.0 (7.6- 8.1)	
	2	20-21	8.8 (7.4-10.3)	3.0 (2.5- 3.5)	
Rhode River Estuary, MD.	3	0-1	60.1 (37.5-76.6)	12.0 (11.3-12.4)	this report
Three high marsh sites, *Spartina*	3	5-6	51.2 (36.9-69.1)	9.5 (7.2-11.0)	
cynusoroides, *S. patens*, *Distichlis*	3	10-11	37.0 (30.8-43.2)	6.1 (4.3- 9.3)	
spicata, *Scirpus* spp., *Hibiscus* spp.	3	20-21	23.8 (19.9-25.9)	3.9 (2.8- 5.1)	
	1	30-31	31.7	3.6	
Great Sippewissett marsh, MA.	12	0-1	38.0 (10.0-64.0)		J. E. Hobbie and
Tall *Spartina alterniflora*	12	2-3	28.0		J. Helfrich,
annual mean.	12	5-6	35.0		pers. comm.
	12	9-10	42.0		
	12	19-20	24.0		
	12	29-30	16.0		

Table 1. (Continued)

Location/site description	*Number of samples*	*Depth (cm)*	*Cell numbers × 10^9*		*Reference*
			per g dry weight	*per cm^3*	
Great Sippewissett marsh, MA. Short *Spartina alterniflora* annual mean.	12	0-1	54.0 (24.0-80.0)		J. E. Hobbie and J. Helfrich, pers. comm.
	12	2-3	29.0		
	12	5-6	29.0		
	12	9-10	28.0		
	12	19-20	18.0		
	12	29-30	12.0		
Great Sippewissett marsh, MA. High marsh, *Spartina patens*, annual mean.	12	0-1	49.0 (16.0-120.0)		J. E. Hobbie and J. E. Helfrich, pers. comm.
	12	2-3	35.0		
	12	5-6	30.0		
	12	9-10	29.0		
	12	19-20	18.0		
	12	29-30	19.0		
Intertidal/Subtidal Mud and Sand Flats					
Sapelo Island, GA. Intertidal sand flat adjacent to *Spartina* marsh, winter samples.	3	0-1		1.0 (0.6- 1.2)	S. Y. Newell, pers. comm.
Newport River Estuary, NC. Subtidal sand, June - November	18	0-1	3.3 (1.6- 5.9)	2.7 (1.2- 4.8)	Shelton (1979)
	18	5-6	2.4 (1.0- 5.9)	1.9 (0.8- 4.8)	
Newport River Estuary, NC. subtidal mud, June - November - "sulfuretum".	18	0-1	8.8 (3.1-22.9)	4.2 (2.0- 8.3)	Shelton (1979)
	18	5-6	5.8 (1.3-20.5)	2.7 (1.1- 7.4)	

Table 1. (Continued)

Location/site description	Number of samples	Depth (cm)	Cell numbers × 10^9 per g dry weight	Cell numbers × 10^9 per cm^3	Reference
Rhode River, MD. subtidal mud flat	3	0-1	17.4 (9.6-26.8)	9.1 (5.0-14.0)	this report
	2	2-3	16.0 (11.7-19.7)	8.4 (6.1-10.3)	
	3	5-6	13.0 (11.8-14.2)	8.7 (7.9- 9.5)	
	3	10-11	8.0 (6.7- 9.1)	6.2 (5.2- 7.1)	
	2	20-21	4.7 (3.3- 6.2)	3.6 (2.5- 4.7)	
Lowe's Cove, Damariscotta Estuary, ME. March - August	6	sfc	2.8 (1.2- 4.7)		M. DeFlaun, pers. comm.
Halifax, Nova Scotia. High intertidal mud flats. April - November	13	sfc	6.0 (2.2- 8.6)	3.5 (2.2- 4.8)	L. M. Cammen, pers. comm.
	13	1	5.7 (3.7- 7.9)	4.5 (3.5- 5.5)	
	13	5	3.5 (2.3- 5.2)	3.6 (2.9- 4.3)	
Halifax, Nova Scotia. Low intertidal mudflat. April - November.	13	sfc	4.2 (1.5- 7.0)	2.8 (1.4- 4.3)	L. M. Cammen, pers. comm.
	13	1	4.0 (2.8- 5.2)	3.5 (2.1- 4.1)	
	13	5	4.0 (2.7- 5.2)	3.8 (2.2- 5.1)	
Petpeswick Inlet, Nova Scotia. Seven intertidal mud and sandflat stations, May - September	17	sfc	2.8 (0.3-10.0)		Dale (1974)
	2	1	2.9 (2.1- 3.6)		
	1	2	6.3		
	16	5	2.5 (0.1- 7.0)		
	14	10	1.8 (0.1- 4.7)		

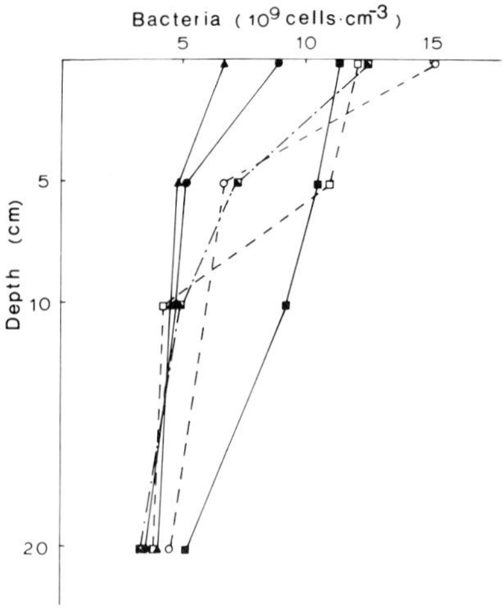

Figure 2. Number of bacteria in sediments of Rhode River Estuary, Maryland. O – *low marsh;* ● – *low marsh creek bed;* ■ – *high marsh (*Distichlis spicata/Spartina patens*);* □ – *high marsh* (Spartina cynusoroides)*;* ◪ – *high marsh (mixed vegetation);* ▲ – *subtidal mud flat.*

plot of his data plus data from six additional studies (Fig. 3) exhibits a correlation and a highly significant regression:

$$N = 1.124 * C + 1.906 \qquad (R^2 = 0.63, F = 195, p<.01, n = 115)$$

where N = number of bacteria (10^9 cells/g dry wt) and C = % organic carbon by weight.

A causal aspect to this relationship is suggested in recent studies. Christian (1976) utilized both *in vitro* and *in situ* experimental approaches in a Georgia marsh and the results indicated that microbial carbon is derived from decaying organic matter. Similarly, Howarth and Teal (1979) suggested that one reason for high levels of sulfate reduction in Great Sippewisset marsh in Massachusetts is large amounts of belowground macrophyte production. Reed (1980) used a bioenergetic model of sulfate reduction to find that availability of dissolved organic matter was a primary factor regulating the rate of sulfate reduction. Similar regulation of other processes is likely, and Howarth and Hobbie (1982) include tidal flushing of sediment, oxygen availability, nutrient quality and supply, predation, and temperature as possible factors regulating anaerobic decomposition processes. Such regulation of processes ultimately affects microbial distribution.

Bacteria-Particle Interactions and Bacterial Transport

The relationship of bacteria to sediment particle surface may be of extreme importance. For example, Costerton et al. (1978) and Marshall (1980)

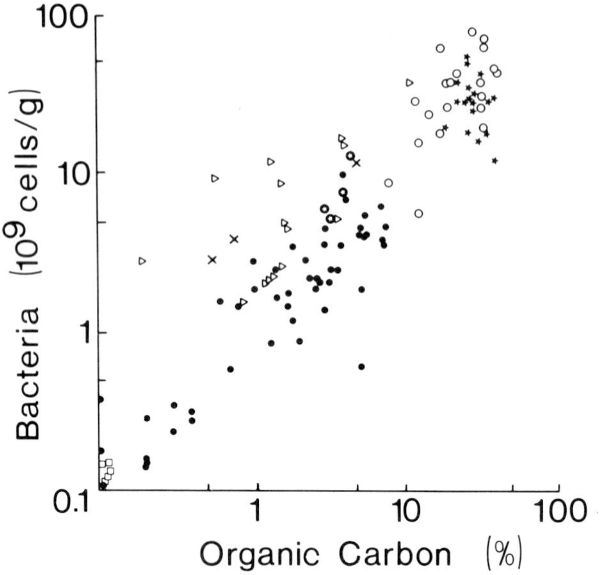

Figure 3. Relationship between bacteria and organic carbon in intertidal and shallow subtidal sediments. ● – intertidal, Nova Scotia (Dale 1974); ○ – marsh intertidal, Maryland (this report); ✪ – subtidal, Maryland (this report); △ – marsh intertidal, North Carolina (Rublee and Dornseif 1978); × – subtidal, North Carolina (Shelton 1979); □ – subtidal North Sea coast (Weise and Rheinheimer 1979); ★ – marsh intertidal, Massachusetts (J. E. Hobbie and J. Helfrich, pers. comm.).

have discussed the reversible attachment process of bacteria to surfaces, and Hobbie and Lee (1980) have pointed out that microbial attachment to sediment surfaces may result in significant quantities of extracellular polysaccharides which in turn could be utilized as a food resource by deposit feeding organisms. Further, recent studies have questioned the paradigm of macrophyte detritus export (Nixon 1980) and one alternative could be decomposition within the marsh and subsequent bacterial export. In fact, Correll et al. (1975) suggested that movement of phosphorus in Rhode River marshes may be in the form of microbial cells.

The relative transport of bacterial cells in sediments in the Rhode River in Maryland has been studied by Rublee et al. (in prep.). They found that the number of bacteria in water seeping from the marshes at low tide and in samples from ground water wells in the marsh were in the same range as that in tidal water (10^6-10^7 cells/ml). Thus, interstitial water contained only 0.1-1% of the total sediment bacteria (Table 1). Second, short (2 cm) sediment cores were taken from intertidal and subtidal surface sediments – both horizontally and vertically – and filtered water was flushed through the soil columns. Counts of the effluent indicated that, in all samples, after an initial "burst" of high bacterial numbers, relatively few bacteria were washed from the column (Fig. 4). An integrated value for the number of bacteria displaced suggests that only a small fraction (<1%) of the total sediment bacteria is washed out under this experimental condition in which flushing far exceeded interstitial water movement. These results suggested that cells adhere to particles with varying degrees of tenacity (see also Fig. 1) and that bacterial cells in sediments are not highly mobile. A third piece of evidence derives from tidal flux studies conducted over one year in both high and low marsh sites in the Rhode River. On eleven tidal cycles, volume-integrated composite samples were taken to determine the mean concentrations of bacteria during flood and ebb periods. Mean concentration of bacteria on flood tides was 7–20% higher than on ebb flows. Thus, marsh sediments did not appear to be a ready source of bacteria for tidal waters, and significant transport of microbial biomass from sediment to water in intertidal and subtidal systems may be restricted to episodes of sediment resuspension.

Size of Sediment Bacteria

Bacterial cell volumes have been determined by direct observation in several studies. Rublee (1982) determined cell sizes over thirteen months in a North Carolina *Spartina alterniflora* marsh. At each sampling date, 100 to 300 cells were observed in samples at each of four depths (surface, 5, 10, and 20 cm). Cells were noted as within one of six size/volume categories dependent upon morphology and dimensions. From this I determined an average volume of 0.20 μm^3 for cells in surface sediments and a slight decrease in size with depth to about 0.18 μm^3 at a depth of 20 cm. Further, I found a positive correlation of cell volume with temperature. Similar cell sizes were found in

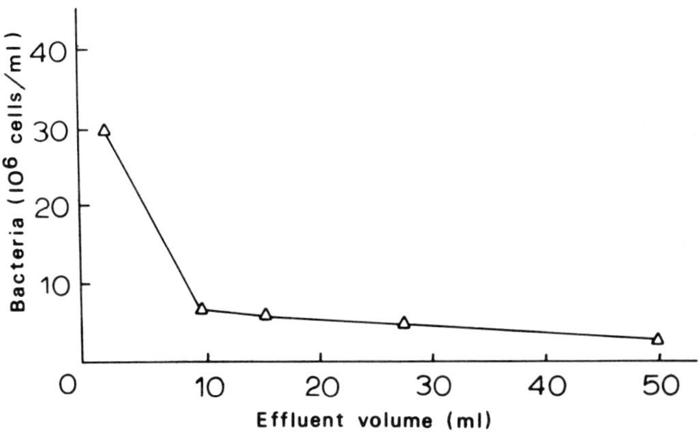

Figure 4. Concentration of bacteria in filtered water flushed through a marsh sediment core.

Massachusetts marshes (J. E. Hobbie and J. Helfrich, MBL, Woods Hole, MA, pers. comm.).

We have conducted a more extensive characterization of sediment bacterial cell volume in the low marsh site of the Rhode River system (Rublee et al. in prep.). Nine sediment samples (two each from depths of 0, 5, 10, 15 cm and one from 20 cm depth) were examined. In each, the individual dimensions of at least 100 cells were recorded and the volume calculated. Duplicate samples were also taken from three ground water wells: one sampled interstitial water from 20-60 cm below the sediment surface; one sampled from 13-17 cm; and one sampled from a depth of 32-38 cm. Predominant interstitial water movement lies in the 20-40 cm depth of these sediments (T. Jordan, CBCES, Edgewater, MD, pers. comm.). Again, 100 cells were sized individually for each sample. To determine average cell size and to compare sizes in sediment vs. interstitial water all samples of each type were pooled. The average cell sizes (0.17 μm^3 in the sediments and 0.18 μm^3 in the interstitital water) were not significantly different since the range of sizes spanned three orders of magnitude (Fig. 5), and therefore the average cell size has wide statistical limits about its mean (coefficient of variation ≈ 100%). A chi-square test of independence, however, showed that the distribution of cell sizes in these samples was significantly different ($\chi^2 = 145$, $p<0.01$, 26 d.f.). Thus, these samples represent morphologically different assemblages.

Determination of Bacterial Biomass

In order to estimate biomass of bacteria one must have conversion factors for dry weight, density, organic carbon content, and cell volume. Based on literature values and empirical determinations of cell carbon content of six estuarine bacterial isolates (R. L. Ferguson, NMFS, Beaufort, NC, unpublished data), Ferguson and Rublee (1976) arrived at a conversion factor of 0.087 g C/cm^3 bacterial biovolume. The assumption of an average cell volume of about 0.2 μm^3 is reasonable and results in a conversion factor for sediment bacteria of 1.7×10^{-15} g C/cell. Utilization of this conversion factor and the cell numbers found in North Carolina, Maryland, and Massachusetts marshes yields bacterial standing crop values in the top 20 centimeters on the order of 15-20 g bacterial carbon/m^2. This value is within the ranges found in other studies and reported by Dale (1974) for subtidal sediments. It is also within the range estimated by White et al. (1979) by lipid phosphate analysis for an estuarine sediment sample in Florida.

Biomass of the Total Microbial Community and its Components

With the information on bacteria and results from other studies, can we determine the total microbial community biomass and contribution of various components? A review of the literature demonstrates that this is a difficult task, based on a number of factors. First, methodology currently in use leaves much to be desired in the way of accuracy and frequently in precision. For ex-

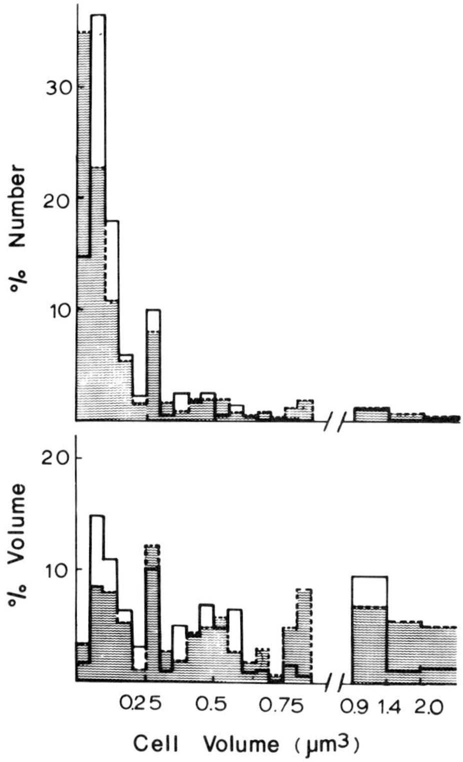

Figure 5. Cell volumes in sediments and interstitial water. Upper graph indicates % of cells in various size classes. Lower figure indicates % of total volume in each size class. Shaded bars represent interstitial water samples; open bars represent total sediment sample.

ample, the bacterial counts discussed previously have a 10% confidence interval around the values for each sample as a result of a compromise in the effort required to count the appropriate number of cells, and coefficients of variation for replicate counts of 10-30% are not unusual. Second, many of the methods are not specific to microbial taxonomic groups; e.g. ATP in samples may derive from larval macrofauna and meiofauna as well as microfaunal organisms. A third problem involves patchiness of the organisms in question. Meiofaunal studies have demonstrated small scale patchiness (e.g. Coull and Bell 1979) and microbial studies frequently rely on small samples and large dilution factors. While small sample sizes may, of necessity, be an acceptable procedure – and probably not too much in error since even small samples (0.1 cm^3) contain many millions of bacteria – there are without doubt microenvironments which are not adequately accounted for (e.g., oxygenated microrhizospheres). Nevertheless, it is instructive to attempt some generalizations regarding the microbial community and its partitioning (c.f. Fenchel 1969; Gerlach 1978).

Total living "microbial" biomass may be estimated by use of ATP assays. A summary of results from various intertidal and subtidal sediments indicates values of ATP in the nanogram to microgram range per cm^3 of sediments (Table 2). Generally, highest values are found in surface sediments and the decrease with depth may be dramatic. Three studies have been conducted in marshes – all being *Spartina alterniflora* marshes of the U.S. east coast. Christian et al. (1975) sieved sediments through a 1 mm mesh, prior to extraction by a bicarbonate buffer, in an attempt to restrict their measure to the microbial and meiofaunal organisms. Rublee (1982) utilized a small sample size (0.2 cm^3) which was probably effective at restricting the smallest macrofaunal organisms. In a South Carolina study, Simons (1976) used 1 cm^3 samples and may have included small macrofauna as well as meiofauna. These studies exhibit ATP values in the range of 0.5 to 13 μ g ATP/cm^3. Estimates of total living "microbial" biomass based on a conversion factor of 250 for the carbon to ATP ratio (but note the reservations on the ratio expressed by Karl [1980]) range from 165-1500 μg C/cm^3 for marsh surface sediments. Integration of values from the deeper layers yields values of about 30 g C/m^2 to a depth of 20 cm in the North Carolina tall and short *Spartina alterniflora* marsh (Rublee 1982) and 13 and 32 g C/m^2 for the same depth in tall and short *S. alterniflora* marshes of Georgia (Christian et al. 1975).

Bacteria contribute significantly to this total microbial biomass. The previous estimates for bacterial biomass (15-20 g C/m^2) represent >50% of the ATP estimates of microbial biomass. While this may seem to be an extremely high value, consideration of the distribution with depth in the sediment suggests that the estimate is not unreasonable. First, most of the ATP is found in the surface sediments. Similarly, most meiofaunal organisms and microbial protozoans are restricted to the upper few centimeters (Fenchel 1969; Coull and Bell 1979), presumably above or near the redox discontinuity layer. As a result, ATP values decrease more rapidly with increasing sediment depth than do bacterial standing crop estimates; Rublee (1982) found that bacteria comprise about 15% of the ATP determined microbial biomass in surface sediments, but at depths of 5 cm or more the bacteria appeared to be the only microbial taxon remaining.

Similar determinations may be made for other microbial groups in the sediments, using direct measurements of individual organisms, or by measuring cell constituents such as chlorophyll *a* in the case of algae. For example, direct microscopic observations have been used in the determination of protozoan communities in some marshes and intertidal sediments. Phleger (1970) reviewed the occurrence of foraminifera in marsh and mudflat systems. In marsh surface sediments, 1 to 150 individuals/cm^2 may be found, and the populations decrease as one proceeds from mudflats to low and finally high marshes. Phleger (1970) gave no estimates of biomass values for this group

Table 2. Recoverable ATP from some salt marsh and subtidal sediment studies. n = number of samples.

Location	*Depth (cm)*	*ATP (μ g/cm³)*	*Reference/extraction method*
Marsh/Intertidal Stations			
Sapelo Island, GA			
Tall *Spartina*	0-1	0.66	Christian et al. (1975);
alterniflora marsh	1-5	0.40	boiling bicarbonate
yearly mean; n = 4	5-10	0.24	
	10-15	0.17	
	15-25	0.14	
	25-35	0.15	
Short *Spartina alterniflora*	0-1	2.11	
marsh yearly mean; n = 4	1-5	1.52	
	5-10	0.54	
	10-15	0.22	
	15-25	0.12	
	25-30	0.05	
North Inlet, SC			
Spartina alterniflora	0-1	5.58*	Simons (1976);
marsh, 8 stations		(1.67-13.63)	cold H_2SO_4
Intertidal creek station	0-1	1.83*	
		(0.90- 3.21)	
Newport River Estuary, NC			
Tall *Spartina*	0-1	4.05	Rublee (1982);
alterniflora annual	5-6	0.32	boiling tris
mean; n = 12	10-11	0.09	
	20-21	0.04	
Subtidal stations			
Appalachicola Bay, FL			
Mud, single station	0-0.5	27.2*	White et al. (1979);
	6-7	11.6*	cold H_2SO_4
	18-19	5.5*	
Newport River Estuary, NC			
Sandy sediments,	0	0.91	Ferguson and Murdoch
annual mean	5	0.24	(1975); boiling tris
	10	0.04	
Sulfuretum,	0	0.38	
annual mean	5	0.06	
	10	0.00	
Newport River Estuary, NC			
Sandy sediments; n = 16	0	0.67	Shelton (1979);
	5	0.20	boiling tris
Sulfuretum; n= 16	0	0.45	
	5	0.08	

*μg/g dry weight

of organisms, however, which may range in size from tens of micrometers to millimeters in length. Fenchel (1969) provided the most complete information regarding subtidal and intertidal protozoans from the Baltic Seas. He sampled the top 20 cm of sediment. In subtidal sandy sediments 5-20 × 10^6 cilliates/m^2 were found; in sulfuretum type sediments the numbers were greater than $10^7/m^2$. These values correspond to a standing crop of 0.2-2.3 g wet weight/m^2. Similar numbers of ciliates have been found in the sediments of marsh pools in Louisiana (Elliott and Bamforth 1975). In the shallow subtidal sediments of a North Carolina estuary Dornseif (1976) found 0-34 ciliates/cm^3 in estuarine sands and 0-11 ciliates/cm^3 in the estuarine sulfuretum community. All of these studies noted the critical role that sediment properties play in limiting the distribution of ciliated protozoans.

Determination of algal biomass from chlorophyll values in marsh and subtidal sediments is characterized by variability in quantitative estimates as a result of the characteristics of the seasonally dominant algal type. For example, Sullivan and Daiber (1975) found filamentous bluegreen algae to be dominant in dwarf *Spartina alterniflora* and *Distichlis spicata* marsh areas in Delaware during summer months, while tall *Spartina alterniflora* sites had no bluegreens. Such bluegreen algal mats may have profound influence on seasonal chlorophyll concentrations. Elliott and Bamforth (1975) found diatoms, bluegreens, and phytoflagellates in sediments of pools within a Louisiana salt marsh. Diatoms are commonly seen in microscopic preparations of marsh sediments and have been the focus of several studies (e.g., Sullivan 1975). The ability of diatoms to migrate vertically in response to tidal inundation and light might also alter distribution of chlorophyll in marsh sediments. Compilation of extant data on chlorophyll content of sediments from east coast salt marshes indicates that there is, as expected, a significant decrease in chlorophyll *a* content with depth (Table 3). The highest chlorophyll *a* values have been found in the Great Sippewisset marsh by Estrada et al. (1974). Lower values have been found in Delaware (Gallagher 1971; Sullivan and Daiber 1975) and in North Carolina marshes (Rublee 1982). Chlorophyll *a* values from intertidal and subtidal mud and sand flats are in the same range (3-50 μg Chl a/cm^3) as those found in salt marsh surface sediments, although bioturbation and daily tidal inundation tend to result in higher subsurface concentrations in mud and sand flats. Estimation of algal biomass using a carbon to chlorophyll *a* ratio of 30 (Ferguson and Murdoch 1975) yields values in the range of 100-3000 μg C/cm^3 in marsh surface sediments.

Fungi may comprise about 10% of the total biomass estimate in marsh surface sediments. There is only one direct count, quantitative study: Hobbie and Helfrich (pers. comm.) used the water soluble aniline blue epifluorescence technique (Paul and Johnson 1977) to measure hyphal lengths in surface sediments of tall and short *Spartina alterniflora* marshes as well as a high marsh site in Massachusetts. Determinations were made in December, May, June, and October, and biomass was estimated by a conversion factor of

Table 3. Chlorophyll *a* concentrations in some intertidal and shallow subtidal sediments of the U.S. east coast. Values are given as mean (and range) of observations.

Location	*Number of samples*	*Depth (cm)*	*Chlorophyll a (μg/cm³)*	*Reference*
Marsh and Intertidal Flats				
Great Sippewissett Marsh, MA				
High marsh (*S. patens* and *D. spicata*)	1	0-0.5	48	Estrada et al. (1974)
	1	0.5-1.0	1	
	1	1.0-1.5	1	
	1	1.5-2.0	1	
	1	2.0-2.5	1	
	1	2.5-3.0	1	
Low marsh (*S. alterniflora*)	1	0.0-0.5	45	
	1	0.5-1.0	5	
	1	1.0-1.5	2	
	1	1.5-2.0	2	
	1	2.0-2.5	2	
	1	2.5-3.0	2	
Pettaquamscutt River, RI Oct. through April				
Bare marsh bank	4	0-0.75	106 (28-183)	Lytle (1969)
S. alterniflora zone	3	0-0.75	114 (78-159)	
S. patens zone	2	0-0.75	57 (38- 76)	
Salt panne	2	0-0.75	221 (185-258)	

Table 3. (Continued)

Location	*Number of samples*	*Depth (cm)*	*Chlorophyll a* ($\mu g/cm^3$)	*Reference*
Canary Creek Marsh, DE				
Yearly mean				
Short *S. alterniflora*	11	0-1	10.3 (7.5-13.3)	Sullivan and Daiber (1975)
Tall *S. alterniflora*	15	0-1	11.0 (1.0-24.8)	Gallagher (1971)
Short *S. alterniflora*	15	0-1	12.7 (5.2-18.3)	
Bare creek bank	15	0-1	3.1 (0.5- 5.3)	
D. spicata	15	0-1	9.2 (3.9-18.0)	
Salt panne	15	0-1	41.6 (24.7-79.8)	
Newport River Estuary, NC				
Tall *S. alterniflora*	10	0-1	2.9 (0.7- 5.2)	Rublee (1982)
August through June	10	5-6	0.7 (0.1- 1.5)	
	10	20-21	0.1 (0.0- 0.2)	
Sapelo Island, GA				
Tall *S. alterniflora*	3	0	10.8 (9.3-12.9)[A]	Pomeroy (1959)
	2	2	4.2 (4.1- 4.2)[A]	
Med. *S. alterniflora*	4	0	7.5 (4.5-10.7)[A]	
	2	2	3.1 (2.5- 3.6)[A]	
Short *S. alterniflora*	1	1	6.0[A]	
Bare marsh/creek surfaces	7	0	7.6 (2.5-25.5)[A]	
	3	2	2.9 (1.7- 3.6)[A]	

Table 3. (Continued)

Location	Number of samples	Depth (cm)	Chlorophyll *a* (μg/cm³)	Reference
Subtidal Estuarine Sediments				
Niantic River Estuary, CT				
Zostera bed	13	0-3	15.8 (7.8-31.0)	Tietjen (1968)
	13	3-5	11.5 (5.6-18.8)	
Bare sand	13	0-3	12.8 (5.9-21.1)	
	13	3-5	9.1 (3.5-15.4)	
Pettaquamscutt River, RI				
Zostera bed	13	0-3	15.6 (6.6-23.9)	Tietjen (1968)
	13	3-5	8.7 (6.3-11.3)	
Bare sandy bottom	13	0-3	11.9 (7.1-18.0)	
	12	3-5	5.5 (3.9-12.7)	
Newport River Estuary, NC				
Subtidal sandy sediments		0	3.77, 7.45[B]	Ferguson and Murdoch (1975)
		5	2.00, 2.01[B]	
		15	0.60 , -[B]	
Sulfuretum sediments		0	3.02, 1.49[B]	
		5	1.21, 0.67[B]	
		15	0.38 , -[B]	
Newport River Estuary, NC				
Subtidal sandy sediments June through December	25	0-1	3.3 (0.5-10.5)	Bigelow (1977)
Sulfuretum sediments June through November	20	0-1	2.9 (0.8- 8.8)	

[A]values are sedimentary chlorophyll units/g dry weight; [B]mean values for summer (June-July) and winter (Nov.-March) samples, respectively.

Table 4. Generalized biomass estimates for microbial groups in estuarine marsh systems. Entries represent means of determinations presented in the text and in tables 1, 2, and 3. Ranges are given in parentheses, where multiple determinations were available. 0-20 cm values for total microbial, bacterial, and algal biomass were determined by summation of linearly interpolated values from surface to 20 cm depth. Values are μ g carbon/cm^3. See text for further description of results and conversion factors.

Depth (cm)	*Total Microbial Biomass*	*Bacterial Biomass*	*Fungal Biomass*	*Algal Biomass*	*Protozoan Biomass*
0- 1	775 (175-1400)	198 (92- 354)	130 (63-228)	1170 (87-3420)	4-46[1]
5- 6	140 (80- 258)	127 (57- 226)	–	21	
10-11	56 (23- 95)	80 (29- 162)	–	3	–
20-21	25 (10- 35)	49 (17- 88)	–	3	–
0-20	3183 (1061-5678)	1975 (818-3670)	130 (63-228)	1255 (198-5756)	4-46

[1]0-5 cm estimate based on range given by Fenchel (1969), assuming dry weight = 20% of wet weight and carbon = 50% dry weight.

1000 m of fungal hyphae = 1 mg carbon (radius of hyphae = 2 μm; density = 1.1 g/cm³; dry to wet weight ratio = 0.15:1; carbon = 50% of dry weight). Mean values in μg C/cm³ of sediment were 50, 39, and 157 for the tall *S. alterniflora,* short *S. alterniflora,* and high marsh sites, respectively. Fungi are not expected to play a significant role below the surface sediments due to the reducing conditions found there. One exception, however, may be fungal involvement with oxygenated rhizospheres of marsh plants (S. Y. Newell, U. Georgia, Sapelo Island, GA, pers. comm.). Qualitative studies have also documented the presence of yeasts in marsh sediments (e.g., Meyers and Ahearn 1974).

A summary of these biomass estimates suggests that algae are the dominant contributors to total "microbial" biomass (as determined by ATP) in marsh surface sediments (Table 4). At greater depths, bacteria generally comprise most of this biomass. The remaining biomass in marsh sediments is comprised primarily of meiofauna, which in some cases may contribute significantly to the sediment community (Sikora et al. 1977; Coull and Bell 1979). Certainly these values are generalizations at best. Each site will be different, due to chemical and physical factors regulating activity and distribution not only of the microorganisms but also of predator populations which in turn influence prey populations.

Acknowledgments

This report stems from research efforts supported variously by the National Science Foundation, the University of North Carolina Sea Grant Program, and the Knapp Foundation. Leon Cammen, John Helfrich, John E. Hobbie, Mary DeFlaun, Steven Y. Newell, and John Paul generously shared unpublished data for inclusion in this report. Susan Merkel, Vivian Vankai, David Light and Maria Faust aided in sample collection and analysis. Earlier versions of the manuscript were improved by Bob Christian, Randy Ferguson, Maria Faust, Tom Jordan, Dennis Whigham, and an anonymous reviewer.

References Cited

Bigelow, G. W. 1977. Primary productivity of benthic microalgae in the Newport River Estuary. M.S. Thesis, North Carolina State U., Raleigh, NC 44p.

Bobbie, R. J., J. S. Nickels, G. A. Smith, S. D. Fazio, R. H. Findlay, W. M. Davis and D. C. White. 1981. Effect of light on biomass and community structure of estuarine detrital microbiota. *Appl. Environ. Microbiol.* 42:150-158.

Christian, R. R. 1976. Regulation of a salt marsh soil microbial community: A field experiment approach. Ph.D. Thesis, U. Georgia, Athens, GA 132 p.

Christian, R. R., K. Bancroft and W. J. Wiebe. 1975. Distribution of microbial adenosine triphophosphate in salt marsh sediments at Sapelo Island, Georgia. *Soil Science* 119:89-97.

Correll, D. L., M. A. Faust and D. J. Severn. 1975. Phosphorus flux and cycling in estuaries, pp. 108-136. *In:* L. E. Cronin (ed.), *Estuarine Research, Vol. I.* Academic Press, New York.

Costerton, J. W. and R. R. Colwell. 1979. *Native Aquatic Bacteria: Enumeration, Activity, and Ecology.* Amer. Society for Testing and Materials, Philadelphia, PA 214 pp.

Costerton, J. W., G. G. Geesey and K. J. Cheng. 1978. How bacteria stick. *Sci. Amer.* 239:86-95.

Coull, B. C. and S. S. Bell. 1979. Perspectives of marine meiofaunal ecology, pp. 189-216. *In:* R. J. Livingstone (ed.), *Ecological Processes in Coastal and Marine Systems.* Plenum Press, New York.

Dale, N. G. 1974. Bacteria in intertidal sediments: Factors related to their distribution. *Limnol. Oceanogr.* 19:509-518.

Daley, R. J. 1979. Direct epifluorescence enumeration of native aquatic bacteria: Uses, limitations, and comparative accuracy, pp. 29-45. *In:* J. W. Costerton and R. R. Colwell (eds.), *Native Aquatic Bacteria: Enumeration, Activity, and Ecology,* ASTM STP 695. Amer. Soc. Testing and Materials, Philadelphia, PA

Davis, W. M. and D. C. White. 1980. Fluorometric determination of adenosine nucleotide derivatives as measures of the microfouling, detrital, and sedimentary microbial biomass and physiological status. *Appl. Environ. Microbiol.* 40:539-548.

Dornseif, B. E. 1976. The distribution of ciliated protozoans in the sediments of the Newport River estuary. M.S. Thesis, North Carolina State U., Raleigh, NC 51 pp.

Elliott, P. B. and S. S. Bamforth. 1975. Interstitial protozoa and algae of Louisiana salt marshes. *J. Protozool.* 22:514-519.

Estrada, M., I. Valiela and J. M. Teal. 1974. Concentration and distribution of chlorophyll in fertilized plots in a Massachusetts salt marsh. *J. Exp. Mar. Biol. Ecol.* 14:47-56.

Fenchel, T. 1969. The ecology of marine microbenthos. IV. Structure and function of the benthic ecosystem, its chemical and physical factors and the microfauna communities with special reference to the ciliated protozoa. *Ophelia* 6:1-182.

Ferguson, R. L. and M. B. Murdoch. 1975. Microbial ATP and organic carbon in the sediments of the Newport River Estuary, North Carolina, pp. 229-250. *In:* L. E. Cronin (ed.), *Estuarine Research, Vol. I.* Academic Press, New York.

Ferguson, R. L. and P. Rublee. 1976. Contribution of bacteria to standing crop of coastal plankton. *Limnol. Oceanogr.* 21:141-145.

Gallagher, J. L. 1971. Algal productivity and some aspects of the ecological physiology of the edaphic communities of Canary Creek tidal marsh. Ph.D. Thesis, U. Delaware, Wilmington, DE 120 p.

Gerlach, S. A. 1978. Food-chain relationships in subtidal silty sand marine sediments and the role of meiofauna in stimulating bacterial productivity. *Oecologia* 33:55-69.

Haines, E. B. 1979. Interactions between Georgia salt marshes and coastal waters: A changing paradigm, pp. 35-46. *In:* R. J. Livingstone (ed.), *Ecological Processes in Coastal and Marine Systems.* Plenum Press, New York.

Hobbie, J. E., R. J. Daley and S. Jasper. 1977. Use of nuclepore filters for counting bacteria by fluorescence microscopy. *Appl. Environ. Microbiol.* 33:1225-1228.

Hobbie, J. E. and C. Lee. 1980. Microbial production of extracellular material: Importance in benthic ecology, pp. 341-346. *In:* K. R. Tenore and B. C. Coull (eds.), *Marine Benthic Dynamics,* U. South Carolina Press, Columbia, SC

Howarth, R. W. and J. E. Hobbie. 1982. The regulation of decomposition and heterotrophic microbial activity in salt marsh soils: A review, pp. 183-207. *In:* V. S. Kennedy (ed.), *Estuarine Comparisons.* Academic Press, New York.

Howarth, R. W. and J. M. Teal. 1979. Sulfate reduction in a New England salt marsh. *Limnol. Oceanogr.* 24:999-1013.

Jones, J. G. 1979. *A Guide to Methods for Estimating Microbial Numbers and Biomass in Fresh Water.* Freshwater Biological Association, Scientific Publication No. 39. 112 p.

Karl, D. M. 1980. Cellular nucleotide measurements and applications in microbial ecology. *Microbiol. Rev.* 44:739-796.

King, J. D. and D. C. White. 1977. Muramic acid as a measure of microbial biomass in estuarine and marine samples. *Appl. Environ. Microbiol.* 33:77-83.

Litchfield, C. D. and P. L. Seyfried (eds.) 1979. *Methodology for Biomass Determinations and Microbial Activities in Sediments.* ASTM STP 673 Amer. Soc. Testing and Materials, Philadelphia, PA 199 p.

Lytle, R. W., Jr. 1969. Primary production of photosynthetic microflora on a tidal marsh in Rhode Island. M.S. Thesis, U. Rhode Island, Kingston, RI 62 p.

Marshall, K. C. 1980. Microorganisms and interfaces. *Bioscience* 30:246-249.

Meyers, S. P. and D. G. Ahearn. 1974. Implication of yeasts and yeast-like fungi in marine processes. *Veroff. Inst. Meeresforsch. Bremerh.* Suppl. 5:321-338.

Moriarty, D. J. W. 1980. Measurement of bacterial biomass in sandy sediments, pp. 131-139. *In:* P. A. Trudinger, M. R.Walter, and B. J. Ralph (eds.), *Biogeochemistry of Ancient and Modern Environments.* Australian Acad. Sci., Canberra and Springer Verlag, Berlin.

Newell, S. Y. and R. E. Hicks. 1982. Direct count estimates of fungal and bacterial biovolume in dead leaves of smooth cordgrass (*Spartina alterniflora* Loisel.). *Estuaries.* 5: In press.

Nixon, S. W. 1980. Between coastal marshes and coastal waters – A review of twenty years of speculation and research on the role of salt marshes in estuarine productivity and water chemistry, pp. 438-525. *In:* P. Hamilton and K. MacDonald (eds.), *Estuarine and Wetland Processes.* Plenum Press, New York.

Paul, E. A. and R. L. Johnson. 1977. Microscopic counting and adenosine 5-triphosphate measurement in determining microbial growth in soils. *Appl. Environ. Microbiol.* 34:263-269.

Paul, J. H. 1982. The use of Hoechst dyes 33258 and 33342 for the enumeration of attached and pelagic bacteria. *Appl. Environ. Microbiol.* 43: In press.

Phelger, F. B. 1970. Foraminiferal populations and marine marsh processes. *Limnol. Oceanogr.* 15:522-534.

Pomeroy, L. R. 1959. Algal productivity of salt marshes of Georgia. *Limnol. Oceanogr.* 4:386-397.

Porter, K. G. and Y. S. Feig. 1980. The use of DAPI for identifying and counting aquatic microflora. *Limnol. Oceanogr.* 25:943-948.

Reed, J. P. 1980. A bioenergetic model of anaerobic decomposition: Sulfate reduction, pp. 173-182. *In:* V. S. Kennedy (ed.), *Estuarine Perspectives.* Academic Press, New York.

Rublee, P. A. 1982. Seasonal distribution of bacteria in salt marsh sediments in North Carolina. *Est. Coast. Shelf Sci.:* In press.

Rublee, P. A. and B. E. Dornseif. 1978. Direct counts of bacteria in the sediments of a North Carolina salt marsh. *Estuaries* 1:188-191.

Rublee, P. A., S. M. Merkel and M. A. Faust. In prep. Bacterial transport in the sediments of a temperate marsh.

Shelton, E. R. 1979. Bacterial distribution and biomass in sediments and water of the Newport River Estuary, North Carolina. M.S. Thesis, North Carolina State U., Raleigh, NC 34 p.

Sikora, J. P., W. B. Sikora, C. W. Erkenbrecher and B. C. Coull. 1977. Significance of ATP, carbon, and caloric content of meiobenthic nematodes in partitioning benthic biomass. *Mar. Biol.* 44:7-14.

Simons, V. M. 1976. Microbial biomass in high marsh sediments of the North Inlet estuary, South Carolina. M.S. Thesis, U. South Carolina, Columbia, SC 80 pp.

Sullivan, M. J. 1975. Diatom communities from a Delaware salt marsh. *J. Phycol* 11:384-390.

Sullivan, M. J. and F. C. Daiber. 1975. Light, nitrogen, and phosphorus limitation of edaphic algae in a Delaware salt marsh. *J. Exp. Mar. Biol. Ecol.* 18:79-88.

Teal, J. M. 1962. Energy flow in the salt marsh ecosystem of Georgia. *Ecology* 42:614-624.

Tietjen, J. H. 1968. Chlorophyll and pheo-pigments in estuarine sediments. *Limnol. Oceanogr.* 13:189-192.

Weise, W. and G. Rheinheimer. 1979. Fluoreszenzmikroskopische Untersuchungen über die

Bakterienbesiedlung mariner Sandsedimente. *Botanica Marina* XXII:99-106.

White, D. C., R. J. Bobbie, J. D. King, J. Nickels and P. Amoe. 1979. Lipid analysis of sediments for microbial biomass and community structure, pp. 87-103. *In:* C. D. Litchfield and P. L. Seyfried (eds.), *Methodology for Biomass Determinations and Microbial Activities in Sediments*, ASTM STP 673. Amer. Soc. Testing and Materials, Philadelphia, PA

White, D. C., R. H. Findlay, S. D. Fazio, R. J. Bobbie, J. S. Nickels, W. M. Davis, G. A. Smith and R. F. Martz. 1980. Effects of bioturbation and predation by *Mellita quinquiesperforata* on sedimentary microbial community structure, pp. 163-171. *In:* V. S. Kennedy (ed.), *Estuarine Perspectives*. Academic Press, New York.

THE REGULATION OF DECOMPOSITION AND HETEROTROPHIC MICROBIAL ACTIVITY IN SALT MARSH SOILS: A REVIEW

Robert W. Howarth and John E. Hobbie
The Ecosystems Center
Marine Biological Laboratory
Woods Hole, Massachusetts

Abstract: Most grass production in salt marshes occurs as the belowground production of roots and rhizomes in soils which are largely anoxic. Although marsh soils accrete organic matter, net rate of accretion is only a small percentage of belowground production. Most organic production is decomposed anoxically within the soil. Rates of decomposition of structural materials such as roots and rhizomes appear to be similar to that found in terrestrial grasslands. Lack of molecular oxygen may limit decomposition of some materials (lignin, aromatic hydrocarbons) but not of most. Anoxia may actually speed decomposition of some materials such as cellulose. Rates of microbial heterotrophy in salt marsh soils are the highest reported for any natural ecosystem. This high activity is probably supported both by slow decomposition of a large pool of dead roots and rhizomes and by rapid use of labile compounds such as those excreted by live roots. Sulfate reduction and related fermentative processes are the major forms of heterotrophy in salt marsh soils. A build up of metabolites and toxins in stagnant pore waters may limit decomposition and microbial heterotrophy, but this has not yet been shown conclusively. Studies of belowground decomposition and microbial activities in salt marsh soils are beset by methodological problems. These problems must be carefully considered if erroneous conclusions are to be avoided. We attempt a comprehensive review of these problems in this paper.

Introduction

Salt marsh grasses annually produce large amounts of structural material such as roots and rhizomes belowground. In all salt marshes studied, belowground production of these components is at least as great as the aboveground production and frequently it is considerably greater (Valiela et al. 1976; de la Cruz and Hackney 1977; Pomeroy and Wiegert 1981; Good et al. 1982). For the common marsh grass *Spartina alterniflora*, estimates for belowground production of roots and rhizomes vary from 220 to 1,680 g C m^{-2} yr^{-1} (calculated from Good et al. 1982, assuming dry weight of organic matter is 48% carbon). Sedimentation rates in marshes are generally estimated to be 1.5 to 6.3 mm yr^{-1}, corresponding to a permanent accretion of carbon of at most 100 to 200 g C m^{-2} yr^{-1} (Redfield and Rubin 1962; Armentano and Woodwell 1975; Howarth and Teal 1980; Pomeroy and Wiegert 1981). Obviously, most of the material of roots and rhizomes produced belowground is decomposed and not merely buried.

ISBN 0-12-404070-5

Decomposition in marsh soils is greater than would be assumed by a comparison of the production of roots and rhizomes with the organic accretion rate because there are other sources of organic carbon in marsh soils. These sources include excretion of dissolved organic carbon from living roots and rhizomes, production by microalgae and photosynthetic bacteria on the marsh surface, production by chemosynthetic bacteria in the soil, and sedimentation of marsh grass leaves and stems and of particulate matter from coastal waters. Thus, most decomposition in salt marshes probably occurs belowground in the soil. Marsh soils are anoxic below the top few millimeters (Teal and Kanwisher 1961), and much of the decomposition is mediated by anaerobic fermenting and sulfate-reducing microbes rather than by aerobic bacteria and fungi. Such decomposition is a major process recycling nutrients to the grasses and controlling export of carbon, nutrients, and energy-rich reduced sulfur compounds from marsh soils to creeks.

Given the potential importance of decomposition and microbial activity in salt marsh soils, surprisingly little is known about rates or controls of these processes. In part, this is due to methodological difficulties, so in this paper we first review the methods used to study decomposition and microbial activity. Then, we summarize pertinent literature and draw tentative, somewhat speculative, conclusions about controls.

Our review concentrates on research in salt marshes and assumes that the reader has a basic knowledge of microbial ecology. Some readers may wish to consult a general text such as the excellent one by Fenchel and Blackburn (1979).

Methods for Studying Decomposition and Heterotrophic Microbial Activity

Four basic approaches have been used to study decomposition and microbial activity in salt marsh soils. Two of these, litterbag studies and seasonal studies of abundance of living or dead roots and rhizomes, measure disappearance of particulate matter. Two other approaches are to measure consumption of electron acceptors (O_2, NO_3^-, $SO_4^=$, CO_2), and turnover and mineralization of specific dissolved organic compounds such as amino acids, glucose, or acetate. The former two approaches give no insight as to the fate of the material lost from the particulate phase (is it mineralized to CO_2 or exported from the soil as dissolved organic carbon?). On the other hand, measures of consumption of electron acceptors or turnover of specific organic compounds give no information on dynamics of the particulate phase.

It is important to remember that these approaches give qualitatively different types of information. Care should be taken in generalizing from any single approach. For instance, a litterbag or root abundance study may suggest that dead roots and rhizomes turn over very slowly, yet heterotrophic activity supported by this decomposition could still be quite large if the standing stock of dead roots and rhizomes is large.

Seasonal Studies of Abundance of Roots and Rhizomes

This approach is reviewed thoroughly by Good et al. (1982) and will only be summarized here. Turnover times for belowground biomass for a variety of species of salt marsh grasses have been found to vary from 1.5 to 10 years. In both Georgia and New Jersey marshes, belowground biomass of *Spartina alterniflora* appears to turn over some 2–3 times faster in stands of the tall form than in stands of the short form.

Studies of turnover of roots and rhizomes are fraught with problems, both with regard to sampling techniques and interpretation (Good et al. 1982). One major problem is that weight loss from winter to summer may reflect translocation of organic matter to aboveground biomass rather than decomposition. Another problem is the differentiation of dead from living roots.

Litter Bag Studies and Related Approaches

We are aware of only two litterbag studies of belowground decomposition in salt marsh soils. Hackney and de la Cruz (1980) placed a mixture of living and dead roots and rhizomes of *Juncus roemerianus* and *Spartina cynosuroides* in litter bags at various depths in marsh soils and followed changes in weight loss and various chemical parameters for one year. Valiela et al. (in press) followed a similar procedure for *Spartina alterniflora* but started with living roots and rhizomes only. Hackney and de la Cruz (1980) found a maximum weight loss of 27% per year for rhizomes of *J. roemerianus* at depths of 0–10 cm. Roots of this species lost only 16% of their weight in a year. Mixed roots and rhizomes of *S. cynosuroides* lost 20% of their weight per year at depths of 0–10 cm and lost no weight below 20 cm. On the other hand, Valiela et al. (in press) found that roots and rhizomes of *S. alterniflora* lost from 40% to 80% of their weight in one year, but from 20% to 50% of the loss occurred in the first week or so, presumably due to leaching. Hackney and de la Cruz (1980) did not find a similar leaching loss in their experiments, probably because they started with a mixture of dead and living roots while Valiela et al. (in press) started with all living material. For belowground litterbag experiments, one ideally would use recently dead material, as pointed out by Hackney and de la Cruz (1980). This would make the procedure comparable to that normally followed in aboveground decomposition studies. In practice, however, it has not been possible to collect recently dead material. Using living material at the start probably overestimates weight loss from recently dead roots and rhizomes. Using a mixture of live and dead material, some of it long dead, may underestimate weight loss of the recently dead material.

Reed (1981) studied the controls on root decomposition in the laboratory. *Spartina alterniflora* roots were labelled with ^{14}C and held under anoxic conditions in containers slowly flushed with water. Decomposition was measured by the amount of ^{14}C which left the chambers. In his study a variety of different incubation conditions caused little apparent change in decomposition. However, addition of natural marsh sediments (or water extracted from such sediments) stimulated the rate of decay by a factor of 2–10 fold. Reed (1981) also demonstrated that decomposition was dramatically decreased at pH = 4 relative to pH = 6 or pH = 8 and that bacteria and not fungi or meiofauna were the primary agents of anoxic decomposition. However, he leached the root material for an extensive amount of time before beginning his experiments which makes it impossible to determine the total amount of material lost from the roots. Even so, there was significant and rapid loss of material; 27% of the ^{14}C was lost in 3 months under anoxic conditions with sediment present. Rate of loss slowed with additional time.

Yet another promising technique for investigating controls has been used by G. Schroeder (Agricultural Research Org., Dor, D.N. Hof-Hacarmel, Israel, pers. comm.). He placed several pieces of standard cotton cloth in various environments of the Great Sippewissett Salt Marsh (Falmouth, MA) in August. Pieces placed on mud flats or in creek waters lost approximately 5% to 10% of their weight in 26 days. Pieces buried in anoxic mud of creek bottoms decomposed considerably faster, losing 20% to 40% or more of their weight during the same time. While one cannot use such data to infer decomposition rates of naturally occurring substrates, this simple experiment clearly demonstrates that the potential rate of cellulose decomposition in this marsh is fast and is greater in anoxic environments than in oxic ones, perhaps because of the greater quantity of nutrients in the anoxic mud.

Turnover and Mineralization of Specific Organic Compounds

Techniques employing radioisotopes are the only methods for studying certain processes and the controls of those processes in sediments of salt marshes. For example, processes such as the breakdown of organic matter into fermentation products can be studied by measuring turnover of added ^{14}C-acetate or ^{14}C-glucose combined with measures of concentrations of these compounds even though the concentrations in the sediments do not change. Yet, one encounters so many problems when working in sediments with radioisotopes that techniques are not secure and practically every study has a number of problems. Many of these problems are easily overcome in the laboratory. But as Christian and Wiebe (1979) point out, *in situ* experiments are necessary and must be used to check extrapolations from laboratory and microcosm studies.

Mixing of sediments: The first experiments carried out in sediments mixed the isotope throughout the sediment sample (Wood 1970; Harrison et al. 1971). In sediments without roots, Hall et al. (1972) found that mixing increased activity of microbes (glucose uptake), but Christensen and Blackburn (1980) found the opposite (alanine turnover reduced by a factor of 2.4). In any case the microstructure of the sediment is affected, and any zonation, such as that suggested by Cappenberg (1974) between sulfate reducing bacteria and methanogens, will be destroyed. Microbes may also respond differently in intact sediments than in artificially slurried sediments. For example, Christian and Wiebe (1979) found no effect of enrichment with glucose on ATP in intact cores but did see an effect in slurries made from the cores. The possibility of exposing anoxic sediments to air while making the slurry is an additional problem

Injection of label: One solution to the problem of mixing isotopes into sediments has been to add small quantities (a few microliters) by injection. Jorgensen (1978) provides a theoretical framework for this method, with particular reference to $^{35}SO_4^=$. The percent of the isotope taken up is the same even though an onion-like sphere of label results from injection, with higher amounts of label in the center and smaller quantities in the outer layers of the "onion". Christensen and Blackburn (1980) calculated that diffusion will reduce the concentration of alanine injected into a 10 μl sphere by a factor of four after 10 min. The injection technique works well for $^{35}SO_4^=$ in marine sediments (including salt marshes) where amount of isotope added does not change the bulk concentration of $SO_4^=$ in the porewater. There can be problems with using this approach for some organic compounds, depending on the concentration of substrate naturally present. Organic substrate concentration can vary tremendously between different marine sediments; Christensen and Blackburn (1980) measured 0.01 μmoles alanine liter^{-1} in porewater of a Danish coastal sediment, while in salt marsh pore waters Hanson and Gardner (1978) found 0.01-0.2 μmoles liter^{-1} (near Skidaway Island, GA) and Henrichs et al. (in press) found about 5 μmoles liter^{-1} (Sippewissett Marsh, MA). Ideally, the amount of radiolabelled compound added should be small relative to the natural concentration, changing this concentration by 10% at most.

The injection technique has been questioned on the grounds that the isotope may be delivered to micropools of high or low substrate concentration relative to the whole sediment (Pomeroy and Wiegert 1981). However, this would only increase the variance of the measure and should not introduce any systematic bias (Howarth and Giblin, in press). Still, the only evidence that the injection technique works is from experiments such as those of Jorgensen and Fenchel (1974) where $SO_4^=$ reduction as measured by injection of $^{35}SO_4^=$ agreed with $SO_4^=$ reduction as measured by changes in the $SO_4^=$ pool after one year. Howarth and Teal (1979) found reasonable agreement between the injection technique and the disappearance of $SO_4^=$ from cores of marsh soils in sealed jars over one week.

Sorption of isotope to sediment: Extensive surface areas of a sediment can sorb added isotope and make it unavailable to microbes. In Christensen and Blackburn's (1980) experiment with ^{14}C-alanine, 30% of the label immediately sorbed to sediment particles. In a salt marsh sediment, we did not observe sorption of ^{14}C-acetate, or at least could not recover any additional isotope by adding additional unlabelled acetate (Hobbie and Howarth, unpublished data). This effect must be examined in every experiment.

Exchange of isotope with roots: In salt marsh soils, roots and rhizomes may make up a significant part of the total volume. When labelled organic compounds such as ^{14}C-alanine are added, much of the disappearance from pore-water solution can be isotopic exchange with large pools of these compounds in the roots (Henrichs et al. in press). In the Sippewissett Salt Marsh, the respiration of ^{14}C-alanine was so low (about 2% of the total uptake) that it is likely that much of the loss of ^{14}C-alanine from the porewater was due to exchange with such internal alanine pools of the roots (Heinrichs et al. in press).

Transformations of added labelled compounds: When labelled compounds are added to sediments they may immediately complex with organic matter in the pore water and become unavailable to microbes. Christensen and Blackburn (Institute of Ecology and Genetics, U. Aarhus, DK-8000 Aarhus C, Denmark, pers. comm.) found that 25% of ^{14}C-acetate added by injection was immediately transformed to a compound of higher molecular weight.

When substrate uptake is calculated from changes in the radiotracer remaining in the porewater, additional problems can occur due to production of soluble metabolic products by microbes. Fleischer (1973) found half of the ^{14}C was no longer in ^{14}C-glucose after 10 min (in activated sludge), while Christensen and Blackburn (1980) found that up to 10% of the ^{14}C was transformed from alanine to volatile fatty acids in coastal Danish sediments. If such transformations are not recognized, consumption of the added tracer is underestimated as it appears that the label is still present in the original substrate.

Isotope dilution: When turnover of labelled compounds is analyzed in biologically rich sediments, a large fraction of the isotope can be taken up even in very short incubations (e.g., 5 min). As incubation continues, a smaller and smaller amount of isotope will be taken up or consumed during each time interval simply because of the reduction in available isotope. The correct model for such uptake is an exponential in which the rate constant of removal is the slope of the relationship between time of incubation and the natural log of the percent isotope remaining in solution. If the uptake is slow and less than 5 or 10% of the labelled substrate is taken up during an incubation, reduction in the amount remaining will be difficult to measure (analyses of the particulate fraction would be better) and a linear model is adequate. If uptake is rapid and a substantial amount of the isotope is taken up, then the exponential formula must be used. For example, Henrichs et al. (in press) found that only 15% of the added alanine was left in the pore water after a 30 min incubation. Calculated as a linear uptake, turnover is 35 min while, calculated as an exponential, turnover is 16 min. Uptake in sediments should always be measured over time to see that large deviations from linearity do not occur.

Concentration of substrate and flux vs. turnover time: Radiotracer techniques measure turnover time of a substrate. To determine actual flux (uptake, consumption, transformation, or respiration) of the substrate, ideally the concentration or pool size of the compound should also be determined. The concentration divided by the turnover time yields the flux. When comparing decomposition rates or microbial activities in different sites or experimental treatments, one should compare fluxes of compounds and not just turnover times. Two sites with similar microbial activities may have different turnover times of substrates merely because of different substrate concentrations (Hanson and Gardner 1978).

Uptake of Electron Acceptors

Processes of mineralization of organic compounds are redox reactions involving transfers of electrons from the organic matter being oxidized to an electron acceptor. Examination of the dynamics of the electron acceptors is a potentially powerful tool for studying decomposition. Electron acceptors usually considered are oxygen, nitrate, sulfate, and carbon dioxide (Fenchel and Blackburn 1979; Sorensen et al. 1979). Oxidized iron compounds may also serve as bacterial electron acceptors (Fenchel and Blackburn 1979), but this has not been studied in salt marshes (Howarth and Teal 1979). Each electron acceptor is studied by a different technique, and the problems of interpreting microbial activities from the electron acceptor dynamics vary with the electron acceptor.

Oxygen: Oxygen uptake rates have been reported for salt marshes from Louisiana to Nova Scotia. Techniques used include measuring change in oxygen tension in gas-filled enclosures in the field (Teal and Kanwisher 1961), in water-filled enclosures in the field (Hopkinson et al. 1978; Smith and Teal as reported in Howarth and Teal 1979), in the gas phase overlying cores (Teal and Kanwisher 1961; Duff and Teal 1965; B. Howes and J. Teal, Boston University Marine Program, Woods Hole, MA, pers. comm.), and in water overlying cores (Hopkinson et al. 1978). Despite differences in temperature, latitude, and marsh types, all of these approaches yielded average estimates of oxygen uptake by marsh muds in the range of 0.014 to 0.20 moles O_2 m^{-2} day^{-1} (Table 1). It is possible that cores with a gas phase produce much higher values (note the two values for Sippewissett Marsh in this table).

Oxygen uptake reflects the sum of respiration by animals, plants, and heterotrophic microorganisms, consumption by chemosynthetic bacteria, and consumption by purely chemical reactions such as the non-biotically catalyzed oxidation of sulfides. For understanding controls on decomposition in marshes, we are interested only in oxygen consumption by heterotrophic microorganisms, yet consumption by other processes is large and cannot be ignored. Antibiotics have been used by Hopkinson et al. (1978) and Smith and Teal (Table 1) to attempt to separate out consumption by heterotrophic microbes. However, the antibiotics used are probably not completely successful in stopping bacterial heterotrophy, do not stop fungal heterotrophy, and probably inhibit bacterial chemosynthetic oxygen consumption.

In addition to oxygen exchange across the mud surface, oxygen also moves down the internal gas spaces of marsh grasses to roots and rhizomes (Teal and Kanwisher 1966). The amount of such movement is not known, but it may be greater than diffusional fluxes calculated by Teal and Kanwisher (1966) if there are major advective fluxes such as occur in water lilies (Dacey and Klug 1979); advective fluxes are virtually unstudied for marsh grasses. The next problem is to determine how much oxygen diffuses from roots and rhizomes into marsh soils. We believe that oxygen diffusion from roots and rhizomes may be significant in wet but unsaturated marsh soils but is probably nonexistent in completely water-saturated soils (Howes et al. 1981; Howes et al. submitted). Thus, oxygen-mediated heterotrophy may play a role in decomposition within unsaturated marsh soils but may be unimportant below the top few millimeters in completely saturated soils. This is an important topic which requires more research.

Nitrate: There are three ways to measure denitrification: collect the pro-

Table 1. Mean values of oxygen uptake by marsh mud.

Location	*Method*	*Oxygen Uptake Rate (moles m^{-2} day^{-1})*	*Investigators*
Louisiana	Cores with water phase	0.014	Hopkinson et al. 1978
	Enclosures with water phase, *Spartina* included	0.072	Hopkinson et al. 1978
Georgia	Cores with water or gas phase	0.055	Teal and Kanwisher 1961
	Enclosures with gas phase, *Spartina* excluded	0.18	Teal and Kanwisher 1961
	Cores with gas phase	0.053	Duff and Teal 1965
Massachusetts	Enclosures with water phase, *Spartina* excluded	0.018	Smith and Teal, (unpublished, as reported in Howarth and Teal 1979, 1980)
	Cores with gas phase	0.20	Howes and Teal (unpublished)
Nova Scotia	Cores with gas phase	0.10	Duff and Teal 1965

ducts (N_2 or N_2O) and analyze the amount of gas produced, add $^{15}NO_3^-$ and measure the isotope changed to a gas, or calculate rates from sediment profiles. When gas analysis is used, the tremendous quantities of N_2 naturally present do not allow any changes to be seen unless the N_2 is first stripped from the sediment or at the very least from the gas phase over a core. Kaplan et al. (1979) measured the rate of production of N_2 in Sippewissett Marsh by placing a bell jar containing helium over the sediment. A problem with this technique is that any diffusion gradient of N_2 in the sediment will be disturbed, and an increased rate of N_2 diffusion will occur until the new equilibrium gradient is established. Thus Kaplan et al.'s measurements are probably maximum rates. When the same method was used in Narragansett Bay, the relatively high rates of N_2 production in the first few days were ignored and the lower rates after two days used instead (S. Seitsinger, Graduate School of Oceanography, U. Rhode Island, Kingston, RI, pers. comm.). When the isotope method is used, there are problems of mixing the sediment and of adding quantities of NO_3^- much in excess of natural levels (DeLaune and Patrick 1980). A third technique (Haines et al. 1977) uses pore water N_2 profiles and a diffusion model, but this is neither precise nor tested.

No matter which techniques are used, rates of denitrification are so low that this process must be insignificant in decomposition (Nixon 1980). Three separate studies using totally different methods all indicate that denitrification is 0.85 moles N m^{-2} yr^{-1} or less in marsh soils (Haines et al. 1977; Kaplan et al. 1979; Delaune and Patrick 1980). Some of this denitrification may represent oxidation of reduced sulfur compounds (Trudinger 1956) rather than microbial heterotrophy. But even if all the denitrification were due to microbial heterotrophy, it would represent a respiration rate of 10 g C m^{-2} yr^{-1} or less. This is trivial relative to the rate of decomposition in marsh soils, and clearly the controls on denitrification are different from those on decomposition or other forms of microbial heterotrophy.

Sulfate: Dissimilatory sulfate reduction has received rather little study. A complete seasonal study of sulfate reduction in vegetated marsh soils has been made for only one marsh, Great Sippewissett Salt Marsh, MA (Howarth and Teal 1979). Measurements have been made in the marshes at Sapelo Island, GA, in June (Skyring et al. 1979) and November (Howarth and Giblin, in press). In both the Massachusetts and Georgia marshes, pyrite (FeS_2) is a major end product of sulfate reduction (Howarth 1979; Howarth and Teal 1979; Howarth and Giblin, in press). If its formation is not considered, the radiotracer method for measuring sulfate reduction can seriously underestimate the actual rate. Thus, the estimate of Skyring et al. (1979) is probably low, but it is impossible to say by how much. Nedwell and Abram (1978) investigated sulfate reduction seasonally in a pan and in a creek bottom in a salt marsh but not in vegetated soils. They too did not include pyrite as a possible end product of sulfate reduction, so their rates also may be low.

Another possible problem, already discussed, is the technique of injecting $^{35}SO_4^=$ into the sediment in microdrops. This avoids problems created by mixing the sediments but raises the possibility that the isotope may be added to areas high or low in natural $SO_4^=$. However, this is unlikely to be an important problem in marsh sediments because diffusion is rapid on the microscale and because grams per liter of $SO_4^=$ are present. Also, when the isotope is injected into many locations in a soil core, the highs and lows of uptake should average out and no systematic bias should result from microscale variation.

Sulfate reduction rates are relatively easy to interpret, and in a system such as a salt marsh they reflect just microbial heterotrophy. Sulfate-reducing bacteria themselves consume only a relatively few number of substrates such as low molecular weight fatty acids, alcohols, and hydrogen (Fenchel and Blackburn 1979; Sorenson et al. 1981). Fermentative bacteria and fungi degrade organic compounds and provide sulfate-reducing bacteria with these substrates. The overall microbial community of sulfate-reducing bacteria and associated fermentative bacteria and fungi oxidize two moles of organic carbon to CO_2 for every mole of sulfate reduced (Jorgensen 1977; Howarth and Teal 1979). Sulfate reduction in stands of short *Spartina alterniflora* in Great Sippewissett Marsh decomposes 1,800 g C m^{-2} yr^{-1}, with daily rates varying from approximately 0.5 g C m^{-2} day^{-1} in January to approximately 10 g C m^{-2} day^{-1} in August and September (Howarth and Teal 1979). Sulfate reduction in stands of short *S. alterniflora* at Sapelo Island, GA, during November decomposes approximately 2 g C m^{-2} day^{-1} (Howarth and Giblin, in press). Thus, sulfate reduction accounts for a major fraction of belowground decomposition in both marshes.

Methane formation: Methanogenesis in salt marsh sediments has also received rather little study. The most comprehensive studies to date have been carried out in Georgia by King (1978) and King and Wiebe (1978, 1980). They measured the net increase of methane in cores incubated under an atmosphere of 80% H_2 and 20% CO_2 or 100% argon. The 80% H_2 and 20% CO_2 atmosphere resulted in higher rates of net methane formation, by a factor of 1.3 to 4.8 in one series of experiments (King and Wiebe 1978) and by a factor of 12 to 860 in a later series (King and Wiebe 1980). These experiments show that bacterial populations respond rapidly to added substrate, but it is difficult to estimate the *in situ* rate of net methane formation since neither of the two atmospheres used in the incubations is similar to the *in situ* atmosphere.

From experiments where $^{14}CH_4$ was added to anoxic cores from salt marsh soils, King (1978) concluded that anaerobic consumption of methane was insignificant. Measurements of methane flux to the atmosphere from marsh soils where the oxic surface layer has been removed (Atkinson and Hall 1976; King and Wiebe 1978), where the surface layer has been poisoned (King and Wiebe 1978), or where helium has replaced air over the marsh soil (Atkinson and Hall 1976) have demonstrated that aerobic consumption of methane at the marsh surface is also insignificant. Thus, measurements of methane flux to the atmosphere appear to reflect methanogenesis rates. Such flux measurements from marsh soils have been reported for marshes in Georgia (Atkinson and Hall 1976; King and Wiebe 1978), Maryland (Lipschultz 1981), and Massachusetts (Howarth and Teal 1979). All of the reported values fall in the range of 0.4 to 40 g CH_4-C m^{-2} yr^{-1} (Table 2). Only the estimates of King and Wiebe (1978) are based on data collected in all seasons of the year. Lipschultz (1981) and Howarth and Teal (1979) measured methane fluxes only from spring through autumn and so probably overestimated the annual flux. Atkinson and Hall (1976) extrapolated an annual flux estimate from measurements in November and thus may have underestimated the annual flux. Additionally, their measurements in November were considerably lower than those reported by King and Wiebe (1978) for very similar sites at the same time of year. Despite this uncertainty, however, it is clear that methane losses are not a major form of decomposition in salt marsh soils. Methane concentrations and fluxes in salt marshes are probably controlled by a variety of factors (Howarth et al., unpub. data). It is only important to note here that changes in concentrations or fluxes in salt marshes do not necessarily reflect changes in rates of decomposition since methanogenesis mediates only a small percentage of the decomposition. The situation would be markedly different in an anoxic system with smaller sulfate concentrations, such as freshwater ecosystems.

Controls of Below-Ground Decomposition

Comparisons Between Salt Marsh and Terrestrial Decomposition

As discussed earlier in this review, the rate of decomposition of roots and rhizomes in salt marsh soils is not known with any certainty. Nonetheless, the rate of weight loss does not seem to differ in any qualitative manner from the rate of weight loss in terrestrial soils. For instance, while belowground bio-

Table 2. Estimates of methane flux from marsh soils to the atmosphere.

Location	*Type of grass*	*Flux (g CH_4-C m^{-2} yr^{-1})*	*Investigators*
Georgia	Tall *Spartina alterniflora*	0.4	King and Wiebe 1978
	Short *Spartina alterniflora*	40.0	King and Wiebe 1978
	Short *Spartina alterniflora*	1.0	Atkinson and Hall 1976
Maryland	Short *Spartina alterniflora*	10.7	Lipschultz 1981
Massachusetts	Short *Spartina alterniflora*	8.0	Howarth and Teal 1979

mass of a variety of salt marsh grasses in a variety of locations turns over once every 1.5 to 10 years (Good et al. 1982), belowground biomass of grass species in a tall-grass prairie turns over once every 4 years (Kucera et al. 1967; Dahlman 1968 as cited in Waid 1974). Weight loss of marsh-grass roots and rhizomes from litterbags buried at 0–10 cm reported by Hackney and de la Cruźz (1980) is slightly lower than that reported for various terrestrial grasses (reviewed by Waid 1974). On the other hand, Valiela et al. (in press) report a weight loss for *Spartina alterniflora* in buried litterbags in anoxic marsh soils which is greater than that reported for timothy grass and fescue grasses and comparable to that for nitrogen-rich red clover decomposing in oxic, terrestrial soils (Waid 1974). Obviously, there is nothing inherent in marsh grass roots and rhizomes or in the anoxic environment of salt marsh soils which prevents or greatly slows the decomposition of these materials.

Rates of Microbial Heterotrophy at Great Sippewissett Marsh

The total amount of microbial heterotrophic activity in salt marsh soils also is not known with any certainty, for reasons discussed earlier, but it is obviously great. Table 3 presents our best estimates for rates of microbial heterotrophy in stands of short-form *Spartina alterniflora* in the Great Sippewissett Marsh. Our biggest gap in understanding microbial heterotrophy is oxygen respiration. Not only are we unsure of the rate of total oxygen consumption (Table 1), but we have no data on what percentage of oxygen consumption is consumption by heterotrophic microbes. The estimate given in Table 3 is little more than a guess, as explained elsewhere (Howarth and Teal 1980). However, it is difficult to see how this estimate could be off by more than a factor of 10 (Table 1), and even if an order of magnitude too low, most of the heterotrophic activity is still anoxic (Table 3).

Table 3. Rates of microbial heterotrophy in stands of short *Spartina alterniflora* in the Great Sippewissett Salt Marsh, Massachusetts.

Form of Activity	*Rate* ($g\ C\ m^{-2}\ yr^{-1}$)	*Adequacy of Estimate*	*Source of Estimate*
Oxygen Respiration	60	Very poor	Howarth and Teal 1980
Denitrification	3	Good	Kaplan et al. 1979
Sulfate reduction	1800	Good	Howarth and Teal 1979
Methanogenesis	16	Fair	Howarth and Teal 1979
Additional fermentative and decompositional activity not accounted for in the above measurements	712	Poor	See text.
Total	2591		

Estimates (Table 3) for carbon consumption by denitrification (Kaplan et al. 1979) and sulfate reduction (Howarth and Teal 1979) are based on seasonally collected data and seem reliable. The estimate for methanogenesis is based on data collected over only half the year and on many assumptions, outlined earlier in this review. However, the rate is clearly low. Note that the estimate of carbon mineralized during methanogenesis (Table 3) is twice as great as the rate of methane formation (Table 2) because of CO_2 formation in methanogenesis. Microbial fermentation is an additional form of microbial heterotrophy in marsh soils which may not be fully accounted for by measures of oxygen respiration, denitrification, sulfate reduction, and methanogenesis. While some of the fermentation products are consumed in sulfate reduction and these other processes, some are probably exported from the marsh soils to creeks before they are consumed. We roughly estimate this otherwise unaccounted for microbial activity as equivalent to the export of dissolved organic carbon from the marsh soil to creeks. We estimate this export as 712 g C m^{-2} yr^{-1} by assuming a pore-water exchange of 13 liters m^{-2} day^{-1} (Howarth et al. in press) and an average concentration of dissolved organic carbon of 150 mg C $liter^{-1}$ (A. Giblin, Woods Hole Oceanographic Institution, Woods Hole, MA, pers. comm.).

Our estimate for total microbial heterotrophy in soils underlying short *Spartina alterniflora* stands at Sippewissett is obviously very rough. It is also very high; higher than any estimates we are aware of for any other natural system. This should not be surprising, however, since the estimate of net production (mostly belowground) for these stands is also higher than that reported for any other natural ecosystem (Valiela et al. 1976) except for some other marshes (Good et al. 1982).

Rates of Microbial Heterotrophy at Sapelo Island

Sapelo Island is the only other salt marsh where enough data have been collected to make even a rough estimate of the rate of microbial heterotrophy (Table 4). The rate of oxygen consumption at Sapelo Island is somewhat better known than at Sippewissett (Table 1), but there are still no data with which to estimate the fraction due to microbial heterotrophy. For the estimate in Table 4, we have merely guessed that 75% of the total oxygen uptake is heterotrophic consumption by microbes. Denitrification at Sapelo Island was calculated from a diffusion model using data collected only from December to June (Haines et al. 1977) and is thus a very rough estimate (Table 4). Carbon consumption by sulfate reduction is calculated from measurements made in November (Howarth and Giblin, in press) by assuming that changes in rate which occur seasonally are similar at Sapelo Island and at Sippewissett Marsh. Methanogenesis is the only satisfactory estimate available for Sapelo Island, and there are problems even with this estimate, as we discussed earlier. To calculate carbon consumption by methanogenesis, we doubled the methane flux estimate of King and Wiebe (1978) to account for CO_2 produced during methanogenesis. Data of Atkinson and Hall (1976)

Table 4. Rates of microbial heterotrophy in stands of short *Spartina alterniflora* at Sapelo Island, Georgia

Form of Activity	*Rate* ($g\ C\ m^{-2}\ yr^{-1}$)	*Adequacy of Estimate*	*Source of Estimate*
Oxygen Respiration	180	Poor	Teal and Kanwisher 1961 and Duff and Teal 1965, assuming 75% of oxygen uptake is microbial heterotrophy
Denitrification	10	Fair	Haines et al. 1977
Sulfate reduction	600	Fair	Howarth and Giblin, in press
Methanogenesis	80	Good	King and Wiebe 1978
Additional fermentative and decompositional activity not accounted for in the above measurements	0	Poor	See text.
Total	870		

would yield a lower estimate (Table 2). We have assumed that these four measures of activity completely account for microbial heterotrophic activity at Sapelo Island (Table 4) since pore-water exchanges, and therefore the export of dissolved organic carbon, are much lower than at Sippewissett (Howarth and Giblin, in press).

Although the estimates are poorly known, the rate of microbial heterotrophy in soils of short *Spartina alterniflora* is apparently greater at Sippewissett Marsh than at Sapelo Island (Tables 3 and 4). Permanent accretion of undecomposed organic carbon is relatively small at both marshes, on the order of 50–100 g C m^{-2} yr^{-1} (Howarth and Teal 1980; Pomeroy and Wiegert 1981). Our analysis suggests, therefore, that the inputs of carbon to the soil are significantly greater at Sippewissett Marsh. This is in agreement with estimates for belowground production of roots and rhizomes in stands of short *Spartina alterniflora* of 1,680 g C m^{-2} yr^{-1} for Sippewissett Marsh (Valiela et al. 1976) and 770 g C m^{-2} yr^{-1} for Sapelo Island (Gallagher and Plumley 1979). For both marshes our estimates of heterotrophic consumption of organic carbon by microbes are greater than these estimates of belowground production of roots and rhizomes. This may merely reflect errors, which are probably considerable. However, it seems likely that there are significant carbon sources in addition to production of roots and rhizomes (see later). Our analysis also indicates that most decomposition is anoxic, the result of sulfate reduction and related fermentative processes.

Oxic Vs. Anoxic Decomposition

Lack of oxygen in marsh sediments does not stop decomposition and in fact can speed decomposition of some organic materials. This is in general agreement with the rapid anoxic decomposition of plant materials in animal rumens. Pasteur demonstrated in the previous century that the consumption of glucose in anaerobic fermentation is much faster than consumption through aerobic respiration. This is because the energy available to microbes mediating the processes is much less anaerobically than aerobically per unit of glucose consumed so microbes must turn over large quantities of substrate when living anaerobically. The same is true of decomposition in marsh soils, where sulfate-reducing bacteria and associated fermentative bacteria can use a maximum of 25% of the potential energy of their organic substrate (Howarth and Teal 1980). Consequently, anaerobic microbes turn over much more substrate to produce or maintain the same biomass, and even though they may be growing on a substrate low in N and P, anaerobic microbes have access to a lot of N and P relative to their needs. They are, therefore, less likely than aerobes to be limited by these nutrients (Fenchel and Blackburn 1979).

Molecular oxygen is required for degradation of some organic compounds such as aromatic hydrocarbons and apparently lignin (Fenchel and Blackburn 1979). These compounds will persist in marsh sediments unless there is oxygen-mediated decomposition. In general, however, lack of oxygen does not inhibit microbial activity or decomposition of most compounds in marsh soils.

Although no general statement can be made about anoxia limiting decomposition, anoxic decomposition under some conditions is undeniably slow. This slowness is often associated with high acidity (low pH's), and of course pickling is an ancient form of food preservation. Decomposition in bogs appears to be slow (Clymo 1965), probably because of their low pH's. As discussed earlier, Reed (1981) showed that roots of *Spartina alterniflora* decompose considerably slower at pH = 4 than at pH = 6 or 8. However, pH's in salt marsh soils are typically greater than 6 although values as low as 4 do occur (Howarth et al. in press; Pomeroy and Wiegert 1981; J. Teal, Woods Hole Oceanographic Institution, Woods Hole, MA, pers. comm.) Low pH seems unlikely to limit decomposition in most salt marshes.

Pore Water Movement and the Build-up of Metabolites

Where pore waters in a marsh are stagnant, decomposition may be slowed because of a build-up of metabolites or toxins. This may explain why Hackney and de la Cruz (1980) found that roots and rhizomes in litterbags buried below 20 cm in the marsh they studied lost no weight during a year.

Turnover of roots and rhizomes is faster in stands of tall-form *Spartina alterniflora* near creeks (where there is good circulation of pore water) than in short-form stands (where pore waters are probably more stagnant) both in New Jersey (Good et al. 1982) and in Georgia (Gallagher and Plumley

1979). In contrast to this finding, however, sulfate reduction rates in Great Sippewissett Marsh are not significantly different in the two types of grass (Howarth and Teal 1979). This apparent discrepancy may just reflect differences in the marshes. But it may also reflect differences in the type of information provided by these approaches. Root and rhizome turnover could be faster in tall sites, yet the sites could still have similar rates of microbial heterotrophic activity if rates of carbon input are similar in the two sites. This would imply that the pool size of dead roots and rhizomes is less in the tall sites. Also, microbial heterotrophy may be fueled by materials arising not only from decomposition of dead roots and rhizomes but also from other sources such as exudates from living roots.

Many workers have argued that microbial activity is greater in tall stands of *Spartina* than in short stands in Georgia marshes (Christian and Wiebe 1978; Skyring et al. 1979; Pomeroy and Wiegert 1981). While this may be true, we do not believe that a difference in total heterotrophic activity has yet been demonstrated. Christian and Wiebe (1978) demonstrated a faster turnover of glucose in tall grass stands by using ^{14}C-labelled glucose. In part this may have been due to a larger pool size of glucose in the short stands. Hanson and Gardner (1978) demonstrated in another Georgia marsh that microbial activities cannot necessarily be inferred from turnover times without pool size information. Glucose concentrations do tend to be higher in the short stands, but the differences in glucose concentration in tall stands and short stands are always less than a factor of two (R. Christian, Biology Dept., East Carolina U., Greenville, NC, pers. comm.). Measured turnover times for glucose, on the other hand, were from 1.5 to 6 times faster in the tall stands (Christian and Wiebe 1978). Thus, the data of Christian and Wiebe (1978) suggest that mineralization of glucose in tall stands is similar to, or may in fact be faster than, that in short stands. However, Christian and Wiebe (1978) screened their mud samples to remove macroscopic organic matter, and this might have prevented them from seeing microbial activity which was tightly coupled to plant roots. Also, they exposed their samples to air while screening this material, which may have inhibited anaerobic microbial activity. And they slurried their samples, possibly affecting microbial activities, as we discussed earlier. But even assuming that glucose mineralization is higher in sediments underlying tall stands, it does not necessarily follow that total heterotrophic activity is greater. As we discuss later, leakage of malate, ethanol, and other plant metabolites from the grass rhizosphere may be important in fueling heterotrophic activity. Glucose mineralization would give no indication of this. The leakage of such compounds is probably greater in stands of short grass (Mendelssohn et al. 1981).

Skyring et al. (1979) concluded that sulfate reduction rates were greater in soils underlying stands of tall marsh grasses than in soils under short grasses in the Georgia marshes. However, their measurements were flawed by failure to consider pyrite as an end product. Their reported rates for both tall and short sites therefore may be underestimated to a significant extent (Howarth 1979; Howarth and Giblin, in press). Our own data for sulfate reduction in Georgia marshes indicated faster rates in short-grass stands rather than in tall-grass ones (Howarth and Giblin, in press). However, we studied only one tall-grass stand and it may not have been representative. Teal and Kanwisher (1961) demonstrated that the redox potential profiles and other sediment characteristics of tall-grass stands in Sapelo Island marshes are quite variable. We urge more research on the rates of sulfate reduction in these stands.

Sources of Organic Matter Fueling Microbial Activity

Christian and his colleagues at Sapelo Island were the first to ask what fuels the heterotrophic activity in salt marsh soils; is it the slow, constant, long-term breakdown of a large mass of dead roots and rhizomes or is it organic matter such as root exudates with a shorter turnover time? (Christian et al. 1978). They manipulated plots or cores of salt marsh by changing water flow, by fertilizing with organic and inorganic nutrients, and by clipping aboveground leaves. By measuring microbial mass and activity, they hoped to find out if there was a close link between these changes and changes in decomposition.

Christian et al. (1978) concluded that long-term decomposition of roots and rhizomes was the main resource for microbes. However, we believe this was not proven because of methodological problems. One method used by Christian et al. (1978) was a measure of microbial biomass and activity by extraction of adenosine triphosphate and other adenylates. They found no changes caused by experimental treatments (clipping, etc.) compared to controls; seasonal changes continued both in controls and experimental treatment plots. The problem with this approach involves the tremendous numbers of microbes in sediments and soils. Most of the direct count measurements made in marsh sediments and soils have found a fairly constant value, 2 to 16 $\times$ 10^9 cells cm^{-3} (Rublee 1982). Why should this be true in arctic and temperate soils as well as in arctic pond sediments and salt marshes? Perhaps space limitations are more important than nutrient supply rates in controlling bacterial numbers (a possibility discussed by Christian et al. 1978). To us, this possibility implies a large population of inactive bacteria. The data of Christian et al. (1978) on community energy charge (a maximum of 0.57) also imply a high percentage of inactive bacteria. Given a large population of bacteria with most inactive and the total number limited by space, is an adenylate measure useful or are any changes in activity overwhelmed by the adenylates of inactive bacteria? We conclude that the community charge measurement must be used with extreme caution.

Christian et al. (1978) also measured turnover of ^{14}C-glucose. They found no major change in the turnover time of glucose (relative to control plots) in clipped plots or in clipped plots which were also enclosed to reduce water flow. This is an interesting finding. But as noted earlier, their method may have artifactually altered microbial activities since they excluded macro-organic matter, they exposed their sample to air while screening out the macro-organic matter, and they used slurries. Also, as discussed earlier, glucose mineralization does not measure all pathways of carbon flow; the technique is biased against some pathways which are particularly tightly coupled to the living grass rhizosphere such as leakage of ethanol from anoxic roots and its subsequent oxidation by microbes.

We believe that belowground processes could be fueled by the slow breakdown of a large mass of dead roots and rhizomes, by a rapid release of organic matter from living roots and rhizomes, by a combination of these (most likely), and possibly by other carbon sources as well. The peak of both $SO_4^=$ reduction (Howarth and Teal 1979) and glucose turnover (Hansen 1979) in the fall, when temperatures are low but when organic materials have been translocated to roots and rhizomes, argues for a close link with the processes of the living plant. Hence the distinctly non-linear trajectory relating sulfate reduction rates to temperature in Great Sippewisset Marsh (Fig. 1). The large mass of dead material and comparisons of production with sedimentation rates argue that slow decomposition of dead matter is also important.

Sources of Organic Matter Belowground in Great Sippewisset Marsh

In Table 5, we have tried to estimate the magnitude of various inputs of

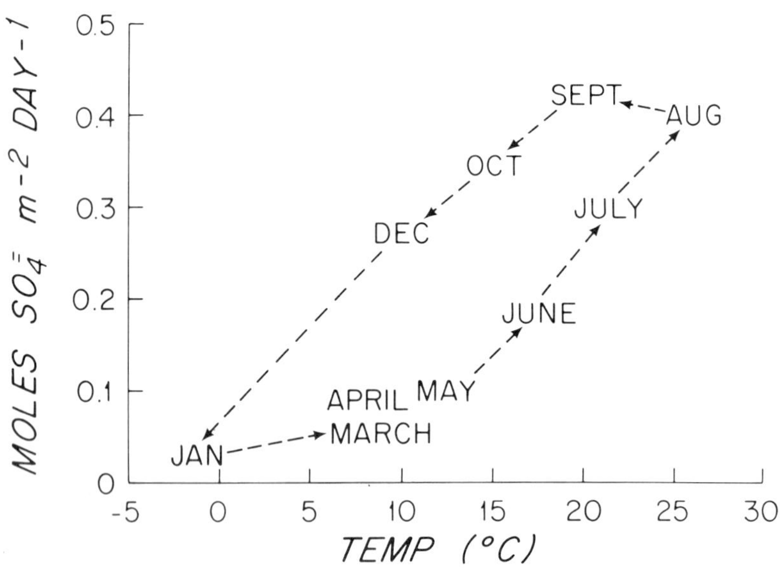

Figure 1. Temperature effects on rates of sulfate reduction in stands of short Spartina alterniflora *in Great Sippewissett Salt Marsh. From Howarth and Teal (1979).*

decomposable organic matter to soils of short *Spartina alterniflora* in Great Sippewissett Marsh. These inputs are roughly divided into two pools; those which we believe would have a short turnover time and those which are likely to have a longer turnover time. Valiela et al. (1976) estimate belowground production of roots and rhizomes to be approximately 1,680 g C m^{-2} yr^{-1}, and Valiela et al. (in press) report that about 20% of the weight of such roots and rhizomes is rapidly leached in buried litterbag studies. Therefore, we estimated that approximately 340 g C m^{-2} yr^{-1} of readily decomposable organic matter are leached from roots and rhizomes as they die. This 20% leaching estimate may be too high because live roots and rhizomes were used in the bags, as discussed earlier. The remainder of the production of roots and rhizomes, minus the permanent organic accretion of 113 g C m^{-2} yr^{-1} (Howarth and Teal 1980), we assume to be slowly decomposable organic matter, 1,230 g C m^{-2} yr^{-1}.

Living roots and rhizomes may contribute additional organic matter to the soil in two ways, by sloughing of material or by excretion of dissolved organic carbon (Table 5). These processes are virtually unstudied for marsh grasses but have been shown to be important for other plants, amounting to a loss of material equivalent to as much as 50% or more of the belowground structural root production (Shamoot et al. 1968; McDougall and Rovira 1970; Head 1973). Excretion of dissolved organic carbon from marsh grass roots may be greater than from most plants. One reason is that these roots often become anoxic (Mendelssohn et al. 1981; Gleason, 1982), and then the plants must carry on anaerobic metabolism (Armstrong 1975). End-

Table 5. Sources of decomposable organic matter in soils of short-form *Spartina alterniflora* in Great Sippewissett Marsh.

Source	*Magnitude of readily decomposable carbon released (g C m^{-2} yr^{-1})*	*Magnitude of slowly decomposable carbon released (g C m^{-2} yr^{-1})*
Production and death of roots and rhizomes:		
– leaching of newly dead material	340	0
– decomposition of organic polymers	0	1230
Sloughing of root material	0	Unknown
Excretion from living roots and rhizomes	Unknown	0
Production and death of chemosynthetic bacteria	7-250 (?)	7-250 (?)
Total	347 to 590 or more	1237 to 1480 or more

products of anaerobic plant metabolism include ethanol (Hook et al. 1971; Mendelssohn et al. 1981), lactic acid (Chirkova 1968; Hook et al. 1971), and malic acid (Crawford and Tyler 1969). Presumably there is some leakage of these organics from roots and rhizomes to the soils, and all are excellent substrates for fermentative and sulfate-reducing bacteria. Mendelssohn et al. (1981) concluded that ethanol produced in *Spartina alterniflora* roots during anaerobic respiration easily diffused into the soils. Such inputs are likely larger in stands of short than tall *Spartina alterniflora* because the soils are more reducing and the short grasses probably carry on more anaerobic respiration (Mendelssohn et al. 1981). We assume that sloughing of root material represents an input of slowly decomposable carbon while excretion of dissolved organic carbon is an input of readily decomposable carbon. We cannot even guess the magnitude of these processes, although we suspect they are large. Both represent net primary production not accounted for by investigations of changes in biomass of belowground components.

Production of chemosynthetic sulfur bacteria represents another input of organic matter to marsh soils (Table 5). This represents secondary production, not primary production, since the energy source is reduced sulfur formed during decomposition of other organic matter (Howarth and Teal 1980). It nevertheless represents production of new organic carbon which can support heterotrophic activity. Assuming that energy released by oxidation of sulfur compounds within the soil is 1.7 watts m^{-2} (13,000 Kcal m^{-2} yr^{-1}) (Howarth and Teal 1980; Howarth et al. in press), that 50% to 100% of this oxidation is biotically catalyzed, and that efficiency of transfer of energy to biomass is 2% to 40% (Kuznetsov 1970), we estimate that the production of chemosynthetic bacteria within the soil might be from 14 to 500 g C m^{-2} yr^{-1}. We guess that half of this would be readily decomposed (Table 5).

Other inputs of organic matter to marsh soil include production by benthic microalgae, input of surface litter not exported or decomposed while standing, and inputs of organic carbon from coastal waters and uplands. In stands of short *Spartina alterniflora* in Great Sippewissett Marsh, these are all fairly small inputs (Howarth and Teal 1980). We therefore estimate that inputs of readily decomposable carbon are in the range of 300 to 600 g C m^{-2} yr^{-1} or more (Table 5), possibly much more if excretion of ethanol and other end products of anaerobic plant respiration are large. Inputs of more slowly decomposed material are in the range of 1,200 to 1,500 g C m^{-2} yr^{-1} or more. Both types of input are important. The total input is in the range of 1,500 to 2,100 g C m^{-2} yr^{-1} or more, again possibly much more. Our estimate of carbon consumption by heterotrophic activity is on the order of 2,600 g C m^{-2} yr^{-1} (Table 3). This is high but is not unreasonable with regard to carbon inputs if excretion and sloughing of materials from roots and rhizomes amounts to a few hundred grams carbon per square meter per year.

Conclusions

It is obvious that little is known about controls of below-ground decomposition in salt marshes. Our review attempts to identify the important processes by setting limits on certain rates. At best, the rates and budgets presented here are good guesses. We continue our guesses with a likely list of controls.

1. *Inherent structural components of roots.* Roots and rhizomes probably would decompose much faster if they had less structural polymeric material and more labile material. They contain much lignin and ligno-celluloses which would decompose slowly even under the best of conditions. Structural arrangement of the plant material limits the rate of fermentation in part through accessibility; the microbes and their enzymes cannot reach all of the material. Although pure cellulose can decompose very rapidly in marsh sediments, much of the cellulose in roots and rhizomes may decompose slowly because of its structural arrangement.

2. *Supply of electron acceptors.* Electron acceptors are always present (e.g., $SO_4^=$, CO_2) in salt marsh soils and their supply rate is not an important control of decomposition. Rate of supply of oxygen (and even more so of nitrate) to marsh sediments is limited. Therefore, $SO_4^=$ is the major electron acceptor consumed in heterotrophic respiration. Sulfate concentrations are apparently always high enough in these soils so that rates of sulfate reduction are not limited. Even if the sulfate concentration were lower, however, it would be unlikely to slow the rate of decomposition or microbial heterotrophy since methanogenesis would probably increase in proportion to a decrease in sulfate reduction. Because most decomposition is anaerobic, nutrients are probably never limiting.

3. *Supply of oxygen.* Although decomposition of most materials is as fast or faster when $SO_4^=$ is the terminal electron acceptor, molecular oxygen is apparently required for decomposition of some materials. Thus, decomposition of lignins, aromatic hydrocarbons, and perhaps some humic substances may be limited by lack of oxygen. Also, if oxygen were present, animals might increase the rate of decomposition and microbial activity by breaking up roots and rhizomes into smaller and smaller pieces which would be more accessible to microbial attack. The presence of oxygen might also allow for more grazing of microbes by animals, and increased grazing might increase the rate of decomposition and microbial activity.

4. *Supply of labile organics from roots.* This route of supply may be large, and if so, the rate of release of substrates from roots would partially control the rate of microbial activity. There are no measurements of this supply rate in salt marshes. However, we can infer a release of plant fermentation products such as ethanol and malate from living roots and rhizomes since these regularly become anoxic and since such products have been found in *Spartina* roots. Such releases are probably greater in shorter stands of grass where soils tend to be more reducing.

5. *Metabolite buildup.* In salt marsh soils, as in other soils, microbes are present in high numbers but many are not very active. It is possible that a

buildup of metabolites and toxins, including H_2S, limits their activity. However, this has not been proven for any salt marsh soil.

Acknowledgments

Preparation of this review was supported by NSF grants DEB-81-04701 and DEB 79-05127. We thank Colleen Cavanaugh, Robert Christian, Courtney Hackney, Brian Howes, Andrew Marinucci, Irvin Mendelssohn, Bruce Peterson, and Park Rublee for reviewing the manuscript and Robert Christian for encouraging us to undertake this effort. We also thank Gerry Schroeder, Anne Giblin, Ivan Valiela, Brian Howes, and John Teal for sharing their unpublished data with us. Figure 1 is used with permission of *Limnology and Oceanography*.

References Cited

Armentano, T. V. and G. M. Woodwell. 1975. Sedimentation rates in a Long Island marsh determined by ^{210}Pb dating. *Limnol. Oceanogr.* 20:452-456.

Armstrong, W. 1975. Waterlogged soils, pp. 181-219. *In:* J. R. Etherington (ed.), *Environment and Plant Ecology*. John Wiley and Sons, London.

Atkinson, L. P. and J. R. Hall. 1976. Methane distribution and production in a Georgia salt marsh. *Est. Coastal Mar. Sci.* 4:677-686.

Cappenberg, T. E. 1974. Interrelationships between sulfate reducing and methane producing bacteria in bottom deposits of a freshwater lake. II. Inhibition experiments. *Anton. van Leeuwenhoek* 40:297-306.

Chirkova, T. V. 1968. Oxygen supply to roots of certain woody plants kept under anaerobic conditions. *Soviet Plant Phys.* 15:475-478.

Christensen, D. and T. H. Blackburn. 1980. Turnover of tracer (^{14}C, ^{3}H labelled) alanine in inshore marine sediments. *Mar. Biol.* 58:97-103.

Christian, R. R., K. Bancroft and W. J. Wiebe. 1978. Resistance of the microbial community within salt marsh soils to selected perturbations. *Ecology* 59:1200-1210.

Christian, R. R. and W. J. Wiebe. 1978. Anaerobic microbial community metabolism in *Spartina alterniflora* soils. *Limnol. Oceanogr.* 23:328-336.

Christian, R. R. and W. J. Wiebe. 1979. Three experimental regimes in the study of sediment microbial ecology. *Special Technical Publication 673, Am. Soc. for Testing and Materials*, pp. 148-155.

Clymo, R. S. 1965. Experiments on breakdown of *Sphagnum* in two bogs. *J. Ecol.* 53:747-758.

Crawford, R. M. M. and P. D. Tyler. 1969. Organic acid metabolism in relation to flooding tolerance in roots. *J. Ecol.* 57:237-246.

Dacey, J. W. H. and M. L. Klug. 1979. Methane efflux from lake sediments through water lilies. *Science* 203:1253-1255.

De la Cruz, A. A. and C. T. Hackney. 1977. Energy value, elemental composition, and productivity of belowground biomass of a *Juncus* tidal marsh. *Ecology* 58:1165-1170.

DeLaune, R. D. and W. H. Patrick. 1980. Nitrogen and phosphorus cycling in a Gulf Coast salt marsh, pp. 143-151. *In:* V. S. Kennedy (ed.), *Estuarine Perspectives*. Academic Press, New York.

Duff, S. and J. M. Teal. 1965. Temperature change and gas exchange in Nova Scotia and Georgia salt marsh muds. *Limnol. Oceanogr.* 10:67-73.

Fenchel, T. and T. H. Blackburn. 1979. *Bacteria and Mineral Cycling*. Academic Press, New York, 225 pp.

Fleischer, S. 1973. A method of studying glucose turnover in wastewater. *Vatten* 3:306-311.

Gallagher, J. L. and F. G. Plumley. 1979. Underground biomass profiles and productivity in Atlantic coastal marshes. *Am. J. Bot.* 66:151-161.

Gleason, M. L. and E. L. Dunn. 1982. Effects on hypoxia on root and shoot respiration of *Spartina alterniflora*, pp. 243-253. *In:* V. S. Kennedy (ed.), *Estuarine Comparisons.* Academic Press, New York.

Good, R. E., N. F. Good and B. R. Frasco. 1982. A review of primary production and decomposition dynamics of the belowground marsh component, pp. 139-157. *In:* V. S. Kennedy (ed.), *Estuarine Comparisons.* Academic Press, New York.

Hackney, C. T. and A. A. de la Cruz. 1980. *In situ* decomposition of roots and rhizomes of two tidal marsh plants. *Ecology* 61:226-231.

Haines, E., A. Chalmers, B. Hanson and B. Sherr. 1977. Nitrogen pools and fluxes in a Georgia salt marsh, pp. 241-254. *In:* M. L. Wiley (ed.), *Estuarine Processes. Vol. II.* Academic Press, New York.

Hall, K. J., P. M. Kleiber and I. Yesaki. 1972. Heterotrophic uptake of organic solutes by microorganisms in the sediment. *Mem. Ist. Ital. Idrobiol. Suppl.* 29:441-471.

Hansen, J. A. 1979. Effects of physical factors on fermentation in salt marsh soils. M.S. Thesis, U. Georgia, Athens, GA 65 pp.

Hanson, R. and W. Gardner. 1978. Uptake and metabolism of two amino acids by anaerobic microorganisms in four diverse salt marsh soils. *Mar. Biol.* 46:101-107.

Harrison, M. J., R. T. Wright and R. Y. Morita. 1971. Methods for measuring mineralization in lake sediments. *Appl. Microbiol.* 21:698-702.

Head, G. C. 1973. Shedding of roots, pp. 237-293. *In:* T. T. Kozlowski (ed.), *The Shedding of Roots.* Academic Press, New York.

Henrichs, S. M., J. E. Hobbie, R. W. Howarth, J. V. K. Helfrich and P. Kilham. In press. Free amino acids in salt marsh sediments: Concentrations and fluxes. *Limnol. Oceanogr.*

Hook, D. D., C. L. Brown and P. P. Kormanik. 1971. Inductive flood tolerance in swamp Tupelo. *J. Exp. Bot.* 22:78-89.

Hopkinson, C. S., J. W. Day and B. T. Gael. 1978. Respiration studies in a Louisiana salt marsh. *An. Centro Cienc. del Mar y Limnol., Univ. Nal. Auton., Mexico* 5:225-237.

Howarth, R. W. 1979. Pyrite: Its rapid formation in a salt marsh and its importance in ecosystem metabolism. *Science* 203:49-51.

Howarth, R. W. and A. Giblin. In press. Sulfate reduction in the marshes at Sapelo Island, Georgia. *Limnol. Oceanogr.*

Howarth, R. W., A. Giblin, J. Gale, B. J. Peterson and G. W. Luther III. In press. Reduced sulfur compounds in the pore waters of a New England salt marsh. *In:* R. O. Hallberg (ed.), *Proceedings of the Fifth International Symposium on Environmental Biogeochemistry*, Stockholm.

Howarth, R. W. and J. M. Teal. 1979. Sulfate reduction in a New England salt marsh. *Limnol. Oceanogr.* 24:999-1013.

Howarth, R. W. and J. M. Teal. 1980. Energy flow in a salt marsh ecosystem: The role of reduced inorganic sulfur compounds. *Am. Nat.* 116:862-872.

Howes, B. L., A. E. Giblin, R. W. Howarth and J. M. Teal. (Submitted). Oxygen loss from roots and rhizomes of *Spartina alterniflora* and its relationship to waterlogging.

Howes, B. L., R. W. Howarth, J. M. Teal and I. Valiela. 1981. Oxidation – reduction potentials in a salt marsh: Spatial patterns and interactions with primary production. *Limnol. Oceanogr.* 26:350-360.

Jorgensen, B. B. 1977. The sulfur cycle of a coastal marine sediment (Limnfjorden, Denmark). *Limnol. Oceanogr.* 22:814-832.

Jorgensen, B. B. 1978. A comparison of methods for the quantification of bacterial sulfate reduction in coastal marine sediments. I. Measurements with radiotracer techniques. *Geomicrobiology J.* 1:11-27.

Jorgensen, B. B. and T. Fenchel. 1974. The sulfur cycle of a marine sediment model system. *Mar. Biol.* 24:189-201.

Kaplan, W., I. Valiela and J. M. Teal. 1979. Denitrification in a salt marsh ecosystem. *Limnol. Oceanogr.* 24:726-734.

King, G. M. 1978. The nature of methanogenesis in soils of a Georgia salt marsh. Ph.D. Thesis, U. Georgia, Athens, GA 194 pp.

King, G. M. and W. J. Wiebe. 1978. Methane release from soils of a Georgia salt marsh. *Geochim. Cosmochim. Acta* 42:343-348.

King, G. M. and W. J. Wiebe. 1980. Regulation of sulfate concentrations and methanogenesis in salt marsh soils. *Est. Coastal Mar. Sci.* 10:215-223.

Kucera, C. L., R. C. Dahlman and M. R. Koelling. 1967. Total net productivity and turnover on an energy basis for tallgrass prairie. *Ecology* 48:536-541.

Kuznetsov, S. I. 1970. *The Microflora of Lakes and its Geochemical Activity*. Univ. Texas Press, Austin, TX 503 pp.

Lipschultz, F. 1981. Methane release from a brackish intertidal salt marsh embayment of Chesapeake Bay, Maryland. *Estuaries* 4:143-145.

McDougall, B. M. and A. D. Rovira. 1970. Movement of ^{14}C-photosynthate into roots of wheat seedlings and exudation of ^{14}C from intact roots. *New Phytol.* 69:37-46.

Mendelssohn, I. A., K. L. McKee and W. H. Patrick, Jr. 1981. Oxygen deficiency in *Spartina alterniflora* roots: Metabolic adaptation to anoxia. *Science* 214:439-441.

Nedwell, D. B. and J. W. Abram. 1978. Bacterial sulphate reduction in relation to sulphur geochemistry in two contrasting areas of saltmarsh sediment. *Est. Coast. Mar. Sci.* 6:341-351.

Nixon, S. W. 1980. Between coastal marshes and coastal waters – a review of twenty years of speculation and research on the role of salt marshes in estuarine productivity and water chemistry, pp. 437-525. *In:* P. Hamilton and K. B. MacDonald (eds.), *Estuarine and Wetland Processes*. Plenum Press, New York.

Pomeroy, L. R. and R. G. Wiegert. 1981. *Ecology of a Salt Marsh*. Springer-Verlag, New York. 271 pp.

Redfield, A. C. and M. Rubin. 1962. The age of salt marsh peat and its relation to recent changes in sea level at Barnstable, Massachusetts. *Proc. Natl. Acad. Sci. U.S.* 48:1728-1735.

Reed, J. P. 1981. Factors influencing the decomposition of *Spartina alterniflora* roots. Ph.D. Thesis, North Carolina State U., Raleigh, NC 72 pp.

Rublee, P. A. 1982. Bacteria and microbial distribution in estuarine sediments, pp. 159-182. *In:* V. S. Kennedy (ed.), *Estuarine Comparisons*. Academic Press, New York.

Shamoot, S., L. McDonald and W. V. Bartholomew. 1968. Rhizo-deposition of organic debris in soil. *Soil Sci. Soc. Am. Proc.* 32:817-820.

Skyring, G. W., R. L. Oshrain and W. J. Wiebe. 1979. Sulfate reduction rates in Georgia marshland soils. *Geomicrobiology J.* 1:389-400.

Sorensen, J., B. B. Jorgensen and N. P. Revsbech. 1979. A comparison of oxygen, nitrate, and sulfate respiration in coastal marine sediments. *Microb. Ecol.* 5:105-115.

Sorensen, J., D. Christensen and B. B. Jorgensen. 1981. Volatile fatty acids and hydrogen as substrates for sulfate-reducing bacteria in anaerobic marine sediment. *Appl. Env. Microb.* 42:5-11.

Teal, J. M. and J. W. Kanwisher. 1961. Gas exchange in a Georgia salt marsh. *Limnol. Oceanogr.* 6:388-399.

Teal, J. M. and J. W. Kanwisher. 1966. Gas transport in the marsh grass, *Spartina alterniflora*. *J. Exp. Bot.* 17:355-361.

Trudinger, P. A. 1956. Fixation of carbon dioxide by extracts of strict autotroph *Thiobacillus denitrificans*. *Biochem. J.* 64:274-286.

Valiela, I., B. L. Howes, R. W. Howarth, A. Giblin, K. Foreman, J. M. Teal and J. E. Hobbie. In Press. The regulation of primary production and decomposition in a salt marsh ecosystem. *In: Proceedings of First International Wetlands Conference,* New Delhi, India.

Valiela, I., J. M. Teal and N. Y. Persson. 1976. Production and dynamics of experimentally enriched salt marsh vegetation: Belowground biomass. *Limnol. Oceanogr.* 21:245-252.

Waid, J. S. 1974. Decomposition of roots, pp. 175-211. *In:* C. H. Dickinson and G. J. F. Pugh (eds.), *Biology of Plant Litter Decomposition.* Academic Press, London.

Wood, L. W. 1970. The role of estuarine sediment microorganisms in the uptake of organic carbon under aerobic conditions. Ph.D. Thesis, North Carolina State U., Raleigh, NC 75 pp.

NITROGEN CYCLING AND ESTUARINE INTERFACES: SOME CURRENT CONCEPTS AND RESEARCH DIRECTIONS

W. Michael Kemp
University of Maryland
Horn Point Environmental Laboratories
Cambridge, Maryland

Richard L. Wetzel
Virginia Institute of Marine Science
Gloucester Pt., Virginia

Walter R. Boynton, Christopher F. D'Elia
University of Maryland
Chesapeake Biological Laboratory
Solomons, Maryland

and

J. Court Stevenson
University of Maryland
Horn Point Environmental Laboratories
Cambridge, Maryland

Abstract: The role of physical interfaces in estuarine nitrogen (N) dynamics is discussed. We consider here: four N-transformation processes (uptake, regeneration, denitrification, nitrification) and five interfaces (water mass fronts and transitions, watershed-estuarine boundaries, the pycnocline of stratified estuaries, the sediment-water boundary, the redox discontinuity layer). Seven examples are presented. First, we show that phytoplankton production and, in turn, NH_4^+ recycling can be stimulated at interfaces where two water masses meet, with one being relatively clear and the other nutrient-rich. Second, data are provided to illustrate that N recycling rates tend to exceed (by 2-8 fold) inputs of "new" N entering across watershed-estuary boundaries, although annual net primary production is more a function of the latter. Third, we argue that periodic occurrences of high NO_2^- concentrations in strongly stratified water columns reflect active NH_4^+ oxidation at the estuarine pynocline. Fourth, evidence is given to indicate that intensive remineralization of NH_4^+ occurs in the uppermost flocculent layer of sediments, and that fluxes estimated from diagenic modeling would tend to overlook this. Next, we show that denitrification, which is concentrated near the sediment redox discontinuity layer (RDL), may be a major component of estuarine N budgets (50-60% of NH_4^+ recycling). Sixth, we indicate that denitrification can be fueled either by sediment nitrification just above the RDL or via NO_3^- diffusion from overlying waters. Seventh, recent experimental results are considered to demonstrate the effects of macrophytic roots enhancing nitrification (and possibly denitrification) by transporting O_2 and deepening the oxidized zone of sediments. Finally, we propose some generic properties of estuarine interfaces which may account for their importance in N cycling.

ISBN 0-12-404070-5

Introduction

Over the last several decades considerable attention has been focused on understanding nutrient cycling in estuarine and coastal waters. Nixon (1981a) has provided an interesting historical account of the changing paradigms of estuarine nutrient dynamics, particularly concerning the perceived importance of various mechanisms of nutrient remineralization. It is evident that since the late 1960's (Pomeroy 1970), major advances have occurred in this field, and these were made possible by the numerous methodological developments for measuring nutrient transformation rates. Many of the key nutrient processes (including phytoplankton uptake, zooplankton excretion, sediment sorption and benthic remineralization, nitrogen-fixation, nitrification and denitrification) have been widely investigated, and numerous reports for these rates are now available in the literature. Only recently has emphasis been placed on factors regulating nutrient processes, and, in particular, several studies have examined the influences of physical and chemical gradients characteristic of estuarine environments (Martens and Goldhaber 1978; Aller and Yingst 1980; Callender and Hammond 1982).

Most of these estuarine gradients are bounded by distinct interfaces that exhibit internal discontinuities with steep rates of change for a particular property. Important interfaces and discontinuities are found throughout many estuaries, including in the vertical direction: 1) the air-water boundary; 2) the pycnocline of stratified water columns; 3) the sediment-water boundary; and 4) the redox discontinuity layer (RDL) in the sediments. In the horizontal direction, interfaces include: 5) the peripheral boundaries at the general watershed shoreline; 6) the edge of the marsh; 7) the transition zones between river and estuary; and 8) the frontal regions between estuary and continental shelf. It appears that such interface zones are often the sites of intensified metabolic and physical-chemical activity and that large fluxes of energy and materials occur across these boundaries. While many investigators have measured nutrient transformations and fluxes at these various interfaces, the potential importance of these sites in overall estuarine nutrient dynamics has been largely unrecognized in the literature.

Our purpose here is to compile and synthesize recent data from a variety of sources to illustrate the spectrum of examples where nutrient cycling rates appear to be accentuated or otherwise influenced by estuarine interfaces. We limit our discussion here to nitrogen, although many of the examples would be equally relevant for other elements of estuarine chemistry. Considerable evidence exists to suggest that nitrogen is the nutrient most often limiting phytoplankton productivity in estuarine and coastal waters (Ryther and Dunstan 1972; Goldman et al. 1973; Thayer 1974; Boynton et al. 1982). We discuss several specific cases in light of current thinking in the literature regarding estuarine nitrogen (N) cycling and present a number of scientific hypotheses which warrant further investigation. Finally, we consider some generic properties of estuarine interfaces which may explain their ecological importance.

Conceptual Framework

In Fig. 1 we present a schematic diagram depicting the relative nature and location of seven specific examples where nitrogen cycling is influenced by one or more estuarine interfaces. Each topic is represented by simplified biochemical reactions circled and labelled, and each circle is positioned in relation to the cross-sectioned, obliquely viewed sketch of an estuarine region. All but one of these processes are examples of intensified N cycling *at* or near an estuarine interface, while the topic of "new" versus "recycled" nitrogen (Example 2, center of Fig. 1) emphasizes the fact that N fluxes *across* such interfaces are crucial to production and regulation of estuarine ecosystems.

Two of the topics presented here are related to interfaces (or discontinuities) encountered along horizontal gradients. Transition zones develop as estuarine waters mix with their riverine and oceanic end members to stimulate phytoplankton production, deposition and benthic remineralization (Example 1, Fig. 1). We consider both the relatively sharp frontal systems occurring on the continental shelf and the more gradual transition zones at the head of the estuary (not shown in Fig. 1). We also discuss, as a kind of interface occurring along a horizontal axis, the watershed-estuarine boundary, across which large fluxes of N enter the estuary via river flow. Here, (Example 2) we consider relative roles of new N inputs across this boundary and N recycling within the estuary to support phytoplankton uptake.

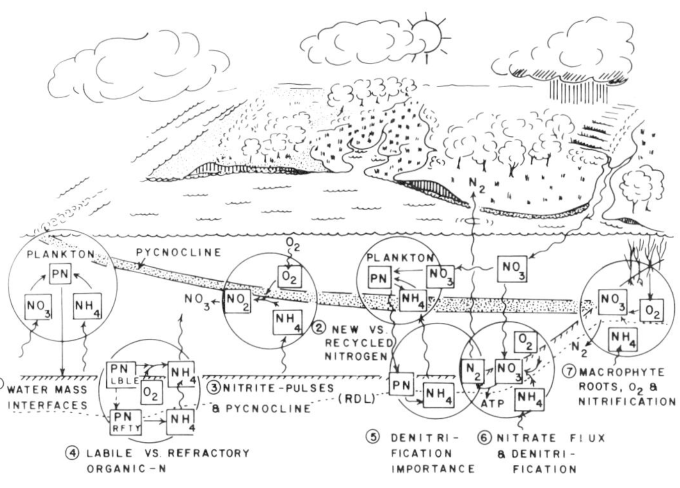

Figure 1. Schematic of seven examples of nitrogen cycling (labeled below circles) at various estuarine interfaces. Small boxes – different species of nitrogen and oxygen; PN – particulate nitrogen (LBLE = labile, RFTY = refractory); DON – dissolved organic nitrogen. Dotted line – redox discontinuity layer (RDL); solid line with arrow – chemical reaction; wavy line with arrow – diffusive or advective transport.

We present five examples of N cycling at estuarine interfaces evident along the vertical dimension. Moving downward from the water surface, we discuss, first the pycnocline of stratified estuaries which may be the site of intense N transformation reactions (Example 3). At the sediment-water interface we consider the uppermost flocculent stratum of sediment where recently deposited labile (LBLE) organic matter is decomposed rapidly to aerobically regenerate NH_4^+ (Example 4). At or below the sediment surface, the RDL demarcates a boundary between aerobic and anaerobic processes. Here (Example 5), we examine the potential importance of denitrification (occurring just below this RDL) in the sediment N budget. The possible role of NO_3^- diffusion from overlying water across the sediment-water interface and the RDL is considered (Example 6) in comparison to *in situ* nitrification as a source of electron-acceptor for denitrification. Finally, the influence of macrophyte roots and rhizomes in enlarging the surface area of this RDL and the resulting effects on nitrification are investigated (Example 7).

Estuarine Interfaces

1. Uptake and Recycling at Water-Mass Interfaces

Physical circulation in estuaries is a complex process, characterized by various discontinuities where different water masses mix along a horizontal gradient. In some cases where estuarine and coastal waters meet, regions of convergence develop characterized by relatively strong vertical motions and sharp frontal boundaries. These estuarine-coastal fronts recently have received considerable attention among oceanographers interested in their physical description (Bowman and Iverson 1978) and ecological significance (Iverson et al. 1979). In particular the influence that such fronts might have on phytoplankton distributions has been widely discussed (Seliger et al. 1981). At the other end of the estuarine gradient, where tidal fresh waters mix with brackish waters, a transitional region develops (particularly in partially stratified estuaries) and is characterized by a broad mixing-zone of highly turbid waters (Schubel 1968) followed by a downstream area of elevated phytoplankton production.

This enhancement of phytoplankton growth which is often found near frontal or transitional zones results from the mixing of a relatively nutrient-rich water mass with a relatively clear, nutrient-depleted water mass (Haines 1979). The characteristically strong vertical circulation provides a mechanism for returning nutrients remineralized by the benthos to the euphotic zone, while deposition of plankton and fecal materials continually supplies the sediment system. This strong relationship between phytoplankton production and benthic processes has been demonstrated for lacustrine (Hargrave 1973) and estuarine (Hargrave 1980; Nixon 1981a) systems.

To illustrate this point, we present summarized data for the riverine-estuarine transition zone in the Patuxent River, MD (Fig. 2). Data demonstrate increasing water clarity (a) and decreasing nutrient concentrations (b)

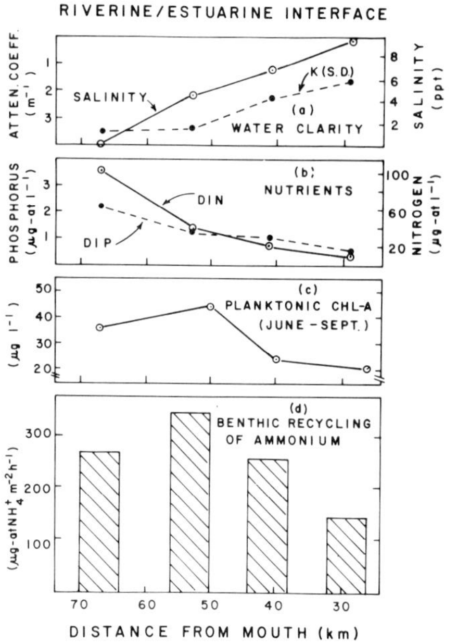

Figure 2. Transects of (a) salinity and water clarity; (K(S. D.) is secchi disk derived attenuation coefficient), (b) nutrients (DIN and DIP are dissolved inorganic nitrogen and phosphorus), (c) planktonic chlorophyll a; *and (d) benthic remineralization of NH_4^+ along the estuarine gradient (through the transition zone) of the Patuxent River, MD. Data, which are annual means except for chlorophyll* a *(summer mean), are adapted from Boynton et al. (1981).*

along the salinity gradient from 0–9 ppt. The turbidity-maximum region extends generally from about 0–4 ppt, followed by a broad peak in phytoplankton chlorophyll *a* between 3 and 6 ppt (c). Coincident with the chlorophyll maximum is a region of rapid NH_4^+ regeneration at about 50–55 km from the estuarine mouth (d). These data are consistent with those reported previously for the middle two stations only (Boynton et al. 1980). Although the generality of these trends awaits further study in other estuaries, Callender (1982) has observed a similar pattern of benthic recycling of phosphorus along the Potomac River estuarine gradient with a local maximum occurring in the transitional zone.

An analogous pattern is shown for a Georgia coastal transect from the Duplin River across the Georgia Bight (Fig. 3). The actual recycling rates beyond the 15 km point are not available (Fig. 3c), but it is likely that rates decrease rapidly with depth beyond this site (Zeitzschel 1979). In fact, Zeitzschel (1979) has shown in his literature review of benthic respiration and NH_4^+ flux, that for all values reported there is a general trend along a coastal depth-gradient with peak values occurring at about 10 m depth (which is consistent

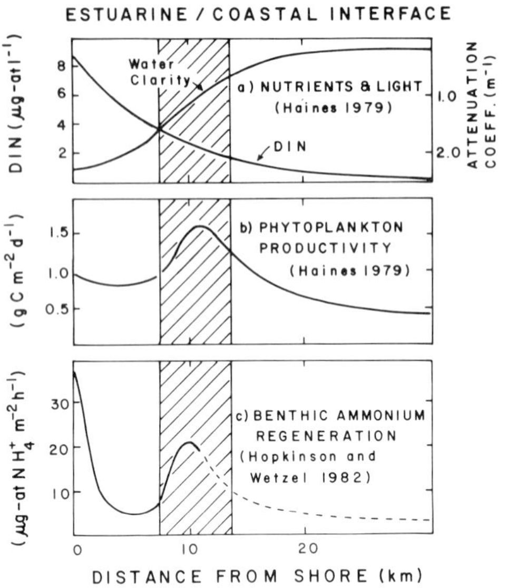

Figure 3. Transects for (a) annual means of nutrients (dissolved inorganic nitrogen, DIN) and water clarity, (b) annual means of phytoplankton productivity and (c) benthic remineralization of NH_4^+ along an inshore-offshore gradient of the Georgia Bight in summer. Dashed line in (c) is based on preliminary observations.

with the depth profiles in Fig. 3). The data compiled by Harrison (1980) suggest similar trends in rates of water-column nitrogen recycling along a coastal shelf transect.

Hence, water mass interfaces which develop at both the riverine and seaward ends of the estuary are characterized by active uptake (and production of phytoplankton) and recycling (benthic and pelagic) of N. However, comprehensive research is needed to define these frontal and transitional regions in estuaries and to document the existence of mechanisms controlling productivity maxima here, as well as the role of sediment-water fluxes.

2. The Watershed-Estuary Interface: Inputs Vs. Recycling

The distinction between "new" and "regenerated" forms of nitrogen for supporting phytoplankton production in coastal and oceanic waters was emphasized by Dugdale and Goering (1967). New nitrogen in their sense refers to allochthonous NO_3^- entering the euphotic zone either from coastal sources or from upwelling. In either case NO_3^- was taken to be the principle form of new nitrogen, whereas NH_4^+ and urea were the predominant forms autochthonously regenerated by zooplankton in the upper portion of the water column. They concluded that net phytoplankton production was controlled largely by inputs of NO_3^-, and the new NO_3^--supported production has been shown to range from 6-66% of the total production in oceanic waters (Dugdale 1976).

What are the implications of this concept for estuarine waters? First, the relatively shallow depths and enclosed periphery which characterize most estuarine systems lead us to conclude that the boundaries for examining this concept should be established somewhat differently than was done for oceanic waters. Here, we can define "new" nitrogen as that entering allochthonously from the watershed or from offshore, and "regenerated" nitrogen as that recycled either within the water column or across the benthic interface. Since the overall importance of N fixation in estuarine N budgets is minor compared to freshwater systems (Nixon 1981b), it is not considered in this discussion. NH_4^+ is the predominant form of recycled nitrogen, and much of the new N is NO_3^-; however, unlike the open ocean, NH_4^+ also constitutes a major source of new N derived from sewage effluents and other sources. Nixon (1981a, b) has argued that NH_4^+ remineralization from estuarine benthos is tightly coupled with autotrophic production, and that because of its inherently slower nature, this heterotrophic recycling may regulate phytoplankton production. We have compiled nitrogen input budgets for eight estuaries (where data were available) and compared these to estimated NH_4^+ regeneration rates within the estuary (Fig. 4). The general pattern of these data agrees reasonably well with those data reported by Nixon (1981a). However, in contrast to Nixon's summary we found no examples where inputs exceeded recycling rates, and the ratio of the latter to the former ranged from 2:1 to 8:1. Regeneration rates appear to increase with increasing inputs, as has been postulated by Eppley et al. (1979) and others for coastal systems. However, it is clear that other factors are involved.

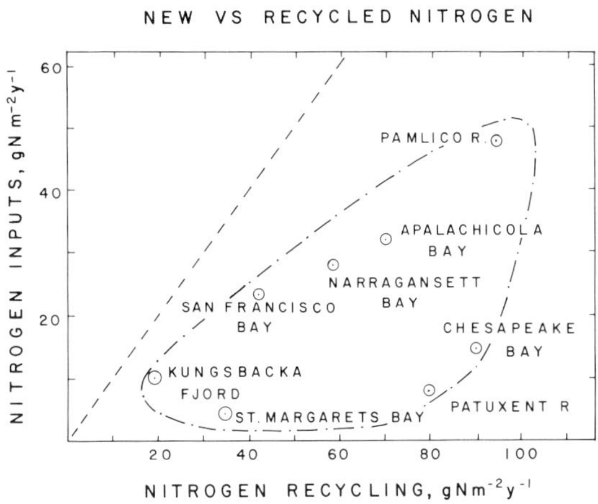

Figure 4. Comparison of nitrogen inputs versus nitrogen recycling for eight estuarine systems. Data are adapted from Boynton et al. (1982).

Were it not for our recent analysis (Boynton et al. 1982), Fig. 4 might lead us to conclude (as did Nixon 1982a) that the effect on productivity from nitrogen inputs via riverflow or other watershed sources is dwarfed by internally regenerated sources. However, the correlative analyses of Boynton et al. (1982) strongly indicate that peak plankton production may be significantly influenced by watershed loading of nitrogen, a conclusion which is consistent with the original concept of Dugdale and Goering (1967). Thus, while estuarine productivity tends to be regulated by recycled NH_4^+ on a day-to-day basis, the levels of peak and total annual production are apparently enhanced by increased inputs of new nitrogen. Experimental evidence and well-constrained nutrient budgets are needed to demonstrate the interrelationships among nutrient loading, nutrient recycling, and phytoplankton productivity.

3. Nitrite Pulses and the Pycnoclinal Interface

Nitrogen occurs in estuarine waters in numerous forms, including the inorganic molecules of nitrate, nitrite, and ammonium, as well as a wide assortment of organic species such as urea and amino acids. Two redox states of inorganic nitrogen salts (NO_3^- and NH_4^+) tend to predominate, and the intermediate NO_2^- is generally encountered only at low (and often undetectable) concentrations. However, recent reports have indicated that periodic pulses of elevated NO_2^- concentrations occur in Chesapeake Bay and other estuaries largely during the late summer and early fall (McCarthy et al. 1977; Webb and D'Elia 1980; D'Elia et al. 1981; Webb 1981). The actual source of this NO_2^- has been a matter of some speculation among these investigators, where nitrification, denitrification, and runoff have all been suggested.

The recent work of Webb and D'Elia (1980) and D'Elia et al. (1981) has indicated that the occurrence of these nitrite peaks appears to be related to the existence of a strong pycnocline separating upper and lower layers of the water column by density stratification. Such stratification is characteristic of coastal plain estuaries such as Chesapeake Bay, but the degree of vertical density difference is variable, being a function of seasonal freshwater input (Seitz 1971), wind conditions (Carter et al. 1978) and spring-neap tidal cycle (Haas 1977). In Fig. 5, we present late summer vertical profiles of dissolved oxygen (O_2), NO_2^- and NH_4^+ across the pycnocline interface for two stations, one in the lower York River (a) and the other in the Chesapeake Bay proper near the mouth of the Patuxent River (b). In both cases the estuary was strongly stratified with vertical salinity gradients of about 0.7 Kg m^{-4}. In the marine environment periodic occurrences of vertical nitrite maxima have been observed near the deep-sea pycnocline. This phenomenon has been often attributed to nitrate reduction (Thomas 1966; Carlucci and Schubert 1969; Wada and Hattori 1972); however, recent studies by Olson (1981a, b) suggest nitrification may be involved.

We hypothesize that in both cases depicted in Fig. 5 the high concentrations of nitrite were attributable to ammonium oxidation, the first step in the

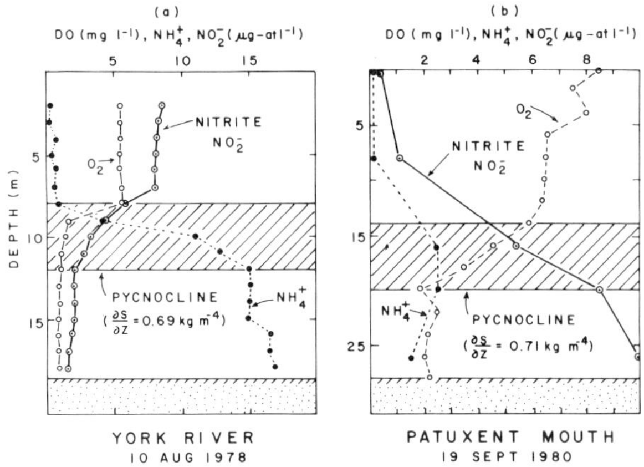

Figure 5. Vertical profiles of ammonium (NH_4^+), dissolved oxygen (O_2) and nitrite (NO_2^-) for water columns of two Chesapeake Bay stations in late summer. (a) lower York River, (b) Chesapeake Bay proper off the mouth of the Patuxent River. Data are adapted from (a) Webb and D'Elia (1980) and (b) D'Elia (unpublished).

nitrification process, and that the second step in the nitrification process, nitrite oxidation, was occurring at an insufficient rate to use up the nitrite produced. Clearly, the pycnocline represents an important boundary vertically separating water masses and is a factor in the accumulation of nitrite in the water column. It may also be a localized site of nitrification activity (Indebø et al. 1977).

The contrast between the York River station in which nitrite accumulated above the pycnocline and the mid-Chesapeake station in which NO_2^- accumulated below the pycnocline suggests to us that NH_4^+ oxidation in estuaries can be spatially variable and difficult to evaluate. Because we lack adequate time series data for concentrations and transformation rates, we can only postulate possible explanations. However, there are several similarities and contrasts between water profiles at the two stations that are worth noting. One common feature at both stations is the inverse relationship between NH_4^+ and NO_2^- concentrations across and below the pycnocline; this is evidence for nitrification. Another common feature is the accumulation of NH_4^+ and depletion of oxygen (O_2) below the pycnocline, probably as a consequence of processes active at the sediment-water interface.

In the York River, below the halocline, oxygen concentrations probably limit nitrification. Above the halocline, very low NH_4^+ concentrations probably limit nitrification and NO_2^- accumulation there probably occurred as the result of processes that were most active prior to the observation period. Under such conditions we might expect the greatest potential for nitrification to occur in the pycnocline where both O_2 and NO_2^- levels are adequate to support vigorous NH_4^+ oxidation. In the mid Chesapeake Bay, below the halocline, both NH_4^+ levels and O_2 levels seem high enough to support active NH_4^+ oxidation, whereas above the halocline levels of NH_4^+ are inadequate. High rates could be maintained only if mineralization of NH_4^+ from the sediment or in the lower water column remained high and diffusion of O_2 across the pynocline continue to provide adequate quantities of both substrates. Obviously, further interpretation awaits direct experimental measurements of nitrification and denitrification through the water (and sediment) column(s), particularly in relation to the development of estuarine pycnoclines.

4. Labile Vs. Refractory Organic Matter At the Sediment-Water Interface

In recent years the flux of remineralized nutrients (particularly NH_4^+) across the sediment-water interface has been much studied (see reviews of Zeitzschel 1979; Boynton et al. 1980; Nixon 1981a). These authors have stressed the importance of this source of recycled NH_4^+ for phytoplankton production. It has been generally assumed that the mechanism for generating such NH_4^+ fluxes is the continual decomposition of organic matter deposited over decades, which leads to establishment of vertical gradients of NH_4^+ in sediment pore-waters (Berner 1976, 1980). Gradient-supported diffusive (and advective) flux then accounts for mass transport across this interface, which can be calculated by solving the appropriate diagenic equations (Berner 1980). A number of investigators have recently compared NH_4^+ fluxes calculated in this way from pore-water concentration profiles with direct measurements of flux (obtained *in situ* by change in concentration under bell jars or other benthic chambers). The results of these comparisons have revealed inconsistent relationships between the two estimates.

We present a summary of these results in Fig. 6. Directly measured fluxes tend to be greater than those calculated from pore-water concentration gradients, with ratios of the former to the latter ranging from about 1.5:1 to 15:1. Various authors have suggested that actual transport of pore-water ions across the sediment-water interface can exceed that predicted from molecular diffusion because of one or more factors, including: stirring of sediments from currents or bioturbation; irrigation of sediments by suspension-feeding invertebrates; or ebullition from methane bubbles (Klump and Martens 1981; Callender and Hammond 1982). Presumably, any or all of these mechanisms may be operative at any given time or place in a particular estuary. The extreme ratios in Potomac and Patuxent transitional reaches may be largely attributable (as suggested by Callender and Hammond 1982) to bioturbation. Yet, there is by no means a one-to-one relation between this ratio and benthic infaunal densities in the Patuxent River (Boynton et al. 1981).

We suggest another mechanism to explain this difference between calculated and observed fluxes – namely, that disproportionately rapid rates of organic decomposition occur at the uppermost sediment surface. It appears that such elevated rates would occur in the unconsolidated flocculent layer because recently deposited organic material is far more labile than

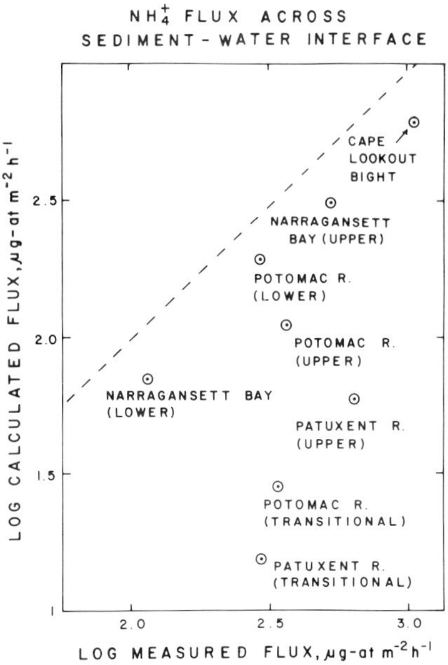

Figure 6. Ammonium flux across the sediment-water interface as calculated from pore-water concentration gradients using a diffusion model versus in situ *measurements of flux into benthic chambers. Note that both axes are logarithmic. Data for Cape Lookout Bight are from Klump and Martens (1981); Narragansett Bay – Elderfield et al. (1981) and McCaffrey et al. (1980); Potomac River – Callender (1979) and Callender and Hammond (1982); Patuxent River – Boynton et al. (1981).*

residual substances which remain at depth, and because oxygen is continuously supplied from the overlying water. Aller and Yingst (1980), in fact, reported exponentially increasing rates of NH_4^+ production in sediment profiles upward from a depth of 10 cm, which predict, by extrapolation, exceedingly high rates at the very surface. Between this floc layer and the RDL, organic material, which is somewhat more refractory than that in the floc above, is decomposed aerobically, while below the RDL the most resistant material degrades slowly under anoxic conditions. The inherently slow nature of anaerobic ammonification by sulfate reduction has been stressed by Vanderborght et al. (1977). This concept of a metabolically active surficial zone has been recently recognized by others, where, for instance, Klump and Martens (1980) and Elderfield et al. (1981) have invoked it to explain anomolously low or high (respectively) *in situ* observed fluxes. In the former case, there appeared to be a net uptake in this flocculent zone while in the latter a net release seemed to be the case.

While net uptake of NH_4^+ may occur in certain seasons and places, vertical profiles of particulate composition in the Patuxent River estuary indicate that, overall, net regeneration is probably occurring in this floc layer. We present nutrient profiles for C, N and P (expressed as % weight) in suspended and deposited particulate material in a water and sediment column sampled during the spring at a lower Patuxent estuarine station (Fig. 7). Both carbon and nitrogen exhibited continual decrease in % composition from 5 m above the sediment surface to 35 cm below it; however, there was a sharp discontinuity in % C and N across the sediment/water interface. While we made no attempt to sample the floc layer separately (our measurements at the sediment surface are for a 0–2 cm homogenized slice), these data suggest that rapid decomposition and remineralization was occurring somewhere between the bottom suspended particles and the upper sediment layer. Phosphorus, on the other hand, exhibited a local maximum % concentration at the sediment surface, indicating perhaps sorption or precipitation had occurred in these oxidized surficial sediments.

Thus, we hypothesize that much of the metabolic activity which yields NH_4^+ regeneration across the sediment-water boundary is, in fact, concentrated in a narrow band at this estuarine interface. While the gradient-supported estimates of NH_4^+ flux may represent more the continued decomposition of organic matter accumulated over decades, direct measurements obtained *in situ* with benthic chambers probably include the additional NH_4^+ recycling of recently deposited organics which rest in the upper, floc layer. All of this, obviously, assumes that these *in situ* chamber measurements represent minimal effect of disturbance during the measurement itself. The perceived importance of these processes occurring during early diagenesis is evi-

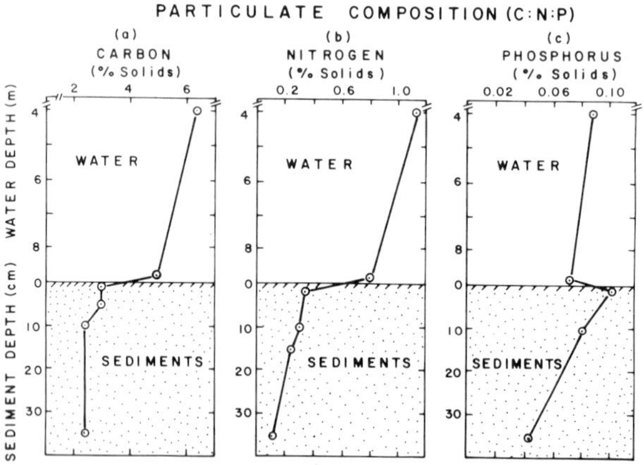

Figure 7. Vertical profiles of particulate matter composition (% weight), as (a) carbon, (b) nitrogen, and (c) phosphorus, from 500 cm above sediment surface to 35 cm below. Data are for station in lower Patuxent River estuary in early spring (from Boynton et al. 1981).

dent in the vast literature which has grown during the last decade (e.g., Berner 1980). Many interesting questions remain concerning the mechanisms of nutrient transformation in estuarine sediments. We need improved techniques for sampling the loosely consolidated surficial sediments, and we need more experimental studies measuring sediment nutrient processes and factors regulating them.

5. Denitrification and the Redox Discontinuity Interface

A process which has been intensively investigated in the last several years is denitrification in estuarine sediments. Limnologists have long been interested in this process (Brezonik 1977); however, the renewed interest in estuarine denitrification is, in part, due to Nixon and his colleagues (Seitzinger et al. 1980; Nixon 1981a) who have postulated that this process plays a crucial role in controlling inorganic N:P ratios in estuarine waters. This dissimilatory NO_3^- reduction requires an organic substrate, anaerobic conditions, and a NO_3^- source. In the general model presented by Vanderborght and Billen (1975) for coastal sediments and by Patrick and Reddy (1976) for flooded soils, NO_3^- is produced (via nitrification) in the upper few cm of sediment or soil and diffuses across the RDL into the anoxic zone where denitrification occurs. This model predicts that the RDL interface is a zone of intense microbial activity – both nitrification and denitrification – with NH_4^+ diffusing upward to support the former process and NO_3^- molecules migrating downward to sites of denitrifying bacteria.

We might expect that estuarine sediment receiving heavy organic loading and rapidly regenerating NH_4^+ would support greater nitrification (if a sufficiently oxidized zone were to exist), which, in turn, would provide NO_3^- for enhanced denitrification. In Fig. 8 we have compiled synoptic measurements of NH_4^+ regeneration and denitrification in nine aquatic sediments (5 estuarine, 2 marsh, and 2 lake sites), and we find that, if we exclude one point, a significant correlation occurs between the two nitrogen transformation processes ($r^2 = 0.85$). In fact, Nixon (pers. comm.) recently has indicated that nutrient enrichment may lead to increased rates of both regeneration and denitrification in the MERL microcosms at University of Rhode Island. However, Jenkins (1981) has found that on a seasonal basis in the Patuxent River, the lack of suitably oxidized sediments can limit the coupled nitrification-denitrification reaction. In general, this regression predicts that denitrification is a substantial component in nitrogen budgets of many aquatic systems, occurring at levels of about 50-60% NH_4^+ recycling rates in aquatic sediments. This again is consistent with the MERL experimental results (Nixon, pers. comm.). However, further experimental evidence is needed to demonstrate the temporal and spatial limits of this relationship which may be a function of sediment porosity, O_2 diffusion, and O_2 demand as suggested by the work of Jenkins (1981).

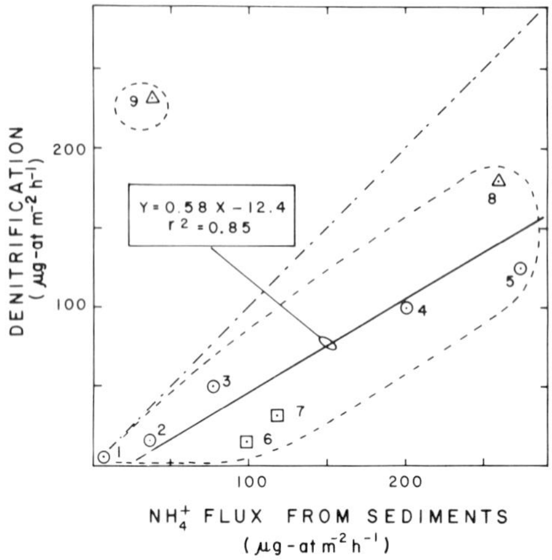

Figure 8. *Comparison of denitrification rates versus NH_4^+ fluxes across the sediment/water interface for nine aquatic systems. Circles indicate estuarine systems, squares are for marshes, and triangles are for lake sediments. Regression line does not include point number 9. Sources of data: 1) North Sea (Vanderborght et al. 1977); 2), 3) North Sea (Billen 1978); 4) Narragansett Bay (Seitzinger et al. 1980); 5) Patuxent Estuary (Boynton et al. 1980); 6), 7) Colne Pt. Marsh (Aziz and Nedwell 1979); 8), 9) Norrviken and Ramsjon Lakes (Tirén et al. 1976).*

6. Denitrification and NO_3^- Flux Across Multiple Interfaces

While the model of Vanderborght and Billen (1975) and Patrick and Reddy (1976) has the intuitive appeal of concentrating the two coupled processes near the RDL, Vanderborght and Billen have also shown that redox conditions in muddy sediments do not always allow for nitrification. However, NO_3^- can be supplied to denitrifiers via diffusion from the overlying water. Vanderborght and Billen (1975) were able to reproduce observed NO_3^- profiles using a one-dimensional diagenic model. They solved for denitrification rates through a trial-and-error manipulation until good agreement between model and data was achieved, and thus inferred rates which were not inconsistent with limited reports of direct flux measurements. The question remains, however, as to what level of denitrification could be supported by NO_3^- flux from overlying waters under a range of redox conditions.

Boynton et al. (1980) previously reported direct *in situ* observations of significant rates of NO_3^- flux into Patuxent estuarine sediments, which were directly proportional to NO_3^- concentrations in the overlying water ($r^2 = 0.75$ and 0.81, Boynton et al. 1980). We hypothesized that this flux supported denitrification, averaging about 70 μg-at $m^{-2}h^{-1}$ at about 10–20 μg-at NO_3 ℓ^{-1} over the entire year. We have thus far not been able to test this hypothesis directly, although additional supportive data have been developed (Jenkins

1981). In Fig. 9 we present (a) some historical data on seasonal NO_3^- concentrations between 1963 and 1980 and (b) recent measurements of NO_3^- flux across the sediment-water interface in the upper Patuxent estuary, which are virtually identical to patterns reported previously (Boynton et al. 1980). Clearly, there appears to have been a continued increase in NO_3^- levels over the last two decades in this Chesapeake tributary, presumably due to anthropogenic sources (Kemp and Boynton 1979), and large quantities of this NO_3^- are lost from the water via diffusion into the sediments. Similar trends of increasing river borne loads of NO_3^- have been observed throughout the world (Walsh et al. 1981), although there are few direct measurements of NO_3^- uptake by these estuarine sediments.

Using a simple diffusion model with an experimentally determined diffusion coefficient of about 10^{-5} cm^2 s^{-1} (Jenkins 1981) and assuming that the RDL lies at 1 cm below the sediment surface, we calculate that a NO_3^- concentration in the overlying water of 20 μg-at ℓ^{-1} could support a denitrification rate of about 7 μg-at $m^{-2}h^{-1}$, or about an order-of-magnitude lower than that observed (Fig. 9). This value is similar to that estimated by Vanderborght and Billen (1975) by diagenic modeling, and Boynton et al. (1980) have argued

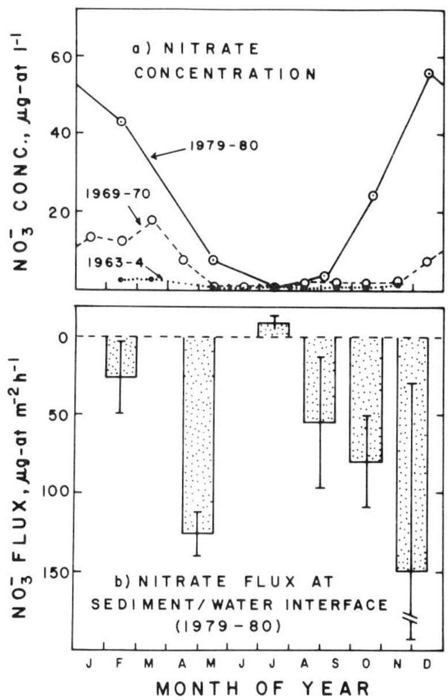

Figure 9. Seasonal patterns of (a) nitrate for three years from 1963 to 1980 and (b) nitrate flux across the sediment-water interface for the 1979-80 year (mean ± standard deviation). Bars pointed downward indicate flux from water to sediments. Data adapted from Mihursky and Boynton (1979) and Boynton et al. (1981).

previously that the NO_3^- uptake is not likely being reduced to NH_4^+. Two possible explanations seem reasonable. First, NO_3^- diffusion may be enhanced by macroinvertebrate irrigation and bioturbation (Callender and Hammond 1982), and secondly, denitrification actually may be occurring above the generalized RDL at anoxic microsites on sediment particles (Jannasch 1960; Brezonik 1977). In fact, Koike et al. (1972) found that denitrification could proceed in lake water samples with up to 1.4 mg ℓ^{-1} O_2. Because of the potential role that denitrification may play as a sink for increased inputs of NO_3^-, this issue deserves further experimental investigation.

7. Effects of Macrophytic Roots on the RDL Interface

Various investigators have shown that aquatic macrophytes, both emergent (Howes et al. 1981) and submerged (Wium-Andersen and Andersen 1972), are capable of oxidizing their sediment environment. The mechanism whereby this oxidation occurs appears to be via passive transport of O_2 through the plant's lacunal system into interstitial waters (Oremland and Taylor 1977; Iizumi et al. 1980). We hypothesize that this translocation of O_2 coupled with the dendritic structure of macrophytic rhizospheres may tend to increase the surface area of the RDL. If O_2 and space are potential limiting factors for nitrification in estuarine sediments, and if denitrification is dependent on availability of NO_3^-, we would expect both processes to be enhanced in the presence of macrophytes.

Marbury et al. (1980) have made some preliminary measurements of these nitrogen transformation reactions in estuarine communities of the submerged macrophyte, *Potamogeton perfoliatus*, and in comparable nonvegetated reference areas. Summertime nitrification rates (measured in triplicate, intact cores using the N-Serve inhibition technique of Henricksen 1980) were about 20 times greater in macrophyte beds (ca 30 μg-at $m^{-2}h^{-1}$) than in unvegetated sediments. Denitrification rates in these experiments were much lower and unrelated to vegetation, while other investigators have reported conflicting relationships between macrophytes and dissimilatory NO_3^- reduction. For example, Iizumi et al. (1980) found rates which were generally greatest in the most abundantly vegetated stations; however, Chan and Knowles (1979) reported higher rates at unvegetated sites in a Canadian lake. Terlizzi et al. (1981) have shown evidence for diel cycles of NO_3^- uptake by macrophytes, which they suggest have competitive advantage over nitrifying bacteria. In general, the impact of macrophyte communities on estuarine nitrogen cycling is poorly understood, although there is sufficient indication that this is an important issue rich in research possibilities. More data are needed quantifying O_2 translocation and oxidation of sediments by estuarine macrophytes, and the relative competition between plants and microbes for nitrogen compounds deserves further attention.

Concluding Remarks

The estuarine environment is characterized by numerous zones where marked gradients exist in the distribution of various substances. We have argued here that these estuarine interfaces are loci of intense metabolic transformation and flux of nitrogen and that the overall N cycling of estuaries is dramatically affected by these zones. What are the unique properties of such interfaces that render them so important in estuarine N dynamics?

Estuarine interfaces are often relatively stable zones with distinct physical structure which may allow for development of rich biotic communities. Within many of these interfaces specific attributes (e.g., chemical compounds) are combined from both systems on either side of a discontinuity region. In addition, the steep concentration (or elevation) gradients that occur across these boundaries provide a mechanism for diffusive (or advective) transport and rapid exchange of materials. Sharp changes in phase (gas, liquid, solid), as well as relatively strong density gradients encountered along the vertical axis provide adequate physical stability needed to concentrate detrital material and living organisms in a narrow band. Hence, for example, much of the life in an estuary occurs in the upper 5 cm of sediment.

The term "ecotone" has been traditionally used among ecologists (particularly for terrestrial environments) to describe transitional zones between two or more ecological communities. Associated with these ecotones is the concept of "edge effect" which means animal and plant populations in these transitional regions are of greater diversity and abundance than in either of the flanking ecosystems (Odum 1971). The mechanism suggested for this increased ecological richness is simply the combination of populations from two adjoining communities and the emergence of new habitats (Odum 1971). However, the effect does not always lead to increased abundance especially in marine-estuarine systems, where the mixing of widely different physical-chemical aqueous environments with the tidal pulse can produce physiologically stressed conditions and reduced biotic populations (Lilly et al. 1953; Odum et al. 1963). Many of the interface systems discussed in this paper seem analogous to ecotone communities. Yet, particularly in the vertical direction, it is more the mixture of chemical (rather than biotic) attributes from above and below which seems to generate metabolically active populations (especially microbes) in these interfaces. For instance, the pycnocline may be a region of intense nitrification because O_2 diffuses down from the epilimnion and NH_4^+ diffuses up from the hypolimnion, while the density discontinuity allows sufficient physical stability to support microbial populations against sinking.

We have presented seven examples of N cycling at estuarine interfaces. Benthic remineralization of NH_4^+ and flux across the RDL or benthic boundary layer are involved in each of these cases, suggesting that it is a key processes in estuarine N dynamics. Undoubtedly, there are numerous examples of interfacial processes which we have not mentioned. The exchange of materials across marsh-estuarine interfaces is certainly one such topic which has been much studied (Haines 1979; Nixon 1980; Odum 1980), although not addressed here. The neuston layer at the air-water interface is another example of interest, since its active chemical and metabolic nature has been demonstrated for various estuarine environments (Gallagher 1975; Pellenbarg and Church 1979), where it may have substantial impact on N cycling.

In this paper, we have alluded to several scientific questions which, we feel, deserve further consideration. We list a few of these below.

- How quickly can benthic processes respond to the development of ephemeral or sporadic fronts?
- How do nitrification, denitrification, and other processes of early diagenesis vary along estuarine gradients through the transition zone?
- What are the biogeochemical properties of the sediment flocculent layer? How does it vary in time and space?
- What is the source of NO_2^- at the estuarine pycnocline? If it is due to nitrification or denitrification, why is the redox reaction incomplete? Why is its occurrence limited to late summer?
- What is the relationship between new and recycled N? Why does the ratio of the two processes vary so?
- What are the factors which control denitrification as it varies seasonally and spatially (both within and between estuaries).
- To what extent can NO_3^- flux into sediments support denitrification? Can denitrification occur to a significant extent at the sediment surface with oxic overlying water?
- How important are macrophytes (and in particular O_2 translocation) in the chemistry of estuarine sediments?

These and other relevant questions are central to an understanding of N cycling in estuaries, and the answers await careful and imaginative observation and experimentation.

Acknowledgments

Many of the ideas presented in this paper were spawned through conversations with our numerous colleagues. We are particularly indebted to K. Webb, C. Hopkinson, T. Callender, S. Nixon, J. Taft, J. Smullen, O. Bricker, J. Means, T. Fisher, R. Twilley and M. Sagoff. Numerous other individuals have contributed to the collection of data upon which we have drawn in this essay, including D. Marbury, C. Keefe, C. Osborne, K. Wood, M. Jenkins, M. Shenton, E. Spalding, S. Demotor, T. Farrell and L. Lane. Without their efforts we would have very little empiricism to ballast our conceptual balloons. As always, the denouement of this endeavor rests on the considerable inputs of A. Hermann and B. Caplins for computation and statistical analyses (which were supported by the University of Maryland Computer Science Center), J. Gilliard for typing and D. Kennedy for drafting. This work was supported in part by grants from the US EPA Nos. R805932010 and R805974, Maryland Sea Grant Program, and the Maryland Department of Natural Resources, Power Plant Siting Program P2-72-02 (78-81). Contribution No. 1313 HPEL from the Center for Environmental and Estuarine Studies; Virginia Institute of Marine Sciences Contribution No. 1047. Permission to use data from Haines (Fig. 3) was given by Plenum Press.

References Cited

Aller, R. C. and J. Y. Yingst. 1980. Relationships between microbial distributions and the anaerobic decomposition of organic matter in surface sediments of Long Island Sound, USA. *Mar. Biol.* 56:29-42.

Aziz, S. A. Abd. and D. B. Nedwell. 1979. Microbial nitrogen transformations in the salt marsh environment, pp. 385-398. *In:* R. L. Jefferies and A. J. Davy (eds.), *Ecological Processes in Coastal Environments.* Blackwell Sci. Publ., Oxford.

Berner, R. A. 1976. The benthic boundary layer from the viewpoint of a geochemist, pp. 33-55. *In:* I. N. McCave (ed.), *The Benthic Boundary Layer.* Plenum Press, New York.

Berner, R. A. 1980. *Early Diagenesis.* Princeton U. Press, Princeton. 241 pp.

Billen, G. 1978. A budget of nitrogen recycling in North Sea sediments off the Belgian coast. *Est. Coastal Marine Sci.* 7:127-146.

Bowman, M. J. and R. L. Iverson. 1978. Estuarine and plume fronts, pp. 87-104. *In:* M. J. Bowman and W. E. Esaias (eds.), *Oceanic Fronts in Coastal Processes.* Spinger-Verlag, Berlin.

Boynton, W. R., W. M. Kemp and C. W. Keefe. 1982. A comparative analysis of nutrients and other factors influencing estuarine phytoplankton production, pp. 69-90. *In:* V. S. Kennedy (ed.), *Estuarine Comparisons.* Academic Press, New York.

Boynton, W. R., W. M. Kemp and C. G. Osborne. 1980. Nutrient fluxes across the sediment-water interface in the turbid zone of a coastal plain estuary, pp. 93-109. *In:* V. S. Kennedy (ed.), *Estuarine Perspectives.* Academic Press, New York.

Boynton, W. R., W. M. Kemp, C. G. Osborne, E. Spalding and C. W. Keefe. 1981. Estuarine community dynamics (Benthic program) Patuxent River estuary preliminary data report. Univ. Maryland Cntr. Environ. Est. Studies Ref. No. 81-52 CBL, Solomons, MD 112 pp.

Brezonik, P. L. 1977. Denitrification in natural waters. *Progr. Water Techn.* 8:373-392.

Callender, E. 1979. Pore-water geochemistry of Potomac riverine and estuarine sediments, pp. 1-15. *In:* Annual Summary of Geochemical Studies in Potomac, U.S. Geological Survey, Reston, VA 256 pp.

Callender, E. 1982. Benthic phosphorus regeneration in the Potomac River estuary. *In:* P. G. Sly (ed.), *Sediment/Freshwater Interactions.* Junk, The Hague (In press).

Callender, E. and D. E. Hammond. 1982. Nutrient exchange across the sediment-water interface in the Potomac River estuary. *Est. Coastal Shelf Sci.* (In press).

Carlucci, A. F. and H. R. Schubert. 1969. Nitrate reduction in seawater of the deep nitrite maximum off Peru. *Limnol. Oceanogr.* 14:197-193.

Carter, H. H., R. J. Regier, E. W. Schiemer and J. A. Michael. 1978. The summertime vertical distribution of dissolved oxygen at the Calvert Cliffs generating station: A physical interpretation. Special Rep. 60 (Ref. 78-1). Chesapeake Bay Instit., Johns Hopkins U., Baltimore, MD 62 pp.

Chan, Y-K. and R. Knowles. 1979. Measurements of denitrification in two freshwater sediments by an *in situ* acetylene inhibition method. *Appl. Environ. Microb.* 37:1067-1072.

D'Elia, C. F., K. L. Webb and R. L. Wetzel. 1981. The effect of the spring-neap tidal cycle on nutrient and oxygen distributions in the lower York River, USA, during August 1978, pp. 597-606. *In:* B. Neilson and L. E. Cronin (eds.), *Nutrient Enrichment in Estuaries.* Humana Press, Clifton, NJ.

Dugdale, R. C. 1976. Nutrient cycles, pp. 141-172. *In:* D. Cushing and J. J. Walsh (eds.), *The Ecology of the Seas.* W. B. Saunders, Philadelphia.

Dugdale, R. C. and J. J. Goering. 1967. Uptake of new and regenerated forms of nitrogen in primary productivity. *Limnol. Oceanogr.* 12:196-206.

Elderfield, H., N. Luedtke, R. J. McCaffrey and M. Bender. 1981. Benthic flux studies in Narragansett Bay. *Am. J. Sci.* 281:768-787.

Eppley, R. W., E. H. Renger and W. G. Harrison. 1979. Nitrate and phytoplankton production in southern California coastal waters. *Limnol. Oceanogr.* 24:483-494.

Gallagher, J. L. 1975. The significance of the surface film in salt marsh plankton metabolism. *Limnol. Oceanogr.* 20:120-123.

Haas, L. W. 1977. The effect of the spring-neap tidal cycle on the vertical salinity structure of the James, York and Rappahannock Rivers, Virginia, USA. *Est. Coastal Mar. Sci.* 5:485-496.

Haines, E. B. 1979. Interactions between Georgia salt marshes and coastal waters: A changing paradigm, pp. 33-46. *In:* R. J. Livingston (ed.), *Ecological Processes in Coastal and Marine Systems.* Plenum Press, New York.

Hargrave, B. T. 1973. Coupling carbon flow through some pelagic and benthic communities. *J. Fish. Res. Bd. Canada* 30:1317-1326.

Hargrave, B. T. 1980. Factors affecting the flux of organic matter to sediments in a marine bay, pp. 243-263. *In:* K. R. Tenore and B. C. Coull (eds.) *Marine Benthic Dynamics.* U. South Carolina Press, Columbia.

Harrison, W. G. 1980. Nutrient regeneration and primary production in the sea, pp. 433-460. *In:* P. G. Falkowski (ed.), *Primary Productivity in the Sea.* Plenum Press, New York.

Henriksen, K. 1980. Measurement of *in situ* rates of nitrification in sediment. *Microb. Ecol.* 6:329-337.

Hopkinson, C. S. and R. L. Wetzel. 1982. Nutrient exchange and oxygen consumption by the nearshore benthic community in the Georgia Bight. VA Inst. Mar. Sci. Contrib. No. 82-37. Gloucester Pt., VA.

Howes, B. L., R. W. Howarth, J. M. Teal and I. Valiela. 1981. Oxidation-reduction potentials in a salt marsh: Spatial pattern and interactions with primary production. *Limnol. Oceanogr.* 26:350-360.

Iizumi, H., A. Hattori and C. P. McRoy. 1980. Nitrate and nitrite in interstitial waters of eelgrass beds in relation to the rhizosphere. *J. exp. mar. Biol. Ecol.* 47:191-201.

Indebø, G., B. Pengerud and I. Dundas. 1979. Microbial activities in a permanently stratified estuary. II. Microbial activities at the oxic-anoxic interface. *Mar. Biol.* 51:305-309.

Iverson, R. L. and nine others. 1979. Ecological significance of fronts in the southeastern Bering Sea, pp. 437-465. *In:* R. J. Livingston (ed.), *Ecological Processes in Coastal and Marine Systems.* Plenum Press, New York.

Jannasch, H. W. 1960. Versuche über denitrifikation und die verfügbarkeit des sauerstoffes in wasser und schlamm. *Arch. Hydrobiol.* 56:355-369.

Jenkins, M. C. 1981. The measurement, occurrence and importance of nitrification and denitrification in Chesapeake Bay. MS Thesis, U. Maryland, College Park, MD 77 pp.

Kemp, W. M. and W. R. Boynton. 1979. Nutrient budgets of the Patuxent estuary: Sources, sinks and internal cycles, pp. 1-42. *In:* W. M. Kemp, J. C. Stevenson and W. R. Boynton (eds.), Submerged Aquatic Vegetation in Chesapeake Bay. Univ. Maryland Centr. Environ. Est. Studies Ref. No. 80-68 HPEL, Cambridge, MD.

Klump, J. V. and C. S. Martens. 1981. Biogeochemical cycling in an organic rich coastal marine basin: II. Nutrient sediment-water exchange processes. *Geochim. Cosmochim.* Acta 45:101-121.

Koike, I., E. Wada, T. Tsuji and A. Hattori. 1972. Studies on denitrification in a brackish lake. *Arch. Hydrobiol.* 69:508-520.

Lilly, S. J., J. F. Sloane, R. Bassindale and J. S. Kitching. 1953. Ecology of Lough Ine rapids with special reference to water current. IV. The sedentary fauna of sublittoral boulders. *J. Animal Ecol.* 22:87-122.

Marbury, D., L. Lane, E. Donaghue, R. Doane, W. M. Kemp and J. C. Stevenson. 1980. Preliminary observations on nitrification, denitrification, and nitrogen-fixation in sediments associated with submerged aquatic macrophytes, pp. 825-841. *In:* W. M. Kemp, W. R. Boynton, J. C. Stevenson and J. C. Means (eds.), Submerged Aquatic Vegetation in Chesapeake Bay.

Univ. Maryland Centr. Environ. Est. Studies Ref. No. 80-68, Cambridge, MD.

Martens, C. S. and M. B. Goldhaber. 1978. Early diagenesis in transitional sedimentary environments of the White Oak River estuary, North Carolina. *Limnol. Oceanogr.* 23:428-441.

McCaffrey, R. J. and eight others. 1980. The relation between pore water chemistry and benthic fluxes of nutrients and manganese in Narragansett Bay, R.I. *Limnol. Oceanogr.* 25:31-44.

McCarthy, J. J., W. R. Taylor and J. L. Taft. 1977. Nitrogenous nutrition of the plankton in the Chesapeake Bay. I. Nutrient availability and phytoplankton preferences. *Limnol. Oceanogr.* 22:996-1011.

Mihursky, J. A. and W. R. Boynton. 1979. Review of the Patuxent Estuary data base. Chesapeake Biological Laboratory, Solomons, MD. Univ. Md. Cntr. Environ. Est. Studies Ref. No. 78-157 CBL. 327 pp.

Nixon, S. W. 1980. Between coastal marshes and coastal waters, pp. 437-525. *In:* P. Hamilton and K. B. MacDonald (eds.), *Estuarine and Wetland Processes.* Plenum Press, New York.

Nixon, S. W. 1981a. Remineralization and nutrient cycling in coastal marine ecosystems, pp. 111-138. *In:* B. Neilson and L. E. Cronin (eds.), *Nutrient Enrichment in Estuaries.* Humana Press, Clifton, NJ.

Nixon, S. W. 1981b. Fresh water inputs and estuarine productivity, pp. 31-57. *In:* Cross, R. and D. Williams (eds.), *Proc. National Sympos. Freshwater Inflow to Estuaries.* U.S. Fish and Wildlife Serv., Office Biol. Services. FWS/OBS-81/04.

Odum, E. P. 1971. *Fundamentals of Ecology.* W. B. Saunders, Philadelphia. 547 pp.

Odum, E. P. 1980. The status of three ecosystem-level hypotheses regarding salt-marsh estuaries: tidal subsidy, outwelling and detritus-based food chains, pp. 485-495. *In:* V. S. Kennedy (ed.), *Estuarine Perspectives.* Academic Press, New York.

Odum, H. T., R. Cuzon, R. J. Beyers and C. Allbaugh. 1963. Diurnal metabolism, total phosphorus, Ohle anomaly, and zooplankton diversity of abnormal marine ecosystems of Texas. *Publ. Inst. Mar. Sci.* 9:404-453.

Olson, R. J. 1981a. ^{15}N tracer studies of the primary nitrite maximum. *J. Mar. Res.* 39:203-226.

Olson, R. J. 1981b. Differential photo-inhibition of marine nitrifying bacteria: A possible mechanism for the formation of the primary nitrite maximum. *J. Mar. Res.* 39:227-238.

Oremland, R. S. and B. F. Taylor. 1977. Diurnal fluctuations of O_2, N_2, and CH_4 in the rhizosphere of *Thalassia testudinum. Limnol. Oceanogr.* 22:566-570.

Patrick, W. H. and K. R. Reddy. 1976. Nitrification-denitrification reactions in flooded soils and water bottoms: Dependence on oxygen supply and ammonium diffusion. *J. Environ. Qual.* 5:469-472.

Pellenbarg, R. E. and T. M. Church. 1979. The estuarine surface microlayer and trace metal cycling in a salt marsh. *Science* 203:1010-1012.

Pomeroy, L. R. 1970. The strategy of mineral cycling. *Ann. Rev. Ecol. Syst.* 2:171-190.

Ryther, J. H. and W. M. Dunstan. 1971. Nitrogen, phosphorus and eutrophication in the coastal marine environment. *Science* 171:1008-1013.

Schubel, J. R. 1968. Turbidity maximum of the Northern Chesapeake Bay. *Science* 161:1013-1015.

Seitz, R. C. 1971. Temperature and salinity distributions in vertical sections along the longitudinal axis and across the entrance of the Chesapeake Bay (April 1968 to March 1969). Graphical Summary Rep. No. 5 (Ref. 71-7), Ches. Bay Instit., Johns Hopkins U., Baltimore, MD 105 pp.

Seitzinger, S., S. Nixon, M. E. Q. Pilson and S. Burke. 1980. Denitrification and N_2O production in near-shore marine sediments. *Geochim. Cosmochim. Acta* 44:1853-1860.

Seliger, H. H., K. R. McKinley, W. H. Biggley, R. B. Rivkin and K. R. H. Aspden. 1981. Phytoplankton patchiness and frontal regions. *Mar. Biol.* 61:119-131.

Taft, J. L. 1976. Phosphorus dynamics in some coastal plain estuaries, pp. 79-89. *In:* M. Wiley (ed.), *Estuarine Processes, Vol. 1*. Academic Press, New York.

Terlizzi, D. E., M. R. Lewis, E. P. Karlander, J. C. Stevenson and W. M. Kemp. 1981. Nitrite accumulation in *Potamogeton perfoliatus:* Evidence for dissimilatory nitrate reduction. Amer. Soc. Limnol. Oceanogr., June 1981, Milwaukee, WI. (ABSTRACT).

Thayer, G. W. 1974. Identity and regulation of nutrients limiting phytoplankton production in the shallow estuaries near Beaufort, North Carolina. *Oecologia* 14:75-92.

Thomas, W. H. 1966. On denitrification in the northeast tropical Pacific Ocean. *Deep-Sea Res.* 13:1109-1114.

Tirén, T., J. Thorin and H. Nômmik. 1976. Denitrification measurements in lakes. *Acta Agric. Scandinavica* 26:175-184.

Vanderborght, J. P. and G. Billen. 1975. Vertical distribution of nitrate concentration in interstitial water of marine sediments with nitrification and denitrification. *Limnol. Oceanogr.* 20:953-961.

Vanderborght, J. P., R. Wollast and G. Billen. 1977. Kinetic models of diagenesis in disturbed sediments: 2. Nitrogen diagenesis. *Limnol. Oceanogr.* 22:794-803.

Wada, E. and A. Hattori. 1972. Nitrite distribution and nitrate reduction in deep sea waters. *Deep-Sea Res.* 19:123-132.

Walsh, J. J., G. T. Rowe, R. L. Iverson and C. P. McRoy. 1981. Biological export of shelf carbon is a sink of the global CO_2 cycle. *Nature* 291:196-201.

Webb, K. L. and C. F. D'Elia. 1980. Nutrient and oxygen redistribution during a spring/neap tidal cycle in a temperate estuary. *Science* 207:983-985.

Webb, K. L. 1981. Conceptual models and processes of nutrient cycling in estuaries, pp. 25-46. *In:* B. J. Neilson and L. E. Cronin (eds.), *Estuaries and Nutrients*. Humana Press, Clifton, NJ.

Wium-Andersen, S. and J. M. Andersen. 1972. The influence of vegetation on the redox profile of the sediment of Grane Langsø, a Danish *Lobelia* Lake. *Limnol. Oceanogr.* 17:948-952.

Zeitzschel, B. 1979. Sediment-water interactions in nutrient dynamics, pp. 195-218. *In:* K. R. Tenore and B. C. Coull (eds.), *Marine Benthic Dynamics*. U. South Carolina Press, Columbia.

SOIL DYNAMICS AND THE PRODUCTIVITY OF *SPARTINA ALTERNIFLORA*

Alice G. Chalmers
University of Georgia Marine Institute
Sapelo Island, Georgia

Abstract: The striking heterogeneity in standing stock and productivity of *Spartina alterniflora* has been explained in a number of ways. Elevation, interstitial salinity, tidal inundation, nitrogen availability and oxygenation of the root zone have each been suggested as the key variable. In fact, there is a great deal of interaction between them. The relationship between elevation and tidal inundation is obvious; both can influence soil salinity. Both salinity and rhizosphere oxygenation can affect nitrogen uptake. There can be a positive feedback between productivity and rhizosphere oxygenation and, conversely, a negative feedback between productivity and soil salinity. Increasing subsurface drainage in an intermediate-height *S. alterniflora* zone can result in increased aboveground growth and can also produce soil conditions similar to those found in the low marsh. No single soil characteristic appears to control *S. alterniflora* productivity; rather, a complex of factors are all affected by subsurface drainage.

Introduction

Spartina alterniflora Loisel. dominates the intertidal zone of Atlantic and Gulf coast salt marshes from Newfoundland to Florida to Texas, occurring in vast, seemingly uniform, monotypic stands. Closer observation reveals a striking heterogeneity in grass height, standing crop and productivity. On the banks of tidal creeks the grass is 1-3 m in height and has an annual aerial productivity of up to 3700 g dry wt/m^2 (Gallagher et al. 1980). Tall *S. alterniflora* plants are robust with relatively wide, dark green leaves and occur in stands of 30-50 stems/m^2. In landward portions of the marsh the grass is shorter (0.5 m or less) and stem density is greater (up to 300 stems/m^2) (Pomeroy and Wiegert 1981). Aerial productivity of short *S. alterniflora* is typically 2-3 times less than that of tall grass in the same marsh and the plants are often chlorotic (Turner 1976; Gallagher et al. 1980). In southern marshes an intermediate height grass (0.5-1 m) is found behind levees and in other low marsh areas (Gallagher et al. 1972).

The causes of this within-marsh heterogeneity in growth and productivity are the subject of a large and growing body of scientific literature. What follows is a review of some of that literature and a discussion of what we know about the causes of the heterogeneity and possible directions for future research.

ISBN 0-12-404070-5

Ecotype or Ecophene?

Some salt marsh investigators have held the view that the tall and short forms are inherently different ecotypes (Chapman 1960; Stalter and Batson 1969), a view supported by some of the taxonomic literature (Church 1940; Fernald 1950). But more recent electrophoretic and reciprocal transplant studies have shown that the heterogeneity is phenotypic rather than genetic (Shea et al. 1975; Valiela et al. 1978). Studies of seed germination and seedling growth suggest that environmental factors are responsible for the variation in growth form within marshes (Mooring et al. 1971), although there is some evidence of latitudinal genetic differences in *S. alterniflora* (Anderson and Treshow 1980).

Field and laboratory studies point to a number of environmental factors which can affect *S. alterniflora* growth, ranging from salinity to nitrogen availability to Eh. Understanding what accounts for the difference in short and tall *S. alterniflora* requires an appreciation of how soil dynamics affect growth. No one variable or mechanism has been identified as the key factor, primarily because of the complex chemical and physical interactions in the soil and physiological responses of the plant.

Salinity

Salinity is cited in a number of studies as being responsible for the differential growth of short and tall *S. alterniflora*. In the laboratory, *S. alterniflora* grows best at low salinities and biomass production decreases with increasing salinity of the growth medium (Mooring et al. 1971; Haines and Dunn 1976; Parrondo et al. 1978; Linthurst and Seneca 1981). In the field, interstitial salinities differ between stands of tall and short *S. alterniflora* (Stalter and Batson 1969; Chalmers 1977; Nestler 1977a).

Nestler (1977a) found a gradient of increasing salinity along transects from the banks of tidal creeks to land that varied inversely with *S. alterniflora* height and biomass in Sapelo Island marshes (Fig. 1). Interstitial salinity in the creekbanks was identical with that in tidal water ($19^{0}/_{00}$), but interstitial salinities in the high marsh were as high as $45^{0}/_{00}$. Elevated salinities in the high marsh were attributed to evaporation when the marsh surface was exposed and to the lack of subsurface water movement (Nestler 1977b). There is little subsurface water flow in the high marsh other than by diffusion, but it increases with proximity to tidal creeks, accounting for the low salinities found in the creek bank soils (Reideburg 1978). The difference in soil water movement was due to the difference in depth of flooding (hydraulic head) and duration of inundation.

In addition to tidal inundation, subsurface water movement, and evaporation, plant growth can itself affect interstitial salinity. In laboratory, Smart and Barko (1978) found an increase in interstitial salinity in containers in which *S. alterniflora* was growing; this they attributed to ion exclusion by plant roots, a common mechanism among halophytes for coping with salt in

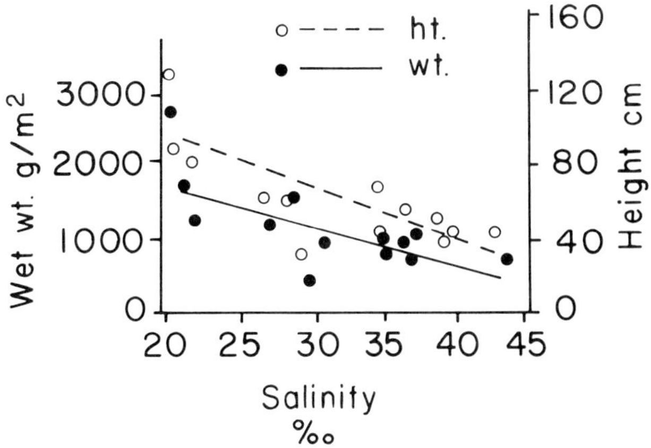

Figure 1. Relationship of shoot height (cm) and biomass (g/m²) of S. alterniflora *to interstitial salinity (⁰/₀₀). Adapted from Nestler 1977a.*

the growth medium (Waisel 1972). A similar increase in soil salinity was observed in two different fertilized plots of short *S. alterniflora* (Christian et al. 1978; Chalmers 1979). In both cases the increases in salinity were attributed to higher rates of transpiration in the fertilized plots resulting from greater plant productivity. This explanation implies that ion exclusion was also taking place.

In spite of the evidence for control of *S. alterniflora* height by salinity, there are marshes in which both tall and short forms occur where salinity stress is not a factor. In Great Sippewissett marsh in Massachusetts there is no consistent pattern of salinity difference in tall and short grass areas and no salinity greater than 33⁰/₀₀ has been measured (Howes et al. 1981). In Louisiana, tall and short forms occur on streamsides and in inland areas, but there are no corresponding differences in interstitial salinity (Buresh et al. 1980).

Nitrogen

It is clear that salinity stress can inhibit *S. alterniflora* growth, but most of the studies cited above have recognized that factors other than salinity are involved in differential growth of tall and short grasses. Many of those studies have investigated the interaction between salinity and other factors, particularly the effect of nitrogen (Haines and Dunn 1976; Linthurst and Seneca 1981).

A number of studies have shown that fertilization with nitrogen increases growth of *S. alterniflora*, but that additions of phosphorus have no stimulating effect (Sullivan and Daiber 1974; Valiela and Teal 1974; Broome et al. 1975; Patrick and DeLaune 1976) (Table 1). Gallagher (1975) found that additions of ammonium nitrate stimulated growth of short *S. alterniflora*

Table 1. Effects of nitrogen fertilization on growth of *Spartina alterniflora* in salt marshes from Massachusetts to Louisiana.

Site	*Fertilizer, Rate of Application*	*Control* g/m^2	*Fertilizer* g/m^2	*Reference*
Massachusetts	urea, 290 g N m^{-2} yr^{-1}	424	834	Valiela et al. 1976
Delaware	ammonium nitrate, 240 g N m^{-2} yr^{-1}	772*	2104*	Sullivan and Daiber 1974
North Carolina	ammonium sulfate, 67.2 g N m^{-2} yr^{-1}	450	1800	Broome et al. 1975
Georgia	ammonium nitrate, 20 g N m^{-2} yr^{-1}	471	803	Gallagher 1975
Georgia	sewage sludge, 100 g N m^{-2} yr^{-1}	396	650	Chalmers et al. 1976
Louisiana	ammonium sulfate, 20 g N m^{-2} yr^{-1}	1666	1916	Patrick and Delaune 1976

* g fresh wt/m^2, all others are g dry wt/m^2

but not of tall, leading to the commonly held view that growth of short *S. alterniflora* but not that of tall is nitrogen limited.

Grass in a short *S. alterniflora* marsh fertilized for four years with urea was almost, but not quite, transformed to the tall form (Valiela et al. 1978). Fertilization increased growth, greenness, height, and stem and leaf width, suggesting nutrient supply (in particular, nitrogen) as an important factor in determining growth. But some secondary factor was also present preventing complete convergence. The implication of these experiments is that either nutrient availability is greater for tall *S. alterniflora* than for short or that the secondary limiting factor is not active for the tall grass, or perhaps some combination of the two. Salinity, iron and water availability were suggested as possible secondary limiting factors.

Chalmers (1979) found that fertilization with dried sewage sludge with a nitrogen content of 2% increased growth of short *S. alterniflora*, interstitial NH_4^+ concentrations and soil total N content. Although the productivity of the grass was increased in the fertilized plots, more inorganic N was available than was utilized. A secondary effect of the fertilization in this experiment was the previously discussed increase in interstitial salinity. Chalmers (1979) believed the stress of increased salinity limited the amount of additional nitrogen that could be utilized.

Soil nitrogen in tall and short *S. alterniflora* areas differs between marshes. In Georgia, soil inorganic nitrogen was the same in both tall and short stands although the annual accumulation of nitrogen in tall *S. alterniflora* was 2.5 times higher than that in the short grass (Chalmers 1977). In this marsh, turnover rates of inorganic N must be greater in the tall zone than in the short *S. alterniflora* zone or there must be some additional source of nitrogen for tall grass. The potential for nitrogen fixation is greater in the tall zone than in the short (Hanson 1977), tidally induced subsurface water flow is greater in the tall zone (Reideburg 1978), and rates of NH_4^+ mineralization are also faster in that zone (E. L. Dunn, Dept. of Biology, Ga. Tech., Atlanta, GA, pers. comm.). Maintenance of low salinities in creek bank soils by tidal flushing helps prevent the salinity stress observed in fertilized high marsh areas.

By measuring tissue concentrations of nitrogen, glutamate dehydrogenase and nitrate reductase, Mendelssohn (1979) found that short *S. alterniflora* in a North Carolina marsh was deficient in nitrogen status relative to the tall form, although interstitial NH_4^+ concentrations were six times higher in the short grass zone, supporting the idea that short *S. alterniflora* was nitrogen limited. Here again, the conclusion was that some factor in addition to nitrogen availability was influencing the production of *S. alterniflora*, but for the first time clear evidence was presented that this secondary factor exerts its influence by preventing uptake and/or assimilation of the available NH_4^+.

The nitrogen uptake kinetics of *S. alterniflora* growing in hydroponic culture are inconsistent with field observations of nitrogen limitation (Morris 1980). Comparisons of Vmax and Km with field measurements of available NH_4^+ suggest no lack of the inorganic nitrogen necessary for *S. alterniflora* to grow well. Morris hypothesized that some edaphic factor is altering nitrogen uptake kinetics and, further, that the observed decrease in productivity with increasing distance from creeks is caused by an increasing alteration of the uptake kinetics by this factor. He suggested that this factor might be salinity, oxygen deficiency in the root zone, hydrogen sulfide, the exchange capacity of the soil, or the diffusion rate in the soil.

Both Haines and Dunn (1976) and Linthurst and Seneca (1981) found that higher nitrogen levels in the growth medium could partially overcome the adverse effects of high salinity on *S. alterniflora* growth. Haines and Dunn (1976) suggested that competitive inhibition of NH_4^+ uptake by Na could result in increased energy expenditure for NH_4^+ uptake at high salinities, thus reducing the energy available for plant growth. This supposes that there is a single cation carrier mechanism for uptake of both Na and NH_4^+. At high salinities the carriers would tend to be occupied by Na and would have to turn over faster to take up the required amount of NH_4^+. Increasing the NH_4^+ concentrations relative to Na, as by fertilization, would decrease the energy expended for NH_4^+ uptake.

Nitrogen is a large component of two compounds used by *S. alterniflora* in osmoregulation, proline and glycinebetaine (Cavalieri and Huang 1981). These compounds are accumulated in the cytoplasm to counteract the low osmotic potential in vacuoles in which the plant sequesters NaCl. Since each of these compounds is 12% nitrogen, accumulation of them by a plant experiencing salinity stress can significantly reduce the nitrogen available for growth. Cavalieri and Huang (1981) found that short *S. alterniflora* growing in an area with high interstitial salinity and low available nitrogen contained higher concentrations of proline and glycinebetaine and lower leaf water potentials than tall grass growing in an area with low salinity and higher N availability. Fertilization with ammonium nitrate resulted in increased growth, higher concentrations of proline and glycinebetaine and lower leaf water potentials in short *S. alterniflora* but had little effect on these parameters in the tall grass. These results imply that tall *S. alterniflora* growing under conditions of relatively low salinity stress has access to sufficient nitrogen for both plant growth and osmoregulation. Short *S. alterniflora*, however, has to utilize a larger proportion of the nitrogen it takes up, as well as a larger proportion of organic carbon and energy, for the production of osmotica. This diversion of resources would result in a reduction of growth relative to the tall form in any case, and if nitrogen uptake were interfered with in any way growth would be further reduced.

Oxygenation of the Root Zone, Eh, and Drainage

Linthurst and Seneca (1981) found that aeration of the root medium in combination with high nitrogen levels was more effective in overcoming the detrimental effect of high salinity than high N alone. Aeration alone also had a beneficial effect on plant growth at all salinities tested. They suggested that aeration enhanced the plant's potential for nutrient uptake. Aeration like that provided in this experiment does not exist in the field, but oxygenation of the root zone does occur via diffusion of O_2 from *S. alterniflora* roots (Teal and Kanwisher 1966; Carlson 1978). Oxygen concentration in the base of *S. alterniflora* shoots declines during tidal submergence and in the dark, and the shoot bases can become anoxic during nighttime high tides (Gleason and Zieman 1981). Gleason and Zieman (1981) also found that *S. alterniflora* shoot height was inversely proportional to the rate and extent of oxygen decline in the shoot bases during laboratory-simulated tidal submergence; the tall shoots had more leaf area out of the water to continue with photosynthesis than the short grass. Anoxia in the roots can cause cellular damage (Gleason and Zieman 1981; Linthurst and Seneca 1981) and reduce the rate of ammonium uptake (Morris 1979). Either of these consequences would reduce plant growth.

Mendelssohn et al. (1981) have found that *S. alterniflora* can respond to oxygen deficiency in its roots by switching from aerobic respiration to fermentation as a means of producing ATP. Alcohol dehydrogenase activity in *S. alterniflora* roots, a measure of alcoholic fermentation, increased along a streamside-landward transect while plant standing crop and culm density decreased. They speculated that the reduced growth may be due to a decrease in carbon available for growth resulting from the increased consumption of glucose during fermentation and subsequent diffusion of ethanol, an end product of fermentation, from the roots. Roots of the highly productive streamside plants primarily use aerobic respiration, although ATP production appears to be oxygen limited. In continuously flooded zones, plants are severely stressed by the low energy status of their roots and the accumulation of toxins in the soil.

Even though *S. alterniflora* can provide some oxygen to the soil via its roots, conditions in the soil and rhizosphere are for the most part highly reducing (Teal and Kanwisher 1961). Breakdown of organic matter in the soil results in a large oxygen demand, high rates of sulfate reduction, and high concentrations of H_2S (Howarth and Teal 1979). Hydrogen sulfide can be toxic to plants by inhibition of oxidase enzymes (Pitts et al. 1972) and can inhibit nutrient uptake (Joshi and Hollis 1977). The filamentous H_2S-oxidizing bacteria, *Beggiatoa,* which has been found associated with *Spartina* roots, detoxifies hydrogen sulfide (Pitts et al. 1972; Joshi and Hollis 1977).

Oxygen diffusion from the roots of *S. alterniflora* can also oxidize reduced sulfur compounds and raise the Eh of the soil. Howes et al. (1981) found that oxidation-reduction potentials were lower in areas of unvegetated peat than in nearby areas newly colonized by *S. alterniflora.* They also found that Eh was higher in the root zone of tall *S. alterniflora* than in the root zone of the short form (Fig. 2). This was due in part to the fact that soil in the short grass zone was more waterlogged than that in the tall area, but they also found direct evidence that *S. alterniflora* itself can alter the Eh of the rhizosphere and that the tall plants oxidize their rhizosphere more than the short plants. Eh was higher in plots which had been fertilized for several years than in adjacent control plots. As discussed earlier, what was originally short *S. alterniflora* was morphologically transformed by long-term fertilization to an intermediate height form (Valiela et al. 1978). Howes et al. (1981) felt that

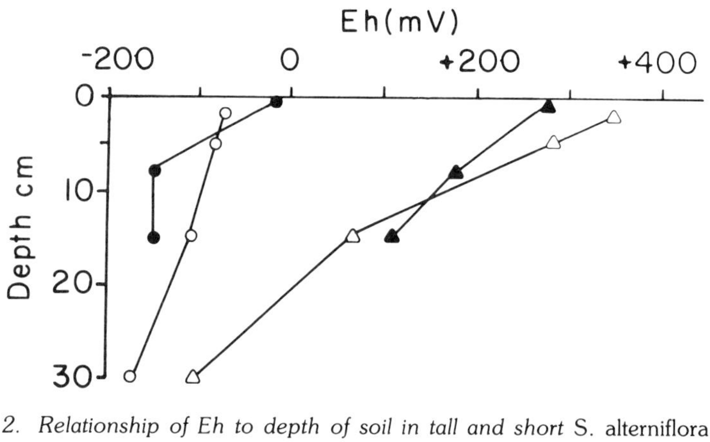

Figure 2. Relationship of Eh to depth of soil in tall and short S. alterniflora. *○,● Short; △,▲ Tall. ○,△ Data from Howes et al. 1981. ●,▲ Data from Linthurst and Seneca 1981.*

the change in Eh was due to the more vigorous growth of *S. alterniflora* rather than to some non-biological effect of the fertilization itself.

Further evidence that Eh is partially controlled by oxygen diffusion from *S. alterniflora* roots is that there are diurnal shifts in Eh in the rhizosphere (Howes et al. 1981) corresponding to the diurnal shifts in oxygen concentration in *Spartina* shoots (Gleason and Zieman 1981). A positive feedback loop exists between rhizosphere oxidation and productivity. Oxidation of the root zone increases nutrient uptake capability and reduces the toxic effect of H_2S, thereby increasing productivity, which in turn increases rhizosphere oxygenation.

Howes et al. (1981) also suggested that increased drainage can aid growth by directly removing toxins from the root zone. Mendelssohn and Seneca (1980) found that plant height increased as soil drainage increased along transects in North Carolina marshes. The redox potential was also highest where drainage potential was highest. They attributed this to drainage and water percolation rather than to any activity of the plants themselves. Soil drainage may determine the extent to which *S. alterniflora* utilizes aerobic respiration or fermentation. Mendelssohn et al. (1981) found that in well drained areas with positive soil Eh, aerobic respiration predominated, but at Eh of -200 mV and below the oxygen in the root tissue became depleted and fermentation was more important.

Wiegert et al. (1982) have found that increasing soil drainage in an intermediate-height *S. alterniflora* marsh increases plant biomass beyond that of an adjacent undisturbed plot and shifts a number of other characteristics toward those typical of tall *S. alterniflora* zones. Drainage was increased by burying drain field pipes in one plot, the tiled plot. The disturbance of burying the pipes was controlled for by digging and refilling similar trenches in an adjacent plot without burying any pipe. A plot next to the tiled plot served as the undisturbed control. Seven months after the perturbation, water infiltration rates were 6 times higher in the tiled plot than in the undisturbed control and 10 times higher than in the disturbed control. At the end of the growing season, standing crop was more than twice as high in the tiled plot than in the undisturbed area and nearly 5 times higher than in the disturbed control. Although this experiment did not elucidate the ultimate mechanisms by which *S. alterniflora* productivity was increased, it is the only field experiment other than fertilization studies that has resulted in increased plant productivity. It confirmed that subsurface drainage is an important factor in determining primary productivity and demonstrated the usefulness of this type of experiment for future investigation of the controls on *S. alterniflora* productivity.

Based in part on measurements made in the same experimental area, King and Klug (1982) have formed the hypothesis that plant height and productivity are a result of the interaction between tidal water movement, dissolved iron and sulfide concentrations in the soil, and sulfate reduction. They found that sulfide concentrations are inversely related to soil water movement and plant productivity, but are not related to rates of sulfate reduction, the major source of sulfide. The concentration of sulfide is also inversely related to that of dissolved iron. The major source of iron in the salt marsh is particulate and colloidal ferric iron suspended in tidal water, so the major input is via sedimentation. King and Klug (1982) proposed that iron input will vary with distance from tidal creeks and frequency of flooding and therefore will be greater in the tall *S. alterniflora* zone than in the short zone. Once on the marsh surface, ferric iron will be reduced to the more soluble ferrous state and can be more easily diffused or transported below the soil surface, where it can precipitate sulfide out of the soil solution. Adams (1963) proposed iron as a limiting nutrient for *S. alterniflora* growth two decades ago, but subsequent studies have shown that the importance of iron as a plant nutrient has been, at best, overemphasized (Broome et al. 1975; Haines and Dunn 1976). According to King and Klug's hypothesis, iron is important for *S. alterniflora* growth, but indirectly rather than directly as a plant nutrient.

Concluding Remarks

Edaphic factors control the heterogeneity in height, biomass, and productivity in *S. alterniflora*. Figure 3 presents a model of the interactions between the factors which affect *S. alterniflora* growth, either directly or indirectly. Field and laboratory studies have shown that salinity is one factor which can influence *S. alterniflora* growth, but there are marshes in which both tall and short forms occur in the absence of salinity gradients.

Fertilization experiments have demonstrated that growth of tall *S. alterniflora* is not nitrogen limited but that productivity of the short form can be increased by nitrogen additions. Other studies have shown that the apparent nitrogen limitation in the short form is not due to a shortage of available nitrogen, but to an alteration in nitrogen uptake kinetics. Salinity stress-

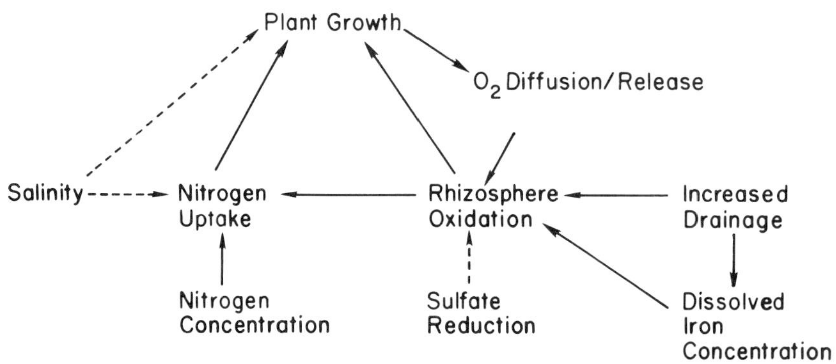

Figure 3. Relationships between factors directly and indirectly affecting growth of S. alterniflora. *Solid lines are positive effects, dashed lines are negative effects.*

caused diversion of nitrogen to the production of osmotica can also reduce the nitrogen available for growth.

High sulfide concentrations and consequent low oxidation-reduction potentials in the rhizosphere can affect nitrogen uptake kinetics. These factors and anoxia can also cause structural damage or alterations in the roots which could affect nutrient uptake. Soil drainage, iron concentrations, oxygen diffusion from *S. alterniflora* roots, and plant productivity itself can all affect sulfide concentrations and redox. In order to understand the controls of *S. alterniflora* productivity, it is essential that the effect of these factors on nitrogen uptake and utilization be demonstrated in the field. Laboratory or hydroponic culture studies are also necessary to elucidate the mechanisms by which nitrogen metabolism is altered. Finally, similar studies in marshes in various geographical areas will be needed to determine if the same factors control the within-marsh heterogeneity in height and productivity of *S. alterniflora.*

Acknowledgments

Preparation of this paper was supported by Grant No. R806592-01 from the Environmental Protection Agency. Contribution No. 449 from the University of Georgia Marine Institute. Figure 1 is published with permission from *Estuarine and Coastal Marine Science.* Copyright Academic Press, Inc. (London) Ltd.

References Cited

Adams, D. A. 1963. Factors influencing vascular plant zonation in North Carolina salt marshes. *Ecology* 44:445-456.

Anderson, C. M. and Treshow. 1980. A review of environmental and genetic factors that affect height in *Spartina alterniflora* Loisel. (salt marsh cord grass). *Estuaries* 3:168-176.

Broome, S. W., W. W. Woodhouse, Jr. and E. D. Seneca. 1975. The relationship of mineral nutrients to growth of *Spartina alterniflora* in North Carolina: II. The effects of N, P, and Fe fertilizers. *Soil. Sci. Soc.* 39:301-307.

Buresh, R. J., R. D. DeLaune and W. H. Patrick, Jr. 1980. Nitrogen and phosphorus distribution and utilization by *Spartina alterniflora* in a Louisiana Gulf Coast marsh. *Estuaries* 3:111-121.

Carlson, P. R. 1978. Radial diffusion of oxygen from the roots of *Spartina alterniflora.* Abstract, 29th Annual Meeting of the Ecological Society of America. *Bull. Ecol. Soc. Amer.* 59:87.

Cavalieri, A. J. and A. H. C. Huang. 1981. Accumulation of proline and glycinebetaine in *Spartina alterniflora* Loisel. in response to NaCl and nitrogen in the marsh. *Oecologia* 49:224-228.

Chalmers, A. G. 1977. Pools of nitrogen in a Georgia salt marsh. Ph. D. Dissertation, U. Georgia, Athens, GA 167 pp.

Chalmers, A. G. 1979. The effects of fertilization on nitrogen distribution in a *Spartina alterniflora* salt marsh. *Est. Coast. Mar. Sci.* 8:327-337.

Chalmers, A. G., E. B. Haines and B. F. Sherr. 1976. Capacity of a *Spartina* salt marsh to assimilate nitrogen from secondarily treated sewage. Tech. Completion Report USDI/OWRT Project No. A-057-GA. 88 pp.

Chapman, V. J. 1960. *Salt Marshes and Salt Deserts of the World.* Interscience Publishers Inc., New York. 392 pp.

Christian, R. R., K. Bancroft and W. J. Wiebe. 1978. Resistance of the microbial community within salt marsh soils to selected perturbations. *Ecology* 59:1200-1210.

Church, G. L. 1940. Cytotaxonomic studies in the Gramineae *Spartina, Andropogon* and *Panicum*. *Amer. J. Bot.* 17:263-271.

Fernald, M. L. 1950. *Gray's Manual of Botany*. American Book Co., New York. 1632 pp.

Gallagher, J. L. 1975. Effect of an ammonium nitrate pulse on the growth and elemental composition of natural stands of *Spartina alterniflora* and *Juncus roemerianus*. *Amer. J. Bot.* 62:644-648.

Gallagher, J. L., R. A. Linthurst and W. J. Pfeiffer. 1980. Aerial production, mortality and mineral accumulation dynamics in *Spartina alterniflora* and *Juncus roemerianus* in a Georgia salt marsh. *Ecology* 61:303-312.

Gallagher, J. L., R. J. Reimold and D. E. Thompson. 1972. Remote sensing and salt marsh productivity, pp. 338-348. *In:* W. J. Kosco (chairman), Proc. 38th Annual Meeting, Amer. Soc. Photogrammetry, Falls Church, VA

Gleason, M. L. and J. C. Zieman. 1981. Influence of tidal inundation on internal oxygen supply of *Spartina alterniflora* and *Spartina patens*. *Est. Coast. Shelf Sci.* 13:47-57.

Haines, B. L. and E. L. Dunn. 1976. Growth and resource allocation responses of *Spartina alterniflora* Loisel. to three levels of NH_4 – N, Fe and NaCl in solution culture. *Bot. Gaz.* 137:224-230.

Hanson, R. B. 1977. Comparison of nitrogen fixation activity in tall and short *Spartina alterniflora*. *Appl. Environ. Microbiol.* 33:569-602.

Howarth, R. W. and J. M. Teal. 1979. Sulfate reduction in a New England salt marsh. *Limnol. Oceanogr.* 24:999-1013.

Howes, B. L., R. W. Howarth, J. M. Teal and I. Valiela. 1981. Oxidation-reduction potentials in a salt marsh: Spatial patterns and interactions with primary production. *Limnol. Oceanogr.* 26:350-360.

Joshi, M. M. and J. P. Hollis. 1977. Interaction of *Beggiatoa* and rice plant: detoxification of hydrogen sulfide in the rice rhizosphere. *Science* 195:179-180.

King, G. M. and M. J. Klug. 1982. The relationship of soil water movement, sulfide concentration and *Spartina alterniflora* production in a Georgia salt marsh. Unpublished manuscript. Kellogg Biological Station, Michigan State Univ., Hickory Corners, MI

Linthurst, R. A. and E. D. Seneca. 1981. Aeration, nitrogen and salinity as determinants of *Spartina alterniflora* Loisel. growth response. *Estuaries* 4:53-63.

Mendelssohn, I. A. 1979. Nitrogen metabolism in the height forms of *Spartina alterniflora* in North Carolina. *Ecology* 60:574-584.

Mendelssohn, I. A., K. L. McKee and W. H. Patrick, Jr. 1981. Oxygen deficiency in *Spartina alterniflora* roots: metabolic adaptation to anoxia. *Science* 214:439-441.

Mendelssohn, I. A. and E. D. Seneca. 1980. The influence of soil drainage on the growth of salt marsh cordgrass *Spartina alterniflora* in North Carolina. *Est. Coast. Mar. Sci.* 11:27-40.

Mooring, M. T., A. W. Cooper and E. D. Seneca. 1971. Seed germination response and evidence for height ecophenes in *Spartina alterniflora* from North Carolina. *Amer. J. Bot.* 58:48-55.

Morris, J. T. 1979. The nitrogen uptake kinetics and growth response of *Spartina alterniflora*. Ph. D. Dissertation, Yale University, New Haven, CT 102 pp.

Morris, J. R. 1980. The nitrogen uptake kinetics of *Spartina alterniflora* in culture. *Ecology* 61:1114-1121.

Nestler, J. 1977a. Interstitial salinity as a cause of ecophenic variation in *Spartina alterniflora*. *Est. Coast. Mar. Sci.* 5:707-714.

Nestler, J. 1977b. A preliminary study of the sediment hydrology of a Georgia salt marsh using rhodamine WT as a tracer. *Southeastern Geol.* 18:265-271.

Parrondo, R. T., J. G. Gosselink and C. S. Hopkinson. 1978. Effects of salinity and drainage on the growth of three salt marsh grasses. *Bot. Gaz.* 139:102-107.

Patrick, W. H., Jr. and R. D. Delaune. 1976. Nitrogen and phosphorus utilization by *Spartina alterniflora* in a salt marsh in Barataria Bay, Louisiana. *Est. Coast. Mar. Sci.* 4:59-64.

Pitts, G., A. I. Allam and J. P. Hollis. 1972. *Beggiatoa:* Occurrence in the rice rhizosphere. *Science* 178:990-992.

Pomeroy, L. R. and R. G. Wiegert. 1981. *The Ecology of a Salt Marsh.* Springer-Verlag, New York. 271 pp.

Reideburg, C. H. 1978. Intertidal pump in a Georgia salt marsh. MS thesis, U. Georgia, Athens, GA 81 pp.

Shea, M. L., R. S. Warren and W. A. Niering. 1975. Biochemical and transplantation studies of the growth form of *Spartina alterniflora* on Connecticut salt marshes. *Ecology* 56:461-466.

Smart, R. M. and J. W. Barko. 1978. Influence of sediment salinity and nutrients on the physiological ecology of selected salt marsh plants. *Est. Coast. Mar. Sci.* 7:487-495.

Stalter, R. and W. T. Batson. 1969. Transplantation of salt marsh vegetation, Georgetown, South Carolina. *Ecology* 50:1087-1089.

Sullivan, M. J. and F. C. Daiber. 1974. Response in production of cord grass, *Spartina alterniflora,* to inorganic nitrogen and phosphorus fertilizer. *Chesapeake Sci.* 15:121-123.

Teal, J. M. and J. W. Kanwisher. 1961. Gas exchange in a Georgia salt marsh. *Limnol. Oceanogr.* 6:388-399.

Teal, J. M. and J. W. Kanwisher. 1966. Gas transport in the marsh grass, *Spartina alterniflora. J. Exp. Bot.* 17:355-361.

Turner, R. E. 1976. Geographic variations in salt marsh macrophyte production: A review. *Contrib. Mar. Sci.* 20:47-68.

Valiela, I. and J. M. Teal. 1974. Nutrient limitation in salt marsh vegetation, pp. 547-563. *In:* R. J. Reimold and W. H. Queen (eds.), *Ecology of Halophytes.* Academic Press, New York.

Valiela, I., J. M. Teal and W. G. Deuser. 1978. The nature of growth forms in the salt marsh grass *Spartina alterniflora. Amer. Natur.* 112:461-470.

Valiela, I., J. M. Teal and N. Y. Persson. 1976. Production and dynamics of experimentally enriched salt marsh vegetation: Belowground biomass. *Limnol. Oceanogr.* 21:245-252.

Waisel, Y. 1972. *Biology of Halophytes.* Academic Press, New York. 395 pp.

Wiegert, R. G., A. G. Chalmers and P. F. Randerson. 1982. Primary production of salt marshes: regulation by soil-water movement. Unpublished manuscript, Department of Zoology, University of Georgia, Athens, GA

EFFECTS OF HYPOXIA ON ROOT AND SHOOT RESPIRATION OF *SPARTINA ALTERNIFLORA*

Mark L. Gleason
Centre College of Kentucky
Danville, Kentucky

and

E. Lloyd Dunn
School of Biology
Georgia Institute of Technology
Atlanta, Georgia

Abstract: Carbon dioxide release (nM $gdw^{-1}s^{-1}$) of shoot and root systems of intact *Spartina alterniflora* was monitored in a 2-chamber laboratory apparatus in darkness while the shoot atmosphere was varied. Whole plant respiration averaged 12.3 nM CO_2 $gdw^{-1}s^{-1}$ at 30°C with shoots in 21% O_2 gas and roots in 0.02% O_2. Shoot CO_2 release exceeded belowground CO_2 release under the conditions tested; however, some of this differential resulted from CO_2 flux from the root system to the shoot through internal gas spaces. Carbon dioxide release of both aerial and belowground tissues declined when oxygen concentration in the shoot chamber was lowered from 21% to 0.02%, but the decline was proportionally greater in shoot tissues. Full submergence of the shoot produced a roughly equivalent decline in CO_2 release. These findings imply that tidal flooding in darkness, by inducing internal hypoxia, can inhibit respiration of *S. alterniflora*.

Introduction

Monotypic stands of *Spartina alterniflora* dominate primary production of tidal marshes along most of the East Coast of the United States. Understanding the ecophysiological interactions that control the productivity of the species is consequently a goal of considerable ecological value.

A growing body of evidence suggests that responses to oxygen scarcity strongly influence the productivity of *Spartina alterniflora*. It is clear that belowground organs of wetland macrophytes, surrounded by anoxic substrate (Teal and Kanwisher 1961), depend on oxygen transferred downward from the shoot through internal gas spaces (Teal and Kanwisher 1966; Armstrong 1975). Several investigators (Linthurst 1979; Mendelssohn and Seneca 1980; Howes et al. 1981) have recently suggested that productivity of *S. alterniflora* is inhibited by reduced, waterlogged substrate and enhanced by substrate aeration. Fertilization experiments (Sullivan and Daiber 1974; Valiela and Teal 1974; Broome et al. 1975; Gallagher 1975; Mendelssohn

ISBN 0-12-404070-5

1978) seem to indicate that nitrogen availability acts as an edaphic limiting factor in most marsh soils; however, culture experiments by Morris (1980) indicate that nitrogen concentrations normally found in interstitial waters of *S. alterniflora* marshes are unlikely to limit productivity. Instead, a physiological inhibitor, such as oxygen insufficiency in the roots, may create an effective nitrogen deficiency by interfering with active uptake (Morris 1980). Nutrient uptake mechanisms of other wetland plants, such as rice, are also sensitive to oxygen supply (Rao and Mikkelsen 1977). In highly reduced substrate, internal oxygen is so severely depleted that anaerobic metabolism can be initiated, accompanied by lower productivity (Mendelssohn et al. 1981).

As the source of energy for uptake mechanisms and of intermediates for synthetic metabolism, respiration is inextricably linked to productivity. However, very little is known about the effects of hypoxia on respiration of *S. alterniflora* (Giurgevich and Dunn 1979). Armstrong and Gaynard (1976) established that respiration of rice and of the bog species *Eriophorum angustifolium* declines when oxygen concentrations in root gas spaces fall below 2.0-2.5%. Oxygen concentrations below 1% in root tips have recently been found to inhibit root growth of rice (W. Armstrong, U. Hull, United Kingdom, 1981, pers. comm.). The thresholds of respiratory and growth inhibition for *S. alterniflora* are unknown. However, Gleason and Zieman (1981) reported prolonged periods of anoxia in the shoot base aerenchyma of *S. alterniflora* during tidal flooding in darkness. This finding, together with the discovery of anaerobic metabolism in roots occupying highly reduced substrates (Mendelssohn et al. 1981), suggests that several types of environmental circumstances may inhibit aerobic respiration of *S. alterniflora*.

The present study was undertaken to determine: 1) the magnitude of aerial and belowground respiration in intact *S. alterniflora*; 2) the effect of hypoxic conditions on respiration; and 3) the effect of tidal submergence on respiration.

Materials and Methods

During July-August 1981, offshoots (20-50 cm tall) from larger *S. alterniflora* culms were removed from mud along a tidal creek bank adjacent to the University of Georgia Marine Institute, Sapelo Island, GA, at the lower margin of the species' vertical distribution. Care was taken to minimize damage to roots and rhizomes. Belowground organs were immersed in $20^{0}/_{00}$ NaCl solution and the plants were placed in darkness for approximately 24 h prior to experimentation. Dry weight of plant tissue was determined by weighing to ± 0.01 g after oven drying at 100°C for 24 h.

Carbon dioxide exchange of aerial and belowground tissues of intact plants was monitored separately using a 2-chamber gas exchange system (Fig. 1). The chamber enclosing a shoot was similar to that described by Mooney et al. (1971) and Giurgevich & Dunn (1978); it was temperature- and humidity-controlled and had a plastic top covered by an opaque light

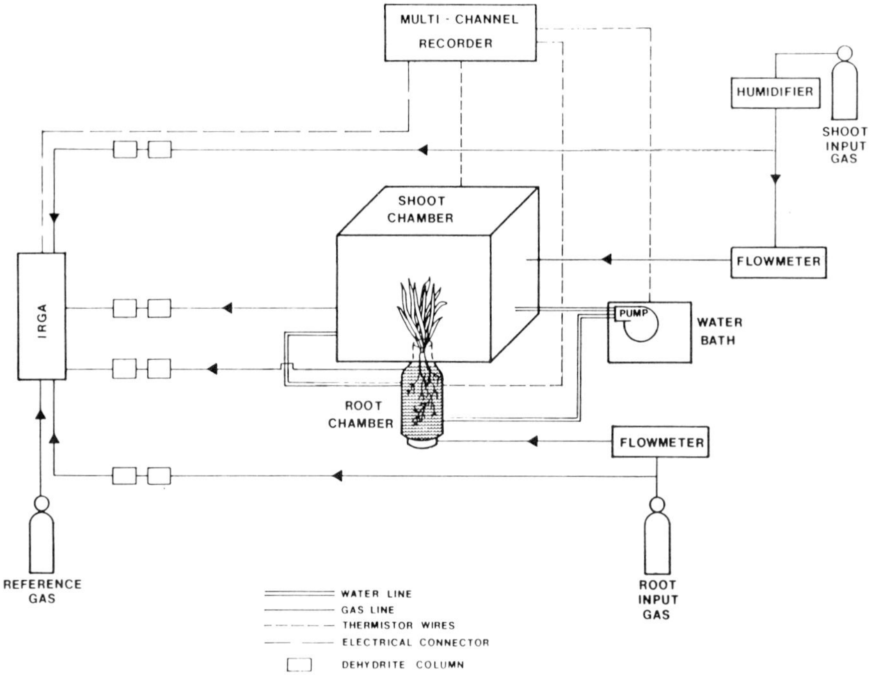

Figure 1. Schematic view of the two-chamber CO_2 exchange apparatus.

shield. The root chamber, modified from the design of Williams and Kemp (1978), consisted of a central glass cylinder (20 cm long × 5 cm I.D.) with a fritted glass disc embedded in its base, fused to an outer glass chamber through which water was circulated for temperature control.

After the inner chamber was filled with 20 $^0/_{00}$ NaCl solution, a plant was secured in the open neck of this chamber by wrapping the shoot base tightly with a length of closed-cell insulating foam, and the neck was sealed with non-hardening rope caulk ("Mortite," Mortell Co.). The shoot and the neck of the root chamber were then fitted through a hole in the base of the shoot chamber and remaining leaks were sealed with Mortite. Temperature in both chambers was controlled by water circulated from a thermostat-regulated water bath. A fan in the shoot chamber aided in temperature control and maintained a well-mixed chamber environment with a constant and small boundary layer. An infrared gas analyzer (Beckman model 865) was used to measure the CO_2 concentration differential between inlet and outlet air streams of both root and shoot chambers.

The shoot environment was varied while the root system remained in 20$^0/_{00}$ NaCl solution through which a gas mixture containing 0.02% O_2 and 310-350 ppm CO_2 was bubbled. Shoot input gases remained within the same range for CO_2, but oxygen was varied between 21% and 0.02%. The oxygen content of all gas mixtures was verified with a Westinghouse model 209 Oxygen Analyzer. In trials simulating tidal flooding, the regular shoot chamber was replaced with a plexiglas tube (60 cm length × 5 cm I.D.) which was attached to the root chamber by a rubber stopper, filled with 20°/₀₀ NaCl solution until the shoot was fully submerged, and sealed at the upper end. Ports near the upper and lower ends permitted gas circulation, and the tube was sheathed with aluminum foil to maintain the shoot in darkness.

Another experiment monitored changes in CO_2 exchange after shoots were excised from root systems at the root-stem junction and sealed within their respective chambers using Mortite.

Chamber temperatures were maintained at 30° ± 1.0°C throughout the CO_2 exchange trials, except during the flooding experiment, in which temperatures in the shoot chamber were 24° ± 1.0°C.

Measurements of CO_2 were determined as concentration differentials between the inlet and outlet gas streams. Corrections for the effects of pressure and temperature on air density and flowmeter calibrations, and for changes in the gas analyzer sensitivity at different CO_2 concentrations were included in the calculations. A correction for microbial activity in the root chamber solution was applied by subtracting the CO_2 exchange of a blank measured when plant tissue was removed from each trial. Microbial CO_2 release averaged 15.4% of total CO_2 release from the root chamber. Carbon dioxide exchange rates are expressed as nM CO_2 $gdw^{-1}s^{-1}$.

Measurements of O_2 uptake by root systems were performed with the shoot uncovered in room air under low illumination (less than 100 microEinsteins $m^{-2}s^{-1}$). Dissolved O_2 concentrations (% volume) were monitored with a YSI model 5331 Clark-type electrode and a YSI model 53 oxygen meter. A magnetic stirrer created water circulation in the root chamber. The solution in the root chamber (20°/₀₀ Instant Ocean) was initially bubbled with compressed air until O_2 saturation occurred, then monitored after the chamber was sealed with a foam plug and petroleum jelly around the stem base. Trials were conducted with intact plants and also with isolated root systems after the shoot was severed and the stub sealed with petroleum jelly. Root chamber temperature was maintained at 22.5° ± 0.5°C throughout the O_2 trials. Oxygen electrode data were converted from percent saturation to ml/l, then changed to nM O_2 $gdw^{-1}s^{-1}$.

Results and Discussion

The mean value of respiration for intact *S. alterniflora* was 12.3 nM CO_2 $gdw^{-1}s^{-1}$ at 30°C with shoots in 21% O_2 and roots in 0.02% O_2. Teal and Kanwisher's (1961; Fig. 6) O_2 uptake data for *S. alterniflora* from Sapelo Island, extrapolated to 30°C, give a higher estimate, approximately 31.0 nM O_2 $gdw^{-1}s^{-1}$. By holding our plants in darkness for about 24 h prior to experimental trials, we reduced or eliminated the portion of respiration known as growth respiration (Hansen and Jensen 1977); therefore, our measurements are estimates of maintenance respiration rather than total respiration.

Carbon dioxide release from shoots clearly exceeded rates from root systems under all conditions tested (Table 1). In intact plants, however, measured CO_2 release rates underestimate belowground respiration and overestimate shoot respiration because of diffusion through internal gas spaces from belowground tissues to shoots. A noticeable rise in CO_2 release by roots and decline in shoot CO_2 release followed separation at the root-stem junction and isolation in separate chambers (e.g., Fig. 2); these results indicate that some CO_2 is normally translocated upward from the roots and is measured as release by the shoots (Teal and Kanwisher 1966; Giurgevich and Dunn 1979). Consequently, respiration estimates for intact *S. alterniflora* based solely on CO_2 exchange of aboveground tissues are liable to overestimate shoot respiration and underestimate respiration for the whole plant (Giurgevich and Dunn, in press).

In a related experiment, belowground O_2 uptake rose sharply after root systems of intact *S. alterniflora* were severed and sealed in chambers (Fig. 3). This finding reaffirms that O_2 is transferred from shoots to roots of *S. alterniflora* through internal gas spaces (Teal and Kanwisher 1966; Gleason and Zieman 1981). This downward flux forms a countercurrent to upward movement of CO_2 from the roots. As with CO_2, accurate O_2 exchange estimates of respiration should take into account internal movement of O_2.

Table 1. Mean rates of CO_2 release from intact *Spartina alterniflora* and from isolated shoot and root systems under two experimental treatments. *Significant at 0.05 level; **Significant at 0.01 level.

	% O_2 in Inflow Gas		*Mean CO_2 Release ($nM\ gdw^{-1}s^{-1}$)*		*F Ratio (3,57) (Aerial vs. Below-ground)*
	Shoot Chamber	*Root Chamber*	*Aerial Tissue*	*Below-ground Tissue*	
	Intact Plants				
Treatment 1	21	0.02	10.01[a]	2.75[a]	169.7**
Treatment 2	0.02	0.02	4.54[a]	1.88[a]	66.0**
	F Ratio (3,57) (treatment 1 vs. treatment 2)		49.7**	3.61*	
	Isolated Shoot and Root Systems[d]				
Treatment 1a	21		8.05[b]		
		0.02		3.11[b]	
Treatment 2a	0.02		5.70[c]		
		0.02		2.51[c]	

[a] n = 30; [b] n = 2; [c] n = 6.

[d] Data taken 1 hour after shoot and root systems separated and isolated.

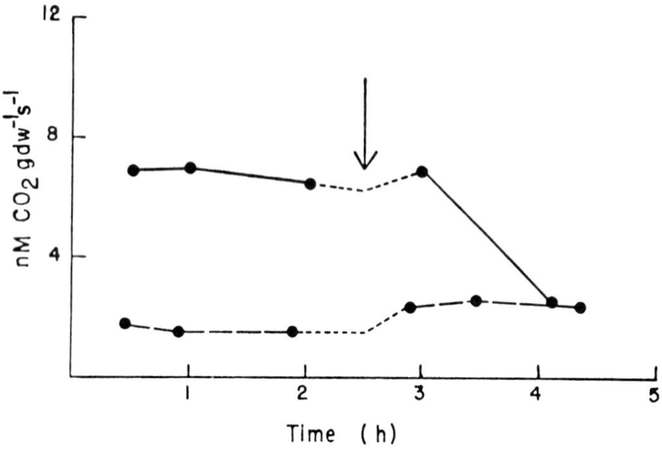

Figure 2. *Carbon dioxide release (nM gdw^{-1}s^{-1}) by* S. alterniflora *shoot (——) and root system (– – –) before and after they were severed and isolated (arrow) in separate chambers. (------) indicates extrapolated lines.*

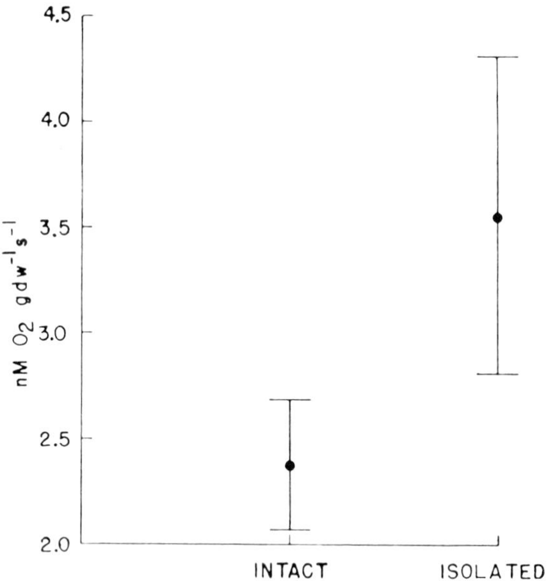

Figure 3. *Oxygen uptake (nM gdw^{-1}s^{-1}) by root systems of intact* S. alterniflora *with shoots in air and by the same root systems after the shoots were excised and the stubs sealed. n = 8; error bars = ± 1 standard error.*

Nevertheless, differences in CO_2 release rates between roots and shoots of intact plants were highly significant ($p<0.01$) under conditions (21% O_2 shoot atmosphere, 0.02% O_2 input to root chamber) superficially resembling those in natural *S. alterniflora* marshes with emergent shoots and deoxygenated substrate (Teal and Kanwisher 1961). Unless upward CO_2 flux from root systems was more than twice as great as the measured rate of CO_2 release by the roots, shoot respiration per gdw must have exceeded root system respiration under these conditions. Indeed, shoot respiration continued to exceed root respiration even after shoot and root systems were isolated (Table 1). In a marsh, atmospheric O_2 supplied to shoot tissues is depleted by respiration as it diffuses downward to the roots, which lack a significant external O_2 supply. So much of this oxygen is apparently consumed enroute that belowground respiration may be inhibited, particularly in tissues (such as root meristems) that are located far from the shoot.

When the oxygen concentration in the shoot chamber was changed from 21% to 0.02%, CO_2 release of aerial and belowground tissues declined (Table 1). The F ratios between treatments indicate that the decline was more significant in aerial tissues. Newman-Kuels tests (Winer 1971) showed a significant (0.01 level) decline for aerial tissues, but the decline for the root system was not significant at the 0.05 level. The external oxygen supply to the shoot was drastically curtailed while the belowground organs remained in a hypoxic environment; consequently, shoot respiration fell off sharply while belowground respiration declined to a lesser degree (Fig. 4). Even with identical 0.02% O_2 input gas, CO_2 release of isolated shoots exceeded that of root systems for at least one hour following separation (Table 1). However, more data and longer monitoring times are needed before conclusions can be drawn about the relative magnitude of respiration in these tissues under identical conditions.

Substantial fractions of shoot and root CO_2 release remained several hours after O_2 in the shoot atmosphere was lowered to 0.02% (Fig. 4), indicating that some respiration persists even when external O_2 sources are nearly eliminated. Available evidence suggests that residual oxygen in internal gas spaces of *S. alterniflora* (Gleason and Zieman 1981) and *E. angustifolium* (Armstrong and Gaynard 1976) is rapidly depleted in darkness when external O_2 sources are cut off; we assume, therefore, that internal O_2 concentrations in our trials approximated the external concentration of 0.02% within several hours after the trial started. Armstrong and Gaynard (1976) established that root respiration of *E. angustifolium* and of rice declines when internal oxygen falls below 2-3%; this decline was believed to coincide with development of anoxic centers in meristematic and stelar tissues. However, some respiration continued in *E. angustifolium* roots until internal O_2 concentrations fell much lower, presumably spreading anoxia to cortical tissues. To extend these assumptions to *S. alterniflora*, our experiments suggest that shoot and root tissues adjacent to internal gas spaces continue to respire as O_2 in internal gas

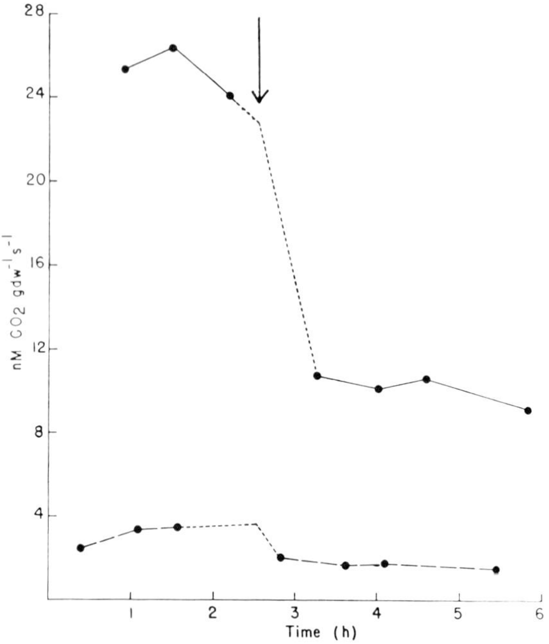

Figure 4. *Carbon dioxide release (nM gdw*$^{-1}$*s*$^{-1}$*) by* S. alterniflora *shoot (——) and root system (– – –) before and after shoot input gas was changed (arrow) from 21%* O_2 *to 0.02%* O_2*. (------) indicates extrapolated lines.*

spaces approaches 0.02%, while more massive tissues have become anoxic and ceased or severely curtailed respiration. *Spartina alterniflora* thus resembles other wetland macrophytes in maintaining respiration under conditions of very limited oxygen availability, although at a reduced rate. Reducing respiration as oxygen becomes scarce may be adaptively advantageous in helping to conserve the remaining oxygen. Because our input gas always contained at least 0.02% O_2, aerobic respiration probably persisted in well-ventilated tissues; however, anaerobic pathways may have been activated in other tissues. Mendelssohn et al. (1981) established that *S. alterniflora* roots in highly reduced substrate respired anaerobically.

Fully submerging a shoot inhibited respiration approximately as much as changing the shoot atmosphere from 21% to 0.02% O_2 (Fig. 5). In both treatments, shoot respiration was much more sharply curtailed than root respiration. Previous work (Armstrong and Gaynard 1976; Gleason and Zieman 1981) has shown that internal oxygen is quickly depleted during submergence of wetland macrophytes in darkness. The present experiments thus indicate that a significant respiratory decline can accompany tidal flooding in darkness, and that internal hypoxia is the probable cause of the decline. Therefore, it appears that at least two common circumstances in *S. alterni-*

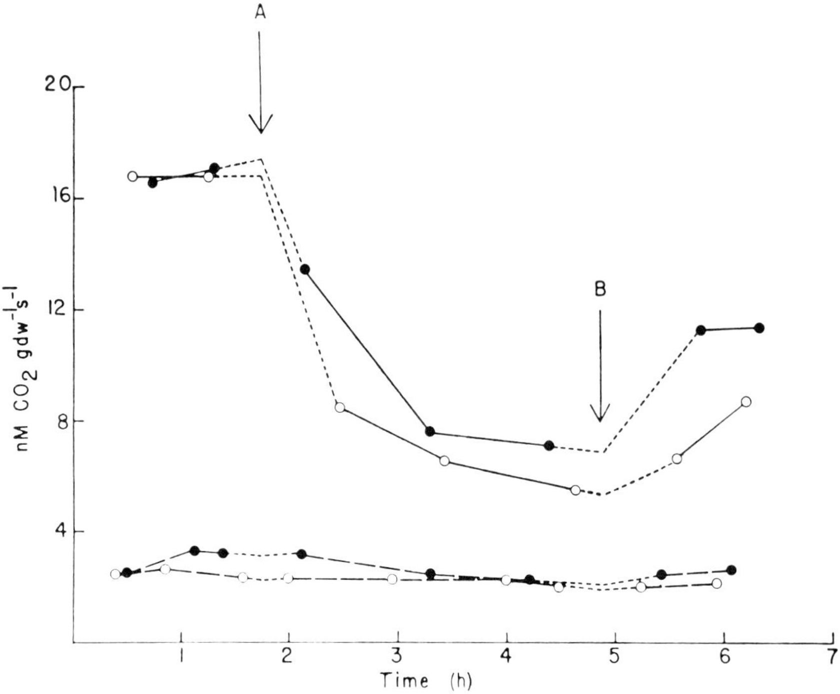

Figure 5. *Carbon dioxide release (nM gdw* $^{-1}s^{-1}$*) by* S. alterniflora *shoots (——) and root systems (– – –) while the shoot environment was manipulated. At time "A," atmosphere of one shoot (—●—) was changed from 21% O_2 to 0.02% O_2, and a second shoot in 21% O_2 (—O—) was fully submerged by 20‰ NaCl solution through which 21% O_2 was bubbled. At time "B," both shoots were returned to 21% O_2 gaseous atmosphere. (------) indicates extrapolated lines.*

flora's marsh habitat – highly anoxic substrates (Mendelssohn et al. 1981) and tidal flooding in darkness – inhibit aerobic respiration. The significance of respiratory inhibition by tidal submergence – including its effect on plant growth – is probably a function of several factors, including the depth, duration, and timing of flooding, shoot height, and temperature (Gleason and Zieman 1981). This inhibition would presumably be greatest at the seaward extreme of *S. alterniflora*'s distribution, where submergence is greatest and internal gases of the shoot base can become anoxic during high tides at night (Gleason and Zieman 1981). It seems probable, then, that flooding-induced respiratory inhibition helps determine the flooding tolerance of *S. alterniflora* and thereby influences the seaward extent of its distribution.

Acknowledgments

This work was funded by grants from the Biomedical Research Committee of the Georgia

Institute of Technology, Atlanta, Georgia, and from Centre College of Kentucky. M. Barton critically reviewed the manuscript and V. Walker and C. Armstrong assisted in its preparation. Contribution number 454 from University of Georgia Marine Institute, Sapelo Island, Georgia 31327.

References Çited

Armstrong, W. 1975. Waterlogged soils, pp. 181-218. *In*: J. R. Etherington (ed.), *Environment and Plant Ecology*. John Wiley, New York.

Armstrong, W. and T. J. Gaynard. 1976. The critical oxygen pressures for respiration in intact plants. *Physiol. Plant.* 37:200-206.

Broome, S. W., W. W. Woodhouse and E. D. Seneca. 1975. The relationship of mineral nutrients to growth of *Spartina alterniflora* in North Carolina: II. The effects of N, P, and Fe fertilizers. *Soil Sci. Soc. Am. Proc.* 39:301-307.

Gallagher, J. L. 1975. Effect of an ammonium pulse on the growth and elemental composition of natural stands of *Spartina alterniflora* and *Juncus roemerianus*. *Am. J. Bot.* 62:644-648.

Giurgevich, J. R. and E. L. Dunn. 1978. Seasonal patterns of CO_2 and water vapor exchange of *Juncus roemerianus* Scheele in a Georgia salt marsh. *Am. J. Bot.* 65:502-510.

Giurgevich, J. R. and E. L. Dunn. 1979. Seasonal patterns of CO_2 and water vapor exchange of the tall and short height forms of *Spartina alterniflora* Loisel in a Georgia salt marsh. *Oecologia* 43:139-156.

Giurgevich, J. R. and E. L. Dunn. (In press). Seasonal patterns of daily net photosynthesis, transpiration and net primary productivity of *Juncus roemerianus* (Scheele) and *Spartina alterniflora* (Loisel) in a Georgia salt marsh. *Oecologia*.

Gleason, M. L. and J. C. Zieman. 1981. Influence of tidal inundation on internal oxygen supply of *Spartina alterniflora* and *Spartina patens*. *Est. Coastal Shelf Sci.* 13:47-57.

Hansen, G. K. and C. R. Jensen. 1977. Growth and maintenance respiration on whole plants, tops, and roots of *Lolium multiflorum*. *Physiol. Plant.* 39:155-164.

Howes, B. L., R. W. Howarth, J. M. Teal and I. Valiela. 1981. Oxidation-reduction potentials in a salt marsh: Spatial patterns and interactions with primary production. *Limnol. Oceanogr.* 26:350-360.

Linthurst, R. A. 1979. The effect of aeration on the growth of *Spartina alterniflora* Loisel. *Am. J. Bot.* 66:685-691.

Mendelssohn, I. A. 1978. Ecophysiological aspects of the height forms of *Spartina alterniflora* in North Carolina. Nitrogen nutrition and soil waterlogging. Ph.D. Thesis, North Carolina State U., Raleigh, NC 142 pp.

Mendelssohn, I. A., K. L. McKee and W. H. Patrick, Jr. 1981. Oxygen deficiency in *Spartina alterniflora* roots: metabolic adaptation to anoxia. *Science* 214:439-441.

Mendelssohn, I. A. and E. D. Seneca. 1980. The influence of soil drainage on the growth of salt marsh cordgrass *Spartina alterniflora* in North Carolina. *Est. Coastal Mar. Sci.* 11:27-40.

Mooney, H. A., E. L. Dunn, A. T. Harrison, P. A. Morrow, B. Bartholemew and R. L. Hayes. 1971. A mobile laboratory for gas exchange measurements. *Photosynthetica* 5:128-132.

Morris, J. T. 1980. The nitrogen uptake kinetics of *Spartina alterniflora* in culture. *Ecology* 61:1114-1121.

Rao, D. N. and D. S. Mikkelsen. 1977. Effects of CO_2, CH_4 and N_2 on growth and nutrition of rice seedlings. *Plant and Soil* 47:313-332.

Sullivan, M. J. and F. C. Daiber. 1974. Response in production of cordgrass, *Spartina alterniflora,* to inorganic nitrogen and phosphorous fertilizer. *Chesapeake Sci.* 15:121-123.

Teal, J. M. and J. W. Kanwisher. 1961. Gas exchange in a Georgia salt marsh. *Limnol. Oceanogr.* 6:388-399.

Teal, J. M. and J. W. Kanwisher. 1966. Gas transport in the marsh grass, *Spartina alterniflora*. *J. Exp. Bot.* 17:355-361.

Valiela, I. and J. M. Teal. 1974. Nutrient limitation in salt marsh vegetation, pp. 547-563. *In:* R. J. Reimold and W. H. Queen (eds.), *Ecology of Halophytes*. Academic Press, New York.

Williams, G. J. and P. R. Kemp. 1978. Simultaneous measurement of leaf and root gas exchange of shortgrass prairie species. *Bot. Gaz.* 139:150-157.

Winer, B. J. 1971. *Statistical Principles in Experimental Design*. 2nd ed. McGraw Hill, New York. 897 pp.

SEDIMENTATION, NUTRIENT ACCUMULATION, AND EARLY DIAGENESIS IN LOUISIANA BARATARIA BASIN COASTAL MARSHES

R. S. Hatton, W. H. Patrick, Jr., and R. D. DeLaune
Laboratory for Wetland Soils and Sediments
Center for Wetland Resources
Louisiana State University
Baton Rouge, Louisiana

Abstract: Vertical marsh accretion, along with sediment and nutrient accumulation, were examined using ^{137}Cs dating and chemical analysis of soil cores taken from various marshes in Louisiana's Barataria Basin. Vertical accretion averaged 1.3 cm yr^{-1} and 0.7 cm yr^{-1} in levee and backmarsh areas respectively. Inorganic sedimentation increased fourfold from freshwater through salt marshes along a transect perpendicular to the Gulf Coast. The accretionary role of inorganic sediments is variable according to marsh type. Mineral sediment being deposited in the marsh is the principle source of iron, manganese, and phosphorus, but not of organic carbon or nitrogen. Manganese and phosphorus appear to be subject to substantial post-depositional remobilization.

Introduction

The controls and consequences of material exchange between salt marshes and estuarine waters have been the subject of considerable interest in recent years. The flux of sediments and associated nutrients to the marsh surface from estuarine waters is an important determinant of fertility and macrophyte production (DeLaune et al. 1979). But marshes themselves appear to perform concurrently as a source of biologically accessible nutrients released from inorganic sediments by virtue of chemical and biochemical transformations below the marsh surface (Nixon 1980). Evidence of this aspect of the accretionary development (Letzsch and Frey 1980) and early diagenesis of marsh soils (McCaffrey 1977; Lord 1980; DeLaune et al. 1981) is limited primarily to saline marshes. The biogeochemical consequences of depositional and other physical processes which distinguish the spatially predominant brackish, intermediate, and freshwater marshes have been the subject of only limited study (Kolb and Van Lopik 1966; Chabreck 1972). Our purpose is to discuss vertical accretion, nutrient accumulation, and early diagenesis in each of four marsh types in the Barataria Basin, Louisiana.

Setting

The unique physiography of Louisiana marshlands is a direct result of

ISBN 0-12-404070-5

the cyclic nature of deltaic sedimentation of the Mississippi River. Due to the process of primary channel switching, periods of fluvial dominance have alternated locally with periods of land loss and marine transgression (Coleman and Gagliano 1964), giving rise to a seaward thickening sequence of overlapping and contemporaneous delta lobes. The modern deltaic plain is thus characterized superficially by a series of interdistributary wetland basins of varying age which are separated by active and relict distributary channels. Barataria Basin is the youngest such interdistributary basin, bounded by the present course of the Mississippi River and by its most recently abandoned channel to the west, Bayou Lafourche (Fig. 1).

Historically, flooding during high river stage has maintained a supply of fluvial sediments to the interdistributary marshes, contributing structurally and chemically to their stability. Progressive channelization of the Mississippi River over the past century for flood control has prematurely terminated the fluvial phase of the Basin's development. Secondary landward redistribution of earlier deltaic, marine, marsh and bay sediments by tides, waves, and wind-induced water movements has replaced fluvial sources of inorganic sediments reaching the marsh surface. Hydraulic energy decreases with distance from the Gulf of Mexico (Byrne et al. 1976). Together with improper land use practices, such hydrologic alterations are generally believed to be responsible for accelerating the natural process of marsh deterioration during local interfluvial periods, resulting in a net loss of land along the historically advancing Louisiana coastline.

The four principal vegetative units we examined in Barataria Basin have been described in detail by Bahr and Hebrard (1976) (Fig. 1). Freshwater marshes (salinity $<1^{0}/_{00}$) which cover roughly 19% of the basin, are characterized by dense stands of *Panicum hemitomon*, *Eleocharis* sp. and

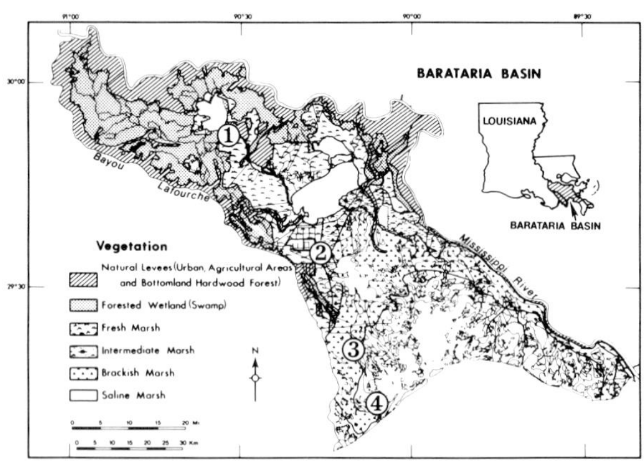

Figure 1. Barataria Basin, its location in Louisiana, vegetation distribution, and location of study sites. (Nos. 1-4).

Sagittaria falcata. Approximately 20% of the basin is termed brackish to intermediate marsh: the brackish system (5-10°/₀₀) is largely vegetated by *Spartina patens* with lesser amounts of *Distichlis spicata* and occasional intrusions of *S. alterniflora*. This predominance of *S. patens* increases into the intermediate marsh (10-15°/₀₀) where *S. alterniflora* is totally absent. Salt marsh vegetation covers some 14% of the basin, and is widely interspersed with open water bodies; *Juncus romerianus* and *Distichlis spicata* are secondary to *S. alterniflora* which represents approximately 60% of cover, and up to 95% locally.

Methods

Representative sites in each marsh type were selected on the basis of vegetation type, soil morphology, and hydrology. Cores for determination of vertical accretion rates were taken at known intervals laterally from the stream banks. Large diameter (15 cm), thin walled (0.2 cm) aluminum coreliners with a sharpened cutting edge were carefully twisted into the substrate to a depth of 50 cm. Cores were capped *in situ* before being raised to the surface, and sealed using specially designed PVC caps and PVC nylon straps. Entire cores were frozen and accurately sectioned into 2 cm horizontal subsamples using a band saw, allowing for blade thickness; freezing did not alter significantly the vertical disposition of sedimentary solids. Samples were weighed wet, and again after drying to constant weight at 100°C. Bulk density was calculated from dry weight of each section of known volume.

Activity profiles of ^{137}Cs were determined using an Ortec (model 8011-1620-S) coaxial, lithium-drifted germanium detector coupled to a (model 6240B) multichannel analyzer. Entire sections were counted for a variable time period determined statistically by ^{137}Cs activity, but not exceeding 2×10^4 sec. Vertical marsh accretion rates were computed directly from depth of burial of the dated (1963) ^{137}Cs activity maximum (DeLaune et al. 1978).

Selected cores were analyzed for total organic carbon (TOC), Kjeldahl nitrogen, and total phosphorus, iron, and manganese as outlined by DeLaune et al. (1981). Organic carbon was determined by dry combustion and organic matter calculated from this value using a multiplier of 1.724 (Wilson and Staker 1932). Kjeldahl nitrogen content of accurately weighed 2-3 g samples was determined by the method of Bremner (1965). Total phosphorus, iron and manganese were extracted by perchloric-nitric acid digestion of 0.5 g soil samples (Standard Methods 1975, pp. 169-170). Iron and manganese analyses were performed on a Fisher (90-750, ICAP) emission spectrometer. Orthophosphate was determined by the standard molybdate blue method (EPA-625-/6-74-003, 1974).

Accumulation rates (A) of organic carbon, nitrogen, iron, manganese, and phosphorus were calculated from mean dry weight concentrations (C, %) and the vertical accretion rate (R, cm yr^{-1}) using the formula

$$A(g\ m^{-2}\ yr^{-1}) = (CRD)10^6$$

where D is the soil bulk density in g cm^{-3}.

Where appropriate, the extent of subsurface depletion (F) and the rate of depletion (LR) were estimated from nutrient profiles using the expressions

$$F = 1 - \frac{C_h}{C_s}$$

and

$$LR\ (g\ m^{-2}\ yr^{-1}) = C_s DFR$$

C_s and C_h denote the surface layer and "steady state" concentrations, respectively (%).

Results

Sedimentation and Vertical Accretion

Accumulation of inorganic sediment increased fourfold from freshwater through salt marsh systems. Organic matter comprised a progressively smaller weight fraction of soil solids seaward in the Basin and towards levee areas but organic matter content per unit volume was remarkably uniform regardless of marsh type or location (Table 1). A predictable curvilinear relationship was found between percentile dry weight organic carbon content (C) and bulk density (D), as

$$D = 0.091 + 1.027 \times e^{-1.78 \times C}$$

This function yielded a mean organic carbon bulk density of approximately 0.09 g cm^{-3} when extrapolated to C = 100.

Geochronologies of ^{137}Cs indicated considerable regional (between marsh types) uniformity in vertical accretion rate, but pronounced variation within specific marshes (Table 1). Accretion rates in the four study areas averaged 1.3 cm yr^{-1} and 0.7 cm yr^{-1} in levee and backmarsh areas respectively.

Total Nutrient Concentrations

Except for particularly high levels of manganese in the freshwater marsh, total iron and manganese concentrations increased progressively in the seaward direction (Table 2). Both metals were consistently more enriched in levee soils than in lower density backmarsh soils. Total iron profiles (Table 3) indicated considerable density-related variability. Correlation analysis yielded a statistically significant positive correlation between iron (Fe) and bulk density (D) over all marsh types and depths such that

$$Fe = 7168 + 4143D \qquad (r = 0.45;\ P<0.01)$$

Table 1. Vertical accretion, bulk density, and sedimentation data. Error limits represent two S.D. (L = levee; B = backmarsh; N = number of cores).

		Vertical Accretion ($mm\ yr^{-1}$)		*Bulk Density*	*Organic Matter Accumulation*		*Inorganic Sediment Accumulation*		
Marsh Type	*Site*	*Mean*	*Range*	*($g\ cm^{-3}$)*	*%*	*($g\ m^{-2}\ yr^{-1}$)*	*%*	*($g\ m^{-2}\ yr^{-1}$)*	*N*
Freshwater	L	10.6	–	0.11 ±0.03	41 ±7	477	59 ±10	689	2
	B	6.5	3.1-6.9	0.09 ±0.01	52 ±6	306	48 ±6	280	7
Intermediate	L	13.5	13.0-14.0	0.18 ±0.04	33 ±6	797	67 ±12	1634	2
	B	6.4	3.8-10.6	0.08 ±0.01	52 ±4	269	48 ±4	243	7
Brackish	L	14.0	10.6-16.9	0.27 ±0.02	22 ±3	826	78 ±11	2954	4
	B	5.9	3.8-8.1	0.14 ±0.01	42 ±3	348	58 ±4	478	8
Salt	B	7.5	5.9-9.4	0.29 ±0.06	20 ±3	435	80 ±12	1740	7

Table 2. Total concentrations (Conc.) and accumulation rates (Acc.) of iron, manganese, phosphorus and nitrogen. Error limits represent two S.D. (L = levee; B = backmarsh; N = number of cores).

Marsh Type	Site	Iron Conc. ($\mu g\ g^{-1}$)	Iron Acc. ($g\ m^{-2}\ yr^{-1}$)	Manganese Conc. ($\mu g\ g^{-1}$)	Manganese Acc. ($g\ m^{-2}\ yr^{-1}$)	Phosphorus Conc. ($\mu g\ g^{-1}$)	Phosphorus Acc. ($g\ m^{-2}\ yr^{-1}$)	Nitrogen Conc. ($\mu g\ g^{-1}$)	Nitrogen Acc. ($g\ m^{-2}\ yr^{-1}$)	N
Freshwater	L	14729 ±3912	19	144 ±52	0.15	927 ±171	1.0	1.50 ±0.30	16	2
	B	9956 ±2007	7	114 ±12	0.07	944 ±82	0.5	1.80 ±0.20	9	7
Intermediate	L	18691 ±2672	46	69 ±12	0.17	648 ±88	1.5	1.23 ±0.18	28	2
	B	10079 ±1590	6	60 ±7	0.03	763 ±34	0.4	2.00 ±0.16	11	7
Brackish	L	20831 ±1251	79	100 ±12	0.38	624 ±135	2.4	0.66 ±0.04	25	2
	B	11830 ±1152	10	77 ±7	0.06	664 ±38	0.5	1.31 ±0.1	10	7
Salt	B	16857 ±854	29	83 ±16	0.14	668 ±112	1.1	0.69 ±0.2	11	7

Table 3. Bulk density and total nutrient profiles. Individual values represent means of two cores in levee areas and seven cores in backmarsh areas.

Depth cm	Bulk Density $g\ cm^{-3}$	Concentration C %	N %	P $\mu g\ g^{-1}$	Fe $\mu g\ g^{-1}$	Mn $\mu g\ g^{-1}$
Freshwater Marsh: Levee						
1	0.07	22.5	1.42	1100	15000	353
5	0.10	21.3	1.31	1196	12500	148
9	0.08	29.7	1.99	1221	7353	94
13	0.11	20.1	1.46	1083	11111	100
17	0.07	24.4	2.22	1063	8594	114
21	0.06	35.8	2.16	1043	9286	118
25	0.09	29.0	1.45	833	14103	115
29	0.18	14.6	0.94	694	20833	125
33	0.16	16.2	0.94	391	24219	129
37	0.14	17.7	1.10	643	24286	
Backmarsh						
1	0.05	33.0	2.00	1386	7678	169
5	0.07	33.6	1.99	1116	5632	82
9	0.07	33.6	2.18	1079	5641	75
13	0.05	37.0	2.14	1095	5310	93
17	0.05	37.4	2.22	990	5520	88
21	0.05	36.1	2.33	806	8049	104
25	0.06	32.2	1.78	838	12104	121
29	0.10	25.0	1.42	674	13653	126
33	0.19	11.0	0.84	714	18137	127
37	0.37	16.2	0.93	745	18623	148
Intermediate Marsh: Levee						
1	0.16	18.6	1.19	880	21760	113
5	0.18	16.2	1.16	769	19231	92
9	0.25	14.6	1.66	538	18396	72
13	0.16	15.1	1.27	664	19711	58
17	0.29	15.0	1.05	387	17453	59
21	0.21	14.5	0.67	510	20192	60
25	0.16	13.9	0.96	548	23039	73
29	0.12	22.7	1.36	720	24152	59
33	0.11	27.3	1.43	721	11441	56
37	0.09	27.9	1.52	702	11538	52
Backmarsh						
1	0.06	24.5	1.67	1012	12325	89
5	0.10	24.1	1.61	764	14421	74
9	0.08	27.9	1.70	726	12567	63
13	0.09	24.6	1.79	684	12675	61
17	0.07	29.5	2.15	725	8430	59
21	0.06	35.6	1.89	694	7616	44
25	0.07	33.6	2.18	744	6186	45
29	0.08	32.5	2.25	784	7177	55
33	0.08	29.6	2.22	755	8229	54
37	0.08	31.3	2.32	736	10434	54
Brackish Marsh: Levee						
1	0.25	14.1	0.73	1184	18275	150
5	0.26	17.6	0.74	759	18989	85
9	0.25	12.1	0.69	570	20313	108
13	0.29	8.8	0.64	604	21615	90
17	0.33	10.0	0.63	611	20779	83
21	0.24	14.7	0.65	544	20192	90
25	0.28	11.0	0.52	549	18359	91
29	0.29	11.9	0.64	520	24230	103
33	0.22	11.4	0.73	529	22449	100
37	0.26	13.4	0.67	369	23113	104
Backmarsh						
1	0.12	17.5	1.35	895	15789	100
5	0.18	18.3	1.07	722	16337	95
9	0.17	18.4	0.97	658	15716	89
13	0.13	24.5	1.26	689	11377	68
17	0.11	26.6	1.42	718	9654	61
21	0.11	25.9	1.30	619	9637	62
25	0.12	26.9	1.36	576	10130	66
29	0.12	22.7	1.24	647	11294	75
33	0.12	25.0	1.43	594	8595	69
37	0.11	32.9	1.81	567	7818	71
Salt Marsh: Backmarsh						
1	0.15	11.3	0.86	833	18666	97
5	0.22	11.1	0.77	733	18879	82
9	0.26	13.2	0.72	695	17303	77
13	0.27	11.4	0.57	642	16629	83
17	0.23	13.3	0.70	665	15442	73
21	0.31	14.3	0.81	671	16408	79
25	0.30	10.7	0.64	657	17169	87
29	0.30	13.5	0.77	631	15238	84
33	0.39	8.1	0.53	640	16898	83
37	0.45	5.4	0.38	511	15930	80

In contrast to iron, manganese profiles were uniformly characterized by well defined surface maxima and inconsistent density dependence (Table 3). Regression analysis of the composite data set, and of subsets according to marsh type and depth, failed to identify significant correlation between manganese and any other variable measured.

In sharp contrast to iron and manganese, phosphorus and nitrogen generally decreased in a seaward direction with higher backmarsh concentrations than in levee soils (Table 2). Surface layer phosphorus maxima characterized soil profiles throughout the study area (Table 3). Correlation analysis indicated positive correlation between phosphorus (P) and organic carbon (C) such that

$$P = 526 + 9.50C \qquad (r = 0.36;\ P<0.01)$$

and an inverse correlation with bulk density ($r = -0.25$; $P<0.01$). Nitrogen (N) and organic carbon were linearly related over all marsh types by the expression

$$N = 0.140 + 0.055C \qquad (r = 0.87;\ P<0.01)$$

which yields a mean carbon:nitrogen ratio of 18.2. Neither carbon nor nitrogen showed signs of progressive depletion with depth (Table 3).

Nutrient Accumulation and Remobilization

Iron, manganese and phosphorus accumulation followed seaward increasing trends. However, whereas iron accumulation increased approximately fourfold from freshwater through salt marshes, manganese varied by a factor of two or less over the corresponding distance. Accumulation of all nutrients was substantially greater in levee marshes than backmarshes, especially in the case of iron.

Subsurface depletion indices (F) calculated from soil profiles (Table 4) indicated that: (1) subsurface depletion of manganese and phosphorus was of the order of 17 to 72% and 24 to 55% respectively; (2) outflux of both manganese and phosphorus was faster in levee than in backmarsh areas; (3) not including the freshwater marsh, a general increase in outflux rate was evident from intermediate to saline marshes.

Discussion

Sedimentation and Vertical Accretion

The data presented quantified the seaward increase in inorganic sediment concentration, bulk density, and accumulation rate in Barataria Basin marshes. In the absence of a significant external source, these inorganic sediments must, by necessity, be of local origin as evidenced by concurrent vertical accretion and lateral erosion in immediately adjacent marshlands (Baumann 1980). Inorganic sediment supply increased seaward in the Basin

Table 4. Remobilization of phosphorus and manganese as indicated by soil profiles. (F = subsurface depletion index; LR = loss rate; L = levee; B = backmarsh).

		Phosphorus		Manganese	
Marsh Type	Site	F (%)	LR ($g\ m^{-2}\ yr^{-1}$)	F (%)	LR ($g\ m^{-2}\ yr^{-1}$)
Freshwater	L	25	0.33	72	0.30
	B	39	0.32	32	0.03
Intermediate	L	31	0.66	33	0.10
	B	28	0.14	41	0.02
Brackish	L	55	2.47	63	0.36
	B	30	0.22	33	0.03
Salt	B	24	0.44	17	0.04

in response to the parallel gradient in hydraulic energy and rate of sediment redistribution. This transgressive process of marsh soil development is characteristic of the cyclic sedimentation of the Mississippi River Delta. Similarly, local gradients in inorganic sedimentation follow directly from lateral hydraulic gradients extending inland from tidal streams and other water bodies.

Nutrient Concentrations and Accumulation Rates

The general parallel between concentrations and accumulation rates of iron and manganese and sedimentation patterns shows that inorganic sediments are the principal carriers of these metals within the Barataria system. This result is consistent with the general geochemistry of these metals (Carroll 1958). Correlation between iron and bulk density over all depths and marsh types is further evidence that iron remains closely associated with inorganic sediments and is not subject to substantial post-depositional mobility within this system. By contrast, lack of statistical correlation and distinct surface maxima observed in the case of manganese indicate substantial mobility, presumably due to release of soluble manganous ions upon reduction of manganic oxides.

Although iron and manganese display some geochemical similarity, their divergent behavior in the Barataria Basin marshes has sound chemical basis. Note that (1) natural oxides of manganese are more readily reduced than iron oxides, resulting in a proportionately greater release of soluble manganous ions into solution upon soil reduction (Gotoh and Patrick 1972, 1974); (2) whereas ferrous ions are subject to significant complexation by organic compounds in soil solution, soluble manganese appears to remain entirely in the unassociated and more mobile ionic form (Stevenson and

Ardakani 1972; Olomu et al. 1973); (3) the solubility of manganese sulfides is far greater than that of iron sulfides (Lisk 1972); (4) free iron but not manganese availability in reduced soils is strongly pH-dependent even under near-neutral conditions (Gambrell et al. 1977). These factors all contribute to an explanation of the lower mobility of iron as compared to manganese in flooded soil systems.

Significant post-depositional remobilization of iron has only been reported from a freshwater lake system (Gorham and Swain 1965), whereas manganese remobilization has been noted in freshwater lake (Robbins and Callendar 1975) and estuarine and tidal marsh deposits (McCaffrey 1977; Lord 1980).

The origin of exceptionally high levels of total manganese (and to a lesser extent, phosphorus) in freshwater marsh soils is not obvious from the data. Agricultural runoff is potentially important in this respect; Kemp (1975) has recorded enrichment of adjacent swamps by agricultural runoff containing phosphorus, although manganese was not examined.

Phosphorus accumulation is qualitatively in accord with sedimentation patterns and the concept that inorganic sediments are the principal carrier of phosphorus to this marsh system. The association of phosphorus with iron and thus the inorganic sediments has been suggested by several workers (Williams et al. 1971; Syers et al. 1973; Patrick and Khalid 1974). However, organic phosphorus may constitute up to 70% of total phosphorus in some lake sediments (Frink 1969; Sommers et al. 1972). It is possible therefore that as a result of plant uptake and slow mineralization under anaerobic conditions, a large proportion of phosphorus in marsh soils may be associated with the organic phase, especially in low density, highly organic substrates of freshwater marshes. Our data are inconclusive in this respect, and it is likely that additional effects, including differential remobilization, are involved.

Almost no data on total concentration and accretionary impoundment of nutrients are available from other marsh systems. McCaffrey (1977) measured iron and manganese flux to a New England brackish marsh of 36 g m^{-2} yr^{-1} and 0.78 g m^{-2} yr^{-1} respectively. These rates are comparable to our own values and those reported previously from Louisiana salt marshes (DeLaune et al. 1981). Organic carbon and total nitrogen concentrations reported for U.S. Atlantic marsh soils (Haines et al. 1977; Lord 1980; Nixon 1980) are generally lower than the Barataria values which we believe to be related to the low inorganic sediment content of these Louisiana marsh soils.

Nutrient Remobilization

Greater diffusive loss of nutrients from the soil surface to flood waters is anticipated in hydrologically dynamic streamside areas, first, because of the greater flow of water over the marsh surface and second, because of the higher concentration of available manganese and phosphorus in levee soils (Brannon 1973). Remobilization from levee soils would be further enhanced by plant uptake in proportion to productivity; not only is productivity greater, but the manganese content of the streamside plants is proportionately higher than the backmarsh form (DeLaune unpublished data). The apparent seaward increase in remobilization rates may be explained similarly by a seaward increase in availability of these nutrients (Hatton 1981) and tidal flushing which would enhance both diffusive loss and loss through detrital export.

Because of the assumptions involved, calculation of remobilization rates from soil nutrient profiles is subject to a large uncertainty. For example, it must be assumed that concentration in incoming sediments has not altered in the time span of the soil section examined; whereas agricultural inputs are potentially important in some areas, the extent of this effect is not obvious from the data. It must be assumed that upper layer concentrations in soil profiles represent incoming sediments, rather than surface accumulations resulting from oxidative re-precipitation. Data from an adjacent Louisiana salt marsh (DeLaune et al. 1981) indicate that whereas this assumption may be true of phosphorus, it may not hold for manganese. The extent of the oxidative re-precipitation is also likely to vary between marsh type according to soil redox condition and flushing time. Finally, selection of a steady state concentration at depth is complicated by significant density-related variability which we have simplified by using a mean subsurface value. But despite these assumptions and the uncertainty that surrounds the calculated remobilization rates, our results appear to be similar to those of McCaffrey (1977) who suggested a manganese export rate of 0.1 ± 0.02 g m^{-2} yr^{-1} from a New England brackish marsh. However, short term direct flux measurements at the marsh surface (Lord 1980) yield values which range from zero to several orders of magnitude greater than indicated by the sedimentary record.

We can compare potential remobilization by plant uptake with export rates indicated by soil profiles. Annual aboveground production of *Spartina alterniflora* in Louisiana salt marshes appears to be of the order of 3×10^3 g m^{-2} yr^{-1} (Hopkinson et al. 1978; White et al. 1978). Together with our preliminary estimates of manganese (1×10^4 g/g) and phosphorus (7×10^{-4} g/g) in aerial components of this species, uptake by *Spartina* is estimated at 0.3 g Mn m^{-2} yr^{-1} and 2.1 g P m^{-2} yr^{-1}. These values are generally higher, but of the same order of magnitude, as the values calculated from soil profiles. This result is interpreted as indicating that there must be considerable recycling *in situ* of manganese remobilized by plants.

In conclusion, despite a pronounced regional landward decrease in inorganic sedimentation in the Barataria Basin, ^{137}Cs geochronologies indicate a remarkable uniformity in vertical accretion rates in analogous marsh areas. This uniformity appears to be sustained by the accumulation of autochthonous organic matter, especially roots. The accretionary role of organic matter increases from saline through freshwater marshes. Vertical marsh accretion in both levee (1.3 cm yr^{-1}) and backmarsh (0.7 cm yr^{-1}) deposits are relatively high compared to estimates from Atlantic coast marshes.

Inorganic sediment is apparently the principal source of iron, manganese, and phosphorus to the Barataria Basin marshes. The general lack of correlation between bulk density and dry weight concentrations of manganese and phosphorus suggest significant influence of other factors which may include modern sediment source, contamination, and chemical and biochemical remobilization with release to estuarine waters. Calculations based on soil profiles and vertical accretion rates indicate (1) that remobilization of phosphorus and manganese from levee marshes exceeds that from backmarshes; and (2) there is a progressive increase from freshwater through salt marshes in phosphorus and manganese outflux rates. Considerable uncertainty surrounds the calculation of nutrient outflux rates from soil profiles; whereas remobilization processes appear to alter significantly the nutrient distribution of marsh soils, the quantitative significance in the fertility of estuarine waters cannot be estimated from the sedimentary record alone.

References Cited

Bahr, L. M. and J. J. Hebrard. 1976. Barataria Basin: biological characterization. Sea Grant Pub. #LSU-T-76-005. Center for Wetland Resources, Louisiana State U., Baton Rouge, LA

Baumann, R. H. 1980. Mechanisms of maintaining marsh elevation in a subsiding environment. M.S. Thesis, Louisiana State U., Baton Rouge, LA 90 p.

Brannon, J. M. 1973. Seasonal variation of nutrients and physicochemical properties in the salt marsh soils of Barataria Bay, Louisiana. M.S. Thesis, Louisiana State U., Baton Rouge, LA

Bremner, J. M. 1965. Inorganic forms of nitrogen, pp. 1179-1237. *I* : C. A. Black (ed.), *Methods of Soil Analysis, Part 2. Amer. Soc. Agronomy*, Madison, WI No. 9.

Byrne, P., M. G. Borengasser, G. Drew, R. A. Muller, B. L. Smith, Jr. and C. Wax. 1976. Barataria Basin: hydrologic and climatologic processes. Final Report to State Planning Office, La. Dept. of Transportation and Development, by Center for Wetland Resources, Louisiana State U., Baton Rouge, LA 175 p.

Carroll, D. 1958. Role of clay minerals in the transportation of iron. *Geochim. Cosmochim. Acta.* 14:1-28.

Chabreck, R. H. 1972. Vegetation, water, and soil characteristics of the Louisiana coastal region. Louisiana State U., Baton Rouge, LA, Agricultural Experiment Station, Bulletin #644.

Coleman, J. M. and S. M. Gagliano. 1964. Cyclic sedimentation in the Mississippi River deltaic plain. Louisiana State U., Baton Rouge, LA, Coastal Studies Institute, Tech. Rept. #16.

DeLaune, R. D., R. J. Buresh and W. H. Patrick, Jr. 1979. Relationship of soil properties to standing crop biomass of *Spartina alterniflora* in a Louisiana marsh. *Est. Coastal Mar. Sci.* 8:477-487.

DeLaune, R. D., W. H. Patrick, Jr. and R. J. Buresh. 1978. Sedimentation rates determined by ^{137}Cs dating in a rapidly accreting salt marsh. *Nature* 275:532-533.

DeLaune, R. D., C. N. Reddy and W. H. Patrick, Jr. 1981. Accumulation of plant nutrients and heavy metals through sedimentation processes and accretion in a Louisiana salt marsh. *Estuaries* 4:328-334.

Frink, C. R. 1969. Fractionation of phosphorus in lake sediments: analytical evaluation. *Soil Sci. Soc. Am. Proc.* 33:326-328.

Gambrell, R. P., R. A. Khalid, M. G. Verloo and W. H. Patrick, Jr. 1977. Transformation of heavy metals and plant nutrients in dredged sediments as affected by oxidation-reduction potential and pH. Vol. II: Materials and Methods/Results and Discussion. U.S. Army Corps of Engineers Contract Rept. D-77-4, 309 p. Waterways Experiment Station, Vicksburg, MS

Gorham, E. and D. J. Swain. 1965. The influence of oxidizing and reducing conditions upon the distribution of some elements in lake sediments. *Limnol. Oceanogr.* 10:268-279.

Gotoh, S. and W. H. Patrick, Jr. 1972. Transformations of manganese in a waterlogged soil as influenced by Eh and pH. *Soil Sci. Soc. Am. Proc.* 36:738-742.

Gotoh, S. and W. H. Patrick, Jr. 1974. Transformation of iron in a waterlogged soil as influenced by redox potential and pH. *Soil Sci. Soc. Am. Proc.* 38:66-71.

Haines, E. A., R. Chalmers and B. Sherr. 1977. Nitrogen pools and fluxes on a Georgia salt marsh, pp. 241-254. *In:* M. L. Wiley (ed.), *Estuarine Processes, 2.* Academic Press, New York.

Hatton, R. S. 1981. Aspects of marsh accretion and geochemistry: Barataria Basin, Louisiana. M.S. Thesis, Louisiana State U., Baton Rouge, LA 116 p.

Hopkinson, C. S., J. G. Gosselink and R. T. Parrondo. 1978. Aboveground production of seven marsh plant species in coastal Louisiana. *Ecology* 59:760-769.

Kemp, G. P. 1975. Agricultural runoff and nutrient dynamics of a swamp forest in Louisiana. M.S. Thesis, Louisiana State U., Baton Rouge, LA

Kolb, C. R. and J. R. Van Lopik. 1966. Depositional environments of the Mississippi River deltaic plain-Southeastern Louisiana, pp. 17-61. *In:* M. L. Shirley (ed.), *Deltas in Their Geologic Framework.* Houston Geologic Society, Houston, TX

Letzsch, W. S., and R. W. Frey. 1980. Deposition and erosion in a Holocene salt marsh, Sapelo Island, Georgia. *J. Sed. Pet.* 50:529-542.

Lisk, D. J. 1972. Trace metals in soils, plants and animals. *Adv. Agron.* 24:267-325.

Lord, J. C. 1980. The chemistry and cycling of iron, manganese, and sulfur in salt marsh sediments. Ph.D. Dissertation, U. Delaware, Newark, DE

McCaffrey, R. J. 1977. A record of the accumulation of sediment and trace metals in a Connecticut, USA, salt marsh. Ph.D. Dissertation, Yale U., New Haven, CT 156 p.

Nixon, S. W. 1980. Between coastal marshes and coastal waters – a review of twenty years of speculation and research on the role of salt marshes in estuarine productivity and water chemistry, pp. 437-525. *In:* P. Hamilton and K. B. MacDonald (eds.), *Estuarine and Wetland Processes.* Plenum Press, New York.

Olomu, M. O., G. J. Racz and C. M. Cho. 1973. Effect of flooding on the Eh, pH, and concentrations of Fe and Mn in several Manitoba soils. *Soil Sci. Soc. Am. Proc.* 37:220-224.

Patrick, W. H., Jr. and R. A. Khalid. 1974. Phosphate release and sorption by soils and sediments: effect of aerobic and anaerobic conditions. *Science* 186:53-55.

Robbins, J. A. and E. Callendar. 1975. Diagenesis of manganese in Lake Michigan sediments. *Am. J. Sci.* 275:512-533.

Sommers, L. E., R. F. Harris, J. D. H. Williams, D. E. Armstrong and J. K. Syers. 1972. Determination of total organic phosphorus in lake sediments. *Limnol. Oceanogr.* 15:301-304.

Stevenson, F. J. and M. S. Ardakani. 1972. Organic matter reactions involving micronutrients in soils, pp. 79-114. *In:* R. D. Dinauer (ed.), *Micronutrients in Agriculture.* Soil Sci. Soc. Amer., Inc., Madison, WI

Syers, J. K., R. F. Harris and D. E. Armstrong. 1973. Phosphate chemistry in lake sediments. *J. Env. Qual.* 2:1-14.

White, D. A., T. E. Weiss, J. M. Trapani and L. B. Thien. 1978. Productivity and decomposition of the dominant salt marsh plants in Louisiana. *Ecology* 59:751-759.

Williams, J. D. J., J. K. Syers, S. S. Shukla, R. F. Harris and D. E. Armstrong. 1971. Levels of inorganic and total phosphorus in lake sediments as related to other sediment parameters. *Env. Sci. Technol.* 5:1113-1120.

Wilson, B. D. and E. V. Staker. 1932. Relation of organic matter to organic carbon in the peat soils of New York. *J. Am. Soc. Agron.* 24:477-481.

ECOLOGICAL IMPLICATIONS OF THE VERTICAL DISTRIBUTION OF MEIOFAUNA IN SALT MARSH SEDIMENTS

Walter B. Sikora and Jean Pantell Sikora
Coastal Ecology Laboratory
Center for Wetland Resources
Louisiana State University
Baton Rouge, Louisiana

Abstract: We have examined the ecological implications of the vertical distribution of nematodes in anaerobic salt marsh sediments. High belowground productivity of *Spartina alterniflora* results in a high detrital component in the anaerobic marsh soils. Nematodes are evolutionarily adapted to these anaerobic soils and important in the transfer of energy from the detritus to other consumers. The energy pathway in anaerobic sediments, however, is not the same as the carbon cycle in aerobic sediments. A new concept of energy pathways involving nematodes is presented, which links the aerobic and anaerobic components of the marsh ecosystem. Export from the marsh which links the *S. alterniflora* marsh with the pelagic food web is discussed.

Introduction

The holistic approach to the analysis of ecosystems attempts to identify and to examine the emergent properties of the system (Kerr 1976). These include such aspects of ecosystem structure and function as community diversity and niche breadth (Lane et al. 1975), energy flow (Teal 1962; Day et al. 1973), and nutrient cycling (Pomeroy 1970; O'Neill et al. 1975). The stabilizing influence of heterotrophs on ecosystem processes has been recognized (Wiegert and Owen 1971; May 1974) and defined as the rate regulation of internal ecosystem function (Reichle et al. 1975). We will, within this framework of heterotrophic control of ecosystem function, discuss the ecological implications of the vertical distribution of meiofauna in salt marsh sediments.

The *Spartina alterniflora* Marsh

Large expanses of estuarine and coastal salt marshes occur along the mid-Atlantic and Gulf coasts of the United States. The coastal zone of the state of Louisiana comprises more than 28,000 km^2 of marshes and estuaries, or about 40% of the total marsh area of the United States (Day et al. 1973). A direct relationship has been postulated between the vast salt marshes and Louisiana's important fisheries (Turner 1977). This paper will examine some of the recent evidence from salt marsh studies that adds to the impetus for

ISBN 0-12-404070-5

change in the paradigm of salt marsh energy flow, relating this dynamic marsh benthic habitat to the pelagic shelf waters.

Coastal salt marshes are dominated by the cordgrass *Spartina alterniflora*. This is a well-studied, highly productive grass, with estimates of aboveground production ranging from 0.5 to 3 Kg m^{-2} yr^{-1} (Day et al. 1973; Hardisky 1980). Belowground production of rhizomes and roots is from 1 to 4 times the aboveground production (Valiela et al. 1976; Gallagher et al. 1977). In addition to the belowground production, a considerable fraction of photosynthate is potentially translocated from aerial culms to the roots during culm senescence (Turner and Gosselink 1975). These plants are therefore allocating a major fraction of the photosynthate, both during active growth and during senescence, to the underground portion of the plant and the foodweb dominated by the sediment microbes and the meiobenthos (Gallagher et al. 1977).

Sediments of *Spartina* Marshes

The typical *Spartina alterniflora* marsh is characterized by finer sediments than *Spartina patens* marshes or marshes with mixed stands of vegetation (Gallagher et al. 1977). In those Louisiana marshes that no longer receive direct sediment inputs from the Mississippi River because of leveeing over the past century, the probable source of sediments is from transport of resuspended sediments from other areas within the drainage basin (DeLaune and Patrick 1980). Marsh soils are characterized by a high peat content, high water content, and relatively low bulk densities because of the high peat and water content.

Because salt marsh sediments are sometimes submerged and are usually waterlogged, they are anaerobic, with only a very thin surface zone becoming oxidized, or aerobic, at times. No free oxygen is found below +350 mV (Patrick and DeLaune 1977), with the consequence that all sediments examined are anaerobic and from moderately to highly reducing, which is typical of coastal marsh sediments (Whelan 1974; Mendelssohn et al. 1981).

Meiofauna of the Salt Marsh

Meiofauna are operationally defined as those benthic invertebrates that pass through a sieve with 500 μm openings and are retained on a sieve with smaller openings, usually 63 μm. The salt marsh meiofauna community is usually dominated by nematodes (Day et al. 1973). Nematodes comprised 87.3% (S. E. 1.7%; N = 168) of the total meiofauna found during a recent study in an estuarine salt marsh in Barataria Bay, near Leeville, Louisiana (DeLaune et al. 1981). Total numbers of meiofauna were highest in early March and lowest in October, with a mean of $2.9 \times 10^6/m^2$. Copepods were usually second in abundance, with small polychaetes, oligochaetes, and turbellarians comprising most of the rest. Kinorhynchs, ostracods, and rotifers were found occasionally. Numbers of minor taxa agree well with the numbers reported by Bell (1980) for samples from a South Carolina salt marsh.

Samples taken in the low marsh in the same South Carolina estuarine marsh complex from which Bell's samples were taken had nematode numbers comparable to Louisiana samples, i.e. $3.0 \times 10^6/m^2$ (Sikora 1977). Abundance and distribution of the meiofauna in the *S. alterniflora* coastal marshes of Louisiana and South Carolina appear to be basically similar.

Methods and Materials

In order to examine the distribution of nematodes with depth in anaerobic salt marsh sediments, we took a series of cores to a depth of 12 cm from a *S. alterniflora* marsh in Barataria Bay, Louisiana (previously described by Day et al. 1973) during October of 1981. Samples used for obtaining the seasonal changes in density had been taken from the same site in an earlier (September 1979 to October 1980) study (DeLaune et al. 1981). The intact cores were frozen in the core tubes in a vertical position with dry ice in the field immediately after removal from the marsh. Measurements of redox potential in the sediments were made using a platinum electrode-calomel half cell and a millivolt meter at the same sampling site (within 50 cm) after a one hour equilibration period. Frozen cores were returned to the laboratory and sectioned while frozen into 1 cm sections that were placed in vials with 10% buffered formalin and stained with Rose Bengal. After fixation and staining each section was sieved and the contents of the sieves were counted under a dissecting microscope.

Redox potential (Eh) was calculated by adding the potential of a calomel electrode against a standard hydrogen electrode (+250 mV) to the measured potential.

Results

Eh for the uppermost section was 78.5 ± 46.9 mV (Fig. 1), indicating that no free oxygen was present (Patrick and DeLaune 1977). Redox potential measured at 4 cm was 90.0 ± 63.2 mV, not significantly different from the uppermost reading. Eh dropped rapidly to -110.7 ± 18.9 mV at 7 cm, and to -146.7 ± 17.1 mV at 10 cm. We found 45.6% of the total nematodes in the uppermost sections (0-1 cm). The next sections (1-2 cm) contained 39.9%. Numbers dropped rapidly below this depth, with 12.3% found in the 2-3 cm sections, and 1.0% found in the 3-4 cm sections. Below this level we found a few nematodes in sections down to the 9-10 cm sections comprising only an additional 2%.

Discussion

Nematode Distribution

Before discussing the vertical distribution of nematodes, we will comment on nematode distribution through geologic time. The first ecosystem to evolve was probably a benthic, detritus-based system (Sibert and Naiman 1980). Contemporary views on the origin of the metazoa acknowledge the

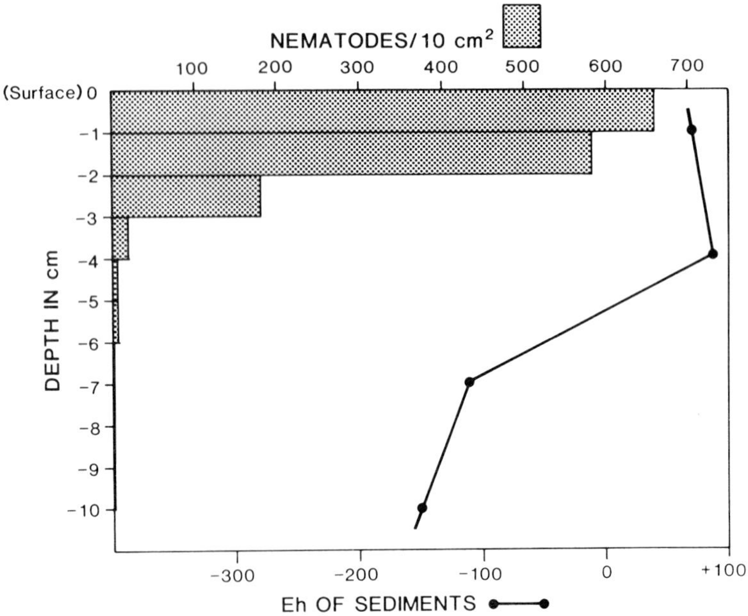

Figure 1. Distribution of nematodes with depth, and Eh of sediments in Barataria Bay, Louisiana.

benthic beginnings of multicellular life (Dahl 1963; Barnes 1974; Boaden 1975) during the Precambrian era. There is controversy over whether meiofauna are primitive groups, primarily anaerobic (Boaden 1975; Platt 1981), or whether their small size and simple structure is the result of secondarily evolved, progenetic adaptations (Swedmark 1964; Gould 1977). What is important, however, is that all of the taxa commonly found in the meiofauna of the marsh, except the copepods, are evolutionarily well adapted for existence in the anaerobic marsh sediments.

It has been known for a century that nematodes inhabit anaerobic muds (Bunge 1888; Juday 1908; Moore 1931; Lindeman 1942). Moore (1931) described their distribution with depth, commenting that the copepods are found only at the surface. Lindeman (1942) reported on the ability of benthic organisms to withstand anaerobiosis and included data on nematodes.

The presence of large numbers of nematodes in anaerobic sediments is not an enigma. Many of the benthic infauna have evolved very efficient anaerobic pathways yielding as much net energy as does aerobic respiration (Pamatmat 1980). Some of these organisms even continue the use of anaerobic metabolism to satisfy their energy requirements when oxygen is available (Chen and Awapara 1969).

The point of controversy is not whether there are large numbers of nematodes in the sediments, but whether the sediments are anaerobic. The cause of this controversy seems to be the unfamiliarity of some biologists with the concept of redox couples and their role in the chemistry of sediments. A number of comprehensive reviews of redox equilibria in freshwater sediments (Golterman 1969), marine waters and sediments (Morris and Stumm 1967), and marsh sediments (Patrick and DeLaune 1977) have demonstrated that no free oxygen occurs below +350 mV and that all other redox couples are at equilibrium in their most highly oxidized states when free oxygen is present. The dark layer in some sediments (as illustrated in Coull and Bell 1979), which is commonly called "the redox layer," is the zone (≅ + 125 mV) where ferric iron is reduced to ferrous iron (Patrick and DeLaune 1977), and is not the dividing line between "oxic" and "anoxic" sediments as Coull and Bell (1979) show it. Reise and Ax (1980) apparently are making a similar assumption when they state ".... 0 mV .... is the approximate border between aerobic and anaerobic conditions."

We have measured Eh levels indicating that no free oxygen is present in the sediments at our study site in the Louisiana salt marsh below 0.5 cm and have found nematodes present in these anaerobic sediments (Fig. 1) in numbers quite similar to those we found in previous studies in a South Carolina salt marsh (Sikora 1977).

New Concepts of Energy Pathways

A series of recently published papers on sulfate reduction and pyrite formation in salt marsh sediments (Howarth 1979; Howarth and Teal 1979, 1980; Peterson et al. 1980) has led to a revision of the concept of energy flow through salt marsh ecosystems. An appreciation of the complexity and an indication of the importance of the anaerobic microbial processes in salt marsh sediments has been emphasized previously (Christian et al. 1975; Pomeroy et al. 1976; Christian and Hall 1977). The importance of the sulfur cycle in anaerobic sediments to energy flow in benthic detrital food webs has also been reported previously (Jørgenson 1977). However, the decoupling of the surface community metabolism from that of the anaerobic community below, the fact that carbon cycling alone is not proportional to total energy flow, and the failure of a carbon budget to represent energy flow in a salt marsh ecosystem did not occur until Howarth and Teal's landmark paper (1979).

Sulfate reduction is many times more important in the degradation of organic matter in the salt marsh than oxygen respiration and denitrification combined (Howarth and Teal 1980). The immediate energy yield is significantly less for anaerobic sulfate reduction than for aerobic respiration. The difference in energy yield is stored in reduced compounds that may be transported by subsurface movement of pore waters and utilized by microbial populations as an energy source in the creek banks and in creeks draining the marsh. These compounds may also be used at a later time. There is a net oxidation of pyrite in the summer and a net build-up during the rest of the year (Howarth and Teal 1979).

If we compare the numbers of nematodes (Fig. 2) in a subtidal salt

NEMATODE STANDING STOCK

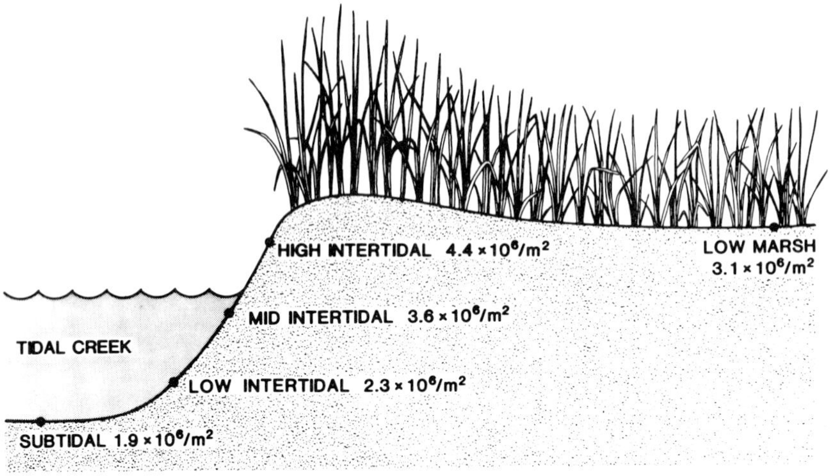

Figure 2. Distribution of nematodes along a transect from the subtidal creek bottom to the low marsh in North Inlet, South Carolina.

marsh creek with the numbers along a transect through low intertidal, mid-intertidal, high intertidal creek bank to the low marsh, we find the annual mean number of subtidal nematodes is 45% of the mean number of high intertidal creek bank nematodes (Sikora 1977). In the high marsh, or short *S. alterniflora* zone, nematode numbers are an order of magnitude lower (Bell 1980). Microbial biomass, as calculated from sediment ATP determinations, is higher in the tall *S. alterniflora* zone just above the high intertidal creek bank, than in the short *S. alterniflora* zone in the marsh (Christian et al. 1975).

Sulfate reduction in the low marsh and high marsh is quite different (King and Wiebe 1980), with higher activity in the tall *S. alterniflora* zones. Activity is also lower in the subtidal sediments (Wiebe 1979). It is probable that the higher number of nematodes in the high intertidal creek bank and the higher levels of microbial biomass in the tall *Spartina* zone are related either directly or indirectly through some aspect of subsurface energy flow in the salt marsh and the sulfur cycle.

Concomitant with the use of organic matter as an energy source and sulfate as an electron acceptor, the anaerobic microbial populations cause the solubilization of that phosphate that has precipitated as iron compounds on clay particles (Patrick and DeLaune 1977). This soluble phosphate is rapidly taken up or immobilized by other microorganisms (Ehrlich 1981). These insoluble stores of phosphates are not released until the death of the cell. Release of this temporarily immobilized phosphate is usually accomplished by grazers: the protozoa and the small metazoa such as nematodes and oligochaetes (Pomeroy and Bush 1959; Johannes 1965; Barsdate et al. 1974; Gardner et al. 1981). Nitrogen products are probably similarly regenerated (Pomeroy 1970).

Regulation of Energy Flow and Nutrient Cycles

The regulation of the rate of energy flow in the salt marsh sediments is accomplished by control of the rate of increase of microbial populations. Ecosystem theory postulates that a persistent, stable ecosystem will have as one of its state variables or essential components an organic energy pool as an alternative to its autotrophic energy base (O'Neill et al. 1975). An example from terrestrial ecosystems would be the litter and soil organic matter and structural biomass of the trees in a forest ecosystem. In the salt marsh, the detritus component of the sediments plays this role. One characteristic of the energy reservoir of an ecosystem is its relative unavailability. High levels of microbial activity on detritus depend on the comminution of larger macrophyte fragments to expose greater surface area and on the grazing of the microbial population to prevent senescence (Reichle et al. 1975). Bacterial numbers and rate of detritus decomposition increase in the presence of grazers (Fenchel 1970, 1972; Hargrave 1970; Mann 1972; Fenchel and Harrison 1976), as does nutrient regeneration (Johannes 1965; Pomeroy 1970). The presence of nematodes significantly increases the amounts of nitrogen and phosphorus in circulation at any given time (Anderson et al. 1981).

Microbial production in the anaerobic salt marsh sediments is as high as it would be in aerobic sediments. The efficiency of substrate utilization is the same (about 60%) whether aerobic or anaerobic pathways are utilized (Payne 1970). This production is not easily available to consumers. After metazoans were shown to possess neither sufficient levels of cellulase nor sufficient gut retention time to permit utilization of detritus (Kristensen 1972), some researchers assumed that the microbial populations on detritus particles were what was being assimilated. Many researchers have found this diffuse organic source to be insufficient to maintain macrofauna (Tenore et al. 1977; Wetzel 1977; Cammen 1980). In addition to the detritus particles and their associated bacteria and fungi, large numbers of small flagellates and ciliates, microalgae including diatoms, and the meiofauna are found in the sediments, which provide the majority of the nutrition for the deposit feeder (Whitlatch 1981). The importance of meiofauna in the nutrition of the "non-selective deposit feeder" or the "detritivore" has been described in some detail (Thorson 1966; Wolff 1973; Elmgren 1977; Tenore et al. 1977; Platt and Warwick 1980; Platt 1981). The importance of the nematodes, particularly, has been suggested (Coull 1973; Platt 1981). The use of ATP analysis of sediments showed 90% of the total sediment ATP to be contributed by nematodes (Sikora et al. 1977). Biomass estimates from actual counts of diatoms, ciliates, flagellates, and nematodes associated with detritus gave similar results (Fenchel and Harrison 1976). Nematodes are not only present in the marsh in high percentages, but they reach the highest density found in any of the habitats they occupy in the salt marsh-estuarine complex (Platt 1981). From the point of view of system function, the grazing of microbial populations by the nematodes regulates the rate of energy transfer from *S. alterniflora* detritus to the food web in the marsh and the rates of nutrient cycles within the marsh sediments (Kitchell et al. 1979).

Ecological Implications of Nematode Distribution

Recognition of the fact that nematodes " are of major energetic importance in littoral ecosystems, forming a significant part of the diet of many other animals" (Platt and Warwick 1980) has been hampered by a unique characteristic of the nematodes themselves. They have no internal or external skeleton but maintain their body form with high hydrostatic pressure. Any mechanical abrasion or rapid pH change, such as that which occurs during ingestion, will cause the nematode to rupture into 2 to 8 "strings", separating along the margins of the hypodermal cords (Bird 1971). These "strings" are usually not recognizable as nematode fragments, and are much more susceptible to rapid digestion than intact organisms. Although ingestion of nematodes by macrofauna and juvenile shrimp and fishes has been demonstrated (Mann 1972; Chao and Musick 1977; Sikora 1977; Condrey 1979; Sheridan and Livingston 1979; Levine 1980), numbers derived from gut content analysis probably underestimate the true ingestion rate.

The feeding of young fish and shrimp on the nematodes forms a link between the anaerobic and the aerobic food webs. It also forms a transport mechanism for the movement of energy produced in the marsh to the offshore waters. Primary production in the *S. alterniflora* salt marsh results in high belowground productivity. Populations of anaerobic bacteria utilize the energy from the detritus for rapid microbial growth, which is grazed by nematodes. Grazing of microbial populations causes increased rates of nutrient regeneration in addition to energy flow to the grazers. Nematode numbers are reduced by the predation of the post-larval shrimp and fish which utilize the salt marsh and the creeks as a nursery area. These species leave the marsh and the creeks for deeper, offshore waters as the temperature falls. Gravid females of these fish species will usually release more than one half million eggs each (Perret et al. 1971; Gaidry and White 1973) during the fall and winter months. Pelagic eggs and yolk-sac larvae are exceedingly vulnerable to predators. No quantitative estimates of predation on eggs or larvae is found in the literature (Dahlberg 1963), but it is probably that no more than 1% of the yolk-sac larvae from each female's spawn survive to return to the estuary and the marshes (Gunter 1956). The adult fish and shrimp, with at least 95% of the carbon, nitrogen, and phosphorus content of their bodies obtained from feeding in the marsh, are also a link from the marsh to the pelagic zones. Considering that 97.5% of the commercial catch in the Gulf consists of estuarine-dependent species (Gunter 1967), this represents a considerable export overlooked by those researchers looking for carbon export in particulate detritus in the water column.

Measurement of Energy Flow

Primary production of the dominant macrophyte, *S. alterniflora*, is not directly linked to the food webs within the marsh-estuarine complex (Odum and Smalley 1959). Decoupling of the sediment microbial community and plant growth (Christian et al. 1978) introduces a time lag between primary production and detritus utilization. Microbial populations are regulated by grazing of nematodes and other consumers (Sorokin 1978). This acts as a feedback mechanism to maintain a rapid transfer of energy when system demand is high, regardless of levels of primary productivity. The role of the consumer complex does not provide stability in the true system sense. It does, however, through its regulatory role, dampen deviations from equilibrium and thus performs a homeostatic function (Lee and Inman 1975).

The further decoupling of the surface community metabolism from that of the anaerobic community beneath the surface (Howarth and Teal 1979) leaves us with no easy method of assessing energy flow. Sulfate reduction appears to be as important in subtidal soft sediments adjacent to marshes and in the very nearshore sediments as it is in marsh sediments (Wiebe 1979). If, because of sulfate reduction and other anaerobic metabolic processes, carbon cycling is no longer proportional to energy flow, and the "conventional use of benthic O_2 uptake data and 'ideal' RQ values (e.g., 0.85) – sweeps an important category of energy under a conceptual rug" (Rich 1979), then we cannot use oxygen uptake measurements as a measure of energy flow in the marshes, the marsh creeks, or the nearshore soft sediments. The few studies that have successfully used benthic oxygen uptake measurements as a measure of benthic community metabolism to relate to other system functions such as nutrient cycling have all been done in sandy sediments (Sørensen et al. 1979; Kemp and Boynton 1980; Hargrave and Phillips 1981). Because of larger pore spaces between the sediment particles, coarser grained sediments have a deeper oxygenated layer (Delamare Deboutteville 1960), extending to as deep as 30 cm (Coull and Bell 1979). Measurement of oxygen uptake over sandy sediments does yield a measure of benthic community metabolism when the benthic community is aerobic.

The uncoupling of energy flow and carbon in anaerobic marsh sediments has another important consequence. A significant fraction of the energy fixed by *S. alterniflora*, anaerobically decomposed, and stored in reduced inorganic sulfur compounds (Howarth 1979; Howarth and Teal 1979, 1980) is transported from the marsh peat to the marsh creeks. There it is used as an energy source by chemoautotrophic bacteria to fix dissolved CO_2 (Peterson et al. 1980). This would result in the bacteria having a $\delta\,^{13}C$ value resembling that of phytoplankton, rather than that of the *S. alterniflora* from which the energy was derived. Organisms feeding on nematodes that have grazed on these bacteria could have $\delta\ ^{13}C$ values different from that of *S. alterniflora*. The linking of the anaerobic portion of the food web to the aerobic portion by nematodes has led to a conflation of the concepts of energy flow with $\delta\ ^{13}C$ values that can only be separated by further research.

Acknowledgments

We thank R. E. Condrey, W. B. Johnson, and I. A. Mendelssohn for their thoughtful comments on an earlier draft, and R. D. DeLaune for providing assistance for the Eh measurements. Publication No. LSU-CEL-81-35. This work was supported in part by the Louisiana Sea Grant College Program, NOAA grant NA81AA-D-00103.

References Cited

Anderson, R. V., D. C. Coleman, C. V. Cole and E. T. Elliott. 1981. Effect of the nematodes *Acrobeloides* sp. and *Mesodiplogaster lheritieri* on substrate utilization and nitrogen and phosphorus mineralization in soil. *Ecology* 62:549-555.

Barnes, R. D. 1974. *Invertebrate Zoology* 3rd ed. W. B. Saunders Co., Philadelphia, PA 870 pp.

Barsdate, R. J., R. T. Prentki and T. Fenchel. 1974. Phosphorus cycle of model ecosystems: significance for decomposer food chains and effect of bacterial grazers. *Oikos* 25:239-251.

Bell, S. S. 1980. Meiofauna-macrofauna interactions in a high marsh habitat. *Ecol. Monogr.* 50:487-505.

Bird, A. F. 1971. *The Structure of Nematodes.* Academic Press, New York. 318 pp.

Boaden, P. J. S. 1975. Anaerobiosis, meiofauna and early metazoan evolution. *Zool. Scr.* 4:21-24.

Bunge, G. 1888. Über das Sauerstoffbedurfnis der Schlambewohner. *Zeitschr. Physiol. Chem.* 12:565-567.

Cammen, L. M. 1980. The significance of microbial carbon in the nutrition of the deposit feeding polychaete *Nereis succinea. Mar. Biol.* 61:9-20.

Chao, L. N. and J. A. Musick. 1977. Life history, feeding habits, and functional morphology of juvenile scianid fishes in the York River estuary, Virginia. *Fish. Bull.* 75:657-702.

Chen, C. and J. Awapara. 1969. Effect of oxygen on the end-products of glycolysis in *Rangia cuneata. Comp. Biochem. Physiol.* 31:395-401.

Christian, R. R., K. Bancroft and W. J. Wiebe. 1975. Distribution of microbial adenosine triphosphate in salt marsh sediments at Sapelo Island, Georgia. *Soil Sci.* 119:89-97.

Christian, R. R., K. Bancroft and W. J. Wiebe. 1978. Resistance of the microbial community within salt marsh soils to selected perturbations. *Ecology* 59:1200-1210.

Christian, R. R. and J. R. Hall. 1977. Experimental trends in sediment microbial heterotrophy: Radioisotopic techniques and analysis, pp. 67-88. *In:* B. C. Coull (ed.), *Ecology of Marine Benthos.* U. South Carolina Press, Columbia, SC

Condrey, R. E. 1979. Draft environmental impact statement and fishery management plan for the shrimp fishery of the Gulf of Mexico, United States waters. Gulf of Mexico Fishery Management Council, Tampa, FL 220 pp.

Coull, B. C. 1973. Estuarine meiofauna: A review: Trophic relationships and microbial interactions, pp. 499-512. *In:* L. H. Stevenson and R. R. Colwell (eds.), *Estuarine Microbial Ecology.* U. South Carolina Press, Columbia, SC

Coull, B. C. and S. S. Bell. 1979. Perspectives of marine meiofaunal ecology, pp. 189-216. *In:* R. J. Livingston (ed.), *Ecological Processes in Coastal and Marine Systems.* Plenum Press, New York.

Dahl, E. 1963. Main evolutionary lines among recent crustacea, pp. 1-26. *In:* H. B. Whittington and W. D. I. Rolfe (eds.), *Phylogeny and Evolution of Crustacea.* Spec. Publ., Museum of Comparative Zoology, Cambridge, MA

Dahlberg, M. D. 1963. A review of survival rates of fish eggs and larvae in relation to impact assessment. *Mar. Fish. Rev.* 41:1-12.

Day, J. W., W. Smith, P. Wagner and W. Stowe. 1973. Community structure and carbon budget of a salt marsh and shallow bay estuarine system in Louisiana. Louisiana State U., Center for Wetland Resources, Sea Grant Publ. No. LSU-SG-72-04. 79 pp.

Delamare Deboutteville, C. 1960. *Biologie des Eaux Souterraines Littorales et Continentales.* Hermann, Paris. 740 pp.

DeLaune, R. D. and W. H. Patrick. 1980. Nitrogen and phosphorus cycling in a Gulf Coast salt marsh, pp. 143-151. *In:* V. S. Kennedy (ed.), *Estuarine Perspectives.* Academic Press, New York.

DeLaune, R. D., C. J. Smith, W. H. Patrick, Jr., W. B. Sikora, J. P. Sikora and G. A. Hambrick, III. 1981. The use of chemical dispersants as a method of marsh restoration following an oil spill. Rept. Exxon Research and Engineering. 64 pp.

Ehrlich, H. L. 1981. *Geomicrobiology.* Marcel Dekker, Inc., New York. 393 pp.

Elmgren, R. 1977. Baltic benthos communities and the role of the meiofauna. *Contr. Askö Lab.* 14:1-31.

Fenchel, T. 1970. Studies on the decomposition of organic detritus derived from the turtle grass *Thalassia testudinum. Limnol. Oceanogr.* 15:14-20.

Fenchel, T. 1972. Aspects of decomposer food chains in marine benthos. *Verh. d. Deutsch. Zool. Ges. 65 Jahresversamml.* 14:14-22.

Fenchel, T. and P. Harrison. 1976. The significance of bacterial grazing and mineral cycling for the decomposition of particulate detritus, pp. 285-299. *In:* J. M. Anderson and A. MacFadyen (eds.), *The Role of Terrestrial and Aquatic Organisms in Decomposition Processes.* Blackwell, Oxford.

Gaidry, W. J. and C. J. White. 1973. Investigations of commercially important penaeid shrimp in Louisiana estuaries. Louisiana Wild Life and Fisheries Commission, New Orleans. Tech. Bull. No. 8. 154 pp.

Gallagher, J. L., F. G. Plumley and P. L. Wolf. 1977. Underground biomass dynamics and substrate selective properties of Atlantic coastal salt marsh plants. U.S. Army Engineer Waterways Exp. Sta., Vicksburg, MS Tech. Rep. D-77-28. 141 pp.

Gardner, W. S., T. F. Nalepa, M. A. Quigley and J. M. Malczyk. 1981. Release of phosphorus by certain benthic invertebrates. *Can. J. Fish. Aquat. Sci.* 38:978-981.

Golterman, H. L. 1969. *Methods for Chemical Analysis of Fresh Waters.* IBP Handbook No. 8. Blackwell, Oxford. 172 pp.

Gould, S. J. 1977. *Ontogeny and Phylogeny.* Belknap Press, Cambridge. 501 pp.

Gunter, G. 1956. Principles of shrimp fishery management. Proc. Gulf Coast Carib. Fish. Inst. 8th Ann. Sess. pp. 99-106.

Gunter, G. 1967. Some relationships of estuaries to the fisheries of the Gulf of Mexico, pp. 621-638. *In:* G. H. Lauff (ed.), *Estuaries.* Publ. No. 83. Amer. Assoc. Adv. Science, Washington, DC

Hardisky, M. A. 1980. A comparison of *Spartina alterniflora* primary production estimated by destructive and nondestructive techniques, pp. 223-234. *In:* V. S. Kennedy (ed.), *Estuarine Perspectives.* Academic Press, New York.

Hargrave, B. T. 1970. The effect of a deposit-feeding amphipod on the metabolism of benthic microflora. *Limnol. Oceanogr.* 15:21-30.

Hargrave, B. T. and G. A. Phillips. 1981. Annual *in situ* carbon dioxide and oxygen flux across a subtidal marine sediment. *Est. Coastal Shelf Sci.* 12:725-737.

Howarth, R. W. 1979. Pyrite: Its rapid formation in a salt marsh and its importance in ecosystem metabolism. *Science* 203:49-51.

Howarth, R. W. and J. M. Teal. 1979. Sulfate reduction in a New England salt marsh. *Limnol. Oceanogr.* 24:999-1013.

Howarth, R. W. and J. M. Teal. 1980. Energy flow in a salt marsh ecosystem: The role of reduced inorganic sulfur compounds. *Amer. Nat.* 116:862-872.

Johannes, R. E. 1965. Influence of marine protozoa on nutrient regeneration. *Limnol. Oceanogr.* 3:434-442.

Jørgenson, B. B. 1977. The sulfur cycle of a coastal marine sediment (Limnfjorden, Denmark). *Limnol. Oceanogr.* 22:814-832.

Juday, C. 1908. Some aquatic invertebrates that live under anaerobic conditions. *Trans. Wisconsin Acad. Sci., Arts, Lett.* 16:10-16.

Kemp, W. B. and W. R. Boynton. 1980. Influence of biological and physical processes on dissolved oxygen dynamics in an estuarine system: implications for measurement of community metabolism. *Est. Coastal Mar. Sci.* 11:407-431.

Kerr, S. R. 1976. Ecological analysis and the Fry paradigm. *J. Fish. Res. Board Can.* 33:329-332.

King, G. M. and W. J. Wiebe. 1980. Regulation of sulfate concentrations and methanogenesis in salt marsh soils. *Est. Coastal Mar. Sci.* 10:215-223.

Kitchell, J. F., R. O. Neill, D. Webb, G. Gallepp, S. Bartell, J. Koonce and B. Ausmus. 1979. Consumer regulation of nutrient cycling. *BioScience* 29:28-33.

Kristensen, J. H. 1972. Carbohydrases of some marine invertebrates with notes on their food and on the natural occurrences of the carbohydrates studied. *Mar. Biol.* 14:130-142.

Lane, P. A., G. H. Lauff and R. Levins. 1975. The feasibility of using a holistic approach in ecosystem analysis, pp. 111-128. *In:* S. Levin (ed.), *Ecosystem Analysis and Prediction.* SIAM, Philadelphia, PA

Lee, J. J. and D. L. Inman. 1975. The ecological role of consumers – an aggregated systems view. *Ecology* 56:1455-1458.

Levine, S. 1980. Gut contents of forty-four Lake Pontchartrain fish species, 1977-1978, pp. 899-1029. *In:* J. H. Stone (ed.), Environmental analysis of Lake Pontchartrain, Louisiana, its surrounding wetlands, and selected land uses. CEL, CWR, LSU, BR, LA 70803. Prepared for U.S. Army Engineer District, New Orleans. Contract No. DACW 29-77-C-0253.

Lindeman, R. L. 1942. Experimental simulation of winter anaerobiosis in a senescent lake. *Ecology* 23:1-13.

Mann, K. H. 1972. Macrophyte production and detritus food chains in coastal waters, pp. 353-383. *In:* U. Melchiorri-Santolini and J. W. Hopton (eds.), *Detritus and its Role in Aquatic Ecosystems.* Mem. Ist. Ital. Idrobiol. Suppl. 29. Pallanza.

May, R. M. 1974. *Stability and Complexity in Model Ecosystems*, 2nd ed. Princeton University Press, Princeton. 265 pp.

Mendelssohn, I. A., K. L. McKee and W. H. Patrick. 1981. Oxygen deficiency in *Spartina alterniflora* roots: Metabolic adaptation to anoxia. *Science* 214:439-441.

Moore, H. 1931. The muds of the Clyde Sea area. III. Chemical and physical conditions; rate and nature of sedimentation; and fauna. *J. Mar. Biol. Ass. U. K.* 17:325-358.

Morris, J. C. and W. Stumm. 1967. Redox equilibria and measurements of potentials in the aquatic environment, pp. 270-285. *In:* R. F. Gould (ed.), *Equilibrium Concepts in Natural Water Systems.* Am. Chem. Soc. Pub.

Odum, E. P. and A. E. Smalley. 1959. Comparison of population energy flow in a herbivorous and a deposit-feeding invertebrate in a salt marsh ecosystem. *Proc. Nat. Acad. Sci.* 45:617-622.

O'Neill, R. V., W. F. Harris, B. S. Ausmus and D. E. Reichle. 1975. A theoretical basis for ecosystem analysis with particular reference to element cycling, pp. 28-40. *In:* F. G. Howell, J. B.

Gentry, and M. H. Smith (eds.), *Mineral Cycling in Southeastern Ecosystems.* ERDA Symposium Series (Conf. 740513). Washington, DC

Pamatmat, M. M. 1980. Facultative anaerobiosis of benthos, pp. 69-90. *In:* Tenore, K. R. and B. C. Coull (eds.), *Marine Benthic Dynamics.* U. South Carolina Press, Columbia, SC

Patrick, W. H. and R. D. DeLaune. 1977. Chemical and biological redox systems affecting nutrient availability in the coastal wetlands. *Geoscience and Man* 18:131-137.

Payne, W. J. 1970. Energy yields and growth of heterotrophs. *Ann. Rev. Microbiol.* 24:17-52.

Perret, W. S., B. Barrett, W. Latapie, J. Pollard, W. Mock, B. Adkins, W. Gaidry and C. White. 1971. *Cooperative Gulf of Mexico Estuarine Inventory and Study,* Louisiana Wild Life and Fisheries Commission, New Orleans. 175 pp.

Peterson, B. J., R. W. Howarth, F. Lipschultz and D. Ashendorf. 1980. Salt marsh detritus: an alternative interpretation of stable carbon isotope ratios and the fate of *Spartina alterniflora. Oikos* 34:173-177.

Platt, H. M. 1981. Meiofauna dynamics and the origin of the metazoa, pp. 207-216. *In:* P. L. Forey (ed.), *The Evolving Biosphere.* Cambridge University Press.

Platt, H. M. and R. M. Warwick. 1981. The significance of freeliving nematodes to the littoral ecosystem, pp. 729-759. *In:* J. H. Price, D. E. G Irvine and W. H. Farnham (eds.), *The Shore Environment: Vol. II. Ecosystems.* Academic Press, London.

Pomeroy, L. R. 1970. The strategy of mineral cycling. *Ann. Rev. Ecol. System.* 1:171-190.

Pomeroy, L. R., K. Bancroft, J. Breed, R. Christian, D. Frankenburg, J. Hall, L. Maurer, W. Wiebe, R. Wiegert and R. L. Wetzel. 1976. Flux of organic matter through a salt marsh, pp. 270-279. *In:* M. L. Wiley (ed.), *Estuarine Processes, Vol. II.* Academic Press, New York.

Pomeroy, L. R. and F. M. Bush. 1959. Regeneration of phosphate by marine animals, pp. 893-895. *In:* M. Sears (ed.), *Int. Oceanogr. Congr. Reprints.* Amer. Assoc. Adv. Science, Washington, DC

Reichle, D. E., R. V. O'Neill and W. F. Harris. 1975. Principles of energy and material exchange in ecosystems, pp. 27-43. *In:* W. H. van Dobben and R. H. Lowe-Connell (eds.), *Unifying Concepts in Ecology.* D. W. Junk, The Hague.

Reise, K. and P. Ax. 1980. Statement on the thiobios-hypothesis. *Mar. Biol.* 58:31-32.

Rich, P. M. 1979. Differential CO_2 and O_2 benthic community metabolism in a soft-water lake. *J. Fish. Res. Board Can.* 36:1377-1389.

Sheridan, P. F. and R. J. Livingston. 1979. Cyclic trophic relationships of fishes in an unpolluted, river-dominated estuary in north Florida, pp. 143-161. *In:* R. J. Livingston (ed.), *Ecological Processes in Coastal and Marine Systems.* Plenum Press, New York.

Sibert, J. R. and R. J. Naiman. 1980. The role of detritus and the nature of estuarine ecosystems, pp. 311-323. *In:* K. R. Tenore and B. C. Coull (eds.), *Marine Benthic Dynamics.* U. South Carolina Press, Columbia, SC

Sikora, J. P., W. B. Sikora, C. W. Erkenbrecher and B. C. Coull. 1977. Significance of ATP, carbon, and caloric content of meiobenthic nematodes in partitioning benthic biomass. *Mar. Biol.* 44:7-14.

Sikora, W. B. 1977. The ecology of *Palaemonetes pugio* in a southeastern salt marsh ecosystem with particular emphasis on production and trophic relationships. Ph.D. Thesis. U. South Carolina, Columbia, SC 122 pp.

Sørensen, J., B. B. Jørgensen and N. P. Revsbech. 1979. A comparison of oxygen, nitrate, and sulfate respiration in coastal marine sediments. *Microb. Ecol.* 5:105-115.

Sorokin, Yu. I. 1978. Decomposition of organic matter and nutrient regeneration, pp. 501-616. *In:* O. Kinne (ed.), *Marine Ecology Vol. 4, Dynamics.* John Wiley and Sons, New York.

Swedmark, B. 1964. The interstitial fauna of marine sand. *Biol. Rev.* 39:1-42.

Teal, J. M. 1962. Energy flow in the salt marsh ecosystem of Georgia. *Ecology* 43:614-624.

Tenore, K. R., J. H. Tietjen and J. J. Lee. 1977. Effect of meiofauna on incorporation of aged eelgrass, *Zostera marina*, detritus by the polychaete *Nephthys incisa*. *J. Fish. Res. Board Can.* 34:563-567.

Thorson, G. 1966. Some factors influencing the recruitment and establishment of marine benthic communities. *Neth. J. Sea Res.* 3:267-293.

Turner, R. E. 1977. Intertidal vegetation and commercial yields of penaeid shrimp. *Trans. Am. Fish. Soc.* 106:411-416.

Turner, R. E. and J. G. Gosselink. 1975. A note on standing crops of *Spartina alterniflora* in Texas and Florida. *Contrib. Mar. Sci.* 19:113-119.

Valiela, I., J. M. Teal and N. Y. Persson. 1976. Production and dynamics of experimentally enriched salt marsh vegetation: Below ground biomass. *Limnol. Oceanogr.* 21:245-252.

Wetzel, R. L. 1977. Carbon resources of a benthic salt marsh invertebrate *Nassarius obsoletus* Say (Mollusca: Nassariidae), pp. 293-308. *In:* M. L. Wiley (ed.), *Estuarine Processes, Vol. II.* Academic Press, New York.

Whelan, T. 1974. Methane, carbon dioxide, and dissolved sulfate from interstitial water of coastal marsh sediments. *Est. Coastal Mar. Sci.* 2:407-415.

Whitlatch, R. B. 1981. Animal-sediment relationships in intertidal marine benthic habitats: Some determinants of deposit-feeding species diversity. *J. exp. mar. Biol. Ecol.* 53:31-45.

Wiebe, W. J. 1979. Anaerobic benthic microbial processes: Changes from the estuary to the continental shelf, pp. 469-485. *In:* R. J. Livingston (ed.), *Ecological Processes in Coastal and Marine Systems.* Plenum Press, New York.

Wiegert, R. G. and D. F. Owen. 1971. Trophic structure, available resources and population density in terrestrial vs. aquatic ecosystems. *J. Theoret. Biol.* 30:69-81.

Wolff, W. J. 1973. The estuary as a habitat: An analysis of data on the soft-bottom macrofauna of the estuarine area of the rivers Rhine, Meuse, and Scheldt. *Zool. Verh.* 126:1-242.

THE INFLUENCE OF FIDDLER CRAB BURROWS AND BURROWING ON METABOLIC PROCESSES IN SALT MARSH SEDIMENTS

Clay L. Montague
Department of Environmental Engineering Sciences
University of Florida
Gainesville, Florida

Abstract: Effects of burrows and burrowing by fiddler crabs on production and decomposition of organic carbon in a salt marsh on Sapelo Island, GA, were measured. Field experiments determined rate of excavation of organic carbon from belowground; efflux of carbon dioxide from burrows; effect of burrows on growth of *Spartina alterniflora;* density of roots near burrows; and salinity, pH, phosphate concentration, and ammonium concentration in burrow water. Fiddler crabs transported 26 g organic C m^{-2} from belowground in July 1979. Annual excavation was estimated at 157 g C m^{-2}, or 20% of the belowground production of *S. alterniflora*. Mean burrow respiration was 2.1 mg CO_2 h^{-1}, accounting for 20% to 90% of salt marsh sediment respiration, depending on marsh wetness. Burrows increased *S. alterniflora* standing stocks by 23% in high marsh. Chemistry of burrow water was much different from interstitial water. Salinity of burrow water was 20-23‰ while that of interstitial water was 37-45‰. Burrows also contained high levels of ammonium. The combined effects of added nutrients, lower salinity, greater oxygen, lower sulfide, and greater flow of water through sediments as a result of fiddler crab burrowing are suggested as reasons for the increased standing stock of short *S. alterniflora*.

Introduction

The environment within which important producers and decomposers grow is in part determined by the activities of animals (Kitchell et al. 1979; Montague et al. 1981). In salt marshes, fiddler crabs (*Uca sp.*) influence the physical environment of both salt marsh cordgrass (*Spartina alterniflora*) and microbial decomposers of the aerobic-anaerobic interface in sediments (Montague 1980a). Feeding by fiddler crabs mixes the top few millimeters of surface sediment at every low tide during all but the coldest months (activity declines when temperature is <10°C [Montague 1980a]). More noticeable than feeding marks are numerous burrows excavated by fiddler crabs. Burrow density ranges from 50 to 700 m^{-2} with density highest near creek banks (Basan and Frey 1977; Krebs and Valiela 1978; Katz 1980).

Burrows can affect production and decomposition in salt marshes in several ways. They increase aerobic surface area of salt marshes by 20% to 60% or more (Teal and Kanwisher 1961; Edwards and Frey 1977; Katz 1980). Water within burrows contains some oxygen and the walls of burrows are aerobic, similar to the sediment-water interface at the marsh surface (Teal

ISBN 0-12-404070-5

1959; Wiedemann 1972; Edwards and Frey 1977). Subsurface aerobic zones provided by burrows should increase aerobic decomposition (Teal and Kanwisher 1961; Aller and Yingst 1978; Katz 1980), neutralize toxic end-products of anaerobic metabolism such as hydrogen sulfide (Morris 1980), and perhaps aerate roots of *S. alterniflora*, which cannot always provide adequate oxygen to its roots via aerenchyma tissue (Mendelssohn et al. 1981). Burrows may improve water circulation through subsurface sediment (Teal and Kanwisher 1961; Kraeuter and Wolf 1974), and this should dilute salts that accumulate in sediments from cordgrass transpiration (Smart and Barko 1980). Lower soil salinity can improve nutrient uptake and growth of *S. alterniflora* (Nestler 1977a; Smart and Barko 1978, 1980; Morris 1980; Linthurst and Seneca 1981). In excavating burrows, fiddler crabs also transport organic carbon from subsurface anaerobic sediments to the surface where more rapid aerobic decomposition occurs (de la Cruz and Hackney 1977; Hackney and de la Cruz 1980).

The objectives of this study were to measure some influences of burrows on production and decomposition in salt marshes. Although several authors have postulated such effects (Kraeuter and Wolf 1974; Katz 1980; Montague 1980a), documentation is lacking. Aller and Yingst (1978) demonstrated the influence of burrows of a polychaete *Amphitrite ornata* on the biogeochemistry and metabolism of water and sediments in an intertidal mudflat. Fiddler crabs are likely to have an even greater effect on ecosystem metabolism because of their great density, burrowing activity, and proximity to the roots of *S. alterniflora*.

Methods

Study Site

All experiments were performed in Airport Marsh, Sapelo Island, GA. This marsh is a 25–30 ha portion of the Duplin River watershed and consists primarily of very uniform stands of short to mid-height *S. alterniflora*. For more descriptive information of this area see Pomeroy and Wiegert (1981).

1. Rate of Excavation of Organic Carbon

Rate of excavation of organic carbon by burrowing fiddler crabs was estimated in a short *S. alterniflora* high marsh (approximately 0.5 ha of Airport Marsh) that contained about 58 burrows m^{-2}. Within this area, ten 0.25 m^2 quadrats were located haphazardly. All sediment excavated by crabs was initially removed by hand on 3 July 1979. Excavated sediment that subsequently appeared in the quadrats was removed 21 times between 3 July and 2 August 1979 and was recorded separately for each quadrat. Some of the excavated sediment was examined under a dissection microscope.

Sediment was dried at 70°C and weighed. Organic carbon of dried sediment samples was determined with a Coleman Model 33 Carbon-

Hydrogen Analyzer. Carbon analysis of ashed sediment (450°C) provided a correction for inorganic carbon.

From 5 July to 27 July 1979, the study site was regularly flooded by tides. During this time, new or resuspended sediment could fall into burrows. To estimate this sediment influx, ten 15 mm diameter polystyrene test tubes were inserted into the marsh surface along a transect through the study site to trap sediments. Trapped sediment was analyzed for carbon to correct the values determined above to net excavation. As maximum daily tidal heights decreased after July spring tide, small differences in quadrat elevation were detected by noting differences in soil moisture among quadrats on each sample day. Marsh coverage was compared to NOAA tide chart predictions to determine tide level required to cover the marsh. Weather was noted during collecting trips.

2. Rate of Outflow of Carbon Dioxide

Effect of fiddler crab burrows on outflow of CO_2 from the salt marsh was measured by infrared gas analysis of air samples removed from a 1 liter opaque bell jar placed directly over a burrow. Diameter of burrows was measured (all were >10 mm). Air samples from bell jars placed over an adjacent area without a burrow served as controls. Each bell jar covered 102 cm^2. Grass under all experimental and control jars was clipped. Clipping removed *S. alterniflora* shoot respiration and its accompanying variability. By comparably clipping, Teal and Kanwisher (1961) found *S. alterniflora* shoot respiration accounted for 40% of the gas exchange of intact marsh.

Air over five burrows and five paired controls was analyzed on each of 7 days in July and August 1979. Measurements began at 0900 EDT to avoid overheating the air in the bell jar. Temperature was monitored in another bell jar placed on the marsh. Syringes were used both to mix the air (by pumping) and to remove 3 ml air samples from each bell jar (all were fitted with a serum stopper). Air samples were removed initially and every 15 min for 45 min. After the measurement period, volume of air space in each burrow was determined by measuring the volume of water required to fill the burrow.

3. Effect of Inhabitated Burrows on Growth of *S. alterniflora*

To evaluate the impact of mud fiddler crabs (*U. pugnax*) on growth of *S. alterniflora*, a 16 mm diameter auger bit and cordless electric drill were used to excavate artificial burrows to a depth of 30 cm in a short *S. alterniflora* high marsh. Because of a dense root mat, few natural burrows existed in the immediate vicinity of the 25 m^2 study site (Ringold 1979). Fiddler crabs routinely fed in the site, however, except when the marsh surface dried during summer neap tides without rain.

Nine burrows were drilled in a grid pattern in each of three 0.44 m^2 experimental plots, 16 burrows in three other plots, and 36 burrows in three additional plots. Locations of these plots within the study site were randomly selected. A paired 0.44 m^2 control plot adjoined each burrowed plot. Within a day or two, all artificial burrows were occupied by fiddler crabs (*U. pugnax*). (Crabs have occupied and maintained these burrows continuously for 46 mo to the time of writing.)

On 28 and 29 November 1979 (after 18 mo), grass was harvested in each of the 18 plots. To avoid any edge effect, only the center 0.25 m^2 from each plot was harvested. Harvested grass was frozen, which intensified the green color of the living portion of plants. Live grass was separated from dead grass on the basis of color, dried at 70°C, and weighed.

4. Comparative Density of Roots and Rhizomes

To compare density of roots and rhizomes around burrows with that of unburrowed sediment, cores of sediment (7.5 cm diameter, 25 cm long) were removed from an area with uniformly short *S. alterniflora*. Five cores with burrows and five without were removed in March 1981. Cores with burrows were examined for burrow configuration. The cores were washed in a solution of sodium hexametaphosphate to remove sediment and leave roots and rhizomes. The solution was sieved through a 1.4 mm sieve to save any loose roots or rhizomes. The clean root mat obtained from each core was dried at 70°C and weighed.

5. Comparative Chemistry of Burrow and Interstitial Water

To compare water standing in burrows with interstitial water, samples were withdrawn from burrows and from small burrow-size wells (1.5–2 cm diameter, 30 cm deep) made with an auger bit or a piece of PVC pipe. The wells were capped with a short piece of PVC pipe so that surface water would not drain into them. Water was sampled both from natural burrows and from the artificial burrows that were drilled for the cordgrass growth experiment (No. 3 above). Interstitial water was sampled from an area within 10 m around the outside of the artificial burrow test plots. Most samples from natural burrows also came from this area, although, because the burrow density was very low, burrows away from this area were also sampled. All burrows sampled were in short to mid-height *S. alterniflora*. Test plots were in an area of uniformly short *S. alterniflora*. Burrow and interstitial water were sampled at neap tides both in March 1981 and again in September 1981 using a 25-cm length of Tygon tubing attached to a 50-ml plastic syringe. Interstitial and burrow water samples were collected and treated identically. Care was taken not to aerate samples, though samples were not collected anaerobically (as in an argon atmosphere) and some exposure to air was inevitable.

Water samples were spun for 10 min in a clinical centrifuge to remove suspended particles. Supernatant was removed for analysis. Salinity of a few drops of each sample was measured with a refractometer. Reactive phosphate and ammonium were measured colorimetrically (Strickland and Parsons 1972). The pH of March samples was measured with a portable Markson Model 88 pH meter; pH was not measured in September.

6. Density of Burrows >10 mm Diameter

Burrows involved in all of the previous experiments were >10 mm diameter. To obtain an estimate of normal density, burrows >10 mm diameter were counted in thirty-nine 0.25 m^2 quadrats along two parallel 300 m transects selected haphazardly and spaced 20 m apart. Transects extended inland from just marshward of the levee near the Duplin River. Thus all samples were taken from the back-levee meadow marsh, as defined by Basan and Frey (1977). This region contains short to mid-height *S. alterniflora*.

Results

1. Rate of Excavation of Organic Carbon

Between 3 July and 2 August 1979, fiddler crabs excavated 182 ± 5 (SE) g of dry sediment m^{-2} with a mean of 3.75 ± 0.57 g of sediment per burrow. The 182 g m^{-2} of excavated sediment contained an average of 19.5 ± 0.5 g organic C. Based on sediment trapped in test tubes, 2.3 ± 0.3 g C m^{-2} may be from new or resuspended sediment that fell into the burrows when the marsh was flooded. Therefore, net transport during the sampling period from belowground to the marsh surface was 17.2 g C m^{-2}.

Plant roots and rhizomes, a few small animals, and feces were visible under a dissecting microscope. Sediment from the test-tube sediment traps contained many small snails and crabs. Carbon values (above) for trapped sediments may be overestimates, but the quantity of trapped sediments was small.

Burrow density and marsh wetness affected rate of excavation. Quantity of excavated sediment was strongly correlated with burrow density. In Fig. 1, dry weight of excavated sediment is plotted against number of burrows in each of 10 experimental plots and three unburrowed control plots. Eighty-eight percent of the excavation occurred during neap tides when the high marsh was dry. Figure 2 shows how organic carbon excavation, a proportion of sediment excavation (roughly 11.1%), changed with marsh wetness. Tides below 2.13 m did not flood the study site. On the first day of drought following a long wet period, a mean of 5.0 ± 0.9 g C m^{-2} was excavated in one 24 h period (28 July).

As the water table falls during neap tides, fiddler crabs must dig deeper to reach water that they need for feeding, temperature regulation, and other physiological processes (Montague 1980a). Although neap tides were to continue for several days, burrowing subsided after 30 July because rain fell on the afternoons of 30 and 31 July and 1 August. The marsh became wetter, the water table rose, and fiddler crabs stopped burrowing.

2. Rate of Outflow of Carbon Dioxide

Outflow of carbon dioxide from 102 cm^2 plots that contained one large burrow was significantly greater than CO_2 outflow from paired, adjacent 102 cm^2 plots without a burrow (Table 1). "Burrow respiration" is CO_2 efflux from burrowed plots minus that from unburrowed plots; it is the difference between

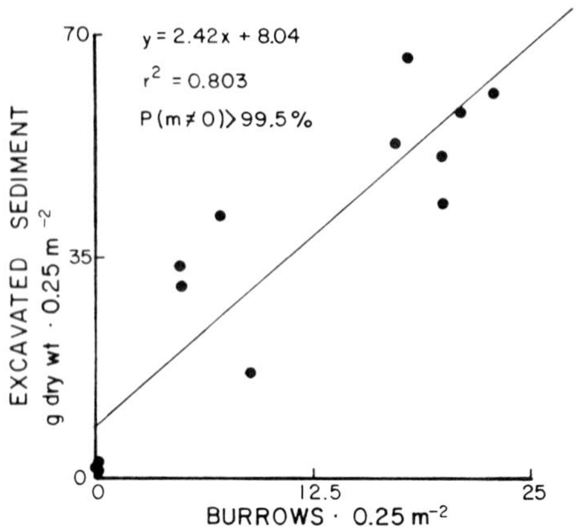

Figure 1. Sediment excavated (4 July to 2 August 1979) per 0.25 m^2 as a function of burrow density. P(m≠0) is the probability that the slope of the regression line is not equal to zero (test from Hines and Montgomery 1972).

each pair of bars in Figure 3. Burrow respiration rates ranged from 0.35 mg CO_2 h^{-1} on 2 August to 7.49 mg CO_2 h^{-1} on 3 July, with an overall mean of 2.09 ± 0.37 (SE) mg CO_2 h^{-1} $burrow^{-1}$ (Fig. 3). The burrows measured were 13.83 ± 0.28 (SE) mm in diameter. Measured burrows may or may not have contained fiddler crabs. Fiddler crab respiration, however, makes a negligible

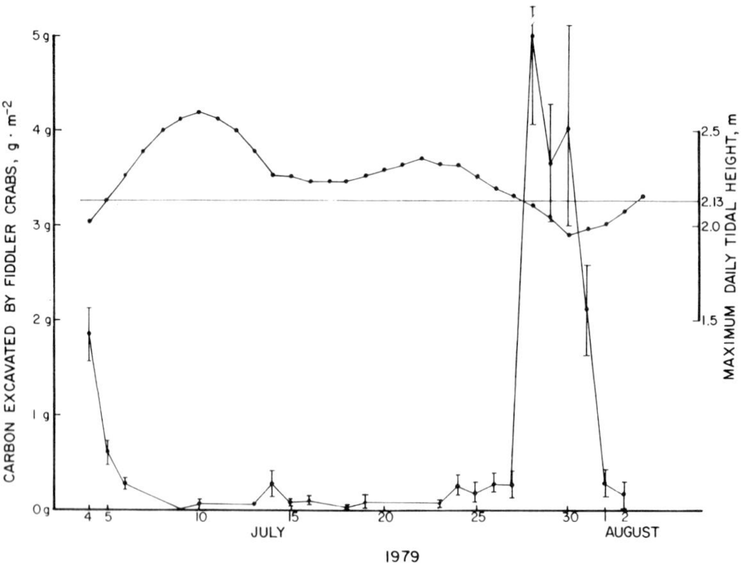

Figure 2. Maximum daily tidal height and carbon excavated by burrowing fiddler crabs (4 July to 2 August 1979). Vertical bars are ± 1 standard error.

Table 1. Mean burrow respiration (mg CO_2 h^{-1}), mean burrow diameter (mm), mean burrow volume (ml), and results of a paired t-test. P(BR>0) is the probability that CO_2 from the burrow is greater than would be expected without a burrow; % Δ is percent increase in respiration over unburrowed, paired control plots when one burrow exists per 102 cm^2 plot.

	3 Jul	*4 Jul*	*29 Jul*	*30 Jul*	*31 Jul*	*1 Aug*	*2 Aug*
Burrow Respiration[1]	7.492	2.106	1.922	3.482	1.888	0.596	0.351
Paired t; df	2.58;1	13.01;4	4.03;4	4.29;4	3.84;4	4.75;4	1.31;4
Mean Burrow Diameter	– –	14.6	14.2	15.0	12.2	14.0	13.0
Mean Burrow Volume[2]	118.5	50.0	49.0	110.0	43.0	12.4	1.0
P(BR>0)[3]	>87.5%	>99.75%	>98.75%	>98.75%	>98.75%	>99.5%	87.5%
% Δ[4]	149	45	39	98	79	24	36

[1]Mean burrow respiration for 32 observations: 2.085 ± 0.371 mg CO_2 h^{-1}.

[2]Burrow volumes were not used in the calculation of CO_2 efflux under the bell jars.

[3]P(BR>O)>99.75% for 32 observations from 7 dates.

[4]Mean % Δ caused by one burrow per 102 cm^2 is 60%.

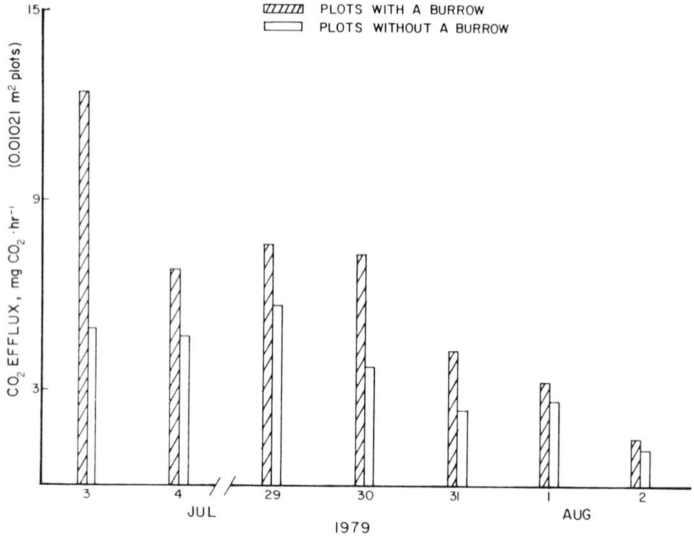

Figure 3. *Carbon dioxide efflux from five 102 cm² plots with one burrow and over five adjacent plots with no burrows, measured from 4 July to 2 August 1979 (two plots each on 3 July). The difference between the bars is called "burrow respiration." See also data in Table 1.*

contribution to this efflux because it is at maximum an order of magnitude lower than the measured CO_2 efflux from burrows (see Montague [1980a] for respiration references). The large burrow respiration value for 3 July is the mean of only two paired observations, whereas values for the other dates are means of five observations.

The contribution of burrows to salt marsh respiration was measured under a variety of moisture conditions. The marsh was dry on the mornings of 3, 4, 29, and 30 July because of low neap tides and no rain. Estimates of air space in measured burrows (burrow volume in Table 1) indicate a falling water table between 29 and 30 July. Rainfall on the afternoons of 30 and 31 July and 1 August caused the marsh surface to become wetter and the water table to rise. As the marsh became wetter, CO_2 efflux from the whole marsh (excluding *S. alterniflora* shoots) generally dropped, perhaps because the marsh sediment became more anaerobic.

Respiration of burrows was also lower after the rains, primarily because of reduced air space in the burrows. Ninety-seven percent of the variation among the daily mean burrow respiration rates in Table 1 can be explained by differences in mean burrow volumes (Fig. 4). The 3 July mean value (circled in Fig. 4) is not included in the regression line because it consisted of only two observations. The regression line indicates a contribution of burrows to salt marsh respiration even when they are completely filled with water. Burrows

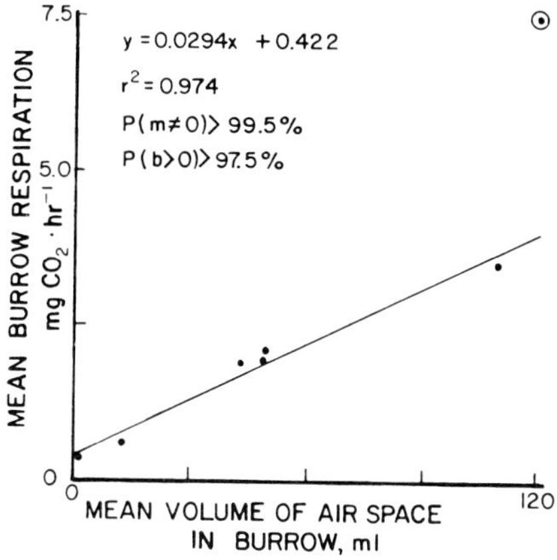

Figure 4. *Mean burrow respiration as a function of mean volume of burrow air space. The 3 July measurement (circled) was not included in the regression (see text). P(m≠0) is the probability that the slope of the regression line is not zero. P(b>0) is the probability that the intercept is greater than zero. Statistical tests are from Hines and Montgomery (1972).*

were nearly full on 2 August (see Fig. 3). At zero burrow volume, the respiration rate of 0.422 mg CO_2 h^{-1} $burrow^{-1}$ is significant ($P[b>0] > 95\%$; Fig. 4).

3. Effect of Inhabited Burrows on Growth of *S. alterniflora*

Routinely occupied fiddler crab burrows increased aboveground standing stock of *S. alterniflora* in short *S. alterniflora* high marsh (Table 2). Compared with unburrowed plots or plots with only 9 burrows per 0.44 m^2, significantly more live *S. alterniflora* occurred in plots with 16 or 36 burrows per 0.44 m^2. Sixteen burrows produced a 21.4% increase and 36 burrows produced a 31.7% increase (Table 2A). These percentages are mean percent increases for the three paired plots. The 6.6% increase with nine burrows was not significant by a paired t-test with controls. A linear regression through the differences between the burrowed plots and their paired controls illustrated the trend of increased live *S. alterniflora* at higher burrow density, yielding a regression of $Y = 0.25X + 0.35$ where Y = burrowed plots − control plots in grams dry weight per 0.25 m^2 and X = burrows per 0.44 m^2 ($r^2 = 0.54$). The probability that the apparent trend is due to chance was less than 2.5% ($P[m \neq 0] > 97.5\%$). A similar trend occurred for standing dead *S. alterniflora* ($Y = 0.36X - 5.49$; $r^2 = 0.31$; $P[m \neq 0] \sim 87\%$) (Table 2B). Thus, the total dry weight of *S. alterniflora* was greater in plots with more burrows ($Y = 0.61X - 5.14$; $r^2 = 0.43$; $P[m \neq 0] \sim 95\%$) (Table 2C). Nine burrows per 0.44 m^2, however, were not sufficient to produce a significant effect, whereas 16 routinely occupied burrows per 0.44 m^2 increased aboveground standing stock by 13.5% and 36 burrows per 0.44 m^2 increased it by 21.6%.

Table 2. Dry weight of A (live), B (dead), and C (live and dead) *S. alterniflora* (g per 0.25 m^2) harvested from plots that contained 9, 16, or 36 burrows per 0.44 m^2 compared to adjacent, unburrowed control plots. P(E>C) is the probability that weight in the burrowed plots (Exp) is greater than that in unburrowed plots (Con); % Δ is percent difference between experimental and control plots.

9		*16*		*36*	
Exp	*Con*	*Exp*	*Con*	*Exp*	*Con*
A. *LIVE WEIGHT*					
33.90	28.11	34.62	27.41	37.80	26.91
32.02	34.38	34.76	31.40	39.93	33.99
30.58	28.82	28.78	22.63	38.73	28.27
% Δ = 6.6		% Δ = 21.4		% Δ = 31.7	
P(E>C)>70%		P(E>C)>97.5%		P(E>C)>97.5%	
All burrowed plots vs all control plots: P(E>C)>99.75%					
B. *DEAD WEIGHT*					
41.71	39.48	52.01	42.90	44.48	45.08
44.99	52.60	44.07	48.55	58.31	52.86
42.82	51.26	47.40	41.49	55.97	40.65
% Δ = −8.4		% Δ = 8.8		% Δ = 15.6	
P(E<C)>75%		P(E>C)>75%		P(E>C)∼85%	
All burrowed plots vs all control plots: P(E>C)>75%					
C. *TOTAL WEIGHT*					
75.61	67.59	86.63	70.31	82.28	71.99
77.01	86.98	78.83	79.95	98.24	86.85
73.40	80.08	76.18	64.12	94.70	68.92
% Δ = −2.7		% Δ = 13.5		% Δ = 21.6	
P(E<C)>65%		P(E>C)>87.5%		P(E>C)>95%	
All burrowed plots vs all control plots: P(E>C)>95%					

4. Comparative Density of Roots and Rhizomes

Density of roots and rhizomes was significantly lower in cores that contained a burrow (16.6 ± 1.7 g DW $core^{-1}$) than in unburrowed cores (21.9 ± 0.8 g DW $core^{-1}$). The difference of 5.3 g is not primarily due to presence of actual burrow space in the core. The root mat was in fact not as dense and cores tended to fall apart on handling unlike unburrowed cores. Total core volume was 1105 ml, but the upper limit to the volume occupied by the burrow was only 7.1% of this volume, assuming a very large burrow 2 cm in diameter and 25 cm deep.* This percentage can account for no more than 1.6 g of the 5.3 g difference.

*In fact, this is an overestimate for the measured burrows, but unfortunately volume estimates were lost. Inspection of the cores revealed vertical burrows without side branches, typical of short *S. alterniflora* marshes with dense root mats (Basan and Frey 1977).

5. Comparative Chemistry of Burrow and Interstitial Water

Salinity of burrow water was significantly lower (probabiltiy >99.5%) than that of interstitial water. Mean salinity of burrow water was 23.6 ± 0.23‰ in March and 20.1 ± 1.01‰ in September. No difference in salinity was detected between artificial and natural burrows (ANOVA F-ratio = 0.22 at 1 and 24 df). Mean interstitial salinity was 37.1 ± 0.35‰ in March and 45.0 ± 1.15‰ in September. Greater percolation of surface water through sediments via burrows could cause burrow water to have the lower salinity more characteristic of surface water.

The pH of burrow and interstitial water was measured in March only. No difference was detected between water from artificial and natural burrows, but burrow water was slightly lower in pH than interstitial water (6.78 ± 0.04 and 6.94 ± 0.03, respectively). This difference perhaps reflects greater oxygen concentrations and hence greater sulfate ion concentration in burrow water as opposed to more sulfide and less oxygen in interstitial water (Teal and Kanwisher 1961; Wiedemann 1972; Edwards and Frey 1977).

Phosphate and ammonium concentrations in burrow water were significantly different from those of interstitial water; these differences were not consistent between the March and September samples (Fig. 5). In addition, phosphate and ammonium in natural burrow water (B_N) was not similar to that of the water from artificial burrows (B_A).

In March, interstitial water was very high in phosphate (20.2 ± 0.9 μmol P l^{-1}) and low in ammonium (6.3 ± 0.5 μmol N l^{-1}). In September, phosphate in interstitial water was only 1.1 ± 0.3 μmol P l^{-1}, but ammonium was 41% higher than it was in March (8.8 ± 1.1 μmol N l^{-1}).

In March, the water from natural burrows contained barely detectable phosphate (<0.1 μmol P l^{-1}) and contained less than half the ammonium concentration of interstitial water (2.3 ± 0.4 μmol N l^{-1}). By comparison, water from artificial burrows in March contained 20 times more phosphate than natural burrow water (0.94 ± 0.23 μmol P l^{-1}), but this value was still only 1/22 of the interstitial phosphate concentration. Ammonium, on the other hand, was 5 times higher than that in natural burrow water and was twice that of interstitial water (11.8 ± 3.2 μmol N l^{-1}).

In September, water from natural burrows contained 19 times more phosphate (0.86 ± 0.23 μmol P l^{-1}) and 32 times more ammonium (72.8 ± 15.9 μmol N l^{-1}) than it contained in March. Compared with natural burrow water in September, interstitial water contained about the same phosphate (1.1 ± 0.27 μmol P l^{-1}) but much less ammonium (8.8 ± 1.1 μmol N l^{-1}). By contrast, water from artificial burrows in September contained 2.8 times the phosphate (2.7 ± 0.6 μmol P l^{-1}) of water from natural burrows, but only half the ammonium (35.6 ± 4.5 μmol N l^{-1}) of natural burrow water (4 times the ammonium of September interstitial water). In September, water from artificial burrows contained 2.9 times the phosphate and 3 times the ammonium that it contained in March.

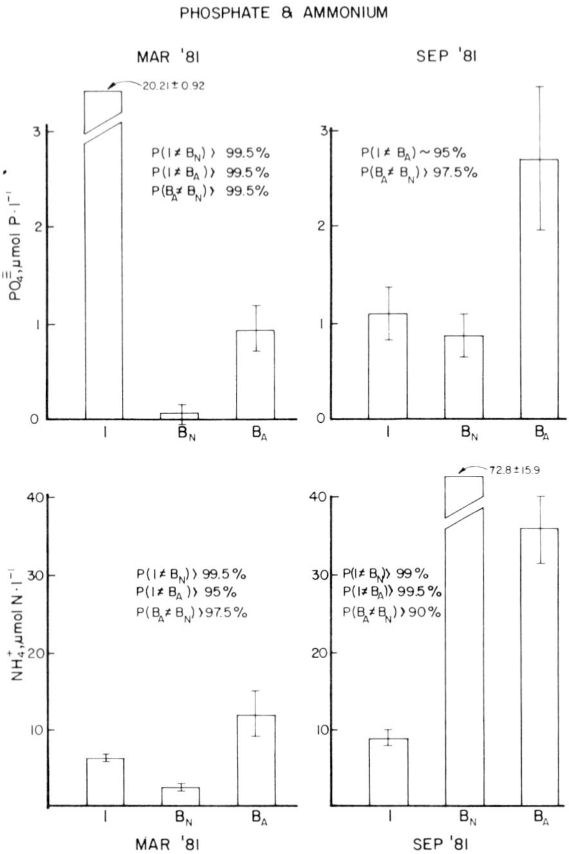

Figure 5. *Phosphate and ammonium concentrations in interstitial water, I; water from natural burrows, B_N, and water from artificial burrows, B_A; in or near cordgrass growth experimental test plots. Standard errors are indicated on each bar.*

6. Density of Burrows >10 mm Diameter

Mean density of burrows >10 mm diameter in the 39 quadrats was 88 ± 5 (SE) m^{-2}. Burrow density along the two transects was 81.5 ± 4.7 and 95.2 ± 5.1 m^{-2}.

Discussion

Changes in salt marsh production, respiration, and biogeochemistry measured in response to the presence of fiddler crab burrows demonstrated the considerable impact of burrows on belowground metabolism in salt marshes. The exact pathways of influence are yet to be identified though the response of the measured variables is clear. Excavation of burrows transports organic carbon from belowground to the surface where it can decompose more rapidly (Hackney and de la Cruz 1980). Excavation of burrows extends and enlarges the aerobic-anaerobic interface vertically into the salt marsh and a measurable result is increased outflow of carbon dioxide. Inhabited fiddler crab burrows measurably enhance the growth of *S. alterniflora*, alter subsurface biogeochemistry, and probably increase the drainage of tidal water through sediments.

Standard error bars for excavated carbon in Fig. 2 are based on the error estimate for sediment dry weight excavated each day. The standard errors increase as the means increase during the dry period at the end of July. Error was higher during intense burrowing because slight differences in elevation between plots created different moisture conditions on the same day. Higher plots dried out first, which caused earlier intense burrowing than in wetter plots. Peak excavation occurred on 28 July for six of the 10 plots, but on 29 July for one plot, and on 30 July for three plots. The latter three plots were slightly lower in elevation, and were closer to the headwaters of a creek. Therefore, burrowing lagged in these plots, which created greater error among all plots lumped together on each day of intense burrowing.

De la Cruz and Hackney (1977) suggested that excavation of roots and rhizomes by fiddler crabs could be a significant transport mechanism for the aerobic decomposition of organic carbon in a salt marsh dominated by *Juncus roemerianus*. Hackney and de la Cruz (1980) demonstrated that decomposition of organic carbon beneath the salt marsh surface was extremely slow when compared to decomposition at the surface. For Airport Marsh, with 88 burrows m^{-2} (>10 mm diameter), the measured July excavation (measured in an area with only 58 burrows m^{-2}) corresponds to 26 g C m^{-2}. For the last 5 years July has been a very hot, dry month on Sapelo Island and burrow excavation was very visible. Burrow excavation should decline in cooler as well as wetter months (Knopf 1966; Katz 1980). If excavation is maximal in July, but declines as a sine wave to zero in January, then fiddler crabs annually excavate 157 g organic C m^{-2}, or 20% of the belowground biomass annually produced by *S. alterniflora* in Georgia (belowground production estimate from Pomeroy et al. 1981).

Rate of excavation evidently declines with increasing latitude. Katz (1980) found fewer and more shallow burrows in a Massachusetts salt marsh, and estimated that crabs turn over 18% of the top 15 cm of sediment annually. If the turnover time of annual belowground *S. alterniflora* production is roughly 1 year, then 18% of the top 15 cm probably represents less than 10% of belowground production (assuming the majority of roots and rhizomes are in the top 30 cm). Cooler temperatures, shorter active season, and lower burrow densities result in less excavation of organic matter by fiddler crabs in more northerly environments. In Georgia salt marshes, extended excavation of organic matter by fiddler crabs over the longer warmer seasons may partially explain the absence of distinct layers of peat (Frey and Basan 1978).

Teal and Kanwisher (1961) calculated that burrows in Georgia salt marshes increase oxygen and carbon dioxide exchange in sediments by about 10% because of increased aerobic surface area extending into the subsurface sediments (see Montague [1980b] for more detail). My direct measurements demonstrated a much greater contribution of burrows to sediment metabolism. For a marsh with 88 burrows m^{-2} (>10 mm diameter), increased carbon dioxide efflux caused by burrows was at least 20% and would be as much as 90% during neap tides when the marsh is dry. If so, estimates for total sediment respiration that do not include the contribution of burrows will be serious underestimates.

Increased sediment respiration reflects increased aerobic surface area created by burrows. By examining plaster casts, Katz (1980) found a surface area increase of 59% caused by fiddler crab burrows in a Massachusetts salt marsh. Surface area increase caused by burrows in Georgia salt marshes should be several times greater than that found by Katz because of deeper and denser burrows in Georgia (Basan and Frey 1977).

Lower root density adjacent to burrows may reflect not only root clipping and high sediment turnover associated with burrow excavation, but also improved nutrient availability. Fertilized plots in a Massachusetts salt marsh contained fewer *S. alterniflora* roots than did unfertilized plots (Valiela et al. 1976). Also, root density is lower both where fiddler crabs burrow and where *S. alterniflora* grows taller (Basan and Frey 1977; Ringold 1979; Pomeroy et al. 1981).

Airport Marsh, with 88 burrows m^{-2} (>10 mm diameter) contains 23% more aboveground *S. alterniflora* than it would without burrows, according to Table 2. In other words, burrows may account for 19% of *S. alterniflora* production. Both burrow density and aboveground growth of *S. alterniflora* increase sharply near creek banks (Basan and Frey 1977; Wiegert 1979; Pomeroy et al. 1981). The greater burrow density near creek banks contributes to an explanation of increased cordgrass growth.

Burrow water is not similar to interstitial water. Teal and Kanwisher (1961) measured oxygen concentration and redox potential of surface water, burrow water, and sediments. Burrow water rarely contained less than 5-8 ppm O_2 (75%-80% saturation). This oxygen concentration is similar to that of the flooding water of Duplin River, to which the study area was adjacent (Frankenberg 1976). Several researchers have shown that salt marsh sediments in Georgia generally have very low redox potentials (Eh<0) except near burrows, or where soils are sandy such as in the landward reaches of the intertidal zone (Teal and Kanwisher 1961; Weidemann 1972; Edwards and Frey 1977).

Burrow water salinity was high, but within the range reported by Pomeroy et al. (1972) for the Duplin River (15-25‰), whereas that of interstitial water was 1.6 to 2.2 times higher (see also Nestler 1977a). Salts should diffuse from interstitial water into burrow water. With additional salinity data for burrows, perhaps the difference between flooding water salinity, burrow salinity, and interstitial salinity could be used to estimate exchange rate of burrow water.

Phosphate concentrations of burrow water were also in the range reported by Pomeroy et al. (1972) (1-6 μmol l^{-1}). Phosphate in the upper reaches of Duplin River was highest in September and lowest in January, March, and April. Burrow water, but not interstitial water, exhibited similar changes.

In September, burrow water contained nearly 3 times the phosphate of interstitial water. The surprisingly low interstitial concentration of approximately 1 μmol P l^{-1} is the equilibrium value for phosphate in aerated water exposed to salt marsh sediments (Pomeroy et al. 1965). Interstitial water, however, is generally anoxic, so a much higher proportion of the phosphate in sediments should be free to move into water (Patrick and Khalid 1974). Interstitial water concentrations should be more similar to the March value of 20 μmol P l^{-1}. Because the samples were treated identically, this difference is probably not due to aeration of samples, though in both months actual concentrations may be higher than reported (see Khalid et al. 1978).

Although salt marsh sediments store much phosphate (Pomeroy et al. 1965; Pomeroy et al. 1972), phosphate extractable from sediment decreased with depth in a Louisiana salt marsh (DeLaune and Patrick 1980). The latter investigators suggest that uptake by *S. alterniflora* causes the depth profile. Perhaps uptake by cordgrass throughout its growing season accounts for the lower September level in interstitial water. Whether greater concentration of phosphate in the test plot burrows can improve growth, however, seems unlikely in light of evidence against phosphate limitation in salt marshes (Buresh et al. 1980; DeLaune and Patrick 1980).

Ammonium concentrations in flooding waters ranged from 1 to 10 μmol N l^{-1} in Georgia and was usually 2-3 μmol N l^{-1} (Haines 1979; A. G. Chalmers, Marine Institute, pers. comm.). Burrow water, however, ranged from 2.3 to 72.8 μmol N l^{-1}.

Fiddler crabs excrete ammonium, and nitrogen additions to salt marshes increased *S. alterniflora* growth in Georgia, Massachusetts, and Louisiana (Gallagher 1975; Valiela et al. 1976; Valiela and Teal 1979; Buresh et al. 1980). From data reported by Nelson et al. (1979), a middle value estimate of ammonium excretion by fed and unfed large crustacea is 3.6 mg NH_4^+ g^{-1} d^{-1}. Thus 10 g of fiddler crabs per m^2 (Wolf et al. 1975) probably regenerate much more than the 4.2 to 5.7 g N m^{-2} yr^{-1} required to supply 19% of the production of *S. alterniflora* (Gallagher 1975; Whitney et al. 1981).

Both in March and September, burrow water from the artificially burrowed marsh contained more ammonium than did interstitial water. The active growing season for *S. alterniflora* in Georgia begins in March; rapid growth of flower culms occurs in autumn (Gallagher et al. 1980). At both these times nitrogen can be depleted because of elevated protein synthesis during periods of rapid growth (Squiers and Good 1974; Mendelssohn 1979;

Buresh et al. 1980; Mattson 1980). Greater ammonium supply together with other features of burrow water (see below) should explain the improved *S. alterniflora* growth in burrowed plots.

Burrow water chemistry, therefore, is similar to published accounts of flooding surface water chemistry except that burrow water contains higher levels of ammonium, as does ebbing water (Haines 1979; Chalmers pers. comm.).

Burrows provide probably five of six factors reviewed by Morris (1980) that improve growth of short *S. alterniflora*. These factors are 1) more oxygen, even though *S. alterniflora* supplies some oxygen internally to its roots (Mendelssohn et al. 1981); 2) less hydrogen sulfide, as shown for rice (also influenced by oxygen); 3) lower salinity (see also Nestler 1977a; Smart and Barko 1978, 1980); 4) increased rates of nutrient diffusion through pore water (see also Nestler 1977b); 5) added nitrogen (see also Gallagher 1975; Smart and Barko 1978, 1980; Buresh et al. 1980; Linthurst and Seneca 1981); and 6) greater exchange capacity of sediments.

Muddy subsurface salt marsh sediments are anoxic everywhere except adjacent to burrows (Teal and Kanwisher 1961; Wiedemann 1972; Edwards and Frey 1977). Hydrogen sulfide should be converted to sulfate in the presence of oxygen, thus reducing any cordgrass growth inhibition it may cause (Howarth 1979; Howarth and Teal 1979, 1980; Wiebe et al. 1981). Salinity of burrow water was significantly lower than interstitial salinity in Airport Marsh, which may indicate increased flow through sediments and increased rates of advective diffusion of nutrients to adjacent plant roots (Nestler 1977b).

In conclusion, burrows of fiddler crabs profoundly influence subsurface metabolism in salt marshes. Transport of organic carbon to the surface, outflow of carbon dioxide, enhanced growth of *S. alterniflora*, and altered biogeochemistry of subsurface water indicate the considerable effects of burrows on subsurface metabolism. As demonstrated not only by these experiments, but also by those of Aller and Yingst (1978), burrowing macrofauna actually define an environment within which live microbes that are so important to subsurface energy utilization and mineral cycling in salt marshes.

Acknowledgments

I sincerely thank Charles Hopkinson and Susan Vince for critical readings of earlier drafts. The Department of Environmental Engineering Sciences supplied travel funds for the presentation of this work at the Estuarine Research Federation Biennial Meeting. This work was partially supported by NSF grants OCE 75-20842 (L. R. Pomeroy and R. G. Wiegert, principal investigators) and DEB 77-20359 (E. B. Haines, principal investigator). Jenny Carter operated the word processor and Randy Koper drafted the figures. My appreciation also to my students William Coggins, John Higman, and Arthur Watson for accompanying me in the collection of water chemistry and root density data. Contribution 453 from the University of Georgia Marine Institute, Sapelo Island, Georgia.

References Cited

Aller, R. C. and J. Y. Yingst. 1978. Biogeochemistry of tube-dwellings: A study of the sedentary polychaete *Amphitrite ornata* (Leidy). *J. Mar. Res.* 35:201-254.

Basan, P. B. and R. W. Frey. 1977. Actual-palaeontology and neoichnology of salt marshes near Sapelo Island, Georgia, pp. 41-70. *In:* T. P. Crimes and J. C. Harper (eds.), *Trace Fossils 2. Geol. J.*, Special Issue 9.

Buresh, R. J., R. D. DeLaune and W. H. Patrick, Jr. 1980. Nitrogen and phosphorus distribution and utilization by *Spartina alterniflora* in a Louisiana Gulf Coast marsh. *Estuaries* 3:111-121.

de la Cruz, A. A. and C. T. Hackney. 1977. Energy value, elemental composition, and productivity of belowground biomass of a *Juncus* tidal marsh. *Ecology* 58:1165-1170.

DeLaune, R. D. and W. H. Patrick, Jr. 1980. Nitrogen and phosphorus cycling in a Gulf Coast salt marsh, pp. 143-151. *In:* V. S. Kennedy (ed.), *Estuarine Perspectives.* Academic Press, New York.

Edwards, J. M. and R. W. Frey. 1977. Substrate characteristics within a holocene salt marsh, Sapelo Island, Georgia. *Senckenbergiana Marit.* 9:215-259.

Frankenberg, D. 1976. Oxygen in a tidal river: Low tide concentration with location. *Estuarine Coastal Mar. Sci.* 4:455-460.

Frey, R. W. and P. B. Basan. 1978. Coastal salt marshes, pp. 101-169. *In:* R. A. Davis, Jr. (ed.), *Coastal Sedimentary Environments.* Springer-Verlag, New York.

Gallagher, J. L. 1975. Effect of an ammonium nitrate pulse on the growth and elemental composition of natural stands of *Spartina alterniflora* and *Juncus roemerianus*. *Am. J. Bot.* 62:644-648.

Gallagher, J. L., R. J. Reimold, R. A. Linthurst and W. J. Pfeiffer. 1980. Aerial production, mortality, and mineral accumulation-export dynamics in *Spartina alterniflora* and *Juncus roemerianus* plant stands. *Ecology* 61:303-312.

Hackney, C. T. and A. A. de la Cruz. 1980. *In situ* decomposition of roots and rhizomes of two tidal marsh plants. *Ecology* 61:226-231.

Haines, E. B. 1979. Nitrogen pools in Georgia coastal waters. *Estuaries* 2:34-39.

Hines, W. W. and D. C. Montgomery. 1972. *Probability and Statistics in Engineering and Management Science.* Ronald Press, New York. 509 pp.

Howarth, R. W. 1979. Pyrite: Its rapid formation in a salt marsh and its importance in ecosystem metabolism. *Science* 203:49-51.

Howarth, R. W. and J. M. Teal. 1979. Sulfate reduction in a New England salt marsh. *Limnol. Oceanogr.* 24:999-1013.

Howarth, R. W. and J. M. Teal. 1980. Energy flow in a salt marsh ecosystem: The role of reduced inorganic sulfur compounds. *Am. Nat.* 116:862-872.

Katz, L. C. 1980. Effects of burrowing by the fiddler crab, *Uca pugnax* (Smith). *Estuarine Coastal Mar. Sci.* 11:233-237.

Khalid, R. A., W. H. Patrick, Jr. and R. P. Gambrell. 1978. Effect of dissolved oxygen on chemical transformations of heavy metals, phosphorus, and nitrogen in an estuarine sediment. *Estuarine Coastal Mar. Sci.* 6:21-35.

Kitchell, J. F., R. V. O'Neill, D. Webb, G. W. Gallepp, S. M. Bartell, J. F. Koonce and B. S. Ausmus. 1979. Consumer regulation of nutrient cycling. *BioScience* 29:28-34.

Knopf, G. N. 1966. Observations on behavioral ecology of the fiddler crab, *Uca pugilator* (Bosc). *Crustaceana* 11:302-306.

Kraeuter, J. N. and P. L. Wolf. 1974. The relationship of marine macroinvertebrates to salt marsh plants, pp. 449-62 *In:* R. J. Reimold and W. H. Queen (eds.), *Ecology of Halophytes.* Academic Press, New York.

Krebs, C. T. and I. Valiela. 1978. Effect of experimentally applied chlorinated hydrocarbons on the biomass of the fiddler crab, *Uca pugnax* (Smith). *Estuarine Coastal Mar. Sci.* 6:375-386.

Linthurst, R. A. and E. D. Seneca. 1981. Aeration, nitrogen, and salinity as determinants of *Spartina alterniflora* Loisel. growth response. *Estuaries* 4:53-63.

Mattson, W. J., Jr. 1980. Herbivory in relation to plant nitrogen content. *Ann. Rev. Ecol. Syst.* 11:119-161.

Mendelssohn, J. A. 1979. Nitrogen metabolism in the height forms of *Spartina alterniflora* in North Carolina. *Ecology* 60:574-584.

Mendelssohn, J. A., K. L. McKee and W. H. Patrick, Jr. 1981. Oxygen deficiency in *Spartina alterniflora* roots: Metabolic adaptation to anoxia. *Science* 214:439-441.

Montague, C. L. 1980a. A natural history of temperate western Atlantic fiddler crabs (Genus *Uca*) with reference to their impact on the salt marsh. *Contrib. Mar. Sci.* 23:25-55.

Montague, C. L. 1980b. The net influence of the mud fiddler crab, *Uca pugnax*, on carbon flow through a Georgia salt marsh: The importance of work by macroorganisms to the metabolism of ecosystems. Ph.D. Dissertation, U. Georgia, Athens, GA 157 pp.

Montague, C. L., S. M. Bunker, E. B. Haines, M. L. Pace and R. L. Wetzel. 1981. Aquatic macroconsumers, pp. 69-85. *In:* L. R. Pomeroy and R. G. Wiegert (eds.), *The Ecology of a Salt Marsh*. Springer-Verlag, New York.

Morris, J. T. 1980. The nitrogen uptake kinetics of *Spartina alterniflora* in culture. *Ecology* 61:1114-1121.

Nelson, S. G., M. A. Simmons and A. W. Knight. 1979. Ammonia excretion by the benthic estuarine shrimp *Crangon franciscorum* (Crustacea: Crangonidae) in relation to diet. *Mar. Biol.* 54:25-31.

Nestler, J. 1977a. Interstitial salinity as a cause of ecophenic variation in *Spartina alterniflora*. *Estuarine Coastal Mar. Sci.* 5:707-714.

Nestler, J. 1977b. A preliminary study of the sediment hydrology of a Georgia salt marsh using Rhodamine WT as a tracer. *Southeastern Geol.* 18:265-271.

Patrick, W. H., Jr. and R. A. Khalid. 1974. Phosphate release and sorption by soils and sediments: Effects of aerobic and anaerobic conditions. *Science* 186:53-55.

Pomeroy, L. R., W. M. Darley, E. L. Dunn, J. L. Gallagher, E. B. Haines and D. M. Whitney. 1981. Primary production, pp. 39-67. *In:* L. R. Pomeroy and R. G. Wiegert (eds.), *The Ecology of a Salt Marsh*. Springer-Verlag, New York.

Pomeroy, L. R., L. R. Shenton, R. D. H. Jones and R. J. Reimold 1972. Nutrient flux in estuaries. *Nutrients and Eutrophication Special Symposia,* vol. 1. pp. 274-291. *Am. Soc. Limnol. Oceanogr.*

Pomeroy, L. R., E. E. Smith and C. M. Grant. 1965. The exchange of phosphate between estuarine water and sediments. *Limnol. Oceanogr.* 10:167-172.

Pomeroy, L. R. and R. G. Wiegert (eds.), 1981. *The Ecology of a Salt Marsh*. Springer-Verlag, New York. 271 pp.

Ringold, P. 1979. Burrowing, root mat density, and the distribution of fiddler crabs in the eastern United States. *J. Exp. Mar. Biol. Ecol.* 36:11-21.

Smart, R. M. and J. W. Barko. 1978. Influence of sediment salinity and nutrients on the physiological ecology of selected salt marsh plants. *Estuarine Coastal Mar. Sci.* 7:487-495.

Smart, R. M. and J. W. Barko. 1980. Nitrogen nutrition and salinity tolerance of *Distichlis spicata* and *Spartina alterniflora*. *Ecology* 61:620-638.

Squiers, E. R. and R. E. Good. 1974. Seasonal changes in the productivity, caloric content, and chemical composition of a population of salt marsh cordgrass *(Spartina alterniflora)*. *Chesapeake Sci.* 15:63-71.

Strickland, J. D. H. and T. R. Parsons. 1972. A Practical Handbook of Seawater Analysis. *Fish. Res. Board Can., Bull. 167,* Ottawa. 310 pp.

Teal, J. M. 1959. Respiration of crabs in Georgia salt marsh and its relation to their ecology. *Physiol. Zool.* 32:1-14.

Teal, J. M. and J. Kanwisher. 1961. Gas exchange in a Georgia salt marsh. *Limnol. Oceanogr.* 6:388-399.

Valiela, I. and J. M. Teal. 1979. The nitrogen budget of a salt marsh ecosystem. *Nature* 280:652-656.

Valiela, I., J. M. Teal and N. Y. Persson. 1976. Production and dynamics of experimentally enriched salt marsh vegetation: Belowground biomass. *Limnol. Oceanogr.* 21:245-252.

Whitney, D. E., A. G. Chalmers, E. B. Haines, R. B. Hanson, L. R. Pomeroy and B. Sherr. 1981. The cycles of nitrogen and phosphorus, pp. 163-181. *In:* L. R. Pomeroy and R. G. Wiegert (eds.), *The Ecology of a Salt Marsh.* Springer-Verlag, New York.

Wiebe, W. J., R. R. Christian, J. A. Hansen, G. King, B. Sherr and G. Skyring. 1981. Anaerobic respiration and fermentation, pp. 137-159. *In:* L. R. Pomeroy and R. G. Wiegert (eds.), *The Ecology of a Salt Marsh.* Springer-Verlag, New York.

Wiedemann, H. U. 1972. Application of red-lead to the detection of dissolved sulfide in water-logged soils. *Z. Pflanzenernaehr. Bodenkd.* 133:73-81.

Wiegert, R. G. 1979. Ecological processes characteristic of coastal *Spartina* marshes of the south-eastern USA, pp. 467-490. *In:* R. L. Jeffries and A. J. Davy (eds.), *Ecological Processes in Coastal Environments.* Blackwell Scientific Publications, Oxford.

Wolf, P. L., S. F. Shanholtzer and R. J. Reimold. 1975. Population estimates for *Uca pugnax* (Smith 1870) on the Duplin estuary marsh, Georgia, USA. (Decapoda Brachyura, Ocypodidae). *Crustaceana* 29:79-91.

EMISSIONS OF SULFUR GASES TO THE ATMOSPHERE FROM ESTUARIES AND COASTAL AREAS

A. R. Bandy, P. J. Maroulis, B. Bonsang, and C. A. Brown

Department of Chemistry
Drexel University
Philadelphia, Pennsylvania

Abstract: Based on recent measurements of the atmospheric concentrations and predicted chemistry of the most abundant atmospheric sulfur gases, an hypothesis is put forward that carbon disulfide (CS_2) is the primary precursor of background sulfur dioxide (SO_2). Preliminary measurements indicating that bay and coastal waters were greatly supersaturated with CS_2 are presented to support this hypothesis. The coastal and bay water samples also were supersaturated with respect to carbonyl sulfide (OCS). These results are the basis for another hypothesis that the oceans were supersaturated with OCS and thus are an important if not dominant source of this material. A volatile sulfur compound, yet unidentified, was also found in large amounts in the water samples analyzed. Some speculation is presented as to the identity of this compound.

Introduction

The ocean's potential importance as a source of sulfur gases has been recognized from the beginning of serious study of the atmospheric part of the global sulfur cycle. Early on in this study, saline marsh and near coastal waters were predicted to be relatively strong sources of these substances compared to land and open ocean areas (Eriksson 1959). This expectation stemmed from the observation that the naturally emitted gases were primarily reduced sulfur gases having oxidation states lower than that of sulfur dioxide (SO_2). These substances were known to be formed in large quantities in the sediments by sulfate-reducing micro-organisms (Eriksson 1959). Further, because of the ease with which some of these reduced sulfur gases are oxidized in water containing oxygen or ozone (O_3), the intensity of the sources should be correlated with the shallowness of the water. These hypotheses have been substantially verified by experiment (Ostlund and Alexander 1963).

Although we have evidence that sulfur gases emitted to the atmosphere by saline water areas, especially shallow areas, are important in atmospheric chemistry, the details and magnitude of their contribution are still relatively unknown. The objective of this paper is to outline the research activity at the frontier of this field and then speculate on where future research may lead.

ISBN 0-12-404070-5

Rationale for Studying the Atmospheric Sulfur Cycle

In proportion to the other elemental cycles, the atmospheric sulfur cycle has received considerable attention in the past twenty years. This interest stems primarily from the fact that reduced sulfur gases are oxidized in both the troposphere and the stratosphere to produce sulfuric acid. This sulfuric acid combines with water vapor to produce very small aerosol particles of highly concentrated sulfuric acid. The particles are strongly acidic and hygroscopic and have mass median diameters of less than about 0.3 μm.

In the troposphere, sulfuric acid aerosol is thought to play an important role in cloud, haze, and fog microphysics which, through perturbations of the earth's radiation balance, can be manifested as a long term influence on weather and climatic processes. In the stratosphere the sulfuric acid aerosol scatters the sun's radiation as well as absorbs the earth's radiation. Therefore, like tropospheric aerosol, it can have direct influence on the earth's radiation budget and climate.

Because sulfuric acid is probably the primary acid produced in the earth's boundary layer it can have profound effects on the other elemental cycles. Consider for example the inorganic halogen cycle in the marine boundary layer. As sea salt aerosol rises in the atmosphere it meets and combines with the sulfuric acid aerosol. Prior to this reaction the sulfuric acid aerosol was very acidic, having a pH below one, whereas the sea salt aerosol was basic, originally having a pH of about 8.3. The wet sea salt aerosol is in effect titrated by the sulfuric acid aerosol. Eventually, the rising sea salt aerosol becomes acidic enough to liberate the hydrogen halides through acid-base reactions involving the alkali metal halides in the sea salt: $Cl^- + H^+ \rightarrow HCl$. It is by this mechanism that the sea salt aerosol becomes depleted in the halides as has been confirmed by many observations (Duce et al. 1965; Winchester and Duce 1966; Duce and Woodcock 1971).

Precursors of Stratospheric Sulfuric Acid Aerosols

There seems to be little doubt that the precursors of stratospheric aerosol are transported to the stratosphere from the troposphere. There is also little doubt that this transport takes place in very intense thunderstorms occurring in equatorial regions. These precursors are probably entrained in the boundary layer air which then is transported rapidly from the base of the cloud up to the tropopause. By a mechanism not fully understood, these gases are transported with a very small efficiency to the stratosphere.

Because the transport is almost totally within the cloud, only relatively insoluble and inert gases can be transported to the stratosphere. The only sulfur gas having these properties as well as a significant concentration in the troposphere is carbonyl sulfide, OCS (Crutzen 1976). Sulfur dioxide is much too reactive and soluble in the water droplet to survive vertical transport within the cloud.

The case for OCS being the primary precursor of background stratospheric aerosol is strengthened by considering the fact that it is the most abundant sulfur gas in the troposphere. Its tropospheric concentration of about 500 pptv (Sandalls and Penkett 1977; Maroulis et al. 1977; Torres et al. 1980) is much greater than that of the next most abundant gas, SO_2, which has a concentration of about 100 pptv (Maroulis et al. 1980). The case for OCS is further enhanced by the observation that its tropospheric residence time is long, probably greater than a few years (Bandy et al. 1981; Johnson 1981). This should be contrasted with the relatively short residence time of SO_2 which is only a few days (Friend 1973; Graedel 1977). Importantly, short atmospheric residence times are manifested as a large variability in the concentration, whereas a long residence time is manifested as a small variability in the concentration.

Sources of OCS

There has been considerable speculation concerning the source of OCS. Presently, we believe that natural sources dominate anthropogenic sources. However, the source strength is small and the ambient concentration is large, thus the sources are hard to detect even when measurements are made close by. Some limited measurements have indicated that saline marshes may be sources of this substance (Adams et al. 1980). However, these measurements were essentially point measurements in a very nonhomogeneous source and therefore cannot be considered definitive.

Oceans as a Source of OCS

Because the hydrolysis of OCS in basic solutions by hydroxide is relatively rapid, several theoretical studies have indicated that ocean waters, including intertidal areas, would be sinks rather than sources of OCS (Johnson 1981). In order to obtain information bearing on this question we have made some preliminary measurements of the sulfur gas concentrations in near shore and bay areas on the east coast of the United States.

The apparatus used is shown in Fig. 1. The system is composed of a cylindrical extraction vessel and a reflux condenser type drying assembly operated at about $-5°$ C. The water to be analyzed is added to the extraction vessel and purged with helium at a flow rate of about 50 ml per min for approximately 40 min. After passing through the cryogenic predryer condenser, the OCS is cryogenically trapped in a 50 cm length of 3.2 mm o.d. teflon tubing cooled in a small dewar of liquid argon. After the trapping phase the condensed material is volatilized with a small dewar of warm water and analyzed by gas chromatography. Chromosil 330 and Carbopak B/XE60 columns were used for separation of the trapped gases. A flame photometric detector operating in the sulfur mode was used as a detector.

These samples were collected near Wallops Island in early summer. Wallops Island is a barrier island on the ocean side of the Virginia Eastern Shore near the Virginia-Maryland border. The locations of the sampling sites

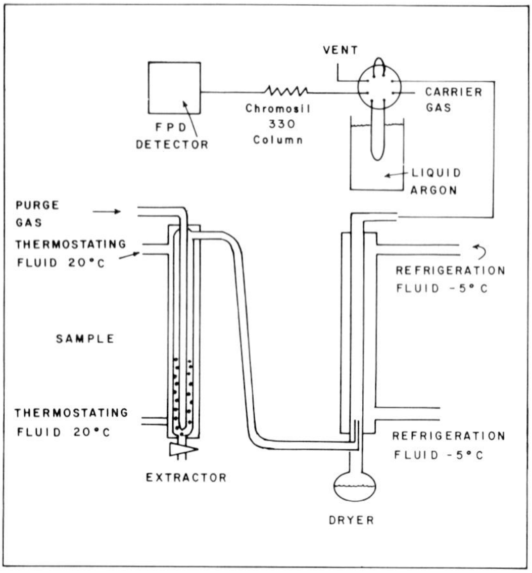

Figure 1. Apparatus for making water measurements – sulfur gas composition of H_2O.

are shown in Fig. 2. Samples were taken both in the bay and the Atlantic Ocean near the beach. The results of the chemical analysis of these sample are shown in Table 1. We have included in this table a calculation of the water

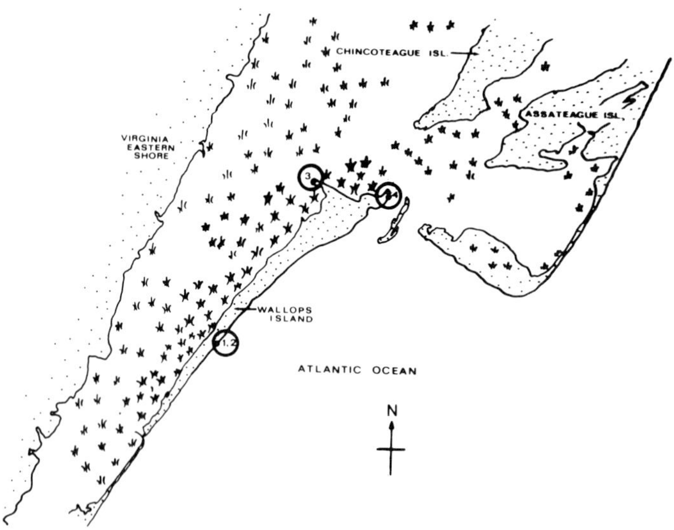

Figure 2. Wallops Island area, showing sampling sites for water samples analyzed.

Table 1. Results (ng S l^{-1}) of carbonyl sulfide (OCS) and carbon disulfide (CS_2) water analysis of samples from Wallops Island, VA.

Sample site	*Measured Concentrations* OCS	CS_2	*Site Number (Fig. 2)*
Atlantic coastal water, east side of Wallops Island	5.9	7.8	1
Atlantic coastal water, east side of Wallops Island	4.0	–	2
Saline marsh water, Wallops Island	6.1	61	3
Coastal water, north end of Wallops Island	6.4	–	4
Mean and Standard Deviation:	5.6±1.1	–	
Calculated Equilibrium Constants[1]:	0.33[2]	0.046[3]	

[1]International Critical Tables 1928; units are ng/ml; [2]At 25°C, $[OCS]_{ambient} = 500$ pptv; [3]At 25°C, $[CS_2]_{ambient} = 30$ pptv.

concentration which should exist under equilibrium conditions for OCS in the air and OCS in the water (International Critical Tables 1928).

The concentration of OCS observed in the water samples was about one order of magnitude greater than the calculated equilibrium value. Although there could be errors as large as a factor of two in our measurements, it is extremely unlikely that the errors are as large as a factor of ten.

Whether the production of the OCS occurs in the water column and/or in the sediments is an interesting question. We can speculate that OCS may play a role in the sulfur redox cycle analogous to that played by nitrous oxide (N_2O) in the nitrogen redox cycle. These roles are shown schematically in Fig. 3. Like N_2O, OCS may be produced in both the oxidation and the reduction branches of the cycle.

Tropospheric Sulfuric Acid Precursors

Until a few years ago many doubted whether or not there existed a background level of SO_2 from which a background level of tropospheric sulfate aerosol could be formed. This issue was substantially resolved by our measurements of the background concentration of SO_2 in the Global Atmospheric Measurements of Trace Aerosols and Gases Program sponsored by the National Science Foundation (Maroulis et al. 1980).

The measurements were made by aircraft over the region shown in Fig. 4. The flags represent the locations where the aircraft landed. Measurements were obtained over a latitude range of 56°S to 70°N over a period of about two weeks in the spring of 1978.

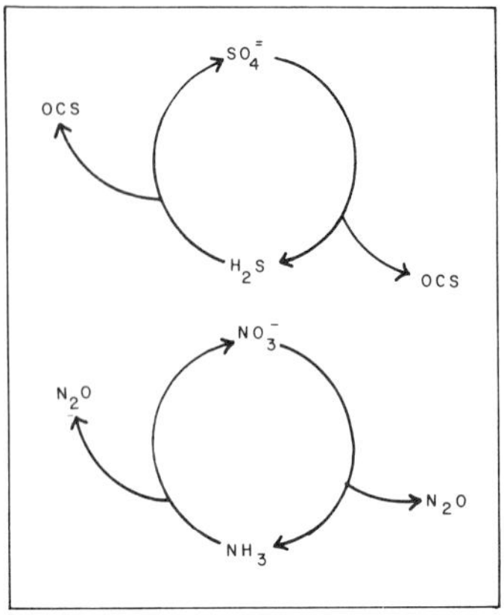

Figure 3. Schematic comparison of production of N_2O in the nitrogen redox cycle and the production of OCS in the sulfur redox cycle.

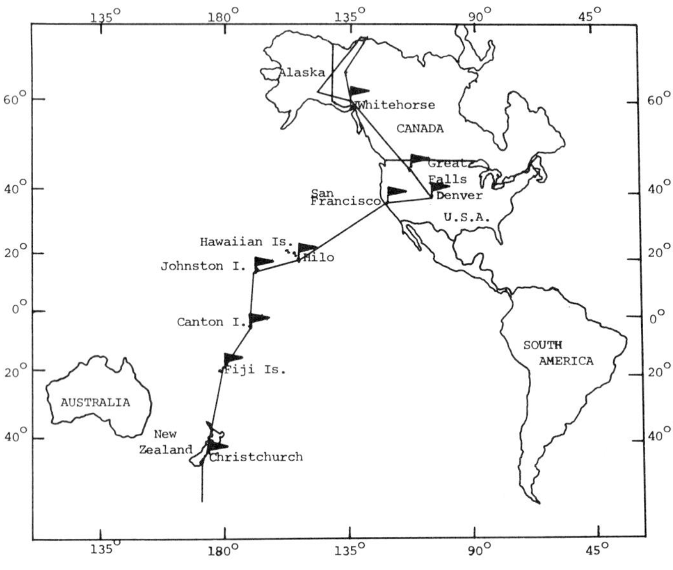

Figure 4. 1978 Gametag field-sampling program flight path.

The SO_2 data obtained in this program are shown in Fig. 5 plotted as a function of latitude. Here we shall focus on the measurements made south of 20°N latitude since these measurements were made over the clean Pacific Ocean. Although the fluctuations are large as expected from the short residence time of SO_2, an almost constant non-zero average SO_2 concentra-

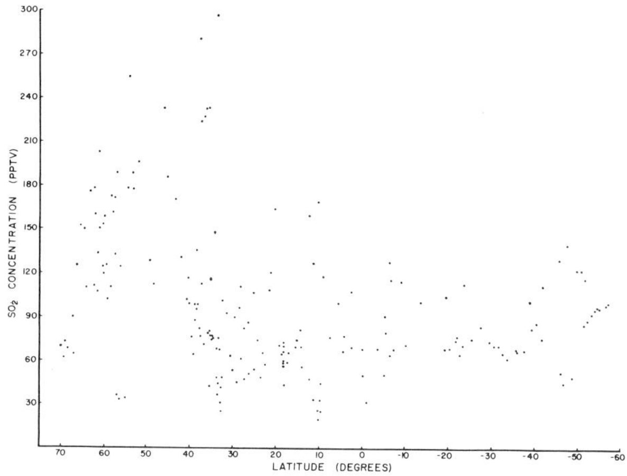

Figure 5. Plot of all the individual SO_2 data as a function of latitude.

tion is observed. Clearly there is a background level of SO_2 which must be formed from a precursor apparently emitted from the ocean.

Further analysis of these data shows that the SO_2 concentration is about a factor of two higher in the free troposphere than in the boundary layer. Thus in our search for a precursor for SO_2, a compound must be found which has very special chemical properties. These properties must include a relatively low reactivity in the boundary layer and somewhat greater reactivity in the mid and upper troposphere.

Substantial speculation has existed about the nature and origin of the precursors of SO_2 (Logan et al. 1979; Sze and Ko 1980). Dimethyl sulfide ($(CH_3)_2S$), hydrogen sulfide (H_2S), carbon disulfide (CS_2), dimethyl sulfoxide ($(CH_3)_2SO$), and OCS have been seriously considered. Extensive literature exists on this subject (e.g., Graedel 1977), yet great uncertainty remains regarding the importance of any of these gases as tropospheric SO_2 precursors. The lack of progress is related to the small amount of data which is available concerning the concentrations of these materials in background air and water.

Carbon Disulfide

Our most recent work has centered on CS_2 primarily because this material may react slowly enough with OH radical to allow a significant fraction of it to be transported from the boundary layer to the free troposphere where its subsequent conversion to SO_2 could contribute to the positive vertical gradient observed for SO_2. Materials such as methyl mercaptan (CH_3SH), $(CH_3)_2S$ and dimethyl disulfide ($(CH_3)_2S_2$) are potential precursors of SO_2; however, these materials react rapidly with OH radical and thus are converted to SO_2 in the boundary layer where the OH radical concentration is relatively high (Graedel 1977). Since the conversion of these reactive gases to SO_2 probably occurs mainly in the boundary layer, they are likely to contribute only to boundary layer SO_2 levels which would tend to make the vertical gradient of SO_2 more negative. One preliminary measurement (Lovelock 1974) suggested that the surface ocean waters may be saturated with CS_2 which at least'is the first prerequisite for the existence of a significant flux of this material from the ocean to the atmosphere.

Sources of Carbon Disulfide

Our boundary layer measurements of CS_2 made in Philadelphia, PA and Wallops Island, VA in relatively unpolluted air indicated that the mean concentration of this sulfur gas in the boundary layer was approximately 30 pptv (Maroulis and Bandy 1980; Bandy et al. 1981). In Philadelphia the air masses considered to be relatively unpolluted originated from the west and north. The Wallops Island data used to obtain the average boundary layer concentration of CS_2 in marine air were made in air that had been over the open ocean for several days. In air arriving from the west and thus having passed directly over the Wallops Island Marsh, the CS_2 concentration was about 80 pptv.

Recently we made CS_2 measurements in free tropospheric air during flights across the continental United States (Bandy et al. 1981). The typical concentration observed was less than 3 pptv. If these values are assumed to be representative of the entire troposphere, an average tropospheric CS_2 concentration of 6 pptv is calculated.

Anthropogenic emissions of CS_2 may also be important. The CS_2 concentration measured in Philadelphia was 5 to 10 times higher for south and southwest winds. Under these conditions the air had passed over the highly industrialized areas in the vicinity of Chester, PA and Wilmington, DE (Maroulis and Bandy 1980).

Adams et al. (1980) have detected CS_2 emissions from marine marshes and swamps as well as cultivated histosol soils. They found CS_2 emission rates of 13 mg S m^{-2} yr^{-1} for an intertidal marsh and 6 mg S m^{-2} yr^{-1} for a stagnant swamp. Emission rates from infrequently flooded marshes and cultivated histosol soils were 125 mg S m^{-2} yr^{-1}.

In our analysis of coastal and saline marsh waters we detected significant concentrations of CS_2 at both locations as shown in Table 1. However, the marsh sample was eight times more concentrated in CS_2 than the ocean sample. Both samples contained much more material than that expected for equilibrium conditions.

Other Sulfur Gases in Water

Recently during our analysis of coastal waters, we obtained some preliminary information that suggests the existence of another sulfur gas in sea water which might be an important precursor of atmospheric SO_2. This

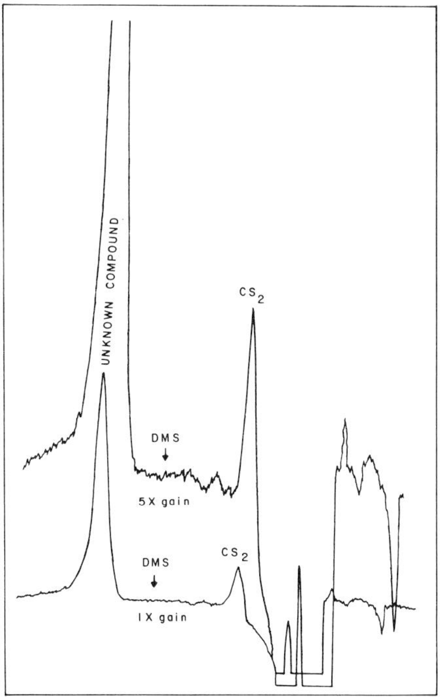

Figure 6. Chromatogram of sulfur gases extracted from marine water samples.

was determined from our chromatograms in which the largest peak was not one of the sulfur gases commonly found in nearshore waters (Fig. 6). A sulfur containing compound appears whose retention time is 130 sec longer than CS_2, 60 sec greater than $(CH_3)_2S$, and which has a concentration of about one order of magnitude higher than CS_2. It should have a molecular weight similar to that of $(CH_3)_2S$ and CS_2. Ethyl mercaptan (CH_3CH_2SH) would elute very close to this compound and should be considered a possibility. However, to our knowledge CH_3CH_2SH has not been observed by other scientists. Irreversible losses may explain the fact that it was not detected in earlier studies. Because of the preliminary nature of these data, we shall not speculate here about the relative importance of this compound to the atmospheric sulfur cycle or about the chemistry and biology of the water column and the sediments which may be responsible for its production.

References Cited

Adams, D. F., F. O. Farwell, M. R. Pack and E. Robinson. 1980. Estimates of natural sulfur source strengths, pp. 35-45. *In:* D. S. Shriner et al. (eds.), *Atmospheric Sulfur Deposition*. Ann Arbor Science Publishers, Ann Arbor, MI.

Bandy, A. R., P. J. Maroulis, L. Shalaby and L. A. Wilner. 1981. Evidence for a short tropospheric residence time for carbon disulfide. *Geophys. Res. Lett.* 8:1180-1183.

Crutzen, P. J. 1976. The possible importance of OCS for the sulfate layer of the stratosphere. *Geophys. Res. Lett.* 3:73-76.

Duce, R. A., J. W. Winchester and T. W. Van Nahl. 1965. Iodine, bromine and chlorine in the Hawaiian marine atmosphere. *J. Geophys. Res.* 70:1775-1799.

Duce, R. A. and A. H. Woodcock. 1971. Difference in chemical composition of atmospheric sea salt particles produced in the surf zone and on the open sea in Hawaii. *Tellus* 23:427-435.

Eriksson, E. 1959. The yearly circulation of chloride and sulfur in nature; meterological, geochemical and pedological implications. *Tellus* 11:375-403 and 12:63-109.

Friend, J. P. 1973. The global sulfur cycle, pp. 177-201. *In:* S. I. Rasool (ed.), *Chemistry of the Lower Atmosphere.* Plenum Press, New York.

Graedel, T. E. 1977. The homogeneous chemistry of atmospheric sulfur. *Rev. Geophys. Space Phys.* 15:421-428.

International Critical Tables 1928. Volume III. McGraw-Hill Book Co., New York. p. 261.

Johnson, J. E. 1981. The lifetime of carbonyl sulfide in the troposphere. *Geophys. Res. Lett.* 8:938-940.

Logan, J. A., M. B. McElroy, S. C. Wofsy and M. J. Prather. 1979. Oxidation of CS_2 and OCS: sources for atmospheric SO_2. *Nature* 281:185-188.

Lovelock, J. E. 1974. CS_2 and natural sulfur cycle. *Nature* 248:625-626.

Maroulis, P. J. and A. R. Bandy. 1980. Measurements of atmospheric concentrations of CS_2 in the eastern United States. *Geophys. Res. Lett.* 7:681-684.

Maroulis, P. J., A. L. Torres and A. R. Bandy. 1977. Atmospheric concentration of carbonyl sulfide in the southwestern and eastern United States. *Geophys. Res. Lett.* 4:510-512.

Maroulis, P. J., A. L. Torres, A. B. Goldberg and A. R. Bandy. 1980. Atmospheric SO_2 measurements on project Gametag. *J. Geophys. Res.* 85:7345-7349.

Ostlund, G. H. and J. Alexander. 1963. Oxidation rate of sulfide in sea water, a preliminary study. *J. Geophys. Res.* 68:3995-3997.

Sandalls, F. J. and S. A. Penkett. 1977. Measurements of carbonyl sulfide and carbon disulfide in the atmosphere. *Atmos. Environ.* 11:197-199.

Sze, N. D. and M. K. W. Ko. 1980. Photochemistry of COS, CS_2, CH_3SCH_3 and H_2S: implication for the atmospheric sulfur cycle. *Atmos. Environ.* 14:1223-1239.

Torres, A. L., P. J. Maroulis, A. B. Goldberg and A. R. Bandy. 1980. Atmospheric OCS measurements on project Gametag. *J. Geophys. Res.* 85:7357-7360.

Winchester, J. W. and R. A. Duce. 1966. Coherence of iodine and bromine in the atmosphere of Hawaii, northern Alaska, and Massachusetts. *Tellus* 18:287-291.

COMPARISONS OF ANADROMOUS FISHES IN ESTUARIES

Convenor: *W. J. Ebel*
Northwest and Alaska Fisheries Center
Seattle, Washington

JUVENILE PACIFIC SALMON IN ESTUARIES: THE LIFE SUPPORT SYSTEM

M. C. Healey
Department of Fisheries and Oceans
Pacific Biological Station
Nanaimo, British Columbia

Abstract: This report summarizes information on the abundance, food requirements and production of five species of Pacific salmon (genus *Oncorhynchus*) in two estuaries in southern British Columbia and compares some of these features among seven other estuaries. Chum salmon are abundant in estuaries for 2 months in early spring, coho for 2 months in late spring, and chinook throughout the spring, summer and autumn. Pink and sockeye spend little time in estuaries. Habitats occupied by each species vary with fish size, tidal stage, and time of year. Tidal creeks through marshes, the junction of major and minor distributaries in the intertidal zone, and the delta front are favoured habitats. Many taxa contribute to the diet of juvenile salmon in different estuaries but relatively few taxa generally constitute the bulk of the diet. Food requirements of juvenile salmon are generally a small percentage of the total standing crop and annual production of prey, but a high proportion of the production of preferred prey is sometimes taken. Major prey tend to be detritus feeders, indicating that the food web supporting juvenile salmon is detritus based. The value of an estuary as rearing habitat for juvenile salmon appears to be influenced by delta configuration, and by the efficiency with which allochthonous organic carbon is trapped.

Introduction

Five species of Pacific salmon (*Oncorhynchus spp.*) constitute the principal fishery resource of the Pacific northwest. Commercial catches are on the order of 170,000 metric tons with a landed value of $360 million. The recreational catch of salmon is also high and is worth hundreds of millions of dollars. Because of their anadromous habit, all Pacific salmon pass through an estuary twice during their lives; first as juveniles migrating from freshwater nursery areas to the ocean, and second as maturing adults returning from the ocean to freshwater to spawn. Urban and industrial expansion in estuaries often conflicts with conservation of salmon habitat. The importance of particular features of natural estuaries to the continued productivity of Pacific salmon has been hotly debated throughout the Pacific northwest. Understanding how salmon utilize estuaries has, therefore, far-reaching practical implications for resource and environmental management.

This report focuses on juvenile salmon in estuaries and addresses several topics: Spatial and temporal utilization of estuaries by each species; food chain relationships of salmon in estuaries; food resource requirements; and the probable life support system for juvenile salmon in estuaries.

ISBN 0-12-404070-5

Sources of Information

Sufficient information exists for a variety of estuaries on which to base a composite description of the spatial and temporal occurrence of juvenile salmon in estuaries. The empirical examples that I shall use to illustrate the behaviour of each species are from my research, much of which is previously unpublished, on the Nanaimo and Nitinat estuaries (Fig. 1) in southern British Columbia (B. C.) (Healey 1979, 1980a, b, unpub. data). Other estuaries considered here include (Fig. 1) the Fraser, Squamish, Cowichan, Courtenay, and Somass in southern B.C., and the Sixes and Yaquina in Oregon (Kask and Parker 1971, 1972a, b, 1974; Reimers 1971; Barraclough and Phillips 1978; Birtwell 1978; Levy and Levings 1978; Levy et al. 1979; Morris and Leaney 1980; Myers 1980; Levy and Northcote 1981). The general behaviour of each species appears quite consistent from estuary to estuary, although there are differences in detail.

Specific information on food chain relationships, food requirements, and the life support system is from my research on the Nanaimo and Nitinat estuaries, again, much of it unpublished. I shall, however, cross reference this information to other estuaries where possible.

The methods by which the unpublished data presented in this report

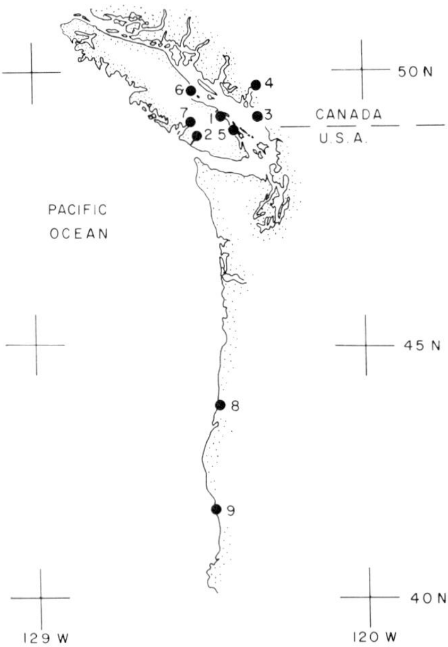

Figure 1. Map showing the location of 9 estuaries referred to in the text: 1, Nanaimo; 2, Nitinat; 3, Fraser; 4, Squamish; 5, Cowichan; 6, Courtenay; 7, Somass; 8, Yaquina; 9, Sixes.

were gathered are the same as those described in my previous publications (Healey et al. 1977; Healey 1979, 1980a).

Description of the Estuaries

Since the empirical examples given in this report are from the Nanaimo and Nitinat estuaries, which are structurally quite different, I shall describe these estuaries in some detail and follow with a brief description of the other estuaries.

The Nanaimo River flows into the Strait of Georgia on the east coast of Vancouver Island (Fig. 1). The area of the estuary is about 11.5 km^2 comprised of 2 km^2 of intertidal marshes, 6.5 km^2 of intertidal sand flat, both dissected by numerous major and minor river channels and tidal creeks, and 3 km^2 of subtidal basin which forms Nanaimo Harbour (Fig. 2). The estuary is sheltered to the North by Protection and Newcastle Islands.

River discharge is characterized by winter freshets of 170-280 $m^3\ s^{-1}$ and summer low flows of 2.8-5.5 $m^3\ s^{-1}$. Surface waters are seldom significantly diluted beyond the harbour entrance except during high winter discharge. River flow is normally dispersed as a low salinity layer a few cm deep over the delta and parts of the harbour area. Typical spring and summer surface salinities in the inner estuary at high tide range over 10-25 ppt with a halocline between 0-1 m. In the outer estuary surface salinities range over 15-29 ppt with the halocline, when present, between 0-2 m. Salinity below

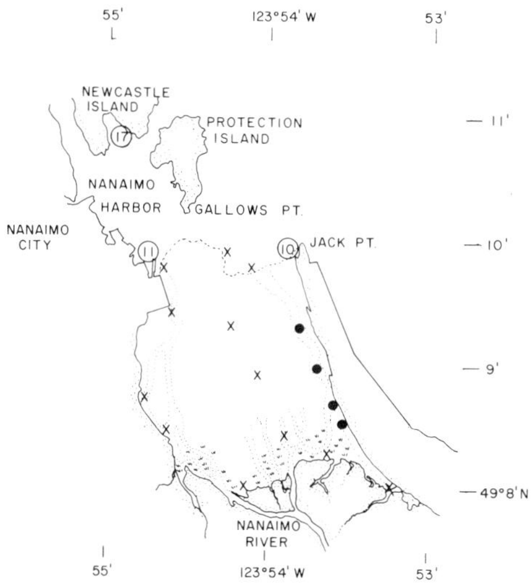

Figure 2. Map of Nanaimo estuary showing: principal physical features; sites sampled weekly for juvenile salmon in the inner estuary (●) and the outer estuary (10, 11, 17); and sites sampled irregularly (X) during 1975-1980.

the halocline is normally 28–30 ppt. During spring and summer, salinity at low tide in river channels crossing the intertidal sand flat is often greater than 24 ppt (Sibert and Brown 1976; Healey et al. 1977; Healey 1980a). Tides in the Nanaimo area are of the mixed, semi-diurnal type and maximum tidal amplitude is 2 (neap) to 3.5 (spring) m.

Annual primary production in the inner estuary ranges from 0.9 g carbon $m^{-2}yr^{-1}$ for macrophytic algae to 73 g carbon $m^{-2}yr^{-1}$ for *Zostera marina* beds. Marsh production, mainly *Carex*, is about 560 g carbon $m^{-2}yr^{-1}$. By far the greatest source of organic carbon to the estuary is the river, which discharges over 2000 g carbon $m^{-2}yr^{-1}$ (Naiman and Sibert 1979). The importance of the various sources of organic carbon to the food chain supporting juvenile salmon is still a subject of debate. As will be discussed later, however, the food chain supporting juvenile salmon in the intertidal appears to be detritus based.

The Nitinat River flows into the head of Nitinat Lake on the west coast of Vancouver Island (Fig. 1). The lake is deep (>200 m), saline, and is 23 km long by 1.2 km wide (area 27.6 km²). It is separated from the open ocean by a narrow (250 m) rocky channel 3 km long and only a few metres deep (Fig. 3). Tidal rise and fall at the river mouth is only about 30 cm, and the shores of Nitinat Lake are steep, so that there is little intertidal zone and no significant

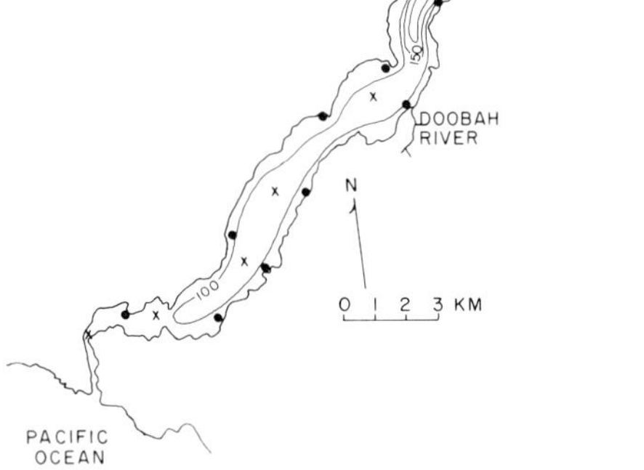

Figure 3. Map of Nitinat estuary showing: principal physical features (depth contours in meters); sites sampled weekly for juvenile salmon throughout the estuary (●), and sites sampled irregularly (X) during 1979 and 1980. The inner Nitinat estuary comprises the deeper north basin; the outer estuary comprises the shallower south basin of the estuary.

river delta. The river runs across a gravel bar and into a bay that shelves quickly to over 100 m deep (Fig. 3).

The Nitinat River is also characterized by winter freshets and summer low flows. Since there are no sizeable lakes in the Nitinat drainage to buffer runoff, and rainfall in the watershed is greater than in the Nanaimo, extremes in discharge are greater at Nitinat. Unfortunately, river discharge is not gauged, so absolute flows cannot be given. During low flow periods river influence hardly extends beyond the bay into which the river empties, while during freshets the whole surface of the lake may fall to less than 1 ppt salinity. Three other rivers (Caycuse, Doobah, Hobiton) as well as numerous seasonal creeks also contribute to the low salinity surface layer of the lake. During the spring and summer, surface salinity in the inner estuary ranges over 1-20 ppt, and in the outer estuary over 10-30 ppt. The halocline normally falls between 0-5 m and salinity below 5 m is usually >30 ppt (Northcote et al. 1964; Fedorenko et al. 1979; Healey unpub. data).

Nitinat lake is vertically stratified throughout the year, and below about 20 m is anoxic (Northcote et al. 1964). Occasional upwelling of H_2S from the anoxic layer causes fish and invertebrate kills in the lake (Fedorenko et al. 1979; Healey unpub. data.).

The smaller estuaries in southern B.C. (Fig. 1), from which comparative observations will be drawn, are structurally similar to the Nanaimo but vary in the relative areas of marsh, intertidal flats and subtidal basin, in the degree to which the marshes and intertidal flats are dissected by river channels and creeks, and in the degree of industrialization. The Fraser estuary is, superficially, a larger version of the Nanaimo estuary, and many of the observations on juvenile salmon in the Nanaimo have been applicable to the Fraser. The Fraser River, however, is fed by snow melt from the interior of B.C. and, therefore, has summer freshets and winter low flows. Also its discharge is so large that the estuary has a much lower salinity than the other estuaries considered here and the discharge lowers surface salinity in most of the southern half of the Strait of Georgia. The Oregon estuaries are drowned river valleys separated from the open sea by coastal sand bars. These estuaries have features in common with both the Nanaimo and Nitinat, but lack the special feature of permanent stratification found in the Nitinat estuary.

Spatial and Temporal Utilization of Estuaries by Juvenile Salmon

As juveniles, the five species of Pacific salmon have different behaviour patterns and this affects their occurrence in estuaries. Pink and chum salmon (*O. gorbuscha, O. keta*) migrate seaward as fry, 30-40 mm in length (0.25-0.4 g), and enter the estuary of their natal stream within a few days after emerging from the gravel spawning beds. In southern B.C., Washington and Oregon this migration occurs during March-May, becoming earlier to the south. With a few exceptions in this geographic area, sockeye and coho (*O.*

nerka, O. kisutch) spend a year in fresh water before migrating seaward, and enter the estuary during April and May of their second year of life when 60-100 mm in length (2-10 g). Chinook (*O. tshawytscha*) enter the estuary either as fry (40 mm, 0.5 g) within a few days after emergence from the gravel in March-May, as fingerling smolts (60-80 mm, 2-5 g) in May or June of their first year, or as yearling smolts (80-110 mm, 5-18 g) in April and May of their second year.

Pink salmon

Among the river systems considered in this report, only the Fraser has abundant runs of pink salmon. Pink fry are present in the tidal marshes of the Fraser delta only during their downstream migration. Pink fry invade the tidal marshes at high tide and leave on the first of the ebb. The residence time of fry arriving in the marsh area is no more than a day or two (Levy et al. 1979). Most of the fry migrating seaward each day move directly through the delta

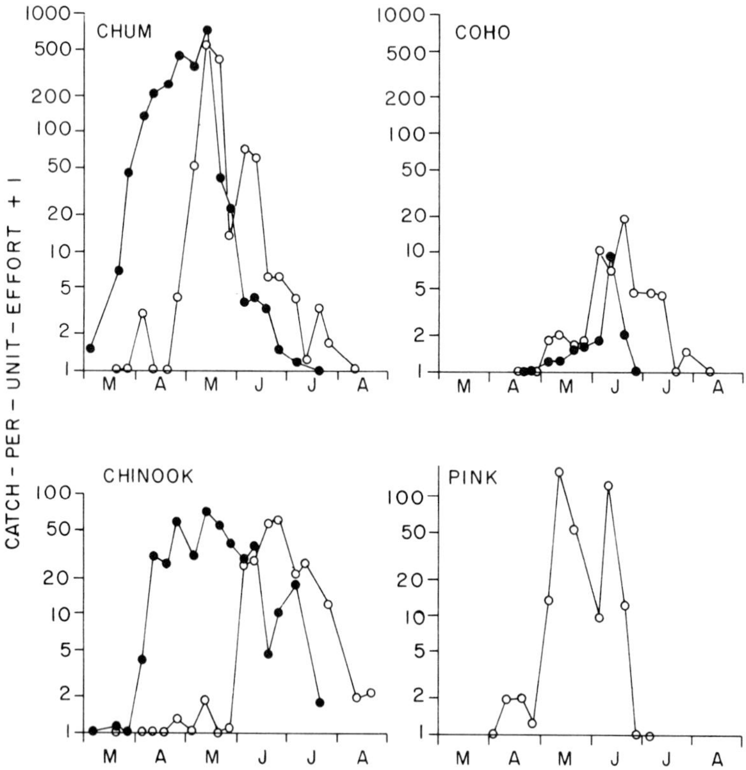

Figure 4. Seasonal changes in catch of four species of Pacific salmon in the inner (●) and outer (○) Nanaimo estuary. Data for 1975-80 averaged. (Healey 1979, 1980a, unpub. data.)

area and into the river plume (Barraclough and Phillips 1978). Pink fry are swept across the southern Strait of Georgia by the river plume and congregate in dense schools near to shore along the east coast of Vancouver Island (Healey 1980b, unpub. data).

Some pink fry, most likely from the Fraser River, find their way into the outer Nanaimo estuary and rear there during May and June (Fig. 4). I captured pink fry mainly at station 17 (Fig. 2), however, where salinity was usually above 27 ppt. Juvenile pinks appear to leave the Nanaimo estuary (Fig. 4) and other habitats in the southern Strait of Georgia in late June or July at which time they are about 80 mm long (Phillips and Barraclough 1978; Healey 1980b, unpub. data).

Chum salmon

Chum salmon occur in both the Nanaimo and Nitinat, and most of the other river systems considered in this report. Their behaviour is quite consistent among estuaries (Healey 1979, unpub. data; Levy et al. 1979; Myers 1980; Levy and Northcote 1981). On arrival at the river mouth chum fry may disperse several km within a few hours. Some, however, remain in the estuary and rear there (Figs. 4, 5). The first habitats occupied are tidal creeks and sloughs high in the delta area, but other marsh habitats and intertidal areas are also quickly colonized. During high tide chum fry congregate in the upper intertidal at the fringe of marshes, and penetrate deep into the marshes along tidal creeks. At low tide the fry retreat into tidal creeks that retain flowing water at low tide and into delta channels. Preferred low tide habitats appear to be at the junction of major and minor distributaries in the delta (Healey unpub. obs. on Nanaimo, Cowichan and Courtenay estuaries).

Release of marked fry in the Nanaimo and Fraser estuaries has demonstrated that chum may spend up to three weeks rearing in the inner estuary, and are quite localized in their movement patterns, even in large estuaries like the Fraser (Healey 1979; Levy et al. 1979).

Superimposed on the twice-daily tidal migration from low tide refugia to the marshes and back again is a seasonal migration seaward in the estuary. This is well demonstrated in the Nanaimo estuary where chum fry are abundant in the inner estuary during April and May and in the outer estuary during May and June (Fig. 4); it is also apparent in the Yaquina, Cowichan and Courtenay estuaries (Myers 1980; Healey unpub. data). The Nitinat estuary differs from the general pattern for southern B.C. in that chum are abundant in both the inner and outer estuary only during April and May, and there is no evidence of a seaward progression through the lake (Fig. 5).

The seaward migration is size related as the larger fry move seaward first (Table 1). Chum are gone from most estuarine habitats in southern B.C. by mid July. In the Yaquina estuary the timing of events is earlier with most chum gone from the estuary by mid May (Myers 1980).

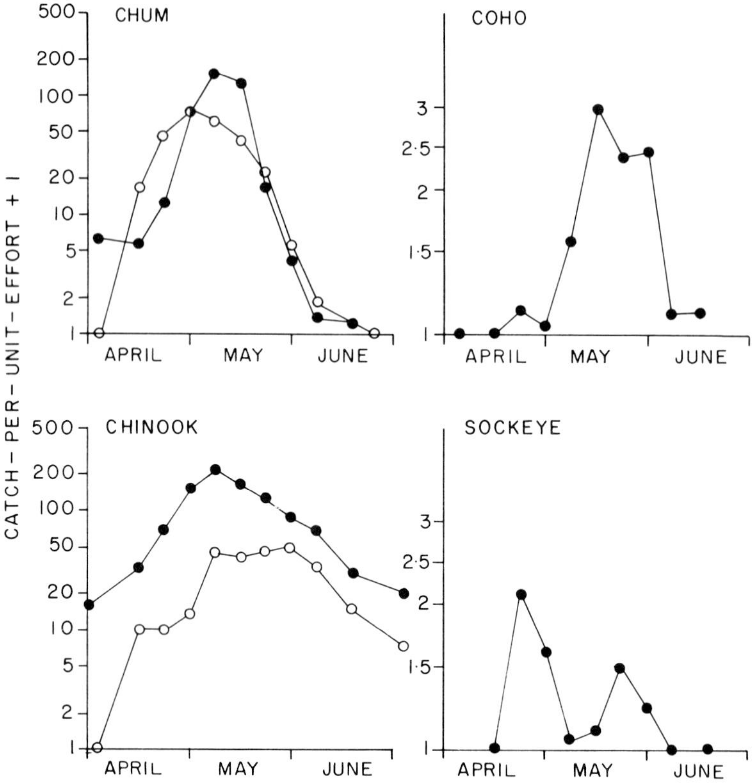

Figure 5. *Seasonal changes in catch of four species of Pacific salmon in Nitinat estuary. Catch of chum and chinook are shown separately for the inner (●) and outer (○) estuary. Catch of coho and sockeye are for the whole estuary. (Healey unpub. data.)*

Sockeye salmon

Among river systems considered in this report, the Fraser and Somass systems are major producers of sockeye salmon and the Hobiton system, tributary to Nitinat lake, is a minor producer. Sockeye smolts are seldom captured in the inner estuary and delta habitats of the Fraser and Somass rivers (Levy et al. 1979; Morris and Leaney 1980). Sockeye are present in the outer Nitinat estuary (Fig. 5) and the outer regions of other estuaries during their downstream migration, but catches are small relative to the number of downstream migrants (Kask and Parker 1971; Barraclough and Phillips 1978). Yearling sockeye smolts, therefore, appear to spend little time in their natal estuary, but move quickly to other nursery areas.

Some races of sockeye in the Fraser River migrate seaward as fry rather than yearlings. These fish show a different behaviour, rearing in the inner

Table 1. Average weights (g ± SE) of juvenile salmon captured in the Nanaimo and Nitinat estuaries. Nanaimo data collected at various times from 1972 to 1979. Nitinat data collected 1979, 1980. (Sources: Anon. 1974; Healey 1979, 1980a, unpub. data.)

	Inner estuary		*Outer estuary*			
Date	*Chum*	*Chinook*	*Chum*	*Chinook*	*Coho*	*Pink*
			Nanaimo Estuary			
1-15 Mar	0.46 ± 0.02					
16-31 Mar	0.50 ± 0.06	0.64				
1-15 April	0.64 ± 0.08	0.73 ± 0.05	0.35			
16-30 April	0.68 ± 0.08	0.86 ± 0.05	0.54 ± 0.26			0.26 ± 0.04
1-15 May	0.66 ± 0.04	0.96 ± 0.09	0.89 ± 0.25	0.53	12.90	0.77 ± 0.39
16-31 May	0.70 ± 0.03	1.01 ± 0.15	1.79 ± 0.50	5.43	23.70 ± 3.36	1.36 ± 0.40
1-15 June	1.28 ± 0.16	2.56 ± 0.07	2.80 ± 0.62	8.08 ± 3.7	17.67 ± 2.33	2.64 ± 0.20
16-30 June	1.67 ± 0.17	3.72 ± 0.36	2.93 ± 0.40	10.14 ± 2.3	19.23 ± 3.64	2.57 ± 0.40
1-15 July		4.32	4.32 ± 0.80	10.35 ± 2.0	42.85 ± 2.33	
16-31 July		4.16 ± 0.28		13.43 ± 1.5		
			Nitinat Estuary			
1-15 April	0.58 ± 0.05	0.68 ± 0.01	0.52 ± 0.02	0.69 ± 0.03		
16-30 April	0.49 ± 0.02	0.67 ± 0.03	0.56 ± 0.03	0.71 ± 0.04		
1-15 May	0.78 ± 0.04	0.74 ± 0.05	0.76 ± 0.07	0.93 ± 0.14		
16-31 May	1.22 ± 0.10	1.11 ± 0.10	1.47 ± 0.38	1.50 ± 0.16		
1-15 June	1.67 ± 0.45	1.72 ± 0.17		2.38 ± 0.33		
16-30 June		2.45 ± 0.41		2.63 ± 0.23		

estuary during April-June, and then migrating into the river plume where at least some remain until September (Healey 1980b).

Coho salmon

Coho smolts migrate seaward from April to June, usually slightly later than pink, chum and chinook fry. Most coho apparently disperse away from their natural estuary, but some remain there (Goodman 1975; Myers 1980). A few were caught in the inner Nanaimo estuary, but most were caught in the outer estuary (Fig. 4) where they were most abundant along the delta front, but particularly at station 10 (Fig. 2) in May and June. Small catches of coho were made throughout Nitinat estuary in May (Fig. 5). In the Yaquina estuary, wild coho were abundant only during April and May. Hatchery coho, however, were resident for about 70 days in the Yaquina following release in June and July (Myers 1980).

Chinook salmon

Chinook, because of their many juvenile life history patterns, have the most varied pattern of estuary utilization. Chinook which migrate seaward as fry colonize the estuary in much the same way as chum fry, first occupying tidal creeks high in the marsh area, and later the outer estuary. Unlike chum, however, chinook fry appear not to occupy high salinity nursery areas. Even though many chinook fry may be swept beyond the river mouth during downstream migration, the estuary appears to be the only habitat suitable for them. In the Nanaimo estuary, chinook fry reside in the inner estuary until late June or July (Fig. 4), and during this time show the same tidal migration pattern from low tide refugia to the marshes and back again as do chum. They also show a clear seaward progression of their low tide distribution with time (Healey 1980a). In the Nitinat estuary, which has no intertidal zone, chinook fry congregate close to shore throughout the lake but are most abundant in the inner estuary (Fig. 5). The peak of abundance in the outer Nitinat estuary is later than that in the inner estuary, indicating a gradual seaward movement (Fig. 5).

When fry migrants reach 70 mm fork length (4.0 g) in the Nanaimo and about 60 mm (2.0 g) in Nitinat they begin to leave the estuary. The dispersal of fry migrant chinook out of these estuaries is coincident with the downstream migration of underyearling chinook smolts. While in southern B.C. large numbers of chinook migrate to the estuary as fry, in the Yaquina and Sixes rivers there are few fry migrants, and most chinook migrate to the estuary as underyearling smolts in May and June (Reimers 1971; Myers 1980). Many underyearling smolts reside in the outer Nanaimo estuary during June and July (Fig. 4) and a few may be found there throughout the year (Healey 1980a). As with coho, greatest numbers of juvenile chinook occurred along the delta front. In the Yaquina and Sixes estuaries underyearling smolts are abundant for a longer time, often into October (Reimers 1971; Myers 1980). On the open Oregon coast, estuaries may provide important sheltered habitat for young chinook throughout the summer and autumn. Earlier disappearance of most chinook from estuaries in southern B.C. may simply reflect the abundance of sheltered coastal habitat in this area.

Chinook which migrate seaward as yearling smolts occur in the outer Nanaimo estuary, and occasionally in the inner estuary, during April-June, after which they migrate away from coastal waters. By contrast, chinook which migrate to sea during their first year of life, often remain in coastal waters throughout their lives (Healey 1982a, unpub. data).

Size and Growth in Estuaries

The size of juvenile salmon captured in estuaries reflects their pattern of migration and residence, and, presumably, the feeding conditions in the estuary. Normally the smallest fish are captured near shore and in the inner estuary, and the larger fish offshore and in the outer estuary. In some cases there appears to be a threshold size governing the movement into deeper or higher salinity water (Healey 1980a). Thus the location from which samples were taken must be considered in interpreting data on the size of juvenile salmon.

Chum salmon

Juvenile chum captured in the inner Nanaimo estuary were heavier than fry captured in the river, and their weight increased with time (Table 1), reflecting the fact that some fry rear for considerable time in the inner estuary. The increase in weight of fish in the inner estuary is not indicative of their true growth rate, however, as this population is continually recruiting recent downstream migrants while its largest fish are the first to move into the outer estuary. After the beginning of May, therefore, when many fry that have reared in the inner estuary begin moving seaward, chum in the outer estuary are significantly heavier ($P<0.05$) than those in the inner estuary (Table 1).

Chum fry dispersed throughout Nitinat Lake within a few hours of downstream migration so that the fish were of similar weight in the inner and outer estuary. Chum captured throughout Nitinat Lake were not significantly different in size from those captured in the inner Nanaimo estuary.

Chum fry captured in Fraser estuary marshes were smaller than those captured in either the Nanaimo or Nitinat estuaries and increased in size more slowly (Levy and Northcote 1981). However, fry are larger further seaward in the Fraser (Goodman 1975) so the small size of fry captured by Levy and Northcote (1981) may not indicate slow growth of chum in the Fraser. Chum in the Squamish and Yaquina estuaries also increased in size more slowly than those in the Nanaimo estuary (Levy and Levings 1978; Myers 1980).

Marked chum captured in the Nanaimo estuary grew at about 5.7% of their body weight per day (Healey 1979). Unmarked chum in Nitinat lake increased in weight about 2.7% per day, but this is probably an underestimate of the true growth rate owing to the emigration of larger chum from the lake.

The true growth is probably closer to 3.5% body weight per day, but still significantly less than in the Nanaimo estuary.

Chinook salmon

Chinook in the Nanaimo and Nitinat estuaries show regional differences in size and seasonal growth similar to chum (Table 1). Interpretation of data for chinook is complicated by the intermixing of fish from freshwater and estuarine nursery areas in June. Recaptures of marked chinook fry in the Nanaimo estuary indicate that their growth rate is about 5.5% per day, similar to chum (Healey 1980a). Marked chinook fry in the Nitinat estuary grew about 3.5% per day in weight, significantly ($P<0.05$) slower than in the Nanaimo (Healey unpub. data). As with chum, chinook throughout the Nitinat estuary were comparable in size with those in the inner Nanaimo estuary.

Chinook in the inner Fraser and Sixes estuaries were intermediate in size between those in the Nanaimo and Nitinat estuaries (Reimers 1971; Levy and Northcote 1981), but chinook in the Yaquina estuary were larger than those in the Nanaimo (Myers 1980). Chinook in the Fraser were fry migrants and their small size may be explained by the fact that the samples were from high in the inner estuary. Chinook from the Sixes estuary were underyearling smolts and their small size was apparently due to poor growth in the river (Reimers 1971). Underyearling smolts from the Nanaimo and Nitinat rivers had grown just as rapidly as fry rearing in the estuary, however, (Healey unpub. data) and underyearling smolts in the Yaquina were also large (Myers 1980). Reimers (1971) noted that chinook in the Sixes estuary grew very little during June and July, and attributed this to high temperature and limited food resources. There was no evidence for a similar cessation in growth of chinook in southern B.C. estuaries (Kask and Parker 1971; Healey 1980a, unpub. data, this study Table 1), except possibly the Squamish (Levy and Levings 1978), or in the Yaquina (Myers 1980).

Coho salmon

Coho captured in the Nanaimo estuary were quite variable in size (Table 1), and probably included fish from a variety of river systems. Tagged juveniles from Big Qualicum hatchery, 60 km to the north, often showed up in the Nanaimo (Healey unpub. data). The seasonal increase in weight observed suggests a growth rate of about 1.5% per day, comparable to the increase of 1.9% per day observed for the Strait of Georgia as a whole (Healey 1980b). No data are available on the size of coho in Nitinat lake. Coho in the Squamish estuary were comparable in size to those in the Nanaimo and increased in size at about the same rate (Levy and Levings 1978). Coho in the Somass and Yaquina estuaries, however, showed no net increase in size over the same period of time, although hatchery coho released in the Yaquina in July grew during August and September (Kask and

Parker 1971; Myers 1980). This suggests that wild coho were resident for only a short period of time in the Somass and Yaquina estuaries.

Pink salmon

Pink fry captured in the outer Nanaimo estuary showed a seasonal increase in weight comparable with chum captured in the same area (Table 1), and comparable with pinks captured elsewhere on the coast (Lebrasseur and Parker 1964). Pinks captured in the inner Fraser estuary were small and showed no increase in size with time, consistent with their brief residence in the estuary (Levy et al. 1979).

Estuary Dependence of Juvenile Pacific Salmon

The species of Pacific salmon clearly utilize the estuary of their natal stream in different ways. Some may spend several months to a year in the estuary; others merely pass through on their way to other nursery areas. Some utilize a variety of habitats in both the inner and outer estuary; others, particularly those which migrate seaward as smolts, occupy mainly the outer estuary. For some, the estuary is virtually their only nursery area for a considerable time after downstream migration; for others the estuary is only one of a number of potential nursery areas. In some estuaries juvenile salmon grow very rapidly (e.g. chum and chinook in the Nanaimo); in others they grow relatively slowly (e.g. chum and chinook in the Nitinat).

Given this variety in patterns of use and apparent benefits to the fish in terms of growth, unequivocal statements about the importance of estuarine habitats to the five species are difficult. I suggest that three criteria be used to evaluate the importance of estuarine habitats: the existence of alternate nursery habitats; the proportion of the population that utilizes alternate habitats as opposed to estuarine habitats; and the length of residence in estuarine habitats. By these criteria chinook are most dependent on estuarine habitat since members of all life history types feed and grow for some time in estuaries, and fry migrants appear totally dependent on the estuary to provide nursery habitat. Chum are second in dependence since a high proportion of the early part of the fry run rears in the estuary, and chum reside and feed in both the inner and outer estuary. Coho are next in dependence as some downstream migrants make use of both the inner and outer estuary. Sockeye and pink salmon appear not to depend on estuaries except, possibly, as a zone of intermediate salinity where acclimatization to sea water can occur. This order of dependence is consistent with that suggested by Dorcey et al. (1978) based on much more limited data.

Food Chain Relationships of Salmon in Estuaries

Salmon which spend more than a few hours in an estuary must derive part of their food resources therein. In fact, all species do some feeding in estuaries. The first few months of sea life are typically a time of rapid growth for young salmon (Lebrasseur and Parker 1964; Healey 1979, 1980a, b, unpub. data) and there is evidence that mortality of small fish may be size dependent (Parker 1971; Healey 1982b). Provision of adequate food resources may, therefore, be one feature of estuaries that makes them attractive to some salmon species.

Salmon diets have been described for a number of estuaries (Lebrasseur et al. 1969; Levy and Levings 1978; Sibert and Kask 1978; Fedorenko et al. 1979; Healey 1979, 1980b; Myers 1980; Levy and Northcote 1981). The principal conclusion from these studies must be that salmon diets vary considerably geographically, seasonally, and annually. Several problems complicate any attempt to compare diets between species and between estuaries. Most important of these are lack of an agreed upon set of parameters that will characterize diet, the abnormal statistical properties of diet data (Sibert and Obrebski 1976), and the difficulty of obtaining samples that are truly comparable. Nevertheless, some generalizations are possible, although these are largely qualitative.

Chum salmon

Chum salmon in the inner Nanaimo estuary in 1975 fed mainly on harpacticoid copepods, decapod larvae, adult insects and calliopiidae (amphipoda) (Healey 1979). In the outer estuary their dominant foods were *Oikopleura spp.*, decapod larvae, and calanoid and harpacticoid copepods (Table 2). In Nitinat Lake in 1980 harpacticoid copepods were absent from their diet. Here the dominant organisms were cladocera, gammarid amphipods, adult insects, and decapod larvae (Table 2). Fedorenko et al. (1979) observed a similar diet of chum in Nitinat Lake, but with greater importance of pelagic zooplankton relative to benthic organisms. In the Fraser estuary the diet of chum was dominated by harpacticoid copepods, chironomid larvae and pupae, adult insects, and amphipods (Levy et al. 1979; Levy and Northcote 1981) while in the Squamish estuary the dominant foods were insect larvae and pupae, mysids, amphipods, and copepods (Levy and Levings 1978).

Seasonal variation in chum diet is considerable (Levy and Levings 1978; Healey 1979). In Nitinat Lake in 1980 gammarid amphipods and adult insects dominated the diet in early April (Fig. 6). In late April and early May diet was much more varied with gammarid amphipods, adult insects, decapod larvae, bivalve molluscs, and *Oikopleura spp.* all contributing substantially. From mid May until June chum diet was dominated by cladocera (Fig. 6). Healey (1979) reported a comparable degree of seasonal variation in the diet of chum from the inner Nanaimo estuary, although involving different taxonomic groups.

Chinook salmon

In the inner Nanaimo estuary in 1978 and 1979 chinook had fed mainly on decapod larvae, mysids, and adult insects, while in the outer estuary

Table 2. Diet items making up 80% of stomach volume ranked by volume contribution (1 = greatest contribution) for juvenile salmon in the Nanaimo and Nitinat estuaries. (Unpub. data except where noted).

	Inner Nanaimo			Outer Nanaimo				Nitinat			
Diet	*Chum*	*Chinook*	*Coho*[1]	*Chum*	*Chinook*	*Coho*[1]	*Pink*	*Chum*	*Chinook*	*Coho*[2]	*Sockeye*[2]
Harpacticoid copepods	1	4		4							
Calanoid copepods				3			1				
Cladocera								1	4		1
Gammarid amphipods							2	2	2	1	5
Caliopiidae (Amphipoda)	4										
Decapod larvae	2	1	2	2	2			4	3		2
Insect larvae			3							3	
Insect adults	3	3			3			3	1	2	4
Oikopleura spp.				1							
Fish larvae (mainly herring)			1		1	1		5		4	
Mysids		2									
Ostracods											3

[1]Data from Anon. 1974; [2]Data from Fedorenko et al. 1979.

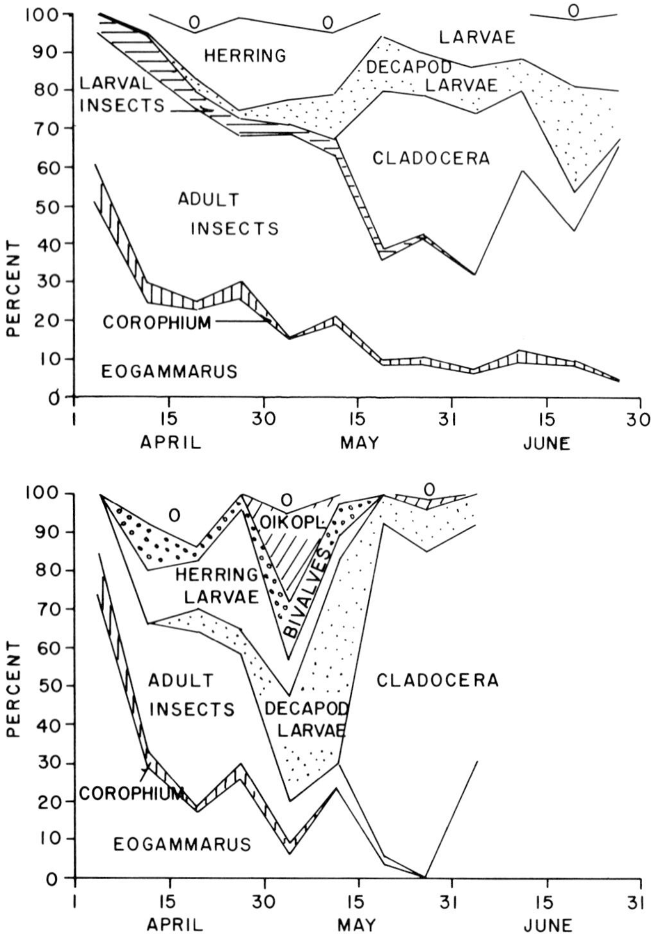

Figure 6. *Diet composition of chinook (upper panel) and chum (lower panel) salmon in Nitinat estuary during April-June, 1980. (Healey unpub. data.)*

they had fed on larval herring (*Clupea harengus pallasi*), decapod larvae, and adult insects (Table 2). Sibert and Kask (1978) described a diet for chinook in the Nanaimo in 1972 that included more amphipods and harpacticoid copepods than I found in 1978 and 1979. Chinook in Nitinat Lake in 1979 and 1980 had fed mainly on adult insects, gammarids, and decapod larvae, with cladocera occasionally important (Table 2). Fedorenko et al. (1979) described essentially the same diet for chinook from Nitinat Lake in 1974 to 1976. Similar organisms appear in the diet of chinook in varying proportions in most estuaries (Table 3).

Seasonal variation in chinook diet is exemplified by data from Nitinat Lake in 1980 (Fig. 6). Amphipods made up 25-50% of the diet in early April

Table 3. Comparison of components making up 80% of diet by volume of chinook salmon in seven estuaries. Components shown in ranked order of contribution to diet, rank 1 being greatest contribution. Where several sources are referenced, ranking was determined by averaging among sources, but account was also taken of the region of the estuary which was sampled, and some ranks were adjusted accordingly.

Diet component	*Nanaimo*[1]	*Nitinat*[1]	*Cowichan*[2]	*Courtenay*[2]	*Squamish*[3]	*Fraser*[4]	*Yaquina*[5]
Harpacticoid copepods	5			4		5	
Cladocera		4				2	
Gammarid amphipods		2	2				
Corophium spp.			5	3			
Decapod larvae	1	3	3	2		4	
Insect larvae			4			3	
Insect adults	2	1	1	1	2	1	2
Mysids	4				1		
Fish larvae	3						1

Sources: [1]This manuscript, and Healey 1980a and unpublished; [2]Healey unpublished; [3]Levy and Levings 1978; [4]Kask and Sibert 1978; Levy and Northcote 1981; Levy et al. 1979; [5]Myers 1980.

but declined to 10% by late June. Cladocera did not appear until mid May, but were about 40% of the diet in late May and early June. Fish larvae and decapod larvae were unimportant until May and June. A similar degree of seasonal variation in diet is evident in other estuaries (Kask and Parker 1971, 1972a, b, 1974; Levy and Levings 1978; Levy et al. 1979; Healey unpub. data).

Coho salmon

Coho in the inner Nanaimo estuary fed on fish larvae, decapod larvae, and larval insects, while those in the outer estuary fed almost exclusively on fish larvae (Table 2) (Anon 1974). In Nitinat Lake coho fed mainly on amphipods and insect larvae and adults. Fish larvae were of secondary importance in Nitinat Lake, although fish did increase in importance with season (Fedorenko et al. 1979). In other estuaries fish larvae commonly dominated coho diet (Myers 1980; Healey unpub. data); however, in the Squamish mysids and amphipods dominated (Levy and Levings 1978).

Pink and Sockeye Salmon

Pink fry captured in the Nanaimo estuary had fed mainly on harpacticoid copepods followed by amphipods (Anon. 1974; Healey unpub. data). Pink fry captured in the Fraser River plume at the same time of year had fed mainly on calanoid copepods (Lebrasseur et al. 1969).

Sockeye captured in Nitinat Lake had fed mainly on cladocera, followed by decapod larvae, ostracods, adult insects, and amphipods (Fedorenko et al. 1979). By comparison, sockeye smolts in the Fraser River plume fed mainly on calanoid copepods and adult insects (Healey 1980b).

Comparison Among Estuaries and Species

Juvenile salmon diets vary considerably among estuaries, habitats within an estuary, and times of sampling (Levy and Levings 1978; Sibert and Kask 1978; Healey 1979, unpub. data; Levy et al. 1979; Levy and Northcote 1981; this study Fig. 6). This indicates that juvenile salmon are opportunistic feeders and are capable of exploiting a variety of taxa. However, some important consistencies are evident in their diets. Insect adults (mainly diptera) and decapod larvae were important in the diet of chinook from seven of the estuaries considered in this report, even though in no two estuaries was 80% of the diet composed of the same suite of organisms (Table 3). Harpacticoid copepods, decapod larvae, and cladocera were common dominants in the diet of chum from several estuaries in southern B.C. (Healey 1979, unpub. data; Levy and Northcote 1981). Sibert and Kask (1978) compared the diets of coho and chinook among estuaries, and among habitats within estuaries, and concluded that each species had a distinct feeding habit within an estuary, but that there was little similarity between estuaries. A relatively small number of taxa consistently appear in juvenile salmon stomachs, however, suggesting that only a few taxa within the community of potential prey animals are at risk. Sibert and Kask (1978) further concluded that coho fed on a narrow spectrum of foods, and that their diet did not overlap with those of other species. Chinook had much more catholic tastes, and their diet overlapped with those of a variety of other species. Levy and Northcote (1981) compared chum and chinook from marsh habitats of the Fraser estuary and concluded that chum diet varied less from place to place than did chinook diet. Thus, the data suggest a range of feeding specificities among juvenile salmon that correlates with the degree of dependence of the species on estuarine habitats. Chinook, having the greatest dependence on estuaries, also appear less restricted in the types of food they can exploit. Chum and coho are less dependent on estuaries and have a more restricted diet.

Whatever the mechanisms dictating their particular diet, all juvenile salmon in estuaries, but particularly chum and chinook, display a considerable dependence on animals living in or associated with the bottom. I include adult insects in this category since, even though they were taken at the surface, almost all were forms with aquatic larvae, and thus were produced from the bottom. The importance of benthic production to the food of juvenile salmon was greatest in the inner estuary. In the outer estuary more pelagic organisms were taken, but benthic foods were often still important. Many of the benthic forms fed upon by juvenile salmon are, themselves, detritivores. Thus, there is an important link between detritus based food webs and juvenile salmon production in estuaries (Sibert et al. 1978).

Population, Biomass, and Food Requirements

I estimated the abundance of chum, coho, chinook, and sockeye in the inner Nanaimo estuary and all of Nitinat estuary by a combination of mark-and-recapture, residence time, and catch-per-unit-effort models (as described in Healey et al. 1977; Healey 1979, 1980a). Data available did not permit me to estimate juvenile salmon abundance for the outer Nanaimo estuary. From the abundance, average weight, and growth rate of the fish I calculated biomass, production, and food requirements for each species.

In the inner Nanaimo estuary chum had the greatest biomass followed by chinook and coho (Fig. 7, Table 4). Although low in abundance coho were much larger than chinook (Fig. 4, Table 1), so that the two species had similar population biomass. Maximum biomass occurred in late April for chum and in June for chinook and coho. Chinook maintained a high biomass throughout May-July (Fig. 7). In the Nitinat estuary chum, chinook and coho all had a similar maximum biomass during mid to late May. Sockeye had a lower maximum biomass at the beginning of May (Fig. 7). Since chinook were resident in the Nitinat estuary longer than either chum or coho, their cumulative biomass (sum of the daily biomass estimates) was greatest (Table 4).

Differences in contribution of the species to total salmon biomass was

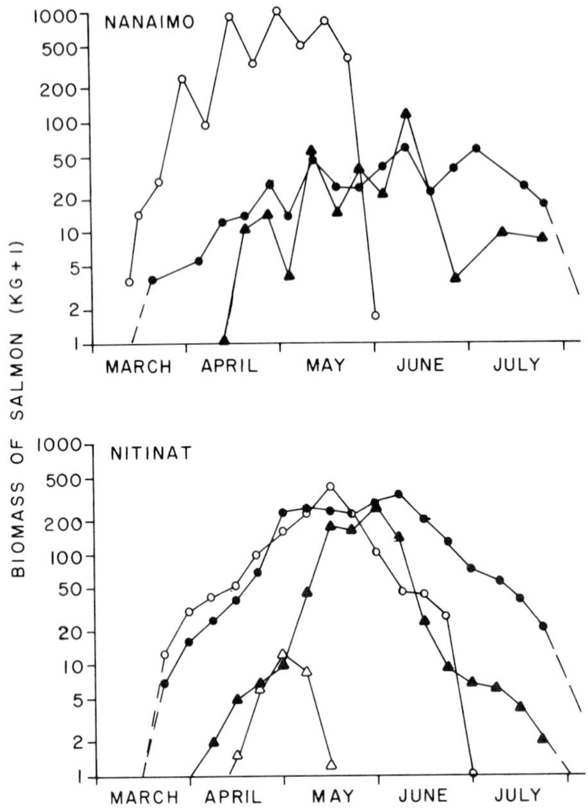

Figure 7. Seasonal changes in biomass of juvenile chum (○), chinook (●), coho (▲)and sockeye (△) salmon in inner Nanaimo and all of Nitinat estuaries.

partly a reflection of the different numbers of juveniles produced by the two river systems. An order of magnitude more chum juveniles are produced in the Nanaimo River than in the Nitinat, while for chinook the reverse is true (Table 4). In spite of the different numerical contributions of the species to total juvenile biomass, cumulative biomass of all species was virtually the same in the two estuaries (Table 4). The whole Nitinat estuary, therefore, appears to be about equivalent to the inner Nanaimo estuary in its capacity to support young salmon.

Annual production (sum of daily biomass × instantaneous growth rate) in the Nanaimo estuary ranged from 1750 kg for chum to 49 kg for coho. Total juvenile salmon production was almost 2000 kg (Table 4). In Nitinat Lake annual production ranged from 734 kg for chinook to 4.8 kg for sockeye. Total production was only 1137 kg (Table 4). While this represents a substantial production over a short time span for such small fish, it is not large relative to the total area of the two estuaries. Production was only 0.67 g/m²

Table 4. Summary of biomass, production and food requirements of juvenile salmon in the inner Nanaimo estuary and Nitinat estuary. (Sources: Healey 1979, 1980a, unpub. data; Canada, Fisheries and Marine Service, unpub. data.)

	Chum	*Chinook*	*Coho*	*Sockeye*	*Total*
		Nanaimo Estuary			
Adult spawners	38.0×10^3	2.1×10^3	2.8×10^3		
Juvenile migrants - fry	40.0×10^6	25.0×10^4			40.2×10^6
- smolt		10.0×10^3	35.0×10^3		45.0×10^3
Cumulative estuary biomass (kg)	30.0×10^3	3.5×10^3	2.6×10^3		36.1×10^3
Maximum 7 day biomass (kg)	7.6×10^3	371	714		7.8×10^3
Growth rate %/day	5.7	5.5	1.9		
Instantaneous rate	0.059	0.057	0.019		
Annual production (kg)	1750	200	49		1999
Maximum 7 day production (kg)	447	21	14		455
Daily ration (%/body wt)	15	15	6		
Annual food requirements (kg)	4500	525	156		5181
Benthic organisms (kg)	3510	383	83		3976
Pelagic organisms (kg)	675	11	73		759
Adult insects (kg)	315	131	0		446
Maximum 7 day food (kg)	1136	56	43		1158
Benthic organisms (kg)	886	41	23		901
Pelagic organisms (kg)	170	1	20		172
Adult insects (kg)	80	14	0		85

Table 4. (Continued).

	Chum	*Chinook*	*Coho*	*Sockeye*	*Total*
		Nitinat Estuary			
Adult spawners	85.0×10^3	1.6×10^3	1.6×10^3	4.6×10^3	
Juvenile migrants - fry	3.2×10^6	2.7×10^6			5.9×10^6
- smolt		4.0×10^4	2.0×10^4	3.5×10^5	4.1×10^5
Cumulative estuary biomass (kg)	9.0×10^3	20.4×10^3	6.6×10^3	2.1×10^2	36.2×10^3
Maximum 7 day biomass (kg)	1.9×10^3	2.7×10^3	1.7×10^3	0.84×10^2	5.5×10^3
Growth rate %/day	3.5	3.5	1.9	2.3	
Instantaneous rate	0.036	0.036	0.019	0.023	
Annual production (kg)	273	734	125	4.8	1137
Maximum 7 day production (kg)	57	97	32	1.9	160
Daily ration (%/body wt)	11	11	6	7	
Annual food requirement (kg)	810	2244	396	15	3465
Benthic organisms (kg)	324	898	277	3	1502
Pelagic organisms (kg)	364	449	79	10	902
Adult insects (kg)	122	898	40	2	1062
Maximum 7 day food (kg)	171	297	102	6	488
Benthic organisms (kg)	68	119	71	1	218
Pelagic organisms (kg)	77	59	20	4	158
Adult insects (kg)	26	119	10	1	112

in Nanaimo and 0.04 g/m^2 in Nitinat. Since the fish are not uniformly distributed in either estuary, however, production per unit area in preferred habitats must be considerably higher.

Daily food requirements for each species were set at 3× the daily growth rate when growth was less than 4%/d. At higher growth rates daily ration was set lower since weights of stomach contents and bioenergetics calculations indicated that 3× the daily growth was too high (Healey 1979). Total food requirements were then calculated as the sum of daily requirements. For the Nanaimo estuary the annual food requirement was 5180 kg, of which over 4000 kg were benthic organisms and insects (Table 4). In terms of the area of the inner estuary this represents about 0.6 g/m^2 taken over a period of 4 months. In the Nitinat estuary, where growth rates and production were lower, the annual food requirement was only 3465 kg, of which 2564 kg were benthic organisms and insects (Table 4). These food requirements represent about 1 g/m^2 taken from the littoral region of Nitinat Lake and about 0.04 g/m^3 taken from the pelagic area during about 4 months.

Since juvenile salmon vary considerably in abundance during their brief residence in estuaries, their biomass and food requirements at peak abundance should also be considered when attempting to gauge their impact on the community of food organisms. I calculated the maximum seven day biomass, production, and food requirements for each species and for the combined salmon population (Table 4). In the Nanaimo estuary, maximum food requirements ranged from 1136 kg for chum to 43 kg for coho, and in the Nitinat from 297 kg for chinook to 6 kg for sockeye. For all species the maximum short term food requirement was 1158 kg in the inner Nanaimo estuary and 488 kg in Nitinat Lake. These food demands represent about 0.15 g/m^2 in Nanaimo estuary and about 0.3 g/m^2 of littoral habitat and 0.012 g/m^3 of pelagic habitat in Nitinat Lake.

Unfortunately, insufficient quantitative sampling of food organisms has been conducted in these two estuaries to permit a precise evaluation of the potential impact of juvenile salmon upon benthic and planktonic communities. Sampling of meiofauna in the inner Nanaimo estuary (Sibert 1979) and of the zooplankton and benthos in Nitinat Lake (Fedorenko et al. 1979) permits a qualitative evaluation of this impact. In the Nanaimo estuary meiobenthos common in the diets of juvenile salmon exceeds 5 g/m^2. The salmon remove about 10% of this standing crop during their four month residence in the estuary, and about 3% during their week of greatest feeding intensity. Since 5 g/m^2 is a minimum estimate of the benthic food resource and since the turnover ratio for small benthic animals is probably on the order of 5-10× per year (Sibert 1979), the proportion of benthic productivity taken by juvenile salmon is very small. The proportion of the annual production of preferred prey taken may, however, be high (Healey 1979; Sibert 1979). Also, the proportion of the meiobenthos actually available to juvenile salmon is unknown. Sibert (1981) reported that benthic organisms commonly fed upon by juvenile salmon occur in dense concentrations in the benthic boundary layer, where they should be readily available to the fish. The relationship of the organisms in the benthic boundary layer to the benthic infauna and their rate of replenishment after exploitation are unknown.

Data on plankton and benthic populations of Nitinat Lake are more limited. Fedorenko et al. (1979) reported numeric densities and size composition of zooplankton from vertical SCORE net hauls and of benthos from stirring and sweep netting within an enclosure. Such density estimates are minimum estimates, especially those for benthos (Fedorenko et al. 1979). I converted the numeric estimates to approximate biomass estimates by the size composition data. Zooplankton populations were $4 g/m^3$ or less during the peak abundance of juvenile salmon, and benthic populations about $3 g/m^2$. These are low standing crops but, provided the organisms are available to the fish, sufficient to support their food demands.

Although food populations are abundant in absolute terms, there are indications that growth and abundance of juvenile salmon in estuaries are influenced by the abundance of food. Food resources were poorer in Nitinat Lake than in the Nanaimo estuary, and growth of chinook and chum was poorer in Nitinat Lake (Table 4). Chinook were most abundant, and had most food in their stomachs, in the Nanaimo estuary in 1975 (Healey 1980a). Reimers (1971) attributed reduced growth of chinook in the Sixes River estuary at high population densities to competition for food.

Definition of adequate food resources and food producing habitat to support juvenile salmon in estuaries involves complicated questions of prey availability, threshold populations for successful feeding, prey composition, and prey aggregation that have yet to be resolved. The abundance of young salmon in a particular habitat and the impact of salmon feeding on the benthic or planktonic communities are probably not simple functions of the total abundance of prey, but involve the response of salmon to particular components of the prey community and the availability of prey in other habitats. For example, chum fry in the inner Nanaimo estuary in 1975 fed very heavily on a single copepod species, *Harpacticus uniremis*, and apparently cropped most of the annual production of this species. The abundance of chum in the estuary was correlated with the abundance of *H. uniremis*, and chum left the estuary when the *H. uniremis* population declined (Healey 1979; Sibert 1979).

Research in the Nanaimo estuary on the relationship among primary production, secondary production, and juvenile salmon production led to the conclusion that juvenile chum salmon were supported by a detritus based food web (Sibert et al. 1978; Healey 1979; Naiman and Sibert 1979; Sibert 1979). While it has not been possible to extend these observations to other salmon species or other estuaries in the same detail, the predominance of detritus feeders in salmon diets suggests that estuarine populations in general are supported by detritus based food webs. The mechanisms for trapping allochthonous organic carbon in estuaries may, therefore, be very important in determining the productivity of an estuary from the point of view of juvenile salmon. Odum et al. (1979) discuss aspects of basin morphometry and transport properties that determine whether an estuary is a net exporter or importer of particulate organic carbon. They argue that retention of organic carbon will be enhanced by restricted exchange with the open ocean and low bed load transport. Similar arguments apply to parts of estuaries, and it may be added that weed beds, and braided and meandering channel morphology, will also enhance the detritus trapping characteristics of an estuary. It has been my experience that juvenile salmon congregate in areas where estuary morphology favours detritus retention to a moderate degree. On the other hand, the small fish in particular seem to avoid both well flushed areas, perhaps because it is too difficult for them to maintain position, and areas of high detritus retention, perhaps because of reduced water quality. Access to marshes appears to be important for young salmon. Marshes are often efficient detritus traps (Odum et al. 1979), and are highly productive of the kinds of organisms on which young salmon feed. The importance of weed beds in the lower intertidal and subtidal is less obvious, however, juvenile salmon were found to be much more abundant in a channel on the foreshore of the Fraser estuary that had eelgrass (*Zostera marina*) than in a similar channel without eelgrass (J. Sibert, pers. comm.) The detritus trapping characteristics of such weed beds probably significantly enhance local food production for young salmon, and may provide shelter from predators. Detritus food chains were the major pathway for the assimilation of eelgrass carbon in an Alaskan lagoon (McConnaughey and McRoy 1979). Furthermore, the ratio of ^{13}C:^{12}C in juvenile salmon from the Nanaimo estuary suggests that eelgrass carbon contributes significantly to the detritus based food web supporting the salmon (J. Sibert, pers. comm.). Thus it may be hypothesized that the complex of intertidal marshes, tidal creeks and secondary river channels, lower intertidal and subtidal weed beds, and basin morphology all contribute to the carrying capacity of the estuary for young salmon, and that the appropriate configurations must be conserved if salmon production is to be maintained.

Acknowledgments

Technical assistance in gathering and analysing the data presented in this report was provided by R. V. Schmidt, F. P. Jordan, and R. M. Hungar. Dr. J. Sibert criticised a draft of the manuscript and permitted me to refer to his unpublished data on the occurrence of juvenile salmon on the Fraser estuary and ^{13}C:^{12}C ratios for juvenile salmon from the Nanaimo estuary.

References Cited

Anon. 1974. An environmental assessment of Nanaimo port alternatives. Lands Directorate, Canada Dep. Environ. Pacific Region. 63 pp. + appendices.

Barraclough, W. E. and A. C. Phillips. 1978. Distribution of juvenile salmon in the southern Strait of Georgia during the period April to July 1966–1969. Canada, Fish. Mar. Serv. Tech. Rep. 826. 47 pp.

Birtwell, I. K. 1978. Studies on the relationship between juvenile chinook salmon and water quality in the industrialized estuary of the Somass River, pp. 58-78. *In:* B. G. Shepherd and R. M. Ginetz (eds.), Proceedings of the 1977 Northeast Pacific Chinook and Coho Salmon Workshop. Canada, Fish. Mar. Serv. Tech. Rep. 759. 164 pp.

Dorcey, A. H. J., T. G. Northcote and D. V. Ward. 1978. "Are the Fraser River Marshes Essential to Salmon?" Westwater Lectures No. 1. Westwater Institute, U. British Columbia, Vancouver, Canada. 29 pp.

Fedorenko, A. Y., F. J. Fraser and D. T. Lightly. 1979. A limnological and salmonid resource study of Nitinat Lake: 1975-1977. Canada, Fish. Mar. Serv. Tech. Rep. 839. 86 pp.

Goodman, D. 1975. A synthesis of the impacts of proposed expansion of the Vancouver international airport and other developments on the fisheries resources of the Fraser River estuary. Canada, Fish. Mar. Serv. Mimeo. Report. 137 pp.

Healey, M. C. 1979. Detritus and juvenile salmon production in the Nanaimo estuary. 1: Production and feeding rates of juvenile chum salmon (*Oncorhynchus keta*). *J. Fish. Res. Board Can.* 36:488-496.

Healey, M. C. 1980a. Utilization of the Nanaimo River estuary by juvenile chinook salmon, *Oncorhynchus tshawytscha. Fish. Bull.* 77:653-668.

Healey, M. C. 1980b. The ecology of juvenile salmon in Georgia Strait, British Columbia, pp. 203-229. *In:* W. J. McNiel and D. C. Himsworth (eds.) *Salmonid Ecosystems of the North Pacific.* Oregon State U. Press, Corvallis, OR.

Healey, M. C. 1982a. Coastwide distribution and ocean migration patterns of stream and ocean type chinook *(Oncorhynchus tshawytscha).* (Submitted to *Can. J. Fish. Aquat. Sci.).*

Healey, M. C. 1982b. Timing and relative intensity of size selective mortality of juvenile chum salmon during early sea life. *Can. J. Fish. Aquat. Sci.* (in press).

Healey, M. C., R. V. Schmidt, F. P. Jordan and R. M. Hungar. 1977. Juvenile salmon in the Nanaimo area 1975: Distribution and abundance. Fish. Res. Board Can. MS Rep. 1369. 161 pp.

Kask, B. A. and R. R. Parker. 1971. Data report. Estuarine pollution Alberni Inlet, B.C. Fish. Res. Board Can. MS Rep. 1161. 1141 pp.

Kask, B. A. and R. R. Parker. 1972a. Data reports. Estuarine pollution Squamish estuary, B.C. 1972. Fish. Res. Board Can. MS Rep. 1029. 135 pp.

Kask, B. A. and R. R. Parker. 1972b. Observations on juvenile chinook salmon in the Somass River estuary Port Alberni, B.C. Fish. Res. Board Can. Tech. Rep. 308. 15 pp.

Kask, B. A. and R. R. Parker. 1974. Data record. Alberni Inlet fishes, April 23 to July 10, 1971. Their distribution, species, size/age composition and stomach content. Fish. Res. Board Can. MS Rep. 1311. 515 pp.

LeBrasseur, R. J., W. E. Barraclough, O. D. Kennedy and T. R. Parsons. 1969. Production studies in the Strait of Georgia. Part III. Observations on the food of larval and juvenile fish in the Fraser River plume, February to May, 1967. *J. Exp. Mar. Biol. Ecol.* 3:51-61.

LeBrasseur, R. J. and R. R. Parker. 1964. Growth rate of central British Columbia pink salmon (*Oncorhynchus gorbuscha*). *J. Fish. Res. Board Can.* 21:1101-1128.

Levy, D. A. and C. D. Levings. 1978. A description of the fish community of the Squamish River estuary, British Columbia: relative abundance, seasonal changes and feeding habits of salmonids. Canada, Fish. Mar. Serv. MS Rep. 1475. 63 pp.

Levy, D. A. and T. G. Northcote. 1981. The distribution and abundance of juvenile salmon in marsh habitats of the Fraser River estuary. University of British Columbia, Westwater Research Centre, Tech. Rep. 25. 117 pp.

Levy, D. A., T. G. Northcote and G. J. Birch. 1979. Juvenile salmon utilization of tidal channels in the Fraser River estuary, British Columbia. University of British Columbia, Westwater Research Centre, Tech. Rep. 23. 70 pp.

McConnaughey, T. and C. P. McRoy. 1979. ^{13}C label identifies eelgrass (*Zostera marina*) carbon in an Alaskan estuarine food web. *Mar. Biol.* 53:263-269.

Morris, S. and A. J. Leaney. 1980. The Somass River estuary, status of environmental knowledge to 1980. Environment Canada and Fisheries and Oceans, Special Estuary Series No. 9, 374 pp.

Myers, K. W. W. 1980. An investigation of the utilization of four study areas in Yaquina Bay, Oregon, by hatchery and wild juvenile salmonids. MS thesis, Oregon State U., Corvallis, OR. 234 pp.

Naiman, R. J. and J. R. Sibert. 1979. Detritus and juvenile salmon production in the Nanaimo estuary: III. Importance of detrital carbon to the estuarine ecosystem. *J. Fish. Res. Board Can.* 36:504-520.

Northcote, T. G., M. S. Wilson and D. R. Hurn. 1964. Some characteristics of Nitinat Lake, an inlet on Vancouver Island, British Columbia. *J. Fish. Res. Board Can.* 21:1069-1081.

Odum, W. E., J. S. Fisher and J. C. Pickral. 1979. Factors controlling the flux of particulate organic carbon from estuarine wetlands, pp.69-80. *In:* R. J. Livingston (ed.), *Ecological Processes in Coastal and Marine Systems.* Plenum Press, New York.

Parker, R. R. 1971. Size selective predation among juvenile salmonid fishes in a British Columbia inlet. *J. Fish. Res. Board Can.* 28:1503-1510.

Phillips, A. C. and W. E. Barraclough. 1978. Early marine growth of juvenile Pacific salmon in the Strait of Georgia and Saanich Inlet, British Columbia. Canada, Fish. Mar. Serv. Tech. Rep. 830. 19 pp.

Reimers, P. E. 1971. The length of residence of juvenile fall chinook salmon in Sixes River, Oregon. Ph.D. Thesis, Oregon State U., Corvallis, OR. 99 pp.

Sibert, J. R. 1979. Detritus and juvenile salmon production in the Nanaimo estuary: II. Meiofauna available as food to juvenile chum salmon (*Oncorhynchus keta*). *J. Fish. Res. Board Can.* 36:497-503.

Sibert, J. R. 1981. Intertidal hyperbenthic populations in the Nanaimo estuary. *Mar. Biol.* 64:259-265.

Sibert, J. R. and T. J. Brown. 1976. The salinity regime of the Nanaimo River estuary. Canada, Fish. Mar. Serv. Data Rec. No. 8. 108 pp.

Sibert, J. R., T. J. Brown, M. C. Healey, B. A. Kask and R. J. Naiman. 1978. Detritus-based food webs: exploitation by juvenile chum salmon (*Oncorhynchus keta*). *Science* 196:649-650.

Sibert, J. R. and B. A. Kask. 1978. Do fish have diets?, pp. 48-57. *In:* B. G. Shepherd and R. M. J. Genetz (eds.), Proceedings of the 1977 Northeast Pacific Chinook and Coho Salmon Workshop. Canada, Fish. Mar. Serv. Tech. Rep. 759. 164 pp.

Sibert, J. R. and S. Obrebski. 1976. Frequency distributions of food counts in individual fish stomachs, pp. 107-114. *In:* C. A. Simenstad and S. Lipovsky (eds.), *Fish Food Habits Studies,* Washington Sea Grant, U. Washington, Seattle, WA.

THE ROLE OF PUGET SOUND AND WASHINGTON COASTAL ESTUARIES IN THE LIFE HISTORY OF PACIFIC SALMON: AN UNAPPRECIATED FUNCTION

Charles A. Simenstad
Fisheries Research Institute
University of Washington
Seattle, Washington

Kurt L. Fresh
Washington Department of Fisheries
Research and Development Division
Olympia, Washington

and

Ernest O. Salo
Fisheries Research Institute
University of Washington
Seattle, Washington

Abstract: Washington State has approximately 100 diverse estuaries, ranging from the more classic coastal estuaries to Puget Sound, a continuum of estuaries with transitional habitats. Of the five Pacific salmon species, chum and chinook utilize these estuaries most extensively. Estimated residence times of individual juvenile salmon range from 4 days (chum salmon) to 6 months (chinook) while individual residence times of adults range from 1-6 weeks. Some salmon populations may, however, remain within Puget Sound until maturity. Juveniles of all species utilize neritic habitats, but chum and chinook also use shallow, sublittoral habitats. Abundant, uniquely estuarine prey organisms are eaten by juveniles of all species, although less so by pink, sockeye, and coho, and contribute to high growth rates in estuaries. Significant predation on juveniles in estuaries has yet to be documented. We hypothesize that Pacific salmon use Washington's estuaries for: 1) productive foraging, 2) physiological transition, and 3) refugia from predators. These functions have probably changed due to salmon culture practices and alterations of estuarine habitat, and it is possible these changes could adversely impact salmon growth and survival. The importance of estuaries to salmon production should be more carefully considered in estuary and salmon management.

Introduction

The numerous, diverse estuaries of Washington State are the result of 40,000 years (50,000-10,000 B.C.) of Wisconsin glaciation and associated sea level changes and the climatology of the following 10,000 years (Crandell 1965). The evolution of Pacific salmon (*Oncorhynchus* spp.), on the other

ISBN 0-12-404070-5

hand, has probably spanned the Pleistocene Epoch, encompassing 0.6-1 million years (Neave 1958). Thus, Pacific salmon have evolved their respective life history patterns as estuaries developed. There has been considerable confusion about, and little documentation on, the importance of estuaries to Pacific salmon, however, and perhaps as a consequence, many estuaries have been substantially and rapidly altered over the last 200 years with uncertain impacts on salmon populations. To address this information gap, we have synthesized published and unpublished information to describe the role of Washington's estuaries (exclusive of the Columbia River) in the life history of five species of Pacific salmon: pink (*Oncorhynchus gorbuscha*), chum (*O. keta*), coho (*O. kisutch*), sockeye (*O. nerka*), and chinook (*O. tshawytscha*).

Our objectives were to:

1. itemize estuaries of Washington utilizable by Pacific salmon;
2. evaluate the utilization of estuaries by each species of salmon and suggest how this utilization may have changed with the alteration of estuaries and the culture of Pacific salmon; and
3. discuss the implications of estuarine utilization by salmon to estuary (shoreline) and salmon management.

For purposes of this paper we have only cited major references; a complete bibliography is available upon request from the second author.

Background

Present and Past Status of Estuaries of Washington State

We utilized Pritchard's (1967) definition of an estuary as "a semi-enclosed coastal body of water, which has a free connection with the open sea, and within which sea water is measurably diluted with fresh water derived from land drainage", to identify 96 coastal and inland estuaries in 14 regions (Table 1). Estuaries within these regions are structurally, hydrologically, and biologically diverse, and range in size from drowned river valleys which characterize major estuaries (e.g., Grays Harbor and Skagit Bay/Port Susan) to the numerous small stream channel estuaries characteristic of Puget Sound and the Strait of Juan de Fuca. However, one region – the island archipelago of northern Puget Sound – has no major estuaries *per se* but is greatly influenced by freshwater outflow of the Fraser (British Columbia, Canada) and Skagit Rivers.

Historical Changes in Estuaries

Extensive changes in Washington's estuaries, either directly to the natural estuarine environment or indirectly through alteration of freshwater habitats in contributing watersheds, have occurred mainly in the last 200 years. Documentation of alterations in freshwater habitat is often extensive and quantitative, especially in the case of modifications associated with major dam construction. However, quantitative information relating changes in

Table 1. Enumeration and characteristics of principal estuaries in Washington State, not including Columbia River (dash indicates no data available; data sources were U.S.G.S. [1978; 1980], Williams et al. [1975], Smith et al. [1977], and U.S. Army Corps Engin., Seattle Dist. [1976]).

Estuarine region	Estuaries	Principal watersheds	Drainage area (km^2)	Average annual discharge $hm^3 yr^{-1}$	Annual maximum discharge $m^3 sec^{-1}$	Extant estuarine area[1] km^2	Water inventory areas[2]
North Sound	Drayton Harbor	Dakota Creek	75	--	--	--	01, 03 (part)
		California Creek	59	--	--	--	
	Bellingham Bay	Nooksack River	2,139	3,520	1,060	13	
	Samish Bay	Samish River	275	--	47	15	
Northern Puget Sound	Port Townsend Bay	Chimacum Creek	87	--	--	--	02, 06 (part), 03 (part)
Skagit Bay/Port Susan	Skagit Bay	Skagit River	8,011	14,900	4,080	67	04, 05, 06(part)
	Port Susan	Stillaguamish River	1,772	2,650	1,785	24	
Possession Sound	Everett Harbor	Snohomish River	4,439	8,890	3,260	19	07
Central/South Puget Sound	Shilshole Bay	Cedar River	487	625	250	--	08, 12, 14, 15 (part)
		Lake Washington	62	50	33		
		Sammamish Lake	472	330	80		
	Chambers Bay	Chambers Creek	48	--	--	--	
	Oyster Bay	Kennedy Creek	53	--	520	--	
	Skookum Inlet	Skookum Creek	n.a.	--	--	--	
	Oakland Bay	Goldsborough Creek	n.a.	--	400	--	
	Hammersley Inlet	Mill Creek	n.a.	--	--	--	
	Case Inlet	Rocky Creek	n.a.	--	--	--	
		Coulter Creek	n.a.	--	--	--	
		Sherwood Creek	n.a.	--	--	--	
	Burley Lagoon	Burley Creek	n.a.	--	--	--	
	Gig Harbor	Crescent Creek	n.a.	--	--	--	
	Olalla Bay	Olalla Creek	n.a.	--	--	--	
	Sinclair Inlet	Gorst Creek	n.a.	--	--	--	
	Dyes Inlet	Clear Creek	n.a.	--	--	--	
		Strawberry Creek	n.a.	--	--	--	
		Chico Creek	n.a.	--	350	--	
	Liberty Bay	Dogfish Creek	n.a.	--	--	--	
	Miller Bay	Grovers Creek	n.a.	--	--	--	
	Port Ludlow	Ludlow Creek	n.a.	--	--	--	
Elliott Bay	Elliott Bay	Duwamish River	1,140	1,370	375	<0.1	09
Commencement Bay	Commencement Bay	Puyallup River	2,455	3,010	1,610	<0.1	10
Nisqually Reach	Nisqually Reach	Nisqually River	1,339	1,630	870	}10	11
		McAllister Creek	--	--	--		
Budd Inlet	Capitol Lake	Deschutes	417	--	3,380	--	13
Hood Canal	Lynch Cove	Union River	61	--	--	--	15 (part), 16, 17
	Big Mission Creek	Big Mission Creek	--	--	--	--	
	Tahuya River	Tahuya River	--	--	--	--	
	Annas Bay	Skokomish River	622	655	610	6.0	
	Dewatto Bay	Dewatto River	43	--	750	--	
	Lilliwaup Bay	Lilliwaup Creek	--	--	--	--	
	Hamma Hamma River	Hamma Hamma River	210	--	2,260	--	
	Anderson Cove	Anderson Creek	--	--	--	--	
	Duckabush River	Duckabush River	172	370	255	--	
	Dosewallips River	Dosewallips River	--	--	--	--	
	Quilcene Bay	Big Quilcene River	--	--	--	--	
		Little Quilcene River	--	--	--	--	
	Jackson Cove	Marple Creek	--	--	--	--	
		Spencer Creek	--	--	--	--	
	Tarboo Bay	Tarboo Creek	32	--	--	--	
	Thorndyke Bay	Thorndyke Creek	31	--	--	--	
	Stavis Bay	Stavis Creek	--	--	--	--	
	Seabeck Bay	Seabeck Creek	--	--	--	--	
	Little Beef Harbor	Little Beef Creek	--	--	--	--	
	Big Beef Harbor	Big Beef Creek	36	35	20	--	
	Port Gamble	Gamble Creek	--	--	--	--	
		Miller Lake	--	--	--	--	
	Squamish Harbor	Shine Creek	--	--	--	--	
Strait of Juan de Fuca	Discovery Bay	Snow Creek	--	--	50	--	18, 19
		Salmon Creek	49	--	--	--	
	Sequim Bay	Jimmycomelately Creek	--	--	--	--	
		Dean Creek	--	--	--	--	
	New Dungeness Bay	Dungeness River	513	350	195	7	
	Freshwater Bay	Elwha River	831	1,340	1,180	--	
	Lyre River	Lyre River	171	--	--	--	
	Pysht River	Pysht River	115	--	--	--	
	Clallam Bay	Clallam River	82	--	--	--	
	Hoko River	Hoko River	113	--	3,620	--	
	Sekiu River	Sekiu River	85	--	--	--	
	Sail River	Sail River	14	--	20	--	
North Coastal	Mukkaw Bay	Waatch River	33	--	35	--	20, 21
		Sooes River	106	--	95	--	
	Ozette River	Ozette Lake/River	229	--	45	--	
	Quillayute River	Quillayute River	1,629	--	--	--	
	Goodman Creek	Goodman Creek	82	--	--	--	
	Hoh River	Hoh River	774	2,240	1,300	--	
	Kalaloch Creek	Kalaloch Creek	--	--	--	--	
	Queets River	Queets River	1,153	3,690	3,690	--	
	Raft River	Raft River	197	--	490	--	
	Quinault River	Lake Quinault, Quinault River	1,124	2,520	1,420	--	
	Moclips River	Moclips River	91	--	120	--	
	Copalis River	Copalis River	--	--	--	--	
Grays Harbor	Grays Harbor	Humptulips River	337	1,190	935	136	22
		Hoquium River	234	--	--		
		Wishkah River	117	--	--		
		Chehalis River	5,358	6,670	1,570		
		Johns River	81	--	--		
		Elk River	47	--	--		

Table 1 cont.

Estuarine region	Estuaries	Principal watersheds	Drainage area (km^2)	Average annual discharge $hm^3 yr^{-1}$	Annual maximum discharge $m^3 sec^{-1}$	Extant estuarine area[1] km^2	Water inventory areas[2]
Willapa Bay	Willapa Bay	Cedar River	33	--	--	167	24
		North River	827	860	990		
		Willapa River	668	590	325		
		Niawiakum River	--	--	--		
		Palix River	95	--	--		
		North Nema River	56	105	55		
		South Nema River	47	--	65		
		Naselle River	142	385	315		
		Bear River	60	--	21		

[1]Combined subaerial and littoral wetlands
[2]See Williams et al. (1975) for description of water inventory areas

estuarine habitat to changes in populations of salmonids and other estuarine fishes is distinctly lacking.

Some of the more adverse impacts on estuaries have resulted from logging practices, particularly during early periods of logging when streams and estuaries were used extensively for log storage and transportation. For example, Wendler and Deschamps (1955) identified the occurrence of over 120 logging dams on rivers emptying into Grays Harbor and Willapa Bay. Dredging and filling have also substantially altered estuarine habitats. Currently, the U.S. Army Corps of Engineers removes 2.3 million m^3 of sediments annually from Washington estuaries as part of maintenance dredging operations, nearly half of this in Grays Harbor (L. Juhnke, U.S. Corps Engin., Seattle Dist., pers. comm.).

Recent information on actual changes in estuarine habitat in Washington is primarily descriptive (Salo and Stober 1971). However, Bortleson et al. (1980) quantified changes which have occurred in some of Washington's major estuaries (Table 2) such as the Duwamish River and Puyallup River deltas, which have lost essentially all their original wetland habitat. Although most smaller estuaries escaped such devastation, many now have road causeways or dikes which have often dramatically altered natural estuarine hydraulics.

Past and Present Status of Relevant Pacific Salmon Stocks

Five species of Pacific salmon are present in Washington coastal and Puget Sound waters, with coho, chinook, and chum salmon being more prevalent than pink and sockeye populations.

Natural Production

Although wild populations of salmon in Washington are assumed to have decreased over the last 100 years, accurate and quantitative information substantiating this is lacking. Only Somerton et al. (unpubl.) have attempted to quantify historical trends in Washington's wild salmon runs. Their analysis (using Puget Sound catch data) suggests that wild runs of chinook, coho, and

Table 2. Estimated historical changes in natural habitat of principal estuaries of Washington state (data from Bortleson et al. 1980, unless otherwise noted).

Estuary	*Estimated (km^2) subaerial wetland*		
	Historical	*Present*	*% change*
Nooksack	4.5	4.6	+0.2
Lummi	5.8	0.3	−89.7
Samish	11.0	0.4	−96.4
Skagit	29.0	12.0	−58.6
Stillaguamish	10.0	3.6	−64.0
Snohomish	39.0	10.0	−74.4
Duwamish	2.6	0.1	−99.2
Puyallup	10.0	0	−100.0
Nisqually	5.7	4.1	−28.1
Skokomish	2.1	1.4	−33.3
Dungeness	0.5	0.5	0
Grays Harbor[1]	195.0	136.0	−30.3
Willapa Harbor[2]	?	167.0	?

[1]Data from Smith et al. (1977).
[2]Data from U.S. Army Corps of Engineers, Seattle District (1976).

chum salmon in Puget Sound may have declined from 1900 to 1935 (Fig. 1); such factors as changes in fisheries and regulations and increasing contributions by hatchery fish after 1935 prohibit interpretation of population trends thereafter. Trends in pink and sockeye salmon run size could not be deduced due to the complex nature of interception fisheries. Current levels of natural production are highest in regions of Puget Sound south of Tacoma and from the Skagit River north. It is an impossible task to define and separate out specific causes of declines of wild salmon populations from general overfishing, but losses and alterations of freshwater and estuarine habitat have likely contributed significantly to the decline of many runs.

Artificial Production

Even though it cannot be shown quantitatively that wild salmon runs have decreased, since the turn of the century wild salmon runs have been continuously enhanced with artificially propagated fish. Some early hatcheries were established to mitigate such modifications to watersheds as logging and hydroelectric dams and direct estuarine alterations. In the early 1900's, coho, chum, and chinook salmon were primarily propagated (Fig. 1), while during the 1950's and 1960's chum salmon production was de-emphasized. Currently, numbers of chinook and chum released are highest, followed by coho, pink and sockeye (Fig. 1), although total production (biomass) of coho is higher than any of the other four species. Since the early 1900's, artificially produced salmon have been released into nearly all river systems of Puget Sound and the Washington coast, and currently much of the total hatchery production is released into systems that have heavily impacted estuaries.

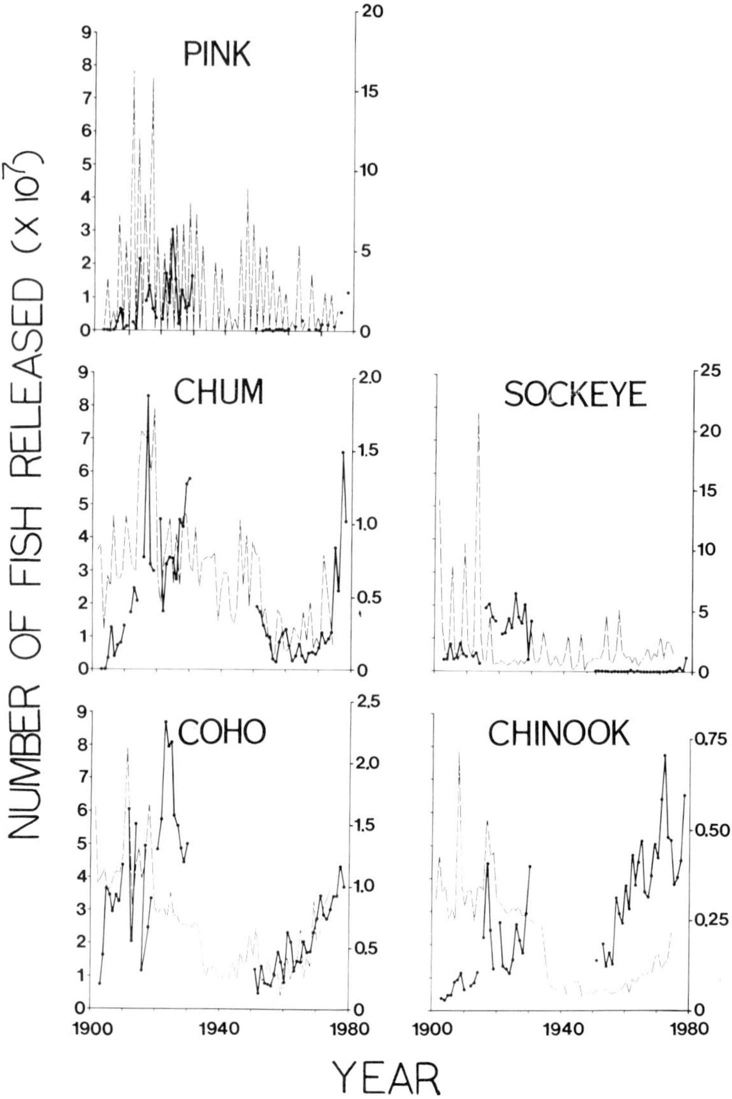

Figure 1. Releases of artificially-propagated salmon in Washington State (——) and Puget Sound salmon catches (– – –), 1901-1979. Release data from various unpublished sources; catch data from Somerton et al. (unpubl.).

Results

Utilization of Estuaries by Salmon

Estuarine utilization by Pacific salmon varies as a function of the diverse life history patterns exhibited by the five species. There are three basic patterns of time and size of entry into the estuary (see Hart 1973 for general review); these are the principal determinants of utilization: (1) upon emergence from the gravel, pink and chum salmon migrate almost immediately to the estuary; (2) coho and chinook rear in rivers for up to 2 years before entering the estuary; and (3) sockeye rear in freshwater lakes until migrating to sea. The following summarizes our knowledge concerning utilization of Washington's estuaries by salmon.

Migrations of Juveniles into the Estuary

The timing of and size at entry of juvenile salmon into estuarine habitats are a function of numerous abiotic and biotic factors, including time of adult spawning, stream temperatures during and after egg incubation, fry (newly emerged from the gravel) size and condition, population density in the stream, food resources, stream discharge and turbidity, physiological changes, tidal (lunar) cycles, and photoperiod. The effect of these factors varies according to species and life history stage (i.e., fry and smolt) (Iwamoto and Salo 1977).

Juvenile pink and chum salmon migrate downstream between early February and late May, with peak migration typically between late March and mid-May for pinks and late March to early May for chum (Wendler et al. 1954; Bostick 1955; Stober et al. 1973; Salo et al. 1980; Seiler et al. 1981). Downstream migration begins soon after emergence, although Salo and Noble (1953), Bostick (1955), and Beall (1972) found that some chum may reside in freshwater for as long as a month. Extended rearing for longer than 30 days occurs in several rivers, principally mainstem reaches of large systems such as the Skagit (Dames and Moore 1976) and Nooksack (Tyler 1964). As downstream movement of both species is almost entirely nocturnal (Salo and Bayliff 1958; Beall 1972; Koski 1975; Seiler et al. 1981), fish likely enter estuaries at night.

Juvenile coho usually migrate as smolts after rearing in freshwater habitats for about one year, although a limited number may migrate as fry after only a short residence (Iwamoto and Salo 1977); however, the ultimate fate or survival of these fry is unknown. Coho smolts usually move into estuarine habitats between mid-April and mid-June, with the peak in early to mid-May (Salo and Bayliff 1958; Sjolseth 1969) while fry migrate somewhat earlier in March and April (Seiler et al. 1981). Most (>90%) coho smolts and fry migrate at night (Seiler et al. 1981).

Sockeye smolts emigrate primarily between early April and early June in both the Cedar River/Lake Washington (Woodey 1970; Traynor 1971) and Quinault River/Lake Quinault populations (Tyler and Wright 1974). While sockeye migrate from Lake Quinault to the ocean almost entirely at night (Tyler and Wright 1974), they migrate from Lake Washington into Puget Sound during both day and night (Woodey 1970).

The diverse racial structure of chinook salmon populations results in migrations of juveniles into estuaries throughout much of the year (Iwamoto and Salo 1977). Most outmigrating juvenile chinook are subyearlings (primarily from fall spawners) and yearlings (from spring spawners) and initiate their migration in early March. Peak migration occurs in late April to early June although some fish continue to move into estuaries through August and September. Fry tend to migrate slightly earlier, from April to late May (Wendler et al. 1954; Sjolseth 1969; Stober et al. 1973; Seiler et al. 1981). Chinook migrate primarily at night, although slightly less so (85-90%) than juvenile coho (Seiler et al. 1981).

Duration of Estuarine Residence

As with factors influencing migration into the estuary, species residence time[1] is determined by biotic (e.g., size at entry and availability of preferred prey) and abiotic factors (e.g., river discharge and estuary topography) (Iwamoto and Salo 1977). While species residence times of juvenile salmonids in estuaries has been studied since the early 1940's, individual residence time is known for only a few estuaries (Table 3).

Juvenile pink and chum salmon are captured in estuaries between January and July for periods ranging from 5-18 weeks. Individual residence time has not been documented for pinks. Estimates of individual residence times of juvenile chum range from 4-32 days, with 30-32 days most common (Table 3). Studies in Hood Canal indicate that estuarine residence of juvenile chum varies seasonally (Salo et al. 1980; Bax and Whitmus 1981). Juvenile chum migrate 8-14 km day^{-1} and spend one week in the estuary in February-March but slow down to 3-7 km day^{-1} and reside up to four weeks in April-June. Several untested hypotheses have been proposed to explain these changes in residence time; Simenstad et al. (1980) and Simenstad and Salo (1982) have proposed that the availability of preferred prey organisms influences migration rate while Bax (in press) suggests that river discharge and surface outflow are responsible.

Large (>200 mm FL) pink and chum salmon have been caught in Puget Sound through winter months during sampling with fine-mesh purse seines (Hartt and Dell unpubl.; Fresh et al. 1981), suggesting that extended residence exists for some populations. Whether these residual pinks and chum are of any particular stock or whether they entered Puget Sound later or at a larger size than normal migrants is unknown.

Up to 3-5% of naturally produced yearling coho entering Puget Sound and the Hood Canal may reside inside the Strait of Juan de Fuca for as long as 600 days after entry into the estuary (essentially until maturity) without migrating to the Pacific Ocean (Rasch and O'Connor 1979). Other estimates of individual estuarine residence range from 6-40 days (Table 3).

[1] Species residence time is here defined as the total time that juvenile salmon of a particular species occur in estuarine habitats. Individual residence time is the maximum estimated time individual fish occur in the estuary (this has been typically estimated from mark-recapture studies).

Table 3. Species residence times (weeks)[1] of juvenile salmon in Washington State estuaries; maximum individual residence time (days) is indicated in parentheses.

Estuary	*Pink*	*Chum*	*Coho*	*Sockeye*	*Chinook*	*Source*
Northern Puget Sound (nearshore)	4	6	12	12	6	Miller et al. 1978
North Sound (Bellingham Bay)		11+	11+		11+	Tyler 1964
(Bellingham Bay)		7+	7		6+, (~20)	Sjolseth 1969
(nearshore)	4	6	6		6	Miller et al. 1978; Fresh 1979
Skagit Bay/Port Susan (Kiket Island)	13	13	12		15, (50)	Stober et al. 1973
(saltmarsh)		14+, (4)			16+, (6)	Congleton et al. in press
Elliott Bay (lower Duwamish)		9	5		8, (42)	Bostik 1955
(estuary)					8	Salo 1969
(lower Duwamish)		12	14+		16+	Meyer et al. 1981b
Commencement Bay (estuary)	16+	8	9		9+	Puyallup Indian Nation, unpubl.
(Hylebos Waterway)		9			8+	Meyer et al. 1981a
Nisqually Reach	12	17+	15, (~40)		11+	Fresh et al. 1979
Hood Canal	18	23, (32)	15, (6)		13	Salo et al. 1980
Strait of Juan de Fuca	14	14	14		16+	Simenstad, unpubl.
Quillayute River estuary		5, (32)			18+	Chitwood 1981
Grays Harbor		10				Wendler et al. 1954
		10+, (~30)	12		29+, (~189)	Simenstad and Eggers 1981

[1] + indicates greater than the values given.

No information is available on the duration of residence of juvenile sockeye in estuaries. Only small numbers have been caught and the short period of capture in northern Puget Sound suggests residence is extremely short. Although similar to coho in their general life history patterns, juvenile chinook salmon reside much longer in estuaries. Chinook may be found in some estuaries for as long as 29+ weeks and individuals may reside for up to 189 days in larger systems such as Grays Harbor (Table 3). Smaller fish tend to reside within the estuary longer, indicating that a threshold emigration size exists (Salo 1969). As with coho, some naturally produced chinook reside in Puget Sound for long periods of time (often until maturity). Recoveries of coded-wire tagged chinook indicate that the probability of their recapture within Puget Sound is much higher with extended rearing in hatcheries (Rasch and O'Connor 1979). This suggests that those juvenile chinook which entered estuaries at a larger size or attained such a threshold size within the estuary may have traditionally been the principal contributors to resident chinook populations in Puget Sound. There are no indications that the large coastal estuaries have ever maintained such resident populations.

Habitat Utilization

Occurrence of juvenile salmon within different estuarine habitats varies as a function of species and size class. Only a few studies have been comprehensive enough to contribute information on utilization of diverse estuarine habitats (Miller et al. 1978, 1980; Fresh et al. 1979, 1981; Salo et al. 1980; Bax and Whitmus 1981).

Upon entry to the estuary, juvenile chum and pink salmon usually occupy shallow sublittoral habitats before moving into neritic habitats. Chum remain primarily in these shallow areas until they are 50-60 mm FL, when the fish become more common in neritic habitats. Eelgrass (*Zostera* spp.) habitats, especially within contained embayments, may be particularly preferred. Juvenile chum usually school in shallow habitats during daylight but disperse into smaller groups at night (Salo et al. 1980). Juvenile chum catches in northern Puget Sound and the Strait of Juan de Fuca indicate that schooling is maintained during migration out of Puget Sound, even in neritic habitats (Fresh 1979). Juvenile coho move directly into neritic waters upon entering the estuary and school much less than pinks and chums. Exposed, cobble, or gravel beaches appear to be preferred nearshore habitats within Puget Sound's fjords (Miller et al. 1978, 1980).

Consistent with their short estuarine residence, juvenile sockeye move directly into neritic habitats, and exhibit no affinity for nearshore habitats. Juvenile chinook salmon of different sizes utilize a number of estuarine habitats, however, during their lengthy estuarine residence. Subyearlings and fry occur mainly in salt marshes where these habitats are available. However, mudflat, foreshore areas can be utilized for some time by larger subyearlings before they move into neritic habitats (Stober et al. 1973; Simenstad and

Eggers 1981; Congelton et al. in press). Yearling chinook move directly into neritic habitats without much utilization of salt marsh or other shallow habitats, although neritic areas associated with contained embayments may be preferred.

Food Habits

Prey spectra of juvenile salmonids in 16 separate estuaries (Fresh et al. 1979, 1981; Simenstad et al. 1979, 1980; Meyer et al. 1981a, b; Simenstad and Buechner 1981; Simenstad and Eggers 1981; Congleton et al. in press) have been summarized in Table 4. The relative importance of prey taxa is based upon the percent of the total Index of Relative Importance (I.R.I.) (Pinkas et al. 1971) modified by substituting prey biomass for volume (Simenstad and Kinney 1978).

Juvenile pink salmon feed almost exclusively upon neritic zooplankton (calanoid copepods, copepod nauplii, and larvaceans) even when found in shallow sublittoral habitats. Juvenile chum, on the other hand, exhibit more complex feeding behavior than pinks. Epibenthic crustaceans (harpacticoid copepods, gammarid amphipods, and isopods) are the principal prey of small (≤50-60 mm FL) chum in shallow habitats, although in salt marshes emergent insects such as chironomids are also eaten. In neritic habitats, larger (>50-60 mm FL) chum feed on planktonic prey similar to those eaten by pinks (calanoid copepods, larvaceans, and hyperiid amphipods) and on some neustonic organisms such as drift insects. Larger organisms such as decapod larvae and fish larvae enter the diet with continued growth and residence in offshore habitats. Compared to the array of zooplankton potentially available as prey, juvenile chum are highly taxa and size selective

Table 4. General prey taxa of juvenile Pacific salmon in Washington estuaries, expressed as range of % total Index of Relative Importance (IRI) and mean IRI in parentheses.

Salmon Species (n=data sets)	Prey Taxa: HARPACTICOID COPEPODS	GAMMARID AMPHIPODS	CUMACEANS	ISOPODS	EMERGENT INSECTS	DRIFT INSECTS	COPEPOD NAUPLII	JUVENILE SHRIMP	LARVACEANS	CALANOID COPEPODS	DECAPOD LARVAE	EUPHAUSIIDS	FISH LARVAE
Pink (n=5)	0.5-15.8 (4.4)	0.3-19.7 (5.0)	0.1-15.7 (3.6)	0.2 (0.1)	-- (0)	0.1-2.2 (0.5)	0.1-21.9 (5.4)	0.1-1.2 (0.3)	2.6-79.5 (26.2)	19.2-90.0 (51.0)	0.1-0.9 (0.6)	0.2 (0.1)	0.1-1.7 (0.4)
Chum (n=15)	0.1-84.7 (27.8)	0.1-17.3 (5.9)	0.1-0.9 (0.8)	8.5 (0.6)	0.4-61.5 (4.7)	0.1-51.8 (4.3)	0.1-5.6 (0.5)	0.1-1.4 (0.5)	0.1-47.1 (7.6)	0.4-80.4 (19.2)	0.1-96.9 (11.1)	0.1-39.4 (5.8)	0.1-51.0 (4.5)
Coho (n=13)	0.1-18.7 (2.3)	0.1-80.4 (22.3)	0.1-8.1 (1.1)	0.1-23.5 (3.4)	0.3-8.2 (0.7)	0.4-15.3 (3.8)	-- (0)	1.0-2.3 (0.7)	-- (0)	0.1-19.7 (1.9)	0.1-99.6 (37.4)	0.1-51.4 (10.6)	0.1-55.2 (9.5)
Sockeye (n=1)	--	2.7	--	--	--	--	2.1	28.5	--	0.7	11.0	48.6	4.4
Chinook (n=16)	0.1-3.0 (0.5)	0.1-40.2 (10.9)	0.1-83.3 (6.3)	0.1-4.6 (0.3)	0.6-52.4 (6.5)	0.5-52.9 (18.2)	0.2- (0.1)	0.2-17.9 (3.6)	0.1-6.0 (0.4)	0.1-25.8 (2.7)	0.2-68.6 (20.7)	0.1-48.3 (4.1)	0.1-53.9 (18.1)

(Simenstad et al. 1980; Simenstad and Eggers 1981), preferentially consuming large, relatively rare harpacticoid (*Harpacticus* spp.) and calanoid (*Calanus* spp.) copepods from available epibenthic and neritic zooplankton, respectively.

Indicative of their large size upon entry into the estuary and occupation of neritic habitats, juvenile coho feed primarily upon large planktonic or small nektonic organisms, including decapod larvae, larval and juvenile fish, and euphausiids. Drift insects and epibenthic gammarid amphipods are also eaten by coho, especially in large turbid estuaries such as Grays Harbor. The prey of juvenile sockeye in estuarine habitats have been documented only from northern Puget Sound, however, and consisted largely of euphausiids, juvenile shrimp, and decapod larvae. Juvenile chinook salmon have the most diverse prey spectra, reflecting their extended estuarine residence, diversity of size classes, and different estuarine habitats utilized. Fry and subyearlings in salt marsh and other shallow habitats prey principally upon emergent insects and epibenthic crustaceans such as gammarid amphipods, mysids, and cumaceans. In neritic habitats, they feed upon small nekton (decapod larvae, larval and juvenile fish, and euphausiids) and neustonic drift insects.

Growth

Juvenile salmon (especially pink and chum) generally have high growth rates in estuaries, in many cases the highest rates in their life histories. Juvenile chum growth estimates by Salo et al. (1980) and Bax and Whitmus (1981) in Hood Canal were between 5.7% and 8.6% body weight day^{-1} and were as high as 10.1% for the first 4 days of estuarine residence. In the Skagit salt marsh, chum grew at a rate of 6% body weight day^{-1} (Congleton et al. in press). Salo (1969) derived a general linear growth relationship for juvenile chinook in the lower Duwamish River and estuary which indicated a growth rate of approximately 2.6 mm week^{-1}.

Predation

While predation upon juvenile salmonids in freshwater habitats is severe under certain conditions, evidence for significant predation in estuarine and nearshore marine habitats in Washington is lacking (Fresh et al. 1979, 1981, 1982; Simenstad et al. 1979; Cardwell and Fresh in prep). Relatively low occurrences of juvenile salmonids have been documented in the stomachs of such piscivores as Pacific cod (*Gadus macrocephalus*), Pacific staghorn sculpin (*Leptocottus armatus*), cutthroat (*Salmo clarki*) and rainbow-steelhead (*S. gairdneri*) trout, and chinook and coho smolts. Given large standing stocks of predators (and size classes large enough to prey upon juvenile salmon), mortality rates due to predation might be significant. Of the potential predators, the two trout species typically have the highest occurrence of juvenile salmonids in their diets.

Marine birds and mammals also represent potentially significant sources of predation mortality. One of the few quantitative studies involved rhinoceros auklets (*Cerorhinca monocerata*) at Protection Island in the Strait of Juan de Fuca (Manuwal 1977); juveniles of four species of Pacific salmon were common prey delivered to auklet chicks during the nesting season. Simenstad et al. (1979) also documented juvenile salmon as prey of double-crested cormorant (*Phalacrocorax auritus*) and common merganser (*Mergus merganser*), although these birds are generally less abundant then rhinoceros auklets. Low incidences of Pacific salmon have been reported in the stomach contents of Pacific harbor seals (*Phoca vitulina richardsi*) and orca (*Orcinus orca*), although the few accounts in Puget Sound give no indication of predation upon juvenile migrants but, rather, on subadults and adults (Simenstad et al. 1979).

Adult Utilization of Estuaries

Adult salmon returning to streams and hatcheries pass through estuaries essentially year-round, although most return between July and September (Table 5). Factors influencing estuarine arrival are poorly understood, but ocean temperature, food supplies, and photoperiod are probably important (e.g., Helle 1979; Burgner 1980). Considerable differences in run timing occur between and within species, even in the same river system.

The behavior of adult salmon in estuaries and factors influencing this behavior are not well documented. Tagging of adult chinook in the Duwamish River estuary (Stauffer 1970) suggested adults aggregated in rather specific staging areas in the estuary before moving upstream. Upriver movements of fish are probably stimulated by such factors as freshets and

Table 5. General estuarine run timing (from various WDF [Harvest Management Division] data sources) and estimated individual residence times (days) for adult Pacific salmon in Washington.

	Run timing		*Est. indiv. residence time*		
Species	*Peak*	*Range*	*Min*	*Max*	*Data source*
Pink	Late July-early Aug.	June-Sept.	6.9	13.7	Barker 1979
(odd years only)			8.1	32.2	Barker 1979
Sockeye	July	June-Aug.	–	–	
Coho	Late Oct.	Sept.-Nov.	13.0	40.0	Eames et al. 1981; in press
			9.2	14.7	Barker 1979
Chinook	Late Aug.-early Sept.	July-Sept.	27.4	39.6	Stauffer 1970
Chum	Mid-Oct.-early Nov.	Sept.-Dec.	11.0	40.0	Eames et al. 1981; in press
			3.0	6.8	Barker 1979
			13.0	20.6	Olney 1976

changing air pressure. We estimated the individual residence times of adults in estuaries (Table 5) from several tagging studies. Given limitations associated with using these studies to determine residence times (e.g., between-estuary variability and unknown prior history of the fish), residence time was estimated to range from 1-6 weeks (Table 5).

Discussion

Importance of Estuaries to Pacific Salmon

Although juvenile and adult Pacific salmon utilize Puget Sound and Washington coastal estuaries to different degrees, the real question is whether such utilization enhances growth and survival. From existing information, we discuss several hypotheses about the significance of estuarine utilization by salmon.

Physiological Transition Zone

Considering the magnitude of the physiological change required to adapt to saltwater, juvenile salmonids may well benefit from the gradual transition from freshwater to marine habitats offered by estuaries (Baggerman 1960; Hoar 1976; Wedemeyer et al. 1980). Moreover, estuarine salinity gradients may be used by juvenile salmon as a directive cue in their seaward migration (McInerney 1964).

Predator Avoidance

While the potential for predation upon juvenile salmon in estuaries may be high, there are a number of factors which we believe may act to limit predation by piscivorous fishes. The behavior of juvenile salmonids, particularly pink and chum, in estuarine habitats (e.g., schooling in shallow eelgrass habitats during the day and dispersion at night by chum 45-55 mm FL) suggests that they are spatially and temporally distributed in a manner that minimizes exposure to predators. In addition, euryhaline conditions may place constraints upon piscivore efficiency. Since predation is basically a visual foraging process, high water turbidity shortens the reactive distance of predators and thereby can reduce the frequency of prey encounter. Because many of the state's estuaries have relatively turbid waters during the period of salmon outmigration, these systems could provide valuable refugia while the juvenile salmon "grow out of" their vulnerability to predators.

Optimum Foraging Conditions

Epibenthic, neritic, and neustonic organisms in estuaries apparently provide optimal prey resources for juvenile salmon from the standpoint of prey size, distribution, and density. Small epibenthic crustaceans are abundant in shallow habitats which juvenile chum salmon utilize. The size, density, and minimal avoidance capabilities of these zooplankters may provide optimum food organisms for the small salmon, which are morphologically and energetically limited in their foraging at this early stage in their estuarine residence. Neritic zooplankton is characteristically more concentrated in estuarine systems than in oceanic habitats (Bowman and Esaias 1978) due to frontal processes or water column stability (depending upon the organism). Although unsubstantiated, estuaries also appear to maintain extensive communities of neustonic organisms such as certain calanoid copepods and exogenous organisms such as drift insects. Sources of insects include both riverine outflow and fallout from adjacent wetlands and uplands. An estuary bordered by a relatively natural terrestrial environment, therefore, acts as a giant insect trap, thus contributing to salmon prey resources within the system.

Estuarine Development and Pacific Salmon

Estuarine development, particularly in larger estuaries, has been historically unregulated and uncontrolled such that virtually all major estuaries in Washington have been affected. Alterations to estuaries have likely contributed to the decline of some salmon populations by changing utilization patterns of salmon in estuaries. For one, reductions in salmon prey resources have probably resulted from: 1) destruction of invertebrate habitat by dredging and filling activities; 2) watershed activities (e.g., logging practices) that increase sediment transported into the estuary, thus reducing the net amount of salt marshes and other shallow habitats available to foraging fish; 3) changes in the amount and sources of organic carbon entering detritus-based food webs; and 4) introductions of toxic substances at levels which kill food organisms. Primary production can be affected, especially due to increases in turbidity from suspended sediment loads reducing the effective euphotic zone. Such reductions in availability of estuarine food organisms may reduce growth of juvenile salmon or modify normal estuarine residence times. Fish would have to forage over a wider area or leave the estuary prematurely in order to locate sufficient densities of prey. Stresses to fish in the estuary such as during saltwater transition or from disease may be exacerbated by reduced fish condition due to inadequate rations. Ultimately, the potential for the estuary to support a given species assemblage and standing stock, i.e. its carrying capacity, could be reduced to the detriment of salmon populations.

In addition, suspended solids from dredging and filling activities or riverine sources can also directly kill or injure fish by clogging and coating gill filaments, by causing abrasive injuries, or by being a source of toxic components (Mortenson et al. 1976). Other degradations in estuarine water quality may force juveniles and adults to avoid areas of low oxygen or high waste levels and move into habitats where they are subjected to increased predation.

Fortunately, over the last two decades a number of new state and federal laws have combined with several older (but recently enforced) laws to provide one of the strongest estuarine (and overall environmental) protection packages in the country, thereby reducing the potential for estuarine alterations harmful to salmon. In addition, under Phase II of the 1974 Boldt Decision, certain Native American Indian tribes of the Pacific Northwest have obtained litigation rights in environmental issues that can be used to challenge, block, or alter developments considered by tribes to be potentially harmful to traditionally fished species, primarily Pacific salmon.

Effects of Artificial Propagation on Estuarine Utilization by Pacific Salmon

The high demand for salmon and the depressed condition of many wild salmon populations necessitate continued enhancement of salmon runs. Hatcheries have long been the most attractive enhancement measure because more juveniles are produced per female spawner than in a natural stream environment, allowing harvest of a greater fraction of returning hatchery fish than is possible for wild fish. But, while egg-to-migrant survival is substantially higher in the hatchery environment than in the stream, migrant-to-adult survival is often lower for artificially produced fish. There are numerous explanations for this, many of which relate to the culture techniques employed (e.g., Salo and Bayliff 1958; Beall 1972; Wedemeyer et al. 1980; Fresh et al. 1982).

Historically, salmon hatcheries usually attempted to mimic the outmigration patterns of fish in the recipient stream, but numerous salmon management and cultural needs (e.g., smolt quality, disease, and allocation of fish to various user groups) have made this difficult and, at times, impossible to accomplish. Thus, hatchery releases often differ from the outmigration of naturally produced fish in terms of density (high-density hatchery releases vs. smaller, naturally produced densities), of size and timing (discrete pulses of hatchery fish vs. a wide time interval of wild fish outmigration). This can result in altered estuarine utilization patterns of the juvenile salmon. For instance, hatchery fish entering estuaries may be either too large or small, too numerous, or too early or late to exploit food resources in the estuary optimally (Simenstad et al. 1980; Simenstad and Salo 1982). Rasch and O'Connor (1979) clearly demonstrated how estuarine utilization by hatchery reared coho and chinook was altered by time and size of release. Estuarine behavior of artificially propagated juvenile salmon may also be modified by "domestication" in the hatchery, resulting in increased vulnerability of juveniles to predators (Thompson 1966; Kanayama 1968) and in non-adaptive feeding behaviors resulting from food delivery (Volovik and Gritsenko 1970; Fresh et al. 1982).

Interspecific interactions (predation or competition) of juvenile salmon in estuaries may also be influenced significantly by culture techniques. Estuarine predation by larger, piscivorous salmon (coho and chinook smolts) on pink and chum of either wild or hatchery origin may be the cause of declines in some Puget Sound pink and chum salmon stocks (J. Ames, WDF, unpubl.; Johnson 1973). Gallagher (1980) and Ames (WDF, unpubl.) hypothesized that competition for food between pink and chum juveniles in estuary and nearshore marine habitats may be a cause of distinct odd-even year cyclicity in natural chum salmon populations in Puget Sound; chum returns tend to be lower from odd-year broods, when pink salmon juveniles coexist with chum in estuaries the following spring. Simenstad et al. (1980) documented dietary overlaps up to 84% for juvenile pink and chum salmon in Hood Canal, suggesting that potential for food competition exists.

Defining and employing the proper culture techniques, however, will help optimize survival and migratory behavior of juvenile salmon. Salmon have evolved diverse life history patterns within and between species which maximize fitness and survival under a wide range of conditions. Thus, while culture and management policies typically seek universal criteria dictating time, size, and density of hatchery releases, a diversity of methods may be desirable in salmon enhancement. A system-by-system approach may be best since culture and release strategies can be tailored to take fullest advantage of system-specific freshwater, estuarine, and marine conditions the fish encounter. For instance, timing, size, and density of hatchery releases should be spatially and temporally structured to allow fish to exploit habitats and food resources optimally (Eggers 1980) and minimize predation and competition. Similarly, selection of enhancement stocks and locations of enhancement projects should include careful consideration of the natural distribution and abundance of salmonids in order to optimize natural food conditions and minimize competition or predation involving naturally-or artificially-produced fish.

Summary and Conclusions

In this paper, we have synthesized available information on utilization of Washington's estuaries by Pacific salmon and concluded that Pacific salmon are adapted to various characteristics of estuaries. All juvenile salmonids utilize estuaries to varying degrees during their initial migratory transition into the north Pacific Ocean. We hypothesize that salmon use estuaries as refugia from predators, for optimum availability of preferred food organisms promoting rapid growth, and as a physiological transition area. What has not been established, however, is the quantitative significance of such estuarine factors to the ultimate survival of adults.

There is also no doubt that substantial alterations have occurred in many major estuaries and some modifications have occurred in almost all estuaries in the region, although deterioration of natural estuarine habitats has decreased over the last two decades. There exists, however, no information which directly ties changes in quantity and quality of Washington's estuaries to changes in abundance of Pacific salmon because there are too many confounding factors. For whatever reasons, however, managers have often had to take substantial steps to augment salmon runs through artificial propagation. But, ironically, use of hatchery salmon has probably led to further changes in estuarine utilization by salmon, some of which may be maladaptive in terms of optimum production of the salmon resource. This is probably more a result of confusing and incomplete information on quantitative advantages of estuarine residence to salmon and other constraints in culture operations than on intentional disregard for the potential benefits of utilizing estuaries to maintain or enhance salmon production. Thus, more emphasis could be given to ecological and economic advantages of managing estuaries and Pacific salmon to optimize beneficial effects of salmon utilization of natural estuarine habitats.

Acknowledgments

Numerous people provided considerable information for this synthesis. We would like to acknowledge the contributions of W. J. Kinney, M. L. Mills, J. Ames, D. Geist, T. Flint, D. Phinney, P. Coleman, and S. Schroder, all of WDF; G. Graves and R. Morinaka of the U.S. Fish and Wildlife Service; R. Smith of the National Marine Fisheries Service; J. Congleton of the University of Idaho Cooperative Fisheries Unit; L. Lestelle of the Quinault Indian Tribe; J. Armstrong of the U.S. Army Corps of Engineers, Seattle District; and D. Somerton and T. Jagielo of the University of Washington. We are grateful for the critical reviews of the manuscript provided by M. Carr of WDF, R. Cardwell of Envirosphere Co., and three anonymous reviewers. Contribution No. 579, School of Fisheries, University of Washington.

References Cited

Baggerman, B. 1960. Salinity preference, thyroid activity and the seaward migration of four species of Pacific salmon (*Oncorhynchus*). *J. Fish. Res. Bd. Canada* 17:295-322.

Barker, M. 1979. Summary of travel time from tagging studies, 1950-1974. Wash. Dept. Fish., Tech. Rep. No. 48. 30 p.

Bax, N. J. (In Press). Seasonal and annual variations in the movement of juvenile chum salmon through Hood Canal, Washington. *In:* E. L. Brannon (ed.), *Proc. Salmon and Trout Migratory Behavior Symp.*, U. Washington, Seattle, WA

Bax, N. J. and C. J. Whitmus. 1981. Early marine survival and migratory behavior of juvenile chum salmon released from the Enetai Hatchery, Washington in 1980. Final Rep., Fish. Res. Inst., U. Washington, Seattle, WA. FRI-UW-8109. 48 p.

Beall, E. P. 1972. The use of predator-prey tests to assess the quality of chum salmon (*Oncorhynchus keta*) fry. M.S. Thesis, U. Washington, Seattle, WA 105 p.

Bortleson, G. C., M. J. Chrzastowski and A. K. Helgerson. 1980. Historical changes of shoreline and wetland at eleven major deltas in the Puget Sound region, Washington. Hydrologic Investigations Atlas HA-617, U.S. Geological Survey, Denver, CO

Bostick, W. E. 1955. Duwamish River seining studies, pp. 5-6. *In:* Puget Sound Stream Studies, Wash. Dept. Fish., Olympia, WA

Bowman, M. J. and W. E. Esaias (eds.) 1978. Oceanic fronts in coastal processes. Workshop Proceedings, Mar. Sci. Res. Center, Stony Brook, NY. Springer-Verlag, Berlin.

Burgner, R. L. 1980. Some features of ocean migrations and timing of Pacific salmon, pp. 153-165. *In:* W. J. McNeil and D. C. Himsworth (eds.), *Salmonid Ecosystems of the North Pacific.* Oregon State U. Press, Corvallis, OR

Cardwell, R. D. and K. L. Fresh. (In Prep.). Predation upon juvenile salmon. Wash. Dept. Fish.

Chitwood, S. A. 1981. Water quality, salmonid fish, smelt, crab and subtidal studies at the Quillayute River project. Rep. to Seattle Dist., U.S. Army Corp Engin., Quileute Fish. Dept., Quileute Indian Tribe, LaPush, WA 92 p.

Congleton, J. L., S. K. Davis and S. R. Foley. (In Press). Distribution, abundance, and outmigration timing of chum and chinook salmon fry in the Skagit salt marsh. *In:* E. L. Brannon (ed.), *Proc. Salmon and Trout Migratory Behavior Symp.*, U. Washington, Seattle, WA

Crandell, D. R. 1965. The glacial history of Western Washington and Oregon, pp. 341-353. *In:* H. E. Wright and D. G. Frey (eds.), *The Quaternary of the United States*. Princeton University Press, Princeton, NJ

Dames and Moore. 1976. Environmental studies report, January-December 1975, Skagit nuclear power project, Northwestern Washington. Rep. to Puget Sound Power and Light Company, Seattle, WA 376 p.

Eames, M., T. Quinn and M. Hino. (In Press). Northern Puget Sound 1977 adult coho and chum tagging studies. Wash. Dept. Fish., Tech. Rep.

Eames, M., T. Quinn, K. Reidinger and D. Haring. 1981. Northern Puget Sound 1976 adult coho and chum tagging studies. Wash. Dept. Fish., Tech. Rep. No. 64. 217 p.

Eggers, D. M. 1980. Feeding ecology of Lake Washington sockeye salmon and the salmon enhancement problem, pp. 165-171. *In:* W. J. McNeil and D. C. Himsworth (eds.), *Salmonid Ecosystems of the North Pacific*. Oregon State U. Press, Corvallis, OR

Fresh, K. L. 1979. Distribution and abundance of fishes occurring in the nearshore surface waters of northern Puget Sound, Washington. M.S. Thesis, U. Washington, Seattle, WA 120 p.

Fresh, K. L., R. D. Cardwell and R. R. Koons. 1981. Food habits of Pacific salmon, baitfish, and their potential competitors and predators in the marine waters of Washington, August 1978 to September 1979. Wash. Dept. Fish., Prog. Rep. No. 145. 58 p.

Fresh, K. L., R. D. Cardwell, B. P. Snyder and E. O. Salo. 1982. Some hatchery strategies for reducing predation upon juvenile chum salmon (*Oncorhynchus keta*) in freshwater, pp. 79-89. *In:* B. R. Melteff and R. A. Nevé (eds.), *Proc. North Pacific Aquaculture Symp.* Alaska Sea Grant Program, U. Alaska, Fairbanks, AL

Fresh, K. L., D. Rabin, C. Simenstad, E. O. Salo, K. Garrison and L. Mathesen. 1979. Fish ecology studies in the Nisqually Reach area of southern Puget Sound, Washington. Final Rep., Fish. Res. Inst., U. Washington, Seattle, WA. FRI-UW-7904. 229 p.

Gallagher, A. F. 1980. An analysis of factors affecting brood year returns of the wild stocks of Puget Sound chum (*Oncorhynchus keta*) and pink salmon (*Oncorhynchus gorbuscha*). M.S. Thesis, U. Washington, Seattle, WA 152 p.

Hart, J. L. 1973. Pacific Fishes of Canada. *Fish. Res. Bd. Canada. Bull.* 180. 740 pp.

Hartt, A. C. and M. B. Dell. (Unpubl.). Early oceanic migrations and growth of juvenile Pacific salmon and steelhead trout. Inter. Pac. Fish. Comm., Fish. Res. Inst., U. Washington, Seattle, WA 249 p. (manuscript)

Helle, J. H. 1979. Influence of marine environment on age and size at maturity, growth, and abundance of chum salmon, *Oncorhynchus keta* (Walbaum), from Olsen Creek, Prince William Sound, Alaska. Ph.D. Dissertation, Oregon State U., Corvallis, OR 118 p.

Hoar, W. S. 1976. Smolt transformation: evolution, behavior, and physiology. *J. Fish. Res. Bd. Canada* 33:1233-1252.

Iwamoto, R. N. and E. O. Salo. 1977. Estuarine survival of juvenile salmonids: a review of the literature. Rep. to Wash. Dept. Fish., Fish. Res. Inst., U. Washington, Seattle, WA 64 p.

Johnson, R. 1973. Potential interspecific problems between hatchery coho smolts and juvenile pink and chum salmon. Puget Sound Stream Studies, Pink and Chum Salmon Investigations. Wash. Dept. Fish., Olympia, WA 40 p.

Kanayama, Y. 1968. Studies of the conditioned reflex in lower vertebrates. X. Defensive conditioned reflex of chum salmon fry in group. *Mar. Biol.* 2:77-87.

Koski, K. V. 1975. The survival and fitness of two stocks of chum salmon (*Oncorhynchus keta*) from egg deposition to emergence in a controlled-stream environment at Big Beef Creek. Ph.D. Dissertation, U. Washington, Seattle, WA 212 p.

Manuwal, D. A. 1977. Marine bird populations in Washington state. Final Rep. to National Wildlife Federation, Washington, DC 116 p.

McInerney, J. E. 1964. Salinity preference: an orientation mechanism in salmon migration. *J. Fish. Res. Bd. Canada* 21:996-1018.

Meyer, J. H., T. A. Pearce and R. S. Boomer. 1981a. An examination of the food habits of juvenile chum and chinook salmon in Hylebos Waterway. Fish. Assist. Office, U.S. Fish & Wildl. Serv., Olympia, WA 13 p. (manuscript)

Meyer, J. H., T. A. Pearce and S. B. Patlan. 1981b. Distribution and food habits of juvenile salmonids in the Duwamish Estuary, Washington, 1980. Fish. Asst. Office, U.S. Fish and Wildl. Serv., Olympia, WA 42 p.

Miller, B. S., C. A. Simenstad, J. N. Cross, K. L. Fresh and S. N. Steinfort. 1980. Nearshore fish and macroinvertebrate assemblages along the Strait of Juan de Fuca, including food habits of the common nearshore fish: final report of three years' sampling, 1976-1979. Final Rep. to NOAA/MESA Puget Sound Project, EPA DOC Res. Rep. EPA-600/7-79-259. 335 p.

Miller, B. S., C. A. Simenstad, L. L. Moulton, K. L. Fresh, F. C. Funk, W. A. Karp and S. F. Borton. 1978. Puget Sound baseline program: nearshore fish survey. Final Rep., July 1974-June 1977, to Wash. Dept. Ecol. Appendix D to Baseline Study Report 10. Lacey, WA 220 p.

Mortensen, D. G., B. P. Snyder and E. O. Salo. 1976. An analysis of the literature on the effects of dredging on juvenile salmonids. Spec. Rep. to U.S. Dept. Navy, Fish. Res. Inst., U. Washington, Seattle, WA. FRI-UW-7605. 37 p.

Neave, F. 1958. The origin and speciation of *Oncorhynchus*. *Trans. Royal Soc. Canada*. 51:25-39.

Olney, F. E. 1976. Population estimation of the 1974-75 Nisqually River chum salmon run. U.S. Fish & Wildl. Serv., Northwest Fish Prog. Spec. Rep. 36 p.

Pinkas, L., M. S. Oliphant and I. L. K. Iverson. 1971. Food habits of albacore, bluefin tuna and bonito in California waters. Calif. Fish Game., Fish. Bull. 151:1-105.

Pritchard, D. W. 1967. What is an estuary: physical viewpoint, pp. 3-5. *In:* G. H. Lauff (ed.), *Estuaries*. Am. Assoc. Adv. Sci. Pub. 83, Washington, DC.

Rasch, T. and D. O'Connor. 1979. 1977 results from micro-tagged salmon experimental groups. Wash. Dept. Fish., Prog. Rep. No. 91. 562 p.

Salo, E. O. 1969. Final Report for the period June 1, 1965-September 30, 1968, estuarine ecology research project. U. Washington, Fish Res. Inst. 80 p.

Salo, E. O., N. J. Bax, T. E. Prinslow, C. J. Whitmus, B. P. Snyder and C. A. Simenstad. 1980. The effects of construction of naval facilities on the outmigration of juvenile salmonids from Hood Canal, Washington. Final Rep., Fish. Res. Inst., U. Washington, Seattle, WA. FRI-UW-8006. 159 p.

Salo, E. O. and W. H. Bayliff. 1958. Artificial and natural production of silver salmon (*Oncorhynchus kisutch*) at Minter Creek, Washington. Wash. Dept. Fish., Res. Bull. No. 4, Olympia, WA 76 p.

Salo, E. O. and R. E. Noble. 1953. Minter Creek biological station, April-June 1953. Wash. Dept. Fish., Olympia. WA 18 p.

Salo, E. O. and Q. J. Stober. 1971. The effects of pollution on the estuaries of the Northwest Pacific Coast, pp. 86-95. *In:* P. Douglas and R. Stroud (eds.), *A Symposium on the Biological Significance of Estuaries*. Spec. Pub. Sport Fishing Inst., Washington, DC

Seiler, D., S. Neuhauser and M. Ackley. 1981. Upstream/downstream salmonid trapping project, 1977-1980. Wash. Dept. Fish., Prog. Rep. No. 144. 197 p.

Simenstad, C. A. and H. Buechner. 1981. Food habits analysis of juvenile chinook (*Oncorhynchus tshawytscha*) and coho salmon (*O. kisutch*) from the Quillayute River and estuary. Appendix F, *In:* S. A. Chitwood, Water quality, salmonid fish, smelt, crab and subtidal studies at the Quillayute River project. Rep. to Seattle Dist., U.S. Army Corps of Engin., Quileute Fish. Dept., Quileute Indian Tribe, LaPush, WA 92 p.

Simenstad, C. A. and D. M. Eggers (eds.). 1981. Juvenile salmonid and baitfish distribution, abundance, and prey resources in selected areas of Grays Harbor, Washington. Final Rep. to

Seattle Dist., U.S. Army Corps of Engin., Fish. Res. Inst., U. Washington, Seattle, WA. FRI-UW-8116. 205 p.

Simenstad, C. A. and W. J. Kinney. 1978. Trophic relationships of outmigrating chum salmon in Hood Canal, 1977. Fish. Res. Inst., U. Washington, Seattle, WA. FRI-UW-7810. 75 p.

Simenstad, C. A., W. J. Kinney, S. S. Parker, E. O. Salo, J. R. Cordell and H. Buechner. 1980. Prey community structure and trophic ecology of outmigrating juvenile chum and pink salmon in Hood Canal, Washington: A synthesis of three years' studies, 1977-1979. Final Rep., Fish. Res. Inst., U. Washington, Seattle, WA. FRI-UW-8026. 113 p.

Simenstad, C. A., B. S. Miller, C. F. Nyblade, K. Thornburgh and L. J. Bledsoe. 1979. Food web relationships of northern Puget Sound and the Strait of Juan de Fuca: a synthesis of the available knowledge. MESA Puget Sound Project, Environ. Res. Labs. DOC/EPA Inter-agency Energy/Environment R&D Program Rep., EPA-600/7-79-259. 335 p.

Simenstad, C. A. and E. O. Salo. 1982. Foraging success as a determinant of estuarine and nearshore carrying capacity of juvenile chum salmon (*Oncorhynchus keta*) in Hood Canal, Washington, pp. 21-37. *In:* B. R. Melteff and R. A. Nevé (eds.), *Proc. North Pacific Aquaculture Symp.* Rep. 82-2. Alaska Sea Grant Program, U. Alaska, Fairbanks, AL

Sjolseth, D. 1969. Studies of juvenile salmon in the Nooksack River system and Bellingham Bay. M.S. Thesis, U. Washington, Seattle, WA 96 p.

Smith, J. M., J. B. Phipps, E. D. Schermer and D. F. Samuelson. 1977. Impact of dredging on the water quality in Grays Harbor. Appendix K, *In* Maintenance dredging and the environment of Grays Harbor, Washington. U.S. Army Corps of Engineers, Seattle District, WA 221 p.

Somerton, D., M. E. Crow, C. M. Lynde and L. J. Bledsoe. (Unpubl.) The Puget Sound runs of salmon: an examination of the changes in run size since 1896. NORFISH, Center for Quant. Sci., U. Washington, Seattle, WA 89 p. (manuscript).

Stauffer, G. D. 1970. Estimates of population parameters of the 1965 and 1966 adult chinook salmon runs in the Green-Duwamish River. M.S. Thesis, U. Washington, Seattle, WA 155 p.

Stober, Q. J., S. J. Walden and D. T. Griggs. 1973. Juvenile salmonid migration through Skagit Bay, pp. 35-70. *In:* Q. J. Stober and E. O. Salo (eds.), *Ecological studies of the proposed Kiket Island nuclear power site.* Final Rep., Fish. Res. Inst., U. Washington, Seattle, WA. FRI-UW-7304. 537 p.

Thompson, R. B. 1966. Effects of predator avoidance conditioning on the post-release survival rate of artificially propagated salmon. Ph.D. Dissertation, U. Washington, Seattle, WA 155 p.

Traynor, J. 1971. Lake Washington sockeye smolt studies, 1971. Coll. Fish., U. Washington, Seattle, WA 10 p. (manuscript).

Tyler, R. W. 1964. Distribution and migration of young salmon in Bellingham Bay, Washington. Circ. 212, Fish. Res. Inst., U. Washington, Seattle, WA 26 p.

Tyler, R. W. and T. E. Wright. 1974. A method of enumerating blueback salmon smolts from Quinault Lake and biological parameters of the 1974 outmigration. Final Rep., Fish. Res. Inst., U. Washington, Seattle, WA. FRI-UW-7414. 29 p.

U.S. Army Corps of Engineers. 1976. Willapa River and harbor navigation projects, Washington. Final Environmental Impact Statement. Seattle, WA 66 p. plus Appendices.

U.S. Geological Survey. 1978. Water resources data for Washington, water year 1977.
U.S. Geol. Surv. Water-Data Rep. WA-77-1, Vol. 1, Western Washington. Tacoma, WA 433 p.

U.S. Geological Survey. 1980. Water resources data for Washington, water year 1979. U.S. Geol. Surv. Water-Data Rep. WA-79-1, Vol. 1, Western Washington. Tacoma, WA 433 p.

Volovik, S. P. and O. F. Gritsenko. 1970. Effect of predation on the survival of young salmon in rivers of Sakhalin. V.N.I.R.O. 71:193-209. (Fish. Res. Bd. Canada Transl. No. 1716).

Wedemeyer, G. A., R. L. Saunders and W. C. Clarke. 1980. Environmental factors affecting smoltification and early marine survival of anadromous salmonids. *Mar. Fish. Rev.* 42:1-14.

Wendler, H. O. and G. Deschamps. 1955. Logging dams on coastal Washington streams. Wash. Dept. Fish., Fish. Res. Paper: 27-28.

Wendler, H. O., G. Deschamps and M. H. Amos. 1954. Downstream migrant studies, pp. 8-14. *In:* Coastal Investigations, March-August 1954. Wash. Dept. Fish., Olympia, WA

Williams, R. W., R. M. Laramie and J. H. Ames. 1975. A catalog of Washington streams. Vols. 1 and 2. Wash. Dept. Fish., Olympia, WA

Woodey, J. C. 1970. Lake Washington sockeye smolt studies, 1970. Coll. Fish., U. Washington, Seattle, WA 7 p. (manuscript).

MIGRATION CHARACTERISTICS OF COHO SALMON (*ONCORHYNCHUS KISUTCH*) SMOLTS IN THE COLUMBIA RIVER AND ITS ESTUARY

Joseph T. Durkin
National Marine Fisheries Service
Hammond Biological Field Station
Hammond, Oregon

Abstract: Beach seine catches indicated a majority of migrating coho salmon (*Oncorhynchus kisutch*) smolts reached the upper Columbia River estuary between late April and early June, with peak numbers occurring between 6 and 17 May. Length measurements revealed that large individuals tended to migrate downstream before small individuals. Based on size, most Columbia River smolts were of hatchery origin rather than of natural spawning origin. Recovery of tagged fish showed rates of downstream movement increased from late March to mid-May. Beach seine and purse seine catches suggest greater migratory movement with daylight. The principal prey organisms utilized by smolts were insects in the river and amphipods in the middle estuary and near the mouth. Predation on other fish was minor.

Introduction

Columbia River coho salmon (*Oncorhynchus kisutch*) originate from a number of hatcheries and tributaries which annually produce millions of smolts. According to Fulton (1970), there were at least 16 coho hatcheries and 72 streams or tributaries with natural spawning runs. The annual smolt migration is a composite of hatchery releases and wild migrants from various geographic areas. Therefore, estuarine samples of juveniles reflect the presence of all Columbia River smolts and not a particular stock of fish.

Coho salmon are one of several economically important species of Pacific salmon found from central California north to Alaska (Haw et al. 1967). They have a three-year life cycle with approximately 18 months spent in fresh water and 18 months in the ocean (Clemens and Wilby 1961; Hart 1973; Wydocki and Whitney 1979). Sexually mature coho salmon average 70 cm in length and weigh 4 kg, though there is local and annual size variation. Adult returns to the Columbia River exceeded 1,050,000 individuals as recently as 1970, but have since declined to 316,000 fish in 1979 (Bohn and Stockley 1981). There is regional concern about the decline in adult returns, and some believe the juvenile smolt migration from river to ocean may be a period of high mortality. Possible causes for the decline may include competition for food (Myers 1980); disease (Iwama and Greer 1980); parasites (Males 1980); nitrogen supersaturation (Ebel 1969); stress (Chiweyite and

ISBN 0-12-404070-5

Schreck 1980); starvation (Snyder 1980); temperature (Zaugg and McLain 1976); herbicides (Miller et al. 1979); hatchery medication (Bouck and Johnson 1979); and descaling caused by man (Bouck and Smith 1979). Since any factor or combination of factors could have a deleterious effect on coho salmon smolts, researchers should have information on the migrational characteristics of coho salmon smolts in riverine, estuarine, and marine habitats in order to understand and correctly assess problem causes.

This paper deals with the results of many published and unpublished National Marine Fisheries Service (NMFS) studies on juvenile coho salmon migrants in the Columbia River estuary between 1966 and 1981. Some of the investigations were complex and conducted over a long period of time; others were limited in scope and short term in duration.

Methods

Nearly all the work described in this report took place in the Columbia River estuary from Rkm (River kilometer) 5 near Sand Island to Rkm 75 east of Puget Island (Fig. 1). Coho salmon catches were made primarily with purse and beach seines.

Purse seines (200 m long; 10 m deep; 18-mm stretch mesh) were set and fished between a seine skiff and seine vessel in the manner described by Johnsen and Sims (1973). The net was held open for 5 min into the current (usually an easterly direction), in a 'U' shaped conformation. The net was pursed and brought aboard the seine vessel after each set. Purse seine use was limited to riverine and estuarine open areas where depths exceeded 9 m.

Beach seines were used in shallow depths in the manner described by Sims and Johnsen (1974). The nets were either 50 m long by 3 m deep or 100 m long by 4 m deep, of 18-mm mesh. The bunt of the beach seine and purse seine was made of knotless netting 12-mm stretch mesh to minimize salmonid descaling. Sets were made by towing one end of the net in an 180° sweep back to the beach.

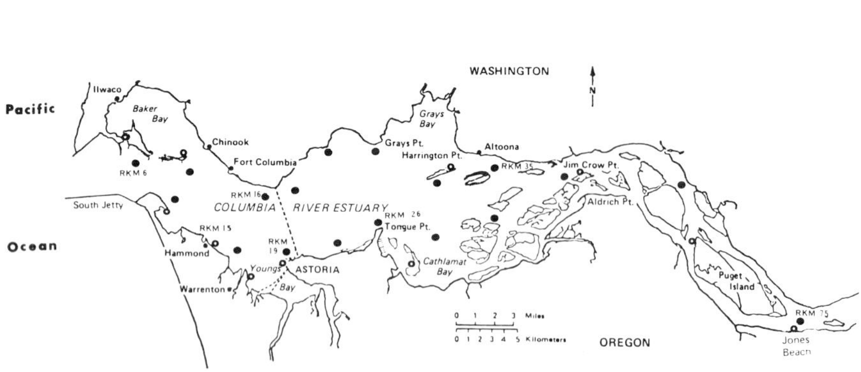

Figure 1. Sampling locations for purse seine (●) and beach seine (O) efforts in the Columbia River estuary. River kilometer (Rkm) locations are marked.

The captured salmon smolts were anesthetized with tricaine methanesulfonate, identified to species, measured (for fork length), and a random subsample removed for food habit determination. Fish stomachs were injected with 10% Formalin on site and removed shortly thereafter in the laboratory. Stomachs were preserved in alcohol and prey items were identified, counted, and weighed. The food consumed by coho smolts at various estuarine locations was grouped to determine the importance of various prey organisms. Use of major food items' proportional numbers, proportional weight, and percent frequency occurrence provided a logical approach to determining prey importance in smolts' diet.

Some juvenile coho were nose tagged at various sites to determine the success of a variety of management-related experiments. Smolts with internal coded wire tags (cwt) were recognized by a clipped adipose fin. Each tag carried identifying marks indicating the fish's origin and the time and place of its release. This made it possible to study the migration route followed by various hatchery or wild smolts.

The coded wire tag most frequently encountered is small (1 mm long × 0.25 mm diameter), has binary exterior code marks, and is made of magnetized stainless steel. Smolts must be sacrificed to obtain the tag. The nose is placed in a potassium hydroxide water bath for 24 h and then a magnetized probe is used to extract the tag. The code is read with a 45 power microscope.

Results and Discussion

Timing of Migration

The majority of coho salmon smolts arrived in the upper Columbia River estuary in early May. Beach seine catch results gathered between 1966 and 1971 indicated the migratory peak occurred between 5 May and 17 May (Fig. 2). Subsequent beach seining in 1977 in the same area confirmed this timing (Dawley et al. 1978). Reports of migratory timing of wild coho salmon in comparatively small streams such as Minter Creek, Washington (from 1938 to 1955) by Salo and Bayliff (1958), and at Waddell Creek, California (from 1933 to 1942) by Shapovalov and Taft (1954), also indicated that peak smolt catches were usually in early May, although there were instances when they occurred in late April or late May. Chamberlain (1907) in Alaska, and Peck (1970) in Lake Superior noted rapid migration of coho salmon smolts in May. Thus, the May timing peak appears to occur irrespective of the size of the stream, the geographic locality, and whether wild or hatchery coho salmon are involved.

While peak movement into the Columbia River estuary may vary by several weeks, the duration of the migration is essentially 6 weeks, from late April to early June. Recently some Columbia River hatcheries have delayed or staggered coho salmon smolt releases and this may result in artificially shifting the migratory peak and duration of the smolt migration.

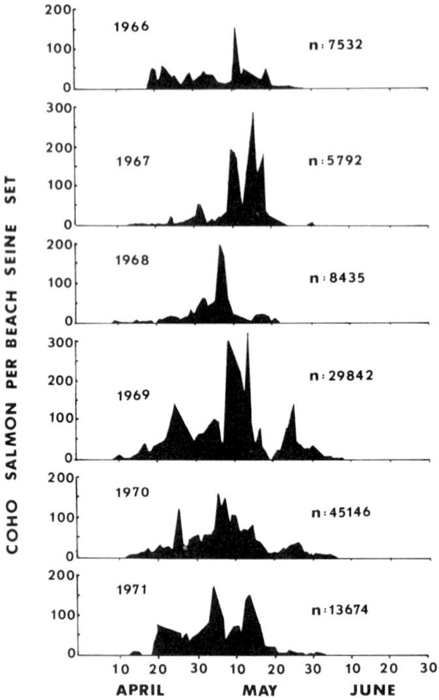

Figure 2. Average daily beach seine catches of juvenile coho salmon in the upper Columbia River estuary (1966–71). n = number of juveniles.

Coho salmon smolts are widely distributed in surface waters of the Columbia River. Substantial numbers of coho salmon were captured both in beach seines and purse seines (Dawley et al. 1981). Purse seines have a substantially greater fishing surface than beach seines and usually averaged more coho smolts.

Size and Time Relationship

Fork lengths of Columbia River coho salmon smolts averaged 12.8 cm to 13.8 cm in the beach seine sampling initiated in 1966. Based on size, coho salmon smolts were primarily hatchery reared fish. Salo and Bayliff (1958) reported that wild coho salmon smolts in Minter Creek, Washington averaged 8.8 to 10.6 cm. Wild coho salmon smolts taken at Waddell Creek, California (Shapovalov and Taft 1954) were also smaller than Columbia River smolts (average length of 11.5 cm).

The average length of Columbia River smolts was greatest in April and gradually decreased during migration (Fig. 3). The pattern was repeated each year, though in 1966, 1967, and 1969 some early migrants were also small. In May 1969, lengths of purse-seine caught coho salmon (which were several millimeters larger than those taken with beach seines), showed a similar downward trend (Fig. 3), although there were just data for 3 weeks. Both Salo and Bayliff (1958) and Shapovalov and Taft (1954) reported similar trends for decreasing lengths of wild coho salmon in Minter and Waddell Creeks. Salo and Bayliff (1958) also pointed out that the condition factor decreased with time and smaller length.

These trends suggest that larger hatchery and wild coho salmon consistently migrate earlier than smaller coho. It would appear there are physiological mechanisms that cause large fish of any hatchery or wild group to migrate first since late migrating small hatchery coho are still substantially larger than early wild coho migrants. This indicates that a particular length by itself is not a characteristic that controls timing. It is important that more be known about size and migration relationships since future changes in smolt size will tend to be obscured as different hatchery managers delay or stagger times of their coho releases.

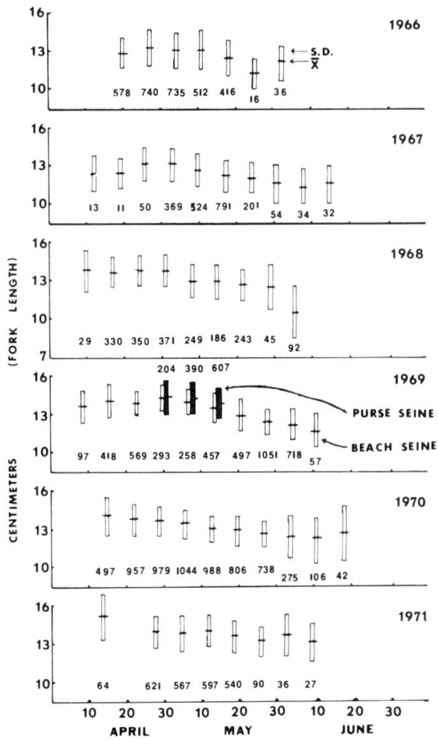

Figure 3. *Mean fork lengths ($\bar{x}$) and standard deviations (S.D.) of juvenile coho salmon captured in the upper Columbia River estuary during each weekly period of the annual smolt migration (1966–71). Numbers are sample sizes.*

Migration Routes

In the estuary, coho salmon smolts appeared to be most easily captured near the surface in open water (Durkin et al. 1981). Approximately 79% of the coho salmon captured in 1980 and 1981 were taken with purse seines, while 18% were taken with beach seines, and 3% with traps and bottom trawls. The large fishing surface of the purse seine undoubtedly influences the proportional catch.

In 1980, 14 tagged (cwt) coho salmon released from three Columbia River hatcheries, located 221 to 268 km upstream of the river mouth, were captured. These smolts were captured with purse and beach seines primarily in or near the open, deepwater parts of the estuary (Fig. 4). Except for a single coho captured just inside Baker Bay, the upriver releases appeared to favor a direct route through the estuary. In contrast, some of 41 tagged coho released from two hatcheries on estuarine tributaries (Grays River Rkm 57, Big Creek Rkm 49) appeared to have moved across tide flats and into bays after entering the estuary (Fig. 5). The Grays River smolts were released on 3 April 1980, while the Big Creek smolts were released on 6 June and 7 July 1980. Both estuarine groups showed a greater tendency to appear in shallow bays than did the upriver smolts. These results may be correct, but more cwt recoveries are needed to understand coho smolt movement in this estuary.

Juvenile coho salmon may have a greater tendency to wander or move upstream when released before mid-April. Between 1967 and 1970 at Rkm 75 (Jones Beach), 355 of a total of 1,297,810 marked fish were recovered that had been released downstream between Rkm 10 and Rkm 68. The majority of these fish were from groups released in early April, though some were released in January, February, and March. Big Creek Hatchery (Rkm 49) contributed the largest number of upstream migrants at Jones Beach, but

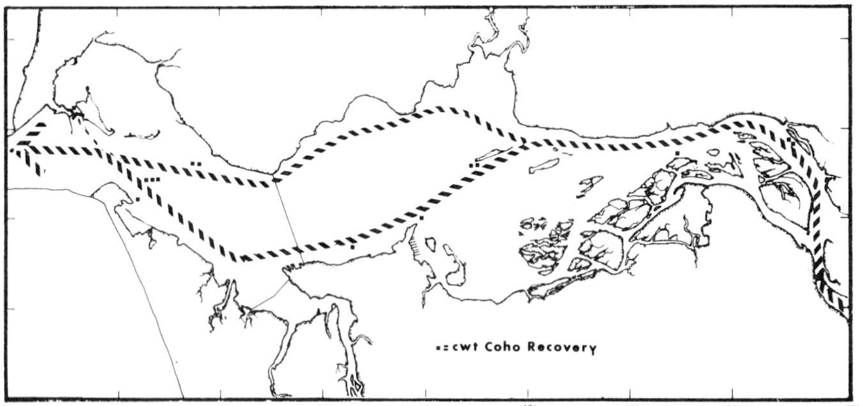

Figure 4. Routes of 14 tagged juvenile coho salmon released from Oregon and Washington hatcheries located over 221 km upstream (right) in 1980.

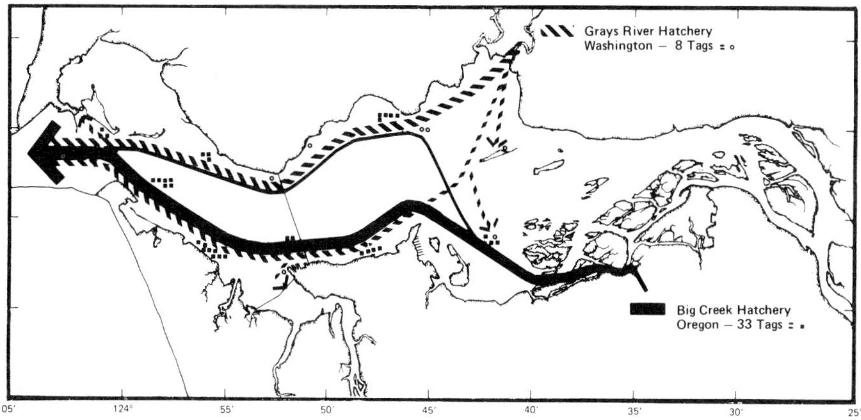

Figure 5. Movement of 41 tagged juvenile coho salmon released from Big Creek, Oregon (Rkm 49) and Grays River, Washington (Rkm 57) hatcheries in 1980.

there also were recoveries of coho salmon released farther downstream at Grays River Hatchery and Youngs Bay.

Migration Rates

In 1967 a coho salmon release experiment was used to test the effects of turbines on smolts at Ice Harbor Dam (Rkm 537). This experiment and another performed in 1968 at the same site provided the means for determining migration rates to the estuary based upon time of release. Recovery data on marked coho salmon released in groups of 64,000 at 8 one-week intervals in 1967 indicated the average number of days to reach the estuary diminished with time (Fig. 6). The average rate of migration to move 461 km to the upper estuary ranged from 40 days for the coho salmon released in mid-March to 12 days for those released in mid-May. The turbine test was conducted at Ice Harbor Dam again in 1968 using fewer smolts (507,000), but the releases were concentrated over two shorter times rather then evenly spaced. The time relationship was less significant in 1968, with releases through 10 April taking 40 days, and releases in late April averaging 23 days.

The increased timing of the 1968 migration may have been due to filling of the 122 km long John Day Dam (Rkm 347) reservoir and the subsequent reduction of water movement. These studies indicated coho smolts have a strong migratory inclination after late April, even though released over 450 km above the estuary, with movements in rivers and reservoirs at rates of 20 to 30 km/day.

Diel Movement

Beach seine sampling at Jones Beach in 1970, during the major part of the May migration, indicated increased catches of coho salmon toward mid-

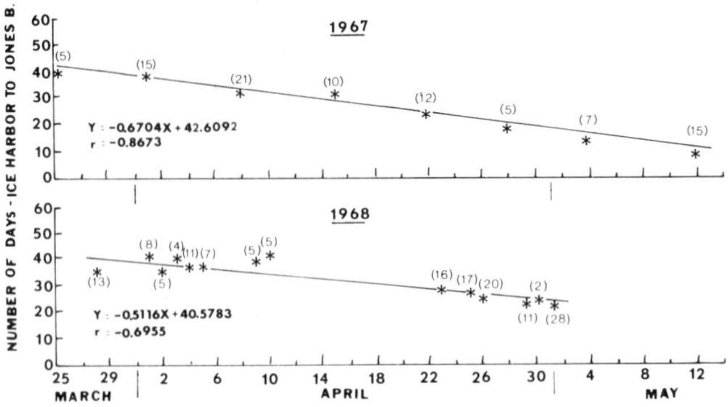

Figure 6. *Relation between migration rate (i.e., the average number of days from place of release at Ice Harbor Dam on the Snake River to time of recovery at Jones Beach) of juvenile coho salmon and the date of their release for 1967 and 1968. Number of coho recovered from each release group is shown in parentheses; average travel time is marked with an asterisk.*

day (Fig. 7) and a subsequent decline in the afternoon. Bauersfeld (1977) made a beach seine set every 2 h for a 24-h period in April and May below Bonneville Dam. His results indicated most coho salmon were caught at night in April, whereas in May the highest catches were in daylight. Dawley et al. (1979) conducted a 24-h beach seine effort at Jones Beach in 1978 and found highest catches occurred in daylight. A 24-h study in the same area was conducted using a purse seine in 1979 (Dawley et al. 1980) and coho salmon catches in open water were also greater in daylight than in darkness. Purse seining in 1980 at Rkm 16 through a 24-h period (Durkin et al. 1981) also indicated coho salmon in the estuary were more frequently captured during daylight rather than in darkness. The results of all these separate studies reveal coho salmon smolts were more frequently captured by day than by night, probably indicating that they swim more rapidly downstream by day.

Prey Selection

Some researchers have suggested that coho salmon may not feed on their migration to the ocean and that starvation, if prolonged, can affect their survival (Snyder 1980). Others are concerned about the possibility of yearling coho smolts consuming subyearling salmonid smolts (chum salmon and chinook salmon) in the river and estuary. Coho smolt stomachs were examined at a number of estuarine sites to determine food consumption. The numerical percent of food items consumed, the percent weight of the food, and the frequency of occurrence are presented in Table 1. The coho smolts captured well above the estuary (Rkm 163) were primarily insect feeders, while in the upper estuary (Rkm 55) they fed on the tube dwelling amphipod *Corophium salmonis*. This amphipod was also a major food item at Rkm 35,

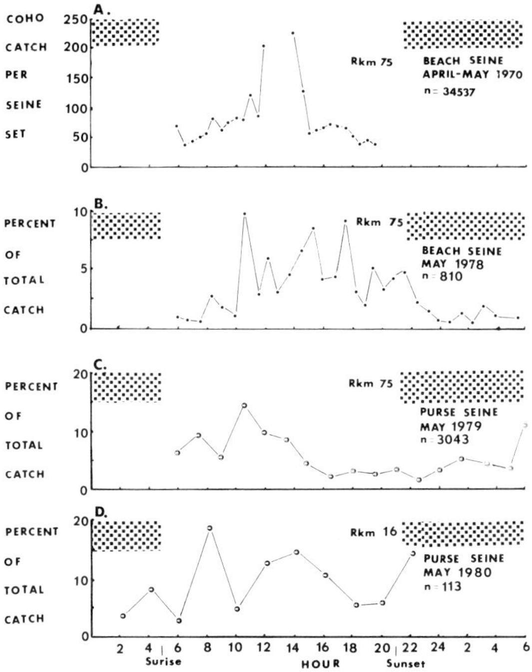

Figure 7. A. Average beach seine catches of coho salmon at Jones Beach per half hour for the period of maximum availability, 26 April to 16 May 1970. B. Percent of total diel beach seine catch at Jones Beach, 18-19 May 1978 (from Dawley et al. 1979). C. Percent of total diel purse seine catch at Jones Beach, 10 and 23 May 1979 (from Dawley et al. 1980). D. Percent of total diel coho salmon catch at the Interstate Bridge, 21-22 May 1980.

26, 22, 19, 17, 15, 11, and 6. Diptera, Homoptera, and *Cancer* megalops were occasionally important diet items. Fish (non-salmonids) were a minor food item only at Rkm 15. At all sites 75% to 95% of the coho salmon stomachs contained food. The results suggest coho salmon smolts were actively feeding, particularly in the estuary, and they rarely consumed fish.

Summary

The seaward migration of coho salmon smolts in the Columbia River and its estuary occurred during a relatively short time (6 weeks), with peak movement in early to mid-May. Rapid downstream movement began in late April and did not slow until the migration was complete. Large smolts tended to migrate earlier or more rapidly than smaller smolts, thereby complicating assessment of growth during their movement to the ocean. The route through the estuary of upriver tagged hatchery smolts appeared to be direct, but the route of lower river hatchery smolts tended toward broad, less direct move-

Table 1. Food consumption of coho salmon smolts captured with purse seines and beach seines between Portland Harbor on the Columbia River and the mouth of the estuary.

Location Upstream from Mouth	Coho Examined	Percent With Food	TYPES OF PREY CONSUMED (IN PERCENT OF TOTAL)					
			NUMBERS		WEIGHT		FREQUENCY OCCURRENCE	
			Primary	Secondary	Primary	Secondary	Primary	Secondary
Portland Harbor, Rkm 163	62	75	Hymenoptera 40%	Diptera 20%	Digested insects 36%	Lepidoptera 36%	Digested insects 53%	Hymenoptera 24%
Jim Crow Pt., Rkm 35-45	17	94	Corophium salmonis 75%	Diptera 8%	Corophium salmonis 41%	Digested material 37%	Digested material 75%	Corophium salmonis 50%
Tongue Pt., Rkm 26	10	80	Corophium salmonis 60%	Diptera 40%	Corophium salmonis 80%	Corophium spinicorne 7%	Corophium salmonis 90%	Diptera 50%
Astoria, Rkm 22	19	95	Corophium salmonis 80%	Calanoid copepods 10%	Digested material 62%	Corophium salmonis 25%	Digested material 90%	Corophium salmonis 75%
Youngs Bay, Rkm 19	10	80	Corophium salmonis 100%		Corophium salmonis 100%		Corophium salmonis 100%	
Skipanon River, Rkm 17	32	75	Corophium salmonis 55%	Corophium spinicorne 23%	Crab megalops 47%	Corophium salmonis 23%	Corophium salmonis 50%	Crab Megalops 39%
Hammond, Rkm 15	5	80	Corophium salmonis 90%	Corbicula manilensis 5%	Corophium salmonis 70%	Fish 21%	Corophium salmonis 100%	Fish 25%
Sand Island, Rkm 6	16	94	Homoptera 50%	Corophium salmonis 40%	Corophium salmonis 40%	Digested material 75%	Digested material 73%	Corophium salmonis 53%

ment. Insects were used as food in riverine areas, while the tube dwelling amphipod *Corophium salmonis* was the primary prey in the estuary. There was little evidence of fish in coho smolts' diet.

Acknowledgments

I express my appreciation to the many National Marine Fisheries Service associates and colleagues who have labored with me on many related projects during the past 14 years. Their assistance in contributing to greater knowledge of aquatic life on the Columbia River estuary is sincerely appreciated. I acknowledge in particular the determination of invertebrate food items by Robert Emmett and the determination of cwt codes by William Muir. Working with dedicated people on a magnificent body of water is a rewarding experience.

References Cited

Bauersfeld, K. 1977. Effects of peaking (stranding) of Columbia River dams on juvenile anadromous fishes below the Dalles Dam, 1974 and 1975. Wash. Dept. Fish. Tech. Rept. 31. 117 pp.

Bohn, B. R. and C. E. Stockley. 1981. Columbia River fish runs and fisheries 1957-1979. Ore. Dept. Fish. Wild., Wash. Dept. Fish. Vol 2; No. 5. 73 pp.

Bouck, G. R. and D. A. Johnson. 1979. Medication inhibits tolerance to seawater in coho salmon smolts. *Trans. Amer. Fish. Soc.* 108:63-66.

Bouck, G. R. and S. D. Smith. 1979. Mortality of experimentally descaled smolts of coho salmon *(Oncorhynchus kisutch)* in fresh and saltwater. *Trans. Amer. Fish. Soc.* 108:67-69.

Chamberlain, F. M. 1907. Some observations on salmon and trout in Alaska. Rep. U.S. Comm. Fish. 1906: 112 pp.

Chiweyite, E. and C. B. Schreck. 1980. Stress and social heirarchy rank in coho salmon. *Trans. Amer. Fish. Soc.* 109:423-426.

Clemens, W. A. and G. V. Wilby. 1961. *Fishes of the Pacific Coast of Canada.* Fish. Res. Bd. Can., Bull. 68. 443 pp.

Dawley, E. M., C. W. Sims and R. D. Ledgerwood. 1978. A study to define the migrational characteristics of chinook and coho salmon and steelhead trout in the Columbia River estuary. Annual Report 1977. Project 712. NOAA, NMFS, Seattle, Washington. 23 pp.

Dawley, E. M., C. W. Sims, R. D. Ledgerwood, D. R. Miller and F. P. Thrower. 1979. A study to define the migrational characteristics of chinook and coho salmon and steelhead trout in the Columbia River estuary. Annual Report 1978. Project 712. NOAA, NMFS, Seattle, Washington. 35 pp.

Dawley, E. M., C. W. Sims, R. D. Ledgerwood, D. R. Miller and F. P. Thrower. 1980. A study to define the migration characteristics of chinook and coho salmon and steelhead in the Columbia River estuary. Annual Report 1979. Project 10990061. NOAA, NMFS, Seattle, Washington. 34 pp.

Dawley, E. M., C. W. Sims, R. D. Ledgerwood, D. R. Miller and J. G. Williams. 1981. A study to define the migrational characteristics of chinook and coho salmon in the Columbia River estuary and associated marine waters. NOAA, NMFS, Seattle, Washington. 68 pp.

Durkin, J. T., T. H. Blahm, G. T. McCabe, T. C. Coley, R. J. McConnell, R. L. Emmett and W. D. Muir. 1981. Salmonid and Non-Salmonid Fish. NMFS Annual Report to Columbia River Data Development Program, Vancouver, Washington. 146 pp.

Ebel, W. J. 1969. Supersaturation of nitrogen in the Columbia River and its effect on salmon and steelhead trout. *Fish. Bull.* 68:1-11.

Fulton, L. A. 1970. Spawning areas and abundance of steelhead trout and coho, sockeye, and chum salmon in the Columbia River basin – past and present. Nat. Mar. Fish. Serv., Spec. Sci. Rep. 618. 37 pp.

Hart, J. L. 1973. *Pacific Fishes of Canada*. Fish. Res. Bd. Can., Bull. 180. 740 pp.

Haw, F., H. O. Wendler and G. Deschamps. 1967. Development of Washington state salmon sport fishery through 1964. Wash. Dept. Fish. Res. Bull. 7. 192 pp.

Iwama, G. K. and G. L. Greer. 1980. Effect of a bacterial infection on the toxicity of sodium pentachlorophenate to juvenile coho salmon. *Trans. Amer. Fish. Soc.* 109:290-292.

Johnsen, R. C. and C. W. Sims. 1973. Purse seining for juvenile salmon and trout in the Columbia River estuary. *Trans. Amer. Fish. Soc.* 102:341-345.

Males, A. 1980. Sensitivity of parasitized coho salmon to crude oil, toluene, and napthalene. *Trans. Amer. Fish. Soc.* 109:293-297.

Miller, R. A., L. A. Norris and B. R. Loper. 1979. The response of coho salmon and guppies to 2,3,7,8 – tetrachlorodibenzo-p-dioxin (TCDD) in water. *Trans. Amer. Fish. Soc.* 108:401-407.

Myers, K. W. 1980. An investigation of the utilization of four study areas in Yaquina Bay, Oregon by hatchery and wild juvenile salmonids. MS Thesis, Oregon State U., Corvallis, OR 234 pp.

Peck, J. W. 1970. Straying and reproduction of coho salmon , *Oncorhynchus kisutch*, planted in a Lake Superior tributary. *Trans. Amer. Fish. Soc.* 99:591-595.

Salo, E. O. and W. H. Bayliff. 1958. Artificial and natural production of silver salmon, *Oncorhynchus kisutch*, at Minter Creek, Washington. Wash. Dept. Res. Bull. 4. 76 pp.

Shapovalov, L. and A. C. Taft. 1954. The life histories of the steelhead rainbow trout *(Salmo gairdneri gairdneri)* and silver salmon *(Oncorhynchus kisutch)* with special reference to Waddell Creek, California and recommendations regarding their management. *Calif. Fish and Game Fish. Bull.* 98. 375 pp.

Sims, C. W. and R. H. Johnsen. 1974. Variable-mesh beach seine for sampling juvenile salmon in the Columbia River estuary. *Mar. Fish. Rev.* 36:23-26.

Snyder, G. R. 1980. Effects of starvation on presmolt coho salmon. Coastal Zone and Estuarine Report. NOAA, NMFS, Seattle, Washington. 55 pp.

Wydocki, R. S. and R. R. Whitney. 1979. *Inland Fishes of Washington*. U. Washington Press, Seattle, WA 220 pp.

Zaugg, W. S. and L. R. McLain. 1976. Influence of water temperature on gill sodium, potassium – stimulated ATPase activity in juvenile coho salmon *(Oncorhynchus kisutch)*. *Comp. Biochem. Physiol.* 54:419-421.

TEMPORAL USE OF AN OREGON ESTUARY BY HATCHERY AND WILD JUVENILE SALMON

Katherine W. Myers
Fisheries Research Institute
University of Washington
Seattle, Washington
and

Howard F. Horton
Department of Fisheries and Wildlife
Oregon State University
Corvallis, Oregon

Abstract: Temporal use of Yaquina Bay, Oregon, (July 1977-December 1978) by accelerated-growth (age-0) hatchery coho (*Oncorhynchus kisutch*) and wild salmon (*Oncorhynchus* spp.) juveniles was determined by periodic sampling of nearshore and channel sites in the estuary. "Residency half-life" ranged from two to nine days for different release groups of hatchery coho juveniles, and was longer for fish released earlier in the year (June) than for groups released later (September-October). Wild populations of chum (*O. keta*) and coho were present at the sample sites for 2-3 mo (March-June), and wild chinook (*O. tshawytscha*) were present during 9 mo (January, April-November). Lack of overlap in peak abundances of wild chum (early April), coho (mid May), and chinook (mid July-early August), suggests interspecific temporal partitioning. Releasing age-0 hatchery coho after peak abundances of wild chinook should be considered as a means of decreasing overlap in temporal use of the estuary.

Introduction

In 1971, the Oregon Legislature authorized licensing of private commercial hatcheries to rear, release, and recapture chum salmon (*Oncorhynchus keta*) at tidewater facilities on coastal Oregon rivers. More recently, coho (*O. kisutch*), chinook (*O. tshawytscha*), and pink (*O. gorbuscha*) salmon were authorized for release. To date, the Oregon Department of Fish and Wildlife has issued permits to 12 private commercial operators for a total annual release of 100.5 million chum, 42.0 million chinook, and 37.8 million coho salmon into Oregon estuaries, or approximately three times the current annual production of Oregon public hatcheries.

Since the issuance of these permits, many biologists, legislators, and fishermen have expressed concern about the impact of large numbers of privately cultured salmon on wild fish and on salmon released from public hatcheries. Some suggested that accelerated-growth (age-0) hatchery coho,

ISBN 0-12-404070-5

released 10-12 mo younger than the usual age of coho at ocean emigration, might remain in estuaries and compete with wild fish (Cummings 1979). Overlap in estuarine utilization is a particular concern, since wild juvenile chinook are known to rear in some Oregon estuaries, and growth may be limited when natural population levels exceed estuarine carrying capacity (Reimers 1973, 1978; Reimers and Concannon 1977).

We present information on length of estuarine residence and size of age-0 hatchery coho in Yaquina Bay, OR, in 1977-78. We examine periods of abundance and size of wild salmon to determine if they use Yaquina Bay as a rearing area, and recommend release strategies that might decrease overlap in temporal use of the estuary by hatchery and wild salmon.

Study Area

Yaquina Bay (Fig. 1) is the fifth largest estuary in Oregon, covering 15.8 km² of land at mean high tide (Oregon Division of State Lands 1973). The river channel and sloughs from the mouth of the Bay to the head of tidewater at River km 42 account for about 65% or the total area; the remaining 35% of the area is tidelands (Percy et al. 1974).

Since 1973, Yaquina Bay has been the location of a release-recapture facility for the largest private commercial salmon hatchery in Oregon, Oregon Aqua Foods (OAF). Most OAF coho salmon released into Yaquina Bay during our study were reared in heated water (11°C) at an inland hatchery at Springfield, OR. With this technique, coho salmon reach migratory size in 5-6 mo; wild coho normally rear for 16-18 mo in freshwater. When the salmon were ready for release, they were trucked to the OAF site at Yaquina Bay (Fig. 1), held and fed in saltwater ponds for 2-3 wk, and then released into the estuary at night, at flood tide.

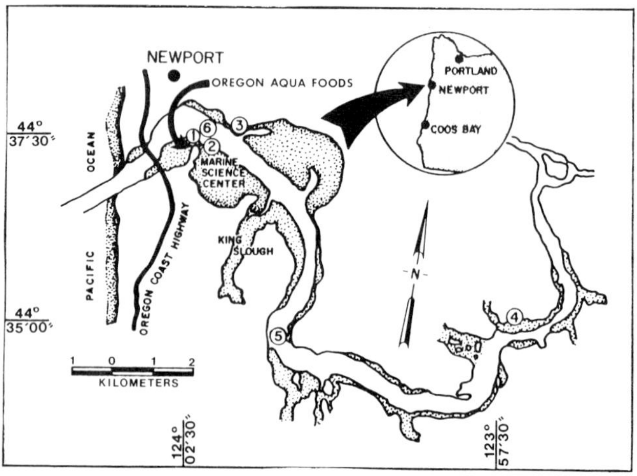

Figure 1. Map of Yaquina Bay, Oregon, showing location of sample sites (1-6).

Yaquina watershed has 66 tributary streams that are utilized by wild anadromous salmonids (Smith and Lauman 1972). Species of Pacific salmon in the watershed include chum, fall chinook, coho, and pink salmon. Smith and Lauman (1972) estimated an annual spawning escapement of approximately 12,600 coho and 2,100 fall chinook salmon. No population estimates of chum salmon are available, and pink salmon spawners are thought to be limited to occasional strays from other river systems (Herrmann 1959). Spawning ground surveys indicate a significant decline in spawning escapement of wild adult salmon in recent years (Scarnecchia and Wagner 1980).

Methods

Sampling was conducted at four nearshore sites (Sites 1-4) and two channel sites (Sites 5-6) (Fig. 1). The nearshore sites were sampled with a 100 m × 3 m varied mesh beach seine, similar to that described by Sims and Johnsen (1974). The net was set in the direction of the tidal current and both intertidal and subtidal areas to a depth of 3 m were sampled. The two channel sites were sampled with a 222 m lampara net, constructed of two 102 m lead wings with 10 cm stretched mesh and an 18 m bag with 0.95 cm stretched mesh. The net, capable of fishing to a depth of approximately 21 m, sampled the entire water column to the bottom.

The four nearshore sites were sampled from July 1977 through December 1978. Samples were taken twice a week during periods of large releases of hatchery fish in July-September of 1977 and 1978, weekly in June 1978 and October 1977 and 1978, and twice a month the rest of the time. The two channel sites were sampled twice a month from March through October 1978.

On a particular sample date, one set was made with the net at each site. Total number of salmon of each species was recorded after each set of the net. Samples or subsamples of juvenile salmon were preserved in 10% buffered Formalin for laboratory determination of fork lengths (FL) and hatchery or wild origin. In 1977 and 1978, OAF released coded wire tagged (CWT) fish marked by removal of the adipose fin. About 60% of CWT coho released in 1978 were part of a genetic experiment conducted by OAF. To maximize survival of fish for this experiment, subsamples of CWT coho were taken when catches of fish with missing adipose fins were large. In 1978 OAF also released fish marked with pressure sprayed fluorescent pigments. Subsamples of up to 100 juvenile coho from each set of the net were examined for dye marks under an ultraviolet light in the laboratory.

Hatchery or wild origin of preserved specimens was determined by tags, fin clips, dye marks, scale characteristics, external parasites, visceral fat, size, fin erosion, and time of capture (Myers 1980). When subsamples were taken, the number of hatchery and wild salmon in the catch was estimated from proportions in the subsamples.

Length of estuarine residence of hatchery coho was determined by capture of CWT and fluorescent dye marked (FDM) fish with known release dates, and by capture of hatchery coho at sample sites after the last hatchery releases in 1977 and 1978. On the basis of the form of the 1977 catch data, length of residence of hatchery coho at Sites 1-4 was modeled by equations of the form: $N = N_o e^{-kt}$ or $\ln N = -kt + \ln N_o$, where N was the number of coho present at the sites at time t, N_o was the number present at $t = 0$, t was the number of days elapsed since the date of release, and k was a rate constant. "Residency half-life" ($t_{1/2}$), or the length of a time interval in which the number of coho at the sites was reduced by 50%, was calculated from the equation: $N/N_o = 1/2 = e^{-kt1/2}$ or $t_{1/2} = \ln(1/2)/-k = 0.693/k$. The rate constant k was estimated by a logarithmic regression of the catch (lnN) on days elapsed since release (t) for individual groups of tagged coho, or for a combination of groups after the last releases in 1977 and 1978. Value of N in the model was the mean daily catch per sample week (MDC/SW). This was the total weekly catch at Sites 1-4 divided by the number of days sampled during that week. If zero catches occurred at all sites for three or more weeks in a row, the first zero catch was assumed to be a small value (0.1), and points after this were not included in the calculation of the regression equations.

Periods of abundance of wild salmon were represented by the catch per unit effort (CPUE) of each species. This was the pooled catch of each species during 2 wk intervals divided by the number of sets of the net made during the same interval. In this paper, "lower estuary" refers to pooled samples taken at Sites 1-3, "channel" refers to pooled samples taken at Sites 5-6, and "upper estuary" refers to Site 4 samples.

Results

Hatchery coho salmon

Approximately 1.3 million age-0 hatchery coho were released on seven dates from May 26 through July 18, 1977. Mean weight at release ranged from 15.5 to 21.3 g. The MDC/SW of hatchery coho at the nearshore sites decreased exponentially after the last release of hatchery fish on July 18. The logarithmic regression (ln) of MDC/SW on time was $\ln N = -0.084t + 5.82$ ($r^2 = 0.89$; SE = 0.01; $H_o{:}\beta = 0$, $p<0.01$; $n = 11$). Residency half-life ($t_{1/2}$) was 8.25 days. Therefore, in three half-lives, or about 25 days, the number of juvenile hatchery coho originally present at the nearshore sites on July 18 was reduced by approximately 90%.

Mean fork lengths of hatchery coho captured at the four nearshore sites in 1977 are shown in Fig. 2. Mean fork length increased from 11.5 cm (mid-July) to 21.5 cm (October). Most of these fish were captured in the lower estuary. The last hatchery coho to occur in upper estuary samples were captured only 7 days after the last release ($\overline{X} = 11.7$ cm FL; $n = 3$). The last juvenile hatchery coho (26.8 cm FL) caught in the lower estuary in 1977 was captured 5 mo after the last release. Five immature hatchery coho ($\overline{X} = 36.0$

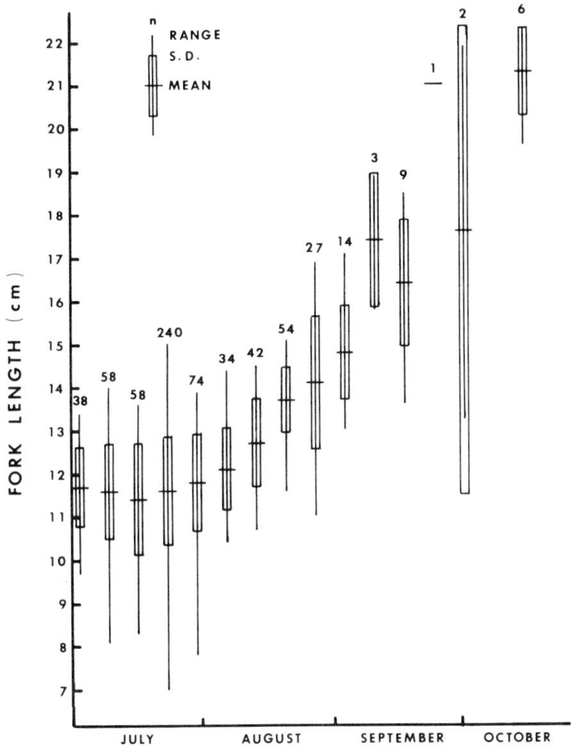

*Figure 2. Weekly mean fork lengths of juvenile hatchery coho salmon (*Oncorhynchus kisutch*) captured by beach seine at four nearshore sites in Yaquina Bay, OR, 1977. S. D. = standard deviation; n = number in sample.*

cm FL) were captured at nearshore sites in the lower estuary in January, March, and April, 1978, prior to the first 1978 release. These fish had, apparently, over-wintered in the estuary or in the nearby coastal waters.

Approximately 8.9 million juvenile hatchery coho were released into Yaquina Bay on 25 dates from April 6 through October 20, 1978. These were all age-0 fish (mean weight ≅ 10-27 g) except for two groups of yearling coho (mean weight ≅ 65 g) released on April 6.

Release and recovery data on CWT and FDM hatchery coho captured at the four nearshore sites in 1978 are shown in Tables 1 and 2, respectively. The CWT coho were recovered up to 3 mo after release. Fish from some CWT releases in June, July, and August were captured in mid-September. With the exception of October releases, no CWT or FDM coho were recovered after mid-September. The number of recoveries of CWT coho was usually low for releases under 10,000 fish (Table 1).

All coho from the FDM releases were recovered at the nearshore sites within three weeks after release (Table 2). Recovery periods for FDM releases were shorter than those of CWT releases of similar number and release date,

Table 1. Release and recovery information for juvenile hatchery coho salmon *(Oncorhynchus kisutch)* marked with coded wire tags and recovered at four nearshore sites in Yaquina Bay, Oregon, during 1978. Estimated number recovered was based on the proportion of tagged fish in subsamples. $\bar{X}$ = mean; s = standard deviation.

Release Date	*No. Released*	*Est. No Recovered (Actual No.)*		*Recovery Period*	*No. Recovery Dates*	*Recovery Sites*	*Recovery Fork Length (cm)* $\bar{X}$	*s*
4/06	10,000	0	(0)	–	–	–	–	–
6/13	9,355	4	(4)	6/15	1	1, 2	11.8	0.62
6/19	203,788	743	(109)	6/22-9/18	19	1, 2, 3, 4	10.9	1.51
6/24	28,918	7	(3)	7/19-9/07	3	1, 2	10.6	1.13
6/30	14,553	14	(2)	7/06, 7/17	2	1, 2	9.8	0.85
7/06	13,595	6	(3)	7/13-8/21	3	2, 3	9.9	2.15
7/07	63,627	288	(37)	7/10-9/18	11	1, 2, 3	10.3	0.97
7/11	14,240	23	(3)	7/13-8/21	3	1, 2, 3	11.1	0.40
7/27	9,372	0	(0)	–	–	–	–	–
8/01	8,231	0	(0)	–	–	–	–	–
8/07	25,310	3	(3)	8/10, 8/21	2	2, 3	10.6	0.26
8/16	35,932	97	(20)	8/17-9/14	6	1, 2, 3	12.3	1.36
9/07	8,419	65	(22)	9/07-9/14	3	1, 2, 3	11.4	1.20

Table 2. Release and recovery information for juvenile hatchery coho salmon *(Oncorhynchus kisutch)* marked with fluorescent dyes and recovered at four nearshore sites in Yaquina Bay, Oregon, during 1978. Estimated number recovered was based on the proportion of dye marked fish in subsamples. $\overline{X}$ = mean; s = standard deviation.

Release Date	*No. Released*	*Dye*	*Est. No. Recovered (Actual No.)*		*Recovery Period*	*No. Recovery Dates*	*Recovery Sites*	*Recovery Fork Length (cm)*	
								$\overline{X}$	*s*
4/06	36,367	Red	17	(15)	4/15, 4/29	2	1, 2, 3, 4	17.2	4.44
4/06	14,480	Yellow	1	(1)	5/13	1	1	14.9	–
6/13	44,399	Green	105	(32)	6/15-7/06	4	1, 2, 3	11.2	0.68
6/29	56,150	Yellow	0	(0)	–	–	–	–	–
7/11	50,988	Red	107	(26)	7/13-8/03	6	1, 2, 3	10.5	1.19
7/27	57,654	Orange/Green	36	(25)	7/31-8/10	4	1, 2, 3	10.4	1.40
8/10	51,089	Red/Green	33	(12)	8/14-8/28	3	1, 2, 3	12.2	1.88
8/22	50,421	Red	146	(52)	8/24-9/14	5	1, 2, 3, 4	11.9	1.67
10/07	55,733	Yellow	1	(1)	10/07	1	1	10.2	–

indicating that our subsamples of up to 100 coho from each set of the net were adequate only for detecting FDM coho when they were most abundant in the estuary. The number of recoveries of yellow FDM coho was much lower than for other dye groups, and was probably due to poor dye retention or our inability to detect this color under the ultraviolet light.

Most CWT and FDM coho were recovered at nearshore sites in the lower estuary (Tables 1 and 2). Only nine marked fish were recovered in the upper estuary. Most of these fish were from spring and fall releases, and were recovered within 1 mo after release. Only 18 CWT and 4 FDM coho were recovered at the channel sites.

Statistics of the residency model calculated for the three largest CWT release groups in 1978 are shown in Table 3. The regressions were statistically significant (H_o: $\beta = 0$, $p<0.01$). Similar to hatchery coho released in 1977, the majority of the coho from the June 19 release group left the sample sites in less than 1 mo after release; a small proportion remained for an extended period of time. The residency half-life of the June 19 release group was over twice as long as the half-life calculated for the July 7 and August 16 release groups. Although these last two groups were released 18 and 58 days after the June 19 release group, respectively, tagged fish from all three release groups were last captured at the study areas in mid-September (Table 1).

Approximately 730,000 juvenile coho were released on September 3 and September 7, 1978. The MDC/SW at the four nearshore sites after the last date decreased exponentially ($\ln N = -0.297t + 7.91$; $r^2 = 0.89$; SE = 0.06; H_o:$\beta = 0$, $p<0.05$; n = 5), and the residency half-life was only 2.3 days.

Over 1.6 million hatchery coho were released into the estuary on October 7 and October 20, 1978, but the largest catch was 62 coho captured at Site 1 on October 7. Although 23 sets of the beach seine were made on six dates after October 7, only 25 juvenile hatchery coho were recovered, and only two hatchery coho were recovered at the channel sites in late October.

Size of CWT coho from the June 19 release group (Table 1) recovered at the nearshore sites in 1978 is shown in Fig. 3. The mean size of coho from

Table 3. Statistics of a model used to describe length of residence of juvenile hatchery coho salmon *(Oncorhynchus kisutch)* in Yaquina Bay, Oregon, calculated for three groups of coded wire tagged fish captured at four nearshore sites in Yaquina Bay during 1978. $t_{1/2}$ = residency half-life = 0.693/k (in days).

Release	*Regression Coefficients*					
Date	*k*	*$\ln N_0$*	*r^2*	*SE*	*n*	*$t_{1/2}$*
6/19	0.077	5.60	0.92	0.0067	14	9.0
7/07	0.186	5.88	0.98	0.0145	6	3.7
8/16	0.164	4.23	0.93	0.0224	6	4.2

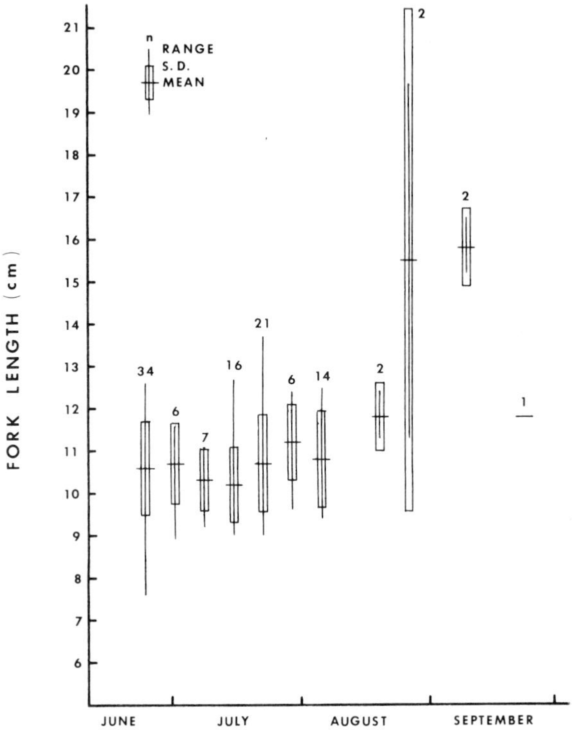

*Figure 3. Weekly mean fork lengths at capture of marked juvenile hatchery coho salmon (*Oncorhynchus kisutch*) released into Yaquina Bay, OR, on June 19, 1978, and captured by beach seine at four nearshore sites in Yaquina Bay during 1978. S. D. = standard deviation; n = number in sample.*

this release group was less than 12 cm FL until the end of August. With the exception of yearling coho released on April 6 and two August release groups, mean size of all CWT and FDM release groups recovered at the nearshore sites in 1978 was less than 12 cm FL (Tables 1 and 2).

Wild salmon

In 1977, juvenile wild chinook salmon were present at nearshore sites in the upper and lower estuary from the start of sampling in July through November. Mean bimonthly CPUE's of wild chinook in the upper ($\overline{X}$ = 0.9, s = 0.6, n = 13) and lower ($\overline{X}$ = 3.2, s = 3.8, n = 13) estuary were low. Maximum CPUE of wild chinook occurred in the upper estuary during late October ($\overline{X}$ = 2.0, s = 1.4, n = 2) and in the lower estuary in mid- to late July ($\overline{X}$ = 14.3, s = 24.2, n = 16). However, these maxima may not represent peak abundances of wild chinook at the nearshore sites in 1977 as sampling did not begin until July. No chinook were captured in samples taken in December 1977. No other species of juvenile wild salmon were captured at the nearshore sites from July through December 1977.

Relative abundance of wild juvenile chum, coho, and chinook salmon at nearshore and channel sample sites during 1978 is shown in Figs. 4 and 5, respectively. The first wild salmon captured at the nearshore sites in 1978 were three yearling chinook (12.2-15.0 cm FL) captured at Site 2 on January 21. The next species to occur was chum salmon juveniles which were present at the nearshore sites during March and April. Catch of chum peaked in the upper estuary during the third week in March and in the lower estuary during the first week in April. No chum were present in samples taken in the upper estuary after April 1 or in the lower estuary after April 29. No chum were captured at the channel sites.

Wild yearling coho salmon did not occur until after the peak of the chum migration and were present at nearshore and channel sites for a period of about 3 mo during April-June (Figs. 4 and 5). Catch of wild coho at the nearshore sites peaked in the upper estuary during the last week in April and in the lower estuary during the second week in May. Catch of wild coho at the channel sites peaked about two weeks later during the last week in May, although this peak represents a CPUE of only two fish (Fig. 5). No wild coho were present in samples taken in the upper estuary after May 27 or in the lower estuary after June 15.

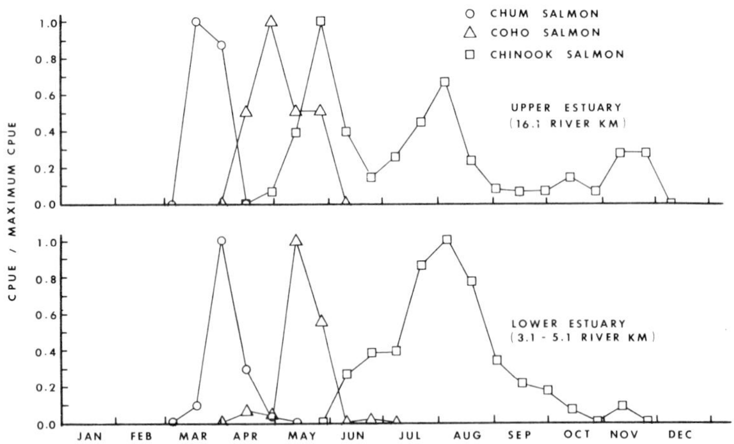

*Figure 4. Relative abundance of wild juvenile chum (*Oncorhynchus keta*), coho (*O. kisutch*), and chinook (*O. tshawytscha*) salmon captured at four nearshore sites in Yaquina Bay, OR, 1978. CPUE = bimonthly catch of each species divided by number of sets of the beach seine. Maximum CPUE, upper estuary: chum = 30.0; coho = 4.0; chinook = 18.0. Maximum CPUE, lower estuary: chum = 18.7; coho = 17.5; chinook = 72.3.*

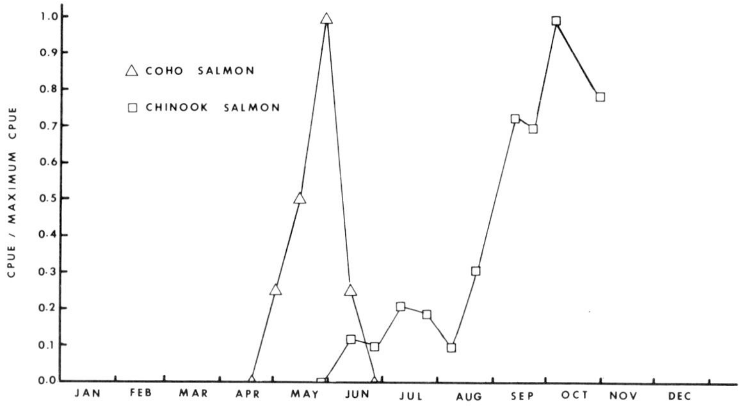

*Figure 5. Relative abundance of wild juvenile coho (*Oncorhynchus kisutch*) and chinook (*O. tshawytscha*) salmon captured at two channel sites in Yaquina Bay, OR, from March through October 1978. CPUE = pooled catch of each species at the two sites divided by the number of sets of the lampara net. Maximum CPUE: coho = 2.0; chinook = 33.5.*

Wild underyearling chinook salmon were not caught at the nearshore sites until the peak of the coho migration and were present for a period of about 7 mo from late April through November. Wild chinook were first captured in the upper estuary during the last week in April but did not occur in samples taken in the lower estuary until the second week in June (Fig. 4). Timing of apparent migrations of chinook through the upper estuary in 1978 was trimodal (late May, early August, November) (Fig. 4). The CPUE of chinook at the nearshore sites in the lower estuary peaked during the first week in August coincident with the lowest CPUE of chinook at the channel sites (Figs. 4 and 5). As the catch of chinook decreased at the nearshore sites after the first week in August (Fig. 4), the catch increased at the channel sites to a peak during the first week in October (Fig. 5). No wild juvenile chinook were captured at the nearshore sites after November 25, 1978. Because sampling at the channel sites was terminated at the end of October 1978, we do not know the length of time that wild chinook remained there.

Mean fork lengths of wild chum and coho salmon sampled at the nearshore sites in 1978 are shown in Fig. 6. Although sample sizes were small, mean lengths of chum increased and mean lengths of coho appeared to decrease over the sample period. During the peak period of wild coho migration in May the mean size was 12.6 cm FL ($s = 1.2$; $n = 29$). Very few wild coho were captured in samples taken at the channel sites; however, their mean size ($\overline{X} = 12.5$ cm FL; $s = 2.1$; $n = 8$) was similar to coho at the nearshore sites.

Figure 7 shows monthly length frequencies of wild juvenile chinook salmon captured at nearshore and channel sites during 1978. In general, size

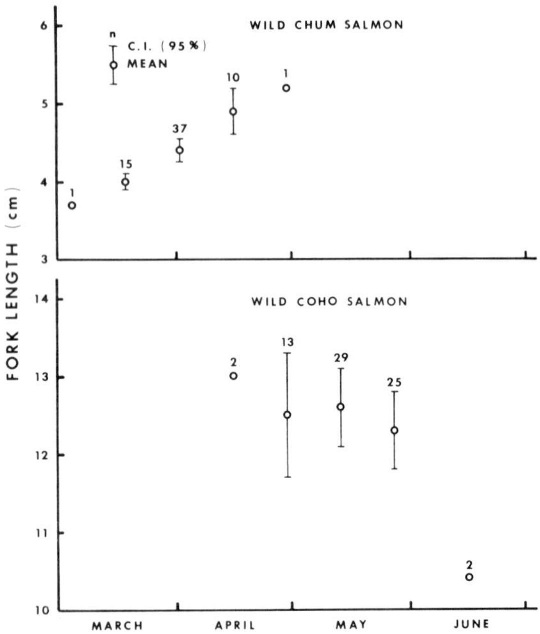

Figure 6. *Mean fork lengths of wild juvenile chum (*Oncorhynchus keta*) and coho (*O. kisutch*) salmon captured by beach seine at four nearshore sites in Yaquina Bay, OR, 1978.*

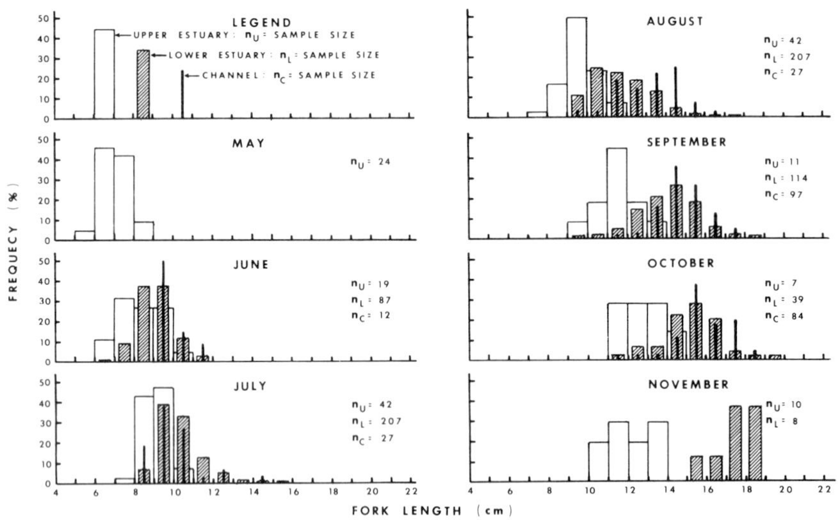

Figure 7. *Monthly length frequencies of wild juvenile chinook salmon (*Oncorhynchus tshawytscha*) captured at four nearshore (upper estuary and lower estuary) and two channel sites in Yaquina Bay, OR, 1978.*

of juvenile chinook increased from the upper estuary through the lower estuary to the channel; however, a large range of sizes was usually present at all the sites (Fig. 7).

Discussion

Age-0 hatchery coho appeared to leave Yaquina Bay, either by death or emigration, at an exponential rate. Residency half-life ranged from approximately two to nine days for different release groups of hatchery coho in 1977 and 1978. Therefore, most individuals within a particular release group had left the study area within 1 mo after release and a smaller portion remained for an extended period of time (>1 mo).

Juvenile hatchery coho that remained in Yaquina Bay after release in 1977 appeared to grow during their extended period of residence (Fig. 2). However, mean fork lengths of a group of CWT hatchery coho released into Yaquina Bay during 1978 remained, for the most part, below 12.0 cm FL (Fig. 3). The lack of increase in mean lengths of coho salmon from this release group indicates emigration to the ocean of fish larger than 12.0 cm FL, or possibly, a suppression of growth due to high population densities of juvenile hatchery coho salmon in the estuary in 1978. Reimers (1973) hypothesized that high population densities were responsible for depressed growth rates of wild juvenile chinook salmon in Sixes River estuary in Oregon in summer.

Time of release appears to be an important factor in determining length of residence of age-0 coho in Yaquina Bay. Regression models of length of residence calculated for different releases of hatchery coho in 1978 (Table 3) and low catches of coho at the nearshore and channel sites in October 1978 indicate that age-0 coho released earlier in the year (June) remained in the estuary for longer periods of time than fish released later (September-October).

The over-wintering of a few immature hatchery coho in Yaquina Bay in 1978 shows that some coho remain in the estuary or in the nearby coastal waters during the entire saltwater phase of their life history. Populations of coho salmon are known to rear for extended periods or the duration of their marine lives in Puget Sound (Allen 1956) and in the Strait of Georgia (Healey 1980a). Fish resident in nearshore ocean waters may, on occasion, enter estuaries to feed. However, we do not know if hatchery coho salmon resident in nearshore ocean or estuarine waters constitute a number sufficient to warrant concern about competition with wild juvenile salmon in estuaries.

Our results suggest that Yaquina Bay is important to wild juvenile salmon as both a staging area for entrance to the ocean and as a rearing area. During 1978, wild populations of chum and coho were present at the sample sites for 2-3 mo (March-June) and wild chinook were present during 9 mo (January, April-November). The short interval of time between peak abundances of wild coho in the upper and lower estuary (Fig. 4) and the lack of increase in mean lengths (Fig. 6) indicate that wild coho use the estuary primarily as a staging area for entrance to the ocean. As with coho, peaks in abundance of wild chum in the upper and lower estuary were offset by only one sample date (Fig. 4), indicating rapid movement of chum populations through the estuary. However, unless smaller individuals emigrate to the estuary before larger individuals, increase in mean fork lengths (Fig. 6) suggests that wild chum use Yaquina Bay as a rearing area prior to ocean entrance. Sibert et al. (1977) demonstrated that individual chum use the Nanaimo estuary in British Columbia as a rearing area.

Extensive temporal use of estuaries by juvenile chinook has been known since the early 1900's (Rich 1922), and recent investigators have shown that river estuaries are important rearing areas for wild chinook salmon (Reimers 1973; Healey 1980b). Both upper and lower Yaquina Bay appear to be important rearing areas for wild chinook. The 1 mo delay in capture of wild chinook between the upper and lower estuary (Fig. 4) and the apparent increase in mean fork lengths during this interval (Fig. 7) indicate that populations of wild chinook entering the estuary in the spring utilize the upper estuary as a rearing area before entrance into the lower estuary. The importance of the lower estuary as a rearing area in summer and fall is indicated by the gradual increase in CPUE of chinook at nearshore sites in the lower estuary from June to August (Fig. 4) and at the channel sites from August through October (Fig. 5). Increase in mean length frequencies at the nearshore sites from June to November and the channel sites from June to October (Fig. 7) indicate growth of individuals within populations of wild chinook in the lower estuary. The 2 mo offset between peak abundances of chinook at nearshore sites in the lower estuary and channel sites, as well as the generally larger mean length frequencies of chinook at the channel areas, suggests that smaller juveniles entering the lower estuary rear in nearshore areas, and move offshore into channel areas as size and abundance increases. Similar spatial-temporal movements were reported for chinook in Sixes estuary (Reimers 1973).

The most striking feature of the temporal use of Yaquina Bay by juvenile wild salmon was the separation of peak abundances of chum, coho, and chinook in time and space so that little overlap in utilization of estuarine sample sites by different species of wild salmon occurred (Figs. 4 and 5). Although we do not know the biological mechanisms involved in this separation, lack of overlap in periods of abundance of different species of juvenile salmon in Yaquina Bay is indicative of resource partitioning (Schoener 1974), and suggests the need for managers to minimize interspecific interactions between hatchery and wild juvenile salmon in the estuary.

Because wild chinook juveniles make extensive temporal use of Yaquina Bay as a rearing area, we think that a greater potential for competition exists between hatchery coho and wild chinook salmon than for other species of wild salmon in the Yaquina system. Therefore, we recommend releasing age-0 hatchery coho after peak abundances of wild chinook salmon in Yaquina Bay. We emphasize that this recommendation applies only to the type of hatchery releases that we studied, i.e. age-0 coho released after the normal peak migration period of wild coho in May. If survival of late releases of age-0 hatchery coho is low, this strategy may not be biologically or economically feasible for private salmon hatcheries.

Acknowledgments

We gratefully acknowledge the valuable field assistance of Alvaro Tresierra Aguilar, Nahanni Boling, Bruce Britton, Delores Eisele, Captain Al Fox and the crew of the *Little Dipper*, Darrel Hand, Gary Hewitt, Dana Horton, Maja Laird, Mac McConnell, John Myers, Chris Strickland, and numerous Oregon State University student volunteers. We also express our gratitude to Dr. Brian J. Allee of Weyerhaeuser Company, Vern Jackson, Norm Moe, and Richard Severson of Oregon Aqua Foods, and Ed Cummings of the Oregon Department of Fish and Wildlife for their cooperation, advice, and technical assistance during this investigation. This research was conducted at Oregon State University with funding provided by Weyerhaeuser Company and the Oregon Agricultural Experiment Station.

References Cited

Allen, G. H. 1956. Migration, distribution and movement of Puget Sound silver salmon. Ph. D. Thesis, U. Washington, Seattle, WA 295 pp.

Cummings, T. E. 1979. Private salmon hatcheries in Oregon. Oreg. Dept. Fish Wildl., Fish Div., Portland, OR 10 pp.

Healey, M. C. 1980a. The ecology of juvenile salmon in Georgia Strait, British Columbia, pp. 203-229. *In:* W. J. McNeil and D. C. Himsworth (eds.), *Salmonid Ecosystems of the North Pacific.* Oregon State U. Press, Corvallis, OR.

Healey, M. C. 1980b. Utilization of the Nanaimo River estuary by juvenile chinook salmon *Oncorhynchus tshawytscha. Fish. Bull.* 77:653-668.

Herrmann, R. B. 1959. Occurrence of juvenile pink salmon in a coastal stream south of the Columbia River. Fish Comm. Oreg., Res. Briefs 7:81.

Myers, K. W. 1980. An investigation of the utilization of four study areas in Yaquina Bay, Oregon, by hatchery and wild juvenile salmonids. M. S. Thesis, Oregon State U., Corvallis, OR 234 pp.

Oregon Division of State Lands. 1973. Oregon estuaries. Oreg. Div. State Lands, Salem, OR 50 pp.

Percy, K. L., C. Sutterlin, D. A. Bella and P. C. Klingeman. 1974. Descriptions and information sources for Oregon estuaries. Sea Grant Coll. Prog., Oregon State U., Corvallis, OR 294 pp.

Reimers, P. E. 1973. The length of residence of juvenile fall chinook salmon in Sixes River, Oregon. Fish Comm. Oreg., Res. Rep. 4. 43 pp.

Reimers, P. E. 1978. The need for research on the estuarine ecology of juvenile fall chinook salmon. Oreg. Dept. Fish Wild., Res. Sec., Info. Rept. Ser., Fish. 78-4, Corvallis, OR 10 pp.

Reimers, P. E. and G. I. Concannon. 1977. Extended residence of hatchery-released juvenile fall chinook salmon in Elk River, Oregon. Oreg. Dept. Fish Wildl., Res. Sec., Info. Rep. Ser., Fish. 77-2, Corvallis, OR 17 pp.

Rich, W. H. 1922. Early history and seaward migration of chinook salmon in the Columbia and Sacramento rivers. U. S. Bur. Fish., Bull. 37 (Doc. 887):1-73.

Scarnecchia, D. L. and H. H. Wagner. 1980. Contribution of wild and hatchery-reared coho salmon, *Oncorhynchus kisutch,* to the Oregon ocean sport fishery. *Fish. Bull.* 77:617-623.

Schoener, T. W. 1974. Resource partitioning in ecological communities. *Science* 185:27-39.

Sibert, J., T. J. Brown, M. C. Healey, B. A. Kask and R. J. Naiman. 1977. Detritus-based food webs: exploitation by juvenile chum salmon (*Oncorhynchus keta*). *Science* 196:649-650.

Sims, C. W. and R. H. Johnsen. 1974. Variable-mesh beach seine for sampling juvenile salmon in Columbia River estuary. *Mar. Fish. Rev.* 36:23-26.

Smith, A. K. and J. E. Lauman. 1972. Fish and wildlife resources of the middle coast basin, Oregon, and their water requirements (revised). Oreg. State Game Comm., Fed. Aid Fish Restor. Comp.,Rep., Proj. F-69-R-8, Job 15. 98 pp.

LIFE HISTORY OF FALL-RUN JUVENILE CHINOOK SALMON, *ONCORHYNCHUS TSHAWYTSCHA*, IN THE SACRAMENTO-SAN JOAQUIN ESTUARY, CALIFORNIA

Martin A. Kjelson, Paul F. Raquel
U. S. Fish and Wildlife Service
4001 North Wilson Way
Stockton, California

and

Frank W. Fisher
California Department of Fish and Game
4001 North Wilson Way
Stockton, California

Abstract: The Sacramento-San Joaquin Estuary is one of the largest on the Pacific Coast and supports a variety of anadromous species. Fall-run chinook salmon, the dominant race in the system, utilize the estuary for rearing and migration. Fry (<70 mm) rear in the estuary for about 2 months, primarily in the upper freshwater Delta. Brackish water bays are used primarily as a migration corridor by smolts (> 70 mm). Chinook in more northern estuaries appear to make greater use of brackish water for rearing. Peak fry rearing (February to March) and smolt migration (April to June) occurs two to three months earlier in the Sacramento-San Joaquin than in most northern estuaries. This reflects earlier spawning and high summer temperatures that force juveniles from the lower river and Delta. Fry abundance and distribution in the estuary are influenced by the magnitude and timing of river flows. Growth rates during estuarine rearing range from 0.4 to 1.2 mm day^{-1} and are similar to other Pacific Coast estuaries. Chinook diet also is similar to other west coast estuaries and is dominated by dipterans, cladocerans, copepods, and amphipods. Survival during smolt outmigration is greater in the lower bays than in the Delta. Survival through the Delta in June is inversely related to water temperature and directly related to river flow as suggested for some northern systems. Alteration of the timing, magnitude, and distribution of flow in the Sacramento-San Joaquin Estuary has a major impact on juvenile chinook survival. Hatcheries produce about 26 million fall-run smolts annually with most released during May and June in upstream and estuarine waters.

Introduction

The Sacramento-San Joaquin River system has one of the largest estuaries along the Pacific Coast (Fig. 1). The major anadromous species it supports include striped bass, *Morone saxatilis*; American shad, *Alosa sapidissima*; white sturgeon, *Acipenser transmontanus*; steelhead trout, *Salmo gairdneri*; and chinook salmon, *Oncorhynchus tshawytscha* (Kelley 1966).

ISBN 0-12-404070-5

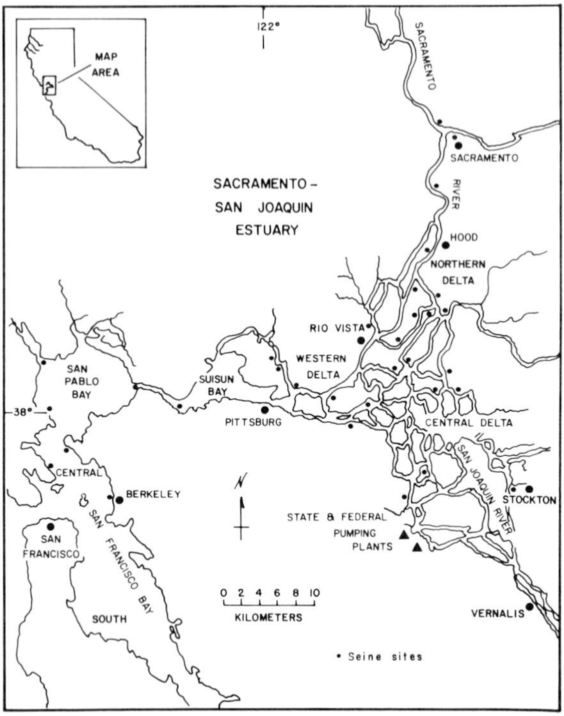

Figure 1. The Sacramento-San Joaquin Estuary, California.

Chinook salmon in this system are at the southern limit of their range. Only the more northern Columbia River system produces more chinook than California's Central Valley streams. Over 90% of the Valley's chinook salmon spawn in the Sacramento River system.

Four major runs (fall, late fall, winter, and spring), identified by the season in which upstream migration occurs, spawn in the Sacramento drainage. The fall-run is the largest numerically (the 1968 to 1978 average number of spawners was 199,000), comprising about 80% of the Sacramento drainage stock. The San Joaquin system supports a fall-run that has averaged less than 4,000 spawners in recent years (1973 to 1978). Fall-run chinook have received the most study.

Chinook populations in the Sacramento-San Joaquin drainages have declined by about 70% since 1953 (Hoopaugh and Knutson 1979). This decline is attributed primarily to the development of water resources through the construction of dams and water diversion projects during the last 40 years. Such development has caused extensive loss and alteration of spawning, rearing, and migration habitat in upstream and estuarine environments. The influence of harvest rates on the decline of chinook in California is unknown.

Three salmon hatcheries located on the American and Feather rivers (State) and on Battle Creek (Federal) were built to mitigate losses of habitat due to water development. All three emphasize production of fall-run chinook. The two State hatcheries release their total fall-run production of about 12 million salmon/year directly into the estuary, while the Federal production (about 14 million annually) is released in the upper Sacramento River and Battle Creek. Hatchery releases are made from mid-May through June. The relative contribution of hatchery produced young to adult stocks is presently unknown.

The objective of this paper is to describe the rearing of fry in the estuary and the migration of smolts through the estuary to the ocean and to compare our observations, where possible, with observations on fall-run chinook in other estuaries from northern California to British Columbia. Most of the data presented in this paper were collected as part of a cooperative study of the Sacramento-San Joaquin Estuary conducted by the California Departments of Fish and Game and Water Resources, the U. S. Fish and Wildlife Service, and the U. S. Bureau of Reclamation. Our chinook salmon studies were designed to assess the impacts of water diversions from the estuary upon chinook salmon survival and to provide information to help correct negative impacts.

Description of Study Area

The Sacramento-San Joaquin Estuary consists of the tidally influenced, freshwater Delta where the Sacramento and San Joaquin rivers join, and a series of downstream embayments, Suisun, San Pablo, and San Francisco Bays (Fig. 1).

The Delta covers about 300,000 ha of land and water. It consists of 1130 km of navigable, freshwater channels and 30 large leveed islands used for farming. Some channels are edged by narrow stretches of freshwater marsh, but most have steep banks of mud or riprap. Delta levees are generally covered by riparian vegetation. Channel widths vary from 50 m to 1.5 km and generally are less than 15 m deep. Annual Delta water temperatures range from 6 to 28°C. Turbidity is high, particularly in the spring, with secchi readings usually less than 0.5 m. Tidal action reaches to the upstream limits of the Delta but the upper edge of the salinity gradient only extends to the western Delta-Suisun Bay area.

The bays downstream from the Delta cover 124,000 ha. Their average depth is 6.1 m although most are less than 2 m deep at low tide. Much of the shore-line is industrialized, but considerable salt and brackish marshland still remains in Suisun, northern San Pablo, and southern San Francisco Bays. Annual surface water temperatures range from 9 to 21°C and salinities range from 28 to 34‰ at the entrance to the ocean and from 0 to 10‰ in Suisun Bay. Kelley (1966), Turner and Kelley (1966), and Conomos (1979) provide more detailed descriptions of the physical and biological characteristics of the estuary.

Water resource project operations in California have altered the distribution, seasonality, and magnitude of flows in the estuary. The historical annual flow passing through the estuary from its 163,000 km^2 drainage basin averaged about 41,000 hectometers^3 in 1900, but consumptive uses upstream and diversions from the Delta by 1960 reduced that flow by about one-half (Kelley 1966).

The Delta is the pivot point in the transfer of water from northern California to southern California. The major out-of-basin diversions (exports) are made via the Federal and State Water Project pumping plants in the southern Delta (Fig. 1). Typical export rates substantially exceed the flow in the San Joaquin River, hence most of its flow goes to the pumps. Also, the San Joaquin River flow to the Delta has been decreased greatly by upstream dam construction. Remaining export needs are met by diversions of Sacramento River water. Part of the flow passes through the central Delta but, due to channel volume limitations, at higher export rates water is drawn upstream to the pumps via the western San Joaquin River. Such net upstream flows are typical in the spring, except in wet years, and in the summer and fall of all years (Chadwick et al. 1977). Future water development plans include construction of additional upstream storage reservoirs and additional export from the Delta.

Definitions and Methods

Fry

We define "fry" as the life stage of salmon (<70 mm) between emergence from the spawning gravel to the completion of upstream or estuarine rearing.

Seasonal abundance and spatial distribution of fry in the estuary were measured by weekly seine surveys at 36 stations during daylight hours (Fig. 1). Delta stations were sampled from 1977 to 1981 while the bays were sampled in 1980 and 1981. The bag seine used was 15.2 by 1.2 m (3.2 mm mesh). Salmon captured were measured to the nearest millimeter fork length (FL). Water temperature and salinities were recorded at all stations. Flow records were provided by the State Department of Water Resources. Each year was classified relative to the amount of unimpaired river runoff that occurred in the Central Valley drainage. These classifications vary from critical, dry, below normal, above normal, to wet, and are determined by the State Water Resources Control Board.

Residence times and growth rates of fry were determined from mark-recapture data in 1980 and 1981. Hatchery fry (40 to 50 mm) were marked with half-sized coded wire nose tags (Jefferts et al. 1963) and by removing the adipose fin. Size variation within each marked release group was small; the standard deviation ranged from 0.05 to 0.1 times that of mean length. One hundred thousand fish per site were released each year in February below Red Bluff (302 km upstream of Sacramento) in the upper Sacramento River, at two sites in the northern Delta, and at Berkeley in San Francisco Bay (Fig. 1). Recoveries of marked fish were made in our seine and trawl surveys (see below).

We performed stomach analyses of 540 chinook fry collected in the Delta in 1979 and 189 fish collected in San Pablo and San Francisco Bays in 1980.

Smolt

Smolts are older salmon (≥70 mm) that are physiologically adapted to enter salt water and are migrating to the ocean.

The timing of smolt migration was estimated by sampling twice weekly during daylight near Pittsburg (Fig. 1) with a 9.1 by 7.9 m (3.2 mm mesh, cod end) midwater trawl. Samples were taken in April through June from 1978 to 1981 and from October, 1979, to March, 1980.

Survival and migration rates through the Delta were estimated from mark-recapture data. Smolts were marked by fin clips in 1969 to 1971, spray dye in 1976 and 1977, and coded wire nose tags in 1978 to 1981. Releases were made in June in the Sacramento River at Sacramento and Suisun Bay (Fig. 1). Recovery of marked smolts from the Sacramento release site in trawl catches at Hood and at Pittsburg from 1976 to 1981 provided estimates of smolt migration rate and smolt survival in the Delta. Survival between Hood and Pittsburg was defined as the ratio of the recovery rate of marked smolts at Pittsburg divided by the recovery rate at Hood. The recovery rates were corrected for sampling effort in time and space.

Recovery of marked adults from the ocean sport and commercial fisheries from 1969 to 1971 and 1978 to 1980 provided an independent estimate of smolt survival through the Delta characterized by larger sample sizes than from Delta trawling. Survival through the Delta between Sacramento and Suisun Bay from adult tag returns was defined as the ratio of the recovery rate of marked salmon from the Sacramento release divided by the recovery rate of marked salmon released in Suisun Bay. We assumed marked salmon behaved similarly to unmarked salmon and that they had similar catchability.

Results and Discussion

Fry Rearing

a. Seasonality and Distribution

Fry abundance in the estuary was usually greatest between February and March (Table 1). Previous salmon surveys in the Delta using fyke nets and seines to capture downstream migrants also showed similar peaks in fry abundance (Rutter 1903; Hatton 1940).

Recruitment of small fry to the Delta occurred during January to March of each year and was reflected by the small increase in mean length and

Table 1. Mean monthly catch (fish per haul) of chinook salmon fry collected by seine in the northern Sacramento-San Joaquin Delta during a variety of water years between 1977 and 1981. Average coefficient of variation (C.V.) for mean monthly catch was 208%.

	Water year classification				
Month	*1977 Critical*	*1978 Wet*	*1979 Dry*	*1980 Wet*	*1981 Dry*
Jan	0.1	15.3	23.5	13.7	5.7
Feb	0.0	25.3	50.8	19.9	20.5
Mar	0.2	22.4	45.6	20.1	10.8
Apr	1.2	7.5	12.8	8.0	11.1
May	6.6	1.3	6.9	1.2	1.2
Jun	3.1	0.7	1.1	1.2	0.6

relatively constant minimum length of fry in the samples (Fig. 2). Similar curves were observed from 1978 to 1980. Emergence of fall-run fry from upstream spawning gravel extended for about three months, which supports this interpretation.

Number of fry in the estuary was influenced by egg deposition and environmental conditions during spawning, incubation, and rearing. Peak catches of fry in the Delta followed major runoff periods (Fig. 3). Low catches in 1977 reflected drought conditions that provided poor spawning and incubation conditions and low runoff after fry emergence. In contrast, 1979 and 1980

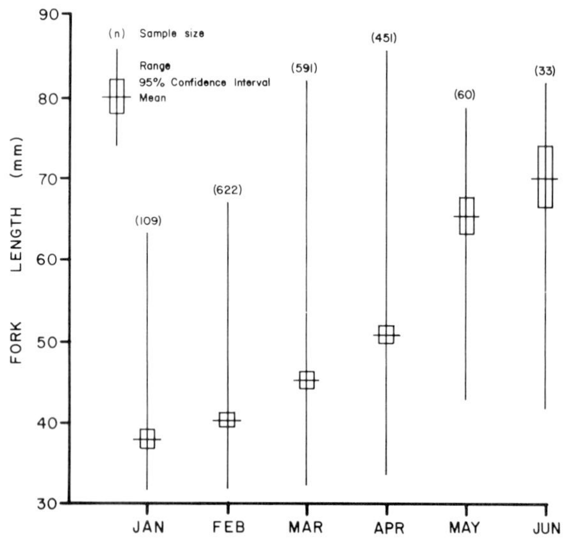

Figure 2. Average size of juvenile chinook salmon collected by seine in the Sacramento-San Joaquin Delta between 1978 and 1981.

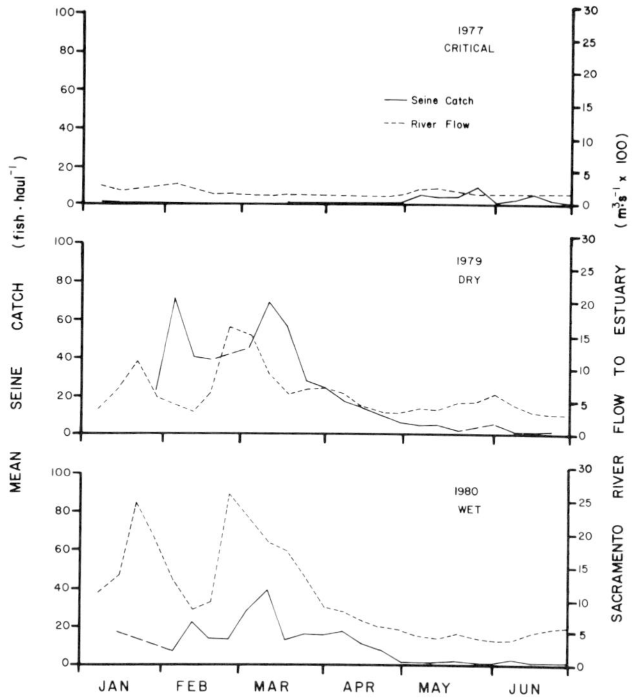

Figure 3. Sacramento River inflow to the Sacramento-San Joaquin Delta at Sacramento and mean weekly seine catches of chinook salmon fry for a variety of water years.

had several large storm runoffs that were followed by high catches of fry after a delay of about two weeks. While the fry may actually enter the estuary with the peak storm flows, their availability to be sampled along the shore may increase as flows subside and the fish concentrate along the shoreline. Even though 1979 was a drier year than 1980, influence of the storm flows on peak fry abundance in the Delta was apparent.

Distribution of fry in the estuary also appears related to storm runoff. In 1980 and 1981 fry were most numerous in the northern Delta (Table 2). However, during 1980 (a wet year), some fry (1.15 fish per haul) were collected as far downstream as central San Francisco Bay, while in 1981 (a dry year), only a few individual fry were seen in the brackish bays (0.01 fish per haul).

Low numbers of fry were observed in the central or western Delta in both years (Table 2). Low numbers in the central Delta suggest that fry moving into the northern Delta via the Sacramento River tend to remain in its channel as opposed to moving off into the smaller channels leading to the central Delta. Some of the fry caught in the central Delta probably represent small populations moving out of the San Joaquin. Lack of many fry in the

Table 2. Mean number of fish per haul of chinook salmon fry collected by seine at varied locations in the Sacramento-San Joaquin Estuary from January to April during 1980 and 1981.

Year	*Northern Delta*	*Central Delta*	*Western Delta*	*San Francisco and San Pablo Bays*
1980	14.37	2.11	3.0	1.15
1981	12.44	2.41	0.5	0.01

western Delta in 1981 probably indicated a lack of sufficient flow to transport them to that location.

The spatial distribution of fry in the Delta channels changes from day to night and with an increase in body size. Seining has revealed that smaller fry are found in shallow water near the shoreline during daylight, but at night the number caught by seine decreases suggesting that they move offshore (Schaffter 1980). Trawling studies have revealed that the size of juvenile salmon increases toward the center of the channel (Wickwire and Stevens 1971). Wickwire and Stevens' studies also indicated that larger fry and smolts were concentrated in the upper 3 m of the water column by day and were more randomly distributed vertically at night.

b. Residence Time

The residence time of marked fry rearing in the Delta was as long as 64 days in 1980 and 52 days in 1981, based on 32 fry recovered in 1980 and 51 recoveries in 1981. Some fish remained in the immediate vicinity of the release site for up to 36 days in 1980 and 49 days in 1981. For comparison, the residence time of marked fry rearing in the upper Sacramento River below Red Bluff was as long as 78 days, based on 11 recoveries.

Four marked fry of the group released at Berkeley in San Francisco Bay during 1980 were recovered 14 days later at a point 7 km west of Berkeley. The small number of recoveries and short residence time may reflect the large amount of shoreline and relatively small sampling effort we expended in San Francisco Bay. No recoveries were made in the Bay after the 1981 release, nor were any wild fry collected. Possibly Bay salinities were adverse during 1981. Low freshwater runoff resulted in salinities of about $24^{0}/_{00}$ at Berkeley that year compared to $18^{0}/_{00}$ in 1980. Wagner et al. (1969) found that chinook fry of the size we released are not able to survive immediate release into water of $25^{0}/_{00}$ salinity, but could withstand salinities up to $20^{0}/_{00}$.

c. Growth

Fry growth rates in the Delta were estimated from marked fry recoveries. Average growth was 0.86 mm day^{-1} (range, 0.57 to 1.23) for 1980, and 0.53 mm day^{-1} (range, 0.40 to 0.69) for 1981 (Fig. 4). Average growth

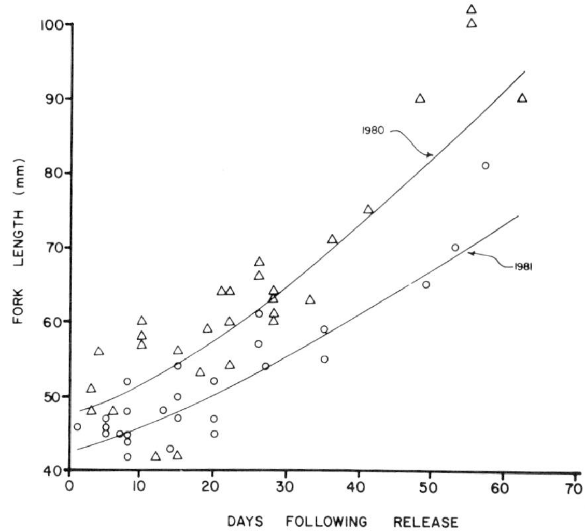

Figure 4. *Growth curves for coded-wire tagged chinook salmon fry released and recaptured in the Sacramento-San Joaquin Estuary between February and April during 1980 (triangles) and 1981 (circles). The curvilinear regression equations for 1980 and 1981 are:* $\log Y = 1.662 + 0.0054X$, $r = 0.89$ *and* $\log Y = 1.629 + 0.0041X$, $r = 0.89$, *respectively.*

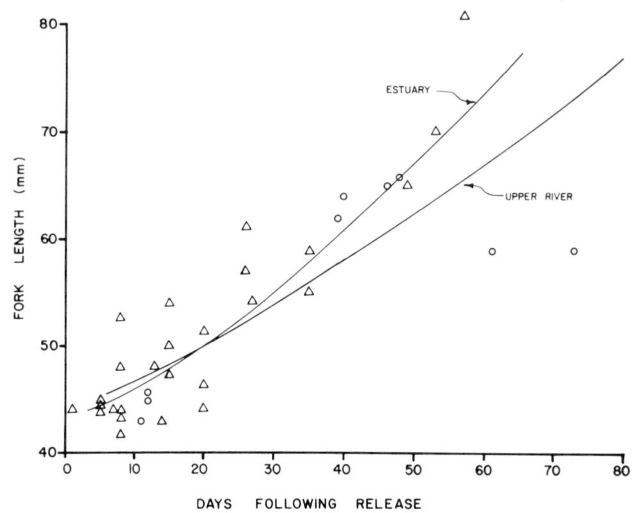

Figure 5. *Growth curves for coded-wire tagged chinook salmon fry released and recaptured in the Sacramento-San Joaquin Estuary (triangles) and the upper Sacramento River (circles), respectively, between February 6 (first date of release) and April 26 (last date of recapture), 1981. The curvilinear regression equations for the Sacramento-San Joaquin Estuary and upper Sacramento River are:* $\log Y = 1.629 + 0.0041X$, $r = 0.89$ *and* $\log Y = 1.647 + 0.0026X$, $r = 0.78$, *respectively.*

rates for fry rearing in the upper Sacramento River in 1981 (Fig. 5) were estimated at 0.33 mm day^{-1} (range, 0.26 to 0.40). Growth curves in Fig. 4 did not have slopes that were significantly different ($T = 0.13$, $df = 59$), while those in Fig. 5 did ($T = -2.37$, $df = 35$, $p<0.05$), suggesting that fry reared in the estuary grew faster than those reared upriver. The curves in Fig. 4 are separated slightly since mean length at release in 1981 was about 5 mm less than that in 1980. Growth rate of marked fry from our 1980 San Francisco Bay release was about 1.01 mm day^{-1} during the two weeks between release and recovery.

d. Diet

Crustacea and insects dominated chinook fry stomach contents, with an increase in crustacea ingestion downstream (Table 3). Cladocera and diptera were consumed frequently in the Delta, while in brackish San Pablo and San Francisco Bays, consumption of copepods, amphipods, and fish larvae increased. Similar food habits were described for older fry and smolts in Delta studies by Rutter (1903), Ganssle (1966), and Sasaki (1966).

Table 3. Taxonomic composition, percent numerical occurrence and percent frequency of occurrence of prey items in the stomachs of juvenile chinook salmon from upper (1979) and lower (1980) Sacramento-San Joaquin Estuary.

Taxa	*Upper Estuary (Delta)*		*Lower Estuary (San Pablo and San Francisco Bays)*	
	% numerical occurrence	*% frequency of occurrence*	*% numerical occurrence*	*% frequency of occurrence*
Crustacea	75	89	88	85
Cladocera	66	76	62	59
Copepoda	8	49	20	34
Amphipoda	1	22	4	41
Other (Mysidacea)	1	<1	2	7
Insecta	24	93	11	65
Diptera	17	89	4	44
Homoptera	3	23	5	33
Plecoptera	<1	6	–	–
Other (Coleoptera)	3	<1	2	5
Miscellaneous (Fish Larvae)	<1	<1	1	13

Smolt Migration

a. Timing and Size

Peak abundance of smolts in the Delta and Bays occurred between April and mid-June (Fig. 6; Ganssle 1966; Messersmith 1966; Aplin 1967). The average size of smolts passing through the Delta and Bays from April through June remained very constant in each year, although mean size varied somewhat between years (Fig. 6; Ganssle 1966; Messersmith 1966; Schaffter 1980).

Other investigators have reported identical results for peak smolt outmigration, with few young salmon seen in the Delta or Bays during the summer (Rutter 1903; Sasaki 1966). Lack of young salmon in the estuary during summer may be caused by greater than optimum water temperatures in the lower Sacramento and San Joaquin rivers and in the Delta from mid-June to September during most years. Summer water temperatures in the main Delta channels usually peak around 22°C. Young chinook prefer temperatures of 12 to 14°C (Brett 1952).

A small secondary peak in smolt abundance occurred in the Delta and Bays during fall months. Midwater trawling has yielded a few salmon around

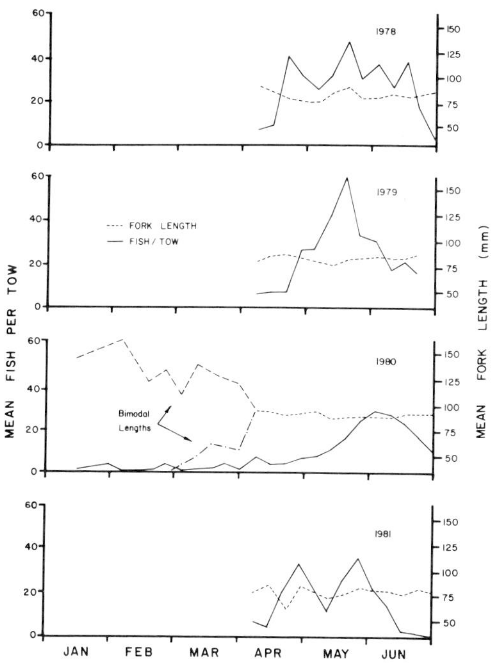

Figure 6. Mean weekly catch and size of chinook salmon for 1978 and 1981 based on midwater trawls at Pittsburg, California in the Sacramento-San Joaquin Estuary. Average coefficient of variation equals 70% and 13% for mean weekly catch and size, respectively.

90 to 200 mm long, apparently from fall and late fall runs that have over-summered in cooler waters upstream from the Delta. Some of these fish also represent yearling hatchery releases.

b. Rate of Migration

The constant size of smolts throughout the Delta and lower bays and mark-recapture data indicate that smolts move through the estuary quickly. Similar sized smolts (94 to 100 mm) were observed in the ocean by Snyder (1924) which further supports this conclusion. Weisbart (1968) and Wagner et al. (1969) found that juvenile chinook could tolerate full sea water upon reaching a size of 70 mm, slightly smaller than the mean sizes we observed during the peak out-migration period (Fig. 6).

Our 1976 to 1980 mark-recapture data based on smolt released at Sacramento and recovered at Pittsburg indicate that smolt migrate through the Delta from 10 to 18 km day^{-1}. At this rate smolts could pass through the estuary (about 172 km from Sacramento to San Francisco) within 10 to 17 days. Salmon trawling planned for 1982 in San Francisco Bay will attempt to recover marked smolts released that year at Sacramento so that migration and growth rates can be estimated for the brackish portion of the estuary. Wickwire and Stevens (1971) estimated smolt migration at 8, 12 and 24 km day^{-1} in April, May, and June, respectively, as water temperatures rose.

c. Survival

Our mark-recapture studies have shown that survival is lower for smolts migrating through the Delta than through the lower estuary. Ocean recoveries ($n = 718$) of marked smolts released in Suisun Bay in June of 1978 were about 80 times greater than those of the smolts released at Sacramento ($n = 9$), yet the migration distance for the Sacramento released fish was only twice that traveled by those released in Suisun Bay. We assume that ocean survival was the same for both release groups.

Survival in the Delta appears to be influenced by water temperatures and/or river flow rate. Temperature of water entering the Delta at Sacramento is influenced by the magnitude of flow and ambient air temperature (Rowell 1972). Smolt survival (based on our ocean recovery data on marked adults) decreased as flow rates decreased and temperatures increased for marked fish released in June ($p<0.01$, $R^2 = 0.97$, $df = 4$; Fig. 7). Temperatures and flow rates were so closely related during the release periods that it was not possible to separate their individual impacts on smolt survival. However, using our trawl recovery data of marked smolts at Pittsburg, temperature alone was related to smolt survival during June ($p<0.05$, $r = -0.86$, $df = 5$; Fig. 8). Our ocean and trawl recovery data indicated that almost total mortality was experienced by migrant smolts in June of 1978 (Figs. 7 and 8) and 1981 (Fig. 8) when temperatures were about 23°C.

Operations of State and Federal pumping plants in the south Delta influence survival of smolts during their migration. Records of smolts collected

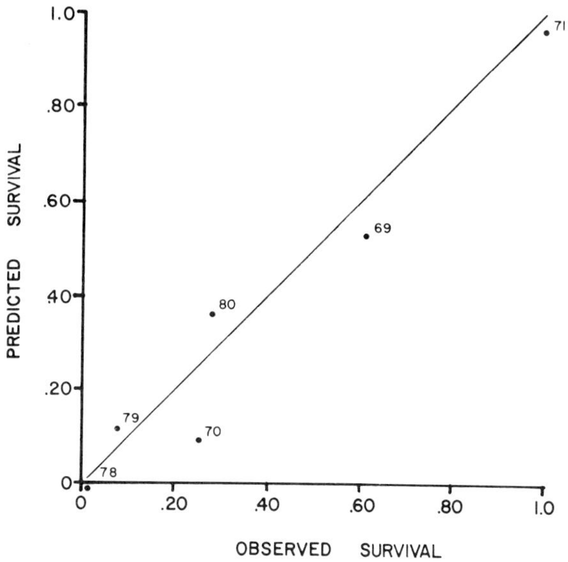

Figure 7. *Relation between observed survival of chinook salmon smolts and survival predicted from June river flows and June water temperatures during June migration through the estuary. The regression equation used to obtain the Y-axis coordinates is* $Y = 1.81 + 0.0008$ *(river flow,* $m^3\ s^{-1}$*)* $- 0.09$ *(water temperature, °C);* $R^2 = 0.95$*. Survival estimates are based on ocean recoveries of marked adults.*

at pumping plant fish screens indicate that as water exports increase more smolts are drawn to the screens (Kjelson et al. 1981). Salmon collected at these screens are returned to the western Delta out of the influence of the pumps, but handling and transport mortality average 58%. An additional 10 to 35% (dependent on size) are lost through the screens (Skinner 1974). While the number of salmon observed at the pumps each year (about 200,000 fish) represent a small fraction of the estimated total smolt production, more fish probably are drawn out of their normal migration path and die. Our earlier mark-recapture studies in 1976 indicated that smolts released in northern Delta channels on a direct path to the south Delta pumps had a lower survival than those released in the Sacramento River which represented the most direct route to the ocean (Kjelson et al. 1981). Additional losses of smolts have occurred from the numerous agricultural, industrial, and municipal diversions in the estuary and upper river systems. Most of these remain unscreened and in total account for considerable losses (Hallock and Van Woert 1959).

Estuarine Comparisons and Summary

We now review some of the readily available literature on life histories

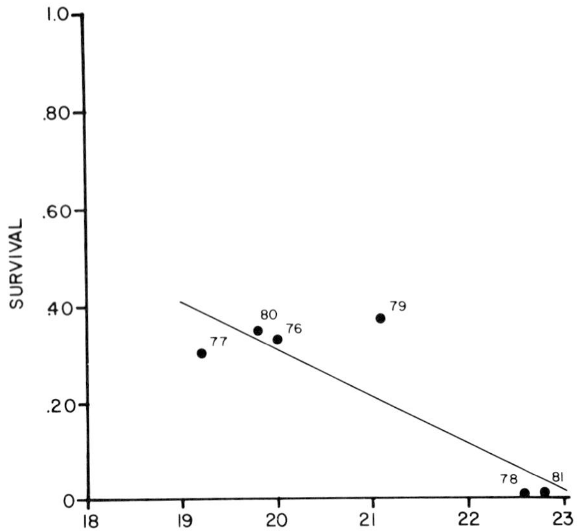

Figure 8. Linear regression between chinook salmon June smolt survival during migration through the Sacramento-San Joaquin Delta and mean June water temperature at Sacramento. Survival estimates are based on mid-water trawl recoveries in the Delta of marked smolts released at Sacramento. Regression equation is: $\hat{S} = 2.28 - 0.10$ (water temperature, °C), $r = -0.86$.

of juvenile fall run chinook salmon in estuaries from northern California to British Columbia. There are a few unique differences and some similarities in salmon life history and their habitats between the Sacramento-San Joaquin Estuary and those further north. Chinook salmon life histories between estuarine systems are highly variable.

Physical Environment

Most estuaries further north which have been studied are much smaller than the Sacramento-San Joaquin Estuary (Reimers 1973; Healey 1980). In absolute terms, most of these smaller systems have little or no tidally influenced delta and limited tidal marshes and mudflats (Healey 1980; Fish and Wildlife Service 1981). Salinity gradients are narrow and river gradients steep making transition from fresh to salt water relatively abrupt compared to that seen in the Sacramento-San Joaquin system and to some extent the Columbia River Estuary. Some estuaries such as in the Mad and Sixes Rivers in California and Oregon have a sill or sand bar that forms across the mouth and restricts exchange with the ocean during low runoff periods (Taniguchi 1970; Reimers 1973). Lower river and estuarine water temperatures can be limiting to chinook residence in summer in estuaries from British Columbia to California.

Seasonal Abundance

Estuarine fry rearing and smolt outmigration by fall-run chinook is about two to three months earlier in the Sacramento-San Joaquin Estuary than further north. Chinook fry appear to be most abundant in the more northern estuaries in April and May (Healey 1980; Puckett 1977) while smolt numbers are highest between June and August (Reimers 1973; Fish and Wildlife Service 1981). Earlier use of the Sacramento-San Joaquin Estuary by salmon appears related to spawning times and the fact they are at the southernmost limit of the chinook range where warmer summer water temperatures in the upper estuary force the salmon to leave. Fall-run spawning in the Sacramento-San Joaquin peaks in October; in some systems further north, it is later by about one month (Reimers 1973). Warm summer water temperature also causes salmon to move out of the main river channel into the estuary in Oregon and British Columbia (Reimers 1973; Healey 1980).

Rearing Environment

Information from the Pacific northwest indicates that estuarine rearing by chinook occurs in both fresh (tidally influenced) and brackish water habitat (Reimers 1973; Healey 1980). In the Sacramento-San Joaquin system most estuarine rearing occurs in the freshwater Delta supplemented by fry rearing in the brackish water bays during high runoff years. The number of fry present in the estuary is influenced, in part, by river flow magnitude both in the Sacramento-San Joaquin system, as well as in Washington (Wetherall 1970) and British Columbia (Healey 1980). Young chinook fry appear to be most abundant in the shallow water, shoreline habitat, while older fry and outmigrant smolts are found further offshore in most Pacific estuarine systems.

Residence Time and Migration Rate

Chinook fry in the Sacramento-San Joaquin Estuary reside there for up to two months which may be slightly longer than in systems further north. Healey (1980) estimated an average residence time of 25 days in Nanaimo Estuary (British Columbia), but some fish remained much longer.

Smolt migration rates appear similar for most estuaries. We estimated rates ranging from 6 to 24 km day^{-1} indicating that movement through the estuary was fairly rapid. Dawley et al. (1981) found an average rate of 6 to 21 km day^{-1} in Columbia River Estuary and also found, as we did in the Sacramento-San Joaquin, that migration rates were higher in the upper river and that estuarine migration rates increased as water temperatures rose.

Growth

Observations of estuarine growth rates varied between different systems. Our measurements in the Sacramento-San Joaquin Estuary ranged from 0.40 to 1.23 mm day^{-1} while Reimers (1973) reported 1.0 mm day^{-1} for Sixes River (Oregon). Sibert (1975) and Healey (1980) estimated growth of

0.71 (smolts) to 1.32 mm day^{-1} (fry) for the Nanaimo River Estuary (British Columbia).

Intermediate circuli spacing on chinook scales, attributed to brackish water estuarine growth by older juveniles (>100 mm), was observed in Columbia River (Rich 1922), in Klamath River (Snyder 1931), and in Sixes River (Reimers 1973). We did not see such a pattern on juvenile and adult scales from chinook in the Sacramento-San Joaquin system. This apparently reflects either lack of significant brackish water rearing by older juveniles in the lower Sacramento-San Joaquin Estuary or similar river and estuarine growth rates.

Diet

Young chinook in the estuaries from northern California to British Columbia appear to consume prey similar to those in the Sacramento-San Joaquin (Herrmann 1971; Lipovsky 1977; Healey 1980). Stomach contents were dominated by copepods, cladocera, amphipods, mysids, insects, and some fish larvae.

Survival

Wetherall (1970) found juvenile chinook survival rates in Green River (Washington) between 37 and 99%; they increased as river flow rates increased. Reimers (1973) provided evidence that there was a survival advantage for extended juvenile chinook residence in Sixes River Estuary. Argue et al. (1979) found that a greater fraction of the chinook smolt population in the Cowichan-Koksilah Estuary (British Columbia) originated from fry that reared exclusively in the estuary. Van Hyning (1973) suggested that the number of adult chinook returning to spawn in Columbia River increased as the May estuarine water temperature decreased during juvenile outmigration. We found that smolt survival during outmigration was better in the lower estuary than in the Delta and that temperature and possibly flow rate influenced survival in the Delta.

Physical and Biological Stresses

Juvenile chinook in more northern estuaries are exposed to somewhat different environmental alterations than those in the Sacramento-San Joaquin Estuary. Salmon in the Pacific northwest estuaries experience the impacts of the lumbering industry, while in the Sacramento-San Joaquin Delta there are no lumbering activities and the greater impacts come from water diversion activities.

Potential problems of exceeding estuarine rearing capacity through hatchery releases have been emphasized by various investigators in the northwest (Reimers 1973; Iwamoto and Salo 1977; Meyer 1979). Hatchery production of multi-races of salmon in Columbia River and Puget Sound systems has been much larger than that in the Sacramento-San Joaquin drainages. Dawley et al. (1981) reports that Columbia River has approximately 150 million young salmon released each year that must pass through the estuary. The problem of exceeding estuarine rearing capacity is of some concern in the Sacramento-San Joaquin (hatchery releases total about 26 million smolts annually), but as yet has not been studied.

Acknowledgments

We wish to thank the Young Adult Conservation Corps enrollees of the U. S. Fish and Wildlife Service and seasonal aides of the California Department of Fish and Game for their assistance in field sampling and laboratory analysis. We are grateful to R. Adair, P. Herrgesell, D. Stevens, D. Vogel, and T. Yocom for their comments on the manuscript.

References Cited

Aplin, J. S. 1967. Biological survey of San Francisco Bay 1963-1966. Calif. Dept. Fish Game, Mar. Res. Oper. Br. Rept. 67(4):1-131.

Argue, A. W., L. M. Patterson and R. W. Armstrong. 1979. Trapping and coded-wire tagging of wild coho, chinook, and steelhead juveniles from the Cowichan-Koksilah River system, 1976. Fish. Mar. Serv. Dev. Tech. Rept. No. 850. Fish. and Mar. Serv., Vancouver, B.C. 117 pp.

Brett, J. R. 1952. Temperature tolerance in young Pacific salmon, genus *Oncorhynchus*. *J. Fish. Res. Board Can.* 9:265-323.

Chadwick, H. K., D. E. Stevens and L. W. Miller. 1977. Some factors regulating the striped bass population in the Sacramento-San Joaquin Estuary, California, pp. 18-35. *In:* W. Van Winkle (ed.), *Proceedings of the Conference on Assessing the Effects of Power-Plant-Induced Mortality on Fish Populations.* Pergamon Press, New York.

Conomos, T. J. (ed.). 1979. *San Francisco Bay: the Urbanized Estuary.* Pacific Division, Amer. Assoc. Advanc. Sci., San Francisco, CA 493 pp.

Dawley, E. M., C. W. Sims, R. D. Ledgerwood, D. R. Miller and J. G. Williams. 1981. A study to define the migrational characteristics of chinook and coho salmon in the Columbia River Estuary and associated marine waters. Coastal Zone and Estuarine Studies Division. Nat. Mar. Fish. Serv., Seattle, WA 68 pp. and Appendix.

Fish and Wildlife Service. 1981. Annual Report – Klamath River fisheries investigation program, 1980. U. S. Dept. Interior. Fish and Wildlife Service, Fisheries Assistance Office, Arcata, CA 107 pp.

Ganssle, D. 1966. Fishes and decapods of San Pablo and Suisun bays. *In:* D. W. Kelley (compiler). Ecological studies of the Sacramento-San Joaquin Estuary. Part I. *Calif. Dept. Fish Game, Fish. Bull.* 133:64-94.

Hallock, R. J. and W. F. Van Woert. 1959. A survey of anadromous fish losses in irrigation diversions from the Sacramento and San Joaquin rivers. *Calif. Fish Game* 45(4):227-266.

Hatton, S. R. 1940. Progress report on the Central Valley fisheries investigations, 1939. *Calif. Fish Game* 26(4):334-373.

Healey, M. C. 1980. Utilization of the Nanaimo River estuary by juvenile chinook salmon, *Oncorhynchus tshawytscha. Fish. Bull.* 77(3):653-668.

Herrmann, R. B. 1971. Food of juvenile chinook and chum salmon in the lower Chehalis River and Upper Grays Harbor, pp. 57-114. *In:* Grays Harbor Cooperative Water Quality Study, 1964-1966. Wash. Dept. Fish., Tech. Rept. No. 7. Grays Harbor, WA

Hoopaugh, D. A. and A. C. Knutson, Jr. (eds.) 1979. Chinook (king) salmon spawning stocks in California's Central Valley, 1977. Calif. Dept. Fish Game, Anad. Fish. Br. Admin. Rept. No. 79-11. 36 pp.

Iwamoto, R. N. and E. O. Salo. 1977. Estuarine survival of juvenile salmonids: A review of literature. U. Washington, Fish. Res. Inst., Seattle, WA 64 pp.

Jefferts, K. B., P. K. Bergman and N. F. Fiscus. 1963. A coded wire tagging system for macro-organisms. *Nature* 198(4877):460-462.

Kelley, D. W. (compiler). 1966. Ecological studies of the Sacramento-San Joaquin Estuary. Part I. *Calif. Dept. Fish Game, Fish. Bull.* 133. 133 pp.

Kjelson, M. A., P. F. Raquel and F. W. Fisher. 1981. Influences of freshwater inflow on chinook salmon *(Oncorhynchus tshawytscha)* in the Sacramento-San Joaquin Estuary, pp. 88-108. *In:* R. D. Cross and D. L. Williams (eds.), *Proceedings of the National Symposium Freshwater Inflow to Estuaries.* U.S. Dept. of Interior, Fish and Wildlife Service, FWS/OBS-81/04. Vol. 2.

Lipovsky, S. J. 1977. Food habits of juvenile salmon in the Columbia River Estuary, pp. 14-21. *In:* R. J. Wahle (convener), *Juvenile salmonids in the estuary: Proceedings of a technical workshop.* Nat. Mar. Fish. Serv., Hammond, OR

Messersmith, J. 1966. Fishes collected in Carquinez Strait in 1961-62. *In:* D. W. Kelley (compiler), Ecological studies of the Sacramento-San Joaquin Estuary. Part I. *Calif. Dept. Fish Game, Fish. Bull.* 133:57-63.

Meyer, J. H. 1979. A review of the literature on the value of estuarine and shoreline areas to juvenile salmonids in Puget Sound, Washington. U. S. Dept. of Interior, Fish and Wildlife Service, Fisheries Assistance Office, Olympia, WA 24 pp.

Puckett, L. K. 1977. The Eel River Estuary – Observations on morphometry, fishes, water quality and invertebrates. Memo. Rept. to Calif. Dept. Fish Game. Sacramento, CA 52 pp.

Reimers, P. E. 1973. The length of residence of juvenile fall chinook salmon in Sixes River, Oregon. Oregon Fish Comm. Research Report, Vol. 4, No. 2. 43 pp.

Rich, W. S. 1922. Early history and seaward migration of chinook salmon in the Columbia and Sacramento rivers. *Bull., U. S. Bur. Fish. Comm.*, Vol. 37:1-77.

Rowell, J. H. 1972. Reservoir and stream temperature predictions in the upper Sacramento River Basin. M.S. Thesis, California State U., Sacramento, CA 151 pp.

Rutter, C. 1903. Natural history of the Quinnat salmon, 1896-1901. *Bull., U. S. Bur. Fish. Comm.*, Vol. 22:65-141.

Sasaki, S. 1966. Distribution and food habits of king salmon, *Oncorhynchus tshawytscha,* and steelhead rainbow trout, *Salmo gairdnerii,* in the Sacramento-San Joaquin Delta. *In:* J. L. Turner and D. W. Kelley (eds.). Ecological studies of the Sacramento-San Joaquin Delta. Part II. *Calif. Dept. Fish Game, Fish. Bull.* 136:108-114.

Schaffter, R. G. 1980. Fish occurrence, size, and distribution in the Sacramento River near Hood, California during 1973 and 1974. Calif. Dept. Fish Game, Anad. Fish. Br. Admin. Rept. No. 80-3. 76 pp.

Sibert, J. 1975. Residence of juvenile salmonids in the Nanaimo River Estuary. Fish. Mar. Serv. Res. Dev. Tech. Rept. No. 537. Fish. Mar. Serv. Pacific Biological Station. Nanaimo, B.C. 23 pp.

Skinner, J. E. 1974. A functional evaluation of a large louver screen installation and fish facilities research on California water diversions project, pp. 225-249. *In:* L. D. Jensen (ed.), *Proceedings of the Second Entrainment and Intake Screening Workshop.* The Johns Hopkins University Cooling Water Research Project, Report No. 15, Baltimore, MD

Snyder, J. O. 1924. Young salmon taken at sea. *Calif. Fish Game* 10(2):63-64.

Snyder, J. O. 1931. Salmon of the Klamath River, California. *Calif. Dept. Fish Game, Fish. Bull.* 35:4-130.

Taniguchi, K. A. 1970. Interval of estuarine residence and outmigration of juvenile chinook salmon in the Mad River, California. M.S. Thesis, Humboldt State U., Arcata, CA 87 pp.

Turner, J. L. and D. W. Kelley (compilers). 1966. Ecological studies of the Sacramento-San Joaquin Delta. Part II. *Calif. Dept. Fish Game, Fish. Bull.* 136. 168 pp.

Van Hyning, J. M. 1973. Factors affecting the abundance of fall chinook salmon in the Columbia River. Res. Rept. Oreg. Fish. Comm., 4(1):1-87.

Wagner, H. H., F. B. Conte and J. L. Fessler. 1969. Development of osmotic and ionic regulation in two races of chinook salmon *(Oncorhynchus tshawytscha). Comp. Biochem. Physiol.* 29(1):325-341.

Weisbart, M. 1968. Osmotic and ionic regulation in embryos, alevins and fry of the five species of Pacific salmon. *Can. J. Zool.* 46:385-397.

Wetherall, J. A. 1970. Estimation of survival rates for chinook salmon during their downstream migration in the Green River, Washington. Ph.D. Dissertation. U. Washington, Seattle, WA 170 pp.

Wickwire, R. H. and D. E. Stevens. 1971. Migration and distribution of young king salmon, *Oncorhynchus tshawytscha,* in the Sacramento River near Collinsville. Calif. Dept. Fish Game, Anad. Fish. Br. Admin. Rept. No. 71-4. 18 pp.

STATUS OF ANADROMOUS FISHES IN SOUTHEASTERN U.S. ESTUARIES

R. A. Rulifson
Center of Environmental Sciences
Unity College, Unity, Maine

M. T. Huish
North Carolina Cooperative
Fishery Research Unit,
Box 5577, NCSU
Raleigh, North Carolina

and

R. W. Thoesen
Fishery Resources, U.S. Fish and Wildlife Service,
Region 4, Atlanta, Georgia

Abstract: Population status of 11 species or races of anadromous fish was assessed for 85 riverine areas in seven Southern coastal States by asking State, Federal, and other agencies to respond to a questionnaire developed for the U.S. Fish and Wildlife Service "Anadromous Species Management Plan for the Southeast." Atlantic and shortnose sturgeons, striped bass (Atlantic race), American and hickory shads, alewife, and blueback herring utilize Southeast Atlantic coast estuaries as spawning grounds or nursery areas. Atlantic sturgeon (Gulf race), striped bass (Gulf race), Alabama shad, and skipjack herring utilize estuaries along the Gulf of Mexico. Some of these populations appear to be stable, but many are declining or threatened and several have been extirpated. Atlantic races of striped bass have been introduced into Gulf coastal waters, and some populations along both coasts are increasing due to years of restocking and restoration efforts. Questionnaire responses suggest that overfishing, poor water quality, low oxygen levels, location of industrial discharges, chemical pollution, dams and impoundments, and inadequate fishway facilities have reduced spawning habitat and nursery areas in many river systems, contributing to the decline of anadromous stocks in Southeastern U.S. waters.

Introduction

State and Federal agencies have monitored the status of anadromous fish stocks in river systems of the Great Lakes and U.S. coastlines for decades. Traditionally these monitoring efforts were associated with commercial catch statistics of species with great socio-economic importance such as striped bass (*Morone saxatilis*), American shad (*Alosa sapidissima*), and the Salmonidae. Along the eastern seaboard, anadromous populations have received greatest attention in the mid-Atlantic and Northeast, where large

ISBN 0-12-404070-5

commercial fisheries have operated since the early 1800's (Beal 1980). Although Southeastern U.S. stocks historically have been of social and economic importance in the seven South Atlantic and Gulf coast States (North Carolina through Louisiana), they have received little attention on a national scale.

Results of recent studies by State agencies and State-Federal cooperative projects suggest that certain anadromous populations have declined in areas along the Southeast Atlantic and Gulf coasts. In response, the U.S. Fish and Wildlife Service designated the status of estuarine and anadromous fishes in the seven Southern coastal States an "Important Resource Problem (IRP)." The IRP system is designed to promote greater understanding of important fish and wildlife resources and to identify where problems affecting these resources are most acute. The Southern coastal fishes problem received a National Priority Ranking Number of 49 out of 78 within the Fish and Wildlife Service. One IRP objective was to submit a document to the U.S. Congress (Rulifson et al. 1982) that 1) summarized the status of anadromous populations, angler demand, and the history of stock restoration efforts, and 2) recommended the development of interagency plans for major rivers and estuarine systems of the Southeast. This paper summarizes a portion of the compiled information. Emphasis is placed on the distribution of all Southern anadromous species and the status of specific stocks within the seven Southern coastal States (with occasional mention of data from Texas). Throughout this paper we distinguish between South Atlantic States (North and South Carolina, Georgia, and Florida's Atlantic coast) and Gulf States (Florida's Gulf coast, Alabama, Mississippi, and Louisiana).

Methods

To aid the development of an anadromous species management plan for the Southeastern U.S., a detailed questionnaire was developed to ascertain the present status of, management practices for, and attitudes toward anadromous species in Texas and in the seven coastal States in the Southeast. Questions and format were reviewed and critiqued by State and Federal agencies before the final questionnaire was completed. Topics addressed included the presence of anadromous species in river systems, status of the stocks, problems of the fisheries, management practices in operation, and future fishery needs. A portion of the questionnaire asked each State agency to identify the river systems important to anadromous fish that are situated within or border their State. Agencies were instructed to indicate the condition of each river by evaluating the relative importance of 21 physical and chemical disturbances that possibly affect anadromous fish stocks inhabiting the river.

An historical background detailing socio-economic importance and life history aspects of all anadromous fish species within South Atlantic and Gulf coast waters was compiled by review of published literature, Federal and

State project reports, memoranda, and verbal communications. Computerized searches of published literature and Federal-State project reports were conducted specifically for completion of the Rulifson et al. (1982) document by the Library Reference Service, U.S. Fish and Wildlife Service, Denver, Colorado, and by Computerized Search Service, Natural Resources Library, Research Services Branch, Washington, D.C. Memoranda and Dingell-Johnson reports were obtained from the Region 4 office, U.S. Fish and Wildlife Service, Atlanta, Georgia.

Seventeen questionnaires were distributed to and completed by freshwater and marine fisheries divisions of the seven Southern coastal States, the freshwater fisheries division of Texas, the Regional office of the National Marine Fisheries Service (NMFS), and the NMFS Laboratory at Beaufort, North Carolina. We assumed that 1) the respondents represented the people most qualified to complete the questionnaire, and 2) lack of response to some questions or parts of questions was due to lack of information. Thus the completed questionnaires and the extensive literature review represent the best information available on Southern anadromous species. The questionnaire, responses by State and Federal agencies, and detailed life histories of each anadromous species are presented in Rulifson et al. (1982).

Results and Discussion

Questionnaire responses indicated the presence of 11 species or races of anadromous fishes and 85 river systems within the seven coastal States. The Percichthyidae are represented by two races of striped bass (*Morone saxatilis*) on the Atlantic and Gulf coasts. The Acipenseridae are represented on the Atlantic coast by the Atlantic sturgeon (*Acipenser oxyrhynchus oxyrhynchus* Mitchell) and the shortnose sturgeon (*A. brevirostrum* LeSeuer), and in the Gulf of Mexico by the Gulf coast sturgeon (*A. o. deSotoi* Vladykov). Atlantic coast clupeids include American shad (*Alosa sapidissima* (Wilson)), hickory shad (*A. mediocris* Mitchell), blueback herring (*A. aestivalis* Mitchell), and alewife (*A. pseudoharengus* Wilson). Gulf coast clupeids include Alabama shad (*A. alabamae* Jordan and Evermann) and skipjack herring (*A. chrysochloris* (Rafinesque)). Of the 85 sounds, rivers, and tributaries with anadromous populations, 29% are located (all or in part) in North Carolina, 15% in South Carolina, 9% in Georgia, 12% in Florida, 14% in Alabama, 6% in Mississippi, and 14% in Louisiana (Fig. 1).

Striped bass

Atlantic race striped bass (*Morone saxatilis*) ranged historically from the St. Lawrence River, Canada, to the St. Johns River, Florida. Their life history is well-documented in Southeast waters. Eggs are found in freshwater tributaries up to $11^{0}/_{00}$ and require current for suspension. Larvae float near the surface one to two days after hatching and form schools within four to five days in shore zones of fresh or brackish waters. Juveniles are most abundant

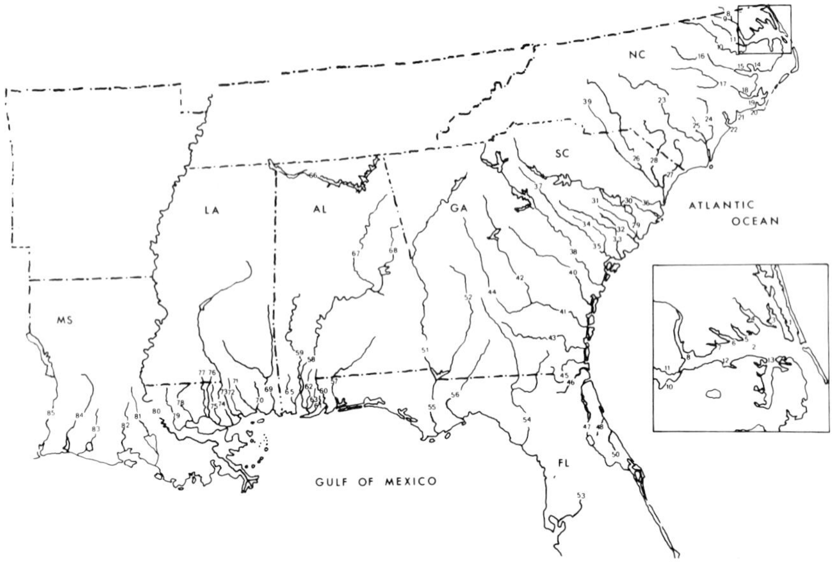

Figure 1. Rivers and sounds of the South Atlantic and Gulf coasts that contain populations of anadromous fish. Numbers refer to the listing of rivers in Table 1.

in shallow, pronounced-current areas of streams, rivers, bays, and sounds. Apparently, adult populations south of North Carolina are endemic and riverine and are rarely caught in offshore ocean waters. Adults overwinter in bays, estuaries, and delta areas. Striped bass populations of the South Atlantic coast are comprised primarily of age classes III (26%), IV (27%), and V (17%) (sexes combined); less than 0.1% attain age XV. Minimum age to maturity is II for males and III for females; minimum size to maturity is 174 mm total length (TL) for males and 432 mm TL for females. Spawning season is delayed with increasing latitude, beginning in mid-February to early April in the St. Johns River, Florida, and in mid-April to early June in the Roanoke River, North Carolina. Spawning occurs in fast-water, rock-strewn areas associated with the Fall Line at temperatures ranging from 13 to 24 C and peaks between 18 and 21.7 C.

The status of striped bass stocks was reported for only 36 of the 50 sounds, rivers, or tributaries in South Atlantic States listed by State and Federal agencies (Table 1). Of these populations, 41% were listed as "stable" and 31% as "declining." Increasing populations occur in the Tar-Pamlico River system, North Carolina, and the St. Johns River, Florida. The New River, North Carolina, population was believed "threatened" or "extripated." Population status was unknown for 8% of the areas.

The presence of Atlantic race striped bass in coastal waters of the Gulf of Mexico is a result of stocking programs conducted by State agencies in

Table 1. Status of anadromous fish stocks in South Atlantic coastal waters of the U.S. I = increasing; S = stable; D = declining; T = threatened; E = extirpated (no longer present); PNP = probably never present; N = not known; dashed line = no response by agencies. Numbers refer to rivers shown in Figure 1.

RIVER SYSTEM	ATLANTIC RACE STRIPED BASS	AMERICAN SHAD	HICKORY SHAD	BLUEBACK HERRING	ALEWIFE	ATLANTIC STURGEON	SHORTNOSE STURGEON
NORTH CAROLINA							
1. Currituck So.	-	-	-	D	D	-	E
2. Albemarle So.	D	D	D	D	D	D	E/T
3. North	-	-	-	D	D	-	E
4. Pasquotank	-	-	-	D	D	-	E
5. Little	-	-	-	D	D	-	E
6. Perquimans	-	-	-	D	D	-	E
7. Yeopim	-	-	-	D	D	-	E
8. Chowan	D	D	D	D	D	D	E/T
9. Meherrin	S/D	D	S/D	D	D	-	E
10. Roanoke	D	D	D	D	D	D	E/T
11. Cashie	-	D	S/D	D	D	-	E
12. Scuppernong	-	-	-	D	D	-	E
13. Alligator	-	-	-	D	D	-	E
14. Pungo	-	-	-	N	N	-	E
15. Pamlico	I	-	-	N	N	S/D	E
16. Tar	I	D	D	D	D	D	E/T
17. Neuse	S/D	D	D	D	D	D	E/T
18. Trent	-	D	S/D	S/D	S/D	-	E
19. North	-	-	-	N	N	-	E
20. Newport	-	-	-	N	N	-	E
21. White Oak	D	N	-	N	-	-	E
22. New	T/E	T	T	T	T	-	E
23. Cape Fear	S/D	D	D	D	D	D	E/T
24. N.E. Cape Fear	S/D	D	D	D	D	D	E/T
25. Black	-	N	-	N	N	-	E
26. Pee Dee	D	D	D	D	D	D	T
SOUTH CAROLINA							
27. Waccamaw	S	S	S	S	-	S	T
28. Little Pee Dee	S	S	S	N	-	S	T
26. Great Pee Dee	S	S	S	S	-	S	T
29. Black	S	S	S	S	-	S	T
30. Santee	S	S/D	S	S	-	S	T
31. Cooper	S	S	S	S/I	-	S	T
32. Ashley	S	S	S	S	-	S	T
33. Edisto	S	S/D	S	S	-	S	T
34. Ashepoo	S	S	S	S	-	S	T
35. Combahee	S	S	S	S	-	S	T
36. Sampit	S	I/S	S	S	-	S	T
37. Salkehatchie	S	S	S	N	-	S	T
38. Savannah	S	S	S	S	-	S	T
39. Lynches	N	S	S	S	-	N	T
GEORGIA							
38. Savannah	D	N	N	N		N	T
40. Ogeechee	D	D	N	N		N	T
41. Altamaha	D	N	N	N		N	T
42. Oconee	N	N	N	N		N	T
43. Satilla	D	D	N	N		N	T
44. Ocmulgee	N	N	N	N		N	T
45. St. Marys	D	D	N	N		N	T
FLORIDA (Atlantic coast)							
45. St. Marys	D/S	S/D	N	N		T/S	N
46. Nassau	D/S	S	N	N		S	N
47. St. Johns	I	D/S	D	D		D	T
48. Pellicer	PNP	D	N	N		PNP	-
49. Moultrie	PNP	PNP	PNP	PNP		PNP	-
50. Tomoka	PNP	D	D	D		PNP	-

order to reestablish the striped bass sportfishery. Of the 33 areas listed by Gulf coast fishery agencies, 88% are believed to contain increasing populations of Atlantic race striped bass (Table 2). Population status was not known for Atlantic race striped bass in the Apalachicola and Vermillion Rivers. Texas currently has an active stocking program for the Atlantic race and has extended its range in coastal waters of the Gulf. This program has established spawning populations in two areas during years of high precipitation. However, since there is little control over water releases from impoundments in the system,

Table 2. Status of anadromous fish stocks in Gulf of Mexico coastal waters of the U.S. Abbreviations as in Table 1.

RIVER SYSTEM	GULF STURGEON	ATL. RACE STRIPED BASS	GULF RACE STRIPED BASS	ALABAMA SHAD	SKIPJACK HERRING
GEORGIA					
51. Chattahoochee	-	I	I	N	S
52. Flint	-	I	I	N	S
FLORIDA (Gulf coast)					
53. Hillsborough	E	PNP	PNP	PNP	-
54. Suwannee	S	PNP	PNP	PNP	-
55. Apalachicola	T/D	N	T/S	D/S	-
56. Ocklockonee	E	I	-	PNP	-
57. Escambia	-	-	-	-	-
ALABAMA					
58. Alabama	N/E	I	D/E	D/S	N/S
59. Tombigbee	N/E	I	D	D/S	N/S
60. Perdido	N	I	N	N	N
61. Bon Secour	N	I	N	N	N
62. Fish	N	I	N	N	N
63. Magnolia	N	I	N	N	N
64. Dog	N	I	N	N	N
65. Fowl	N	I	N	N	N
66. Tennessee	E	I	PNP	S	-
51. Chattahoochee	-	-	-	S	S
67. Coosa	E	I	E	D	D
68. Tallapoosa	E	I	E	D	D
MISSISSIPPI					
69. Pascagoula	N	-	N	-	-
70. Tchouticabouffa	N	I	N	-	-
71. Biloxi	N	I	N	-	-
72. Wolf	N	I	N	-	-
73. Jourdan	N	I	N	-	-
74. Pearl	N	I	N	-	-
LOUISIANA					
74. Pearl	N	I	N	N	S
75. Bayou LaCombe	N	I	N	N	S
76. Tchefuncte	N	I	N	N	S
77. Tangipahoa	N	I	N	N	S
78. Tickfaw	N	I	N	N	S
79. Amite	N	I	N	N	S
80. Mississippi	N	I	N	N	S
81. Atchafalaya	N	I	PNP	N	S
82. Vermillion	N	N	PNP	N	S
83. Mermentau	N	I	PNP	N	S
84. Calcasieu	N	I	PNP	N	S
85. Sabine	N	I	PNP	N	S

spawning is not successful in dry years (E. G. Simmons, Texas Parks and Wildlife Dept., Austin, pers. comm.).

The life history and status of the Gulf race striped bass is not well-documented. Historically they ranged from the Suwannee River, Florida, to Lake Pontchartrain, Louisiana, but their egg, larval, and juvenile distributions are unknown. Adult populations are endemic and riverine; few fish venture into the Gulf of Mexico, though they may enter adjacent river systems. Adults migrate upstream in spring to spawn; however, little is known about local riverine habitats and distributions. The spawning season in the Apalachicola River, Florida, ranges from early February to mid-May. Spawning activity occurs from 15 to 22.3 C and is greatest at approximately 20 C. The stock status of Gulf striped bass was not known for 55% of the riverine areas listed, and another 24% of the areas probably never supported Gulf race populations. Areas in which the Gulf striped bass no longer occur include the Coosa and Tallapoosa Rivers (Table 2).

Sturgeon

The shortnose sturgeon (*Acipenser brevirostrum*) ranged historically from the Saint John River, New Brunswick, to the St. Johns River, Florida, and is considered an endangered species by the four South Atlantic States. Little is known of its migratory patterns or localized distributions and habitats. Adults are frequently observed in large tidal rivers, but are also collected in brackish and salt waters. Shortnose sturgeon may not be as migratory as other sturgeon species. Populations have been extirpated in 46% of the riverine areas listed by State agencies, and 51% of the areas contain threatened populations. Shortnose sturgeon status was unknown for the St. Marys and Nassau Rivers, Florida (Table 1).

The life history and status of Atlantic sturgeon (*A. oxyrhynchus oxyrhynchus*) is not well-documented for Southeastern waters. Atlantic sturgeon range from the Hamilton River Inlet, Labrador, to the St. Johns River, Florida. The adhesive and demersal eggs are found in fresh or brackish waters. Larval distribution is not known. Juveniles are benthic and reside in rivers for one to three years before emigrating to sea. The offshore distributions and spawning migrations of adults are poorly understood. Riverine distributions are a function of fish sex and migration direction (upstream or downstream). Adults frequent cool, deep holes during summer. Most populations in South Atlantic States are stable (42%) or declining (27%). None was listed as increasing or extirpated, and only the St. Marys River population may be threatened. Status was not known for populations in 21% of the areas listed (Table 1).

The condition and life history of Gulf sturgeon (*A. o. deSotoi*) in coastal waters of the Gulf of Mexico is not well-documented. Historically, this subspecies ranged from Tampa Bay, Florida, to Louisiana; presently 19% of the populations are considered extirpated (Table 2). Local habitats and distributions are not known, but Gulf sturgeon occur in the Suwannee River, Florida, and in most major coastal rivers of the Gulf drainage. Minimum age to maturity is VII for males and VIII for females. The spawning season in the Suwannee River ranges from late May through October. None of the areas was reported to have increasing populations, and only the Suwannee River population was considered stable. The Apalachicola River population was thought to be in decline or threatened. Gulf sturgeon status for the remaining 75% of the areas was not known (Table 2).

Clupeids

The life history and status of American shad (*Alosa sapidissima*) in South Atlantic States is fairly well-known. Shad range from the St. Lawrence River, Canada, to the Tomoka River, Florida. The demersal, non-adhesive eggs are dispersed in fresh water by currents. Larvae, most abundant at the surface, drift downstream to disperse in fresh or brackish waters after reaching sufficient size. Juveniles form schools at 20 to 30 mm TL, preferring deep pools away from shorelines in non-tidal areas. Emigrating to estuarine areas by autumn, juveniles enter the ocean for three to five years before returning to natal streams as adult spawners. Shad from U.S. rivers feed in the Bay of Fundy, Canada, during summer and then move south to overwinter, possibly offshore Middle and South Atlantic States. In the spring, adults migrate north or south to spawn in natal rivers. South Atlantic coast stocks are comprised primarily of age classes IV (36%), V (46%), and VI (14%) (sexes combined), and less than 0.1% of the fish are older than age XI. Minimum age to maturity is IV for males and V for females. Spawning, which occurs in tidal freshwater with extensive flats or in shallows near creek mouths, follows a south to north temporal progression beginning as early as December in Florida and extending to June in Canada. Spawning temperatures range from 12 to 20 C. Most American shad populations in the South Atlantic States are stable (32%) or in decline (44%) (Table 1). Only the Sampit River, South Carolina, population is thought to be increasing, and the New River, North Carolina, population is the only one threatened. Population status was unknown for 14% of the habitats; no populations were believed extirpated.

The hickory shad (*A. mediocris*) occurs from the Bay of Fundy to the Tomoka River, Florida, and its life history is poorly documented for Southeastern waters. The adhesive eggs, which are easily dislodged by currents, are semi-demersal in slow waters but become buoyant in turbulent waters. Larvae are found in fresh and brackish rivers and tributaries, though local habitats and distributions are unknown. Juveniles frequent fresh- and brackish-water areas; lower estuaries and ocean nearshore areas are possibly utilized as nursery grounds. The oceanic distributions for adults have not been determined. Adult riverine distribution is similar to that of American shad. Both males and females may be sexually mature at age II. The spawning season probably begins in early March and continues through mid-May;

spawning temperatures range from 9.5 to 22 C. Stable (40%) or decreasing (31%) stocks are present in the majority of Southeast river systems. The New River, North Carolina, population is threatened (Table 1). Population status for 24% of the systems is not known.

Blueback herring (*A. aestivalis*), also known as river herring or glut herring, range from Cape Breton, Nova Scotia, to the Tomoka River, Florida. Eggs are somewhat adhesive, being demersal in still waters and pelagic in lentic waters. Larvae are found throughout river systems, with greatest occurrence in creeks and tributaries. Juveniles congregate in creeks and tributaries during summer months and emigrate in the fall to overwintering areas (sounds, bays, and deep estuarine areas). The parental stocks apparently move to offshore bottom waters during the winter and inhabit a narrow band of coastal water. In the spring these stocks enter fresh and brackish waters to spawn. Southeastern stocks (sexes combined) consist primarily of fish from age classes IV (38%), V (40%), and VI (13%); both sexes attain sexual maturity by age III. The spawning season progresses temporally from south to north, starting in January in Florida and continuing to mid-July in northern latitudes. Spawning occurs in shallow, heavily-vegetated areas, rice-fields, upstream river swamp areas, and small tributaries above the tidal zone; temperatures range from 13 to 27 C. A large proportion (40%) of blueback herring populations in South Atlantic States is declining (Table 1). An additional 22% of the populations are stable. Stock status for 32% of the systems was not known. The New River, North Carolina, population was believed threatened and the Cooper River, South Carolina, population may be increasing.

Although the historical range of alewife (*A. pseudoharengus*) extends from Newfoundland southward to South Carolina, the Lake Mattamuskeet canals bordering the Pamlico and Albemarle Sounds appear to be the southern boundary for alewife populations large enough to support commercial fisheries (M. W. Street, Div. Marine Fish., Morehead City, NC, pers. comm.). Most (74%) North Carolina waters contain declining alewife stocks, and their status in 20% of the areas is unknown (Table 1). The New River population is believed to be threatened.

The Alabama shad (*A. alabamae*) is a little-studied clupeid that inhabits Gulf coastal waters and occurs from the Suwannee River, Florida, west to the Mississippi River. The status of Alabama shad populations is unknown for the majority (66%) of the Gulf coast rivers listed by State agencies (Table 2). Approximately one-fourth of the areas contain stable (12%) or declining (12%) stocks.

Little information has been published concerning the life history and population dynamics of skipjack herring (*A. chrysochloris*) in Gulf coastal waters. Responses by State agencies estimated that 63% of the populations contained in Gulf coastal rivers are stable and 8% may be in decline. Status of populations in the remaining areas (29%) was unknown.

Environmental Factors Influencing Fish Stocks

Many sounds, estuaries, and tributaries of the four South Atlantic States have been altered physically and chemically by industrial, municipal,

Table 3. Factors possibly important in contributing to the decline of certain populations of anadromous fish in coastal waters of South Atlantic and Gulf States.

RIVER SYSTEM	NONE OR INADEQUATE INFORMATION	CHANNELIZATION	DREDGE AND FILL PROJECTS	BULKHEADING	DAMS AND IMPOUNDMENTS	INDUSTRIAL WATER INTAKES	INDUSTRIAL DISCHARGE LOCATIONS	ROAD CONSTRUCTION	ACID RAIN	CHEMICAL POLLUTION (eg. heavy metals)	THERMAL EFFLUENTS	TURBIDITY	LOW OXYGEN LEVELS	SEWERAGE OUTFALLS	INADEQUATE FISHWAY FACILITIES	POOR CONTROL OF WATER RELEASE (dams)	REDUCED FRESHWATER TO ESTUARIES	REDUCED SPAWNING HABITAT	REDUCED NURSERY HABITAT	POOR FOOD AVAILABILITY	SPAWNING GROUNDS TOO ACCESSIBLE	GENERALLY POOR WATER QUALITY	OTHER (See footnotes)
NORTH CAROLINA																							
Currituck So.	X																						
Albemarle So.													X									X	
Chowan		X	X	X		X	X			X		X	X	X			X	X	X	X	X	X	
Roanoke		X	X		X	X	X			X		X	X	X		X		X	X	X	X	X	
Pungo	X																						
Other Albemarle area rivers																						X	
Albemarle So. tributaries		X	X	X						X			X	X				X	X	X	X	X	
Tar		X			X	X	X			X		X	X	X		X		X	X	X	X	X	
Pamlico	X																						
Neuse		X			X	X	X			X		X	X	X	X	X		X	X	X	X	X	X[1]
Trent	X																						
North	X																						
Newport	X																						
White Oak	X																						
New		X																X	X				
Cape Fear		X			X	X	X			X		X	X	X	X	X		X	X	X	X	X	X[1]
N.E. Cape Fear		X								X		X	X	X				X	X	X	X	X	
Pee Dee					X	X	X			X		X	X	X		X						X	
SOUTH CAROLINA																							
Waccamaw	X																						
Pee Dee					X										X			X	X				
Black	X																						
Sampit										X												X	
Santee					X										X			X	X				
Cooper					X													X	X				
Ashley	X																						
Edisto	X																						
Ashepoo	X																						
Combahee	X																						
Lynches	X																						
Savannah					X										X			X	X			X	
GEORGIA																							
Savannah			X		X		X			X								X	X	X		X	
Ogeechee		X	X		X		X											X	X	X		X	
Altamaha		X	X		X		X			X								X	X	X		X	
Satilla		X	X		X		X											X	X	X		X	
St. Marys			X		X		X											X	X	X		X	
Apalachicola					X													X					

Table 3 (cont'd.)

RIVER SYSTEM	NONE OR INADEQUATE INFORMATION	CHANNELIZATION	DREDGE AND FILL PROJECTS	BULKHEADING	DAMS AND IMPOUNDMENTS	INDUSTRIAL WATER INTAKES	INDUSTRIAL DISCHARGE LOCATIONS	ROAD CONSTRUCTION	ACID RAIN	CHEMICAL POLLUTION (eg. heavy metals)	THERMAL EFFLUENTS	TURBIDITY	LOW OXYGEN LEVELS	SEWERAGE OUTFALLS	INADEQUATE FISHWAY FACILITIES	POOR CONTROL OF WATER RELEASE (dams)	REDUCED FRESHWATER TO ESTUARIES	REDUCED SPAWNING HABITAT	REDUCED NURSERY HABITAT	POOR FOOD AVAILABILITY	SPAWNING GROUNDS TOO ACCESSIBLE	GENERALLY POOR WATER QUALITY	OTHER (See footnotes)
FLORIDA (Atlantic coast)																							
St. Marys		X		X		X	X			X									X			X	
Nassau				X		X	X			X									X			X	
St. Johns		X	X	X	X	X	X			X	X		X	X			X	X	X		X	X	X[2]
Tomoka										X												X	X[3]
FLORIDA (Gulf coast)																							
Hillsborough					X		X							X	X			X	X			X	X[4]
Suwannee							X																X[5]
Ocklockonee	X																						
Apalachicola		X	X	X	X		X			X		X		X	X			X	X			X	X[6]
ALABAMA																							
Perdido							X			X			X									X	
Alabama				X	X		X			X			X		X	X		X	X			X	
Tennessee				X	X		X			X					X	X		X	X				
Tombigbee					X		X			X			X		X	X		X	X			X	
Coosa					X		X			X					X	X		X	X			X	
Tallapoosa					X		X			X					X	X		X	X			X	
Fish	X																						
Magnolia	X																						
Bon Secour							X																
Dog	X																						
Fowl	X																						
MISSISSIPPI																							
Pascagoula	X																						
Tchouticabouffa	X																						
Biloxi	X																						
Wolf	X																						
Jourdan	X																						
Pearl	X																						
LOUISIANA																							
Pearl		X			X																		
Bayou LaCombe		X			X																		
Tchefuncte		X																					
Tangipahoa		X																					
Tickfaw		X																					
Amite		X																					
Mississippi		X	X																				
Vermillion	X																						
Mermentau		X	X		X																		
Calcasieu	X																						
Sabine	X																						
Atchafalaya		X	X																				

[1] Locks and barriers

[2] Agricultural drainage; agricultural and domestic non-point source pollution; overfishing

[3] Agricultural and domestic non-point source pollution

[4] Agricultural and domestic non-point source pollutants; overfishing; proposed phosphate strip-mining

[5] Agricultural and domestic non-point source pollutants; proposed phosphate strip-mining

[6] Overfishing

and agricultural practices. Generally poor water quality is characteristic of 53% of the South Atlantic coastal areas listed by State and Federal agencies (Table 3). Reduced spawning habitat for anadromous stocks is a problem in 50% of the areas, and 47% have reduced nursery habitat. Chemical pollution, dams and impoundments, industrial discharge locations, and channelization each affect over a third of the systems. Other problems identified as contributors to stock decline in South Atlantic States included poor food availability, low levels of dissolved oxygen, sewerage outfalls, dredge and fill projects, industrial water intakes, turbidity, and exploitation of stocks on spawning grounds. Less than 20% of South Atlantic coastal areas with anadromous stocks are affected by bulkheading, inadequate fishway facilities, or poor control of water releases. Other contributing factors listed by agencies included locks and barriers, agricultural drainage, domestic non-point source pollution, and overfishing. Reduced freshwater input to estuaries, thermal effluents, acid rain, and road construction were not considered important to the decline of anadromous stocks in South Atlantic States (Table 3).

Coastal waters of the Gulf of Mexico have also been physically and chemically altered. Many of these areas (44%) listed by State agencies may not be affected or were not assessed due to lack of information (Table 3). Factors possibly important in contributing to the decline of anadromous stocks in Gulf coastal waters include channelization (31%), dams and impoundments (31%), chemical pollution (30%), inadequate fishway facilities (22%), poor control of water releases (16%), and dredge and fill projects (12%). Additional factors affecting less than 10% of the river systems include bulkheading, low oxygen levels, sewerage outfalls, and turbidity. Anadromous stocks in Florida are also affected by overfishing and agricultural and domestic non-point source pollutants. Proposed phosphate strip mining in Florida may also affect certain anadromous stocks. Factors that apparently do not contribute to the decline of stocks in Gulf coastal waters are industrial water intakes, road construction, acid rain, thermal effluents, poor food availability, reduced freshwater input to estuaries, and fisherman accessibility to spawning grounds (Table 3).

In conclusion, the following summary was developed from the information provided by responses to our questionnaire:

1) Stable populations were reported for over 40% of all Southeastern stocks of Atlantic race striped bass, Atlantic sturgeon, hickory shad, and skipjack herring.

2) Stock status for over 50% of the Gulf race striped bass, Gulf sturgeon, and Alabama shad populations is not well-documented.

3) Declining populations characterize over 40% of all Southeastern stocks of American shad, blueback herring, and alewife.

4) Most populations (88%) of Atlantic race striped bass in Gulf coastal waters have increased due to intensive stocking efforts.

5) The only species considered endangered by Southeast State and Federal fisheries agencies is the shortnose sturgeon, *Acipenser brevirostrum*.

6) Nearly one-half of the South Atlantic coastal areas listed have poor water quality and reduced spawning and nursery habitat for anadromous fishes. A variety of physical and chemical alterations to estuarine areas apparently contribute to the decline of specific stocks along the South Atlantic coast.

7) Coastal waters of the Gulf of Mexico have also been altered physically and chemically, but many (44%) could not be assessed by the agencies due to lack of information.

8) Extensive additional information is needed before intelligent management decisions can be developed concerning anadromous stocks in the Southeastern United States. Future research is needed regarding life histories, population dynamics, and environmental factors that influence status of the stocks.

Acknowledgments

We thank all of the Southeastern State and Federal fisheries agencies and their personnel who responded to our questionnaire and supplied supplemental life history information on Southeastern anadromous species. R. T. Bowyer, K. L. Beal, G. G. Rulifson, and an anonymous reviewer gave valuable comments which improved the final manuscript. The work was sponsored by the U.S. Fish and Wildlife Service and accomplished under the supervision of the North Carolina Cooperative Fishery Research Unit, which is jointly supported by North Carolina State University, North Carolina Wildlife Resources Commission, and the Fish and Wildlife Service, U.S. Department of the Interior. The views expressed are those of the authors and do not necessarily represent the official policy of Southeastern State agencies or of the U.S. Fish and Wildlife Service, Department of the Interior, or any of its agencies.

References Cited

Beal, K. L. 1980. Territorial sea fisheries management and estuarine dependence, pp. 67-77. *In:* V. S. Kennedy (ed.), *Estuarine Perspectives*, Academic Press, New York.

Rulifson, R. A., M. T. Huish and R. W. Thoesen. 1982. Anadromous fish in the Southeastern United States and recommendations for development of a management plan. U.S. Fish and Wildlife Service, Fishery Resources, Region 4, Atlanta, GA, 525 p. (Available from Fishery Resources, USFWS, 75 Spring St., S. W., Atlanta, GA 30303.)

FACTORS AFFECTING RECRUITMENT OF POTOMAC RIVER STRIPED BASS AND RESULTING IMPLICATIONS FOR MANAGEMENT

Tibor T. Polgar
Martin Marietta Corporation
Environmental Center
1450 South Rolling Road
Baltimore, Maryland

Abstract: Studies of the Potomac River estuary's striped bass *(Morone saxatilis* Walbaum) population have shown that year-class strengths are largely determined by density-independent environmental factors. The level of recruitment of mature adults into the estuary's fishery follows trends in year-class strength and is also modified by fluctuations in fishing effort in and out of the estuary. Ichthyoplankton distributions, together with information on environmental variability, provide a basis for determining year-class strength prior to the juvenile stage. Feeding habits of larvae and juveniles and food availability are important links between abiotic environmental factors and eventual year-class strength. Given the resulting dominant-year-class nature of the population (resource), in which excess, harvestable production is provided by the occasionally successful year-class, management should focus on conserving yields from large cohorts. My analysis demonstrates that this strategy would increase yields over time because it increases net biomass by lowering mortalities during the early phases of recruitment.

Introduction

"Under constant environmental conditions, a population whose size remains bounded usually has some kind of density dependent regulation . . . . However, there is no guarantee that all populations remain within finite bounds or that density-dependent factors predominate. Environmental conditions vary in time, and the organism involved may be sufficiently sensitive to environmental changes so that the population size is determined largely by extrinsic factors." (Sawyer and Slatkin 1981)

The authors of the quote above pose a very important generalization about the factors controlling biological populations. The question of how the combination of environmental and intrinsic biological population controls affect the dynamics of fish stocks is an extremely practical issue for fisheries managers, whose concerns are optimization of yields, stabilization of catches, and persistence of stocks. Proper strategies for reaching management objectives, however, will have to account for the characteristics of natural population-control mechanisms, be they density-dependent, environmentally regulated, or combinations of these.

ISBN 0-12-404070-5

This paper discusses an analysis of population-control factors for the stock of striped bass *(Morone saxatilis* Walbaum) that spawns in the Potomac River (hereafter called Potomac striped bass) that emphasizes inferences drawn from a four-year field study and from historical catch records. Specifically, evidence for the dependence of spawning success on environment is examined through the various life stages, which will shed some light on the relationship between the size of spawning stock and recruitment. Population modeling, incorporating the concepts of population control derived from field data and literature information, allows quantification of relationships among recruitment level, spawning-stock size, and environmental variables affecting the persistence of the population. The results form the basis for a discussion of management strategies that are consistent with factors controlling population renewal.

Although the Potomac striped bass population is only a single, indigenous subpopulation of the Atlantic striped bass stock, the population-control mechanisms affecting it can be generalized to subpopulations in all the other coastal estuaries. The transferability of findings is supported by historical catch records and measures of year-class strength. Recruitment strength into the entire Chesapeake Bay fishery can be forecast from seining index values for juveniles (Chadwick 1964; Boone 1978; Cooper and Polgar 1981). The index values are specific to all major tributary systems used for spawning in the Chesapeake Bay; while the values vary consistently in magnitude among subsystems, their temporal variability has common short- and long-term trends. The parallel temporal trends among regions, in both the measures of recruitment strength and catches, are due to two principal causes. First, the migratory stocks of Chesapeake Bay striped bass are the major contributors to the Atlantic coastal fishery (Clark 1968; Berggren and Lieberman 1978; Van Winkle et al. 1979); and second, as evidenced by parallel trends in indices of spawning success, the subpopulations in the Bay tributaries seem to respond to common factors controlling population size.

The primary hypothesis underlying the concepts presented in this paper is that the mutually shared factors that control the sizes of striped bass populations in various estuaries where they spawn are climatological. In other words, striped bass is a species that “may be sufficiently sensitive to environmental changes so that the population size is determined largely by extrinsic factors.” (Sawyer and Slatkin 1981)

Population Control Mechanisms

Classical approaches to analyzing the regulation of population size in fisheries were derived primarily for self-regulating populations that exist under relatively constant environmental conditions. The formulations most commonly used are those of Beverton and Holt (1957), Cushing (1971), and Ricker (1975). All three rely on density-dependent, spawner-recruit functions to stabilize population sizes around an equilibrium value, whether the population is fished or not. The principal difference among the three formulations is in the severity of prerecruitment mortality associated with increases in parent stock. This population-stabilizing dynamic behavior is generally referred to as compensation, or compensatory mortality. It is achieved only by the dependence of progeny on parent-stock density in a limited habitat. The literature suggests that the major biological mechanism involved in density-dependent compensation is larval survival, which depends on competition, cannibalism, predation, and behavior of the spawning stock. Thus, the density-dependent processes regulating population size are thought to be determined during an early, critical period in the life history, especially at the time when larvae first feed (Hjort 1926; Lasker 1965; Jones and Hall 1973; May 1974). Clark (1974) suggested modifications in spawner-recruit formulations to include depensatory mortality, or an increase in early mortality when the biomass of the parent stock is low and decreases further. This is also a density-dependent process, thought to operate on schooling pelagic fishes, and both kinds of density dependence have been applied simultaneously to some stocks (e.g., Parrish and MacCall 1978).

Some of the population properties reflected in stock data that can be used to diagnose density dependence are: relatively constant age structure, little change in catch per unit of effort, predictability of immature abundances by the size of the spawning stock, and no significant correlations with environmental variables. The specific biotic mechanisms that bring about this kind of dynamic behavior include the responses in growth, fecundity, and mortality to existing population density, with negligible environmental effects on the survival of life stages. The major consequence of density-dependent self-regulation is population stability, which provides for predictability and manageability of stocks. Therefore, yield-per-recruit, surplus-production, and stock-production models, which all rely heavily on population levels being at equilibrium, properly reflect the dynamics of self-regulating populations.

When the effects of extrinsic environmental factors become significant, population levels begin to manifest large temporal variations. However, the fact that recruitment depends heavily on environmental variation does not preclude the possibility that it may depend on the size of the parent stock. Density dependence may simply be masked by the effects of environmental variation, while still operating on the population during the critical period of early development. The necessity for demonstrating that both density-dependent and density-independent components of control coexist in a single year-class population is clear. If recruitment of a single year class of species depended entirely on randomly acting environmental factors, then the species could quickly become extinct. However, assessing the density dependence of multiple-age populations, in which several age classes of repeat spawners contribute to any given year's recruitment, is much more difficult. Conceivably, minimum reproductive potential could be maintained through many years of unfavorable (environmentally affected) recruitment, with negligible or nonexistent density dependence. Nonetheless, to test hypotheses regarding the relative contributions of density-dependent and environmental factors on population regulation, both kinds of mechanisms must be analyzed.

The usual symptom of weakening of density-dependent regulation by environmental factors is that no obvious empirical relationships can be found between the spawning stock and the immature progeny it produces. However, adult stock levels can be forecast reliably from juvenile indices. Environmentally induced, dominant, year classes produce unstable age structures in the adult population over time. Thus, equilibrium-based, optimum yields cannot be defined by applying yield models that assume biologically regulated, stable population levels. Surplus production in populations that are affected significantly by environmental variation depends principally on the size of occasionally successful or dominant year-classes, and not on compensatory (density-dependent) reserves.

Early investigators of striped bass population dynamics recognized the great importance of environmental population controls (Pearson 1938; Merriman 1941; Vladykov and Wallace 1952). So, many fishery managers have been reluctant to apply standard management techniques to striped bass for fear of endangering population persistence in favor of optimizing yields (Polgar 1980; Cooper and Polgar 1981). This is partly because of the uncertainty about the relative strength of stabilizing, self-regulating, density-dependent, population-control mechanisms. If their programs are to be effective, the managers of striped bass resources need to assess the relative importance of environmental control in population persistence.

Classical spawner-recruit functions may be multiplied by functions of environmental variables to improve the amount of variability in recruitment that is explained by density-dependent mechanisms alone. With this approach, density-dependent mechanisms are only apparently independent of the effects of the environment, or in other words, compensatory or depensatory responses are still only explicit nonlinear functions of parent-stock density. However, there may be statistical interactions among density-dependent mortalities and the stochastic environmental components of mortality. The choice of environmental variables to consider in such treatment is only clear when environmentally induced variation in recruitment is strong, and correlations exist among some set of variables and recruitment levels and/or catch patterns. The selection of environmental influences for consideration may involve single variables or combinations of variables, based upon known requirements of the organism through its critical period of development.

Some successful attempts to elucidate mechanisms of environmental population control have been made for widely different types of fisheries. Parrish and MacCall (1978) demonstrated that when environmental dependence is considered, the fit for the spawner-recruit relationship for Pacific mackerel improves significantly. The maximum variability in recruitment explained by density-dependent formulations alone was approximately 30%, whereas as much as 76% could be explained when larval transport and upwelling variables were also considered. The spawner-recruit relationship thus derived can be interpreted in terms of larval feeding requirements and other processes during the critical developmental period. It is worth noting that although environmental effects are considered to be explicitly density-independent, Parrish and MacCall (1978) entertain the idea that environmental effects may act on populations in both density-dependent and independent modes, due to possible interactions of abiotic and biotic processes that affect density. An important conclusion of their study is that "the recovery of the stock is more likely to be related to environmental conditions than to resiliency of a population that is under carrying capacity." Thus, for Pacific mackerel, a multiple age-class fishery, it seems that a minimum reproductive potential is always maintained to cope with periods of adverse environmental conditions.

Nelson et al. (1977) explained variations in Atlantic menhaden recruitment by using a survival index to account for the deviations from a density-dependent Ricker function. The most significant environmental variable incorporated into the index was zonal Ekman transport, which carries larvae from offshore spawning areas to inshore nursery grounds. Although the authors said that environmental variables act independently of stock density, they conceded the possibility that changes induced in larval density by abiotic mechanisms may become virtually density-dependent, and would consequently appear in the density-dependent portion of the spawner-recruit relationship.

Complete analyses of population-control mechanisms are hardly possible for most species, either because of the quantity or the quality of data required. It is difficult at the inception of most population studies to anticipate the potentially important environmental data to collect, or the combinations of variables to scrutinize as partial predictors of population size. Thus, most quantitative studies are by necessity pieced together through hindsight from various sources of primary data. The Potomac Striped Bass Study suffered from the same drawbacks. It covered only a few years, less than a generation time for striped bass, yet its results confirmed some suspected hypotheses on environmental dependence. As will be seen, supplemental historical data were indispensable to the analyses.

The Potomac Striped Bass Study

The Maryland Power Plant Siting Program undertook a four-year, intensive study of striped bass, beginning in 1974, to determine the potential impacts of power plants on the Potomac River subpopulation. The study objectives recognized the importance of treating an entire regional spawning population in detail to assess the impact of local perturbation. Further, the possible uniqueness of the subpopulation was recognized, so population-control mechanisms were not assumed from previous striped bass investigations, but an understanding of them was an objective of the study. This philosophy resulted in the following general approach. Impacts on vulnerable life stages could be put into context only if their natural dynamics were understood. Thus, early life stages in the entire system were followed until year-class strength was determined, past the critical stages of development. Abundance, distribution, and structure of the spawner population were also investigated with the hope of establishing relationships among reproductive potential, eventual spawning success, and distributional mechanisms acting on ichthyoplankton. To relate distributions of early life stages to environmental and biotic factors affecting survival of those stages, the program developed a large set of data from concurrent physical, chemical, and biological measurements. (Details of the various sampling programs are in several references cited below.)

Striped bass ichthyoplankton were sampled weekly for abundance measurements during spawning and development. These measurements were made generally at twelve transects with four stations each and encompassed 100 km of the Potomac estuary within the salinity range from low mesohaline to tidal fresh water. Ichthyoplankton were classified into four life stages: eggs, yolk-sac larvae, finfold larvae, and post-finfold larvae. Data on larval length distributions and stomach contents from the 1976 and 1977 collections were related to zooplankton abundances (Setzler-Hamilton et al. 1981).

Juvenile striped bass abundances were determined biweekly. The start of the collection program overlapped the period when the striped bass develop into the post-finfold larval stage. Collections by beach seines and by bottom sleds used standardized effort. Stomach contents of juvenile striped bass were also analyzed. Boynton et al. (1981) give details of the juvenile investigation.

Over the four study years, seven gill-net sampling stations provided several types of information on the adult spawning stock (details are given by Jones et al. 1977). First, the experimental catch data described variations in age, sex, fecundity, and abundance among years and during each year's spawning run. Second, the data allowed some generalizations to be made between the spatial and temporal distributions of spawned eggs and the origin of these eggs in various classes of the spawning stock. Third, the population fecundity data provided a benchmark for comparing the egg estimates from ichthyoplankton sampling.

Results of the Ichthyoplankton Studies

Many of the important observations relating to spatial and temporal distributions of striped bass ichthyoplankton have been summarized by Setzler-Hamilton et al. (1981). The temperature-triggered onset of spawning, upstream progression of spawning activity in time, and appearances of multiple spawning peaks are the common characteristics found in the four years of data.

The most noteworthy observations relating to ichthyoplankton are the spatial and temporal changes in distribution through the developmental stages. Interpretation of these distributional changes establishes the connections between a stock's reproductive potential (observed egg abundances or population fecundity) and eventual measures of year-class strength (juvenile abundances). The spawning areas and nursery areas for all the larval stages of striped bass are in tidal fresh and oligohaline regions of the estuary that at most have weak, two-layered estuarine circulation. There is generally a marked and continuous downstream net flow through the areas where striped bass ichthyoplankton are found. Yet, the presumably planktonic larval stages seem to have been abundant consistently in the region of peak spawning, or well upstream of it. Early hypotheses explaining this phenomenon (Polgar et al. 1976) have been borne out by the consistency in distributional patterns over the four-year study and by rigorous modeling of the coupled spawning-developmental-mortality-transport process (Ulanowicz and Polgar 1980). In a gross sense, the virtual retention of successive larval stages within a limited segment of the estuary results from the sum of three competing processes: upstream movement of spawning adults in time, continual downstream transport of eggs and larvae, and significantly higher mortalities of larvae toward the downstream nursery area.

Modeling of the developmental process, using measured transport properties of the estuary and observed ichthyoplankton abundances (Ulanowicz and Polgar 1980), allowed the mapping of the egg distributions that produced the distributions of the surviving, last larval stage. From this analysis, we concluded that a small fraction of late, upstream-spawning adults produced the successful late-stage larvae, whereas almost all of the progeny of the peak spawn, which was earlier in the season and farther downstream, died before the finfold stage. While the spawning stock size declined continually over the four years, spawning success, as measured by the ratio of post-finfold larval abundance and egg production, varied irregularly and by a factor of 35. The trend in the variation of spawning success was paralleled, however, by the trend in the juvenile index.

These results suggest that environmental factors dominate the survival of larval stages so greatly that year-class strength is determined independently of parent-stock potential. Analyses by Beaven and Mihursky (1980), Setzler-Hamilton et al. (1981), and myself indicate that food availability is the direct cause of differential mortality of larvae with distance downstream and among years. This conclusion was reached by relating stomach contents in all larval stages to microzooplankton densities in the same areas. The data show that larvae are feeding on a limited group of species and stages of zooplankton from the available assemblages. The striped bass larvae seem to be preferring the less abundant adult and copepodite life stages of copepods and cladocerans.

The timing of spawning and larval development with respect to the abundance of several preferred larval prey also had various degrees of coincidence, as reflected in larval stomach contents. Analysis of differences in length between feeding and nonfeeding larvae indicates that much of the critical period encompasses the life span for the yolk-sac (first-feeding) stage and some of the finfold stage. Preferred prey are abundant in the upstream regions of the nursery area, and their declines seem to be associated with calculated increases in larval mortality in the downstream direction (Ulanowicz and Polgar 1980; Setzler-Hamilton et al. 1981).

Several potentially complex interactions affect the evolution of larval distributions. The rate and extent of upstream movement by spawning adults could be influenced by their density and sex ratio (Jones et al. 1977) and by the rate of the spring temperature rise. The latter factor may be related to river flow, which also delivers nutrients for microzooplankton production. High spring river flow and an associated slower increase in temperature, which could cause farther and later penetration of adults upstream, may increase not only the proportion of eggs orginating upstream, but would tend to increase the residence time of devloping larvae under more favorable nutritional conditions. High flows would tend to increase the size of the suitable nursery area. Adequate data describing the complex variations of these factors are not available for a long enough period for rigorous analysis, so the consistency of results among the various life stages still must be examined.

Results of Juvenile Studies

Analyses of historical index records for the Potomac (Boone 1978) show that juvenile abundances are good indices of year-class success because they are significantly related to subsequent commercial catches (Polgar 1980; Boynton et al. 1981; Kohlenstein 1981). Although the trend between measures of spawning success and juvenile indices seems consistent for the four-year study, no relationship is apparent between juvenile abundances and reproductive potentials, or between juvenile and egg abundances. The ichthyoplankton studies, which indicate that the yolk-sac and finfold larval stages are critical, underscore the fact that year-class strength is largely determined by the advent of the juvenile stage.

A study of juvenile feeding habitats (Boynton et al. 1981) showed that the composition of the diet merely reflects natural estuarine gradients in prey abundances. The totally nonselective, opportunistic, feeding strategy of juveniles points to easy adaptation to alternative food items. This may be an important mechanism for stabilizing and lowering juvenile mortality rates, thereby consistently establishing year-class strength following the larval period. The consequence of this, of course, is that juvenile abundances can be used to forecast the relative abundances of future adult stocks (Polgar 1980).

The early studies of striped bass development (Merriman 1941; Vladykov and Wallace 1952) recognized that cold winters prior to the spawning period and large river discharges into the estuary during spawning may have had something to do with spawning success. Heinle et al. (1977) observed that following early winter freezes, greater amounts of carbon appeared in the estuary during the spring. The association between carbon availability and zooplankton production as a function of winter temperature is clear, especially in coastal estuaries with extensive rooted aquatic vegetation. Heinle et al. (1977) also observed some weak correlations between low winter temperatures and higher striped bass catches.

Since juvenile abundances bore no correlation to striped bass landings, and had no parallel trends with egg abundances, Walter Boynton (of Chesapeake Biological Laboratory) and I started to search for an explanation of juvenile index variability using temperature and river-flow variables, which might influence spawning success. After extensive trial and error using stepwise regressions and combinations of variables, we found that variability in the Potomac juvenile index (j) between 1954 and 1978 was best explained by the following regression:

$$j = -1.73 * DT + 0.21 * AT + 2.31 * AF$$

with $R^2 = 0.73$ and F (24 d.f.) $= 19.91$ $(P<0.01)$.

The predicted variable is the juvenile index, with the predictors being December air temperature (DT) and April air temperature (AT) measured in °C, and late April river flow (AF) measured in cubic meters per second. Air temperatures were used because of the lack of consistent, historical water temperature data. In light of the discussion on the processes affecting ichthyoplankton survival, the dependence of the juvenile index on these environmental variables can be interpreted in terms of factors determining spawning success. The cold December temperatures in the winter prior to spawning provide a carbon source for spring microzooplankton production, because early icing freezes in place portions of the previous year's carbon production for spring use. High April flows provide the delivery system for the stored carbon into the nursery area. The relatively small, but significant contribution by the April air temperature has no obvious explanation. However, solar heating and the quantity of cold river runoff compete in controlling spring water temperature, and thus, air temperature may explain only a flow-independent mechanism for modifying water temperature, and the rate of spawner migration. The April flow variable probably also contains temperature effects.

The most important result of the clear-cut dependence of the juvenile index (year-class strength) on environmental variables is that these variables are consistent with limited observations on larval mortality patterns, estuarine transport, spawning-stock behavior, and preferred larval food distributions, as discussed previously. Thus, the end point in the measurement of spawning success is juvenile abundance, which integrates all processes through the critical larval period, and ends up being independent of reproductive potential.

Results of Spawning Stock Studies

Jones et al. (1977) and Setzler-Hamilton et al. (1981) summarized the results of experimental gill-net catches of the adult spawning stock. I will only review some major points relating to the broadcasting of eggs by the spawning population.

The absolute peak abundances of spawning adults consistently occurred in the downstream portion of the nursery area, coincidentally with the clearly dominant peak in egg abundance early in the spawning season. The subsequent peak abundances of spawning adults progressed upstream (the progression was only interrupted and dispersed temporarily in 1976 by a regional drop in temperature). As upstream migration progressed, the sex ratio reversed, favoring females, so that at the end of the spawning period, catches were composed entirely of females. Sizes (and ages) also increased coincidentally with upstream migration. Thus, the apparent origin of the majority of successful, surviving, late larvae (Ulanowicz and Polgar 1980) is in the highly fecund, older female portion of the stock that arrives late at the upstream end of the spawning grounds. This provides for a larval subpopulation that develops under favorable nutritional conditions. Hence, the structure and movements of the spawning population are consistent with the origin of successful larvae in the upstream portion of the spawning grounds.

The Stock-Recruitment Relationship and Management

The results of the four-year Potomac River field study only defined semiquantitative relationships for the predominantly environmental mechanisms controlling striped bass population size. These relationships were supported by relationships derived from longer-term juvenile abundance data. However, the juvenile data only reflect year-class strength, and not recruitment of spawners. Males could be sexually mature at age 2; females start spawning in considerable numbers at ages 4 and 5. The fishery is heavily exploited, and the age-selective harvest pattern is strongly affected by variations in abundance (Polgar 1980). Therefore, recruitment of sexually mature fish in the years following the juvenile stage may become significantly uncorrelated with the initial juvenile index measure of year-class strength because there may be significant differential exploitation by sex.

I developed a model from historical Potomac commercial-catch and juvenile-index records (covering 1954–1978) that allows calculations of the relationship between sexually mature female parents and sexually mature female progeny. This model takes into account the time-variable rates of exploitation that affect initial year-class strength.

The details of the model will only be sketched here, but it is similar in approach to Deriso's (1980) autoregressive population model. His model achieved significant fits for southern New England yellowtail flounder, Georges Bank haddock, and Pacific halibut, all without compensatory terms. The model uses statistical estimates of strength of the relationship between reported commercial landings (Potomac River Fisheries Commission, Colonial Beach, VA, personal communication) and measured year-class strength to estimate sex-specific and age-class-specific commercial fishing rates and other mortality rates (e.g., from natural causes and fishing away from the spawning ground). This relationship is expressed as a multiple linear regression between commercial catch and lagged year-class strength (a time-delayed model):

$$\sum_{i=1}^{8} Y_i(t) = \hat{\beta}_1 N_{01}(t-1) + \hat{\beta}_2 N_{02}(t-2) + \ldots + \hat{\beta}_i N_{0i}(t-i) + \ldots + \hat{\beta}_8 N_{08}(t-8)$$

where

$\sum_{i=1}^{8} Y_i(t)$ = the reported total commercial catch of all age-classes during the spawning period in year t,

$\hat{\beta}_i$ = the estimates of regression coefficients, and

$N_{0i}(t-i)$ = the initial year-class strength of i-year-old fish. The measure of year-class strength used in these analyses is the juvenile index.

The regression coefficients estimated from the data records are interpreted in terms of a cohort model to estimate migration rates of spawning females, fishing rates, and other sources of mortality, assuming constant growth rates and a constant fishing effort through the period of the available fishing record.

These rate estimates are then used in the model to reconstruct the composition of stocks of striped bass on the spawning grounds for any given year. Age and sex characteristics of the model-generated stock, total commercial catch, relative catch on the spawning grounds, measured weight of individuals in the commercial catch, etc., are compared to parameters measured for the same periods in the four-year and other independent studies. When the characteristics of the model-predicted stocks concurred statistically with the characteristics of the observed stocks, the model was judged to have provided acceptable estimates of the average fishing and mortality rates during the period of historical record.

To generate time- and stock-dependent fishing rates, some of the basic model assumptions (e.g., constant fishing effort over the period of record) are relaxed and replaced with alternate assumptions. For example, for a fishable stock of some given size and age-sex composition, the model can maximize numbers of fish in the commercial catch. (Fishermen on the Potomac do this by fishing a gill net with a mesh size that catches the most abundant size class.)

The fishing rates thus obtained assume variable catchability for each age-sex class, and, again, the parent stock on the spawning grounds can be reconstructed. The characteristics of these stock estimates are compared to the entire historical record to determine the reliability of the new estimates.

At the end of the process, the reconstructed age classes are used to estimate all recruitments of sexually mature females into the fishery, with appropriate delays to sexual maturity, taking into account fishing and mortality rates, participation of females in the spawning process (migration rates), sexual maturity of males and females, and mean relative fecundity of each age class. The strength of relationships among time-delayed recruits and parent stock and physical environmental variables (temperature, freshwater flow, etc.) is measured using linear and nonlinear multiple regression techniques.

An entire suite of nonlinear regressions was performed for fitting the relationship of derived stock to progeny plotted in Fig. 1. Based on the fit of the juvenile index to environmental variables, the most general functional form assumed was an environmentally modified Ricker function:

$$R = aP * \exp[bP + c(DT) + d(AT) + e(AF)]$$

where R = computed mature female recruits; P = computed mature female

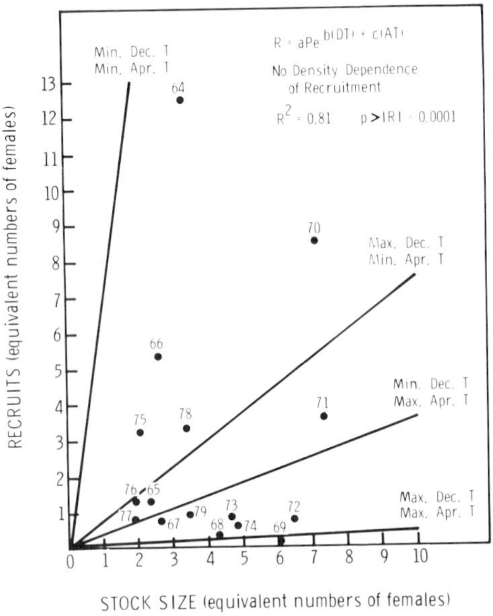

Figure 1. *Derived, mature, female-equivalent estimates of spawners and recruits for the Potomac striped bass population. Lines show the boundaries of stock-recruitment response with respect to extrema in December and April temperatures (see text). The equation represents the best fit of recruitment to independent variables.*

spawning stock; and DT, AT, and AF are the environmental variables previously defined. The comparisons of models, which forced in the independent variables in different sequences, are too lengthy to evaluate here. However, the model that explains most of the variance with the best fit-statistic is:

$$R = 1239 \, P* \, \exp \, [-0.26 \, DT \; - \; 0.59 \, AT].$$

This density-independent model explains 81% of the variability in recruitment, using a record length of 25 years. Linearity tests between observed and predicted values show a fit that has a probability of the correlation being zero of 0.0001. The classical, density-dependent Ricker model for these data explains only 0.12% of the variance. Intermediate model forms, in which the compensatory term is significant, explain only 75% of the variation, with a poorer fit.

The dependence of recruitment on December temperatures in the year previous to spawning has been discussed in connection with the variations of juvenile abundances. This dependence is not changed by the inclusion of fishing history. However, the inclusion of the linear parent-stock term, together with December temperatures, makes the April temperature term significant. Examination of the significance of various environmental terms in other models and the negative sign of the April temperature term indicates that this dependence likely reflects a significant negative correlation of April flow with April temperature. Indeed, if the boundaries of regions with extreme December and April temperatures are plotted in Fig. 1, dominant and poor year classes fall where they ought to according to the environmental dependence of the juvenile index alone.

Thus, the stock-recruitment model confirms the qualitative hypotheses for mechanisms of successful year classes indicated by ichthyoplankton distributions and adult-stock behavior. The explicit compensatory term is absent, indicating the overwhelming environmental dependence of recruitment. It cannot be assumed that the environmental mechanisms either concentrate or dilute larvae, which may be construed as an apparent density dependence in a fixed-size nursery habitat. But the magnitude of river flow does change the habitat size for striped bass larvae, and in the Sacramento-San Joaquin system, this is clearly related to spawning success (Stevens 1977). We do not know whether larval density responds to flow changes independently of changes in habitat size.

The strong dominance of environmental factors on recruitment and the lack of explicit compensation suggest that the striped bass demonstrates a strategy of persistence in a fluctuating environment, rather than saturating carrying capacity (Ulanowicz and Polgar 1980). As is obvious from Fig. 1, fairly small stock sizes may produce dominant year classes. This, in turn, ensures that the population will survive conditions that are adverse for spawning for periods comparable to a large fraction of the effective maximum life span (approximately 10 years under heavy exploitation). Catastrophically small parent stocks will not be likely to develop over such long periods in an exploited stock because fishing pressure will be reduced due to scarcity or regulation.

In addition to the question of persistence, the environmentally-controlled population regulation also has ramifications for optimizing the benefits of exploitation. As alluded to earlier, yields cannot be optimized in the same way as in populations that have density-dependent compensation, mostly due to the lack of a stable equilibrium level. Polgar (1980) and Cooper and Polgar (1981) suggested that since surplus production is contained, for the most part, in the moderately successful or occasionally dominant year classes, management strategies should focus on conserving yields from these over time. This could be implemented most effectively by relaxing fishing pressure on 2- and 3-year-old fish. This strategy takes advantage of the obvious tradeoff between long-term, adult mortalities and individual growth. In a statistical sense, there is some theoretical optimum in biomass yield, although this cannot be determined due to long-term fluctuations in environmental conditions. Polgar (1980) has shown that by halving fishing pressure on 2- and 3-year-old fish, biomass yield from the population could be increased by 13%, averaged over a long period. The calculation, however, assumed that environmental variability and its effects on recruitment will remain unchanged. The general features of this strategy could be retained, however, even under changing environmental conditions. Reduction of fishing pressure on early, exploitable year classes could be keyed to low juvenile abundances, since they forecast landings by an average of 3 years.

Whether management of biomass yields by year class, as outlined above, conserves reproductive potential and thereby increases spawning success is an open question. For the Pacific mackerel population, which is controlled by environmental factors to a lesser extent than striped bass, Parrish and MacCall (1978) concluded that "maintenance of a larger spawning biomass might have helped the stock survive [the] poor spawning successes . . . ." The strengths of dominant year classes in striped bass subpopulations could improve; however, increased reproductive potentials would not be likely to have much effect on environmentally-induced mediocre spawning levels.

Acknowledgments

This work was supported by the Maryland Power Plant Siting Program, Department of Natural Resources. I would like to thank my colleagues at the Chesapeake Biological Laboratory for their dedication and hard work in the field studies and in analyses that made this integrated assessment possible. I am particularly indebted to Drs. J. A. Mihursky, W. R. Boynton, and E. M. Setzler-Hamilton, and to my co-workers G. M. Krainak and P. A. Souza, for their assistance on various aspects of the Potomac striped bass study.

References Cited

Beaven, M. and J. A. Mihursky. 1980. Food and feeding habits of larval striped bass: an analysis of larval striped bass stomachs from the 1976 Potomac estuary collection. Final report to Maryland Power Plant Siting Program. University of Maryland, Chesapeake Biological Laboratory, Solomons, MD, USA. Rept. No. UMCEES 79-45-CBL. 92 pp.

Berggren, T. J. and J. T. Lieberman. 1978. Relative contribution of Hudson, Chesapeake and Roanoke striped bass, *Morone saxatilis*, stocks to the Atlantic Coast fishery. *Fish. Bull.* 76:335-345.

Beverton, R. J. H. and S. J. Holt. 1957. On the dynamics of exploited fish populations. *U.K. Minist. Agric. Fish. Food, Fish. Invest.* Ser. II, 19. 533 pp.

Boone, J. G. 1978. Performance report, estuarine fish recruitment survey: shore zone seining for fish distribution and abundance. Federal Aid in Fish Restoration Project F-27-P5, Maryland Fisheries Administration, Annapolis, MD, USA.

Boynton, W. R., T. T. Polgar and H. H. Zion. 1981. Importance of juvenile striped bass food habits in the Potomac estuary. *Trans. Am. Fish. Soc.* 110:56-63.

Chadwick, H. K. 1964. Annual abundance of young striped bass, *Roccus saxatilis*, in the Sacramento-San Joaquin Delta, California. *Calif. Fish Game* 50:69-99.

Clark, C. W. 1974. Possible effects of schooling on the dynamics of exploited fish populations. *J. Cons., Cons. Int. Explor. Mer* 36:7-4.

Clark, J. R. 1968. Seasonal movements of striped bass contingents of Long Island Sound and the New York Bight. *Trans. Am. Fish. Soc.* 97:320-343.

Cooper, J. C. and T. T. Polgar. 1981. Recognition of year-class dominance in striped bass management. *Trans. Am. Fish. Soc.* 110:180-187.

Cushing, D. H. 1971. The dependence of recruitment on parent stock in different groups of fishes. *J. Cons., Cons. Int. Explor. Mer* 33:340-362.

Deriso, R. B. 1980. Harvesting strategies and parameter estimation for an age-structured model. *Can. J. Fish. Aquatic Sci.* 37(2):268-282.

Heinle, D. R., D. A. Flemer and J. F. Ustach. 1977. Contribution of tidal marshlands to mid-Atlantic estuarine food chains, pp. 309-320. *In:* M. L. Wiley (ed.), *Estuarine Processes, Vol. II: Circulation, Sediments and Transfer of Materials in the Estuary.* Academic Press, New York.

Hjort, J. 1926. Fluctuations in the year-classes of important food fishes. *J. Cons., Cons. Int. Explor. Mer* 1:5-38.

Jones, P. W., J. S. Wilson, R. P. Morgan, II, H. R. Lunsford, Jr. and J. Lawson. 1977. Potomac River fisheries study; striped bass spawning stock assessment. Interpretive report 1974-1976. University of Maryland, Chesapeake Biological Laboratory, Solomons, MD, USA. Rept. No. UMCEES 77-56-CBL. 391 pp.

Jones, R. and W. B. Hall. 1973. A simulation model for studying the population dynamics of some fish species, pp. 35-39. *In:* M. S. Bartlett and R. W. Hiorns (eds.), *The Mathematical Theory of the Dynamics of Biological Populations.* Academic Press, London and New York.

Kohlenstein, L. C. 1981. On the proportion of the Chesapeake Bay stock of striped bass that migrate into the coastal fishery. *Trans. Am. Fish. Soc.* 110:168-179.

Lasker, R. 1965. The physiology of Pacific sardine embryos and larvae. *Mar. Res. Comm., Calif. Coop. Ocean. Fish. Invest. Rept.* 10:96-101.

May, R. C. 1974. Larval mortality in marine fishes and the critical period concept, pp. 3-19. *In:* J. H. S. Blaxter (ed.), *The Early Life History of Fish.* Springer-Verlag, New York.

Merriman, D. 1941. Studies on the striped bass *Roccus saxatilis* of the Atlantic Coast. *Fish. Bull.* 50(35):1-77.

Nelson, W. R., M. C. Ingham and W. E. Schaaf. 1977. Larval transport and year-class strength of Atlantic menhaden, *Brevoortia tyrannus. Fish. Bull.* 75(1):23-41.

Parrish, R. H. and A. D. MacCall. 1978. Climatic variation and exploitation in the Pacific mackerel fishery. *Calif. Fish Game* 167:1-110.

Pearson, J. C. 1938. The life history of striped bass, or rockfish, *Roccus saxatilis* (Walbaum). *Fish. Bull.* 49:825-851.

Polgar, T. T. 1980. Is striped bass a manageable resource?, pp. 189-202. *In:* H. Clepper (ed.), *Proceedings of the Fifth Annual Marine Recreational Fisheries Symposium.* Sport Fishing Institute, Washington, DC.

Polgar, T. T., J. A. Mihursky, R. E. Ulanowicz, R. P. Morgan and J. S. Wilson. 1976. An analysis of 1974 striped bass spawning success in the Potomac estuary, pp. 151-165. *In:* M. L. Wiley (ed.), *Estuarine Processes, Vol. 1: Uses, Stresses and Adaptation to the Estuary.* Academic Press, New York.

Ricker, W. E. 1975. Computation and interpretation of biological statistics of fish populations. *Fish. Res. Board Can. Bull.* 191:1-382.

Sawyer, S. and M. Slatkin. 1981. Density independent fluctuations of population size. *Theoretical Population Biology* 19:37-57.

Setzler-Hamilton, E. M., W. R. Boynton, J. A. Mihursky, T. T. Polgar and K. V. Wood. 1981. Spatial and temporal distribution of striped bass eggs, larvae and juveniles in the Potomac estuary. *Trans. Am. Fish. Soc.* 110:121-136.

Stevens, D. E. 1977. Striped bass *(Morone saxatilis)* year-class strength in relation to river flow on the Sacramento-San Joaquin estuary, California. *Trans. Am. Fish. Soc.* 106:34-42.

Ulanowicz, R. E. and T. T. Polgar. 1980. Influences of anadromous spawning behavior and optimal environmental conditions upon striped bass *(Morone saxatilis)* year-class success. *Can. J. Fish. Aquatic Sci.* 37(2):143-154.

Van Winkle, W., B. L. Kirk and B. W. Rust. 1979. Periodicities in Atlantic Coast striped bass *(Morone saxatilis)* commercial fisheries data. *J. Fish. Res. Board Can.* 36:54-62.

Vladykov, V. D. and D. H. Wallace. 1952. Studies of striped bass *Roccus saxatilis* (Walbaum), with special reference to the Chesapeake Bay region during 1936-1938. *Bull. Bingham Oceanogr. Collect.* 14(1):132-177.

LARVAL RETENTION IN ESTUARIES

Convenor: *S. D. Sulkin*
Horn Point Environmental Laboratories
Cambridge, Maryland

ESTUARINE LARVAL RETENTION MECHANISMS ON TWO SCALES

W. C. Boicourt
Chesapeake Bay Institute
The Johns Hopkins University
Shadyside, Maryland

Abstract: Circulation processes can explain the observational evidence of estuarine larval retention, but knowledge of water motion is required on the small spatial and temporal scales appropriate to larval transport. Two cases are presented – the oyster *Crassostrea virginica* and the blue crab *Callinectes sapidus* – in which the operative mechanism is recruitment or return rather than retention in the strict sense. In the case of the oyster, marked differences in spat settlement success between two adjacent tributaries in the lower Choptank River (Chesapeake Bay) are explained by subtle differences in the exchange mechanisms between the tributaries and the lower Choptank proper. In the case of the blue crab, the return of later stage larvae to Chesapeake Bay from the continental shelf can be provided by the northward flowing inner shelf band which is driven by the wind against the prevailing southward flow.

Introduction

The estuarine larval retention problem has often seemed intractable to both biologists and physical oceanographers. On one hand, biologists are faced with difficulties in achieving sufficient sampling in time and space to allow a description of the time history of larval distributions. On the other hand, physical oceanographers have difficulty providing circulation details on scales relevant to larval transport. The classical two-layer estuarine circulation has been proffered as the most likely mechanism for retaining larvae, but many questions remain, such as the role of larval motility and behaviour, and the timing and distribution of larval sources. In addition, larval retention must be explained in estuaries with low stratification and weak gravitational circulations. The difficulty of the problem probably has discouraged the dedication of the observational resources that are deemed sufficient from either the biological or physical perspective.

Recent advances in the understanding of the smaller scales of water motion in the estuary and on the continental shelf has placed physical oceanographers in a position where they can begin to address the larval transport problem. The motivation provided by the clear need to explain the mechanisms of observed larval retention has fostered research that is beneficial to both disciplines: the attempt to address the smaller space scales

ISBN 0-12-404070-5

(which are also appropriate scales for the transport of water, salt, heat, and momentum) of larval transport has proved instructive with regard to the physical oceanography alone.

Two studies of water motion in Chesapeake Bay and the adjacent continental shelf are presented herein. They differ in scale and intent, but they are both relevant to the larval retention problem, and the results from both stand in contrast to expectations derived from conventional circulation knowledge. One study in the lower Choptank River on the eastern shore of Chesapeake Bay was designed to address specifically the larval transport and retention problem. Here, two tributary estuaries with weak density-driven circulations, Broad Creek and the Tred Avon River (Fig. 1), show markedly different oyster larvae settlement (spatfall) success. A field program of current measurement and dye dispersion was carried out to examine the hypothesis that circulation processes were responsible for this difference. The other study was designed to examine the circulation at the mouth of Chesapeake Bay and on the adjacent inner continental shelf. While the goals were to examine the estuary-continental shelf interaction and to provide a proper formulation for the seaward boundary conditions for an estuarine numerical model, the results are relevant to blue crab larval history.

Background

If a wealth of knowledge were available on the life history of an organism with a planktonic larval stage, then the construction of a larval transport

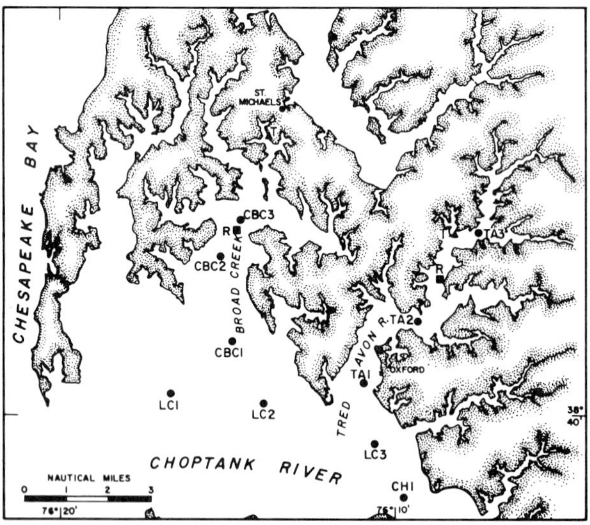

Figure 1. Map of lower Choptank River, showing Broad Creek and Tred Avon River tributaries. ●-Current-meter moorings; ■-dye release points; ▲-tide gauge emplacements (T).

model would be the obvious procedure for predicting the success of the later life stages. The information that would be brought to bear in constructing the model would be: 1. Water Circulation; 2. Larval Motility and Behaviour; 3. Larval Sources and Time Variability; 4. Larval Mortality; and 5. Verification Data. Water circulation is listed first, not to reflect the bias of a physical oceanographer, but to remind us that water motion is the crucial determinant of larval transport, regardless of the possible role of larval motility. Effective larval motility modes are limited to vertical swimming or sinking. In the horizontal, currents clearly dominate larval transport. On the continental shelf and in the estuary, daily transport distances are on the order 5–10 km. The ability of a larva to swim horizontally, even an order of magnitude less than this distance in one day, is doubtful. Various vertical motility modes have been suggested as effective mechanisms for larval retention in estuaries. Although there are circulation and resuspension mechanisms sufficient to retain larvae and transport them landward of brood stock, there is evidence that the larvae of certain species can take selective advantage of flood tidal currents by timing their vertical swimming activity (Bousfield 1955).

Information on both sources and sinks of larvae is crucial to the modeling effort, but difficult to obtain. The location and timing of a larva's entry into the water column is seldom known, except in a few simple cases. The rate of mortality of larvae, once they are introduced into the water column, is not known to sufficient accuracy for predicting the success of later life stages. At present, proceeding with a model would require a simplifying assumption such as an exponential die-off, and a series of runs with varying coefficients. This procedure obviously assumes spatial uniformity of predation and a mortality directly proportional to the concentration of larvae, which may not prove valid. In addition to the listed input information, a set of independent verification data would be desirable in order to test the variety of assumptions necessary for the construction of the model. Unfortunately, few data sets of sufficient coverage and resolution are available because the larval sampling and counting effort required has been prohibitively large (Wood and Hargis 1971; Andrews 1979).

In the assessment of the suitability of available knowledge for constructing a model, water circulation should not be exempt from scrutiny. The resolution and coverage necessary for describing circulation on scales appropriate to larval transport are seldom known for a given body of water, and if known, are not easy to achieve. An estimate of larval transport derived from archived data or from classical knowledge may not suffice, or may even mislead, as will be seen from two examples discussed herein.

Oyster Larvae Retention and Recruitment: a Comparison of Two Adjacent Tributaries

On the eastern shore of Chesapeake Bay there are two estuaries tributary to the lower Choptank River (Fig. 1) with oyster spatfall records extending back over 40 years (Meritt 1977; Kennedy 1980). One tributary, Broad Creek, has shown consistently greater (one or two orders of magnitude) oyster larvae settlement than the adjacent Tred Avon River. The geometry, fresh water inflow, temperature, salinity, and biology appear so similar in each estuary that circulation processes seem the next reasonable suggestion as to the cause of the marked difference in spat settlement. The records begin sufficiently long ago that differential mortality due to toxic pollutants seems unlikely in this predominantly rural region.

A current measurement and dye dispersion program was carried out in the summer of 1979 in the two tributaries and in the connecting waters (Boicourt et al. 1982). The program was designed to test the hypothesis that circulation dictated spat settlement success. The first model of the spat settlement was that larvae were retained in Broad Creek and flushed out of the Tred Avon River.

Ten current-meter moorings were placed in the two adjacent tributaries and in the lower Choptank River (Fig. 1). In addition, a raft containing a reservoir of Rhodamine-WT dye and a precision release pump was located in the middle reaches of Broad Creek and the Tred Avon. While the larval transport pathways could not be directly simulated with dye because the source locations were not known, the intent was to take advantage of the complementary information from Eulerian and Lagrangian flow measurement techniques in order to provide sufficient spatial resolution to infer possible larval pathways.

Details and results of the experiment are reported elsewhere (Boicourt et al. 1982) so only a summary is given here. The primary finding is that circulation processes can account for the observed difference in oyster larvae settlement. Both the wind-driven circulation and the density-driven circulation favor Broad Creek over the Tred Avon River for recruitment of larvae from outside the tributary. Broad Creek circulation consists of a simple, wind-driven, two-layer circulation superimposed on a weaker, classically two-layer estuarine circulation. Transient northerly winds produce a strong inflow in the lower layer, augmenting the slower gravitational flow. The Tred Avon River, in contrast, shows both a weak wind-driven flow, and a weak gravitational flow in the lower reaches. The reason for the lack of strong response to the wind is that the curvature of the main stem (Fig. 1) does not allow a consistent axial pressure gradient to be built up over long reaches of the River. The weak gravitational flow in the lower reaches appears to be the result of the low-salinity water in the upper layer of the Choptank River estuary flowing seaward past the mouth of the Tred Avon River. The salinity of this outflowing layer is sufficiently low that the horizontal salinity gradient (and therefore density gradient) in the upper layers of the Tred Avon is low or even directed into the River. Current measurements and dye dispersion experiments (Fig. 2) show that the exchange between the lower Tred Avon River and the lower Choptank is slow. While Broad Creek shows a strong two-layer flow (both wind and density driven), the Tred Avon shows a progression of weak one-layer, two-layer, and three-layer flow patterns. Not only does the Tred Avon River have a weak wind-driven component of the circulation, but also the density-driven component acts to prevent a significant amount of larvae from entering the tributary from the lower Choptank River. The Tred Avon, then, appears to be dependent on local production of larvae for its spat settlement, while Broad Creek has a significantly greater ability to augment its supply with recruitment from the productive bars in the lower Choptank River. In addition, the greater transport and dispersion observed within Broad Creek aids the larval search for a suitable substrate upon which to settle.

Neither the Eulerian nor the Lagrangian measurement technique suffices alone to allow the deciphering of the circulation processes. The simplest interpretation of the Eulerian mean currents even appear contradictory to the first interpretation of the dye dispersion patterns. As in the three-layer circulation profile of Baltimore Harbor, the vertical scales appear remarkably small. The upper-layer transport for the wind-driven circulation occurs within < 2 m of the surface in both Broad Creek and the Tred Avon River. The density-driven upper layer appears to be on the order 3 m thick. Perhaps these small scales should not be surprising in light of Carter's (1967) results from a similar study of another eastern shore tributary to the south of the Choptank River. Carter found a variety of circulation regimes in various sections of the Manokin River, with either the wind, the river, or the density-driven circulation dominating the flow. Providing an accurate measure of flows in these thin surface layers is difficult with either Eulerian or Lagrangian techniques. Moored current meters are prone to damage and wave contamination at these shallow depths. The physical scale of the Lagrangian drag elements prevents adequate spatial resolution in the vertical. Rhodamine-WT dye does not suffer from this deficiency, but it does diffuse more rapidly than salt in the vertical direction. In addition, the time spent sampling the water column for dye is often time removed from the horizontal tracking of the dye patch or plume if a limited number of vessels is employed. The answer for current measurement in these shallow estuaries is probably profiling sensors from a fixed, rigid platform such as a pile or tower. Profiling from surface vessels with four anchor points may provide accurate current measurements, but the record length dictated by the wind-driven flow variability is on the order of one month. Time series of this length obviously requires unattended measurement.

The results of the current-meter and dye experiments are, in a sense, opposite to the original hypothesis in that the Tred Avon River is more retentive of oyster larvae than Broad Creek. They are satisfying in that they provide a plausible explanation for the difference in oyster larvae settlement success, but they are also revealing as to the circulation scales of shallow tributary estuaries with weak gravitational circulations. Prior to these experiments, two estuaries which appear so similar in physical characteristics would not be ex-

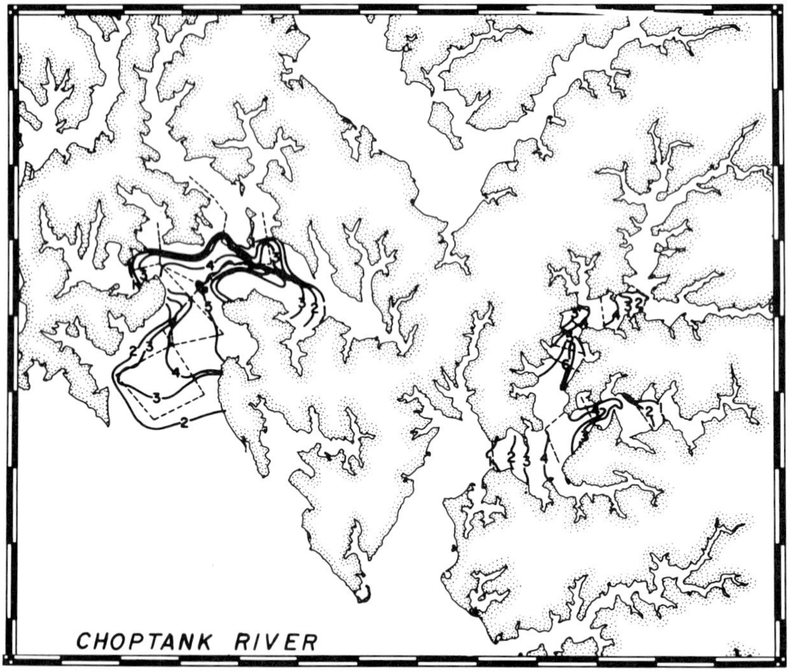
CHOPTANK RIVER

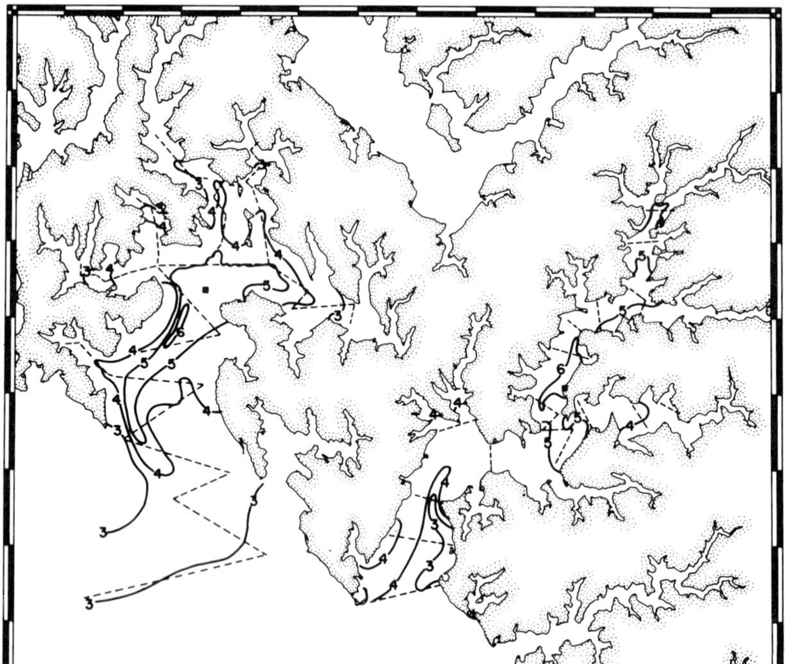
CHOPTANK RIVER

pected to exhibit such a marked difference in the amount of exchange with the larger water body to which they are connected.

Blue Crab Larval Export to the Continental Shelf: is there a Mechanism for Return?

The Chesapeake Bay blue crab life history involves estuarine circulation at the opposite end of the spatial scale from that of Broad Creek and the Tred Avon River. Chesapeake Bay represents the archetypical partially mixed estuary with strong stratification and a well-developed density-driven two-layer circulation. The blue crab spawns in the lower reaches of the Chesapeake Bay, predominantly on the eastern shore near Cape Charles. The available evidence is that the majority of the larvae are subsequently exported to the continental shelf through the nearby Bay entrance (S. Sulkin, pers. comm.). It is at this point that physical oceanographers have presented a dilemma to those studying the crab larval history: if the larvae are exported to the continental shelf, how can the larvae return to the Bay if the mean currents sweep southward past the mouth of the Bay toward Cape Hatteras? Is Chesapeake Bay dependent on recruitment of larvae from estuaries that are upstream in the shelf flow?

The southward mean flow on the Middle Atlantic Bight continental shelf has been long established by the time-series current-pole measurements of Haight (1942) and the extensive drifter observations of Bumpus (1973). Bumpus also documents occasional reversals to this flow during summers of persistent southerly winds, but the preponderance of moored current-meter information (Beardsley et al. 1976) shows a clear mean flow to the south of order 5-10 cm/s. Long-term current meter moorings on the outer shelf also show occasional reversals to this southward mean (Beardsley and Boicourt 1980), but they appear sufficiently rare that they seem unlikely to provide an acceptable mechanism for the return of blue crab larvae to Chesapeake Bay.

The answer to the dilemma may be provided not only by some new current measurements on the inner shelf near the entrance to Chesapeake Bay, but also by a reinterpretation of the available information on the flow on the continental shelf. Recent winter current measurements in the Bay mouth region (Boicourt 1981) are in accordance with conventional expectations in

Figure 2. (opposite page). Examples of Rhodamine-WT dye distribution patterns in Broad Creek and Tred Avon River. Contours are log ratio of observed concentration to concentration of dye at source (■). Dashed lines are tracks of tracer vessels. Figure 2a (upper) shows dye distribution pattern on 7 July, 1 d after continuous release began; Broad Creek shows greater dispersion of dye, especially seaward. Figure 2b (lower) shows dye distribution pattern on 11 July, 6 d after release began. Distribution in lower Tred Avon shows typical tendency of fresher water to enter from Choptank on eastern side of the River, while surface layers on western side are moving seaward.

that the flow on the inner shelf is southward, parallel to isobaths. There is some evidence that there are time intervals wherein the (subtidal) mean flow is into the Bay at all depths in the northern part of the Bay entrance cross-section, but a definitive statement will have to await measurements with increased spatial resolution. Recent summer measurements (Fig. 3) show the upper layers of the inner continental shelf waters to be flowing northward with a 38-day mean of 7 cm/s. The Chesapeake Bay outflow plume, which forms a stable, low-salinity jet along the Virginia-North Carolina coast, is moving southward toward Cape Hatteras. If these measurements stood alone, then the northward-flowing water would probably be interpreted as a manifestation of a particularly strong or persistent southerly wind, causing a reversal of the mean southerly flow across the shelf. However, an examination of earlier measurements (Beardsley et al. 1976; Boicourt and Hacker, 1976) shows the inner shelf moving independently of the outer shelf during the summer (stratified) season. These current meter records are shown as vector time series in Fig. 4. Mooring 408A is located 28 km offshore, slightly south of the mouth of Chesapeake Bay. Moorings 413A, 415A, and 416B are outer shelf moorings, on the order 100 km offshore. All current meter records from the offshore moorings show a mean *southward* flow, while the inner shelf mooring (408A) records show a mean *northward* flow. The primary reason for this behaviour is that the outer shelf mean flow is not driven by the wind, but by pressure gradients associated with the ocean interior circulation. This component of the flow decreases in the onshore direction, so that in the shallow regions of the inner shelf the prevailing southwesterly winds of the summer can dominate the flow. This wind dominance can be seen easily in Fig. 4. The inner shelf currents at station 408A reflect directly the driving wind stress. As

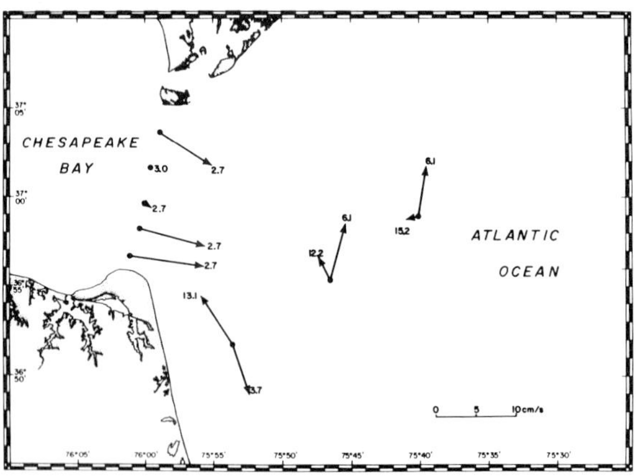

Figure 3. Mean velocities in Chesapeake Bay mouth region for a 38 d interval beginning 23 June 1981. Depth (m) of measurements is indicated near head of velocity arrow. From Boicourt (1981).

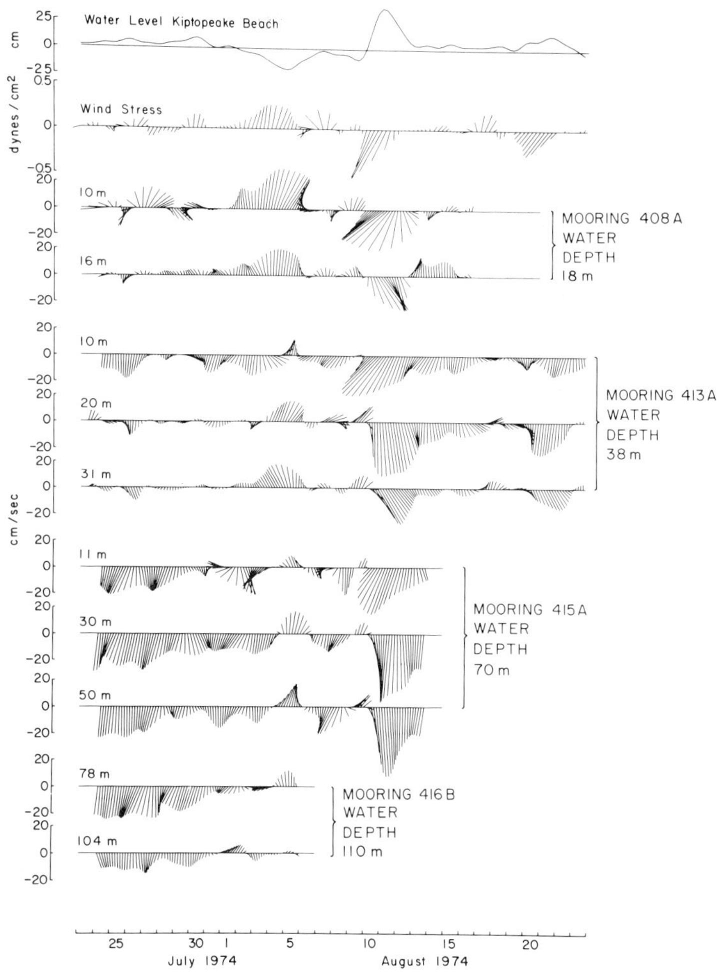

Figure 4. Time series of wind, coastal sea level, and currents for continental shelf off Chesapeake Bay. Offshore distance of current-meter mooring positions are shown in Fig. 5. From Beardsley and Boicourt (1980).

the water deepens in the offshore direction, the wind's dominance of the flow decreases, until the wind-driven component is seen as a modulation of the mean southerly flow. The resultant mean flows for a winter wind event, a summer wind event, and a summer monthly mean are shown in Fig. 5.

While the possibility that the wind can reverse the mean flow on the inner shelf while the outer shelf continues to flow southward was recognized at the time of these measurements, the implications to the transport and mixing along and across the shelf were not. With the overwhelming portion of Middle Atlantic Bight current-meter measurements made either on the outer shelf or

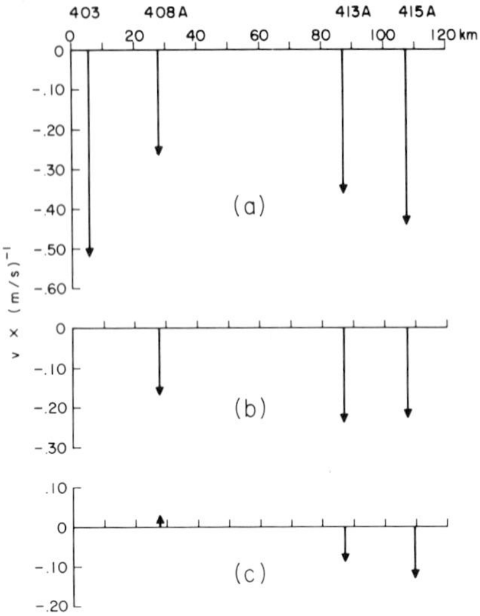

Figure 5. Mean alongshore surface velocities for a winter storm (a, 3 to 6 Feb. 1974), a summer wind event (b, 10 to 13 Aug. 1974), and a summer monthly mean (c, 21 July to 27 Aug. 1974). Moorings are indicated by distance offshore along 36°50′ N.

in winter, the few summer records from the inner shelf showing northerly flow were probably dismissed as a curiosity, easily explained by invoking local wind driving. The suggestion of a banded shelf in the southern half of the Middle Atlantic Bight has profound implications not only to the transport of blue crab larvae, but also to the transport and mixing of water, salt, heat and momentum. Figure 6 contains a schematic summer diagram of the continental shelf in the offing of Chesapeake Bay. The outer shelf moves typically southward, except when occasionally reversed by strong southerly winds. The inner shelf can be reversed by weaker winds. The 15–20 km band near the coast contains the rapidly flowing southward coastal jet set up by the low-salinity outflow from Chesapeake Bay. The strength and spatial extent of this jet depend on: 1) the salinity difference between the outflowing Bay water and the inner shelf waters, 2) the flux of the outflowing water, and 3) the strength of southwesterly winds, which can move the jet offshore to form a broad diffuse plume in mid-shelf. The northward flowing inner shelf waters provide a return mechanism for blue crab larvae that are swept southward by the Bay outflow jet.

Outflow plumes are also developed by the Hudson River (Ketchum et al. 1951) and Delaware Bay (Boicourt 1973). The lower river flow and the smaller amplitude of the circulation in these estuaries, however, produce an outflow plume that does not extend as far from the mouth of the estuary as does the Chesapeake outflow plume. The smaller outflow with a lower salini-

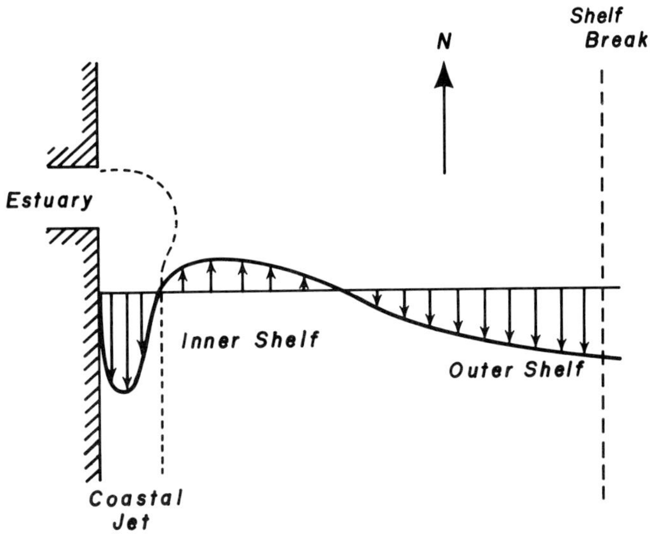

Figure 6. Schematic alongshore surface velocity profile for summer southern Middle Atlantic Bight.

ty difference between the outflow and the ambient shelf waters produces a plume that is less stable than the Chesapeake's plume.

Once the crab larvae are delivered to the mouth region from the inner shelf, the most effective means by which they can be transported into the Bay is via the inflowing lower-layer flow. The blue crab larvae are not required to be near bottom because the depth range of the inflowing water spans a large portion of the water column (Boicourt 1981).

Discussion

Larval transport and retention have been examined from a circulation perspective on two spatial scales – in small shallow tributaries with weak gravitational circulation and in Chesapeake Bay proper with a strong classical estuarine flow pattern. In neither case have particular larval pathways been delineated. Retention mechanisms have been proposed that suffice to explain observed larval distributions. In the process, a few possible hypotheses have been eliminated, such as the suggestion that Broad Creek is more retentive of oyster larvae than the Tred Avon River.

The inferred larval retention mechanisms have been proposed without recourse to larval motility. That circulation alone suffices to explain the observed larval history does not preclude the possibility that vertical larval motion enhances the retention mechanisms for the Chesapeake Bay oyster and blue crab. Behaviour modes wherein the oyster larva can phase vertical swimming activity with tidal currents (or sink to the lower layers and be suspended in the longer and stronger flood currents) or wherein the blue crab larva is positively phototaxic, are beneficial to the observed transport during the larval life history. There is some suggestion from the Tred Avon-Broad Creek study that tidally phased swimming on the part of oyster larvae is not a sufficient mechanism for transport to the head of the estuary because tidal currents and depths are similar in both tributaries. The importance of larval motility will remain unanswered for both the larvae of the oyster and the blue crab, however, until observed vertical migration information can be coupled with circulation information. A numerical circulation model which includes the advective terms in the equations of motion would be a useful tool here. Numerical experiments can be run with a variety of source positions and a variety of larval migration modes to test their effectiveness in transport.

The larval transport and retention problem will require interdisciplinary communication for resolution. This statement is not intended to pay gratuitous homage to the joys of interdisciplinary research, but rather to emphasize that the problem stands at the state of the art in both physical and biological fields. On one side is the difficulty in observing and describing larval distributions over time, and on the other side is the difficulty in describing the circulation details relevant to the transport of these larvae. In both the case of the Choptank tributaries and in the exchange between the Chesapeake Bay with the inner shelf waters the necessary details were not readily available from conventional wisdom.

References Cited

Andrews, J. D. 1979. Pelecypoda: Ostreidae, pp. 293-341. *In:* A. C. Giese and J. S. Pearse (eds.), *Reproduction of Marine Invertebrates. V. Molluscs: Pelecypods and Lesser Classes.* Academic Press, New York.

Beardsley, R. C., W. C. Boicourt and D. V. Hansen. 1976. Physical oceanography of the Middle Atlantic Bight. *Limnol. Oceanogr.*, Spec. Symp. Vol. 2: 20-34.

Beardsley, R. C. and W. C. Boicourt. 1980. On estuarine and continental shelf circulation in the Middle Atlantic Bight, pp. 198-233. *In:* B. A. Warren and C. Wunsch (eds.), *Evolution of Physical Oceanography.* MIT Press, Cambridge, MA

Boicourt, W. C. 1973. The circulation of water on the continental shelf from Chesapeake Bay to Cape Hatteras. Ph.D. Thesis, The Johns Hopkins U., Baltimore, MD 183 pp.

Boicourt, W. C. 1981. Circulation in the Chesapeake Bay entrance region: estuary-shelf interaction, pp. 61-78. *In:* J. W. Campbell and J. P. Thomas (eds.), Chesapeake Bay Plume Study: Superflux 1980. NASA Conference Publication 2188.

Boicourt, W. C. and P. W. Hacker. 1976. Circulation on the Atlantic continental shelf of the United States, Cape May to Cape Hatteras. *Memoires de la Societe Royale des Sciences de Liege, 6e Serie* 10:187-200.

Boicourt, W. C., V. S. Kennedy and S. R. Rives. 1982. Oyster larvae settlement success in the Choptank River tributaries: an example of circulation control of planktonic transport. Unpublished manuscript.

Bousfield, E. L. 1955. Ecological control of the occurrence of barnacles in the Miramichi estuary. *Bull. Nat. Mus. Canada* 137:1-69.

Bumpus, D. W. 1973. A description of the circulation on the continental shelf of the east coast of the United States. *Progress in Oceanography* 6:111-157.

Carter, H. H. 1967. A method of predicting brood stock requirements for oyster *(C. virginica)* producing areas with application to the Manokin River. Chesapeake Bay Institute, The Johns Hopkins U. Spec. Rep. 13, Ref. 67-2, 46 pp.

Haight, F. J. 1942. Coastal currents along the Atlantic coast of the United States. U.S. Dept. of Commerce Coast and Geodetic Survey Special Publication 230, 73 pp.

Kennedy, V. S. 1980. Comparison of recent and past patterns of oyster settlement and seasonal fouling in Broad Creek and Tred Avon River, Maryland. *Proc. Natl. Shellf. Assoc.* 70:36-46.

Ketchum, B. H., A. C. Redfield and J. C. Ayers. 1951. The oceanography of the New York Bight. *Papers in Physical Oceanography and Meteorology* 12: 46 pp.

Meritt, D. W. 1977. Oyster spat set on natural cultch in the Maryland portion of the Chesapeake Bay (1939-1975). U. Maryland, Center for Envir. and Est. Studies, Spec. Rep. No. 7, 111 pp.

Wood, L. and W. J. Hargis, Jr. 1971. Transport of bivalve larvae in a tidal estuary, pp. 29-44. *In:* D. J. Crisp (ed.), *Fourth European Marine Biology Symposium*, Cambridge U. Press, Cambridge.

LARVAL RECRUITMENT IN THE CRAB *CALLINECTES SAPIDUS* RATHBUN: AN AMENDMENT TO THE CONCEPT OF LARVAL RETENTION IN ESTUARIES

Stephen D. Sulkin and William Van Heukelem

Horn Point Environmental Laboratories
Center for Environmental and Estuarine Studies
University of Maryland
Cambridge, Maryland

Abstract: Mechanisms of larval export from estuaries and post-larval recruitment back are examined for the blue crab, *Callinectes sapidus* Rathbun. Zoeal Stages I, IV, and VII do not alter phototaxis or level of activity in response to temperature or salinity change, except in Stage I where reduction in salinity temporarily reduces activity level. In Stage I, inhibition of upward migration occurs at a halocline of 10 ppt salinity. Sensitivity to haloclines increases through zoeal development. A pycnocline in nature will not inhibit movement to the surface of Stage I larvae and thus cannot promote retention. Megalopa are negatively geotactic and exhibit high barokinesis with a low pressure threshold, adaptations which can promote movement to the surface depending on locomotor activity. A conceptual model, based on evidence from the literature and on experimental data presented here, is proposed for recruitment of *Callinectes sapidus* and invokes offshore recruitment as the dominant factor. Coastal circulation can maintain larvae of estuarine invertebrates nearshore and provide the potential for offshore recruitment. This mechanism may be essential to population dynamics in *C. sapidus* and potentially significant in maintaining genetic communication among estuarine populations of other species.

Introduction

Although blue crabs (*Callinectes sapidus* Rathbun) occupy estuaries as adults, larvae typically are found in great abundance in coastal marine waters (Nichols and Keney 1963; Pinschmidt 1963; Tagatz 1968; Dudley and Judy 1971; Sandifer 1975; Smyth 1980). Sulkin et al. (1980) have demonstrated behavioral traits in larvae of various stages which will promote both export of larvae from the estuary and maintenance of later zoeal stages nearshore where they are potentially available for transport back into the estuary as post-larvae or juveniles. Recruitment to the estuarine habitat from larval populations occupying shelf waters has been suggested by Tagatz (1968), Sandifer (1975), and Scheltema (1975).

If larvae of an estuarine species such as *C. sapidus* do indeed possess behavioral adaptations which promote export from estuarine spawning grounds, as indicated by Sulkin et al. (1980), and if mechanisms for recruitment back to the estuary from offshore can be described, then models of

ISBN 0-12-404070-5

recruitment which invoke estuarine retention should be amended to include retention in and recruitment from coastal marine waters.

We present here new information on behavioral traits of *C. sapidus* larvae and post-larvae which relate to estuary-coastal water exchange and we evaluate these traits in terms of life history strategy in *C. sapidus* and the general concept of larval retention within estuaries.

Materials and Methods

Response of Zoeae to Temperature and Salinity Change

Entrapment of crab larvae below the pycnocline which is characteristic of stratified estuaries of the mid-Atlantic Bight is a possible mechanism for substantial estuarine retention in spite of reported adaptations (Sulkin et al. 1980) which promote larval export in surface waters. Experiments were conducted to determine the response of larvae of zoeal Stages I, IV and VII to temperature and salinity changes. Methods of rearing larvae of *C. sapidus* for the purpose of conducting behavioral experiments are discussed in detail in Sulkin et al. (1980).

Influence of temperature and salinity on sign of phototaxis

Sign of phototaxis was determined using the apparatus shown diagrammatically in Fig. 1. Samples of larvae were placed in the ports of the removable bridge, light of controlled quantity and quality was aimed down the axes of the six parallel raceways, larvae were released by slowly raising the bridge, and change in distribution was measured by counting larvae in each segment of the chamber at predetermined time intervals. In experiments designed to test the effect of temperature and salinity change on sign of phototaxis, light was filtered to 500 nm by interference filters and adjusted to

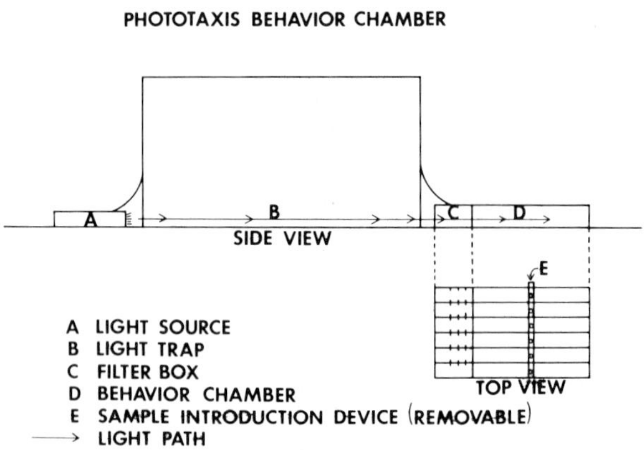

Figure 1. Diagrammatic representation of apparatus used to measure sign of phototaxis.

an intensity of 0.25 W/m^2 using neutral density filters. Light condition was constant and uniform among the six raceways.

Groups of light-adapted larvae were held for 24 h at either 15°C, 20°C, or 25°C for temperature tests (30 ppt salinity) and at 25 ppt, 30 ppt, or 35 ppt salinity (25°C) for salinity tests. In any one experiment, the test temperature (salinity) was uniformly constant among the six raceways and samples of larvae (n = 10) acclimated to each of the three temperatures (salinities) were introduced to two of the six raceways chosen at random. The entire experiment was conducted three times, each time using one of the three test temperatures (salinities) in the observation chambers. Experiments were repeated five times, each time using larvae from a different brood (n = 100 larvae per test condition). Distributions obtained after 5 min were used for interpretation. Larvae were scored as positive, neutral, or negative depending upon their movement relative to the light source. The experiment was repeated with larvae of zoeal Stages I, IV and VII.

Response of zoeae to laboratory-constructed haloclines

A series of haloclines was established in a set of vertically oriented chambers measuring 50 cm long and 5 cm × 5 cm in cross section. Haloclines tested were 2.5 ppt, 5 ppt, 7.5 ppt, and 10 ppt salinity, with the bottom salinity always 30 ppt. A control chamber was set up in identical manner with a vertically homogeneous medium of 30 ppt salinity. Chambers were marked off into three segments, the middle segment containing the salinity gradient. The haloclines were very stable and the gradients thus very steep (instantaneous). Larvae acclimated to 30 ppt salinity were introduced to the bottom of each chamber. A white light (26 W/m^2) at the top of each chamber was used to induce larvae to swim upwards. Distribution among treatments was compared directly. Duration of each experiment was 30 min. Experiments were repeated 6-8 times depending upon stage of development. Larvae from a number of different broods were used.

Numbers of larvae in each segment at t = 30 min were transformed to percentages and plotted as means. Percentages for each of the three segments of the chambers were compared among treatments for zoeal Stages I, IV, and VII using one way analysis-of-variance with arc sine transformation. Differences among means were determined by the Student-Newman-Keuls test at the 5% level of significance.

Depth Regulation in Megalopa

The megalopa stage has been identified as the likely agent of recruitment from offshore to the estuarine habitat (Tagatz 1968; More 1969; Williams 1971; Sandifer 1975). Its dispersal, effected by passive transport, will be mediated by its vertical distribution. Reports of vertical distribution of megalopa vary considerably. Tagatz (1968), Sandifer (1973), and Goy (1976) reported megalopa most common in bottom waters; however, Williams (1971) and both Epifanio (College of Marine Studies, Lewes, DE) and Provenzano (Old Dominion University, Norfolk, VA) (pers. comm.) report them to be common in surface waters. Experiments were conducted to determine behavioral responses of megalopa to conservative environmental parameters which are central to depth regulation in a zooplankter, namely gravity (orientation stimulus) and hydrostatic pressure (rate stimulus).

Geotaxis

Geotaxis is here defined as oriented movement in the vertical plane in the absence of light. Sign of geotaxis was determined by direct observation of individual laboratory-reared megalopa. The room was darkened and a dim red (λ>650 nm) background light was used to silhouette the megalopa. Individual megalopa were dropped gently into a column of water. Vertical movements were characterized as passive sinking, active downward swimming, and active upward swimming.

Direct observation was confirmed by placing a sample of 20 megalopa, freshly captured from the field, into an observation chamber measuring 16 cm wide × 20 cm high × 5 cm deep. A natural photoperiod was established in the room. A series of photographs was taken at one hour intervals for a period of 38 h. A digital watch included in the photographic field marked the time of each exposure. The camera was activated with a timing device which triggered the shutter and winder. A deep red flash (peak output at wavelengths greater than 600 nm) was used to reduce impact on normal behavior.

Megalopa were readily visible. Percent of total sample above the bottom was plotted as a function of time. Photographs taken during the natural dark period were examined for geotaxis. All megalopa oriented at least 30° from the horizontal plane were scored as negative (up) or positive (down).

Barokinesis

Two separate experiments were used to test the effect of hydrostatic pressure on locomotory response in megalopa. In one experiment which utilized laboratory-reared megalopa, two identical observation chambers measuring 46 cm long and 5 cm in diameter were set up with the long axis oriented vertically. Each chamber was marked off into seven equal segments. One of the chambers was connected to a mercury manometer. A sample of the megalopa was placed in the bottom of each tank and the tanks were sealed. In the experimental chamber, the pressure was increased to 2.5 psi above ambient by raising the distal end of the manometer. One minute later, the distribution of megalopa among the seven segments in each chamber was noted. The pressure in the experimental chamber was then raised to a total increment of 5 psi, followed one minute later by notation of distribution of megalopa as before. This sequence was repeated for 6 min until a total pressure increment of 15 psi was obtained. The experiment was repeated twelve times with megalopa from seven different broods. Percent of each sample above the bottom segment was calculated and plotted against time (pressure). For each time interval (pressure increment), the percent above the bottom was compared between experimental and control chambers using a paired t-test (arc sine transformation). Pressure acclimation effect was determined by repeating the experiment, but raising the pressure immediately to an increment of 15 psi. Distributions were determined as before at 1 min and at 6 min after the pressure increase. Percentages above the bottom for all distributions measured at 15 psi were compared by one-way analysis of variance after arc sine transformation.

In a second experiment, swimming rates of megalopa freshly captured from the field were measured directly in a series of pressure increments. The apparatus and protocol has been described in detail in Sulkin et al. (1980). Swimming rates of 20–29 megalopa were measured at pressure increments of 2, 4.5, 6, 7.5, 9, 11, 13, and 15 psi. Data were tested for linearity (Li 1964). The test of linearity was significant and the data were subjected to regression analysis.

Results

Influence of temperature and salinity on sign of phototaxis

Shown in Fig. 2 are the results of phototaxis tests for larvae acclimated to 15, 20, or 25°C. It is readily apparent that positive phototaxis predominated independent of temperature of acclimation and that a change in temperature from that of acclimation had no effect.

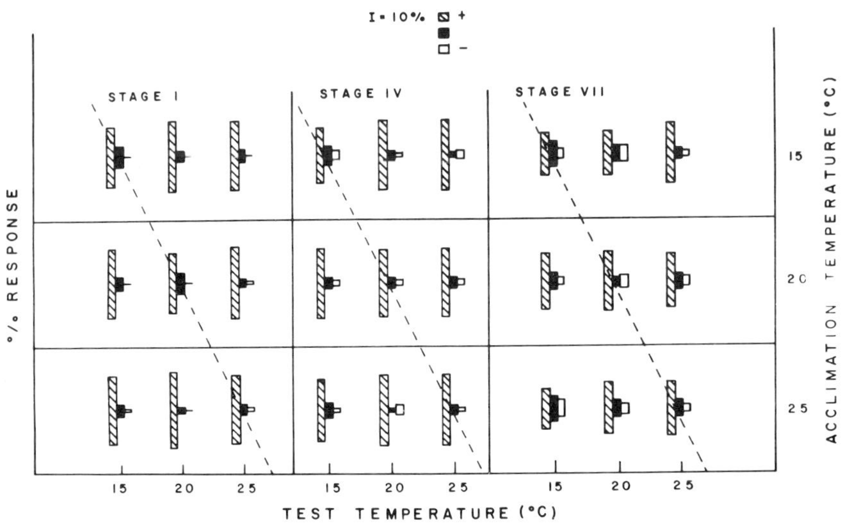

Figure 2. Sign of phototaxis for larvae acclimated to 15, 20, or 25°C. Light conditions described in text. Each kite diagram represents mean of 10 replicates, measured 5 min after all larvae introduced to center of chamber (neutral zone). Dashed line indicates experiments in which larvae were tested at the temperature of acclimation.

Phototaxis responses of larvae acclimated to 25, 30 or 35 ppt salinity and tested at 25, 30 or 35 ppt salinity are shown in Fig. 3. In all three stages, larvae tested at or above the salinity of acclimation showed positive phototaxis in a majority of individuals. In Stages IV and VII, larvae tested at 5 or 10 ppt salinity below that of acclimation showed little change in response. However, in Stage I, a 5 or 10 ppt reduction in salinity induced what appears to be a neutral response. In fact, this result was due to reduction in locomotory activity in most of the larvae. However, larvae acclimated to the reduced salinity swam toward the light, implying resumption of locomotory activity after extended exposure to the reduced salinity.

Response of zoeal Stages I, IV and VII to haloclines

Figure 4 shows the response of Stages I, IV and VII to a series of increasingly strong haloclines. Shown are the mean percents of sample in the top, middle, and bottom segments of the chambers for each stage. Larvae in the top segment have traversed the halocline, while those in the middle segment are in the process of negotiating it. It appears that the proportion of larvae passing through the halocline declined with increase in Δ salinity and the proportion retained in the region of the halocline increased. We define the former case as "inhibition" and the latter as "disruption" of vertical migration.

Statistical analysis confirms this conclusion. Table 1 presents the results of a one-way analysis of variance and Student-Newman-Keuls tests for each stage. In Stage I and IV, disruption occurred at a halocline as low as 2.5 ppt; however, inhibition did not occur in Stage I until a halocline of 10 ppt was

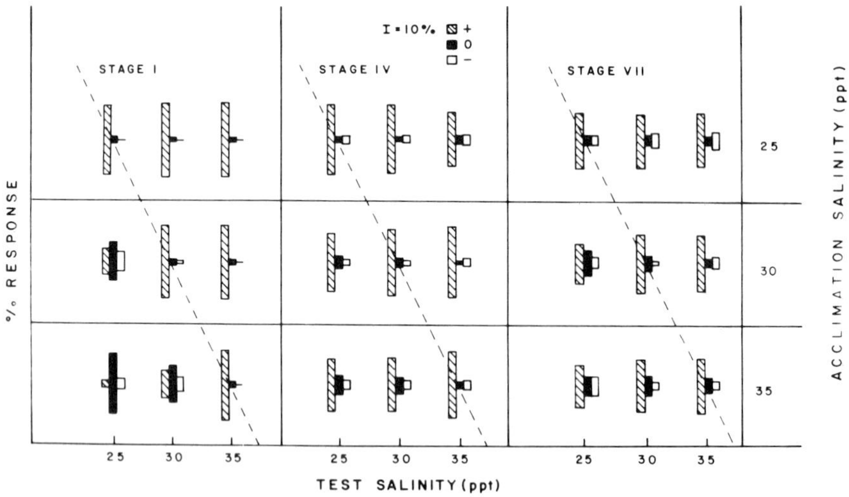

Figure 3. Sign of phototaxis for larvae acclimated to and tested at 25, 30 or 35 ppt salinity. Light conditions described in text. Kite diagrams and dashed lines as in Fig. 2.

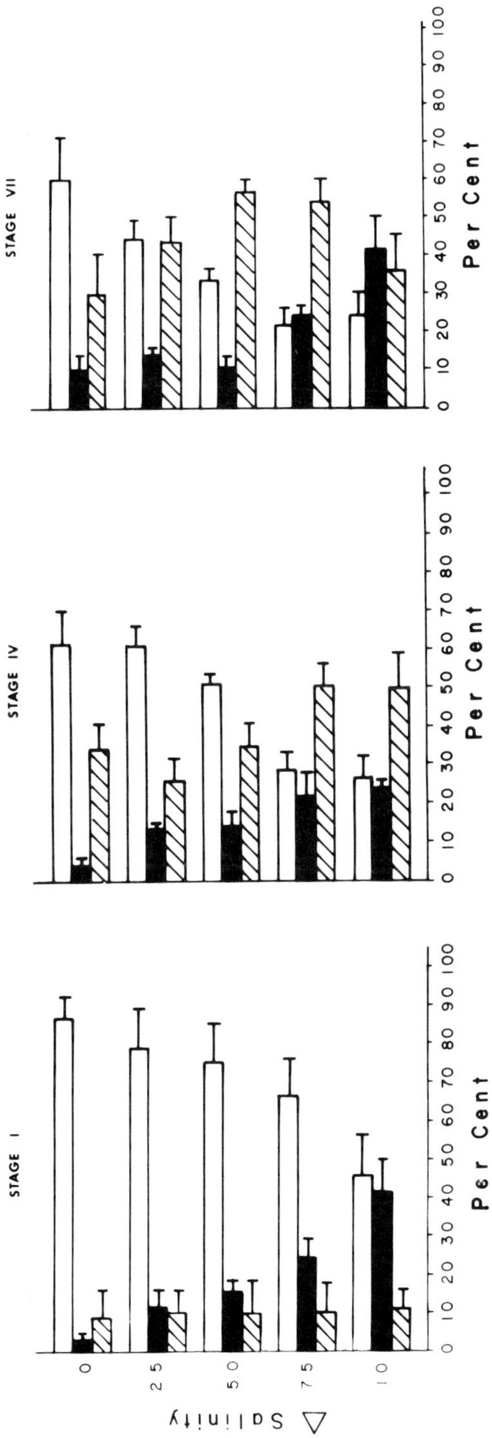

Figure 4. Mean percent of sample (plus one standard error) in the top (clear), middle (solid), or bottom (cross hatched) segment of the observation chamber 30 min after all larvae entered at bottom segment. Data are presented for each halocline tested for indicated zoeal stages. See Table 1 for statistical analysis.

Table 1. Results of analysis of variance (arc sine transformation) and Student-Newman-Keuls test (5%) for halocline experiments. For each stage, mean percent of sample in each of the three segments of the observation chamber were compared among halocline treatments. Treatments bearing the same letter designation were not significantly different. Dashes indicate no Student-Newman-Keuls test performed.

Zoeal Stage	*Section of Chamber*	*Δ Salinity* 0	*2.5*	*5*	*7.5*	*10*	*ANOVA*
	Top	A	A	A	A	B	$P = 0.05$
I	Middle	A	AB	B	BC	C	$P < 0.001$
	Bottom	–	–	–	–	–	$P > 0.05$
	Top	A	A	A	B	B	$P < 0.001$
IV	Middle	A	AB	B	B	B	$P < 0.005$
	Bottom	–	–	–	–	–	$P > 0.05$
	Top	A	AB	B	B	B	$P < 0.005$
VII	Middle	A	A	A	AB	B	$P < 0.001$
	Bottom	–	–	–	–	–	$P > 0.05$

reached. The Δ salinity which results in inhibition decreased through the course of development.

Geotaxis

Direct observations indicated that downward movement was via passive sinking. When swimming ceased, megalopa assumed a horizontal position, usually dorsal side down. To measure passive sinking rate, megalopa were narcotized by adding a few crystals of ethyl carbamate to the culture medium and were timed as they sank through a column of filtered sea water. Mean sinking rate (± 1 SD), measured for a sample of 30 megalopa, was 1.8 ± 0.55 cm/sec. Active swimming invariably involved upward movement, usually initiated as a broad upward arc, occasionally resulting in an upward spiral, but often continuing in a straight line. This behavior indicates negative geotaxis.

Analysis of the photographs taken during the 38 h study permitted determination of the percent of megalopa off the bottom at hourly intervals. Data for the entire 38 h period are shown in Fig. 5, along with diurnal light pattern and tidal cycle at site of collection. Analysis of the photographs taken during the dark phase showed that 108 megalopa could be scored as moving in the vertical plane. All of these megalopa were swimming upward. This was true independent of vertical distribution of the sample at various times during the dark phase. This analysis confirms the presence of negative geotaxis in the megalopa.

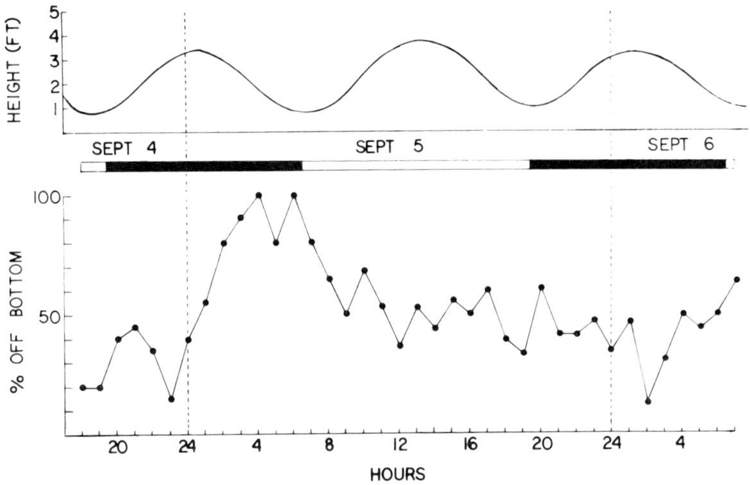

Figure 5. Percent of sample above the bottom of the chamber over a 38 h period. Also shown is natural light: dark cycle (black and white bars) simulated in the laboratory, and tidal cycle from site of capture.

Barokinesis

The graph in Fig. 6 shows the sequence of pressure increments and the percent of megalopa swimming above the bottom at each increment for the control (ambient) and experimental chambers. Also shown are the percent of sample above the bottom of the chamber for the samples in which pressure was initially raised to 15 psi. The mean percentages at each time interval were compared using a paired t-test. Significant differences were found at each time (pressure) interval ($P<0.001$). In a separate analysis, the three distributions measured at 15 psi were compared by one-way analysis of variance. No significant difference was found ($P>0.05$). These results indicate the presence of high barokinesis; that is, an increase in pressure produced an increase in locomotory activity which resulted in movement up from the bottom, the site of initial entry of larvae.

The graph in Fig. 7 shows the mean horizontal swimming rate of megalopa measured directly as a function of pressure, along with the regression line. Table 2 shows the results of the statistical analysis which involves a test of linearity and the regression analysis. The slope of the regression ($b = 0.24$) is significant, confirming the presence of high barokinesis.

Discussion

Sulkin et al. (1980) have shown that in *C. sapidus*, change in salinity affected sign of geotaxis only in zoeal Stage IV. The present study indicates that sign of phototaxis is not influenced by a temperature change of 10°C and that salinity variation of up to 10 ppt has a reversible effect on activity level in

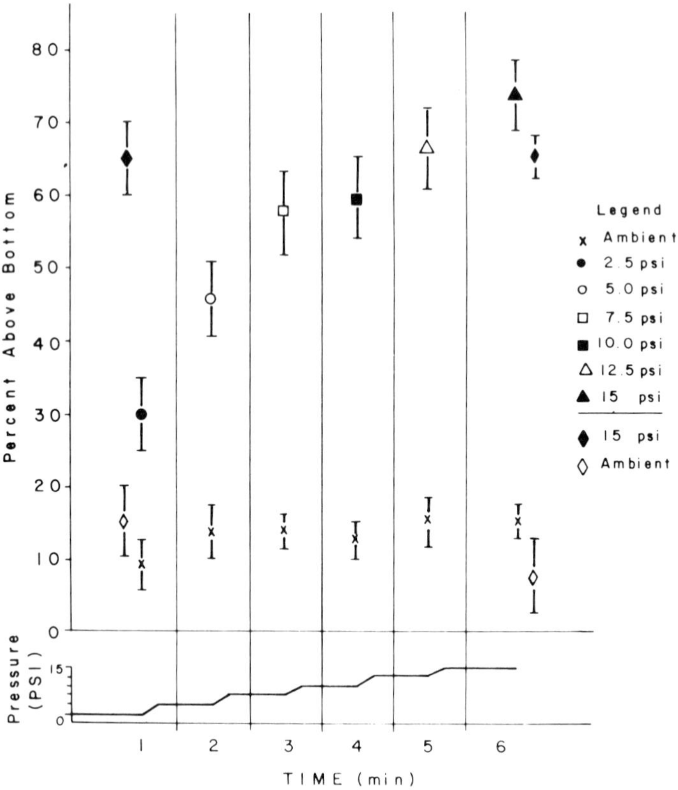

Figure 6. Mean percent (± one standard error) of sample above bottom of experimental and control observation chambers. Also shown is stepwise sequence of pressure increment. The diamond symbols represent mean distributions in which pressure was initially raised by an increment of 15 psi.

zoeal Stage I. The comparatively low sensitivity to temperature change shown by *C. sapidus* larvae is in contrast to that reported for other groups. Thorson (1964) reported that larvae of 14 invertebrate species, which normally show positive phototaxis, become negative with an increase in temperature. Ewald (1912) reported similar behavior in *Balanus perforatus* larvae. On the other hand, Fraenkel (1931) showed that normally negatively phototactic *Squilla* sp. larvae become positive when temperature is reduced. The work of Ott and Forward (1976) on larvae of the brachyuran crab *Rhithropanopeus harrisii* Gould is perhaps more relevant to the present case. Although the sign of geotaxis was dependent upon temperature in both the first and last zoeal stage, sign of phototaxis was temperature-independent, although reduced response occurred at temperature extremes.

Latz and Forward (1977) reported that in *R. harrisii* larvae, a sudden change in salinity of 1–2 ppt can temporarily reverse the sign of phototaxis,

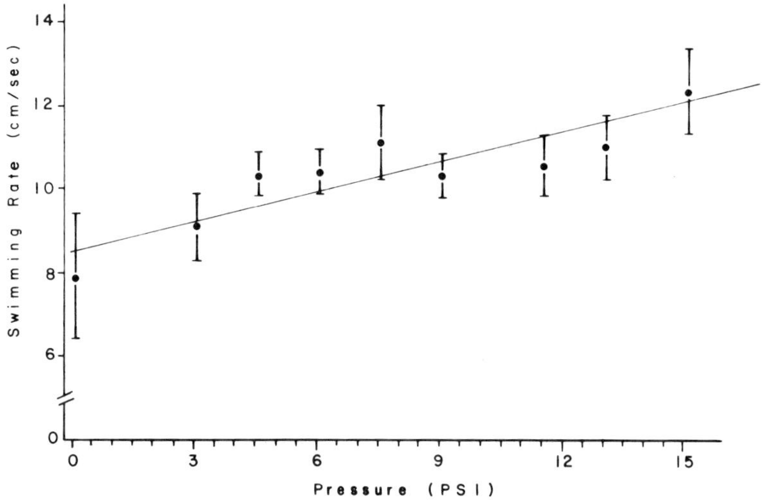

Figure 7. Mean swimming rates (± one standard error) of megalopa exposed to a series of pressure increments.

Table 2. Regression analysis for mean swimming rate of megalopa as a function of pressure increment. *a* indicates significance among means; *b* indicates that the slope of the regression line is greater than zero; *c* indicates that data pass the test of linearity.

Source of Variation	*Degrees of Freedom*	*Sum of Squares*	*Mean Square*	*F Statistic*	*Probability*
Among	8	331.48	41.44	3.90	<0.001[a]
Regression	1	263.25	263.25	24.67	<0.001[b]
Remaining	7	68.23	9.75	0.92	N.S.[c]
Within	212	2259.45	10.65		
Residual	218	2327.68	10.67		
Total	219	2590.93			

although normal phototactic response was regained in as little as 5.5 min. Results reported here are similar except that salinity reduction reduces locomotory activity. Although this response is reversed after a period of acclimation to the lower salinity, even a temporary reduction in activity could disrupt upward migration of Stage I zoeae in nature and entrap larvae at or below the pycnocline.

Response of zooplankters to salinity discontinuities has been studied in detail by Lance (1962), Harder (1968), and Roberts (1971). In general, zooplankters collect at salinity discontinuities, although minimum thresholds vary widely. Larvae of the anomuran crab *Pagurus longicarpus* were capable of detecting ΔS of only 2.5 ppt and would collect at salinity discontinuities due to avoidance of the upper salinity (Roberts 1971).

Similar response is seen in the present study. It is apparent that sharp salinity gradients as low as 2.5 ppt will cause concentration of zoeae at the discontinuity, but that much stronger haloclines are necessary to inhibit upward migration in Stage I blue crab larvae. Furthermore, even the sharpest gradients in nature are unlikely to approach those established in these laboratory studies. Similar results have been obtained for behavior at thermoclines (R. McConnaughey, Horn Point Environmental Laboratories, Cambridge, MD, pers. comm.). This analysis indicates that the presence of a pycnocline in the estuarine spawning grounds of *C. sapidus* is unlikely to inhibit migration to surface waters and will not provide a mechanism for estuarine retention. Substantial changes in behavior occur in *C. sapidus* at the megalopa stage. In contrast to the terminal seventh zoeal stage (Sulkin et al. 1980), the megalopa exhibits pervasive negative geotaxis and high barokinesis and is a much stronger swimmer. The presence of negative geotaxis in the megalopa is in contrast to gravity response reported for megalopa of the crab *Leptodius floridanus* Gibbe (Sulkin 1973). However, it seems likely that behavioral adaptations which influence vertical distribution in the megalopa will vary among species and are likely to be related to adult adaptation and the requirements of recruitment to the adult habitat. Barokinesis results reported here for *C. sapidus* are similar to those reported by Naylor and Isaac (1973), except that the response threshold reported here is lower.

The high degree of pressure sensitivity exhibited by megalopa in combination with negative geotaxis indicates the behavioral basis for high precision depth regulation in the post-larva. This response will promote upward movement or maintenance of position high in the water column and thus provides a mechanism to explain surface swarms of megalopa reported by Dudley and Judy (1971), Williams (1971) and Smyth (1980). However, the exact nature of vertical distribution of *C. sapidus* megalopa will depend upon base activity levels which may be rhythmic. Indeed the data in Fig. 5 imply the presence of a locomotory rhythm although the experimental design is not sufficient for appropriate analysis. Sulkin et al. (1979) reported a diel locomotory rhythm in the first zoeal stage of *C. sapidus*. Periodicity of base activity level under endogenous or exogenous control could explain apparent discrepancies in vertical distribution reported from field studies.

***Larval Dispersal and Recruitment in* C. sapidus**

Early zoeal stages possess a variety of behavioral traits which should result in distribution high in the water column (Sulkin et al. 1980; present study). Proximity of blue crab spawning grounds to the estuary mouth and net non-tidal seaward flow of surface waters in stratified estuaries should result in massive export of larvae to coastal marine waters, a conclusion amply supported by field evidence referenced earlier.

Change in behavioral responses as zoeal development proceeds (Sulkin et al. 1980) suggest adaptations to promote deeper distribution of later zoeal stages with resultant landward transport (Norcross and Stanley 1967; Scheltema 1975). Field evidence does not uniformly support this conclusion, however. While Tagatz (1968), Dudley and Judy (1971), Sandifer (1971), and Goy (1976) report late zoeal stages to be most common in deep water samples, Smyth (1980) and Provenzano (pers. comm.) report large numbers in the neuston. Although early zoeal stages (and megalopa) are common near the mouth of the estuary, late zoeal stages are found there far less frequently, but are widely reported at considerable distance from shore (Nichols and Keney 1963; Dudley and Judy 1971; Sandifer 1975; Smyth 1980). The simplest explanation of these apparently conflicting reports is that intermediate and late zoeal stages tend to be scattered vertically, and hence, geographically. Those that are found in the mouth of the estuary are invariably found at depth and may be a relatively minor source of recruitment to the estuary from offshore.

The major source of recruitment back to the estuary is likely to be the megalopa and, perhaps, juvenile stages (Tagatz 1968; More 1969; Williams 1971; Sandifer 1975; Smyth 1980). There are profound changes in behavior at the megalopa stage which result in a depth regulatory mechanism of high precision. This phenomenon, in combination with appropriate locomotor rhythms, could provide for transport into the estuary based both on net nontidal flow (consistent with reports that megalopa are present in the bottom waters) and on tidal flow (accomplished by periodic movement to surface waters on flood tides).

Recruitment Mechanism and Life History Strategy in **C. sapidus**

The reproductive strategy of *C. sapidus* has retained conservative features (Norse 1977) in the face of substantial adult adaptation. Larval development is characterized by long pelagic duration and a comparatively large and variable number of instars. Furthermore, although adults have successfully invaded the estuarine environment, spawning must take place near the mouth of the estuary because of high salinity requirements of the larvae. In spite of these requirements, larval behavior virtually assures export to coastal marine waters, while the long pelagic period assures wide dispersal. However, the probability of recruitment also appears to be enhanced by behavioral adaptations of late zoeal and megalopa stages and by the capacity of megalopa to delay metamorphosis, particularly at low temperature and high salinity (Costlow 1967). This latter trait is consistent with the model for larval development recently proposed by Jackson and Strathmann (1981) in which a long precompetent phase (zoea) which results in wide dispersal is balanced by a prolonged competent phase (megalopa) which results in recruitment to suitable adult habitats. A life history strategy which promotes wide dispersal as well as providing for recruitment to the adult habitat is well suited to *C. sapidus* whose reproductive strategy is constrained by high salinity requirements (primitive) for larval development and whose distribution is at the geographic limit of the family (Williams 1974). In the mid-Atlantic Bight region for example, the potential for wide dispersal permits recruitment to marginal estuarine habitats at the northern end of the range.

Callinectes sapidus *and the Concept of Estuarine Retention*

There is substantial evidence that larvae of estuarine invertebrates are found in coastal marine waters. In addition to larvae of *C. sapidus*, Nichols and Keney (1963) reported the occurrence of other estuarine decapods in shelf waters off the southeastern U.S. coastline, including species for which estuarine retention mechanisms have been described, such as *Panopeus* sp., *Eurypanopeus* sp., and *Rhithropanopeus* sp. The critical question is whether such larvae are lost from the adult habitat or whether they can be recruited back to the estuarine environment and if so, how significant is this source of recruitment?

A considerable body of literature indicates that recruitment to populations of coastal marine invertebrates is enhanced (in some cases governed) by mechanisms which act either to retain larvae nearshore or to return larvae to the coastal habitat. Perhaps the most generally stated case is that of Makarov (1969) who described "larval belts" along continental margins created by input from large spawning populations of marine invertebrates and maintained by a system of nearshore currents. Efford (1970) proposed a counter-current mechanism to explain recruitment of the sand crab *Emerita analoga* to exposed sandy beaches along the west coast of North America. Lough (1976) postulated a combination of longshore currents and upwelling events to explain coastal retention of larvae of the Dungeness crab *Cancer magister*. In more general terms, Peterson et al. (1979) have described nearshore retention of copepod populations in Oregon effected by upwelling events and Cox and Wiebe (1979) have postulated a recycle mechanism in the mid-Atlantic Bight which can promote retention of meroplankton in the northeastern shelf region in spite of prevailing southwesterly drift of shelf currents. These studies all suggest that hydrographic features characteristic of coastal waters combine with adaptive features of the life history to promote retention of larvae of coastal marine invertebrates.

Indeed, behavioral adaptations which selectively exploit characteristic hydrographic features have been identified with both estuarine retention of estuarine species and nearshore retention of marine species. It is reasonable to speculate that these same phenomena will serve to retain nearshore the larvae of estuarine invertebrates which are transported to shelf waters from estuarine spawning grounds. Retention within waters of the continental shelf and subsequent recruitment from offshore can occur to greater or lesser degrees in this manner for many estuarine species.

Given the vagaries likely associated with recruitment to estuaries from offshore, this model predicts considerable net loss of offspring and creates a distinction between retention within parental populations and retention on a species level. Indeed there is likely to be considerable exchange of offspring among parental populations with coincident gene flow, which will be defined largely by regional hydrographic features. Even in an extreme case such as *C. sapidus*, however, retention within parental populations can occur, depending upon the size of the parent population and its relative contribution to the offshore population of larvae. One would predict, for example, a genetically homogeneous population of *C. sapidus* among estuaries of the mid-Atlantic Bight, a conclusion partially substantiated by Cole and Morgan (1978). Nevertheless, annual recruitment in estuaries of the region is likely to be dominated by larvae produced in Chesapeake Bay, by far the largest source of larvae in the region. Thus, by definition, there will be considerable parental retention in Chesapeake Bay, in spite of the dominance of offshore recruitment.

Clearly, the relative significance of offshore recruitment will vary among estuarine species. *Callinectes sapidus* spawns close to the estuary mouth and, typical of marine invertebrates, migrates to surface waters early in larval development. As a consequence, the potential for retention is reduced and the significance of recruitment from offshore is increased. In other estuarine species in which spawning is remote from the mouth of the estuary or which show specific adaptations to increase retention within the parental estuary, transport of larvae from offshore will be a less significant source of recruitment. Even in these cases, however, a small degree of recruitment from offshore may have significance beyond that relating directly to population dynamics. These emigrants to coastal waters may be important in maintaining genetic cohesion among spatially separated populations. Furthermore, the potential for offshore recruitment, even though of minor significance in some estuarine species, provides a mechanism for recruitment in species which inhabit both stratified and non-stratified estuaries. In so doing, it obviates a paradox implicit in recruitment models which depend strictly on retention within parent estuaries, but which are relevant only to one type of estuary or the other.

Thus, although *C. sapidus* may represent an extreme case, one in which offshore recruitment is central to population dynamics, larvae of other estuarine invertebrates which are exported from estuaries are not necessarily "lost", but may serve important ecological and biological functions for the species. If this is indeed the case, there is no paradox in the fact that estuarine species vary in the efficiency with which their larvae are retained within parental estuaries.

Acknowledgments

We wish to thank Ms. Patti Kelly, Ms. Sharon Van Aller, and Mr. Robert McConnaughey for their technical assistance. We are also indebted to the intellectual stimulation of our colleagues Dr. Charles Epifanio, University of Delaware, and Dr. Tony Provenzano, Old Dominion University. This research has been supported by the National Science Foundation which supported Ms. Van Aller through its Undergraduate Research Participation Program (Grant No. SPI-8025449), and by the Maryland Sea Grant Program (Project R/F-19). The U.S. government is authorized to produce and distribute reprints for governmental purposes notwithstanding any copyright that may appear hereon. This publication is referenced as Contribution No. 1291 from the Center for Environmental and Estuarine Studies.

References Cited

Cole, M. A. and R. P. Morgan, II. 1978. Genetic variation in two populations of blue crab, *Callinectes sapidus*. *Estuaries* 1:202-205.

Costlow, J. D., Jr. 1967. The effect of salinity and temperature on survival and metamorphosis of megalops of the blue crab, *Callinectes sapidus*. *Helgoländer wiss. Meeresunters.* 15:84-97.

Cox, J. and P. H. Wiebe. 1979. Origins of oceanic plankton in the Middle Atlantic Bight. *Est. Coastal Mar. Sci.* 9:509-527.

Dudley, D. L. and M. H. Judy. 1971. Occurrence of larval, juvenile and mature crabs in the vicinity of Beaufort Inlet, North Carolina. *National Marine Fisheries Service Special Scientific Report Fisheries* No. 637, 1-10.

Efford, I. E. 1970. Recruitment to sedentary marine populations as exemplified by sand crab, *Emerita analoga* (Decapoda, Hippidae). *Crustaceana* 18:293-308.

Ewald, E. F. 1912. On artificial modification of light reactions and the influence of electroytes on phototaxis. *J. Exp. Biol.* 13:591-612.

Fraenkel, G. 1931. Die Mechanik der Orientierung der Tiere im Raum. *Biol. Rev.* 6:36-87.

Goy, J. W. 1976. Seasonal distribution and the retention of some decapod crustacean larvae within the Chesapeake Bay, Virginia. M.S. Thesis, Old Dominion Univ., Norfolk, Va. 334 p.

Harder, W. 1968. Reactions of plankton organisms to water stratification. *Limnol. Oceanogr.* 13:156-160.

Jackson, G. A. and R. R. Strathmann. 1981. Larval mortality from offshore mixing as a link between precompetent and competent periods of development. *Amer. Naturalist* 118:16-26.

Lance, J. 1962. Effects of water of reduced salinity on the vertical migration of zooplankton. *J. Mar. Biol. Assoc.* U.K. 42:131-154.

Latz, M. U. and R. B. Forward, Jr. 1977. The effect of salinity upon phototaxis and geotaxis in a larval crustacean. *Biol. Bull.* 153:163-179.

Li, J. C. R. 1964. *Statistical Inference I*, Edwards Bros. Inc., Ann Arbor, MI 658 p.

Lough, R. G. 1976. Larval dynamics of the Dungeness crab, *Cancer magister*, off the central Oregon coast 1969-71. *Fish. Bull.* 74:353-376.

Makarov, R. R. 1969. Transport and distribution of decapoda larvae in the plankton of the Kamchatka shelf. *Oceanology* 9:251-261.

More, W. R. 1969. A contribution to the biology of the blue crab *(Callinectes sapidus* Rathbun) in Texas, with a description of the fishery. *Texas Parks and Wildl. Dept. Serv.* 1:1-31.

Naylor, E. and M. J. Isaac. 1973. Behavioral significance of pressure responses in megalopa larvae of *Callinectes sapidus* and *Macropipus* sp. *Mar. Behav. Physiol.* 1:341-350.

Nichols, P. and P. M. Keney. 1963. Crab larvae *(Callinectes)* in plankton collections from cruises of M/V Theodore N. Gill. South Atlantic coast of the United States 1953-54. *U.S. Fish Wildl. Service. Special Scientific Report, Fisheries* 448:1-14.

Norcross, J. J. and E. M. Stanley. 1967. Part II. Inferred surface and bottom drift. *In:* W. Harrison J. J. Norcross, N. A. Pore and E. M. Stanley. Circulation of shelf waters off the Chesapeake Bight. Environmental Sciences Services Administration Professional Paper 3. U.S. Dept. of Commerce, Environmental Sciences Services Adm. 82 p.

Norse, E. A. 1977. Aspects of the zoographic distribution of *Callinectes* (Brachyura: Portunidae). *Bull. Mar. Sci.* 27:440-447.

Ott, F. S. and R. B. Forward, Jr. 1976. The effect of temperature on phototaxis and geotaxis by larvae of the crab *Rhithropanopeus harrisii* (Gould). *J. Exp. Mar. Biol. Ecol.* 23:97-107.

Peterson, W. T., C. B. Miller and A. Hutchinson. 1979. Zonation and maintenance of copepod populations in the Oregon upwelling zone. *Deep Sea Res.* 26A:467-494.

Pinschmidt, W. C., Jr. 1963. Distribution of crab larvae in relation to some environmental conditions in Newport River estuary, North Carolina. Ph.D. Dissertation, Duke U., Durham, NC 112 p.

Roberts, M. H., Jr. 1971. Larval development of *Pagurus longicarpus* Say reared in the laboratory. III. Behavioral responses to salinity discontinuities. *Biol. Bull.* 140:489-501.

Sandifer, P. A. 1973. Distribution and abundance of decapod crustacean larvae in York River estuary and adjacent lower Chesapeake Bay, Virginia, 1968-69. *Chesapeake Sci.* 14:235-257.

Sandifer, P. A. 1975. The role of pelagic larvae in recruitment to populations of decapod crustaceans in the York River estuary and adjacent lower Chesapeake Bay, Virginia. *Est. Coast Mar. Sci.* 3:269-279.

Scheltema, R. S. 1975. Relationship of larval dispersal, gene-flow and natural selection to geographic variation of benthic invertebrates in estuaries and along coastal regions, pp. 373-391. *In:* L. E. Cronin (ed.), *Estuarine Research, Vol. 1.* Academic Press, New York.

Smyth, P. O. 1980. *Callinectes* (Decapoda: Portunidae) larvae in the middle Atlantic Bight, 1975-77. *Fish. Bull.* 78:251-265.

Sulkin, S. D. 1973. Depth regulation of crab larvae in the absence of light. *J. Exp. Mar. Biol. Ecol.* 13:73-82.

Sulkin, S. D., I. Phillips and W. Van Heukelem. 1979. On the locomotory rhythm of brachyuran crab larvae and its significance in vertical migration. *Mar. Ecol. Progress Series* 1:331-335.

Sulkin, S. D., W. Van Heukelem, P. Kelly and L. Van Heukelem. 1980. The behavioral basis of larval recruitment in the crab, *Callinectes sapidus* Rathbun: A laboratory investigation of ontogenetic changes in geotaxis and barokinesis. *Biol. Bull.* 159:402-417.

Tagatz, M. E. 1968. Biology of the blue crab, *Callinectes sapidus* Rathbun in the St. Johns River, Florida. *Fish. Bull.* 67:17-33.

Thorson, G. 1964. Light as an ecological factor in the dispersal and settlement of larvae of marine bottom invertebrates. *Ophelia* 1:167-208.

Williams, A. B. 1971. A ten-year study of meroplankton in North Carolina estuaries: annual occurrence of some brachyuran development stages. *Chesapeake Sci.* 12:53-61.

Williams, A. B. 1974. The swimming crabs of the genus *Callinectes* (Decapoda: Portunidae). *Fish. Bull.* 72:685-798.

COMPARISON OF DISPERSAL OF CRAB LARVAE IN DELAWARE BAY, USA, AND THE GULF OF NICOYA, CENTRAL AMERICA

C. E. Epifanio
College of Marine Studies
University of Delaware
Lewes, Delaware

and

A. I. Dittel
Escuela de Biologia
Universidad de Costa Rica
San Pedro de Montes de Oca
Costa Rica
America Central

Abstract: We present a comparison of factors affecting dispersal and recruitment of larvae of estuarine species in tropical and temperate regions. Species of the blue crab *Callinectes* spawn in the lower reaches of estuaries in both regions and their larvae are exposed to risk of transport out of the estuary and consequently to habitats that are unsuitable for adult existence. We review available data demonstrating that larvae of *Callinectes sapidus* are indeed carried out of the major estuaries of the Middle-Atlantic Bight, undergo development on the shelf, and are eventually transported back into the estuary. We present data suggesting that larvae of *Callinectes arcuatus* may also be exported from the estuaries of the eastern tropical Pacific, and we discuss turbulent and advective mechanisms of recruitment in the light of known patterns of circulation in both areas.

Introduction

There is convincing evidence that the larvae of many species of estuarine invertebrates are, for the most part, retained in their parent estuary. This may be accomplished by tidally related, vertical migration where larvae rise on flood currents and sink on ebb currents (e.g. Carriker 1951; 1967) or by a general change in vertical position where early stage larvae are transported upstream in bottom water (e.g. Bousfield 1954). Laboratory investigations indicate that changes in salinity or hydrostatic pressure may be important cues for tidally related vertical migration in crustacean larvae (e.g. Latz and Forward 1977; Wheeler and Epifanio 1978; Sulkin et al. 1980) while changes in vertical position with stage of development may be a function of diminished negative geotaxis in later stage larvae (Sulkin 1973).

ISBN 0-12-404070-5

Regardless of the mechanisms involved, retention of larvae of estuarine invertebrates would appear a highly adaptive trait. The limited dispersal that occurs within the parent estuary allows exploitation of newly available habitat while retention of the majority of the larvae assures continuity of the species in the estuary. Additionally, those larvae that are not retained allow gene flow between populations in different estuaries. But in spite of the apparent advantages in retention, there are a number of common estuarine species whose larvae are not retained. Instead, these are routinely flushed from the parent estuary to undergo development in the waters over the continental shelf. This pattern of neritic dispersal appears to carry immensely increased risks of dispersal away from favorable adult habits. Accordingly, the pattern is more common among species that utilize both the lower estuary and the coastal shelf as adult habitat (Sandifer 1975; Dittel and Epifanio 1982).

Larvae in coastal shelf waters are dispersed by both diffusive and advective processes. Diffusive processes involve random or turbulent water motions and can be visualized as the spreading of a dye patch that is introduced at a single point in a water mass. Some of the dye molecules remain near the starting point and others return to that point after having been dispersed, but in general there is a tendency for movement of the molecules away from the release point. Larvae relying on turbulent (eddy) diffusion for dispersal and recruitment need display no special behavioral adaptations; they are widely distributed in the water column and are carried along with the water. Jackson and Strathmann (1981) have presented a detailed analysis of dispersal of coastal larvae in terms of a simplified (two-dimensional) model of turbulent diffusion, but their model ignores the effects of advective processes (mean flow) that are often dominant in estuarine and coastal waters.

For example, in the Middle Atlantic Bight, advective processes involve a strong southerly flow of surface water on the inner shelf and a landward flow of bottom water with convergence at the mouths of major estuaries (Bumpus 1965; Beardsley et al. 1976). The imposition of these advective processes on turbulent diffusion would obviously have large effects on the dispersal of larvae. To use our dye example again, we can conceive of the patch spreading in three dimensions while being carried along in one dimension. In a two-dimensional advective system where surface and bottom waters are not necessarily traveling at the same rate or even in the same direction, the situation becomes more complicated yet.

The important point is that larvae can overcome the effects of diffusion, at least in the vertical plane, through their own swimming activities. Coastal larvae have the potential of taking advantage of two-dimensional advective processes to control the extent of their dispersal. Additionally, there are gyre-like patterns of advection that may also serve to limit the extent of dispersal of larval forms (Cox and Wiebe 1980).

In the present paper we discuss the state of understanding of neritic dispersal as it applies to larvae of the crab genus, *Callinectes*, in a tropical and a temperate estuary. In one case, *Callinectes sapidus* Rathbun, the adults are generally restricted to the estuary while the larvae appear to undergo development almost exclusively on the shelf. In the second case, *Callinectes arcuatus* Ordway, adults are again restricted to the estuary, but we have limited knowledge of the distribution of the larvae.

Oceanographic Comparison of Delaware Bay and Gulf of Nicoya

Our discussion here is not intended as a general comparison of circulation characteristics of tropical and temperate estuaries, but rather as a comparison of investigations by our research group in the Gulf of Nicoya on the Pacific coast of Costa Rica (approximately 10°N 85°W) and Delaware Bay (approximately 39°N 75°W). The Gulf of Nicoya is a tectonic estuary extending approximately 80 km northward from the Pacific Ocean to the mouth of the Rió Tempisque. The estuary is situated in a geographic region that is

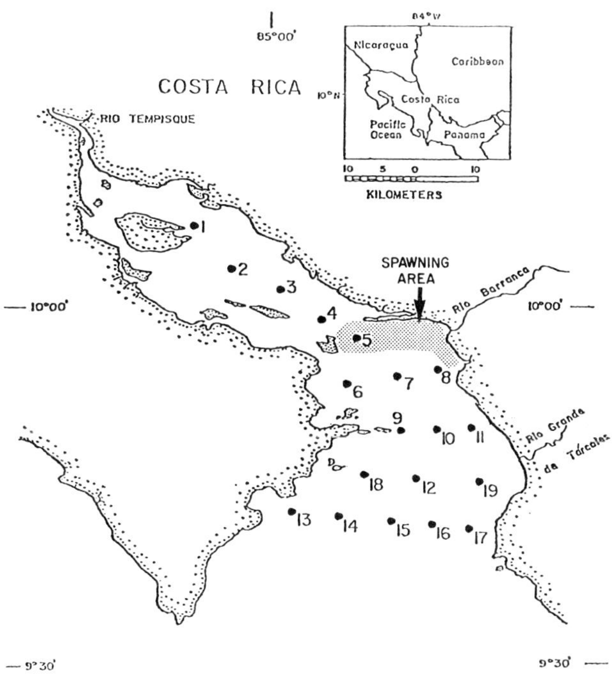

Figure 1. Gulf of Nicoya, Central America. Location of stations where physical data were collected and area where female Callinectes arcuatus *congregate for spawning.*

subjected to extreme seasonal variation in rainfall, and to a large degree it is the consequent seasonal change in the discharge of the rivers entering the Gulf that controls circulation in the estuary. We estimated circulation in the Gulf from vertic profiles of temperature and salinity at 20 stations (Fig. 1) and from hourly measurements of current speed and direction over a tidal cycle at four of those stations. Measurements of temperature and salinity were made *in situ* with a salinity-temperature-conductivity probe and were accurate to $\pm 0.1°C$ and $\pm 0.1°/_{oo}$. Current speed and direction were also measured *in situ* with an electronic current meter. The entire analysis was conducted once during dry season and again during rainy season.

Results indicate that the Gulf is a positive estuary in both seasons with seaward flow concentrated in the surface waters along the eastern shore and compensating landward flow in the deeper waters. The upper Gulf is highly estuarine in character and is vertically stratified during rainy season due to reduced salinity of the surface waters. The lower Gulf is less influenced by seasonal changes in rainfall, but there is a permanent thermocline between 30 and 40 m in depth, and the difference in temperature between surface and bottom water can be as great as 14°C (Fig. 2). Generally the residual flow above the thermocline is seaward while that below is landward (Fig. 3). The circulation time for the Gulf (the average time required to advect a water parcel over the longitudinal dimension of the Gulf by subtidal currents) is approximately one week.

Delaware Bay differs from the Gulf of Nicoya in at least two important respects. First and most obviously, the Bay is subject to large seasonal changes in temperature, and the spawning season of *C. sapidus* is restricted to the warmer portion of the year. Secondly, lower Delaware Bay is relatively unstratified in either temperature or salinity during that spawning season

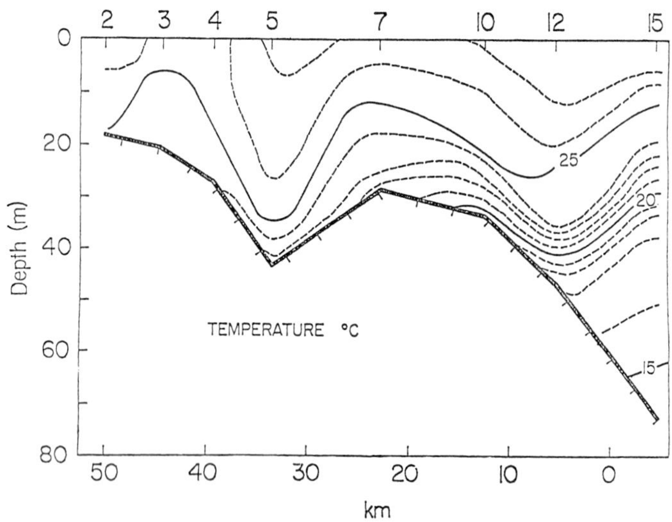

Figure 2. Gulf of Nicoya, Central America. Temperature (°C) isopleths along vertical section down axis of Gulf.

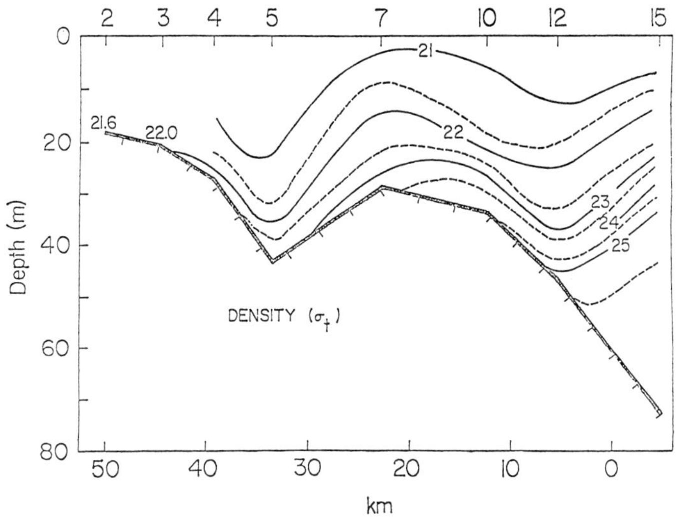

Figure 3(A). Isopleths for density (σ_t) along vertical section down axis of Gulf.

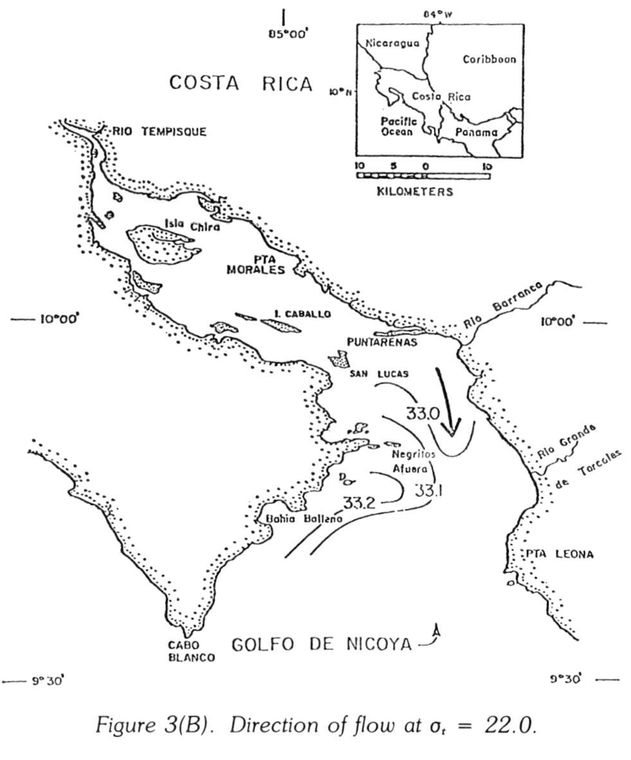

Figure 3(B). Direction of flow at σ_t = 22.0.

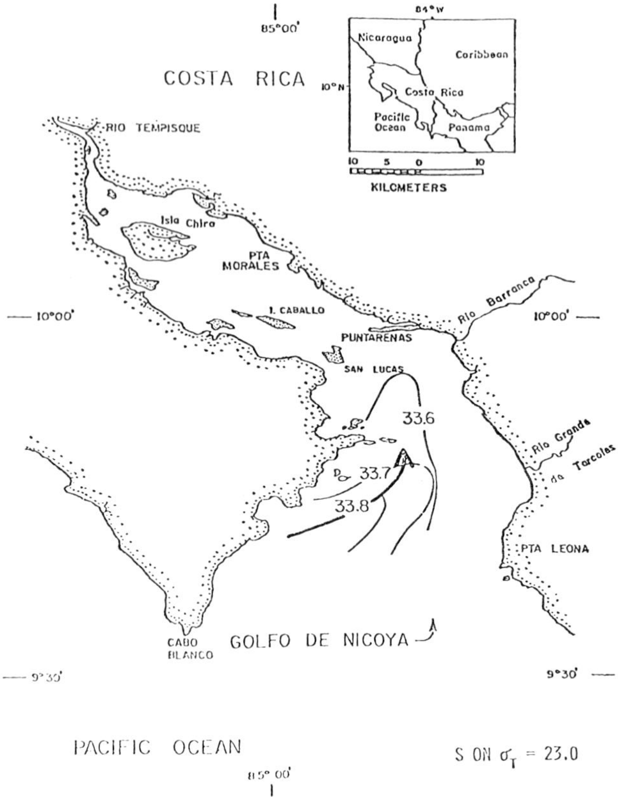

Figure 3(C). Direction of flow at σ_t = 23.0.

(Cronin et al. 1962). However, the general circulation patterns of the two estuaries are grossly similar; Delaware Bay also shows a residual seaward flow of surface waters with a related landward flow of bottom waters (Polis and Kupferman 1973).

Details of residual circulation in the lower Bay have recently been studied as part of our investigation of dispersal of crab larvae in that estuary; that study dealt with Lagrangian circulation, i.e. mean velocity of specific water particles, by means of extensive release of surface and bottom drifters (Pape 1981). Results indicated that estuarine circulation extends at least 40 km onto the adjacent shelf and the velocity of surface water is an order of magnitude greater than that of bottom water. Thus, surface drifters were quickly carried out of the influence of Delaware Bay and were subjected to the general southerly transport in the Middle Atlantic Bight. Surface transport from Delaware Bay to Chesapeake Bay took an average of 26 days. However, advection on the shelf appeared to be greatly influenced by wind patterns and one incidence of reversal of flow was documented.

Comparative Scenarios for Dispersal of Larvae

The life histories of *C. arcuatus* and *C. sapidus* are similar in a number of respects; most importantly, the females of both species migrate to higher salinity portions of estuaries to brood and hatch their eggs (Churchill 1919; Norse and Estevez 1977). This places the resultant larvae in close proximity to the shelf waters and increases the difficulty of retaining larvae in the estuary. The fact that larval development in both species appears to include seven or eight zoeal stages lasting three to five days per stage increases the probability of loss of larvae to the shelf (Sulkin 1978; Dittel and Epifanio unpublished). Clearly, retention in either Delaware Bay or the Gulf of Nicoya could only be effected if larvae spent a considerable portion of time in bottom water. Pape's results (1981) suggest that larvae would have to spend at least an order of magnitude more time in bottom water than surface water to remain in the general circulation of Delaware Bay.

With *C. sapidus*, however, the larvae do not seem to spend such amounts of time on the bottom. For example, results of laboratory experiments have shown a strong negative geotaxis in Stage I zoeae; this would obviously tend to keep the larvae in the near surface waters (Sulkin et al. 1980). Additionally, field data from both the Chesapeake and Delaware estuaries have shown a preponderance of Stage I zoeae in surface water regardless of tidal stage or time of day (Sandifer 1975; Goy 1976; Dittel and Epifanio 1982). These results, in combination with reports of the relatively common occurrence of advanced *C. sapidus* zoeae in shelf waters, strongly suggest that development occurs on the shelf (Nichols and Keney 1963; Smyth 1980).

How, then, are the larvae transported back to the estuary? The least conservative mechanism would involve eddy diffusion. Here larvae are dispersed and concomitantly retained by the turbulent processes discussed in our introduction. Transport back to the parent estuary, or any estuary for that matter, becomes totally a matter of chance with the probabilities highly in favor of loss to unsuitable habitat, e.g. the outer shelf (Jackson and Strathmann 1981). Additionally, in the Middle Atlantic Bight, advective processes would skew this type of dispersal to the south.

A more conservative mechanism would take advantage of advective processes on the shelf. Here larvae would be flushed from the Delaware Estuary, dispersed to the south in surface water, and eventually transported back to shore as they took up deeper positions in the water column during their advanced developmental stages. In this scenario the majority of larvae from the Delaware Estuary are recruited into the Chesapeake population while the Delaware population is dependent on larvae from estuaries farther north. An obvious problem with the scenario is that only during years of protracted reversed flow in the Bight (Bumpus 1969) can there be substantial advective transport of larvae to the northern estuaries. (The only other possibility of advective transport is entrainment of larvae from the Chesapeake into the Gulf Stream and subsequent onshore movement in warm core eddies. However, the necessary time of transport here generally exceeds the period of larval development (Cox and Wiebe 1980)). So in this scheme, the estuaries north of the Delaware are dependent on some type of turbulent process for recruitment during most years.

Another problem with this scenario is the existence of a strong thermocline in the Bight during the summer season. This intersects the bottom at a depth of about 15 m, and waters more than 10 km from shore are often highly stratified (Table 1). Since it is not clear that *C. sapidus* larvae are capable of development at the low bottom temperatures that exist over much of the inner shelf, onshore advection in these bottom waters may not be a major mechanism for recruitment. Additionally, results of the most extensive survey of the vertical distribution of *C.sapidus* larvae on the shelf suggest a maximum occurrence at the surface in the neuston (Smyth 1980). But these data are the result of a comparison of integrated oblique tows from bottom to surface and discrete tows that sampled only the neuston. Therefore the relative abundance in the neuston may have been overestimated relative to other discrete depths in the water column. This and the results of a study of water movement in the Bight by Boicourt and Hacker (1976) suggest a possible answer to our dilemma. Boicourt and Hacker's (1976) findings indicate a strong onshore flow of water immediately above the thermocline and larvae may therefore be able to take advantage of onshore advection without descending through the thermocline. Conceivably larvae are concentrated at discontinuities such as the surface or the thermocline, and careful sampling of the thermocline may show a large number of larvae there as well as in the neuston.

Our knowledge of the biology of *C. arcuatus* larvae is much less complete, hence our scenarios concerning dispersal are even more speculative. The key information that we do have is the fact that eggs are brooded and

Table 1. Middle Atlantic Bight off Delaware Bay. Surface and bottom temperatures as a function of distance from shore. Measurements were made *in situ* with a conductivity-salinity-temperature-depth meter on August 5, 1981.

Distance (km)	*Temperature (°C)* Surface	Bottom
5	22.0	19.9
10	22.3	17.9
15	24.2	12.6
20	24.3	11.4

hatched in the high salinity portions of estuaries. In the Gulf of Nicoya, hatching occurs in the region bordering the upper and lower Gulf (Fig. 1). If larvae are not to be transported to the lower Gulf they must maintain a position deep in the water column. On the other hand, if the larvae behave as do those of *C. sapidus*, they would be rapidly transported to the lower Gulf. Additionally, because circulation time in the lower Gulf is approximately one week, occupancy of the surface waters for more than the first two zoeal stages would result in export from the Gulf and subsequent neritic dispersal. Not enough is known of shelf circulation in the region to predict the extent of this dispersal, but the extremely narrow shelf along the Central American coast would appear to increase the risk of loss to the deep sea.

Some retention in the general estuarine system could be effected by turbulent diffusion, but just as for *C. sapidus*, substantial loss of larvae to unsuitable habitats would occur. The scenario of retention through advective processes is also similar to that for *C. sapidus*. In this model, larvae spend the first two zoeal stages at the surface and are transported into the lower Gulf. With development to the third stage, the larvae sink or migrate into the deeper waters of the lower Gulf and are carried back to the upper Gulf. However, this would require the larvae to cross the thermocline in the lower Gulf and to enter the cold waters below. Because these are larvae of a presumably stenothermal, tropical species, we might expect that they would not be capable of continued development at temperatures as low as 14°C. In addition, the water below the thermocline in the Gulf of Nicoya contains relatively little dissolved oxygen (Peterson 1958), thus imposing further physiological stress on the larvae. Therefore, it would be surprising if recruitment of *C. arcuatus* larvae involved migration of the larvae through the thermocline. Alternative advective mechanisms could take advantage of landward flow along the thermocline or of gyres in the coastal waters adjacent to the Gulf (Efford 1970), but this is total speculation as our knowledge of the circulation of the area is incomplete.

In any event, we find it difficult to believe that recruitment of *Callinectes* larvae is the result of largely turbulent processes such as eddy diffusion when advective transport allows considerable conservation of larvae. Surely selective processes must have led to behavioral adaptation to the flow patterns common to most estuarine-shelf systems. This is not to say that turbulent processes do not allow the retention of some larvae, but advection certainly appears to be the dominant process in systems such as Delaware Bay or the Gulf of Nicoya. We conclude, then, that dispersal of *Callinectes* larvae in the Middle Atlantic Bight largely involves north to south transport with occasional transport to the north in years when wind-induced surface flow is reversed (Bumpus 1969). We would expect especially heavy recruitment in Delaware Bay as well as in the smaller estuaries farther north during these years. Additionally, we would not expect that recruitment in Chesapeake Bay would be diminished during years when there are unusually long espisodes of flow reversal because larvae that are ordinarily lost from that system to the Gulf Stream or to the South Atlantic Bight would be carried back into the parent estuary.

Clearly there are certain advective processes common to most estuarine-shelf systems that would minimize loss of larvae to unsuitable habitat, and we can see no selective advantage in a species' maximizing its loss of larvae. We propose that the conservative nature of natural selection has resulted in the evolution of behavioral traits that allow larvae to take advantage of these advective flows.

Acknowledgments

Our studies in Delaware Bay were supported by funds from the U.S. Department of Commerce, Office of Sea Grant, while support for our work in the Gulf of Nicoya came from the International Cooperative Assistance Program of the Office of Sea Grant, from the Organization for American States, and from the University of Costa Rica. Analyses of physical oceanographic data from the Gulf of Nicoya were performed by Dr. Arthur D. Voorhis of Woods Hole Oceanographic Institution, and he is presently preparing a more detailed treatment of these results for publication. Additionally, we extend our thanks to Dr. Richard W. Garvine of the University of Delaware for his helpful discussions on circulation in Delaware Bay and to the colleagues, students, assistants, and crew members of the research vessels who helped in the work. Finally we extend special thanks to Dr. Manuel M. Murillo of the University of Costa Rica who provided general supervision of the work in the Gulf of Nicoya.

References Cited

Beardsley, R. C., W. C. Boicourt and D. V. Hansen. 1976. Physical oceanography of the Middle Atlantic Bight. *Limnol. Oceanogr. Spec. Symp.* 2:20-34.

Boicourt, W. C. and P. W. Hacker. 1976. Circulation on the Atlantic continental shelf of the United States, Cape May to Cape Hatteras. *Mem de la Soc. Roy. Scie. Liege* 6:187-200.

Bousfield, E. L. 1955. Ecological control of the occurrence of barnacles in the Miramichi Estuary. *Nat. Mus Canada,* Bull. 137, 46:1-69.

Bumpus, D. F. 1965. Residual drift along the bottom on the continental shelf in the Middle Atlantic Bight area. *Limnol. Oceanogr. Suppl.* 10:R50-R53.

Bumpus, D. F. 1969. Reversals in the surface drift in the Middle Atlantic Bight area. *Deep Sea Res. Suppl.* 16:17-23.

Carriker, M. R. 1951. Ecological observations on the distribution of oyster larvae in New Jersey estuaries. *Ecol. Monogr.* 21:19-38.

Carriker, M. R. 1967. Ecology of estuarine benthic invertebrates, pp. 442-487. *In:* G. H. Lauff (ed.), *Estuaries.* Amer. Assoc. Adv. Sci., Washington, DC

Churchill, E. P., Jr. 1919. Life history of the blue crab. *Bull. U.S. Bur. Fish.* 36:91-128.

Cox, J. and P. H. Wiebe. 1980. Origin of oceanic plankton in the Middle Atlantic Bight. *Est. Coast. Mar. Sci.* 9:509-528.

Cronin, L. E., F. C. Daiber and E. M. Hulbert. 1962. Quantitative seasonal aspects of zooplankton in the Delaware River estuary. *Chesapeake Sci.* 3:63-93.

Dittel, A. I. and C. E. Epifanio. 1982. Seasonal abundance and vertical distribution of crab larvae in Delaware Bay. *Estuaries* 5: (in press).

Efford, I. E. 1970. Recruitment to sedentary marine populations as exemplified by the sand crab, *Emerita analoga* (Decapoda, Hippidae). *Crustaceana* 18:293-308.

Goy, J. W. 1976. Seasonal distribution and retention of some decapod crustacean larvae within the Chesapeake Bay, Virginia. M.S. Thesis. Old Dominion U., Norfolk, VA 101 pp.

Jackson, G. A. and R. R. Strathmann. 1981. Larval mortality from offshore mixing as a link between precompetent and competent periods of development. *Am. Nat.* 118:16-26.

Latz, M. I. and R. B. Forward, Jr. 1977. The effect of salinity upon phototaxis and geotaxis in a larval crustacean. *Biol. Bull.* 153:163-179.

Nichols, P. R. and P. M. Keney. 1963. Crab larvae *(Callinectes)* in plankton collections from cruises of R/V *Theodore N. Gill*, South Atlantic coast of the United States, 1953-54. *U.S. Fish Wildl Serv. Spec. Sci. Rept., Fisheries* 448:1-14.

Norse, E. A. and M. Estevez. 1977. Studies on portunid crabs from the eastern Pacific. I. Zonation along environmental stress gradients from the coast of Colombia. *Mar. Biol.* 40:365-373.

Pape, E. H. 1981. A drifter study of the Lagrangian mean circulation of Delaware Bay and adjacent shelf waters. M.S. Thesis. U. Delaware, Newark, DE 148 pp.

Peterson, C. L. 1958. The physical oceanography of the Gulf of Nicoya, Costa Rica, a tropical estuary. *Bull Inter-Am. Trop. Tuna Comm.* 3:139-216.

Polis, D. F. and S. L. Kupferman. 1973. Physical oceanography. *Delaware Bay Report Series Vol. 4.* U. Delaware, Newark, DE 143 pp.

Sandifer, P. A. 1975. The role of pelagic larvae in recruitment to populations in the York River estuary and adjacent lower Chesapeake Bay, Virginia. *Est. Coast. Mar. Sci.* 3:269-279.

Scheltema, R. S. 1975. Relationship of larval dispersal, gene flow, and natural selection to geographic variation of benthic invertebrates in estuaries and along coastal regions, pp. 272-291. *In:* L. E. Cronin (ed.), *Estuarine Research, Vol. 1.* Academic Press, New York.

Smyth, P. O. 1980. *Callinectes* (Decapoda: Portunidae) larvae in the Middle Atlantic Bight, 1975-77. *Fish. Bull.* 78:251-265.

Sulkin, S. D. 1973. Depth regulation of crab larvae in the absence of light. *J. exp. mar. Biol. Ecol.* 13:73-82.

Sulkin, S. D. 1978. Nutritional requirements during larval development of the portunid crab, *Callinectes sapidus* Rathbun. *J. exp. mar. Biol. Ecol.* 34:29-41.

Sulkin, S. D., W. Van Heukelem, P. Kelly and L. Van Heukelem. 1980. The behavioral basis of larval recruitment in the crab *Callinectes sapidus* Rathbun: a laboratory investigation of ontogenetic changes in geotaxis and barokinesis. *Biol. Bull.* 159:402-417.

Wheeler, D. and C. E. Epifanio. 1978. Behavioral responses to hydrostatic pressure in larvae of two species of xanthid crabs. *Mar. Biol.* 46:167-174.

TIMING OF LARVAL PRODUCTION AND FLUX OF INVERTEBRATE LARVAE IN A WELL-MIXED ESTUARY

John H. Christy
Belle W. Baruch Institute for Marine Biology and Coastal Research
University of South Carolina
Columbia, South Carolina

and

Stephen E. Stancyk
Belle W. Baruch Institute for Marine Biology and Coastal Research
and
Department of Biology
University of South Carolina
Columbia, South Carolina

Abstract: Flux of invertebrate larvae between an estuary and coastal waters and the effect of the timing of larval production by estuarine crabs on larval transport were studied in North Inlet estuary, South Carolina. Invertebrate larvae were sampled from three transects across the entrances to this well-mixed estuary during spring and neap tidal periods in each season in 1979. Crab zoeae also were sampled nightly throughout the summer at a single station in the upper reaches of the estuary. The net flux of most larval groups corresponded to the net flux of water and was not different from zero. Six of the 20 larval groups sampled (e.g. echinoderm plutei, barnacle cyprids) showed significant imports to the estuary during one or more sampling periods. Only crab zoeae were exported significantly from the estuary. Such exports occurred during spring tidal periods in the summer when the densities of stage I zoeae of the seven dominant estuarine species were 6 to 57 times greater in the lower estuary and 100 to 10,000 times greater in the upper estuary than during neap tidal periods. Maximum zoeal densities occurred when female crabs released larvae. The timing of larval release may result in rapid seaward transport of newly hatched zoeae on nocturnal ebb tides of greatest amplitude during the semilunar cycle, leading to significant net export from the estuary.

Introduction

Larvae of several taxa of estuarine invertebrates hatch or are liberated only at specific times in the diel, tidal, and semilunar or lunar cycles (Naylor 1976; Saigusa and Hadaka 1978; Clark 1979; DeCoursey 1979). However, most models of mechanisms leading to retention of larvae of estuarine organisms ignore the potentially important effects of the timing of larval entry into the water column on larval dispersal and eventual settlement in ap-

ISBN 0-12-404070-5

propriate habitats (but see Carriker 1961; Russell-Hunter et al. 1972). Instead, such models invoke either larval responses to salinity gradients, patterns of nontidal circulation and tidal currents (Bousfield 1955; Carriker 1967; Sandifer 1975) or the effects of currents and eddy diffusion alone (Ayers 1956; de Wolf 1974) to explain patterns of larval distribution and abundance. In addition, models of retention rarely have been tested directly by measuring the net flux of larvae across the boundaries of an estuary.

This paper presents the results of a study of the seasonal and annual flux of invertebrate larvae between a well-mixed estuary and coastal waters and a concurrent study of the semilunar timing of larval production by brachyuran crabs common in the upper-reaches of this estuary. The major objectives of this research were (1) to determine the possible effects of the timing of larval release by crabs on larval retention as measured by the net flux of crab zoeae, and (2) to compare the signs and magnitudes of the net fluxes of the different larval groups.

Materials and Methods

North Inlet estuary encompasses approximately 3200 ha of salt marsh near Georgetown, South Carolina (Fig. 1). This estuary has a single inlet to the sea at the confluence of three major tidal creeks (Debidue, Town, and Jones Creeks) that average about 3 m deep at mean tide level. The mean tidal range is 1.7 m and peak current velocities of 2.3 m s^{-1} have been recorded in Town Creek. Salinities in the northern portion of the marsh typically are above 32 ppt because the only freshwater input is in the form of runoff, which usually is a very minor component of the tidal prism (Kjerfve and McKeller 1980). In contrast, the southern portions of the marsh receive brackish water from Mud Bay and salinities are generally lower and more variable than in the northern portions (Schwing and Kjerfve 1980).

Larval Flux

Samples were collected in 1979 from anchored boats at two stations on transects near North Inlet across Town and North Jones Creeks, and from one station in South Jones Creek (Fig. 1). Detailed discussions of station placement and materials sampled can be found in Kjerfve et al. (1981) and Chrzanowski et al. (1979). Samples were collected every two hours for 25 h during both neap and spring tide periods in May, July, and October-November, and during mid and neap tide periods in February. Two complete tidal cycles occurred during each 25 h sampling period, but there was no replication of day and night cycles within periods.

Zooplankton were collected with 153 μm mesh, 30 cm mouth diameter nets which had 1 m long cyclindrical sleeves to prevent clogging. Paired, sequential, vertical oblique tows were made from the stationary boats by lowering closed nets to the bottom, opening them, and bringing them to the surface. Vertical oblique tows were taken from a moving boat in South Jones

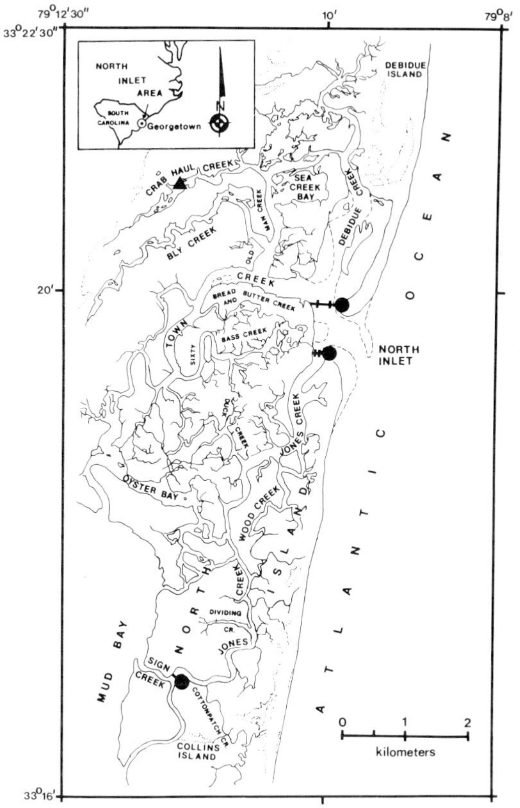

Figure 1. *Map of North Inlet estuary showing locations of transects used in the study of the flux of invertebrate larvae (circles) and the sampling station used in the study of production of brachyuran larvae (triangle).*

Creek because current velocities were too low for stationary sampling. The amount of water filtered was quantified with General Oceanics flowmeters in the net mouths. Between 0.6-7.6 m^3 of water was filtered during tows lasting 1.5-5 min, depending upon current velocity.

Samples were diluted or concentrated so that at least 400 organisms were present in a 2 ml Stemple pipette subsample. Organisms were identified to species (copepods) or larval group and enumerated. All counts were adjusted to yield numbers of organisms per m^3. A residual analysis showed that samples of less than 0.6 m^3 of water did not accurately estimate zooplankton abundance, so such samples were excluded from the analysis of larval flux. In all, 1173 samples were collected over the year, and 92% (1077) were used in flux calculations. Water velocity measurements were taken concurrently at each station with biplane current crosses (Pritchard and Burt 1951).

The instantaneous flux of zooplankton was calculated from:

$$F = \sum_{j=1}^{n} [\overline{V}_j W_j D_j \overline{C}_j] \quad (1)$$

where F is the transport or flux; n is the number of sampling stations; $\overline{V}_j$ is the depth-averaged instantaneous velocity at station j; W_j and D_j are width and depth of the partial cross-section represented by station j; and $\overline{C}_j$ is the concentration of zooplankton at station j. Integration of velocity over depth is possible in these flux calculations because instantaneous velocity varies little with depth (Kjerfve and Proehl 1979; B. Kjerfve, Baruch Institute, pers. comm.). Although samples were collected over two tidal cycles, single tidal cycle transports were calculated by summing the instantaneous fluxes for each tidal cycle to permit comparison with other studies (e.g. Carlson 1978). Tidal cycle estimates calculated in this manner ignore natural rhythms in the data and exaggerate estimates of variance. By our sampling two consecutive tide cycles, sustained rhythms were detected, mathematically defined, and incorporated in an equation to calculate flux. This method supplied an estimate of variance, was not affected by a small number of missing observations, and allowed statistical evaluation of the mean net transport relative to zero transport. Since zooplankton were collected only for two tidal cycles, the method may underestimate variability, but it still provides a reasonable approximation of flux. The net transport calculated for each tidal series (two tidal cycles) was obtained by modelling the instantaneous flux (F) (eq 1) as a function of time (T). The model was

$$\begin{aligned} F_T = \mu &+ \alpha_1 \operatorname{Sin}(2\pi T/24.84) + \beta_1 \operatorname{Cos}(2\pi T/24.84) \\ &+ \alpha_2 \operatorname{Sin}(2\pi T/12.42) + \beta_2 \operatorname{Cos}(2\pi T/12.42) \\ &+ \alpha_3 \operatorname{Sin}(2\pi T/6.21) + \beta_3 \operatorname{Cos}(2\pi T/6.21) \\ &+ \alpha_4 \operatorname{Sin}(2\pi T/24.00) + \beta_4 \operatorname{Cos}(2\pi T/24.00) + \varepsilon \end{aligned} \quad (2)$$

where μ is the mean instantaneous flux; $\alpha_1, \alpha_2, \alpha_3, \alpha_4, \beta_1, \beta_2, \beta_3, \beta_4$ are coefficients; and ε is a random error term. The mean flux per tidal cycle was calculated by multiplying μ by the length of a tidal cycle in the appropriate time units. More detailed discussions of this model may be found in Chrzanowski et al. (in press a, b).

Least squares estimates of $\mu, \alpha_1, \alpha_2, \alpha_3, \alpha_4, \beta_1, \beta_2, \beta_3, \beta_4$ were made using the general linear model procedures of the SAS computing package (Helwig and Council 1979). A complete development of the statistical theory of the general linear model is available in Graybill (1976). The standard t-test from the general linear model was used to test (α = 0.05) the null hypothesis μ = 0 versus the two-sided alternative where μ is the average instantaneous flux (eq 2) for each larval group. About one out of every 20 such tests should be significant at the α = 0.05 level by chance alone; whether such significant fluxes are ecologically meaningful can be judged using other criteria (see below). A more conservative multiple comparisons test would lack the sensitivity to reveal some of the significant fluxes seen in this study.

Brachyuran Larval Production

Brachyuran zoeae were collected nightly near the time of high tide from 18 May to 10 October, 1979, at a station in a shallow (about 1.7 m at mean high tide) tidal creek in the upper marsh (Fig. 1; Miller and Gardner 1981). This sampling time was chosen because the fiddler crabs, *Uca pugilator, Uca pugnax* and *Uca minax,* which are among the numerically dominant brachyurans in the marsh, release larvae only at night near the time of high tide (DeCoursey 1979; Bergin 1981).

Two sampling methods were used. From 18 May to 23 July, three 10-min samples were taken each night with a centrifugal pump powered by a gasoline engine. The intake of the suction line (1.6 cm diameter) was lowered to 10 cm above the creekbed for the first 5 min of each sample and then raised to 10 cm below the surface for the remaining 5 min. The three nightly samples were taken at 30 min intervals, the first beginning (mean ± SD) 1.5 ± 0.56 h after high tide. An average of 0.224 m^3 of water was filtered for each sample. Nightly zoeal densities are the averages of the three nightly samples.

From 24 July to 10 October, one 1.5 h sample was taken each night by drawing water with an electric centrifugal pump through a 153 μm mesh net mounted vertically in a plastic pipe (15.25 cm diameter) fitted to the intake of the suction line of the pump. The net and pipe were bouyed so that the mouth of the net remained about 20 cm below the surface at all tide heights. Sampling began 0.09 ± 0.25 h after high tide. An average of 2.25 ± 0.12 m^3 of water was filtered each 1.5 h sample. Zoeae collected each night were preserved as single samples.

Zoeae in samples were identified by comparison with larvae hatched in the laboratory (*Uca* spp., *Sesarma cinereum, Sesarma reticulatum*) and from published descriptions. All zoeae were counted in samples containing up to about 200 individuals of the most abundant species. At higher densities, three to five aliquots were taken with 5 ml Stemple pipette and density estimates were based on the average count per aliquot when this number exceeded 20. This procedure resulted in counting the number of zoeae of the most abundant species in a volume of sample equal to a minimum of 38 l of water in the marsh and the number of zoeae of the rarest species in a volume of sample equal to at least 114 l of creek water. The coefficient of variation among aliquots with counts of 20 or more individuals was 9.4 ± 1.0%. Sampling error could not be estimated because simultaneous replicate samples were not taken.

Results

Larval Flux

Estimates of mean net flux of twenty different larval categories during spring, mid, or neap tides were calculated for each sampling period (Table 1). Variation about these estimates was large (coefficients of variation up to 200%) because larval densities varied greatly between paired sequential samples and day and night cycles were not replicated. Few of the net fluxes were significantly different from zero. Copepodids and copepod nauplii, including those of the dominant copepod *Acartia tonsa*, showed occasional significant import, especially during neap tides in July and October. There was significant import of barnacle cyprids on spring tides in February and May. Hydrozoan larvae (November), pilidium larvae (May, July) and pluteus larvae (July) also were imported significantly on some spring tides. Had observation of significant fluxes been due to chance alone there should have been an approximately equal number of imports and exports. However, 8 of the 10 significant fluxes were imports even though water was exported from the estuary during all sampling periods (Chrzanowski et al. in press a, b).

Only crab zoeae showed consistent export from North Inlet during times of maximum larval production (Table 1). Exports of crab zoeae were significant during May and July spring tides, and were 6 to 57 times greater than corresponding neap tide exports. Spring tide export of crab zoeae was 2.4 times greater than neap export in October/November as well, but was not significant. During times of significant export, densities of crab zoeae ranged as high as 28,766 m^{-3}, while mean densities ranged from 462 to 9423 m^{-3}, depending upon the creek sampled (Table 2). There was considerable fluctuation in numbers of crab zoeae with time of day, tidal height and velocity, but patterns are difficult to interpret due to the lack of replication and large between-sample variation.

The proportion of the larval population that is entrained in the flow of water through the transects and exported over a tidal cycle can be approximated by comparing the absolute flux (direction ignored) with the net flux. Most of the categories had net fluxes which were 0.3-5.0% of the absolute flux. When crab zoeae were significantly exported, however, they had net fluxes that were 8.3% (May spring tide) and 7% (July spring tide) of the absolute flux. Net flux of crab zoeae on the corresponding neap tides was 2.3 and 3.0%. Absolute flux is equal to two times the population crossing the boundaries (all organisms out plus all organisms in) minus the organisms which do not return (net export). This indicates that about 13-15% of the crab zoeal population was exported during each tidal cycle. Thus, in five to seven tidal cycles, nearly all of the crab zoeae in the water flowing across the boundaries of the estuary could be lost through export.

Brachyuran Larval Production

Stage I zoeae of 14 taxa (16 species) were identified and enumerated in the samples taken each night near the time of high tide in the upper marsh (Table 3). *Uca* spp. clearly dominated all samples, accounting for about 94% of all zoeae counted. Zoeae of the grapsids, *Sesarma cinereum* and *S. reticulatum*, the xanthid, *Panopeus herbstii*, and the pinnotherid, *Pinnixa*

Table 1. Net flux (numbers per tidal cycle × 10^8) of larval forms through the North Inlet estuary, SC, for different tidal periods and different seasons. Negative values are imports; positive values are exports. *$P \leq 0.05$

	Feb 1979		*May 1979*		*July 1979*		*Oct/Nov 1979*	
Larval Category	*Mid*	*Spring*	*Neap*	*Spring*	*Neap*	*Spring*	*Neap*	*Spring*
Acartia copepodids	0.0	0.0	106.90	0.0	−3953.00	129.80	−657.10*	−608.00
Other copepodids	63.30	−25.10	−20.60	−19.00	94.60	24.10	−329.10*	−18.25
Copepod nauplii	−89.50	−13.80	29.50	10.10	−2604.00*	11.52	−0.83	−33.90
Hydrozoan larvae	0.0	0.13	−82.60	4.54	36.90	−8.42	−1.16	−22.70*
Muller's larvae	0.0	0.0	0.0	0.0	0.0	0.0	0.0	0.34
Pilidium larvae	0.19	0.0	−0.17	−1.80*	7.64	−3.54*	−0.12	2.47
Trochophore larvae	−54.67	−27.50	10.83	−0.23	9.83	2.83	0.16	5.00
Polychaete larvae	138.00	14.56	−37.65	13.40	13.20	−119.60	−11.47	−11.03
Bivalve larvae	3.40	1.16	11.68	47.10	−137.50	−106.70	20.50	0.40
Gastropod veligers	−3.93	−0.008	0.36	−6.07	28.93	−3.93	3.84	−5.71
Cyphonautes larvae	0.02	0.05	−24.21	2.31	−173.30	−20.45	−3.14	−42.21
Actinotroch larvae	0.0	0.0	−0.15	0.0	0.0	−0.07	0.03	0.37
Crab zoeae	0.0	−0.05	150.80	919.10*	15.40	876.90*	0.90	2.20
Crab megalopae	0.0	0.0	−0.008	2.62	0.57	−0.85	0.45	0.28
Shrimp zoeae	0.0	0.06	8.62	16.69	−15.91	9.0	0.10	−0.59
Barnacle nauplii	1.29	0.98	−4.25	9.94	−13.78	−38.25	6.50	−24.25
Barnacle cyprids	5.92	−31.60*	−16.20	−443.60*	−206.70	171.10	−0.098	2.68
Pluteus larvae	0.0	0.0	−25.00	2.10	417.00	−25.00*	−1.05	−1.80
Tornaria larvae	0.0	0.0	−0.08	0.0	0.0	0.0	−0.06	7.00
Ascidian larvae	0.0	0.0	0.0	0.0	0.0	0.42	−0.10	1.97

Table 2. Average densities (± 1 standard error of the mean) of crab zoeae over two tidal cycles in the creeks near the boundaries of the North Inlet estuary.

Month, 1979	*Tidal Period*	*Creek*	*Number of Zoeae* m^{-3}
February	Mid	Town	0
		North Jones	0
		South Jones	0
	Spring	Town	0
		North Jones	0.5 ± 0.50
		South Jones	0
May	Neap	Town	973.0 ± 103.58
		North Jones	1133.6 ± 172.46
		South Jones	2939.8 ± 807.40
	Spring	Town	2463.0 ± 551.69
		North Jones	2397.5 ± 473.10
		South Jones	9423.5 ± 1762.36
July	Neap	Town	306.7 ± 38.95
		North Jones	362.8 ± 45.10
		South Jones	1610.8 ± 307.52
	Spring	Town	916.5 ± 150.80
		North Jones	462.1 ± 116.01
		South Jones	1927.7 ± 468.22
October	Neap	Town	4.9 ± 1.86
		North Jones	5.1 ± 1.88
		South Jones	0.7 ± 0.40
November	Spring	Town	5.8 ± 2.18
		North Jones	11.6 ± 4.75
		South Jones	0.3 ± 0.27

chaetopterana, also were frequently observed at lower densities. Together, these seven species comprised 99.5% of all zoeae. Their relative abundance was unaffected by variation in zoeal densities, and did not change significantly between semilunar maxima and minima. Later zoeal stages made up only 0.019% of all zoeae counted (16 stage II and 5 stage V *Uca* spp.; 3 stage II *Eurypanopeus depressus*). Species composition and relative abundance of zoeae in flux samples compared favorably with samples from the upper marsh. The seven species that were dominant in the upper marsh comprised 98% of the zoeae present in eight representative flux samples taken during May and July spring tidal periods, and 83% of the zoeae in six samples collected during the respective neap tidal periods. Later zoeal stages also were rare in flux samples, accounting for only 0.1% of 5066 zoeae identified to species in 14 representative samples from all transects.

The density of *Uca* spp. zoeae varied significantly among the three samples taken each night from 18 May to 23 July (Freidman's two-way

Table 3. Zoeal species composition and abundance in samples taken at night near the time of high tide at a station in the upper reaches of the North Inlet estuary.

Taxon	*Number of zoeae counted*	*% of total*	*Cumulative % of total*
Uca spp.	115090	93.80	93.80
Sesarma cinereum	2749	2.24	96.05
Sesarma reticulatum	1842	1.50	97.55
Panopeus herbstii	1411	1.15	98.69
Pinnixa chaetopterana	978	0.79	99.49
Pinnotheres ostreum	299	0.24	99.74
Eurypanopeus depressus	250	0.20	99.94
Pinnotheres maculatus	31	0.03	99.97
Pinnixa sayana	25	0.02	99.98
Porcellanid	8	0.01	99.99
Neopanope texana	5	<0.01	
Emerita talpoida	2		
Menippe mercenaria	1		
Rhithropanopeus harrisii	1		100.00
Total	122692		

ANOVA; $\chi^2 = 10.897$; 2 df; $P<0.005$). Densities in the third sample were consistently lower than in the first two samples. This suggests that the time of sampling could have affected the daily estimates of zoeal densities significantly. However, there was no correlation between the time a set of three samples was begun and the density estimate (*Uca* spp.) for that night ($r = -0.149$; $N = 57$; $P>0.10$). Errors introduced by variation in sampling time were sufficiently small to be masked by true daily variation in density. Sampling time varied little from 24 July to 10 October.

Densities of stage I zoeae of each of the seven most abundant species varied markedly (Fig. 2). *Uca* spp. were 1000 to 10,000 times more dense when abundant than when rare. Densities of the other species varied over two to three orders of magnitude. All species displayed the same temporal variation; maximum densities occurred during the spring tides near the time of the full and new moons and minimum densities occurred during neap tides near the first and last quarter moons. Semilunar variation in zoeal densities was most pronounced from June through September.

Uca spp., and, to a lesser extent, *Sesarma reticulatum* and *Panopeus herbstii,* displayed a lunar cycle in maximum zoeal densities from June through August. Zoeae were 3 to 5 times more abundant during the new moon spring tides than during full moon spring tides (Fig. 2).

Although all seven dominant species displayed strong semilunar periodicities in zoeal densities, each also exhibited unique short term variation. An exception appears to be the marked and unexpected decline in densities

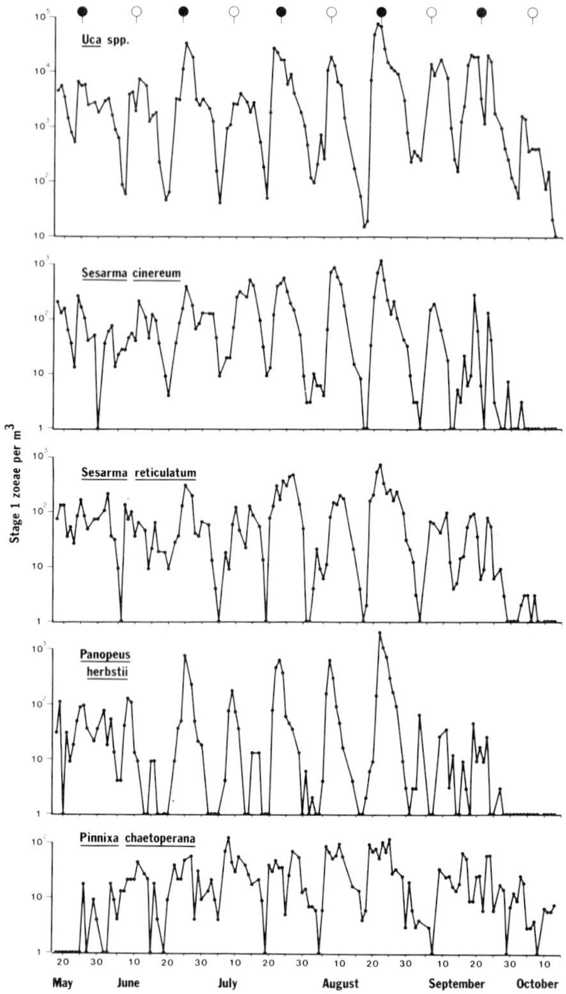

Figure 2. *Densities of common stage I brachyuran zoeae in the samples taken in the upper reaches of the North Inlet estuary. Densities of 0 m^{-3} or 1 m^{-3} are plotted as 1 m^{-3}. The dates of new moons are indicated by closed circles and the dates of full moons are indicated by open circles.*

for a 2-3 d period during the new moon spring tides on 21-23 September, which occurred about 15 d after Hurricane David affected the marsh.

Discussion

Few larval forms showed significant imports or exports from North Inlet estuary. Retention of larvae in this system may have been affected by aggregation of larvae in water masses which maintain their integrity over several tidal cycles (Dyer 1982). Coefficients of variation between sequential samples ranged from 85-200%, whereas simultaneous oblique plankton tows in North Inlet have coefficients of variation of 15% or less (S. E. Stancyk and L. Barker, unpublished). Large variation between sequential samples probably reflects true patchiness in the plankton as has been described in numerous other planktonic environments (Steele 1980). Dye patches placed in tidal creeks adjacent to the transects used during this study remained visible for 3-4 tidal cycles and longer under certain conditions (B. Kjerfve, pers. comm.). Patches of larvae may have been passing back and forth across the inlet boundaries with little net loss or gain.

Alternatively, vertical migration by larvae may have minimized export from the estuary. Trinast (1975) showed that *Acartia (Acanthacartia)* n. sp. tended to congregate at depth during ebb tide, becoming entrained in a gyre system which enhanced retention. Similarly, some copepods *(Pseudodiaptomus hessei, Acartia longipatella, A. natalensis)* and mysids *(Rhopalophthalamus terranatalis, Mesopodopsis slabberi)* aggregate with respect to tidal currents, thereby maintaining position in the estuary (Wooldridge and Erasmus 1980). Whether larvae with clumped distributions in North Inlet behave similarly requires further study.

Although water was exported from the North Inlet estuary during all sampling periods (Chrzanowski et al. in press a, b), some larval forms showed significant imports. High rates of consumption or settlement within the estuary may explain net imports of larvae that may have originated and occurred at high densities in the ocean (e.g. pilidium larvae, plutei). Other larval forms, such as barnacle cyprids (Bousfield 1955) and crab megalopae (Sandifer 1975; Sulkin et al. 1980), may exhibit behavior that enhances import into the estuary.

Only crab zoeae were significantly exported from the North Inlet estuary. Exports were greatest during spring tidal periods in the summer when densities of stage I zoeae of the seven dominant brachyuran crabs reached semilunar maxima in both the upper and the lower estuary. Plankton samples were taken in the upper estuary near the time of high tide at night, the time when *Uca* spp. release larvae (DeCoursey 1979; Bergin 1981), suggesting that semilunar variation in the abundance of stage I zoeae of *Uca* spp. resulted directly from cycles of larval release. This suggestion is supported by the observation that *U. pugilator, U. pugnax,* and *U. minax* all exhibited peaks of larval release in the laboratory near the time of the full and new moons throughout most of the 1979 breeding season (J. Christy, unpublished). Independent evidence of semilunar timing of larval production among the other dominant brachyurans is available only for *Sesarma cinereum* (Dollard 1980); this species also released larvae near the time of the syzygies during the summer of 1979 at North Inlet. If there is a constant proportional relationship between the magnitude of larval release at high tide at night and at other times of the day and phases of the tide, then semilunar variation in the abundance of stage I zoeae of the other species may result from semilunar timing of larval production by these crabs as well. Since stage I zoeae of *Uca* spp. (Christiansen and Yang 1976; Wheeler 1978), *Sesarma cinereum* (Costlow et al. 1960) and *Panopeus herbstii* (Costlow et al. 1962) take 2-4 d to reach stage II at summer temperatures (25-30°C), some stage I zoeae in the upper marsh samples may not have been released the night they were collected. However, larval mortality and seaward transport, which may be rapid in this estuary, will remove older stage I zoeae from the upper estuary, increasing the correlation between nightly densities of stage I zoeae and the magnitude of larval release.

The lunar cycle of variation in the density of stage I zoeae, most evident for *Uca* spp., may result from lunar variation in the tidal prism, not from lunar periodicity in the intensity of larval release. During July and August, 1979, the lunar perigee (when the moon is closest in its orbit to the earth) and the full moon were nearly coincident. Tides were about 0.5 m higher and the volume of water on the marsh correspondingly greater during the full moon spring tides than during the new moon spring tides, which may have produced a relative dilution and concentration of zoeae, respectively.

The unexpected reduction in zoeal densities for all species during the spring tides in late September occurred one semilunar cycle after the unusually severe weather associated with Hurricane David. Males of several species of *Uca* exhibit semilunar cycles of courtship (von Hagen 1970; Zucker 1976, 1978; Christy 1978) which coincide with cycles of female sexual receptivity, mating, and oviposition (Christy 1978). The incubation period for several species of *Uca* is 12-15 d at summer temperatures (Feest 1969; von Hagen 1970; Greenspan 1975; Christy 1978). The Hurricane may have disrupted breeding behavior sufficiently to reduce sharply the number of matings on 4-6 September, resulting in the observed decline in zoeal densities 15 d later.

Peaks of larval production by the seven dominant brachyuran crabs in the upper reaches of the North Inlet estuary occurred near the time of the full and new moons when nocturnal ebb tides were greatest in amplitude. Semilunar cycles of larval production have been described for several species of *Uca* (Christy 1978, and references therein; Wheeler 1978), four grapsid crabs (Warner 1967; Saigusa and Hadaka 1978; Dollard 1980), and two gecarcinids (Gifford 1962; Henning 1975; Klaasen 1975) at sites ranging from the east coast of India to the mid-Atlantic coast of the United States. With one exception, peaks of larval release by these crabs and nocturnal ebb tides of maximum amplitude occur near the time of the full and new moons. At a site in Charlotte Harbor on the west coast of Florida, larval release by *Uca pugilator* and nocturnal ebb tides of maximum amplitude usually occur during the quarter moons (Christy 1978; NOAA Tide Tables). This strongly suggests that larval release is timed to occur during large amplitude nocturnal ebb tides, not during a particular phase of the moon.

The immediate consequence of the timing of larval production seen in the brachyurans will be rapid seaward transport of newly hatched zoeae. Few later stage zoeae were collected in either the upper or lower estuary and up to 15% of the zoeal population passing across the boundaries of the North Inlet estuary was exported each tidal cycle. Rapid export of stage I zoeae into coastal or oceanic waters appears to be an important aspect of the dynamics of the dispersal of brachyuran larvae in this well-mixed estuary. Export of other larvae from the North Inlet estuary was rare and non-significant when it occurred. Large exports of water were not necessary to effect transport of larvae, and imports or zero fluxes occurred despite water exports. Further study may reveal that some larval forms possess behavioral adaptations that aid retention. Models dealing with transport of crab zoeae should be extended to include larvae of other groups only with extreme caution.

Acknowledgments

We would like to thank the many undergraduate and graduate student volunteers who helped with sampling during the flux study, as well as the numerous technicians and staff of the Baruch Institute who made our work possible. T. L. Ferrell particularly is thanked for his tireless zooplankton enumeration, and Dr. J. D. Spurrier and G. O. Osenga for designing the flux model and fitting it to the zooplankton data. This research was supported by Grants DEB 8004275 (SES) and OCE 77-20960 (JHC) from the National Science Foundation. This is contribution number 439 from the Belle W. Baruch Institute for Marine Biology and Coastal Research.

References Cited

Ayers, J. C. 1956. Population dynamics of the marine clam, *Mya arenaria*. *Limnol. Oceanogr.* 1:26-34.

Bergin, M. E. 1981. Hatching rhythms in *Uca pugilator* (Decapoda: Brachuyura). *Mar. Biol.* 63:151-158.

Bousfield, E. L. 1955. Ecological control of the occurrence of barnacles in the Miramichi Estuary. *Bull. Nat. Mus. Can.* 137:1-69.

Carlson, D. M. 1978. The ecological role of zooplankton in a Long Island salt marsh. *Estuaries* 1:85-92.

Carriker, M. R. 1961. Interrelation of functional morphology, behavior and autecology in the early stages of the bivalve *Mercenaria mercenaria*. *J. Elisha Mitchell Sci. Soc.* 77:168-241.

Carriker, M. R. 1967. Ecology of estuarine benthic invertebrates: a perspective, pp. 442-487. *In:* G. H. Lauff (Ed.), *Estuaries*. Am. Assoc. Adv. Sci. Pub. No. 83. Washington, DC

Christiansen, M. E. and W. T. Yang. 1976. Feeding experiments on the larvae of the fiddler crab *Uca pugilator* (Brachyura, Ocypodidae), reared in the laboratory. *Aquaculture* 8:91-98.

Christy, J. H. 1978. Adaptive significance of reproductive cycles in the fiddler crab *Uca pugilator*. *Science* 199:453-455.

Chrzanowski, T. H., L. H. Stevenson and B. Kjerfve. 1979. Adenosine 5'-triphosphate flux through the North Inlet marsh system. *Appl. Environ. Microbiol.* 37:841-848.

Chrzanowski, T. H., L. H. Stevenson and J. D. Spurrier. (in press, a). Seasonal variability and transport of suspended microfungi in a southeastern salt marsh. *Appl. Environ. Microbiol.*

Chrzanowski, T. H., L. H. Stevenson and J. D. Spurrier. (in press, b). Transport of particulate organic carbon through the North Inlet ecosystem. *Marine Ecology Prog. Series.*

Clark, R. B. 1979. Environmental determination of reproduction in polychaetes, pp. 107-121. *In:* S. E. Stancyk (ed.), *Reproductive Ecology of Marine Invertebrates*. U. South Carolina Press, Columbia, SC

Costlow, J. D. Jr., C. G. Bookhout and R. Monroe. 1960. The effect of salinity and temperature on larval development of *Sesarma cinereum* (Bosc) reared in the laboratory. *Biol. Bull.* 118:183-202.

Costlow, J. D. Jr., C. G. Bookhout and R. Monroe. 1962. Salinity and temperature effects on the larval development of the crab *Panopeus herbstii* Milne-Edwards, reared in the laboratory. *Physiol. Zool.* 35:79-93.

DeCoursey, P. J. 1979. Egg-hatching rhythms in three species of fiddler crabs, pp. 399-406. *In:* E. Naylor and R. G. Hartnoll (eds.), *Cyclic Phenomena in Marine Plants and Animals.* Pergamon Press, Oxford.

de Wolf, P. 1974. On the retention of marine larvae in estuaries. *Thalassia Jugoslavica* 10:415-424.

Dollard, H. A. 1980. Larval release patterns in the wharf carb, *Sesarma cinereum,* from North Inlet, South Carolina. MS Thesis, U. South Carolina, Columbia, SC 39 p.

Dyer, K. R. 1982. Localized mixing of low salinity patches in a partially mixed estuary (Southampton Water, England), pp. 21-36. *In:* V. S. Kennedy (ed.), *Estuarine Comparisons.* Academic Press, New York.

Feest, J. 1969. Morphophysiolgische Untersuchungen zur Ontogeneses und Fortpflanzungsbiologie von *Uca annulipes* und *Uca triangularis* mit Vergleichsbefunded an *Illyoplax gangetica. Forma Functio* 1:159-225.

Gifford, C. A. 1962. Some observations on the general biology of the land crab *Cardisoma guanhumi* (Latrielle) in South Florida. *Biol. Bull.* 123:207-223.

Graybill, F. A. 1976. *Theory and Application of the Linear Model.* Duxbury Press, North Scituate, MA

Greenspan, B. N. 1975. Male reproductive strategy in the communal courtship system of the fiddler crab *Uca pugnax.* Ph.D. Thesis, Rockefeller U., New York. 195 pp.

Helwig, J. T. and K. A. Council. 1979. *SAS User's Guide.* SAS Institute Inc., Raleigh, NC

Henning, H. G. 1975. Kampf-, Fortpflanzungs- und häutungsverhalten-Wachstum und Geschlechtsreife von *Cardisoma guanhumi* Latreille (Crustacea, Brachyura). *Forma Functio* 8:463-510.

Kjerfve, B. and H. McKellar, Jr. 1980. Time series measurements of estuarine material fluxes, pp. 341-358. *In:* V. S. Kennedy (ed.), *Estuarine Perspectives.* Academic Press, New York.

Kjerfve, B. and J. A. Proehl. 1979. Velocity variability in a cross-section of a well-mixed estuary. *J. Mar. Res.* 37:409-418.

Kjerfve, B., L. H. Stevenson, J. A. Proehl, T. H. Chrzanowski and W. M. Kitchens. 1981. Estimation of material fluxes in an estuarine cross section: a critical analysis of spatial measurement density and errors. *Limnol. Oceanogr.* 26:325-335.

Klaasen, F. 1975. Oekologische und ethologische Untersuchungen zur Fortpflanzungsbiologie von *Gecarcinus lateralis* (Decapoda, Brachyura). *Forma Functio* 8:101-174.

Miller, J. L. and L. R. Gardner. 1981. Sheet flow in a salt-marsh basin, North Inlet, South Carolina. *Estuaries* 4:234-237.

Naylor, E. 1976. Rhythmic behaviour and reproduction in marine animals, pp. 393-429. *In:* R. C. Newell (ed.), *Adaptation to Environment, Essays on the Physiology of Marine Animals.* Butterworths, London.

Pritchard, D. W. and W. V. Burt. 1951. An inexpensive and rapid technique for obtaining current profiles in estuarine waters. *J. Mar. Res.* 10:180-189.

Russell-Hunter, W. D., M. L. Apley and R. D. Hunter. 1972. Early life-history of *Melampus* and the significance of semilunar synchrony. *Biol. Bull.* 143:623-656.

Saigusa, M. and T. Hadaka. 1978. Semi-lunar rhythm of the zoeae-release activity of the land crabs *Sesarma. Oecologia* 37:163-176.

Sandifer, P. A. 1975. The role of pelagic larvae in recruitment of populations of adult crustaceans in the York River estuary and lower Chesapeake Bay, Virginia. *Est. Coastal Mar. Sci.* 3:269-279.

Schwing, F. B. and B. Kjerfve. 1980. Longitudinal characterization of a tidal marsh creek separating two hydrographically distinct estuaries. *Estuaries* 3:236-241.

Steele, J. H. (ed.). 1980. *Spatial Patterns in Plankton Communities.* Plenum Press, New York. 470 p.

Sulkin, S. D., W. Van Heukelem, P. Kelly and L. Van Heukelem. 1980. The behavioral basis of larval recruitment in the crab *Callinectes sapidus* Rathbun: a laboratory investigation of ontogenetic changes in geotaxis and barokinesis. *Biol. Bull.* 159:402-417.

Trinast, E. M. 1975. Tidal currents and *Acartia* distribution in Newport Bay, California. *Est. Coastal Mar. Sci.* 3:165-176.

von Hagen, H. O. 1970. Anpassungen an das spezielle Gezeitenzonen-Niveau bei Ocypodiden (Decapoda, Barchyura). *Forma Functio* 2:361-413.

Warner, G. F. 1967. The life history of the mangrove tree crab *Aratus pisoni. J. Zool.* 153:321-335.

Wheeler, D. 1978. Semi-lunar hatching periodicity in the mud fiddler crab *Uca pugnax* (Smith). *Estuaries* 1:268-269.

Wooldridge, T. and T. Erasmus. 1980. Utilization of tidal currents by estuarine zooplankton. *Est. Coastal Mar. Sci.* 11:107-114.

Zucker, N. 1976. Behavioral rhythms in the fiddler crab *Uca terpsichores*, pp. 145-159. *In:* P. J. DeCoursey (ed.), *Biological Rhythms in the Marine Environment.* U. South Carolina Press, Columbia, SC

Zucker, N. 1978. Monthly reproductive cycles in three sympatric hoodbuilding tropical fiddler crabs (genus *Uca*). *Biol. Bull.* 155:410-424.

TIDALLY TIMED BEHAVIOR: EFFECTS ON LARVAL DISTRIBUTIONS IN ESTUARIES

Thomas W. Cronin
Department of Biology
Yale University
New Haven, Connecticut

and

Richard B. Forward, Jr.
Duke University Marine Laboratory
Beaufort, North Carolina
and
Department of Zoology
Duke University
Durham, North Carolina

Abstract: Zoeal larvae of the xanthid crab *Rhithropanopeus harrisii* are retained in the upper regions of the Newport River estuary, North Carolina. This retention is effected by rhythmic vertical migrations which are similar for all zoeal stages, ranging above and below the depth of no net flow. Time series analysis of these migrations revealed that their dominant periodic components were usually those of the tidal cycle (period = 12.4 h) and the current flow cycle (period = 6.2 h). Under constant laboratory conditions, larvae taken directly from the estuary continued only the tidal vertical migration, reaching their greatest depth at the time of low tide. In contrast, larvae reared in the laboratory under a 12 h:12 h light:dark cycle had only a weak circadian migration rhythm. The tidal rhythm is also expressed in larvae tested immediately after hatching from Newport River crabs, but the migration amplitude is greater for larvae which have spent at least one day under field conditions. Time of hatching may also play a part in setting the phase of this rhythm. Larvae are released near the time of local high tide; this timing is evidently adapted to take advantage of maximum salinity of the water and to help determine the region of the estuary in which they will subsequently develop. Tidal behavior and tidal rhythms could be widespread among estuarine zooplankton, particularly in regions where the tides exert strong dynamic effects on the physical structure of estuarine waters.

Introduction

Many estuarine species of invertebrates produce planktonic larvae. In those species which have sessile or sedentary adults, the larvae must be present in the proper estuarine regions at the time of metamorphosis or adult populations will not persist. The most adaptive distribution of late-stage larvae can be attained either by the continuous retention of all intervening developmental stages or by an initial downstream migration of larvae into the

ISBN 0-12-404070-5

lower estuary or adjacent coastal waters followed by a return migration of older larvae.

Most estuaries have currents which vary systematically with depth but which are dominated by tidal influences (see reviews in Kjerfve 1978). Thus, there commonly are tidal variations in depth, salinity and other water quality factors, and vertical turbulence-induced mixing (Bowden 1978). The direction of net transport of estuarine plankton depends upon their distribution by depth. Because net flow low in the water column is landward and that near the surface is seaward, larvae can potentially control both the speed and direction of transport by regulating how much time they spend at each depth. Such larval vertical migrations in estuaries could be related to the obvious exogenous cycles of current speed or tidal conditions. This behavior would contrast with that of typical marine zooplankton and larvae, which have diel vertical migrations.

The estuarine xanthid crab *Rhithropanopeus harrisii* (Gould) passes through four planktonic stages and a semi-benthic megalopal stage before molting to the juvenile crab. The larval stages are usually found concentrated in the upper regions of estuaries and are rarely found in coastal waters (Connolly 1925; Bousfield 1955; Pinschmidt 1963; Tagatz 1968; Williams 1971; Sandifer 1973, 1975; Goy 1976; Cronin 1982). Such distributions imply that an effective retention mechanism operates in this species. In this paper, we review the vertical migration behavior of *R. harrisii* larvae under natural estuarine conditions and show that tidal and current flow cycles influence larval migrations. We then discuss experiments in which these larvae were monitored under constant conditions in the laboratory and demonstrate that larvae which have been entrained in their normal environment have tidal rhythms of vertical migration and activity. Finally, we review other known tidal rhythms of estuarine crabs and suggest future directions for research.

Larval Vertical Migrations in the Field

Vertical migrations of zoeal larvae of *R. harrisii* were studied in the Newport River estuary near Beaufort, North Carolina. In this system, tides are semidiurnal with an average period of 12.4 h. The upper regions of the estuary, particularly the sinuous channel known as the Newport Narrows, are populated by adult crabs (unpublished data), and zoeal larvae of *R. harrisii* are abundant in the water column (Pinschmidt 1963; Cronin 1982). To study larval migrations and the environmental cycles associated with them, we set up an anchored sampling station in the Newport Narrows on four occasions during the summers of 1977-1978 (Cronin 1982). Sampling occurred at hourly intervals for at least 2 days on each occasion, with larval samples from more than 1 m^3 of water being taken at each of four depths: surface, 1 m, 2 m, and just above the bottom (2.5-3.5 m). A gasoline-driven pump with its outflow filtered through a no. 2 mesh plankton net was used to sample larvae, and vertical profiles of temperature, salinity, light intensity, and current velocity were obtained simultaneously. Samples were counted for each of the larval stages of *R. harrisii*, and each hour's counts were depth-averaged to obtain the mean depth for each hour. This mean depth was a useful statistic for describing the depth of the larval population, since in most hours the larvae were aggregated near it, not widely dispersed vertically.

Typical vertical migrations of *R. harrisii* zoeal larvae are shown in Fig. 1. The results plotted are for stage I (solid line) and stage IV (dashed line); the intermediate stages generally had mean depths between these two lines. Regular changes in vertical position are evident, and stages I and IV migrated in close unison (along with stages II and III, which are not shown). Stage IV was deeper than stage I a significant number of hours (χ^2 test; $p < 0.05$), but the average difference in depth was only 14 cm. We consistently observed only slight downward movements with age in this larval population (c.f. Sandifer 1975).

The times of low tide at the sampling site, defined as the time of minimum bottom salinity, are also indicated in Fig. 1. The larval migrations were obviously related to the tidal cycle. Larvae rose to a minimum depth soon after low tide, rapidly descended just before high tide, and remained relatively deep until just after the next low tide. The migrations had features which were nontidal as well; note that alternate cycles are different in shape. Figure 2 includes data from the same study as illustrated in Fig. 1, but in this case the proportion of total zoeal larvae in each depth's sample is given. The clumping

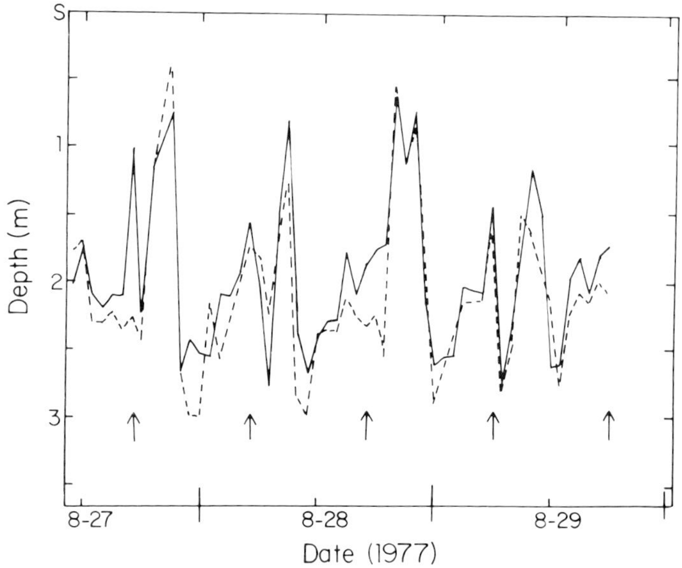

Figure 1. Mean depths (m) in study 2 (August 27–29, 1977) for the stage I zoea (solid line) and the stage IV zoea (dashed line) of R. harrisii. *Arrows indicate the times of low tide (minimum bottom salinity) at the sampling site.*

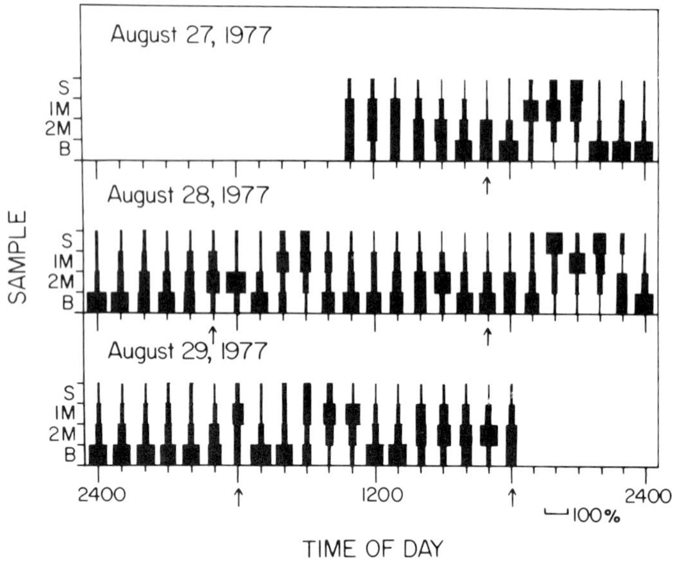

Figure 2. Proportion of total zoeal larvae of R. harrisii *in each of the four hourly samples throughout study 2 (August 27–29, 1977). A 100% scale bar is given below the abscissa. Arrows indicate the times of low tide, defined as in Fig. 1. S = surface, B = bottom.*

of larvae at a single depth in each hour is evident, as is the relationship between tidal phase and larval depth. Larval ascents consistently occurred 2 to 3 h after each low tide.

In each of the field studies, cyclical changes in larval mean depth were noted and all zoeal stages migrated similarly. Therefore, for each study the time series of hourly mean depths was determined for the combined zoeal larvae. The overall average depth was calculated by averaging these hourly means and in every case this "grand mean depth" lay near the depth of no net flow in the Newport Narrows (Table 1). Although the larvae continually migrated vertically, their longitudinal displacement in the estuary was slight since they migrated around the depth of no net flow and thus were retained near the parent populations throughout larval development.

A Fourier power spectrum (Kirk et al. 1979) was calculated for each study's time series (Fig. 3). This allows estimation of the activity at frequencies of interest, especially those of the major environmental cycles. Although such an analysis does not prove a causal relationship between environmental cycles and larval migrations, it is noteworthy that tidal and/or current flow frequencies are dominant in all larval migration time series (Fig. 3).

To define more rigorously the associations between the larval depth series and the environmental series, cross-spectral analysis was also employed (Fuller 1976). These analyses were carried out in a pair-wise

Table 1. Overall average depths of all *R. harrisii* zoeal larvae (calculated from hourly mean depths of each field study) and depth of no net flow (determined by interpolation of tidally time-averaged current data obtained at 0.5 m depth intervals).

Study	*Average larval depth (m)*	*Depth of no net flow (m)*
1 (July 1-3, 1977)	1.60	1.56
2 (Aug 27-29, 1977)	2.01	2.00
3 (July 7-11, 1978)	1.94	2.13
4 (Aug 29-Sept 2, 1978)	2.15	1.45

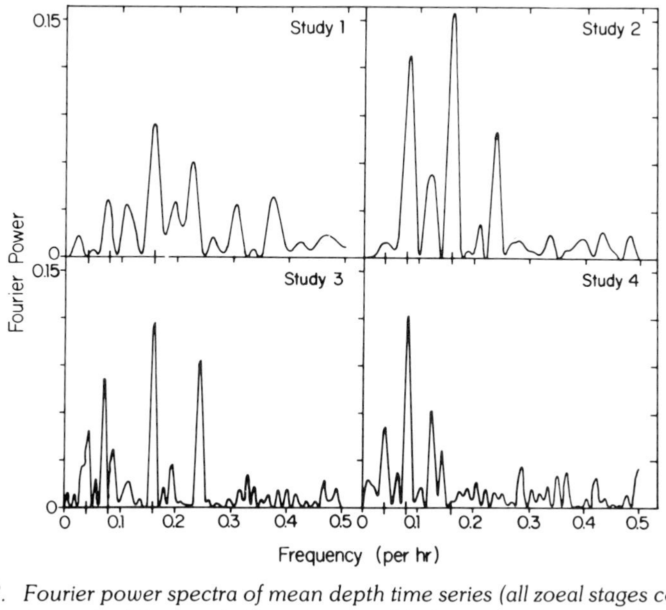

Figure 3. Fourier power spectra of mean depth time series (all zoeal stages combined) for each of the four field study series. Marks on the abscissa at 0.042 (1/24 h), 0.081 (1/12.4 h), and 0.161 (1/6.2 h) indicate the frequencies of the environmental diel, tidal, and current flow cycles, respectively.

fashion, so that the larval migration series was analyzed with each environmental time series in turn. This analytical procedure permits testing only at frequencies which are harmonics of the total length of the time series. Therefore, frequencies were tested at the harmonic nearest to each environmental frequency. Any systematic influence of tidal height on larval mean depth which would be caused by the regular change in depth of the bottom sample was removed by normalizing each mean depth (D) to the depth of the bottom sample (D_{max}), and depths were converted to heights by subtracting the quotient from 1:

$$Y = 1 - \frac{D}{D_{max}}$$

A mean height of 1 would be at the surface and 0 at the bottom. The results of the cross-spectral analyses are given in Table 2. Only in study 4 was there no significant association between the larval depth series and either tidal or current flow cycles. This was also the only field study made during neap tides, so it is likely that the smaller tides at this time of the month had less effect on larval behavior than did the larger tides of the other studies.

The physiology of the larval behavior which leads to migrations that are closely tied to tidal influences in this estuary is not known. The observed crab larval migrations are superficially similar to those claimed for estuarine oyster larvae, which are reported to swim up from the bottom during the rising tide and sink at the commencement of ebb tide, thus moving upstream with each tidal cycle (Nelson 1912; Carriker 1951; Haskin 1964; Wood and Hargis 1971). In the case of oyster larvae, however, the proposed mechanism seems clear; larvae lying on the bottom sense increases in salinity as the tide moves past them and respond by repeatedly swimming into overlying water, but they become inactive in decreasing salinities (Haskin 1964).

Periodic sampling of overlying water for salinity change is not available to wholly planktonic organisms, since they move with the water mass. There is no evidence from the Newport River to suggest that *R. harrisii* zoeal larvae ever leave the water column (Cronin 1982). Therefore, it is most likely that the migrations we observed are responses to mixing events in the water col-

Table 2. Results of cross-spectral analyses of transformed larval mean depth time series (all zoeal larvae of *R. harrisii* combined, see text) and time series of environmental factors related to the tides. "Significance" = statistical significance level of the squared coherency (n.s., not significant). "Phase" = calculated time delay (−) or advance (+) of the larval cycle relative to the environmental cycle.

	Current Cycle (6.2 h)		*Salinity Cycle (12.4 h)*	
Study Number	*Significance*	*Phase (h)*	*Significance*	*Phase (h)*
1	<0.05	−1.45	ns	
2	<0.01	−1.13	<0.001	+1.78
3	<0.01	−1.34	ns[1]	
4	ns		ns	

[1]In study 3, larvae migrated with the salinity cycle only during the first two days, with $p<0.05$, phase = +0.35 h.

umn or to other dynamic tidal features such as flow-induced turbulence, which is maximal at mid-flood and mid-ebb (Cronin 1979). Carriker (1961) suggested that clam veliger larvae are stimulated to swim up in turbulent flows; the current-related migrations of *R. harrisii* larvae could reflect similar behavior, which could also cause the 1.3 h lag of the larval peak after the current maximum (Table 2). Since larval numbers in the water do not change with current speed (Cronin 1979), *R. harrisii* zoeal larvae are not dynamically entrained into tidal currents, as are barnacle cyprid larvae in some locations (deWolf 1974). Zoeal larvae of *R. harrisii* alter their phototactic and geotactic behavior in sudden salinity changes as small as $1^0/_{00}$ (Latz and Forward 1977), changes which they probably experience when mixing of previously stratified water occurs (Cronin 1979). This type of behavior could contribute to tidal vertical migration.

It is unrealistic, however, to invoke salinity changes as the only possible cue for tidal or current-related vertical migrations of estuarine larvae. Larvae of fiddler crabs, *Uca* spp. have regular tidal migrations in North Inlet Estuary, South Carolina (DeCoursey 1976a), a system which has virtually no salinity change over a tidal cycle. The cue for migration is unknown. Whatever might be the driving stimulus for the *R. harrisii* vertical migrations, their functional significance is clear. These larvae prevent potentially disastrous horizontal displacements in the net seaward flow by remaining near the estuary's depth of no net flow (Table 1). *Rhithropanopeus harrisii* larvae were also centered on this depth in the Miramichi estuary (Bousfield 1955); thus, migrations about the depth of no net flow could be common among larval populations of this species and could help explain its widespread success in estuaries.

Endogenous Tidal Rhythms in Estuarine Crabs and Their Larvae

Tidal influences regulate the lives of many intertidal marine and estuarine organisms, and a common adaptation of such organisms is to have endogenous tidal rhythms controlling activity, physiology, and reproduction (for examples, see Palmer 1973; DeCoursey 1976b; and Naylor and Hartnoll 1979). Such tidal rhythms have seemed unlikely in estuarine planktonic organisms, which are buffered from most tidal influences by being contained within a particular water mass. On the other hand, the results of the field studies just described show that zoeal larvae of *R. harrisii* migrate vertically in relation to the tidal cycle, and thus the larvae conceivably have an endogenous tidal clock.

To test for the presence of the postulated clock, we performed experiments under laboratory conditions (Cronin and Forward 1979). We followed larval vertical distributions in a 1.9 m tall lucite column which was maintained in constant darkness and at constant temperature. Each half hour the column was scanned by a moving closed-circuit television camera connected to a videotape recorder, with the column backlit by a dim red light for the 2 minute scan time. Videotapes were later analyzed to determine each hour's mean depth of the larval population. Experimental larvae were fed with *Artemia* nauplii and were monitored in the column over the time period from the third stage zoea until metamorphosis to the juvenile crab 6 to 8 days later.

Each group of zoeal larvae was tested for rhythmic behavior as described above after having lived under either laboratory or field conditions for the first two larval stages. Laboratory-reared larvae were obtained from a crab collected in the Neuse River estuary, North Carolina, a non-tidal system. During embryonic development and after hatching these larvae were maintained on a 12 h:12 h light:dark cycle, and the larvae were provided with new 20°/₀₀ water and food each morning. When added to the experimental column they had a weak circadian rhythm ($p<0.01$, Fisher's periodogram test; Fuller 1976) of period approximately 24.6 h (Enright's periodogram; Enright 1965); no tidal component was apparent (Fig. 4A).

For comparison, larvae were collected on two occasions with a plankton net in the same area of the Newport Narrows where the field studies had been performed, taken directly to the laboratory for sorting of stage III zoeal larvae of *R. harrisii*, and placed under experimental conditions. For larvae collected at the time of spring tides, a striking rhythmic migration was observed which reached its greatest depths coincident with the times of low tide at the collecting site (Fig. 4B). The migration was statistically significant ($p<0.01$) and had a periodogram peak at 12.3 h. Alternate cycles had a slightly different form, which suggested that a circadian rhythm might be superimposed on the tidal one.

To learn whether the migration rhythm of the field-entrained larvae was truly tidal or only a modified circadian rhythm, we repeated the experiment using larvae caught at neap tides, when the tidal and diel cycles are phase shifted by 6 h compared to spring tides. Once more, a statistically significant tidal rhythm was observed ($p<0.01$; period = 12.3 h) with the greatest depths occurring at low tide. In this case the alternation of migration patterns was not apparent (Fig. 4C). The amplitude of this neap tide migration was reduced compared to spring tides. In the field data, no tidal influences on larval vertical migration were seen at the time of neap tides (Table 2).

In summary, larval *R. harrisii* can be internally entrained to natural tidal cycles, and there is probably an endogenous component in the vertical migrations which occur in the field. Laboratory light:dark cycles can entrain a weak circadian migration rhythm as well (Fig. 4A), but it is clear that under field conditions tidal entrainment is much more effective.

Although the form of the laboratory migration superficially resembled the field migration, and phase relationships between larval movements and the tidal cycle were similar in both cases, the field migration pattern on Fig. 1 was much more complex than the laboratory patterns of Fig. 4. Furthermore, tidal migrations were not always observed in the field. We hypothesized that

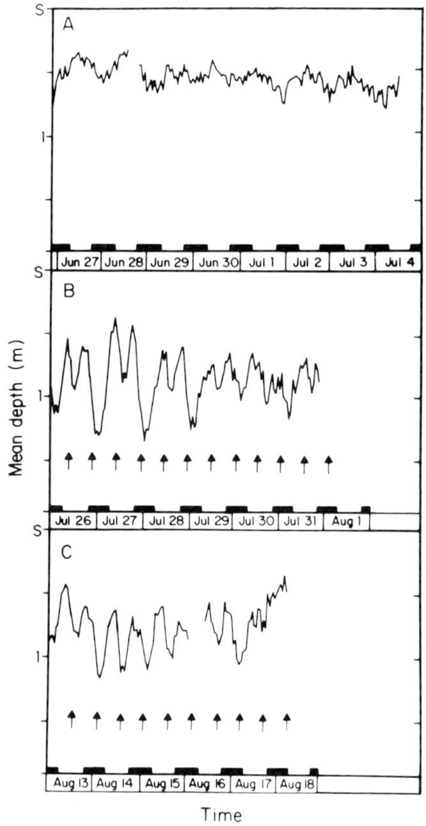

Figure 4. Movement of mean depth of populations of R. harrisii *larvae in a vertical column in constant dark at 25°C. Dates are in 1978; S = surface. Gaps indicate missing data. Dark and light bands indicate the photoperiods to which the larvae were exposed before the experiment began. (A) Laboratory-reared larvae. Initially, 100 larvae were introduced into the column. (B) Field-caught larvae taken at spring tides. Initially, 62 larvae were introduced. Arrows indicate predicted times of low tide at the collection site. (C) Field-caught larvae taken at neap tides. Initially 100 larvae were introduced. Arrows as in (B). (Modified from Cronin and Forward 1979).*

in the field the light:dark cycle could modify the basic pattern, since *R. harrisii* zoeal larvae are very phototactic (Forward and Costlow 1974; Forward 1974, 1976).

To test this hypothesis, we set up a laboratory experiment as before, but imposed a square-wave 12 h light: 12 h dark cycle using a tungsten lamp suspended above the column. The experiment began on September 20, 1978, when the natural photoperiod also contained days and nights of equal length. Intensity of light at the surface of the column was adjusted to equal the noon intensity in the Newport Narrows at a depth of 1 m (equivalent to 0.012 W m^{-2} at 500 nm, measured with a submersible radiometer filtered to have a spectral sensitivity similar to that of *R. harrisii* larvae; see Cronin 1982). Larvae were collected in the Newport Narrows as for the previous experiment and 100 stage III zoeal larvae were added to the column. The light:dark cycle imposed in the laboratory unexpectedly caused an increased departure of the migration sequence from that of the field; larvae moved very near the surface during the day and were about 1 m deeper at night (Fig. 5), whereas in the field larvae were generally deeper during the day (Fig. 1). There was a hint of a residual tidal migration, with descents during the day just after the time of low tide, but the diel migration was dominant, a pattern never seen in the field. This larval population appeared to become entrained to the imposed light:dark cycle, since by the end of the experiment larvae were rising before lights-on and sinking before lights-off (Fig. 5).

As described earlier, circadian entrainment of larvae in the field is less effective, though there is weak evidence for it in Fig. 4B where the larvae were somewhat higher in the column at night. The results of the experiments suggest that complex chemical and physical events of the natural tidal cycle are more important in shaping migrations observed in the field than is the

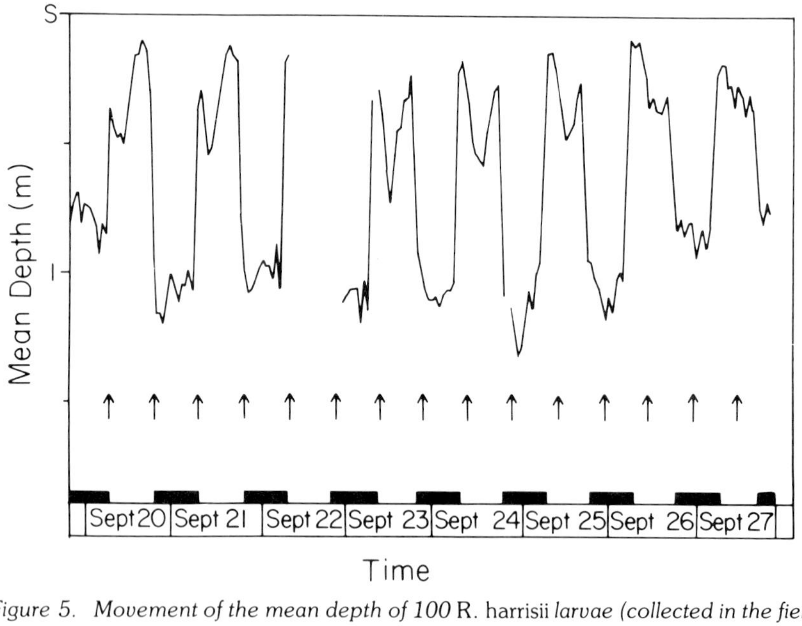

Figure 5. Movement of the mean depth of 100 R. harrisii *larvae (collected in the field) in a vertical column (25°C; 12 h:12h light:dark cycle). Dates are in 1978; S = surface. Gaps indicate missing data. Dark and light bands indicate timing of the imposed photoperiod (identical to the natural photoperiod). The arrows indicate the predicted times of low tide at the collection site.*

daily light:dark cycle; while in the laboratory, the temporal (square-wave) and spatial (highly directional radiance distribution) characteristics of the light stimulus are more effective in altering larval behavior than are field light:dark stimuli.

Tidal vertical migration rhythms exhibited by crab larvae could reflect endogenous circatidal changes in orientation, pressure sensitivity, locomotor activity, or other processes. To test for tidal activity rhythms in these larvae, we collected larvae from the Newport Narrows as before and placed them under constant conditions. Using a closed-circuit television system, we measured their swimming speeds at 2 h intervals for several tidal cycles (Forward and Cronin 1980). Whether larvae were collected at spring or neap tides, they had higher-than-average swimming speeds in the interval from low to high tide, and lower-than-average speeds from high to low tide (Fig. 6). Thus, assuming that swimming results in upward movement, they were more likely to rise as high tide approached and to sink with the coming of low tide. The endogenous tidal vertical migration rhythm most likely results from the activity rhythm.

While entrainment of the tidal rhythms of intertidal animals is usually by

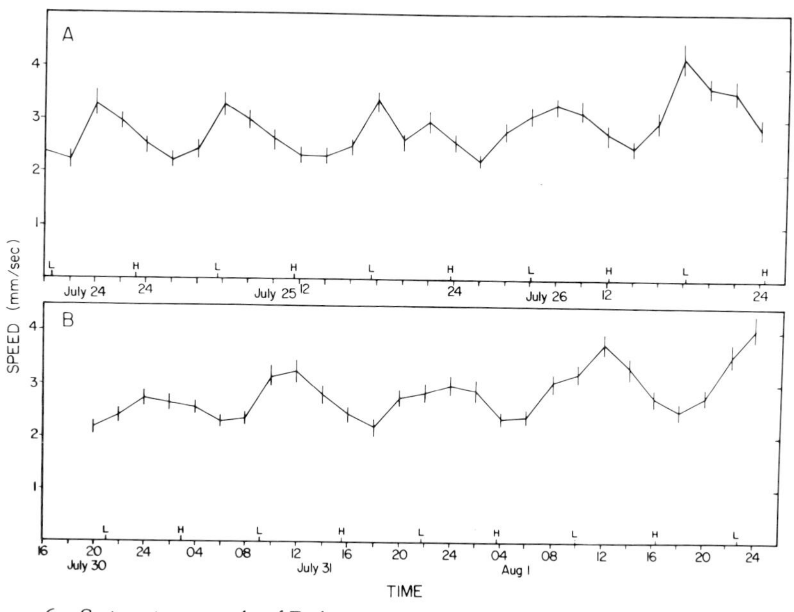

Figure 6. Swimming speeds of R. harrisii *zoeal larvae collected in the Newport River and maintained under constant dim light at 30°C. Dates are in 1979. Means and standard errors of 45 determinations are shown at 2 h intervals. Predicted tide times at the collection site are given; H = high tide, L = low tide. In part A the larvae were collected near spring tides; in part B, near neap tides (from Forward and Cronin 1980).*

obvious stimuli, such as wave agitation or pressure cycles (Palmer 1973; Naylor 1976), the zeitgeber(s) for the larval rhythms are not so evident. Recent work in our laboratory indicated that even newly-hatched stage I larvae from crabs which were collected shortly before larval release and maintained in constant laboratory conditions had tidal vertical migrations (unpublished observations). However, field-caught stage I zoeal larvae had rhythms of greater amplitude; therefore, some experience in the planktonic environment seems necessary for full entrainment.

Ovigerous crabs could also influence larval distributions by releasing their larvae at a particular tidal phase. Semilunar hatching cycles occur in some fiddler crab species (Christy 1978; Wheeler 1978) and in some species of *Sesarma* (Saigusa 1981). Also, in several fiddler crab species, hatching in the laboratory usually occurs very near the time of high tide at the site of collection of crabs (DeCoursey 1979; Bergin 1981). Preliminary work in our laboratory suggests that *R. harrisii* from the Newport River similarly maximize larval release near the time of high tide (Fig. 7). Such timing could be important for releasing larvae into high-salinity water to improve survival, and it also might induce upward swimming to move the larvae up into the water column (Forward, unpublished data). Since both larval release by crabs and vertical migration of newly-hatched larvae include tidal rhythms, there possibly exists some mechanical or chemical communication between the ovigerous crabs

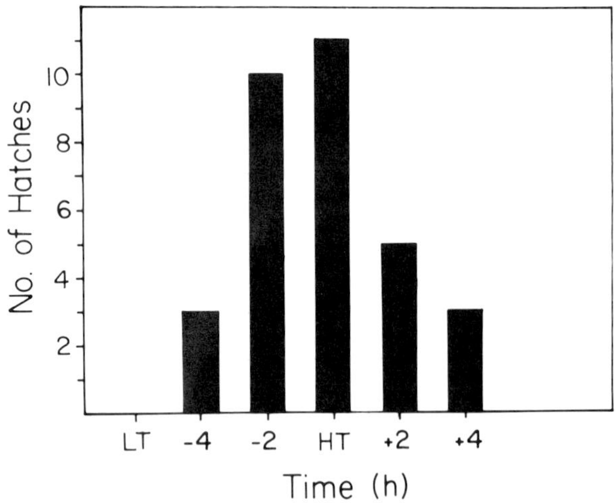

Figure 7. Tidal relationship of hatching of larvae from Newport River crabs maintained in constant dim light at 30°C. Hatching peaked near the predicted time of high tide at the collection site. The hatching frequency distribution is significantly different from a uniform distribution ($p<0.05$; Kolmogorov-Smirnov goodness-of-fit test). HT = high tide, LT = low tide.

and their developing embryos near the end of development which would maintain appropriate phase relationships between their individual tidal clocks.

Tidal Rhythms of Estuarine Larvae: Present Knowledge and Areas for Research

Larvae of *R. harrisii* are well adapted for retention and survival in the Newport River estuary, and indirect evidence suggests that larval populations of this species are equally successful in other estuarine systems (Connolly 1925; Bousfield 1955; Pinschmidt 1963; Tagatz 1968; Williams 1971; Sandifer 1973, 1975; Goy 1976). Although evidence is lacking, it is likely that responsiveness to estuarine tidal cycles is widespread among larvae of this species. In the Newport River, *R. harrisii* zoeal larvae often migrate vertically in close association with the tidal cycle and with the alternating current flows created by the tides, and this is particularly true when the water column alternates between times of strong stratification and mixing to homogeneity (Cronin 1982). These tidal vertical migrations aid retention by moving larvae about the depth of no net flow in the estuary, thus reducing their transport downstream. However, we must emphasize that tidal vertical migration is not essential for retention, since other patterns of vertical migration about the depth of no net flow can achieve the same result (Bosch and Taylor 1973). In fact, *R. harrisii* larvae were retained even when their migrations were not strongly tidal (Tables 1 and 2).

Endogenous tidal rhythms in vertical migration are also characteristic of *R. harrisii* larvae from the Newport River (Cronin and Forward 1979) and probably reflect tidal locomotion rhythms (Forward and Cronin 1980). These tidal rhythms are probably present even in newly-hatched larvae. Tidal rhythms of larval release from ovigerous crabs are observed as well (Forward, unpublished data). The influence of tides on rhythmic larval behavior has been demonstrated solely in this species (though tidal rhythms of adults are also known in fiddler crabs; DeCoursey 1979; Bergin 1981), but it is very unlikely that *R. harrisii* larvae are unique in this regard. Responsiveness to tidal conditions has been noted in *Uca* spp. larvae (DeCoursey 1976a), and could have an endogenous component. Estuarine zooplankton in general and larvae in particular would certainly profit from fine-tuning of numerous aspects of their physiology to the tides, even among those species in which larval retention is incidental.

Among the estuarine crabs which have been studied, only *R. harrisii* exhibits clearly documented retention of larval stages throughout development. Continuous larval retention has several implications. It seems wasteful to produce planktonic larvae if they are simply returned to the region of the parent estuary; direct development would likely be more efficient. The adult crabs, however, are small and probably do not range widely. They are mainly associated with oyster flats and clumps in Newport River estuary (Pinschmidt 1963), which are isolated and ephemeral habitats. Possession of planktonic larvae offers this population the opportunity to exploit all habitats available and to maintain genetic contact with otherwise isolated crab populations in the estuary. Furthermore, both larvae and adults of this species could become genetically adapted to the specific conditions of individual estuaries. There probably are numerous physiological and behavioral adaptations which aid both retention and survival in particular estuarine systems. This matter should be investigated using populations of *R. harrisii* which originate from estuaries offering little opportunity for exchange of larvae.

We feel that there are other areas of research involving both laboratory and field work, preferably in concert, which could prove particularly rewarding. The prevalence of tidal migration or other tidal behavior among estuarine zooplankton certainly demands better understanding. The relationships between the physical hydrography of estuaries and the behavior of their zooplankton are poorly understood at present. Descriptions of larval vertical migrations in estuaries are few. In order properly to understand these migrations, which are often complex (Fig. 1), data collecting programs must include frequent sampling at several depths and must extend well beyond the length of the longest period of interest. Sampling simply at surface and bottom and only on a few occasions for a small number of tidal cycles provides insufficient data for a useful analysis. The converse implies heroic sampling efforts, of course, as well as providing intimidating quantities of samples for counting, but if the sampling program is designed to answer a few specific questions and to test multiple hypotheses, the information that will be gained will be far more valuable than the more diffuse results of traditional sampling sequences.

In the laboratory, responsiveness to tidal variables such as salinity or turbulence demands wider study, notwithstanding the valuable work already done (Haskin 1964; Latz and Forward 1977; Sulkin et al. 1980). The moderating effect of light is also not well understood (Fig. 5). The recently revealed tidal rhythms of estuarine crabs and their larvae provide a host of unanswered questions: What are the zeitgebers? What is the physiology of the tidal clock and where is it located? Are all populations of *R. harrisii* or other estuarine crabs capable of endogenous tidal rhythmicity? Is there communication between mother crabs and their broods with regard to phase setting of rhythms? Are there interacting tidal and circadian rhythms leading to semilunar or lunar rhythms? Studies to date have examined rhythmic behavior of populations of larvae, but it will be necessary to examine rhythmicity in individual larvae to understand the physiology of rhythmicity in this species. Obviously, these aspects of the biology of estuarine larva-producing invertebrates await further imaginative investigation.

Acknowledgments

This material is based on research supported by the National Science Foundation under grants No. OCE 77-26838 and OCE 8007434. Helpful comments on the manuscript were provided by T. Waterman, N. Blackstone, and N. Knowlton. Figure 4 is used with permission and is

References Cited

Bergin, M. E. 1981. Hatching rhythms in *Uca pugilator* (Decapoda: Brachyura). *Mar. Biol.* 63:151-158.

Bosch, H. F. and W. R. Taylor. 1973. Diurnal vertical migration of an estuarine cladoceran, *Podon polyphemoides*, in the Chesapeake Bay. *Mar. Biol.* 19:172-181.

Bousfield, E. L. 1955. Ecological control of the occurrence of barnacles in the Miramichi estuary. *Bull. Natn. Mus. Can. Biol. Ser.* 46(137):1-65.

Bowden, K. F. 1978. Mixing processes in estuaries, pp. 11-36. *In:* B. Kjerfve (ed.), *Estuarine Transport Processes*. U. South Carolina Press, Columbia, SC

Carriker, M. R. 1951. Ecological observations on the distribution of oyster larvae in New Jersey estuaries. *Ecol. Monogr.* 21:19-38.

Carriker, M. R. 1961. Interrelation of functional morphology, behavior, and autecology in early stages of the bivalve *Mercenaria mercenaria*. *J. Elisha Mitchell Sci. Soc.* 77:168-241.

Christy, J. H. 1978. Adaptive significance of reproductive cycles in the fiddler crab *Uca pugilator:* A hypothesis. *Science* 199:453-455.

Connolly, C. J. 1925. The larval stages and megalopa of *Rhithropanopeus harrisii* (Gould). *Contr. Can. Biol.* 2:329-334.

Cronin, T. W. 1979. Factors contributing to the retention of larvae of the crab *Rhithropanopeus harrisii* in the Newport River estuary, North Carolina. Ph.D. thesis, Duke U., Durham, NC 206 pp.

Cronin, T. W. 1982. Estuarine retention of larvae of the crab *Rhithropanopeus harrisii*. *Est. Coastal Shelf Sci.* (In press).

Cronin, T. W. and R. B. Forward, Jr. 1979. Tidal vertical migration: an endogenous rhythm in estuarine crab larvae. *Science* 204:1020-1022.

DeCoursey, P. J. 1976a. Vertical migration of larval *Uca* in a shallow estuary. *Am. Zool.* 16:244.

DeCoursey, P. J. 1976b. *Biological Rhythms in the Marine Environment*. U. South Carolina Press, Columbia, SC 283 pp.

DeCoursey, P. J. 1979. Egg-hatching rhythms in three species of fiddler crabs, pp. 399-406. *In:* E. Naylor and R. G. Hartnoll (eds.), *Cyclic Phenomena in Marine Plants and Animals*. Pergamon Press, New York.

deWolf, P. 1974. On the retention of marine larvae in estuaries. *Thalassia Jugoslavica* 10:415-424.

Enright, J. T. 1965. The search for rhythmicity in biological time series. *J. Theoret. Biol.* 8:428-468.

Forward, R. B., Jr. 1974. Negative phototaxis in crustacean larvae: possible functional significance. *J. Exp. Mar. Biol. Ecol.* 16:11-17.

Forward, R. B., Jr. 1976. A shadow response in a larval crustacean. *Biol. Bull.* 151:126-140.

Forward, R. B., Jr. and J. D. Costlow, Jr. 1974. The ontogeny of phototaxis in the crab *Rhithropanopeus harrisii*. *Mar. Biol.* 26:27-33.

Forward, R. B., Jr. and T. W. Cronin. 1980. Tidal rhythms of activity and phototaxis of an estuarine crab larva. *Biol. Bull.* 158:295-303.

Fuller, W. A. 1976. *Introduction to Statistical Time Series*. J. Wiley and Sons, New York. 470 pp.

Goy, J. W. 1976. Seasonal distribution and the retention of some decapod crustacean larvae within the Chesapeake Bay, Virginia. M.S. thesis, Old Dominion U., Norfolk, VA 334 pp.

Haskin, H. H. 1964. The distribution of oyster larvae, pp. 76-80. *In:* N. Marshall (ed.), *Proc. Symp. Exper. Mar. Ecol.*, U. Rhode Island Occ. Publ. No. 2.

Kirk, B. L., B. W. Rust and W. Van Winkle. 1979. *Time Series Analysis by the Maximum Entropy Method.* Oak Ridge National Laboratory, Oak Ridge, TN 212 pp.

Kjerfve, B. 1978. *Estuarine Transport Processes.* U. South Carolina Press, Columbia, SC 331 pp.

Latz, M. I. and R. B. Forward, Jr. 1977. The effect of salinity upon phototaxis and geotaxis in a larval crustacean. *Biol. Bull.* 153:163-179.

Naylor, E. 1976. Rhythmic behaviour and reproduction in marine animals, pp. 393-429. *In:* R. C. Newell (ed.), *Adaptation to Environment.* Butterworths Press, Boston.

Naylor, E. and R. G. Hartnoll. 1979. *Cyclic Phenomena in Marine Plants and Animals: Proceedings of the 13th European Marine Biology Symposium.* Pergamon Press, New York. 477 pp.

Nelson, J. 1912. Report of the Biologist. Observations of natural propagation of oysters at Barnegat, NJ, 1911. New Jersey Agric. Coll. Expt. Sta., Biol. Dept. Report, pp. 271-309.

Palmer, J. D. 1973. Tidal rhythms: the clock control of the rhythmic physiology of marine organisms. *Biol. Rev.* 48:377-418.

Pinschmidt, W. C., Jr. 1963. Distribution of crab larvae in relation to some environmental conditions in the Newport River estuary, North Carolina. Ph.D. thesis, Duke U., Durham, NC 112 pp.

Saigusa, M. 1981. Adaptive significance of a semilunar rhythm in the terrestrial crab *Sesarma. Biol. Bull.* 160:311-321.

Sandifer, P. A. 1973. Distribution and abundance of decapod crustacean larvae in the York River estuary and adjacent lower Chesapeake Bay, Virginia, 1968-1969. *Chesapeake Sci.* 14:235-257.

Sandifer, P. A. 1975. The role of pelagic larvae in recruitment to populations of adult decapod crustaceans in the York River estuary and adjacent lower Chesapeake Bay, Virginia. *Est. Coastal Mar. Sci.* 3:269-279.

Sulkin, S. D., W. van Heukelem, P. Kelly and L. van Heukelem. 1980. The behavioral basis of larval recruitment in the crab *Callinectes sapidus* Rathbun: a laboratory investigation of ontogenetic changes in geotaxis and barokinesis. *Biol. Bull.* 159:402-417.

Tagatz, M. E. 1968. Biology of the blue crab, *Callinectes sapidus* Rathbun in the St. Johns River, Florida. *Fish. Bull.* 67:17-33.

Wheeler, D. E. 1978. Semilunar hatching periodicity in the mud fiddler crab *Uca pugnax* (Smith). *Estuaries* 1:268-269.

Williams, A. B. 1971. A ten-year study of meroplankton in North Carolina estuaries: annual occurrence of some brachyuran developmental stages. *Chesapeake Sci.* 12:53-61.

Wood, L. and W. J. Hargis. 1971. Transport of bivalve larvae in a tidal estuary, pp. 29-44. *In:* D. J. Crisp (ed.), *Fourth European Marine Biology Symposium.* Cambridge University Press, Cambridge.

SELECTION FOR RETENTION OR EXPORT OF LARVAE IN ESTUARIES

R. R. Strathmann
Zoology Department and Friday Harbor Laboratories
University of Washington
Friday Harbor, Washington

Abstract: Two quite different sorts of selection could favor export of offspring from estuaries: (1) animals which spread offspring over several estuaries increase relative to non-spreaders in the population when favorability of estuaries varies independently; (2) animals which export offspring from an estuary during the larval stage increase relative to non-exporters in the population when survival or growth of larvae is greater outside the estuary. Adaptation to local conditions and the occurrence of homing in many life histories are indirect evidence against the first hypothesis. Conclusions are premature, but predictions are that (1) varying favorability among estuaries rarely favors the spread or dispersal of larvae among estuaries, (2) waters outside estuaries are often safer for larvae than waters inside estuaries, and (3) when there is selection for export of larvae from estuaries, it is selection for migration to areas more favorable to larval development rather than dispersal. In this view, the large scale spread of offspring is an accidental though important consequence of a long planktonic period; the spread of offspring confers no short term advantage.

Introduction

Behavior of some estuarine invertebrates enhances retention of larvae in the estuary; behavior in other species may enhance export of larvae from the estuary (Cronin and Forward 1982; Sulkin and Van Heukelem 1982). Observations of behavior apparently directed toward enhancing or decreasing retention of larvae raise questions about selection for retention or export of offspring. Under what circumstances should we expect selection for retention of larval offspring? When will selection favor export of offspring?

A direct approach to the problem is difficult. One might attempt to introduce estuarine invertebrates to a new environmental regime. Field experiments would show extent and direction of changes in larval behavior or larval stages which can occur over a few generations. Laboratory experiments could measure heritable variation in traits of interest. Studies of species which have been accidentally introduced into new locations may provide selection experiments with a longer time scale. Interpreting results of selection experiments on larval retention would require much additional information, however. A decrease or an increase in retention of larvae could depend on initial traits in

ISBN 0-12-404070-5

the population and on any of several features of the environment. One would want to know how the initial traits influenced the direction of change and which features of the environment were most important. Experiments would complement but not replace studies of patterns of environmental variation, physical constraints on life history traits, and constraints imposed by ancestry. I know of no direct observations on heritable increases or decreases in retention of larvae in a population of estuarine invertebrates which has been introduced to a new environmental regime. I shall therefore take a less direct approach to answering these questions.

First I examine two hypotheses on selection of offspring. 1) Animals which spread offspring among several estuaries increase relative to non-spreaders in the population because of independent variation of estuaries over time. (2) Animals which migrate from the estuary for part or all of the larval phase increase relative to non-exporters in the population because survival or growth of larvae is greater outside the estuary. I shall advance arguments for rejecting the first hypothesis and for the plausibility of the second hypothesis in some cases. Throughout I shall assume that larvae are affected by their environment in a density independent manner.

In the second section of this paper I offer some reasons why the same environmental regime might select for quite different larval traits in different populations of marine invertebrates. The outcome (retention or export of offspring) will depend strongly on traits of the starting population.

I admit at the outset that my arguments are very indirect and highly simplified. The reader is referred to Stearns (1977) for a discussion of limitations inherent in the type of arguments which I advance. Despite numerous potential pitfalls, general models and arguments of this sort are useful preliminary to more detailed studies of specific cases.

Two Hypotheses on Selection for Export of Offspring

(1) Spreading larval offspring among estuaries.

Dispersal of larval offspring on a small scale may offer numerous advantages. Local crowding, local concentrations of predators, and local disturbances favor dispersal of offspring on a small scale (for example, Paine and Levin 1981). Advantages of dispersing offspring among several estuaries are less obvious. In the long run all the offspring of a non-disperser will die when the estuary disappears or is altered radically. For example, in a population of the polychaete *Cirratulus cirratus* in Sussex, England, the larvae are demersal; the population disappeared during unusually cold weather in 1962-63, and the population had not reappeared as of George's (1968) writing. In the short run, however, there may be substantial losses during the spread of part of an individual's offspring to another estuary. Under what circumstances will there be advantages to such spread in the short term? There are a variety of analytical treatments of the effect which spreading offspring in a varying environment has on rate of increase (Cohen 1967; Balkau and Feldman 1973;

Strathmann 1974; Marcus 1980; Frank 1981; and Kuno 1981). These papers provide some insight into the basis of advantages or costs to spreading offspring in a varying environment. Under some circumstances dispersal damps temporal variation in survival and reproduction of an individual's descendents by averaging spatial variation. This can result in higher rates of increase relative to non-dispersers. In other circumstances, such as when sites are consistently ranked, dispersers suffer a net export of offspring from favorable sites to unfavorable sites and thereby decrease relative to non-dispersers. The studies cited do not examine many patterns of variation and only Marcus (1980) compares effects of spreading on different scales. Because the analytical treatments are limited in scope, A. R. Palmer and I employed simulation models of consequences of spread of larvae on different spatial scales and under different patterns of environmental variation. The remainder of this section is a discussion of the application of Palmer and Strathmann's (1981) results to estuarine invertebrates.

Our models are equivalent to a haploid single locus model of selection. Generations do not overlap. We modelled favorability in two ways. For density independent models, environmental favorability is defined as the number of larvae produced per settling larva at a given site and time. For density dependent models, environmental favorability is defined as the total number of larvae which can be produced from a given site at a given time, and this number of offspring is apportioned equally among the larvae that have arrived at the site. Rates of increase of dispersers are measured relative to non-dispersers. Scale of dispersal does not affect recruits per unit of reproductive effort, an assumption which is equivalent to assuming that dispersers can compensate for increased larval mortality by reducing size of eggs and increasing number of eggs. Only a few of the results can be summarized here. I have selected several simulations which examine effects of variance and of spatial and temporal correlation of favorability. These are based on a lognormal distribution of environmental favorability with replicate runs of 100 generations. In the results reported here, edge effects are eliminated by dividing populations into non-overlapping cells; greater spread of larvae is achieved by increasing the number of sites in a cell. We have previously varied a number of these simplifying assumptions with no qualitative change in the results.

The model shows that if the environment varies randomly with no spatial or temporal correlation, then spreading offspring results in a relatively higher rate of increase of dispersers. The result is the same in both density dependent and density independent models (Fig. 1). There is an asymptote of maximum advantage which is approached as scale of spread increases. There are clearly diminishing returns to increasing scale of spread. If the variance in favorability is increased, the asymptote of maximum advantage is increased, and the approach to the asymptote is slowed as scale of spread is increased (Fig. 1), but the big gains from dispersal are obtained with spread

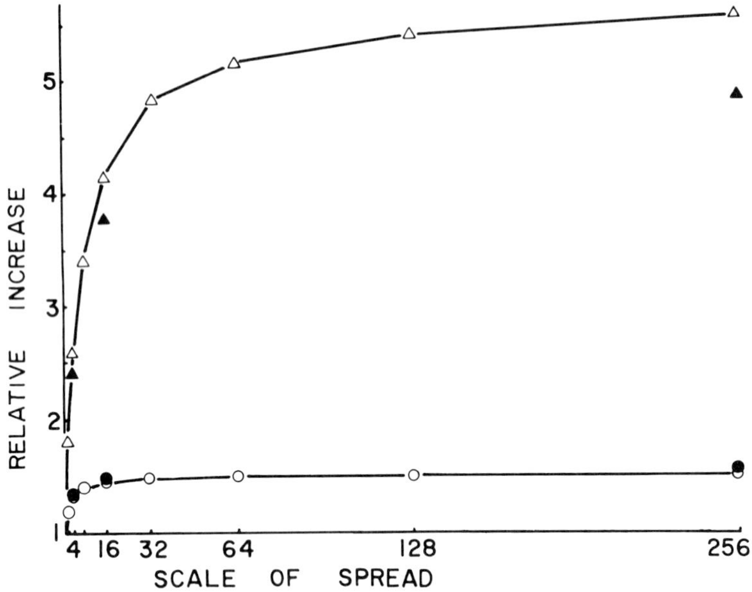

Figure 1. Effect of variance in favorability on relative rates of increase per generation of animals dispersing on different scales (lognormal distribution of favorability with mean of 1 and variance 3 [circles] or 81 [triangles]). Each symbol is the mean of three replicate runs with the density independent model (connected open symbols) or density dependent model (solid symbols) (after Palmer and Strathmann 1981).

over only a few sites. If sites in the model are assumed to correspond to portions of estuaries, then in this sort of world dispersing to the next estuary would add some sites and therefore give a slight advantage as long as leaving the estuary imposed no additional losses on larvae.

One might expect, however, that one estuary would differ from another. What if one estuary is the best spot for settling larvae in one year, another estuary in another year? This sort of situation was simulated by increasing the spatial correlation. In this model a site's favorability is likely to be similar to the neighboring site's favorability, but there is no correlation between favorability in one year and the next. If spatial correlation is increased, the approach to maximum rate of increase with increased dispersal is slower (Fig. 2). In this sort of world, advantages of exporting offspring to the next estuary might outweigh costs of substantial losses of larvae.

In the real world, however, some places are consistently favorable while others are consistently unfavorable. If one estuary is consistently more favorable than its neighbors, then dispersers will export more larvae from the favorable estuary than are returned to it from the less favorable estuaries. We modelled this situation by increasing the temporal correlation over a finite in-

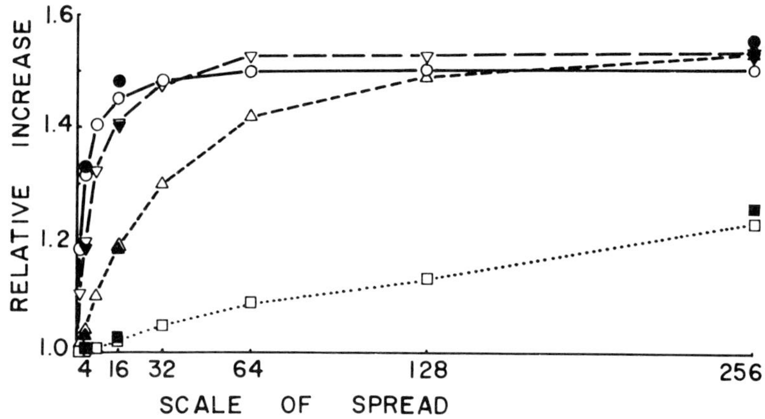

Figure 2. *Effect of increased spatial correlation on relative rates of increase per generation of animals dispersing on different scales (lognormal distribution of favorability with mean of 1, variance of 3, and spatial correlations of about zero [circles], 0.35 [triangles apex down], 0.80 [triangles apex up], and 0.96 [squares]). Open or solid symbols as in Fig. 1 (after Palmer and Strathmann 1981).*

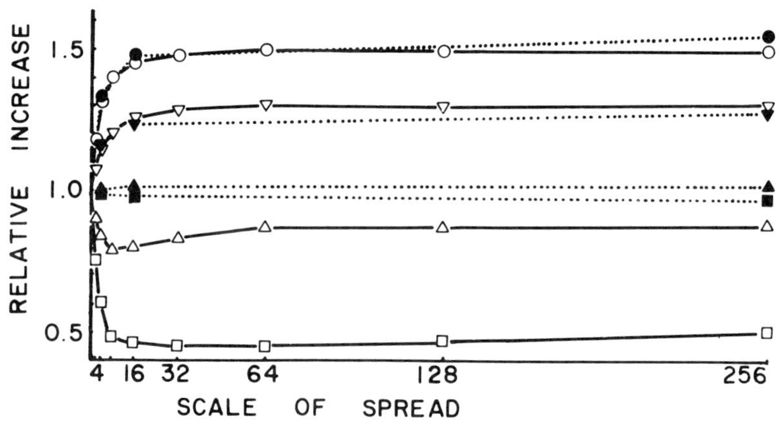

Figure 3. *Effect of increased temporal correlation and decreased temporal variance on relative rates of increase per generation of animals dispersing on different scales (lognormal distribution of favorability with mean of 1, variance of 3, spatial correlation zero, and temporal correlation about zero [circles], 0.34 [triangles apex down], 0.78 [triangles apex up], and 0.93 [squares]). Open or solid symbols as in Fig. 1 (after Palmer and Strathmann 1981).*

terval (100 generations) and thereby also decreasing the temporal variance. In density independent models dispersal is quite costly (Fig. 3). In density dependent models the situation is more complicated. Advantages to dispersal disappear as temporal correlation is increased and temporal variance decreased, but over 100 generations the rate of decrease of dispersers relative to non-dispersers is not great. Some additional simulation showed that dispersers and non-dispersers approach an equilibrium which depends on carrying capacities and also on initial numbers at each site. We can conclude, however, that there is at least no advantage to spreading larvae among estuaries when (1) one estuary has a consistently higher carrying capacity and probabilities of return to each estuary are equal or (2) the estuaries have equal carrying capacities but the probability of entering one estuary is much greater.

How does the real world compare with these simulations? When one considers the spatial scale of several estuaries, or a system of estuaries and coastal lagoons, is it entirely chaotic as in Fig. 1, highly correlated spatially but not temporally as in Fig. 2, or highly correlated temporally as in Fig. 3? An indirect line of evidence suggests that Fig. 3 is closer to reality for most species dispersing on a large geographic scale. The models discussed above assume that the only traits which vary are those affecting dispersal, but animals from different locations often exhibit genetic differences which adapt them to their home location. Then each site becomes the best one for members of the local population. By reasoning backwards, we can take local genetic adaptation as a measure of consistent differences among sites. Individuals spreading their offspring more widely must either be generalists or suffer more selective deaths each generation. Either way, consistent differences among benthic sites are disadvantageous for dispersers.

Scheltema (1975) and Burton and Feldman (1982) reviewed examples of divergence among populations inhabiting different estuaries and concluded that those species with planktonic larvae differ less among local populations for a variety of traits. Nevertheless, selection may sometimes be strong enough to produce local variation in widely dispersing marine and estuarine animals. Divergence among sites has been observed for a variety of traits and for a wide range of spatial scales, from clines over thousands of kilometers of coastline to clines within a few hundred meters (for example Struhsaker 1968; Williams et al. 1973; Koehn et al. 1976; Ament 1979). These differences within populations of wide dispersal have been interpreted as the result of differing selective regimes. If this is so, one result of spreading offspring is substantial selective mortality each generation.

An estuary presents a strong gradient in physical conditions and in biota, and this internal gradient may impose costs on dispersers that do not return to the favorable part of the gradient following the larval stage. A comparison of settling behavior in two species differing in larval dispersal shows that one cost of dispersal may be behavior inappropriate to a part of the gradient. The polychaete *Spirorbis borealis* has a very brief period of larval swimming, and there are heritable differences among local populations in the substratum preferences of the settling larvae. The larvae prefer to settle on *Fucus* at exposed sites (Doyle 1975). At protected sites, however, *Fucus* is scarcer relative to *Ascophyllum*, on which the larvae can settle but with less adhesion than on *Fucus*. From the outer (exposed) to the inner (protected) part of a fjord, MacKay and Doyle (1978) found a cline in substratum preference from a preference for *Fucus* to about equal preference for *Fucus* and *Ascophyllum*. If *S.borealis* dispersed its offspring throughout the fjord, then mortality would increase. Either they would settle on *Ascophyllum* in the outer fjord and be dislodged, or they would delay settlement while waiting to encounter *Fucus* in the inner fjord and thus risk being eaten.

Settling larvae of the barnacle *Balanus glandula* may make fatal errors in settlement because they are dispersed in an estuarine gradient (Strathmann and Branscomb 1979; Strathmann et al. 1981). In the San Juan Islands and on the outer coast of Washington and British Columbia, adults of this species are largely restricted to the upper intertidal zone. In extensive areas of Puget Sound and the Straits of Georgia adults are abundant almost from mean lower low water to mean higher high water. The larval period of two to four weeks is sufficient for exchange of larvae among regions of quite different zonation. The larvae settle in the lower intertidal zone in the San Juan Islands even though at most sites the low settlers will be eaten by snails before they reproduce. Field experiments, in which settling plates were incubated at different tidal heights and then paired at the same height, demonstrated a preference by *B. glandula* larvae for low sites. Why should larvae arriving on shores in the San Juan Islands settle where fitness will be zero? All our evidence suggests that it is because this preference is suitable for larvae from the adjacent regions of Puget Sound and the Straits of Georgia. If *B. glandula* did not disperse so widely, selection might produce different substratum preferences in different regions and post-settlement mortality would be greatly reduced.

In both the above examples, differences along the gradient of the fjords are quite consistent from year to year. If spreading of offspring is costly and keeping them home reduces costs, then why do so many estuarine species have long planktonic larval periods? Perhaps a long planktonic larval period should be considered as a migration into a different habitat for part of the life cycle and dispersal as an accidental consequence.

The numerous cases of fishes that migrate from and to spawning grounds provide an example of a migratory pattern in which dispersal is carefully avoided. If an invertebrate exports its larvae from an estuary, the larvae are likely to be dispersed widely. Though the larvae may be concentrated again when they are returned to the estuary and select preferred settling sites, it is unlikely that they are capable of homing. They may return to a similar site, but it is unlikely to be the parental site. If they did return to the parental site, we would not hesitate to call the process homing and would not be considering hypotheses on dispersal. Salmon have abundant opportunities for dispersal among spawning sites, but instead they return to the parental spawning ground with remarkable precision. Several studies show the advantage of this homing. One example is the appropriateness of upstream or downstream swimming by sockeye fry hatching at the outlet or inlet of a lake.

Fry from outlets and inlets preferentially make the correct choice (Brannon 1967; Raleigh 1967; reviewed by Ricker 1972). If the salmon dispersed widely, the young stages would not be so closely adapted to the spawning site. Attempts to transplant salmon are frequently unsuccessful (Ricker 1972; Altukhov et al. 1980). Evidence for homing is not so direct for marine or estuarine fishes, but indirect evidence supports the hypothesis that homing occurs in numerous stocks (Harden Jones 1968; Leggett 1977).

Life histories of homing fishes are distinctly different from those of marine invertebrates: adults are returning to spawn, not larvae to settle. These life histories of homing fish do carry a moral for invertebrate zoologists, however: many strong swimmers pass up an opportunity to disperse and return instead to a home site to complete part of the life cycle. This suggests that selection for export of larvae from estuaries, when it occurs, could be selection for migration, and that the resulting spread of offspring is an accidental consequence which confers no short term advantage. (The long term consequences for geographic range, speciation rates, and extinction rates may be great, but that is a different argument [Jackson 1974; Scheltema 1977; Hansen 1980; Jablonski 1980].)

(2) The larva as a migrant.

An oocyte, embryo, or larva is released by the mother into the water and later returns to the bottom as a settling larva or juvenile. Several hypotheses on larval adaptation have viewed these events as a migration and ignored dispersal entirely. A recent model is that of Vance (1973) together with suggested modifications (Strathmann 1977; Christiansen and Fenchel 1979). Factors of major importance in this view are (1) differences between the bottom and the plankton as sites for survival and growth of small offspring and (2) trade of reduced fecundity with greater parental care (or larger eggs) against high fecundity with less parental care (or smaller eggs). Export or retention of larvae of estuarine invertebrates is a special case of such migration models. Offspring are retained in an estuary (1) if the larval period is short or absent, (2) if the larvae stay on or in the bottom, or (3) if the swimming larvae exploit estuarine current systems to control their horizontal movements. When could selection favor export of offspring from an estuary for a period before their subsequent return? This question can be divided into two parts. When is it better to leave the bottom, and when is it better to leave the estuary? To answer these questions we must compare physical conditions, food supplies, and predation rates.

Are salinity, temperature, or oxygen concentration in estuaries often ultimately limiting for larval stages? Eggs, embryos, and larvae of some estuarine invertebrates are well adapted to low salinities, others are not. Sometimes tolerances vary even among closely related populations, as in worms classified as *Nereis diversicolor* (Smith 1964). Saigusa (1981) suggests that release of larvae in synchrony with the tides is important for crabs of two species of *Sesarma* because the larvae must get to sea quickly, but that release is less synchronous with the tides in a third species because the larvae are more tolerant of fresh water. Younger or smaller stages are usually but not always less tolerant of low salinities. In *Gammarus palustris* newly hatched juveniles are more tolerant of low salinities than the adults (Gable and Croker 1978). Though generalizations about energetic costs, risks, or obstacles of larval adaptation to low salinity are not obvious, low salinities may impose risks on larvae of some estuarine animals and may be one reason for leaving an estuary. Low salinity is not the only physical hazard associated with estuarine conditions; it exemplifies the difficulties in generalizing about physical hazards for larvae in estuaries.

Food may be a reason for staying in an estuary. Along most coasts, small particulate food is more abundant in shallow estuaries than offshore. It is unlikely that larvae will usually find more food outside a shallow estuary.

Risks from predators are difficult to compare. I will assume that estimates of mortality rates for copepods can be used as a measure with which to compare safety of habitats for small animals. I use these data because there appears to be more data for copepods than for any one type of larva in other taxa of invertebrates, but the estimates of instantaneous mortality rates have been derived in different ways. These methods are subject to different sources of error and include different parts of the life history. Some of the mortality rates that I have tabulated are from graphs or numbers presented for quite different purposes. My aggregation of data by author, site, and species obscures differences in replication among studies. The direct comparisons of estimates offered here should therefore be interpreted with caution.

The estimates in Table 1 are derived in varying ways from estimates of naupliar abundance and developmental rates. Flushing of larvae from the study area is a potential source of bias. I have eliminated estimates from fall and winter because estimates of mortality are often lower and many animals are not producing larvae at those seasons. I have taken arithmetic means where numerous estimates were made. I made no other attempts to adjust the data for differences in season, stage, or size, nor have I thrown out questionable estimates. The highest estimates are generally for estuaries, but there are low mortality rates from Jakle's Lagoon and the CEPEX bag experiment. The estimates from coastal waters are from only two studies and are not from waters near any of the estuaries. If estuarine and coastal mortality rates are compared in a Wilcoxon two sample rank test, they are not significantly different ($p>0.10$); if the estimate from the CEPEX bag is eliminated, then the difference is significant at $0.10>p>0.05$. These data are consistent with the hypothesis that coastal waters are often safer for larvae than estuarine waters, but they do not prove it.

The estimates in Table 2 are death rates which lump all stages. Most are derived from the difference between instantaneous rate of change of numbers in the population and instantaneous rate of birth, with rate of birth derived

Table 1. Estimates of instantaneous mortality rates per day for copepod nauplii in estuarine and coastal waters.

Environment	*Season*	*Rate*	*Source*
ESTUARINE			
North Inlet, South Carolina			D'Apolito and Stancyk 1979
Euterpina acutifrons	Apr.	0.31-0.35	
Patuxent R., Chesapeake Bay			Heinle 1966
Acartia tonsa	July-Sept.	0.82	
Rhode R., Chesapeake Bay			Allan et al. 1976
Acartia tonsa	July-Oct.	0.71 or 0.59	
Eurytemora affinis	Apr-July	1.01	
Jakle's Lagoon, near Strait of Juan de Fuca			Landry 1976, 1978
Acartia clausi	Apr-Aug	0.05 or 0.09	
CEPEX bag, Saanich Inlet (a fjord), British Columbia			Parslow et al. 1979; Grice et al. 1977
Pseudocalanus sp.	May-July	0.05	
COASTAL			
off Villefranche, Mediterranean			Gaudy 1976
Paracalanus parvus	May-June	0.09	
Acartia clausi	May-June	0.22	
Centropages typicus	May-June	0.24	
Oithona helgolandica	May-June	0.12	
Oithona nana	May-June	0.04	
Euterpina acutifrons	May-June	0.23	
off Southern California			Mullin and Brooks 1970
Calanus helgolandicus	Apr-Aug	0.42 or 0.15	

Table 2. Estimates of instantaneous death rates per day for copepods of several estuarine habitats

Habitat	*Season*	*Rate*	*Source*
Infaunal			
North Inlet, South Carolina			Fleeger 1979
Stenhelia bifida	Apr-Sept	0.19	
Mukilteo, Puget Sound			Feller 1980
Huntemannia jadensis	Apr-Sept	0.48	
Epibenthic			
North Inlet, South Carolina			Fleeger 1979
Microarthridion littorale	Apr-Sept	0.75	
North Inlet, South Carolina			M. A. Palmer 1980
Microarthridion littorale			
intertidal	Apr-Sept	0.46	
subtidal	Apr-Sept	0.64	
deeper subtidal	Apr-Sept	0.80	
Planktonic-epibenthic			
Chesapeake Bay			Lonsdale 1981
Scottolana canadensis	June-July	0.11	
Planktonic			
Chesapeake Bay			Lonsdale 1981
Oithona colcarva	Aug-Sept	0.07	

from counts of eggs and of egg-bearing females, but I derived the estimate from Lonsdale (1981) from the difference between rates of population change in the field and rates in the laboratory under similar physical conditions. The data of Table 2 suggest that the bottom is as hazardous for small animals as is the water column in estuaries, though Heinle's (1966) estimate for mortality of planktonic copepodids of *A. tonsa* (mean of 0.50 per day) is about the same as the estimates for all stages of *M. littorale* in Table 2. Fleeger (1979) and Hicks (1977) conclude that epibenthic copepods are subject to higher mortality rates than are infaunal copepods, but Feller's (1980) estimates of mortality of infaunal *H. jadensis* are quite high. There is no indication that life is safer for larvae on or near the bottom of an estuary. Risks of staying near the bottom in order to avoid export from an estuary might be high. (Periodical vertical swimming to avoid export may also involve risks or energetic costs, and these are also difficult to quantify.)

The hypothesis that selection often favors migration out of estuaries during the larval stage because coastal or offshore waters are more favorable for the animals when they are small is plausible, but evidence that estuarine waters are more hazardous for small animals is still weak, and abundance of phytoplankton in many estuaries suggests that estuaries could be more favorable than coastal waters for larvae of some species. A shifting balance of risk from predators and advantage from high food supplies is possible along many estuarine to offshore gradients but cannot yet be evaluated quantitatively.

Varying Adaptations to the Same Environmental Regime

In the previous sections I discussed patterns of environmental variation that could select for or against export of larvae from estuaries. Even under environmental regimes which favor export, selection would not necessarily produce a life history in which larvae are exported. First, larval anatomy in some species may prevent the evolution of a sufficiently prolonged planktonic stage. Second, different complexes of traits could present alternative adaptive solutions to the problems presented by a given environmental regime.

(1) Constraints from ancestral anatomy.

Many traits are quite conservative and not suited to routine large scale dispersal of larvae (Strathmann 1978 a, b). Consider a bryozoan with lecithotrophic development. The larva cannot feed and in most cases lacks a gut entirely. The gut develops only after settlement and metamorphosis. The period of larval swimming is short and there is no basis for extending the larval period by extending the period of larval feeding. Cost of extending the larval period by packing more nutritional material into each egg would be a reduced number of larvae. The postsettling stages are sessile. Such animals inhabit estuaries, but an extended planktonic stage each generation is not one of their evolutionary options. If they are estuarine, their entire life history must be adapted to life within the estuary.

Nereids represent a different sort of case. Nereid larvae are not well equipped for planktonic feeding, so these worms have some of the same limitations as bryozoa. The adults are mobile, however. Estuarine nereids have been observed swimming in large numbers (Clark 1961; Estcourt 1966; Dean 1978). Prespawning migrations out of an estuary are a possibility for many species and occur in some species (Clark 1961).

If an estuarine species lacks a feeding larval stage, the length of the larval period is constrained and thus export of larvae from estuaries is also limited. An adult body plan adapted to a sessile life constrains migration from an estuary as an adult. New types of feeding larval stages have evolved, and some mobile animals are probably derived from sessile ancestors, but these major changes in form are rare events.

(2) Alternative complexes of traits.

Consider a population of invertebrates in which juvenile and adult stages have become established in an estuary. Assume also that predation rates on larvae are lower outside the estuary. One possible result is selection for a longer period of larval feeding, smaller and more numerous eggs, and (through change in timing of release or appropriate vertical swimming by larvae) flushing of larvae out of the estuary as quickly as possible. Once outside the estuary the larvae grow in a relatively safe environment, and when they have reached a larger and safer size the larvae could return to the estuary (again, by appropriate vertical swimming). Another possible result is selection for a short period of larval feeding, larger hatchlings more capable of avoiding predators, and (through timing of release or appropriate vertical swimming of larvae) retention of larvae near the parental area in the estuary so that an extended search for an appropriate area to settle is not necessary. Initial traits which might result in selection in the first directions would be a long larval period and poorer tolerance of low salinities by the larvae. In that case individuals with still smaller eggs, higher fecundity, a faster departure from the estuary, and more growth before becoming competent to settle might initially leave more offspring, grandoffspring, and greatgrandoffspring. If tolerance to low salinity were high, eggs larger, and fecundity lower, selection might proceed in the opposite direction. The direction of selection could depend on the initial traits of species newly invading an estuary.

Vance (1973) has presented an analytical demonstration of alternate stable outcomes of selection in the same environmental regime. Though this model has been criticized as unrealistic for the phenomenon for which it was devised, it is useful for its simplicity and clarity. Vance models benthic versus planktonic development and feeding versus non-feeding development of larvae, but the trade of growth and high fecundity against mortality in a longer larval stage is similar to the possible exchange of risks and benefits when larvae are retained in an estuary or exported from it. To adapt this model to the retention versus export alternative, substitute "estuarine" for "benthic" and "lecithotrophic" as interpretations for the parameters in Vance's equation (14) and let k either equal zero or be extra costs of adapting a larva to estuarine conditions. We then see that, under some conditions, selection might produce either a migrator or non-migrator depending on initial traits (see Vance's equations 16 and 17). The relation between habitat, survival, and fecundity in this model may not describe these relations for development in estuarine and coastal waters, but the general conclusion does not depend on unrealistic features of the model. Under some circumstances, initial traits could affect the direction and outcome of selection for retention or export.

Predictions

Conclusions are premature, but the following predictions may be useful. (1) Varying favorability among estuaries does not select for spread of offspring among estuaries. (2) Waters outside estuaries are often safer for larvae than waters inside estuaries. (3) Selection for export of larvae from estuaries, when it occurs, is selection for migration, and the resulting large scale spread of offspring is an accidental consequence which confers no short term advantage. These hypotheses will not be easy to test, but they are testable for some species and estuaries.

Acknowledgments

NSF grant OCE8008310 and the John Simon Guggenheim Foundation supported this study at the Marine Biological Association of the U.K. and the Friday Harbor Laboratories. D. Allan, R. Harris, M. Landry, and E. K. Roberts provided references on mortality rates of copepods. Megumi Strathmann, V. S. Kennedy, A. R. Palmer, and J. Pearse critized a draft of the manuscript. I thank Springer-Verlag for permission to use three figures from *Oecologia*.

References Cited

Allan, J. D., T. G. Kinsey and M. C. James. 1976. Abundances and production of copepods in the Rhode River subestuary of Chesapeake Bay. *Chesapeake Sci.* 17:86-92.

Altukhov, Y. P., E. A. Salmenkova, G. D. Ryabova and N. I. Kulikova. 1980. Genetic differentiation of chum salmon populations and effectiveness of some acclimatization measures. *Soviet J. Mar. Biol.* 6:125-184.

Ament, A. S. 1979. Geographic variation in relation to life history in three species of the marine gastropod genus *Crepidula:* growth rates of newly hatched larvae and juveniles, pp. 61-76. *In:* S. E. Stancyk (ed.), *Reproductive Ecology of Marine Invertebrates.* U. South Carolina Press, Columbia, SC.

Balkau, B. and M. W. Feldman. 1973. Selection for migration modification. *Genetics* 74:171-174.

Brannon, E. L. 1967. Genetic control of migrating behavior of newly emerged sockeye salmon fry. *Progr. Rept. Intern. Pac. Salmon Fisheries Comm.*, No. 16, 31 p.

Burton, R. S. and M. W. Feldman. 1982. Population genetics of coastal and estuarine invertebrates: does larval behavior influence population structure?, pp. 537-551. *In:* V. S. Kennedy (ed.), *Estuarine Comparisons.* Academic Press, New York.

Christiansen, F. B. and T. M. Fenchel. 1979. Evolution of marine invertebrate reproductive patterns. *Theor. Pop. Biol.* 16:267-282.

Clark, R. B. 1961. The origin and formation of the heteronereis. *Biol. Rev.* 36:199-236.

Cohen, D. 1967. Optimization of seasonal migratory behavior. *Amer. Natur.* 101:5-17.

Cronin, T. W. and R. B. Forward, Jr. 1982. Tidally timed behaviors: their effects on larval distributions in estuaries, pp. 505-520. *In:* V. S. Kennedy (ed.), *Estuarine Comparisons.* Academic Press, New York.

D'Apolito, L. M. and S. E. Stancyk. 1979. Population dynamics of *Euterpina acutifrons* (Copepoda: Harpacticoida) from North Inlet, South Carolina, with reference to dimorphic males. *Mar. Biol.* 54:251-260.

Dean, D. 1978. Migration of the sandworm *Nereis virens* during winter nights. *Mar. Biol.* 45:165-173.

Doyle, R. W. 1975. Settlement of planktonic larvae: a theory of habitat selection in varying environments. *Amer. Natur.* 109:113-126.

Estcourt, I. N. 1966. The life history and breeding biology of *Nicon aestuarensis* Knox (Annelida, Polychaeta). *Trans. Roy. Soc. NZ Zoology* 7:179-194.

Feller, R. J. 1980. Quantitative cohort analysis of a sand dwelling meiobenthic harpacticoid copepod. *Est. Coastal Mar. Sci.* 11:459-476.

Fleeger, J. W. 1979. Population dynamics of three estuarine meiobenthic harpacticoids (Copepoda) in South Carolina. *Mar. Biol.* 52:147-156.

Frank, P. W. 1981. A condition for a sessile strategy. *Amer. Natur.* 118:288-290.

Gable, M. F. and R. A. Croker. 1978. The salt marsh amphipod *Gammarus palustris* Bousfield, 1969 at the northern limit of its distribution. II. Temperature-salinity tolerances. *Est. Coastal Mar. Sci.* 6:225-230.

Gaudy, R. 1976. Étude du plancton de la zone nord de la rade de Villefranche-sur-Mer a la fin du printemps (17 Mai 71 au 16 Juin 71). III. - Production secondaire des copépodes pélagiques. *Vie Milieu Ser. B.* 26:77-106.

George, J. D. 1968. The effect of the 1962-63 winter on the distribution of the cirratulid polychaetes *Cirratulus cirratus* (Müller) and *Cirriformia tentaculata* (Montagu) in the British Isles. *J. Anim. Ecol.* 37:321-337.

Grice, D. G., M. R. Reeve, P. Koeller and D. W. Menzel. 1977. The use of large-volume transparent enclosed sea-surface water columns in the study of stress on plankton ecosystems. *Helgoländer Wiss. Meeresunters.* 30:118-133.

Hansen, T. A. 1980. Influence of larval dispersal and geographic distribution on species longevity in neogastropods. *Paleobiology* 6:193-207.

Harden Jones, F. R. 1968. *Fish Migration*. Edward Arnold. 325 pp.

Heinle, D. R. 1966. Production of a calanoid copepod, *Acartia tonsa*, in the Patuxent River estuary. *Chesapeake Sci.* 7:59-74.

Hicks, G. R. F. 1977. Breeding activity of marine phytal harpacticoid copepods from Cook Strait. *New Zealand J. Mar. Freshwat. Res.* 11:645-666.

Jablonski, D. 1980. Apparent versus real biotic effects of transgressions and regressions. *Paleobiology* 6:397-407.

Jackson, J. B. C. 1974. Biogeographic consequences of eurytopy and stenotopy among marine bivalves and their evolutionary significance. *Amer. Natur.* 108:541-560.

Koehn, R. K., R. Milkman and J. B. Mitton. 1976. Population genetics of marine pelecypods. IV. Selection, migration, and genetic differentiation in the blue mussel *Mytilus edulis*. *Evolution* 30:2-32.

Kuno, E. 1981. Dispersal and the persistence of populations in unstable habitats: a theoretical note. *Oecologia* 49:123-126.

Landry, M. R. 1976. Population dynamics of the planktonic marine copepod, *Acartia clausi* Giesbrecht, in a small temperate lagoon. PhD Thesis, U. Washington, Seattle, WA, 200 pp.

Landry, M. R. 1978. Population dynamics and production of a planktonic marine copepod, *Acartia clausi*, in a small temperate lagoon on San Juan Island, Washington. *Int. Revue ges. Hydrobiol.* 63:77-119.

Leggett, W. C. 1977. The ecology of fish migrations. *Ann. Rev. Ecol. Syst.* 8:285-308.

Lonsdale, D. J. 1981. Regulatory role of physical factors and predation for two Chesapeake Bay copepod species. *Mar. Ecol. Prog. Ser.* 5:341-351.

MacKay, T. F. C. and R. W. Doyle. 1978. An ecological genetic analysis of the settling behaviour of a marine polychaete: I. Probability of settlement and gregarious behaviour. *Heredity* 40:1-12.

Marcus, R. 1980. Optimal diffusion in a random environment. *Mathl. Biosci.* 49:223-233.

Mullin, M. M. and E. R. Brooks. 1970. Production of the planktonic copepod *Calanus helgolandicus*. *Bull. Scripps Inst. Oceanogr.* 17:89-103.

Paine, R. T. and S. A. Levin. 1981. Intertidal landscapes: disturbance and the dynamics of pattern. *Ecol. Monogr.* 51:145-178.

Palmer, A. R. and R. R. Strathmann. 1981. Scale of dispersal in varying environments and its implications for life histories of marine invertebrates. *Oecologia* 48:308-318.

Palmer, M. A. 1980. Variation in life history patterns between intertidal and subtidal populations of the meiobenthic copepod *Microarthridion littorale*. *Mar. Biol.* 60:159-165.

Parslow, J., N. C. Sonntag and J. B. L. Matthews. 1979. Technique of systems identification applied to estimating copepod population parameters. *J. Plankton Res.* 1:137-151.

Raleigh, R. F. 1967. Genetic control in the lakeward migrations of sockeye salmon (*Oncorhynchus nerka*) fry. *J. Fish. Res. Bd. Canada* 24:2613-2622.

Ricker, W. E. 1972. Heredity and environmental factors affecting certain salmonid populations, pp. 27-160. *In*: R. C. Simon (ed.), *The Stock Concept in Pacific Salmon*. H. R. MacMillan Lectures in Fisheries, U. British Columbia, Vancouver, Canada.

Saigusa, M. 1981. Adaptive significance of a semilunar rhythm in the terrestrial crab *Sesarma*. *Biol. Bull.* 160:311-321.

Scheltema, R. S. 1975. Relationship of larval dispersal, gene-flow and natural selection to geographic variation of benthic invertebrates in estuaries and along coastal regions. *Estuarine Research* 1:372-391.

Scheltema, R. S. 1977. Dispersal of marine invertebrate organisms: Paleobiogeographic and biostratigraphic implications, pp. 73-108. *In*: E. G. Kauffman and J. E. Hazel (eds.), *Concepts and Methods of Biostratigraphy*. Dowden, Hutchinson & Ross, Stroudsburg, Pennsylvania.

Smith, R. 1964. On the early development of *Nereis diversicolor* at different salinities. *J. Morph.* 114:437-464.

Stearns, S. C. 1977. The evolution of life history traits: a critique of the theory and a review of the data. *Ann. Rev. Ecol. Syst.* 8:145-171.

Strathmann, R. R. 1974. The spread of sibling larvae of sedentary marine invertebrates. *Amer. Natur.* 108:29-44.

Strathmann, R. R. 1977. Egg size, larval development, and juvenile size in benthic marine invertebrates. *Amer. Natur.* 111:373-376.

Strathmann, R. R. 1978a. The evolution and loss of feeding larval stages of marine invertebrates. *Evolution* 32:894-906.

Strathmann, R. R. 1978b. Progressive vacating of adaptive types during the Phanerozoic. *Evolution* 32:907-914.

Strathmann, R. R. and E. S. Branscomb. 1979. Adequacy of cues to favorable sites used by settling larvae of two intertidal barnacles, pp. 77-89. *In:* S. E. Stancyk (ed.), *Reproductive Ecology of Marine Invertebrates*. U. South Carolina Press, Columbia, SC

Strathmann, R. R., E. S. Branscomb and K. Vedder. 1981. Fatal errors in set as a cost of dispersal and the influence of intertidal flora on set of barnacles. *Oecologia* 48:13-18.

Struhsaker, J. W. 1968. Selection mechanisms associated with intraspecific shell variation in *Littorina picta* (Prosobranchia: Mesogastropoda). *Evolution* 22:459-480.

Sulkin, S. D. and W. VanHeukelem. 1982. Larval recruitment in the crab *Callinectes sapidus* Rathbun: an amendment to the concept of larval retention in estuaries, pp. 459-475. *In:* V. S. Kennedy (ed.), *Estuarine Comparisons*. Academic Press, New York.

Vance, R. R. 1973. On reproductive strategies in marine benthic invertebrates. *Amer. Natur.* 107:339-352.

Williams, G. C., R. K. Koehn and J. B. Mitton. 1973. Genetic differentiation without isolation in the American eel, *Anguilla rostrata*. *Evolution* 27:192-204.

POPULATION GENETICS OF COASTAL AND ESTUARINE INVERTEBRATES: DOES LARVAL BEHAVIOR INFLUENCE POPULATION STRUCTURE?

Ronald S. Burton

Department of Biology
University of Pennsylvania
Philadelphia, Pennsylvania

and

Marcus W. Feldman

Department of Biological Sciences
Stanford University
Stanford, California

Abstract: Numerous investigations of the genetic structure of natural populations of marine and estuarine invertebrates suggest that there is generally an inverse correlation between the length of pelagic larval life and the extent to which geographically separated conspecific populations have diverged. There are, however, exceptions to this observation which may be attributed to at least two factors: (1) Natural selection can act to promote divergence of populations even when there is extensive gene flow, and (2) species which appear to have high capacity for dispersal may not always experience high levels of gene flow. In this paper, we point out that even among species with long-lived pelagic larvae there is considerable variation in population structure. Based on limited data, it appears that species that release eggs or weak-swimming larvae into the water column tend to show little population differentiation over broad geographic ranges; in contrast, there is suggestive evidence that species which liberate strong-swimming larvae (e.g., crustaceans) show significant population differentiation. We propose that the behavioral capacities of strong-swimming larvae allow them to exert some control over their dispersal (for example, via depth-regulatory behavior), and that this control limits the passive dispersal which is manifested in species that liberate eggs. Hence, we hypothesize that larval behavior influences population structure in some species of coastal and estuarine invertebrates.

Introduction

Laboratory experiments have clearly demonstrated that the pelagic larval stages of a variety of invertebrate species show specific behavioral responses to a number of environmental stimuli, including light (e.g., Thorson 1964), temperature (Ott and Forward 1976), salinity (Latz and Forward 1977), and hydrostatic pressure (e.g., Sulkin 1973). It has been widely recognized that these behavioral responses, especially in conjunction with the temporal information provided by biological clocks (Forward and Cronin 1980), could allow larvae actively to navigate coastal or estuarine waters in

ISBN 0-12-404070-5

order to remain in or eventually return to appropriate habitat for settling and adult life. However, the extent to which active behavior, such as depth regulatory mechanisms, actually influences the dispersal of larvae and inter-population recruitment remains in question.

Much of the data bearing on the issue of pelagic larval dispersal is derived from direct oceanic and estuarine plankton sampling studies (e.g., Johnson 1939; Scheltema 1971; Lough 1975). While such studies provide valuable data concerning the geographic distribution of larvae in the water column, they are frequently difficult to interpret since the ultimate fate of larvae in the sampled waters is unknown. Hence, although larvae may be found at great distances from any potential source population, the extent to which these represent larval wastage versus viable recruits is difficult to determine. Furthermore, while non-random larval distributions are suggestive of behaviorally-mediated processes, such patterning may also result from marine and estuarine circulation.

On a smaller geographic scale, there is good evidence that larval behavior influences the microdistribution of individuals within the available habitat. It has long been recognized that many larval species discriminate between substrate types, showing pronounced preferences in their patterns of settlement (e.g., Knight-Jones 1953; Crisp and Barnes 1954). While such observations set a precedent for suggesting that larval behavior can influence dispersal and recruitment, the difference in the scales of the processes suggests that behavioral capacities sufficient to allow local substrate selection may not necessarily be sufficient for controlling the entire dispersal process.

To assess directly the extent to which larval dispersal effectively homogenizes the gene pools of geographically separated populations, it is clear that documentation of the geographic distribution of larvae is insufficient; we must look at the gene pools themselves to quantify levels of divergence among populations. While such an approach was essentially impossible only fifteen years ago (i.e., prior to the work of Lewontin and Hubby 1966), biochemical genetic techniques have made direct genetic comparisons of populations almost routine today. Gel electrophoresis of proteins combined with histochemical techniques for visualization of specific enzyme activities has provided a powerful tool for the investigation of population structure and recruitment patterns in marine and estuarine organisms.

In this paper, we will (1) briefly review some of the available data concerning the relationship between larval dispersal patterns and population genetic differentiation as revealed by electrophoretic studies, (2) point out some circumstances under which genetic data might implicate an important role for larval (or adult) behavior in the control of effective dispersal, and (3) suggest the types of studies that might prove most valuable in determining both the genetic consequences of larval retention in or leakage from estuaries and the relative importance of larval behavior in dispersal and recruitment processes. For a broader discussion of the population genetics of marine and estuarine organisms, the reader is referred to Gooch (1975) and Levinton (1980).

Relationships Between Larval Dispersal and Genetic Differentiation of Geographically Separated Populations

Berger (1973) used electrophoretically-determined esterase polymorphisms to compare the extent of population genetic differentiation in three sympatric littorinid gastropod species with varying capacities for larval dispersal. He found that in *Littorina littorea*, a species with pelagic development and free-swimming larvae, there appeared to be little population differentiation, while in *L. obtusata* and *L. saxatilis*, both of which lack pelagic development and free-swimming larval stages, significantly more differentiation was observed. A number of additional studies (e.g., Gooch et al. 1972; Snyder and Gooch 1973; Campbell 1978) have yielded similar results; Gooch (1975) and Crisp (1978) have drawn the conclusion that there is generally an inverse relationship between dispersal capacity, as measured by the duration of pelagic larval life, and the degree of genetic differentiation between intraspecific populations.

Scheltema (1975) has pointed out that effective dispersal not only reflects the length of pelagic larval life, but also the consequences of larval behavior and oceanic circulation patterns. Such effects are apparent in the data of Scheltema (1971) which were collected to address the question of trans-oceanic transport of the larvae of shallow-water benthic gastropods; these data show a clear lack of correspondence between the length of larval life of a given species and the extent to which the larvae of that species were found dispersed offshore (see Tables I and IV in Scheltema 1971). On the basis of larval distributions in the plankton, Scheltema (1971) concluded that trans-oceanic larval dispersal is not infrequent in a number of gastropod species. While this conclusion gains some support from the rough correspondence between levels of morphological divergence between eastern and western Atlantic populations and the extent to which larvae were observed in the plankton, no genetic evidence has been presented to verify the proposed gene flow. Such data are of particular importance in light of Berger's comparisons of littorinid populations on opposite sides of the Atlantic (Berger 1977); this work unambiguously established that despite its relatively long-lived pelagic larvae, *L. littorea* experiences no trans-Atlantic gene flow (North American and European populations showed strong divergence at all nine polymorphic loci studied).

While the inverse relationship between length of larval life and genetic differentiation of populations generally accounts for much of the data in the literature addressing this issue, there are some notable exceptions (e.g., Struhsaker 1968; Koehn 1978). We have previously suggested (Burton and Feldman 1981) that these exceptions may be attributed to two factors: (1) Strong differences in natural selective forces arising from habitat heterogeneity among populations can result in genetic differentiation even where there is significant inter-population gene flow. (2) Species which appear to have high capacity for dispersal may not always experience high levels of gene flow. Distinguishing between these factors has often proven to be difficult, but specific types of data can sometimes allow strong inferences to be made as discussed below.

Probably the most intensively studied case of genetic differentiation among populations of marine and estuarine invertebrates involves that in the common mussel *Mytilus edulis*. Steep clines in allele frequencies have been observed at the *LAP* locus that correlate with salinity changes in the environment (Boyer 1974; Koehn et al. 1976; Theisen 1978). While such clines might simply reflect restricted larval dispersal between geographically isolated populations, an alternative explanation could be that there is essentially free gene flow over the length of the cline and natural selection subsequently results in differentiation according to prevailing local environments. With respect to the *LAP* cline observed between *M. edulis* populations in Long Island Sound versus those on the Atlantic coast just outside the Sound, Koehn et al. (1980) have presented striking evidence for distinguishing between these mechanisms. By determining size-dependent *LAP* allele frequencies at a series of five sites along the cline in two successive years, an annual influx of recruits from the oceanic populations (genetically marked with a high LAP^{94} allele frequency) reaching up to 20 miles into the Sound was made apparent; hence, despite their having a pelagic larval life of up to 7 weeks, larval dispersal from oceanic populations into Long Island Sound is limited. Lassen and Turano (1978) have discussed the hydrographic features that are likely to restrict larval dispersal into the Sound. Koehn et al. (1980) have also provided strong evidence for a size-dependent decline in the frequency of the LAP^{94} allele with increasing shell size which is also site specific; i.e., the decline of LAP^{94} in larger mussels was only significant at sites that were furthest into Long Island Sound. Hence, the observed pattern of genetic differentiation in these populations is apparently the result of restricted larval dispersal due to hydrographic conditions combined with natural selection acting in a heterogeneous environment.

Theisen (1978) found strong allele frequency clines at three loci (including *LAP*) in populations of *M. edulis* from the Baltic and the Kattegat. The fact that all three clines between the two areas are coincident (as opposed to discordant) strongly suggests that significant gene flow occurs among the sites studied. To test if natural selection was important in maintaining population differentiation in spite of the apparent gene flow, a population sample from a Baltic site was transplanted to the Kattegat; after less than eight months, transplanted adults were sampled and allele frequencies compared to those determined at the start of the experiment. Results showed that allele frequencies at all three loci changed significantly in the direction expected from a natural selection argument: allele frequencies after the experiment were more like those characteristic of Kattegat populations than they were initially. For further discussion of selection in the *M. edulis* system, the reader is referred to Levinton (1980) and Koehn et al. (1980); our purpose here is merely to point out that with appropriate population sampling and experimental studies, some assessment of the relative importance of natural selection versus restricted gene flow can be made from genetic data.

Using Population Genetic Data, Can We Distinguish Between Active and Passive Recruitment Processes?

We have referred to a few investigations which provide what we feel are valuable protocols for data collection and interpretation with regard to the relative importance of gene flow and natural selection in determining patterns of population genetic differentiation. We now turn our attention to a different question: can population genetic studies give us any insight into the mechanisms of gene flow – i.e., are patterns of population differentiation consistent with the passive dispersal of larvae by currents, or is there evidence suggesting that larval (or adult) behavior influences the extent of the dispersal process?

If recruitment processes simply reflect the passive drifting of pelagic larvae in estuarine and oceanic water masses, then population genetic structure will be the result of three factors: (1) when and where larvae are released into the water column, (2) duration of free-swimming larval life, and (3) spatial and temporal patterns of water circulation. While we rarely know any of these parameters with great precision, some approximation of each is usually available for a given system of interest. Hence, we can usually derive some sort of qualitative *a priori* expectation of what the population structure of a particular species should be. Returning to our earlier discussion of the relationship between intraspecific population differentiation and larval dispersal patterns, we predict that species with long-lived free-swimming larvae inhabiting areas with well developed currents would tend to be genetically homogeneous if passive processes dominate recruitment; species lacking pelagic larvae, or inhabiting regions of restricted current flow are expected to show considerably greater inter-population genetic differentiation.

We propose that in some cases, to be considered below, a general lack of correspondence between observed and expected patterns of population genetic differentiation provides suggestive evidence for an active role of behavior in the recruitment process; i.e., discordance between the observed pattern of population differentiation and that expected based on the qualitative predictions of a passive model implies that other forces are involved. By giving larvae the potential to control their dispersal, behavior can clearly influence patterns of gene flow and subsequent genetic differentiation of geographically separated populations. This is not to imply that behavior is the only force that can produce discordance between observed and expected patterns of differentiation; as we have already seen, natural selection will also result in discordance. Similarly, we recognize that an active role of behavior will not always be detected in studies of population structure; if allele frequencies at all loci studied are similar in each population, little inference concerning gene flow mechanisms can be made.

In the following paragraphs, we will first consider a case study of population structure in a marine copepod where behavior may be influencing the extent of differentiation of neighboring populations, then suggest that similar inferences may be made with respect to some other species, and finally propose that there is an underlying pattern in the observations made to date.

Population Structure of the Intertidal Copepod Tigriopus californicus*: A Case Study.*

Since 1976, our laboratory has undertaken extensive studies of the genetic structure of natural populations of *T. californicus*, a common inhabitant of high intertidal and supralittoral rock pools along the Pacific coast of North America. Our interest in this species stemmed from its restriction to rocky habitat which is irregularly distributed along the coast; our studies were aimed at determining the extent of genetic differentiation among populations of *T. californicus* which are geographically separated from one another to varying degrees depending on local, regional, and long-distance patterns of habitat distribution.

The life cycle of *T. californicus* includes six naupliar larval stages and six copepodid stages (the last of which is the adult form), all of which are feeding and free-swimming forms. The total life cycle takes about 24 days in the laboratory at 21° C, but probably takes from 3-6 weeks or longer in nature depending on pool temperature. As all life stages can often be found in a given rock pool, it is clear that populations occupying single pools experience at least some internal recruitment. However, wave activity during high tides or storms frequently washes large numbers of *T. californicus* out of high pools and into the sea where, presumably, they could be transported along shore by prevailing inshore currents. Based on these features of the natural history of this species, it appears that *T. californicus* has considerable capacity for dispersal on at least a local and regional scale; hence, we hypothesized that *T. californicus* populations occupying pools located within several kilometers of one another would be genetically homogeneous. On a larger scale (i.e., hundreds of kilometers), some differentiation might be anticipated especially where these distances include long stretches of sandy beach completely devoid of rock habitat.

Results of some of our population survey work are shown in Table 1 (also see Burton et al. 1979, and Burton and Feldman 1981). The ten sites for which data are presented cover a wide range of inter-site distances as shown in Fig. 1. The major points to be made from these data are: (1) even on a local scale (the distance between the Santa Cruz and West Capitola sites

Table 1. Allele frequencies for five polymorphic gene loci at ten sites along the California coast. Each allele frequency is based on a sample of at least 50 individuals. PGM: phosphoglucomutase; PGI: phosphoglucose isomerase; GOT: glutamate-oxaloacetate transaminase; ME: malic enzyme; EST: esterase.

Site	*PGM*					*PGI*			*GOT*				*ME*			*EST*		
	109	*107*	*105*	*100*	*96*	*105*	*100*	*89*	*107*	*100*	*96*	*84*	*112*	*103*	*100*	*100*	*97*	*Null*
Moss Beach (MB)	–	–	.80	.20	–	–	1.0	–	–	1.0	–	–	–	–	1.0	.40	.30	.30
Pescadero (PO)	–	–	.07	.93	–	.48	.52	–	–	1.0	–	–	–	–	1.0	1.0	–	–
Santa Cruz (SC)	–	–	.60	.30	.10	–	1.0	–	–	.98	.02	–	–	–	1.0	.95	–	.05
West Capitola (WC)	–	.24	.50	.26	–	–	1.0	–	–	1.0	–	–	–	.45	.55	1.0	–	–
Monterey (MY)	–	–	1.0	–	–	–	.98	.02	.30	.70	–	–	–	–	1.0	1.0	–	–
Carmel (CL)	–	–	.90	.10	–	–	.93	.07	–	.90	.10	–	–	–	1.0	1.0	–	–
Carrillo (CO)	–	–	–	1.0	–	–	1.0	–	–	.95	.05	–	–	–	1.0	1.0	–	–
Palos Verdes (PV)	–	–	1.0	–	–	–	1.0	–	–	–	–	1.0	–	–	1.0	1.0	–	–
Newport Beach (NB)	.18	–	.82	–	–	–	.69	.31	–	1.0	–	–	1.0	–	–	1.0	–	–
Laguna Beach (LB)	.43	–	.57	–	–	–	.50	.50	–	1.0	–	–	1.0	–	–	1.0	–	–

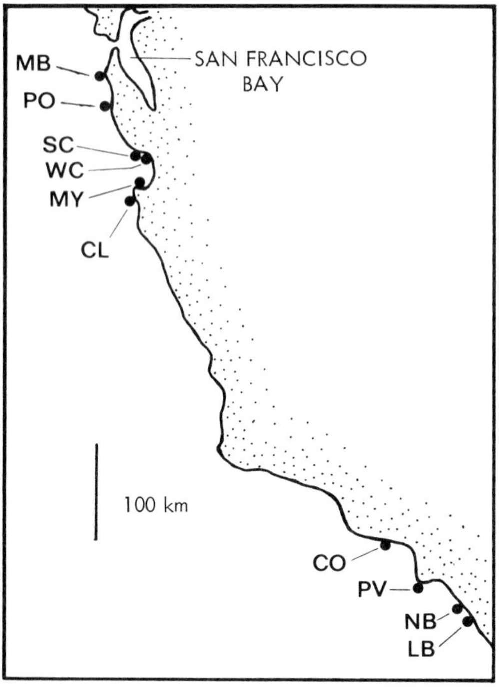

Figure 1. *Locations of* T. californicus *populations sampled along the California coast. Abbreviations as in Table 1.*

is less than 8 km), populations show sharp genetic differentiation, with alleles that are common in one population sometimes being completely absent from a neighboring population, and (2) there is no apparent patterning of allele frequencies in space; i.e., polymorphisms are patchily distributed with many alleles restricted to small portions of coast while the dominant allele remains the same in populations to both the north and south of the polymorphic area. The distribution of unique alleles at the five loci for which data are presented clearly establishes that gene flow among these populations is extremely restricted.

To investigate further the geographic scale of gene flow among *T. californicus* populations, we have carried out detailed studies at three sites (Burton and Feldman 1981). Results from the Pescadero site are shown in Fig. 2. The four sites sampled are all located within a 2 km stretch of coastline and yet are clearly distinguished from one another by their respective frequencies of PGI and PGM alleles. Of particular interest here is the fact that the patterns of differentiation at the two loci are not parallel to one another as would be expected if migration played a dominant role in determining allele frequencies (Christiansen and Simonsen 1978). The apparent restriction of gene flow at even this geographic scale is further supported by the fact that

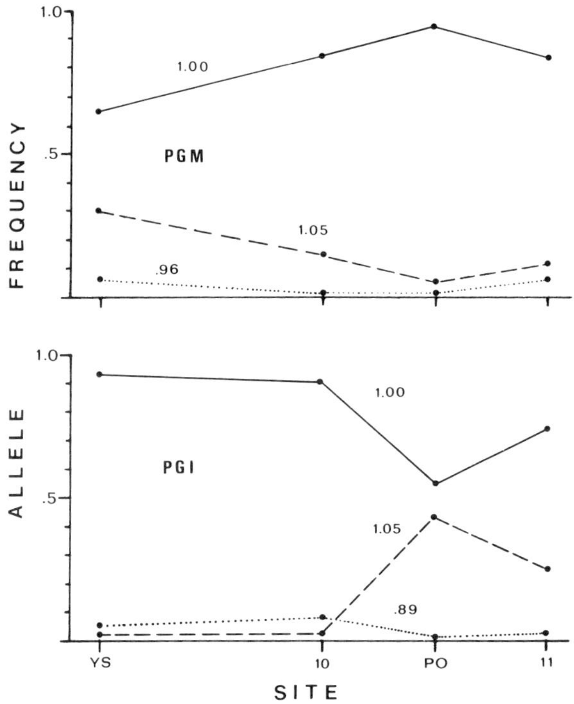

Figure 2. PGI and PGM allele frequencies at four sites near Pescadero State Beach, California. The sites are arranged from south (YS) to north (site 11) along the coast; the total distance between YS and site 11 is approximately 2 km. Sample sizes for each allele frequency exceeded 70 individuals.

the PGI$^{.89}$ allele which has been observed in frequencies of over 13% at site 10 (Burton and Feldman 1981), has never been observed 500 m to the north at the PO site, with over 1000 PO individuals scored. Finally, as we have reported elsewhere (Burton and Feldman 1981), the pattern of differentiation at these sites has not varied significantly over the past three years.

Hence, despite what appears to be a considerable capacity for dispersal in all the life stages of *T. californicus,* our population genetic data show that along-shore transport of this species, as would be expected from passive drifting of individuals washed out of high intertidal habitat, is extremely low. This observation has led us to the hypothesis that active behavioral mechanisms are involved in preventing passive along-shore dispersal. Such mechanisms may include both grasping the substrate to prevent seaward dispersal and active swimming to remain in or return to supralittoral pools after being washed out. We have only anecdotal evidence for the former mechanism: *T. californicus* larvae and adults in the laboratory and in the field are frequently difficult to remove from the substrate with relatively strong pipette suction. With respect to the latter, Bozic (1975) has shown that when presented with a choice, *Tigriopus* sp. preferentially move to vessels that have been previously occupied by adult conspecifics. Dethier (1980) has pointed out that the behaviors evident in Bozic's study might play an important role in the return of copepods to the high intertidal after being washed out by severe wave activity. While we currently lack detailed knowledge of *T. californicus* behavior, our genetic data, combined with qualitative natural history information, strongly suggests that behavioral processes significantly influence population structure in this species.

Some Suggestive Patterns of Differentiation in Other Species.

Detailed studies of population genetic structure of marine invertebrates in which data on more than one locus from a series of geographic sites are reported are not common in the literature. Earlier in this paper we considered a number of the available studies in discussing the inverse relationship between length of larval life and the extent of population differentiation. Interestingly, these studies dealt primarily with a variety of species of molluscs; it was concluded that patterns of differentiation were generally consistent with models of passive larval dispersal combined with hydrographic conditions and the action of natural selection. In contrast, the occurrence of unique alleles in high frequencies in neighboring populations of the copepod *T. californicus* seems inconsistent with any models including passive dispersal (unless we are willing to accept tremendously strong natural selection causing a common allele in one population to be lethal in an environmentally similar habitat as little as 500 m away!). The intriguing possibility that behavior plays an important role in population structure in *T. californicus* has led us to a literature search to consider the following hypothesis: Do species which enter the plankton as relatively strong swimming larval forms show greater population differentiation than species which enter the plankton as either non-motile eggs or poorly-developed weak swimming larvae? As a first approach, we have selected the crustaceans as a group to represent the former condition while the molluscan data discussed above represent the latter. Before proceeding, we note that in each group, species that lack any pelagic life stages tend to show significant genetic differentiation on a small geographic scale (for molluscs, see discussion above; for crustaceans, see e.g., Salemaa 1978; Bulnheim and Scholl 1981).

Though based on generally small sample sizes, the study of Tracey et al. (1975) on the American lobster *Homarus americanus* presents data on population structure in this species. As pointed out by the authors, allele frequencies at the malic enzyme (ME) locus show sharp differentiation between samples obtained from inshore versus offshore populations (Woods Hole and Martha's Vineyard populations compared to Georges Bank populations). Striking differentiation at ME was also apparent in the comparison of a Gulf of St. Lawrence sample to any of the other samples in the study. Aside from the ME comparisons, the authors restricted their discussion of the data to overall genetic distance measures; in doing so, we think they have failed to recognize some other significant facets of the data. In particular, their data indicate that there is significant differentiation at some loci in neighboring populations. For example, populations located quite close together (as the LSB and LSA sites offshore and the MVS and WHP populations inshore) are significantly differentiated at the MDH-2 locus, where the WHP and LSA samples each possess the MDH^{105} allele (at frequencies of 0.20 and 0.11 respectively) which is absent from the MVS and LSB samples (represented by 45 and 60 individuals respectively). In fact, at the only other moderately polymorphic locus for which reasonable sample sizes were analyzed, EST-2, MVS and WHP also appear to be differentiated. Similarly, LSA and LSB are significantly differentiated at the PGM-4 locus. Clearly, while more extensive sampling would be desirable, available data indicate substantial differentiation of local populations of *H. americanus* despite its pelagic larval stages.

With respect to estuarine crustaceans, we have found the data to be particularly lacking. Cole and Morgan (1978) have presented preliminary data comparing populations of the blue crab *Callinectes sapidus* from Chesapeake and Chincoteague Bays. While they found striking similarity of allele frequencies at three polymorphic loci, the data presented pooled results obtained from two sample sites in Chesapeake Bay and involved animals purchased from commercial sources from each bay; the extent to which these factors account for the observed extreme deviations from Hardy-Weinberg proportions is difficult to assess. While we may accept genetic similarity of these populations (since we lack other data), there is good evidence that *C. sapidus* from other locations along the Atlantic coast are significantly differentiated from the Chesapeake and Chincoteague populations: Nelson and Hedgecock (1980) reported that *C. sapidus* from the Wando River in South Carolina were polymorphic at the hexokinase locus (with a heterozygosity of 0.587), while Cole and Morgan (1978) reported that hexokinase was monomorphic in the Chesapeake and Chincoteague populations. Similarly, there are differences at malic dehydrogenase and phosphoglucose isomerase loci, based on reported heterozygosities in Hedgecock and Nelson (1980) versus Cole and Morgan (1978).

There is also evidence for differentiation of *Palaemonetes varians* populations (Christensen and Lomholt 1972) and *Penaeus setiferus* populations (comparing the PGM data in Marvin and Caillouet, 1976, for a South Carolina estuary versus data for a Georgian estuary presented in Nelson and Hedgecock 1980). Despite the scant nature of these data and those cited earlier, it appears that much of the differentiation among crustacean populations involves the patchy occurrence of unique alleles rather than the patterned clines discussed above for *Mytilus edulis* populations. While detailed studies of each set of populations would be needed to assess the relative importance of natural selection versus restricted gene flow in the process of differentiation, we note that the occurrence of unique alleles in conspecific populations is generally indicative of restricted gene flow among those populations.

Differences in Levels of Population Differentiation Among Species with Pelagic Larvae: A Hypothesis.

The above discussion has pointed out that even among species with pelagic larvae, there is considerable variation in the extent to which conspecific populations have diverged genetically. While some of this variation may reflect differences in the current patterns to which larvae of different species are exposed, this is clearly not always the case. The genetically uniform populations of *Nassarius obsoletus* (Gooch et al. 1972) and *Littorina littorea* (Berger 1973) occur over much the same stretch of coastline as the genetically differentiated populations of *Homarus americanus* and *Callinectes sapidus* considered above.

On the basis of admittedly sparse data, we suggest that the differences in population structure among these and other species are the result of differential behavioral capacities of the pelagic developmental stages of the species. In the case of species like *L. littorea* whose egg capsules are shed into the sea, population structure will naturally reflect passive, current-borne dispersal as behavioral influences are unlikely. In contrast, pelagic larvae of crustaceans are typically strong swimmers when first released into the water column and show well developed depth-regulatory behavior at that time (e.g., *Homarus gammarus:* Ennis 1973; *Callinectes sapidus:* Sulkin et al. 1980; *Rhithropanopeus harrisii:* Forward and Cronin 1980; *Emerita analoga:* Burton 1979). Hence, despite their relatively long pelagic larval life, these species have behavioral mechanisms that may act to prevent passive dispersal and thereby influence the extent of inter-population gene flow.

The above hypothesis implies that there is some evolutionary advantage to an individual larva in preventing passive dispersal. The most obvious possibility is that behavioral control over dispersal prevents larvae from being transported out of a region where their parents successfully reproduced and into potentially hostile environments. Furthermore, by providing a mechanism by which successive generations continuously recruit into the same population, behavioral control over dispersal creates a population structure that is conducive to the evolution of population adaptations to local environmental conditions. Such evolution would result from the increased reproductive success of individuals with locally advantageous traits and the reduced cost of dispersal in the sense of Strathmann et al. (1981).

Larval Retention in Estuaries: Addressing the Role of Behavior Via Population Genetic Studies.

Though in the above discussion we have treated coastal and estuarine populations alike in that they both face the problem of recruitment to geographically restricted habitat, the estuarine environment can present special challenges to the recruitment process. In particular, the estuarine environment may impose dramatic barriers to larval dispersal in the form of steep salinity and thermal gradients. To date, population genetic techniques have been largely unexploited in the investigation of estuarine recruitment processes. While we feel that available data are insufficient to prove that larval behavior significantly influences population genetic structure, our discussion illustrates the manner in which genetic data can be used to address this issue. The limiting factor at this time is the availability of large-scale detailed data sets of the genetic structure of natural populations of estuarine invertebrate species with pelagic developmental stages with differing behavioral capacities. Only after such data become available will a critical evaluation of our hypothesis be possible.

Acknowledgments

We thank P. Petraitis for helpful discussions. This work was supported in part by a grant from the Lerner Fund for Marine Research of the American Museum of Natural History, an NIH training grant in integrative biology, and NIH grants GM 28016 and GM 10452. A travel grant from the Office of the Dean, Faculty of Arts and Sciences, University of Pennsylvania, is gratefully acknowledged.

References Cited

Berger, E. M. 1973. Gene-enzyme variation in three sympatric species of *Littorina*. *Biol. Bull.* 145:83-90.

Berger, E. M. 1977. Gene-enzyme variation in three sympatric species of *Littorina*. II. The Roscoff population, with a note on the origin of North American *L. littorea*. *Biol. Bull.* 153: 255-264.

Boyer, J. F. 1974. Clinal and size-dependent variation at the LAP locus in *Mytilus edulis*. *Biol. Bull.* 147:535-549.

Bozic, B. 1975. Détection actométrique d'un facteur d'interattraction chez *Tigriopus* (Crustacés, Copépodes, Harpacticoides). *Bull. Soc. Zool. Fr.* 100:305-311.

Bulnheim, H. -P. and A. Scholl. 1981. Genetic variation between geographic populations of the amphipods *Gammarus zaddachi* and *G. salinus*. *Mar. Biol.* 64:105-115.

Burton, R. S. 1979. Depth regulatory behavior of the first stage zoea larvae of the sand crab *Emerita analoga* Stimpson (Decapoda: Hippidae). *J. Exp. Mar. Biol. Ecol.* 37:255-270.

Burton, R. S. and M. W. Feldman. 1981. Population genetics of *Tigriopus californicus:* II. Differentiation among neighboring populations. *Evolution* 35:1192-1205.

Burton, R. S., M. W. Feldman and J. W. Curtsinger. 1979. Population genetics of *Tigriopus californicus* (Copepoda: Harpacticoida): I. Population structure along the central California coast. *Mar. Ecol. Prog. Ser.* 1:29-39.

Campbell, C. A. 1978. Genetic divergence between populations of *Thais lamellosa* (Gmelin), pp. 157-170. *In:* B. Battaglia and J. Beardmore (eds.), *Marine Organisms, Genetics, Ecology, and Evolution*. Plenum Press, New York.

Christensen, B. and B. Lomholt. 1972. Amylase heterogeneity in *Palaemonetes varians* (Leach) (Crustacea, Decapoda). *Ophelia* 10:63-65.

Christiansen, F. B. and V. Simonsen. 1978. Geographic variation in protein polymorphisms in the eelpout, *Zoarces viviparous* (L.), pp. 171-194. *In:* B. Battaglia and J. Beardmore (eds.), *Marine Organisms, Genetics, Ecology, and Evolution*. Plenum Press, New York.

Cole, M. A. and R. P. Morgan. 1978. Genetic variation in two populations of blue crab, *Callinectes sapidus*. *Estuaries* 1:202-205.

Crisp, D. J. 1978. Genetic consequences of different reproductive strategies in marine invertebrates, pp. 257-273. *In:* B. Battaglia and J. Beardmore (eds.), *Marine Organisms, Genetics, Ecology, and Evolution*. Plenum Press, New York.

Crisp, D. J. and H. Barnes. 1954. The orientation and distribution of barnacles at settlement with particular reference to surface contour. *J. Anim. Ecol.* 23:142-162.

Dethier, M. N. 1980. Tidepools as refuges: predation and the limits of the harpacticoid copepod *Tigriopus californicus* (Baker). *J. Exp. Mar. Biol. Ecol.* 42:99-111.

Ennis, G. P. 1973. Behavioral responses to changes in hydrostatic pressure and light during larval development of the lobster *Homarus gammarus*. *J. Fish. Res. Board Can.* 30:1349-1360.

Forward, R. B. and T. W. Cronin. 1980. Tidal rhythms of activity and phototaxis of an estuarine crab larva. *Biol. Bull.* 158:295-303.

Gooch, J. L. 1975. Mechanisms of evolution and population genetics, pp. 349-409. *In:* O. Kinne (ed.), *Marine Ecology, Vol. II. Physiological Mechanisms*. Wiley, London.

Gooch, J. L., B. S. Smith and D. Knupp. 1972. Regional survey of gene frequencies in the mud snail *Nassarius obsoletus*. *Biol. Bull.* 142:36-48.

Johnson, M. W. 1939. The correlation of water movements and dispersal of pelagic larval stages of certain littoral animals, especially the sand crab, *Emerita*. *J. Mar. Res.* 2:236-245.

Knight-Jones, E. W. 1953. Decreased discrimination during settling after prolonged planktonic life in larvae of *Spirorbis borealis* (Serpulidae). *J. Mar. Biol. Ass. U.K.* 32:337-345.

Koehn, R. K. 1978. Biochemical aspects of genetic variation at the LAP locus in *Mytilus edulis*, pp. 211-227. *In:* B. Battaglia and J. Beardmore (eds.), *Marine Organisms, Genetics, Ecology, and Evolution*. Plenum Press, New York.

Koehn, R. K., R. Milkman and J. B. Mitton. 1976. Population genetics of marine pelecypods. IV. Selection, migration, and genetic differentiation in the blue mussel, *Mytilus edulis*. *Evolution* 30:2-32.

Koehn, R. K., R. I. E. Newell and F. Immermann. 1980. Maintenance of an aminopeptidase allele frequency cline by natural selection. *Proc. Natl. Acad. Sci. USA* 77:5385-5389.

Lassen, H. H. and F. J. Turano. 1978. Clinal variation and heterozygote deficit at the LAP-locus in *Mytilus edulis*. *Mar. Biol.* 49:245-254.

Latz, M. I. and R. B. Forward. 1977. The effect of salinity upon phototaxis and geotaxis in a larval crustacean. *Biol. Bull.* 153:163-179.

Levinton, J. S. 1980. Genetic divergence in estuaries, pp. 509-520. *In:* V. S. Kennedy (ed.), *Estuarine Perspectives*. Academic Press, New York.

Lewontin, R. C. and J. L. Hubby. 1966. A molecular approach to the study of genic heterozygosity in natural populations. II. Amount of variation and degree of heterozygosity in natural populations of *Drosophila pseudoobscura*. *Genetics* 54:595-609.

Lough, R. G. 1975. Larval dynamics of the Dungeness crab, *Cancer magister*, off the central Oregon coast, 1970-71. *Fish. Bull.* 74:353-375.

Marvin, K. T. and C. W. Caillouet. 1976. Phosphoglucomutase polymorphism in white shrimp, *Penaeus setiferus*. *Comp. Biochem. Physiol.* 53B:127-131.

Nelson, K. and D. Hedgecock. 1980. Enzyme polymorphism and adaptive strategy in the decapod crustacea. *Amer. Natur.* 116:238-280.

Ott, F. S. and R. B. Forward. 1976. The effect of temperature on phototaxis and geotaxis by larvae of the crab *Rhithropanopeus harrisi* (Gould). *J. Exp. Mar. Biol. Ecol.* 23:97-107.

Salemaa, H. 1978. Geographic variability in the colour polymorphism of *Idotea baltica* (Isopoda) in the northern Baltic. *Hereditas* 88:165-182.

Scheltema, R. S. 1971. Larval dispersal as a means of genetic exchange between geographically separated populations of shallow-water benthic marine gastropods. *Biol. Bull.* 140:284-322.

Scheltema, R. S. 1975. Relationship of larval dispersal, gene-flow, and natural selection to geographic variation of benthic invertebrates in estuaries and along coastal regions, pp. 372-391. *In:* L. E. Cronin (ed.), *Estuarine Research, Vol. 1: Chemistry, Biology, and the Estuarine System.* Academic Press, New York.

Snyder, T. P. and J. L. Gooch. 1973. Genetic differentiation in *Littorina saxatilis* (Gastropoda). *Mar. Biol.* 22:177-182.

Strathmann, R. R., E. S. Branscomb and K. Vedder. 1981. Fatal errors in set as a cost of dispersal and the influence of intertidal flora on set of barnacles. *Oecologia* 48:13-18.

Struhsaker, J. W. 1968. Selection mechanisms associated with intraspecific shell variation in *Littorina picta* (Prosobranchia; Mesogastropoda). *Evolution* 22:459-480.

Sulkin, S. D. 1973. Depth regulation of crab larvae in the absence of light. *J. Exp. Mar. Biol. Ecol.* 13:73-82.

Sulkin, S. D., W. Van Heukelem, P. Kelly and L. Van Heukelem. 1980. The behavioral basis of larval recruitment in the crab *Callinectes sapidus* Rathbun: A laboratory investigation of ontogenetic changes in geotaxis and barokinesis. *Biol. Bull.* 159:402-417.

Theisen, B. F. 1978. Allozyme clines and evidence of strong selection in three loci in *Mytilus edulis* L. (Bivalvia) from Danish waters. *Ophelia* 17:135-142.

Thorson, G. 1964. Light as an ecological factor in the dispersal and settlement of larvae of marine bottom invertebrates. *Ophelia* 1:167-208.

Tracey, M. L., K. Nelson, D. Hedgecock, R. A. Shleser and M. L. Pressick. 1975. Biochemical genetics of lobsters: Genetic variation and the structure of American lobster *(Homarus americanus)* populations. *J. Fish. Res. Board Can.* 32:2091-2101.

GENETIC CONSEQUENCES OF LARVAL RETENTION: THEORETICAL AND METHODOLOGICAL ASPECTS

Dennis Hedgecock
University of California
Bodega Marine Laboratory
Bodega Bay, California

Abstract: To the extent that larvae of estuarine organisms are retained in their estuary of origin, population genetic theory predicts differences will develop among conspecific populations inhabiting different estuaries. The rate at which these differences evolve depends most importantly on genetically effective population sizes, the intensity of diversifying natural selection, the magnitude of larval "leakage" among estuaries, and the demographic history and stability of extant populations. While electrophoretic techniques are now routinely employed to gather gene frequency data from natural populations, lack of information on these ecological, demographic, and historical factors often renders interpretation of such genetic data equivocal. Larval leakage cannot be inferred from genetic similarities among populations in different estuaries because populations may be large, diversifying selection weak, or time since divergence insufficient for evolution by random genetic drift. Where differences do exist among estuaries, closer inspection of gene frequencies in larval populations and in new recruits, together with experimental manipulation of these, may help narrow the limits on some of the parameters affecting divergence. Genetics is thus a necessary, but not a sufficient, component in the investigation of larval retention in estuaries.

Introduction

"Study of statistical differences among local populations is an important line of attack on the evolutionary problem. While such differences can only rarely represent first steps toward speciation in the sense of the splitting of the species, they are important for the evolution of the species as a whole. They provide a possible basis for intergroup selection of genetic systems, a process that provides a more effective mechanism for adaptive advance of the species as a whole than does the mass selection which is all that can occur under panmixia." (Wright 1943)

Fortunately for purposes of discussing genetical consequences of larval retention in estuaries, a well developed mathematical theory already describes the consequences of subdivision of a species into incompletely isolated local populations (Fisher 1922, 1930; Wright 1931, 1942, 1943). For fifty years the effects upon the distribution of gene frequencies among local populations of population size, migration, mutation, selection, and time since subdivision have been recognized. Yet the essential ecological, demographic and historical parameters of this theory have proven extremely difficult to measure in natural populations. On the other hand, electrophoresis and specific enzyme staining have greatly facilitated observation of gene fre-

ISBN 0-12-404070-5

quencies in natural populations. In turn, this has offered the promise of testing evolutionary and population genetic theory or at least of estimating indirectly some of the ecological or demographic parameters that have been resistant to direct measurement (Lewontin 1974).

In the present context we ask, "What are the genetic consequences of larval retention in estuaries?" In general we expect conspecific populations inhabiting different estuaries to evolve differences over time depending on how well in fact they are isolated (i.e. depending on how much larval "leakage" exists). However, more specific predictions of gene frequency distributions depend on rather precise estimates of gene flow and at least order of magnitude estimates of population sizes. Demographic history and stability of extant populations must also be known. Because it is unlikely that we will ever have this detailed ecological information for most organisms, the tendency is to turn the question around and ask, "Can genetic studies provide information on population structure and help determine rates of larval leakage?" To answer this we will focus not on hydrographic and biological factors that might or might not affect larval dispersal (as previously discussed by Scheltema 1975, for example), but upon what genetic theory and observations can and cannot tell us about the evolution of estuarine populations.

The basic elements of the mathematical theory of population genetic structure will be reviewed briefly in order to identify the parameters of interest and how they interact in the evolution of a subdivided species. Methodological problems arising in the estimation of population parameters will then be discussed in light of the overall problem of identifying causes of observed amounts and patterns of divergence among conspecific populations inhabiting different estuaries. The basic impasse reached in such studies is that lack of knowledge of population parameters more often than not renders interpretation of genetic data equivocal. This is not to say that genetic techniques such as gel-electrophoresis are not useful in the study of population structure, but that limitations on their scope need to be appreciated.

The Theory

Aside from unique mutational events such as polyploidization or chromosomal rearrangements, factors affecting gene frequencies in populations may be classified as deterministic or stochastic. Systematic evolutionary pressures (Wright 1949) are those producing directed changes in gene frequencies; mutation, migration, and selection operating at constant rates or intensity are the primary deterministic factors. A basic equation describing, for example, change in gene frequency as a result of migration is:

$$\Delta p_i = m(p - p_i) \qquad (1)$$

where p_i is the frequency in population i of one of a pair of alleles at a locus, p is the frequency of this allele among migrants (the average frequency in the species if migrants are drawn at random from all conspecific populations), and m is the proportion of the population that has immigrated. At equilibrium, $\Delta p_i = 0 = (p - p_i)$, so that a local population under constant migration pressure becomes indistinguishable from other conspecific populations.

If the allele represented by frequency p is selected either for or against such that it is reproduced in a ratio of $1 - s{:}1$ relative to the alternative allele at the locus, then the change in gene frequency may be written as

$$\Delta p_i = sp_i(1 - p_i) \tag{2}$$

provided s is sufficiently small. Combining these two linear pressures on the gene frequency of population i we obtain

$$\Delta p_i = sp_i(1 - p_i) + m(p - p_i) \tag{3}$$

which when set to 0 may be solved approximately by the quadratic formula (eliminating squared terms) to yield equilibria conditions (Wright 1931). For $s > m$, i.e. local selection on the gene is much stronger than migration pressure, $p_i = 1 - (m(1-p))/s$ or $-mp/s$, depending on the direction of selection. If on the other hand $m > s$, $p_i = p[1 + (s(1 - p_i))/m]$ and there will be little difference between population i and the rest of the species.

These deterministic equilibria, however, ignore one of the most important factors in the evolution of real populations, the stochastic fluctuation of gene frequencies in finite populations known as "random genetic drift." If we consider a population of N monoecious individuals in which a pair of alleles has frequencies of p and $1 - p$ and there is no selection, migration, or mutation, then the probability distribution of the numbers of each allele in a sample of N male and N female gametes is given by the binomial expansion

$$[p + (1 - p)]^{2N} \tag{4}$$

If a random variable x is now used to represent the first generation gene frequency in samples from a collection of such completely isolated, finite populations, the mean and variance of population gene frequencies will be given by

$$\begin{aligned} \bar{x} &= E(x) = p, \\ V_x &= E[(x - p)^2] = p(1 - p)/2N. \end{aligned} \tag{5}$$

If the same pattern of reproduction continues for t generations, it can be shown that

$$\begin{aligned} \bar{x}^t &= p, \\ V_x{}^t &= p(1 - p)\,[1 - (1 - 1/2N)^t] \end{aligned} \tag{6}$$

(see Crow and Kimura 1970). In the limit as t goes to infinity, $V_x = p(1-p)$ while $\bar{x}$ remains p; in other words, after a sufficiently long time, p populations will have become fixed for one allele and $(1-p)$ populations will have become fixed for the other allele. This process of random genetic drift is illustrated in Fig. 1 which shows gene frequency distributions through time for those populations that remain unfixed or polymorphic. It is clear that both population size and the number of generations that a population is isolated are very important determinants of gene frequencies in finite populations.

If a linear evolutionary pressure is now introduced into the above model of population subdivision, an equilibrium distribution of gene frequencies may be achieved such that a proportion of the populations remains polymorphic. For example, the equilibrium distribution of gene frequencies, $\phi(x)$, with constant migration among demes as described in the deterministic case above and with negligible mutation rates is given by (Wright 1931),

$$\phi(x) = \frac{\Gamma(4Nm)}{\Gamma[4Nmx_m]\,\Gamma[4Nm(1-x_m)]}\, x^{4Nmx_m - 1}(1-x)^{4Nm(1-x_m)-1} \tag{7}$$

where x_m is the gene frequency of immigrants. This distribution with initial gene frequency $p = 0.5$ is shown in Fig. 2 for various values of Nm. For migration rates greater than $m = 1/2N$ the curve is bell-shaped indicating the cohesive influence of even very low rates of migration among demes. Only when migration is less than one individual every other generation does the complete divergence of demes become appreciable.

In order to quantify the amount of gene frequency dispersion in a subdivided population, Wright (1943) introduced the measure

$$F_{ST} = V_x/\bar{x}(1-\bar{x}) \tag{8}$$

which gives the gene frequency variance over subpopulations, V_x, as a proportion of maximum variance possible for a given mean gene frequency, $\bar{x}$. We will consider this measure, or the equivalent measure, G_{ST}, of Nei (1973), in further discussion of population structure.

Rearranging equation (6), we see that for an infinite number of subpopulations completely isolated for t generations

$$F_{ST}{}^{t} = 1-(1-1/2N)^{t}. \tag{9}$$

In the limit as t goes to infinity, F_{ST} becomes 1.0. If migration occurs among subpopulations as described above, the standardized gene frequency variance at $t = \infty$ tends towards an approximate maximum value of

$$F_{ST}{}^{\infty} = 1/(4Nm+1). \tag{10}$$

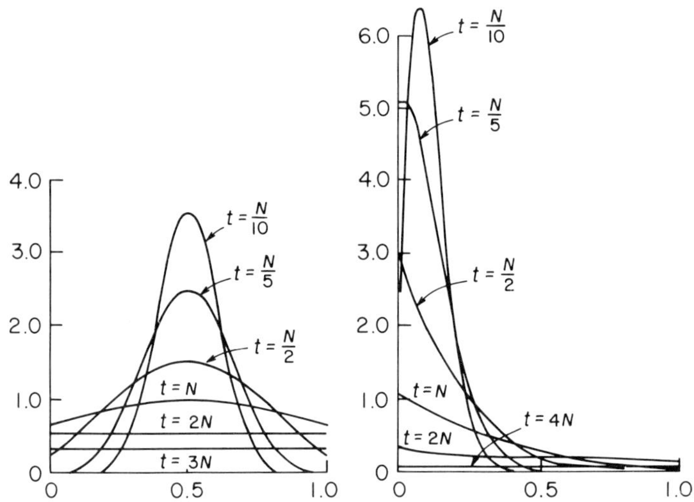

Figure 1. *Gene frequency distributions for subpopulations remaining polymorphic after* t *generations of random genetic drift. Initial frequency for curves on the left was* p = *0.5 and on the right* p = *0.1 (after Kimura 1955).*

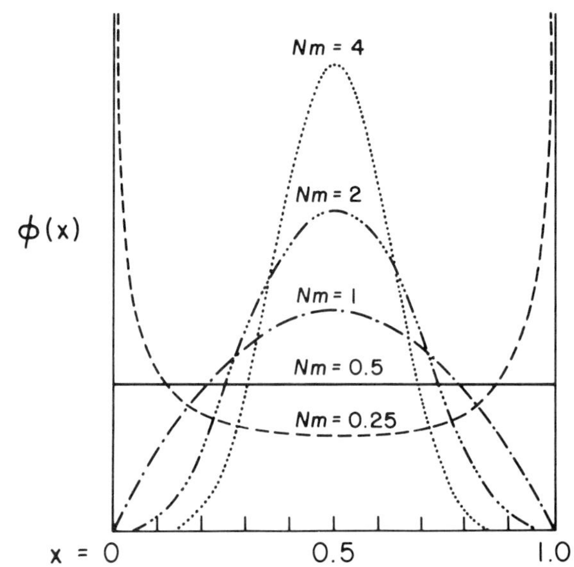

Figure 2. *Gene frequency distribution of polymorphic subpopulations at equilibrium for various values of* Nm *and initial gene frequency,* p = *0.5 (after Wright 1931).*

With this model of population structure there is no variance in F_{ST} by definition. Nei and Chakravarti (1977) and Nei et al. (1977) have pointed out, however, that when the number of subpopulations is finite rather than infinite, F_{ST} estimated from data on multiple loci may have a mean value quite different from that given by the above approximation and will have considerable sampling variance due to random genetic drift. Since the number of estuaries over which a species may range is likely to be limited, this source of sampling error becomes important. Figure 3 compares the mean and variance of F_{ST} over a small number of subpopulations with some expected values of this quantity based on Wright's model; the results give little hope that estimates of mean F_{ST} can be used to calculate the product *Nm*. Also note, again, that the number of generations since divergence is an important parameter.

The model of population structure that we have considered thus far was termed the "island model" by Wright (1943) because immigrants are drawn at random from conspecific populations. Several alternative models of population structures have been described including continuous populations in one or two dimensions (Wright 1943) and stepping-stone models in several

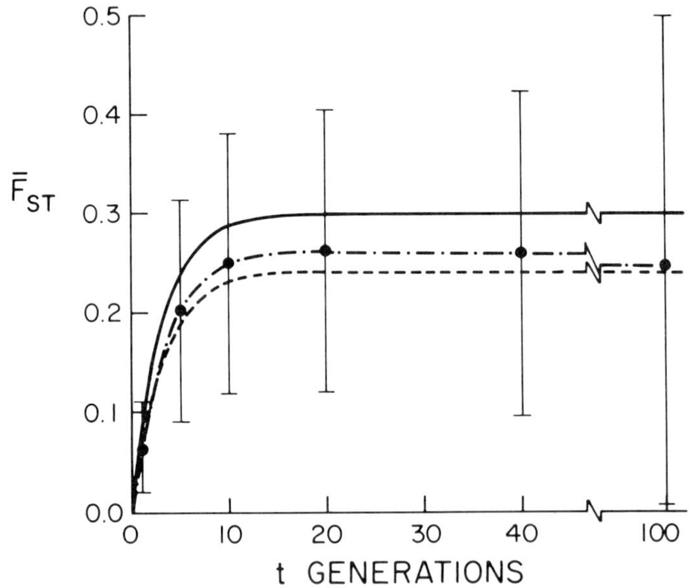

Figure 3. *Expected values of* F_{ST} *over time for an infinite number of subpopulations (solid line),* F_{ST} *values corrected for* $s = 5$ *subpopulations by the factor* $(s - 1/s)$ *(dashed line), and the theoretical expected values of* F_{ST} *for* $s = 5$ *(dashes and dots); data from Nei et al. 1977. Bars on the last curve show one standard deviation about the mean value. In all cases* $N = 5$, $m = 0.1$ *and* $p = 0.5$.

dimensions (Kimura and Weiss 1964; Weiss and Kimura 1965). The latter seem appropriate to estuarine organisms if the number of stepping-stone populations is made finite; actually the island model can be used in such cases if in equation (7) x_m is the actual gene frequency of immigrants and $m(1-r)$, where r is the correlation of gene frequencies among adjacent islands, is substituted for m. In general, the global amount of differentiation among populations is higher in these models since migration takes place only among adjacent demes whose gene frequencies tend to be correlated. This correlation not only affects the amount of differentiation measured by F_{ST} but also introduces spatial patterns in gene frequencies even in the absence of selection (Endler 1977).

We have said little about the effects of selection on gene frequency distributions. From the deterministic equations for linear genic selection (eqs 2 and 3) it is clear that selection can make gene frequencies either more homogeneous or more heterogeneous than expected under the operation of random genetic drift alone. A balance between opposing pressures of gene flow and selection may even appear to be an equilibrium state without selection, but with an Nm product quite different from the actual value. Many types of selection have been explored theoretically, but suffice it to say that selection necessarily complicates an already complex description of selectively neutral genes in finite, subdivided populations. Without *a priori* knowledge of the strength and direction of natural selection on a locus it is impossible to predict a spatial gene frequency distribution; to make a prediction simultaneously for several genes as might be detected in an electrophoretic study is certainly out of the question. Thus, a theory assuming no selection becomes, in almost all cases, the null hypothesis against which observations must be tested.

Methodological Problems in Applying the Theory

From this brief overview of population genetics theory we see that three parameters need to be specified in order to state the genetic consequences of larval retention: N, m, and t. Definition of these parameters and problems in estimating them in natural populations will be discussed. Due to a paucity of data, illustrative examples include some organisms that are not strictly estuarine.

The Population Number

The population size parameter, N, should really be designated N_e because we are talking about the concept of genetically effective population size. In the stable finite population discussed above with N monoecious reproductive individuals it was assumed that the average parent contributed two gametes to the next generation with binomial variance around this mean. N is the effective population number in this ideal case because random genetic drift is directly related to this number. Real populations, of course, contain juvenile and senescent individuals that do not affect the sampling of gametes in each generation; the first problem, then, is to identify and enumerate the number of reproducing adults (the breeding size). Even when this number is known there are several other factors that reduce breeding size to a smaller genetically effective size. Fluctuations in abundance over time reduce N_e to near the minimum number rather than the arithmetic average (because the cumulative sampling variance of gene frequencies over generations is a function of the harmonic mean of *N;* Wright 1939). For dioecious populations, deviations from 1:1 sex ratio similarly reduce N_e to twice the harmonic mean of the numbers of each sex. Inequalities of adult reproductive success – specifically, V_k, the variance in the number of gametes per adult actually contributed to the following generation – reduce N_e to $(4N-4)/(V_k+2)$ at equilibrium. Finally, any systematic deviations from random mating (inbreeding) will enhance the rate at which genetic diversity is lost from a finite population.

Of these factors, variation in breeding success may be the most significant for marine or estuarine organisms. The common life history features of high fecundity and high early mortality of pelagic larvae make conceivable, and at least numerically possible, a reproductive variance several orders of magnitude greater than the "ideal" binomial variance (Hedgecock et al. 1982). If so, the resulting reduction of N_e (among other possibilities) would explain why crustaceans in general maintain much less genetic variation than organisms of their abundance ought to maintain according to models of selectively neutral variation *(op. cit.).* Moreover, to the extent that variance in reproductive success affects sampling characteristics of the larval recruitment process, it is an alternative to natural selection as an explanation for some patterns of geographic variation.

In the Pacific North American barnacle *Balanus glandula,* for example, an electrophoretic survey shows high genetic similarity over much of the species range; indeed, differences among populations account for only 4% of the total genetic diversity (Hedgecock et al., unpublished). A month-long pelagic larval phase presumably ensures sufficient gene flow to prevent divergence on a macrogeographic scale, but statistical comparisons of gene frequency profiles reveal cases of significant differentiation on a microgeographic scale. If gene flow is so high, how do these differences arise? There are two possible explanations: (1) differential survival of individuals after settlement coupled with spatial heterogeneity in the environmental factors causing this selective mortality, and (2) spatial and temporal heterogeneity in the genetic makeup of that fraction of adult reproductive output that by chance survives to settlement. These alternative hypotheses may be tested by analyzing genetic diversity of barnacles recruiting to settling plates and by controlled transplantation of settling plates among different sites. What is germane to our discussion is that (1) differentiation of populations can occur despite very high rates of immigration (additional examples follow in the next section), and (2) population divergence can be due to selection (the most common interpretation) or to sampling artifacts of the reproductive and recruitment processes, or to both. In abundant species having strict larval retention and high V_k, divergence of populations in different estuaries might occur faster or to a greater extent than expected from large populations.

Migration and Gene Flow

In order to estimate rate of gene flow we must know not only the proportion of larvae recruited to an estuary from elsewhere but also the probability that this proportion remains the same until reproduction. Origin and gene frequency of immigrants must also be known. In the island model one migrant every other generation is sufficient to prevent substantial divergence of populations; in models where migration occurs between adjacent populations with correlated gene frequencies, m must be still lower for divergence of neighboring populations. Whatever the model, one must be prepared to estimate very low migration rates if the goal is to predict the consequences of larval retention or leakage. On the other hand, if m is to be estimated *a posteriori* from observed gene frequency distributions, N_e must be known with some precision because $\phi(x)$ depends on the product Nm (eq 7). Often investigators will assume that m is substantially above the level at which random differentiation is possible (as in the case of *Balanus glandula*) or that m is essentially zero because of physical, oceanographic, or biological barriers between populations. Selected examples illustrate these approaches.

Extensive studies have been made of the population structure of mussels, particularly *Mytilus edulis* (Koehn 1975; Koehn et al. 1976; Levinton and Koehn 1976). The most striking findings are widespread deviations from random mating proportions of genotypes in the direction of excess homozygotes, declines in these deviations and changes in gene frequency with age or size, and dramatic clines in allele frequencies both on macrogeographic and microgeographic scales. One main conclusion from these studies has been that migration – the pelagic larvae of *M. edulis* remain in the water column from 21 to 55 days and are thus broadly dispersed – has little effect on the genetic constitution of a local population (Koehn 1975). Rather, differential viability of genotypes as a function of spatial heterogeneity in important environmental parameters such as temperature, salinity, and wave shock is thought to be the major factor controlling divergence of populations. The aminopeptidase polymorphism in *M. edulis,* for example, shows a sharp cline between Atlantic Ocean and Long Island Sound localities that is stable in residents and fluctuates temporally and spatially in immigrants (Koehn et al. 1980). That this cline is maintained by differential viability of genotypes along a salinity-temperature gradient is supported by studies of the biochemical and physiological functions of this enzyme. Genetic differences among localities are thus maintained despite high rates of gene flow.

Although this is the view adopted by most students of mussel population genetics, alternative hypotheses have appeared in the literature to explain certain features of population structure. For example, Tracey et al. (1975) suggest that temporal and spatial variation in the spawning of local populations may give rise to deviations from expected random mating genotypic proportions in the mussel population as a whole. This hypothesis is similar to the recruitment sampling error explanation for differentiation of barnacle populations in that it proposes a non-selective, chance mechanism for population divergence. A tentative explanation for clines in gene frequencies in Long Island Sound is that mussels from the low salinity waters of the Sound and those from the adjacent open ocean are ecologically isolated by differential adaptations to salinity and temperature (Levinton and Lassen 1978). Whether or not these hypotheses will stand close scrutiny, they at least highlight the need for more complete understanding of population structure parameters.

Now consider a case in which gene flow among subpopulations is inferred to be very low or absent. Inhabiting coastal rivers of the Indian subcontinent, southeast Asia, Indonesia, the Philippines, New Guinea, and Australia, the freshwater prawn, *Macrobrachium rosenbergii,* has larvae that require brackish water but have low tolerance for undiluted seawater (Sandifer et al. 1975). Because of its aquacultural importance, efforts have been made to assess genetic variation over the vast range of this species (Malecha 1977). Electrophoretic (Hedgecock et al. 1979) and morphological studies (Lindenfelser 1980) have shown that it comprises at least two well-differentiated and widespread races. The two races are referred to as "east" and "west" types with respect to Wallace's Line, a biogeographic boundary apparently delimiting the Pleistocene continental margins of southeast Asia from oceanic land masses to the east (Mayr 1954). Across this line, more than 20% of the loci studied by electrophoresis show a complete divergence of alleles such that certain enzymes serve as diagnostic markers of race (Fig. 4). Larval intolerance for oceanic waters apparently does act to isolate island populations. This is further illustrated by the divergence of a Sri Lankan sample from the remaining western populations, and by the differentiation of Australian populations from other eastern populations. In both cases the divergence is complete at only a single locus, suggesting much more recent separation of populations possibly as the result of changes in Pleistocene sea levels. The Australian prawns do differ in color pattern and chelae shape from other eastern types (Malecha 1980). While patterns of differentiation are loosely consistent with the biogeography and geological history of the region (see below), the extent to which diversifying natural selection has distorted these patterns is unknown.

Classical studies on physiological variation among American oyster populations along the Atlantic and Gulf coasts (Stauber 1950; Loosanoff and Nomejko 1951) have long been cited as evidence for the genetic differentiation of these populations. In the case of spawning cycle differences that are maintained even after transplantation and acclimation to a new site, the

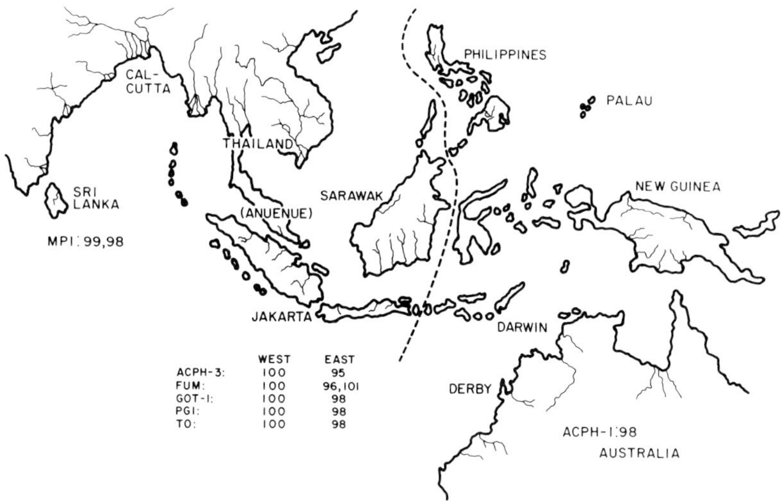

Figure 4. Sampling localities for Macrobrachium rosenbergii *and diagnostic allozyme differences between the major racial groups east and west of Wallace's Line (dotted line). MPI is polymorphic for alleles 99 and 100 in all western populations except Sri Lanka and fixed for allele 97 in all eastern populations. Divergence of Australian population at ACPH-1 is noted also. (after Hedgecock et al. 1979).*

evidence for adaptive divergence of reproductive characters is strong. Nevertheless, electrophoretic studies of oyster populations from throughout the broad range of this species have shown that genetic divergence of populations occurs on a much larger geographic scale than does the physiological differentiation. Gulf of Mexico and Gulf of Campeche populations appear to have diverged from a remarkably homogeneous Atlantic *C. virginica* population distributed from Florida to Canada (Buroker et al. 1979; W. W. Anderson, U. Georgia, Athens, GA, pers. comm.; L. J. Lester, U. Houston, Clear Lake, TX, pers. comm.; Hedgecock and Menzel, unpublished). How do we reconcile these different views of population structure?

Such conflict in information concerning geographic variation is clearly seen in the well-studied species of the *Drosophila willistoni* group. Chromosomal inversions are heterogeneously distributed, for example among Caribbean island populations, while allozyme frequencies are fairly homogeneous over the same populations (Ayala et al. 1971). Ayala and colleagues interpret the chromosomal data as evidence for low gene flow and the allozyme data as evidence for widespread balancing natural selection of enzyme polymorphisms. If, for the sake of discussion, we ignore other evidence that supports this interpretation, an alternative conclusion concerning population structure appears viable. Because inversion polymorphisms are known to be adaptive features under strong natural selection (Dobzhansky 1970), chromosomal differentiation could be due to diversifying selection that overcomes rates of gene flow sufficient to homogenize the frequencies of more weakly-selected, or even selectively-neutral, allozymes. Applying this lesson to oysters, does the existence of physiological races that are electrophoretically indistinguishable point to very low rates of gene flow with uniform selection of enzyme polymorphisms or to substantial rates of gene flow with strong diversifying selection of polygenic, physiological adaptations? Independent information on population structure parameters and a more thorough understanding of the genetic basis of the physiological traits are critical to the resolution of this conflict.

Evolutionary Time and Population History

The amount of time that estuarine populations may have been isolated and whether extant populations have reached demographic and structural equilibria pose even thornier problems than measuring N_e and m. Ideally, geological history can be reconstructed to give at least a rough idea of the number of generations that estuaries have been populated. The western North Atlantic coast, for example, is known to have been drastically affected by glaciation as recently as 11,000 years ago (CLIMAP 1976). Because a severe reduction in population size can have effects that last for thousands of generations (Nei et al. 1975; Chakraborty and Nei 1977), care must be taken in interpreting the gene frequency distributions presently observed in organisms from this coast. Gene frequency clines have commonly been observed along the Atlantic coast (Koehn et al. 1976; Corbin 1977; Powers and Place 1978); a striking number of these clines are step-shaped at the latitude of Cape Cod. Are these clines due strictly to ongoing differential adaptation of organisms across steep environmental gradients in this region, or might they represent residual effects of progressive recolonization of habitat formed as the ice sheet retreated? Sometimes the latter alternative can be ruled out as in the case of *Fundulus heteroclitus* in which north-south clines thought to be adaptive to temperature changes with latitude are consistently repeated in independent north-south drainage basins (D. Powers, John Hopkins U., Baltimore, MD, pers. comm.); such a pattern cannot be accounted for by random genetic drift in a recolonizing species.

Several electrophoretic studies of marine organisms may be cited for demonstrating a plausible concordance between geographic patterns of genetic variation and the presumed geological history of the species or group (Schopf and Murphy 1973; Vawter et al. 1980; cf. Lessios 1981). Genetic distance between east and west races of *Macrobrachium* is consistent not only with low migration across oceanic barriers but also with a separation of several million years. (Different calibrations of the "protein clock" yield, however, estimates of divergence times ranging from 3×10^5 to 8×10^6 years!) It is tempting to speculate that *Macrobrachium*'s colonization of Pacific oceanic land masses may have occurred when tectonic events brought New Guinea and Australia into close proximity with the Indonesian Archipelago about 10 million years ago (Lindenfelser 1980). Nevertheless, the error variance associated with estimating absolute divergence times from genetic distance is substantial. Independent fossil evidence for the ages of populations is essential for testing the hypothesis that history and genetic drift in finite populations are the prime causes of a particular pattern of geographic variation.

Conclusions

Since the mid-1960's the techniques of gel electrophoresis and specific enzyme assay have greatly facilitated observations of gene frequencies in natural populations. (These techniques and their applications to population genetics, systematics and evolution have been reviewed by several authors such as Avise 1974; Lewontin 1974; Ayala 1975; and Selander 1976. Although focused on allozyme data, the previous discussion is applicable to any form of genetic data such as polychromatism frequencies in isopods or copepods.) What can electrophoretic or other genetic studies of estuarine organisms tells us about larval retention?

If populations maintain completely different sets of alleles at a locus for several generations, then gene flow must be zero or nearly so, and retention mechanisms must be quite efficient. Among conspecific populations, however, one generally observes nearly the same sets of alleles at similar frequencies over broad geographic areas (Ayala 1975); while gene frequencies may vary significantly from site to site, the standardized measure of geographic variation, F_{ST}, is usually 10% or less of its maximum value. This poses the problem of interpreting F_{ST} as a function of effective population sizes of demes, the rates of gene flow among them, and the evolutionary history of the extant population structure, assuming for modelling simplicity that selection is negligible. With only one data set and three parameters that are unknown or little better than educated guesses, interpretation of the situation is likely to be speculative. Even if we can assume that time has been sufficient for the population to have reached equilibrium, the sampling variance of F_{ST} makes any estimate of genetic divergence among populations, particularly one derived from data on only a few genes, of limited value in calculating the product Nm. Of course, additional information on the spatial patterning of gene frequencies over several estuaries might also be available, but unequivocal interpretation of spatial patterns requires knowledge of the same ecological parameters.

To capsulize these interpretive problems:

- similarities among populations need not imply high rates of larval leakage;
- differences among populations need not imply strict larval retention mechanisms or diversifying selection overriding the effects of larval leakage;

- spatial patterns such as clines need not imply selection along an environmental gradient.

The conclusion is inescapable that genetic information alone is insufficient to decide upon the magnitude of larval retention. On the other hand, genetics is necessary in the study of larval retention because population genetic structure is what ultimately must be explained when enough is known about the parameters of population history, size, and gene flow. Electrophoretic studies, for example, have revealed a wealth of information about the genetics of natural populations. While unequivocable interpretation of these data is rarely possible, these studies often do allow us to set at least rough limits on unknown parameters and to focus our attention on weaknesses in our knowledge of population biology.

Acknowledgments

Figure 1 is reprinted with permission of M. Kimura; Figure 2 with permission of S. Wright and *Genetics;* Figure 4 with permission of *World Mariculture Society*.

References Cited

Avise, J. C. 1974. Systematic value of electrophoretic data. *Syst. Zool.* 23:465-481.

Ayala, F. J. 1975. Genetic differentiation during the speciation process, pp. 1-78. *In:* T. Dobzhansky, M. K. Hecht and W. C. Steere (eds.), *Evolutionary Biology, Vol. 8.* Plenum Press, New York.

Ayala, F. J., J. R. Powell and T. Dobzhansky. 1971. Polymorphisms in continental and island populations of *Drosophila willistoni*. *Proc. Nat. Acad. Sci. USA* 68:2480-2483.

Buroker, N. E., W. K. Hershberger and K. K. Chew. 1979. Population genetics of the family Ostreidae. II. Interspecific studies of the genera *Crassostrea* and *Saccostrea*. *Mar. Biol.* 54:171-184.

Chakraborty, R. and M. Nei. 1977. Bottleneck effects on average heterozygosity and genetic distance with the stepwise mutation model. *Evolution* 31:347-356.

CLIMAP Project Members. 1976. The surface of the ice-age earth. *Science* 191:1131-1137.

Corbin, K. W. 1977. Phosphoglucose isomerase polymorphism and natural selection in the sand crab, *Emerita talpoida*. *Evolution* 31:331-340.

Dobzhansky, T. 1970. *Genetics of the Evolutionary Process.* Columbia University Press, New York. 505 pp.

Endler, J. A. 1977. *Geographic Variation, Speciation, and Clines.* Princeton University Press, Princeton. 246 pp.

Fisher, R. A. 1922. On the dominance ratio. *Proc. Roy. Soc. Edinb.* 42:321-341.

Fisher, R. A. 1930. *The Genetical Theory of Natural Selection.* Clarendon Press, Oxford. 272 pp.

Hedgecock, D., D. J. Stelmach, K. Nelson, M. E. Lindenfelser and S. R. Malecha. 1979. Genetic divergence and biogeography of natural populations of *Macrobrachium rosenbergii*. *Proc. World Maricul. Soc.* 10:873-879.

Hedgecock, D., M. L. Tracey and K. Nelson. 1982. Genetics, *In:* L. G. Abele (ed.), *Biology of the Crustacea, Vol. 2.* Academic Press, New York.

Kimura, M. 1955. Solution of a process of random genetic drift with a continuous model. *Proc. Nat. Acad. Sci. USA* 41:144-150.

Kimura, M. and G. H. Weiss. 1964. The stepping stone model of population structure and the decrease of genetic correlation with distance. *Genetics* 49:561-576.

Koehn, R. K. 1975. Migration and population structure in the pelagically dispersing marine invertebrate, *Mytilus edulis*, pp. 945-959. *In:* C. L. Markert (ed.), *Isozymes. IV Genetics and Evolution.* Academic Press, New York.

Koehn, R. K., R. Milkman and J. B. Mitton. 1976. Population genetics of marine pelecypods. IV. Selection, migration and genetic differentiation in the blue mussel, *Mytilus edulis*. *Evolution* 30:2-32.

Koehn, R. K., R. I. E. Newell and F. Immermann. 1980. Maintenance of an aminopeptidase allele frequency cline by natural selection. *Proc. Nat. Acad. Sci. USA* 77:5385-5389.

Lessios, H. A. 1981. Divergence in allopatry: Molecular and morphological differentiation between sea urchins separated by the Isthmus of Panama. *Evolution* 35:618-634.

Levinton, J. S. and R. K. Koehn. 1976. Population genetics, pp. 357-384. *In:* B. L. Bayne (ed.), *Marine Mussels: Their Ecology and Physiology*. Cambridge University Press, Cambridge.

Levinton, J. S. and H. H. Lassen. 1978. Experimental mortality studies and adaptation at the *Lap* locus in *Mytilus edulis*, pp. 229-254. *In:* B. Battaglia and J. A. Beardmore (eds.), *Marine Organisms*. Plenum Press, New York.

Lewontin, R. C. 1974. *The Genetic Basis of Evolutionary Change*. Columbia University Press, New York. 346 pp.

Lindenfelser, M. E. 1980. Subspecific variation and evolution in the freshwater prawn *Macrobrachium rosenbergii* (de Man) (Decapoda: Palaemonidae) as reflected in morphometric and allozyme variation. Ph.D. Dissertation. U. Hawaii, Honolulu, HI 232 pp.

Loosanoff, V. L. and C. A. Nomejko. 1951. Existence of physiologically-different races of oysters, *Crassostrea virginica*. *Biol. Bull.* 101:151-156.

Malecha, S. R. 1977. Genetics and selective breeding of *M. rosenbergii*, pp. 328-355. *In:* J. A. Hanson and H. L. Goodwin (eds.), *Shrimp and Prawn Farming in the Western Hemisphere*. Dowden, Hutchinson and Ross, Inc., Stroudsburg, PA

Malecha, S. R. 1980. Development and general characterization of genetic stocks of *Macrobrachium rosenbergii* and their hybrids for domestication. *Sea Grant Quarterly*, Vol. 2, No. 4. U. Hawaii Sea Grant College Program, Honolulu, HI 6 pp.

Mayr, E. 1954. Wallace's Line in the light of recent zoogeographic studies. *Quart. Rev. Biol.* 29:1-14.

Nei, M. 1973. Analysis of gene diversity in subdivided populations. *Proc. Nat. Acad. Sci. USA* 70:3321-3323.

Nei, M. and A. Chakravarti. 1977. Drift variances of F_{ST} and G_{ST} statistics obtained from a finite number of isolated populations. *Theor. Pop. Biol.* 11:307-325.

Nei, M., A. Chakravarti and Y. Tateno. 1977. Mean and variance of F_{ST} in a finite number of incompletely isolated populations. *Theor. Pop. Biol.* 11:291-306.

Nei, M., T. Maruyama and R. Chakraborty. 1975. The bottleneck effect and genetic variability in populations. *Evolution* 29:1-10.

Powers, D. A. and A. R. Place. 1978. Biochemical genetics of *Fundulus heteroclitus* (L.). I. Temporal and spatial variation in gene frequencies of *Ldh-B*, *Mdh-A*, *Gpi-B*, and *Pgm-A*. *Biochemical Genetics* 16:593-607.

Sandifer, P. A., J. S. Hopkins and T. I. J. Smith. 1975. Observations on salinity tolerance and osmoregulation in laboratory-reared *Macrobrachium rosenbergii* postlarvae (Crustacea: Caridea). *Aquaculture* 6:103-114.

Scheltema, R. S. 1975. Relationship of larval dispersal, gene-flow and natural selection to geographic variation of benthic invertebrates in estuaries and along coastal regions, pp. 372-391. *In:* L. E. Cronin (ed.), *Estuarine Research, Vol. 1*, Academic Press, New York.

Schopf, T. J. M. and L. S. Murphy. 1973. Protein polymorphism of the hybridizing seastars *Asterias forbesi* and *Asterias vulgaris* and implications for their evolution. *Biol. Bull.* 145:589-597.

Selander, R. K. 1976. Genic variation in natural populations, pp. 21-45. *In:* F. J. Ayala (ed.), *Molecular Evolution.* Sinauer Assoc., Sunderland, MA

Stauber, L. A. 1950. The problem of physiological species with special reference to oysters and oyster drills. *Ecology* 31:109-118.

Tracey, M. L., N. F. Bellet and C. D. Gravem. 1975. Excess allozyme homozygosity and breeding population structure in the mussel *Mytilus californianus. Mar. Biol.* 32:303-311.

Vawter, A. T., R. Rosenblatt and G. C. Gorman. 1980. Genetic divergence among fishes of the eastern Pacific and the Caribbean: Support for the molecular clock. *Evolution* 34:705-711.

Weiss, G. H. and M. Kimura. 1965. A mathematical analysis of the stepping stone model of genetic correlation. *J. Appl. Prob.* 2:129-149.

Wright, S. 1931. Evolution in Mendelian populations. *Genetics* 16:97-159.

Wright, S. 1939. Statistical genetics in relation to evolution. *Act. Scient. et Indus.* 802:1-63.

Wright, S. 1942. Statistical genetics and evolution. *Bull. Amer. Math Soc.* 48:223-246.

Wright, S. 1943. Isolation by distance. *Genetics* 28:114-138.

Wright, S. 1949. Adaptation and selection, pp. 365-389. *In:* G. L. Jepson, G. G. Simpson and E. Mayr (eds.), *Genetics, Paleontology and Evolution.* Princeton U. Press, Princeton, NJ

ESTUARINE SEDIMENT DYNAMICS AND SEDIMENTATION CONTROL

Convenor: *W. H. McAnally, Jr.*
Waterways Experiment Station
Vicksburg, Mississippi

INTRODUCTION TO SESSION ON ESTUARINE SEDIMENT DYNAMICS AND SEDIMENTATION CONTROL

William H. McAnally, Jr.
USAE Waterways Experiment Station
Vicksburg, Mississippi

These papers address the collision of the inevitable infilling of estuaries with sediment and the irrepressible human determination to conduct navigation in estuaries. The product of this collision is often massive maintenance dredging programs. Recently, the monetary and environmental costs of estuary channel maintenance have led to a renewed search for alternatives to standard dredging methods and for ways of making unavoidable dredging more effective. At about the same time, renewed activity in sediment dynamics research has revealed potential new directions in sedimentation control. The following papers examine both of these phenomena.

Dr. Parker's paper describes detailed field observations of how some estuarine sediment beds form from dense suspensions. This work and subsequent laboratory experiments have revealed much about the mechanics of deposition and have lead to highly successful sediment management techniques. The paper by Dr. Mehta and his colleagues shows how resuspension of deposited sediments is much more complex than previously thought, being dependent on both the deposition process and the stress history of the bed.

The final three papers constitute an excellent review of practical solutions to sedimentation problems. Mr. Luo discusses in detail some of China's efforts to optimize maintenance programs by skillful use of natural forces including agitation dredging. Dr. Hoffman's paper shows the unique sedimentation problems faced by the US Navy and several nonconventional practices that have been applied to solve these problems. Similarities with the Chinese experiences are obvious. Finally, Mr. Mohr gives us an evaluation of standard dredging practice, revealing its strengths, its limitations, and what is needed to make it more effective.

By examining both sediment dynamics and sedimentation control, we hope to accomplish two goals. First, we wish to accelerate the transfer of knowledge from researchers to practitioners. We are witnessing enormous increases in understanding of the basic processes and the time is right for using this developing knowledge to improve sedimentation control practices. Second, we wish to create greater interest in sediment dynamics research so that the renewal of work in this field will not falter, but grow.

ISBN 0-12-404070-5

TIME DEPENDENT PROPERTIES OF COHESIVE SEDIMENT RELEVANT TO SEDIMENTATION MANAGEMENT – EUROPEAN EXPERIENCE

W. R. Parker and R. Kirby
Institute of Oceanographic Sciences
Crossway, Taunton, Somerset
United Kingdom

Abstract: Field observations show fine sediment suspensions often have a mobile, high-concentration layer adjacent to the sea bed. Stationary suspensions which have time-varying acoustic and mechanical properties develop from these layers. In these circumstances, conventional survey echosounders can not provide unambiguous information on sea bed altitude relevant to safe navigation and optimum dredging practices. In Europe, echosounding is being replaced by *in situ* density measurement as the means to define navigable depth and to monitor or control dredging. Repeated *in situ* density surveys allow time series of mass deposition to be constructed, dredging schedules to be optimised, and productivity to be increased. Techniques may then be developed to trap sediment in designated sites where stationary pumping systems remove spoil automatically at an optimum density.

Introduction

Sediment management is needed to reduce the economic, environmental or social consequences of uncontrolled sedimentation, or scour, to acceptable levels. In practice this often means learning to live with the problem.

A dredging requirement arises when sediment accumulates in a navigation channel to a level which obstructs safe navigation. The requirement is detected and quantified by hydrographic surveys using leadlines or echosounders. These data are used to specify dredging need or achievement, define long term sedimentation trends, or calibrate numerical models predicting sedimentation.

In muddy areas the sea bed often consists of a discontinuous transition from muddy water, through watery mud, to firm mud (Delft Hydraulics Laboratory 1962; Migniot 1968; Kirby and Parker 1977). Echosounder records from these areas are often difficult to interpret due to the presence of more than one reflector at the sea bed (Fig. 1). Practical problems arise when dredging returns do not equate with the echosounder measurements of depth gained and further difficulties ensue when comparing the results of echosounder surveys and leadline surveys in an attempt to establish the dredging achieved.

ISBN 0-12-404070-5

The objective of maintaining navigable depth is based on measurements of depth. The inter-relationships between the actual criteria on which depth is generally defined by leadline, echosounder, or numerical model are arbitrary. Leadline results refer to the combined buoyancy and sediment shear strength which prevent further penetration of the lead; echosounder data relate to the acoustic reflection coefficient of the bed, or interfaces within it, and the levels predicted from model studies generally relate to the erodibility of the substrate as influenced by the rate of deposition and consolidation. However, all these data are directed towards navigation and for this purpose the sea bed is more realistically defined in terms of the passage of a ship's hull. Within the zone between the reflectors on the echosounder records in Fig. 1 occur (i) the level which is resistant to a ship, and therefore of principal interest to the surveyor and dredging superintendent and (ii) the level which is resistant to erosion and would constitute the bed in a time-averaged model. Each of these levels is parameterised by different properties and it is highly unlikely that they occur at the same altitude. Thus, where the sea bed is depends on how it is defined, and this is determined by the purpose for which it is defined. It is, therefore, more productive in sediment management terms to ask "what is the sea bed?" rather than "where is the sea bed?" so that when properly defined, the altitude of the sea bed can be correctly determined.

This paper deals with one approach that has been developed for living with the problems produced by rapid fine sediment deposition leading to the formation of fluid mud or fluff (a high-concentration suspension of fine sediment) and the associated mud maintenance dredging.

To trace the development of a practical solution to this problem the sediment properties and processes associated with defining navigable depth in mud areas will be outlined on the basis of a simple conceptual model of fine sediment behaviour. The consequences of this behaviour for various aspects of sediment management will be discussed, a solution identified, and its application in Rotterdam-Europoort described where a new definition of the sea bed has been devised and the technology and methods to measure it developed.

Laboratory Study of the Properties of Fine Sediment Suspensions

Fine sediment suspensions in estuaries consist of aggregates of smaller particles. The size and density of the aggregates depends upon local particle size and a number concentration, local physicochemical and microhydrodynamic conditions, and microbiological activity in the water or at the particle surface (Ives 1978; Yariv and Cross 1979).

The resistance of a mud bed to erosion, or the deposition of mud from suspension, is in part controlled by the response of the mud to shear. This may be evaluated experimentally by studying its rheological behaviour (Fig. 2).

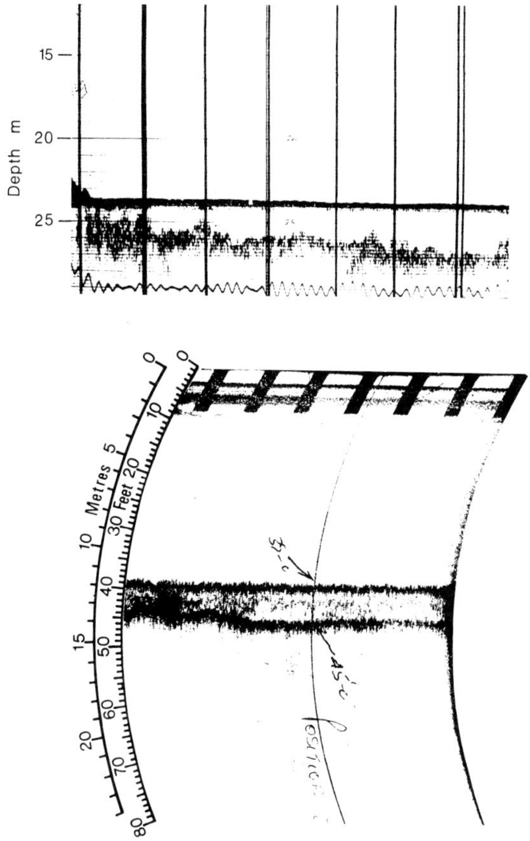

Figure 1. Records from 200 kHz (upper) and 30 kHz (lower) survey echosounders showing typical records from fluff areas. Prudence dictates that the surveyor chooses the top reflector as the sea bed but navigable material may extend for some distance below. This results in unnecessary dredging.

The rheological behaviour of mud suspensions ranges from Newtonian, at low solids volume concentration, to pseudoplastic or dilatant at higher solids content (Yariv and Cross 1979). With increasing solids content a transition from pseudoplastic, through a 'pseudoplastic with yield' to a Bingham plastic behaviour may be observed, although Street (1958) suggests that there is no experimental justification for a yield value with clay suspensions. It is well known (eg. Street 1958) that with these materials the increase in shear stress (τ) per unit area, required to produce a unit increase in the rate of shear depends upon the rate of shear ($\dot{\gamma}$). Thus, experimental measurement of the rheological behaviour requires that both the shear stress and shear rate are accurately determined in apparatus for which a hydrodynamic analysis shows that the sample is subjected to a uniform shear field (eg. Mooney and Ewart

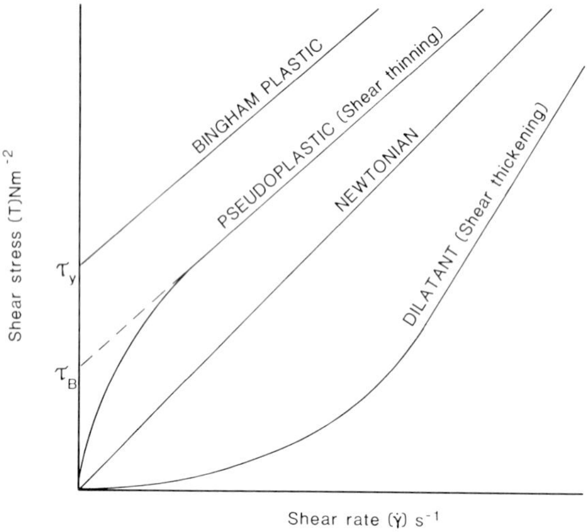

Figure 2. Idealised equilibrium shear stress/shear rate curves for typical materials.

1934). In Fig. 2, the Bingham plastic yield stress (τ_y) or, in the case of the pseudoplastic, the upper Bingham yield stress (τ_B), are generally extrapolated back to zero shear rate from measurements at higher shear rates where the shear stress/shear rate curve is linear.

Although many measurements claim to describe a Bingham plastic behaviour they are often based on measurements in which the sample is subjected to a wide range of shear rate in what is nominally measurement at a single shear rate (eg. Krone 1963; Migniot 1968; Owen 1970; Gularte 1979). Data showing pseudoplastic behaviour is often interpreted as Bingham plastic (eg. Krone 1963; Gularte 1979) and measurements are made on remoulded materials with water content more characteristic of a settled and partly consolidated bed (eg. Gularte 1979).

The principle problem with using non-analytical apparatus is that data can only be presented as a stress (or torque) and some measure of rotation speed. Thus, even if the apparatus can operate at sufficiently low shear rates to minimise extrapolation, the actual shear rate may be unknown. Following Street (1958) it is therefore of little value to quote a shear stress if its dependence on shear rate is unquantified. Although there are very few examples where available hydrodynamic analyses of the apparatus are applied (cf. Bryant et al. 1980), when measurements are made at the low shear rates typical of the benthic boundary layer ($\dot{\gamma} < 10\ sec^{-1}$) a gross non-linearity in the shear stress/shear rate curve is observed. This is consistent with the aggregated character of the mud suspensions (Flegmann et al. 1969) and is observed on materials of widely differing origins and compositions (Bryant et al. 1980).

The detailed behaviour for any particular sediment depends upon its particle size versus particle number distribution and the distribution of mineral species with size. At the solids contents (Ø) typical of moving suspensions or recently deposited beds (Ø <0.15) the behavioural characteristics of particular relevance to sediment management are: (i) the generally low shear stress extrapolated at zero shear (Fig. 3), indicating very low shear resistance. (ii) the measurably non-linear shear stress versus shear rate curves at the shear rates ($\dot{\gamma}$ <10 sec^{-1}), typical of the benthic boundary layer (Fig 3). These two characteristics mean that in natural circumstances these materials can sustain very little shear and will flow, even at high concentration. This contrasts strongly with the Bingham model generally applied to cohesive sediment (eg. Gularte et al. 1979).

Field Behaviour of Fine Sediment Suspensions

As has been reported by many authors (Meade 1972; Allen et al. 1974; Kirby and Parker 1977), the fine sediment of an area passes through repeated cycles of erosion, transport and deposition on tidal (eg. semi-diurnal) and lunar (spring-neap-spring) timescales, superimposed on the seasonal variations due to river discharge, wind stress, etc. As a result of this recycling, fine sediment may exist as either mobile suspensions, stationary suspensions, or settled mud, which can individually be clearly identified (Fig. 4). From the sediment management point of view, the most important features of this behavioural pattern may be summarized as follows: (i) Mobile suspensions tend to have a layered concentration field, characteristically with a high concentration layer adjacent to the bed (Kirby and Parker 1977, 1981). (ii) These high concentration layers have a non-Newtonian flow behaviour (Williams and James 1978; Bryant et al. 1980) with low resistance to movement but high resistance to mixing (Kendrick 1980; Parker et al. 1980). (iii) The mobile high concentration near-bed layers evolve to form stationary suspensions (Kirby and Parker 1977) which have time dependent structural and acoustic properties (Parker and Kirby 1977; Kirby et al. 1980).

This behaviour has many far reaching consequences for sediment control, as will be discussed subsequently, but since the acoustic measurement of depth is central to estimating sedimentation, it is appropriate that the formation and evolution of stationary suspensions are examined in some detail.

Formation of Stationary Suspensions

Sampling reveals that the material detected in Fig. 1, often called fluid mud, fluff, slib, or creme de vase, is a high concentration suspension of fine sediment. These fluid muds may be regarded as grossly under-consolidated soil and they will develop wherever the vertical mass flux of sediment exceeds that which self-weight consolidation can maintain as a normally consolidated bed. Mobile high concentration layers can easily provide such a condition and it is probable that the mobile high concentration layer adjacent to the bed is

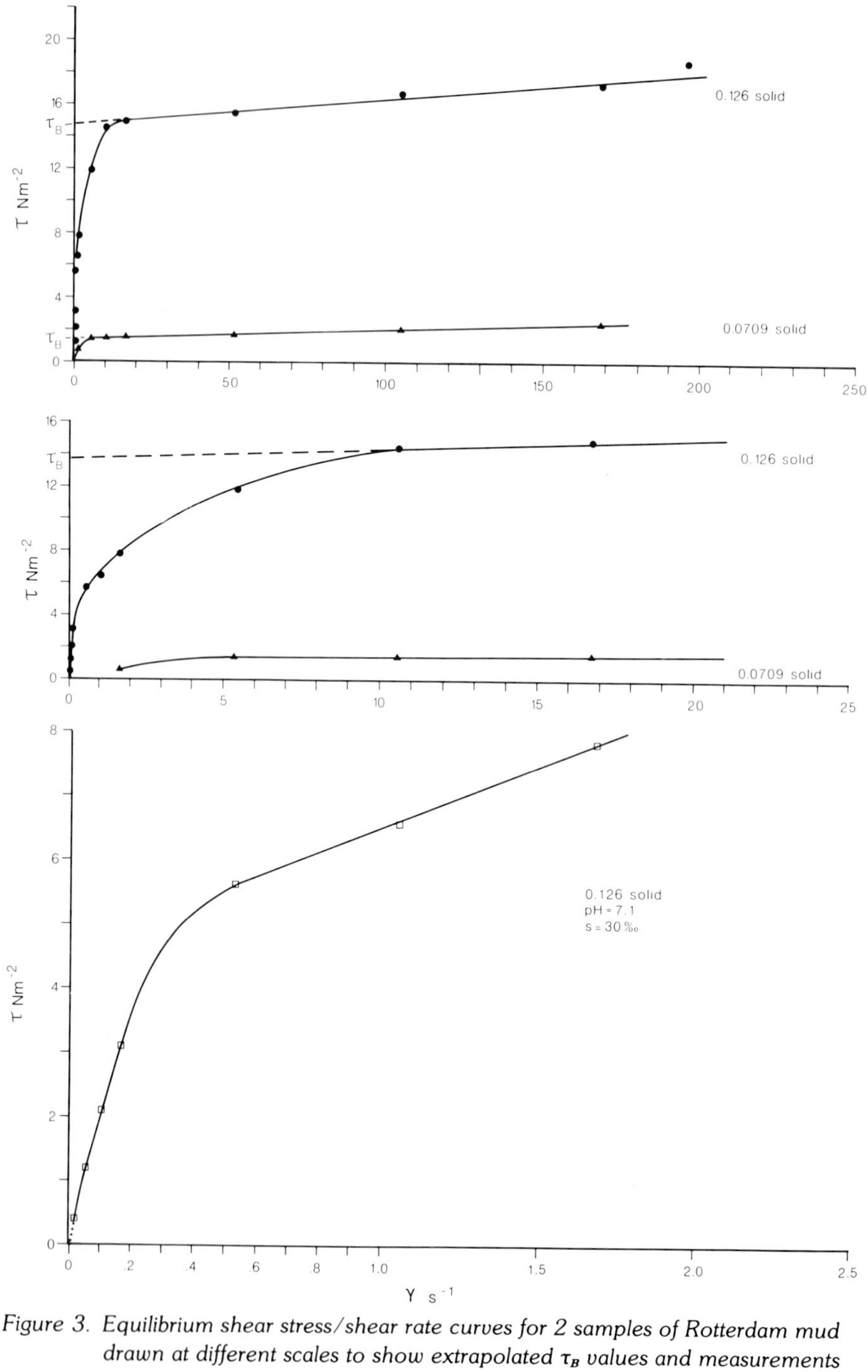

Figure 3. Equilibrium shear stress/shear rate curves for 2 samples of Rotterdam mud drawn at different scales to show extrapolated τ_B values and measurements at low shear rates. Note that at high concentration, shear rates <0.1 sec^{-1} and shear stress 0.4 Nm^{-2} no evidence of a yield point is observed (cf. Krone 1963; Migniot 1968; Gularte 1979). (Data from Williams and James 1978).

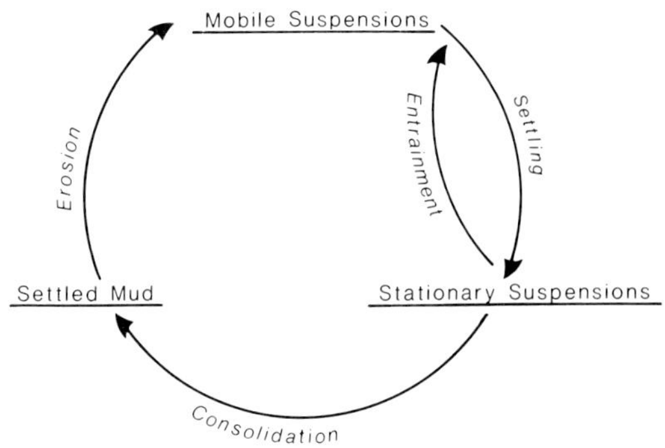

Figure 4. Identifiable states of occurrence of fine sediment and linking processes.

the precursor of the stationary suspension which forms when the layer stops moving along (Kirby and Parker 1977).

Consolidation of Stationary Suspensions

The general structural evolution of a single layer stationary suspension is shown schematically in Fig. 5. Whilst the surface of the suspension settles, zones of higher density build up from the bed until they intercept the falling suspension surface. Owen (1970) and McRoberts and Nixon (1976) have used Kynch's Theory (Kynch 1952) to try to predict the internal structure of these suspensions. However, this theory relates to incompressible particles in 'dilute' dispersions and makes no allowance for compression of the particles which leads to increases of density. Parker and Kirby (1977) interpreted data of Fuerstenau (Michaels and Bolger 1962) to develop a qualitative picture of early consolidation of a stationary suspension and Been (1980, 1982) and Been and Sills (1981) have made comprehensive, fully instrumented studies of the self-weight consolidation of natural muds under laboratory conditions. From these experiments, and also from field observations (Kirby and Parker 1974; British Transport Docks Board 1981) it may be shown that the muddy substrates developing from stationary suspensions have a multi-layered time dependent structure. The evolution of this structure and the attendant changes in parameters of interest (shear strength, acoustic properties, density) depend upon the state of consolidation which, for a given soil, depends mostly on the mass flux onto the bed.

Lee and Sills (1981) show that for deposition of a single layer the state of consolidation improves with time, but for steady sedimentation the state of consolidation becomes worse with time. The state of consolidation of the bed remains constant only when the mass flux is proportional to $t^{1/2}$ (t = time), not a generally anticipated natural short term situation.

Evolution of Acoustic Properties

The evolution of the physical properties of stationary suspensions is ac-

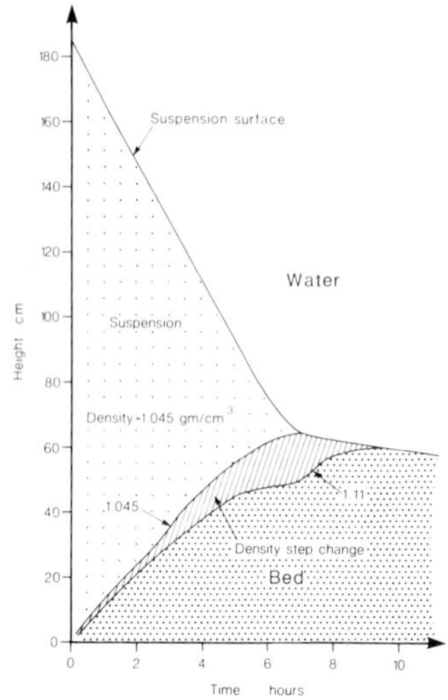

Figure 5. Height/Time/Density plot for consolidation of laboratory suspension of natural mud in tapwater. Initial density was 1.045 gm cm^{-3}. (Redrawn from Been 1980).

companied by changes in their acoustic properties. The properties of principal concern to hydrographic surveys are sound speed, which governs the apparent depth and thickness of the layer, and the target strength of the layer which determines its detectability.

A Hookian model leads to an expression for the propagation speed (c) of a compressional wave as

$$c = \left\{ \frac{(K + 4/3\ G)}{\varrho} \right\}^{1/2}$$

where K is the bulk modulus, G is the shear modulus, and ϱ is the density. For suspensions with zero shear modulus this can be recast in terms of the compressibility and density of the fluid and mineral phases (Wood 1941):

$$c = \left(\frac{1}{\{\varnothing_m \beta_m + (1 - \varnothing_m)\ \beta_w\}\ \{\varnothing_m \varrho_m + (1 - \varnothing_m)\ \varrho_w\}} \right)^{1/2}$$

where β = compressibility, ϱ = density, and subscripts *m* and *w* refer to mineral and water respectively. With increasing solids content (Ø), sound speed decreases by 3% to 4% of the water value at about 20% of volume solids content after which it increases again (Hampton 1967; Akal 1972). As sound speed varies only a little, it is the change in the density of the suspension which produces the significant change in acoustic impedance, $Z(= \varrho c)$, of the suspension as it consolidates (Parker and Kirby 1977).

The detectability of the surface of stationary suspensions depends upon the relationship between the sensitivity of the echosounder and the acoustic energy returned by scattering and reflection. Suspension clouds have been studied by Proni et al. (1976) and Orr and Hess (1978) using back scattering systems operating in the range 20-200 kHz.

Although a scattering model might be more appropriate when considering detection of a suspension, Tsuchiya et al. (1973, 1977) have successfully used a Rayleigh plane wave model when observing the target strengths of natural high concentration suspensions.

The Rayleigh reflection coefficient (R), may, for plane wave normal incidence onto a flat surface, be represented by the Rayleigh equation (Rayleigh 1894):

$$R = \frac{\varrho_2 c_2 - \varrho_1 c_1}{\varrho_2 c_2 + \varrho_1 c_1}$$

Subscript 1 refers to an upper layer (eg. water above the suspension) and subscript 2 refers to a lower layer (eg. the suspension). For a plane 'step' interface the target strength (TS) may be calculated from

$$TS = 20 \log_{10} R$$

The density change (and hence acoustic impedance change) at the surface of the stationary suspension is least when it is forming (Fig. 5). Thus some modified reflection coefficient, depending upon the relationship between wavelength of the incident radiation and impedance gradient, would be more appropriate (Gupta 1965, 1966). An evolving density structure similar to Fig. 5 has been used to calculate profiles of target strength (Fig. 6) using a simple plane wave model. The reflection coefficient of the sediment-water interface is strongly affected by porosity (Faas 1969; Hamilton 1970; Akal 1972) so that as the suspension consolidates the reflection coefficient at the surface changes (Parker and Kirby 1977), as do those related to the density steps within. Thus the echosounder may detect different parts of the consolidating layer depending on the target strength taken as the lower limit for detection and the state of consolidation. As is observed in the field, reflectors may appear at the surface of the suspension, as well as within it, due solely to consolidation. The frequency dependence of suspension detection has been discussed by Tsuchiya et al. (1973, 1977) and Parker and Kirby (1977). Higher frequency systems detect much smaller density changes and thus less well consolidated suspensions. It has also been demonstrated that similar frequency systems give differing results which may be due to some unique character of the equipment and its relation to the state of consolidation of the suspension (Fig. 7).

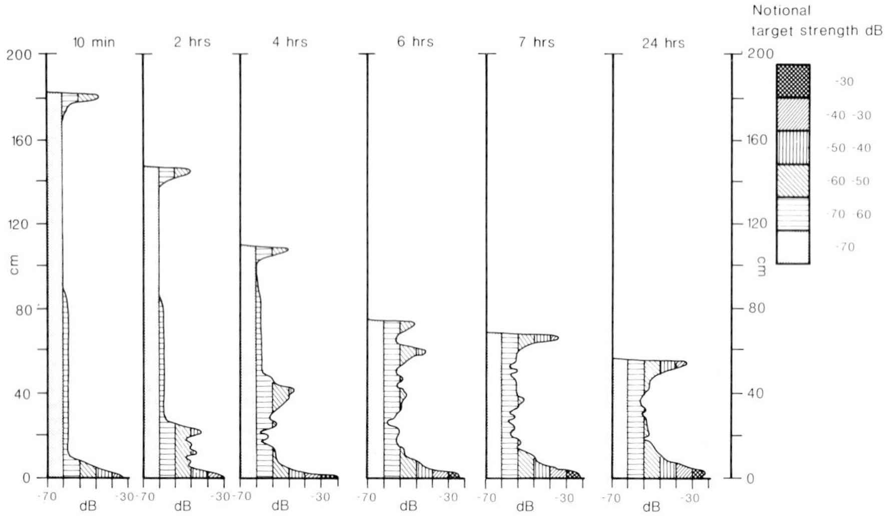

Figure 6. *Time series of profiles of notional target strength through a consolidating suspension. Surface values are steady up to 6 hours and jump by 7 hours as the 'bed' reaches the surface (cf. Fig. 5). Note stronger targets at the top of the bed rising upwards through the consolidating column. (Based on laboratory data of Been [1980] in Fig. 5).*

Thus, reflectors on the echosounder record do not relate to a specific value of density but to a critical density, and hence impedance, gradient as shown in Fig. 8. As the consolidation proceeds the reflectors from lower frequency signals show no consistent relationship to the density contours. This has also been demonstrated in field investigations (Parker and Kirby 1977; Kirby et al. 1980).

Effect of Gas Microbubbles

Although suspended solids will only produce a sound speed deviation of up to 4% below the clear water velocity, the places where high concentration suspensions occur and stationary suspensions develop are also commonly areas of high bacterial activity. In such areas, gas microbubbles, some of bacterial origin, may be expected (Schubel and Schiemer 1973; Whelan et al. 1978; Siddiquie et al. 1981).

Anderson and Hampton (1980) have shown that sound speed is strongly influenced by the ratio of the echosounder frequency to the bubble resonance frequency and the percentage of pore space filled with gas. The resonance frequency of a bubble is inversely proportional to its size and the range of most survey echosounders (20 to 200 kHz) corresponds to a range of bubble diameters between ≅ 0.03 and 0.003 cm. Bubbles observed in sediment range from sub-millimetre to the order of 0.5 cm (Siddiquie et al. 1981) and may thus affect a wide range of acoustic equipment.

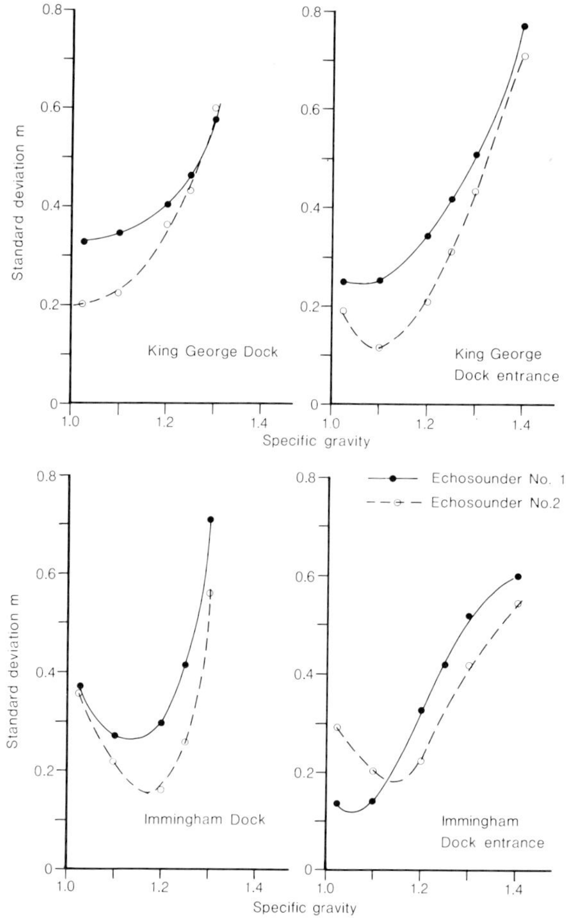

Figure 7. Standard deviation of the distance between the first reflector and specific density values for two nominal 200 kHz survey echosounders. (Redrawn from British Transport Docks Board 1981).

Since natural suspensions and sediments contain gas, some distortion of the dimensional relationships in the record, both above and within the fluff layer, may be expected. Although traditional bar checks avoid this problem in most of the water column, sound propagation properties may be expected to vary, especially in the bottom few metres. Thus local dimensions on the record must be treated with caution.

In terms of defining navigable depth in mud areas, echosounders do not provide unambiguous data on sea bed altitude and leadline data are no more dependable. Watts (1954) reported a spread of more than 0.6 m in leadline depths measured under laboratory conditions and up to 2.0 m in field trials. This imprecision in navigable depth determination has a number of

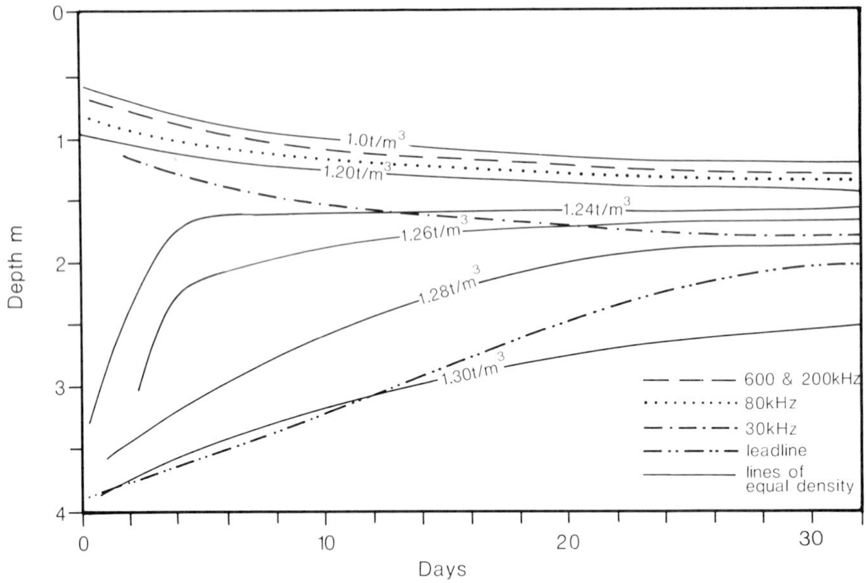

Figure 8. *Comparison of density structure, echosounder reflectors at 600, 200, 80 and 30 kHz, and leadline depth during 32 day laboratory self-weight consolidation of Europoort mud. (Redrawn after Rijkswaterstaat 1979).*

clearly definable consequences for sedimentation management. If sea bed altitude cannot be defined unambiguously, then sediment volumes, dredging need, or dredging achieved cannot be quantified unambiguously and these data are inadequate for numerical model calibration. If the time dependence of bed properties and structure has such a profound effect on acoustic or mechanical evaluation of navigable depth, and its "knock-on" effects on dredging operations and costs, an alternative definition of seabed altitude should be sought.

Sea Bed Defined in Terms of Navigability – the 'Nautical Depth' Concept

In many under-developed countries it is a well known and long accepted practice to allow vessels to navigate with a "negative" underkeel clearance: ie. they sail though the top of the sea bed as defined by an echosounder. In developed countries where ports handle very large crude carrier (VLCC) and hazardous cargo traffic this is an unacceptable practice since it leads to increased hull wear, damage, and occasional groundings. However, if this proven practical principle were placed on a more secure scientific basis, it might provide some advantages in increased navigable depth and reduced dredging.

The passage of a ship's hull in confined waters is influenced by its proximity to the sea bed. If the vessel is to sail through a medium which is more dense and more viscous than water, what parameter best defines the bed quality which will deleteriously affect the ship's behaviour and may therefore be regarded as the sea bed? Intuitively it was proposed that this would be the shear strength of the fluid mud through which the vessel was to be sailed. Although this cannot be measured *in situ*, bulk density, which may empirically relate to shear strength, can (Parker et al. 1975). Thus the concept of 'Nautical Depth' was evolved (Kirby et al. 1980), being defined as "the altitude of the density level within a stationary suspension above which a ship can safely sail" (Fig. 9).

On the basis of physical models, numerical models, and field prototype tests using a 318,000 dwt VLCC, it was established that if the Nautical Depth was defined by the altitude of the 1.2 t/m^3 (74.9 lb/ft^3) density horizon, handling characteristics of the vessel were not significantly impaired and some positive advantages accrued at the cost of increased engine revolutions (Nederlof and van Bochove 1981). This definition of navigable depth has been adopted in Rotterdam-Europoort.

Dredging Control in Rotterdam-Europoort

The combined annual dredging requirement for this port is in excess of 12×10^6 m^3 at an annual cost exceeding \$20 $\times 10^6$. To reduce dredging costs, three aspects of maintenance dredging were examined: technical research into sediment behaviour and properties, methods and plant for surveying and dredging, and contract and management procedures. Only the first two will be mentioned here.

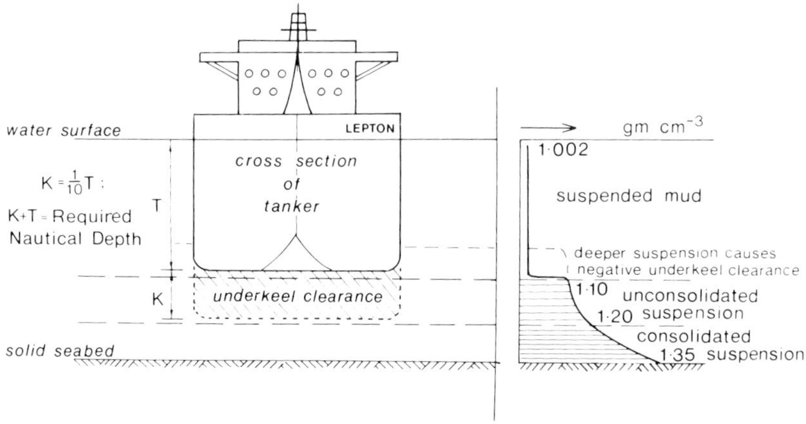

Figure 9. Relationships between draught (T), underkeel clearance (K), and density structure in "Nautical Depth" concept.

Sediment sources and mechanisms

Results of echosounder surveys suggested that the abrupt appearance of fluff was associated with storms in the North Sea, the time lag between the storm and the apparent rapid siltation reflecting the time taken for the material to arrive, consolidate, and become acoustically detectable. Observations of turbidity profiles (Kirby and Parker 1977) showed that below the general low turbidity, stratified water column (solids <0.1 g l^{-1}) high concentration layers (>40 g l^{-1}) were flowing slowly into Europoort. Laboratory measurements suggest that these suspensions are pseudoplastic (Williams and James 1978).

Operational improvements

As a consequence of the ship handling studies, routine echosounder surveys were at first supplemented and subsequently superseded by routine *in situ* density surveys (van Oostrum et al. 1980). From these surveys a number of very specific advantages accrued: (i) Regular density surveys allowed the rapid incursion of material to be monitored and only dredged when it had consolidated. They provided accurate information on the seasonal variation of mass sedimentation rate and thus more accurate scheduling of dredger capacity. (ii) By ignoring the material less than 1.2 t/m^3, the quantity of dredging required was decreased and the dredgers worked to a higher density achieving greater productivity. (iii) Only that material which had to be dredged was moved. Immediate gains in navigable depth were achieved without dredging by devising the correct definition of what the sea bed was in terms of navigability and by locating it. (iv) Correct identification of the mechanism whereby the sediment entered the port, as dense nearbed layers, allowed the design of alternative dredging procedures. A silt trap of 4×10^6 m^3 capacity was dug and the trapped material monitored using density surveys to follow consolidation and achieve optimum dredger production. Other developments include placing stationary pumping facilities into the pit and automated emptying controlled by *in situ* instrumentation.

Cost benefits of this integrated approach to dredging, including revised working procedures, more effective use of plant, revised contract procedures, and controlled scheduling of capacity, have resulted in a real reduction in dredging costs to 50% of the 1975 levels.

Application to Other Areas

In situ density determination to guide dredging is currently in regular use in Rotterdam-Europoort and in some UK ports. It has also been used in Hamburg (Germany), Zeebrugge (Belgium), Bordeaux (France) and several ports in the United States and is also being considered for use in Maracaibo (Venezuela). At present it offers the only unequivocal method of measurement of soil properties for control of navigability and dredging. It does not necessarily replace echosounding but provides an essential explanatory set of data to supplement conventional survey techniques.

Conclusions

Acoustically defined bed levels in fluff areas may give ambiguous or misleading data on quantities of deposited sediment or sedimentation rates. The source of this problem has been identified as mobile, high-concentration, near-bed suspension layers by which most, if not all, of the significant mud transport is achieved.

Evolution of stationary suspensions not only creates surveying problems but may contribute to other difficulties. The relationships between current speed and the rate of entrainment of sediment from the sea bed is often parameterised by erosion experiments using thin, normally consolidated beds and 'equilibrium' flow conditions (eg. Thorne and Parsons 1980). Natural mud beds in fluff areas have been shown to have a layered structure (Kirby and Parker 1974) and a layered bed model may more accurately represent conditions on the estuary bed where the moving pseudoplastic suspension may grade downwards through zones of increasingly structured material to a Bingham-plastic type substrate.

The large discrepancies which frequently occur between prediction of dredging need and hopper capacity required to maintain a given depth cannot be attributed to only one source, eg. 'bulking factors'. Almost every aspect of sediment management, from hydrographic surveys, via models, to dredging operations, has suffered from confusion concerning depth to the sea bed. *In situ* density measurement and a redefinition of navigable depth may provide a sound scientific basis for improved sediment management.

Acknowledgments

We wish to thank the British Transport Docks Board, the Netherlands Rijkswaterstaat (Lower Rivers Division), Dr. K. Been (Golder Associates, Canada), Dr. G. Sills (Oxford) and Dr. D. J. Williams (Swansea) for permission to use data in Figs. 3, 5, 7 and 8 and for their discussion of several aspects of this topic. Dr. K. R. Dyer (IOS) made useful comment on an early draft and Mrs. L. D. M. Marshall (NERC Computing) wrote and ran the programs for target strength.

References Cited

Akal, T. 1972. The relationship between the physical properties of underwater sediments that affect bottom reflection. *Mar. Geol.* 13:251-226.

Allen, G. P., P. Castaing and J. M. Jouanneau. 1977. Méchanisms de remise en suspension et de dispersion des sédiments fins dans l'estuaire de la Gironde. *Bull. Soc. géol. France.* 7. (XIX) No. 2:167-176.

Allen, G. P., P. Castaing and A. Klingebiel. 1974. Suspended sediment transport and deposition in the Gironde estuary and adjacent shelf. *Mem. Inst. Geol. Bassin Aquitaine.* 7:27-36.

Anderson, A. L. and L. D. Hampton. 1980. Acoustics of gas-bearing sediments. (I and II). *Jour. Acoust. Soc. Am.* 67(6):1865-1889; 1890-1903.

Been, K. 1980. Stress-strain behaviour of a cohesive soil deposited underwater. D.Phil. Thesis, Department of Engineering Science. Oxford, U.K. 257 pp.

Been, K. 1982. Non-destructive soil bulk density measurements using X-ray attenuation. In Press. (Dec. 1982). *Geotechnical Testing Journal.* ASTM.

Been, K. and G. C. Sills. 1981. Self weight consolidation of soft soils: An experimental and theoretical study. *Geotechnique.* 31(4):519-535.

British Transport Docks Board. 1981. *Bed density measurements and echo sounding at Humber ports. Report No. 286.* (June 1981), British Transport Docks Board, Southall, Middlesex, U.K. 44 pp.

Bryant, R., D. J. A. Williams and A. E. James. 1980. Rheology of cohesive suspensions, pp. 279-287. *In:* Collins, M. B., F. T. Banner, P. A. Tyler, S. J. Wakefield and A. E. James (eds.), *Industrial Embayments and Their Environmental Problems.* Pergamon Press, New York.

Delft Hydraulics Laboratory. 1962. *Demerara Coastal Investigation.* Delft, The Netherlands. 237 pp.

Faas, R. W. 1969. Analysis of the relationship between acoustic reflectivity and sediment porosity. *Geophysics* 34:546-553.

Flegmann, A. W., J. W. Goodwin and R. H. Ottewil. 1969. Rheological studies on kaolinite suspensions. *Proc. Brit. Ceram. Soc.* 13:31-45.

Gularte, R. C., W. E. Kelly and V. A. Nacci. 1979. Rheological methods for predicting cohesive erosion. *Proceedings 5th Annual Conference of the Marine Technology Society:* 251-258.

Gupta, R. N. 1965. Reflection of plane waves from a linear transition layer in liquid media. *Geophysics* 30:122-132.

Gupta, R. N. 1966. Reflection of sound waves from transition layers. *J. Acoust. Soc. Amer.* 39:255-260.

Hamilton, E. L. 1970. Reflection coefficients and bottom losses at normal incidence computed from Pacific sediment properties. *Geophysics* 35:995-1004.

Hampton, L. D. 1967. Acoustic properties of sediment. *Jour. Acoust. Soc. Amer.* 42(4):882-890.

Ives, K. J. (ed.). 1978. *The Scientific Basis of Flocculation.* N.A.T.O. Advanced Study Institute Series E. No. 27: Sijthoff and Noordhoff. Netherlands. 369 pp.

Kendrick, M. 1980. Port of Belawan, Indonesia. 3rd. Report on hydraulic studies. Report EX879. Part 1 and 2. (June 1980). Hydraulics Research Station, Wallingford, U.K., 103 pp.

Kirby, R. and W. R. Parker. 1974. Seabed density measurements related to echosounder records. *Dock and Harbour Authority* 54:423-424.

Kirby, R. and W. R. Parker. 1977. The physical characteristics and environmental significance of fine sediment suspensions in estuaries, pp. 110-120. *In: Estuaries, Geophysics and The Environment.* National Academy of Sciences, Washington, DC.

Kirby, R. and W. R. Parker. 1981. The distribution and structure of fine sediment phenomena in the Severn Estuary. *Can. J. Fish Aquatic Sci.* (In Press).

Kirby, R., W. R. Parker and W. H. A. van Oostrum. 1980. Definition of the sea-bed in navigation routes through mud areas. *Int. Hydrographic Review.* LVII(1):107-117.

Krone, R. B. 1963. Rheologic properties of estuarial sediments. U. California, Berkeley, CA. Reports S.E.R.L. No. 63-8, 91 pp.

Kynch, G. J. 1952. A theory of sedimentation. *Trans. Faraday Soc.* 48(2):166-176.

Lee, K. and G. C. Sills. 1981. The consolidation of a soil stratum including self-weight effects and large strain. *Numerical and Analytical Methods in Geomechanics.* 5:105-428.

McRoberts, E. C. and J. F. Nixon. 1976. A theory of soil sedimentation. *Canadian Geotechnical Jour.* 13:294-310.

Meade, R. H. 1972. Transport and deposition of sediments in estuaries. *Mem. Geol. Soc. Amer.* 133:91-120.

Michaels, A. J. and J. C. Bolger. 1962. Settling rates and sediment volumes of flocculated kaolin suspensions. *I. & C. E. Fundamentals.* 1(1):24-33.

Migniot, C. 1968. Étude des proprietes physiques de differents sediments très fins et de leur comportement sous des actions hydrodynamique. *La Houille Blanche* 23(7):591-620.

Mooney, M. and R. H. Ewart. 1934. A conicylindrical viscometer. *Physics* 5:350-354.

Nederlof, L. and G. van Bochove. 1981. Manoeuvering behaviour of ships in muddy canals and harbours. *Dock and Harbour Authority*. May 1981. pp. 2-6.

Orr, M. H. and F. R. Hess. 1978. Remote acoustic monitoring of natural suspensate distributions, active suspensate resuspension, and slope/shelf water intrusions. *Jour. Geophys. Res.* 83(C8):4062-4068.

Owen, M. W. 1970. Properties of a consolidating mud. *Report I.N.T. 83.* 51 pp. Hydraulics Research Station, Wallingford, U.K.

Parker, W. R. and R. Kirby. 1977. Fine sediment studies relevant to dredging practice and control, pp. 15-26. *In:* Stephens, H. S., N. G. Coles and J. A. Clarke (eds.). *Second International Symposium on Dredging Technology*. Texas A. and M. University, B.H.R.A., Cranfield, U.K.

Parker, W. R., G. C. Sills and R. E. A. Paske. 1975. In-situ bulk density measurement in dredging practice and control, pp. 25-41. *In:* Hemmings, S. K. (ed.), *First International Symposium on Dredging Technology*. Canterbury, U.K. B.H.R.A., Cranfield, U.K.

Parker, W. R., T. J. Smith and R. Kirby. 1980. Observation of density stratification due to suspended fine sediment, pp. 955-966. *In:* Carstens, T. and T. McClimans (eds.), *Second International Symposium on Stratified Flows*. Tapir. Trondheim, Norway.

Proni, J. R., F. C. Newman, D. C. Rona, D. E. Drake, G. A. Berberian, C. A. Lauter Jr. and R. L. Sellers. 1976. On the use of acoustics for studying suspended oceanic sediment and for determining the onset of the shallow thermocline. *Deep Sea Res.* 23:831-837.

Rayleigh, J. W. S. 1894. *Theory of Sound*. Vol. II. p. 149. MacMillan.

Rijkswaterstaat. 1979. *Density Measurements in Situ*. Report by Lower Rivers Division (Research Section), Public Works Department, Dordrecht, Netherlands, 47 pp.

Schubel, J. R. and E. W. Schiemer. 1973. The cause of the acoustically impenetrable, or turbid, character of Chesapeake Bay sediment. *Marine Geophys. Researches*. 2:61-71.

Siddiquie, H. N., D. Gopala Rao, K. H. Vora and R. S. Topgi. 1981. Acoustic masking in sediments due to gasses on the western continental shelf of India. *Marine Geol.* 39:M27-M37.

Street, N. 1958. Viscosity of clay suspensions. *World Oil*. December 1958:151-156.

Thorn, M. F. C. and J. G. Parsons. 1980. Erosion of cohesive sediments in estuaries: an engineering guide, pp. 349-358. *In:* Stephens, H. S. and C. A. Stapleton (eds.), *Third International Symposium on Dredging Technology*. Bordeaux, France, B.H.R.A., Cranfield, U.K.

Tsuchiya, A., K. Ogawa and M. Nishimura. 1973. Acoustical characteristics of underwater floating mud layer. *J. Fac. Mar. Sci. Technol. Tokai Univ. Japan*. 7:179-191.

Tsuchiya, A., K. Ogawa and M. Nishimura. 1977. Detection of mud sediment or floating mud layer by ultrasound. *Rapp. P-v Reun. Cons. Int. Explor. Mer.* 170:98-104.

van Oostrum, W. H. A., W. R. Parker and R. Kirby. 1980. Maintenance dredging in fluid mud areas, pp. 117-190. *In:* Stephens, H. S. and C. A. Stapleton (eds.), *Third International Symposium on Dredging Technology*. Bordeaux, France, B.H.R.A., Cranfield, U.K.

Watts, G. M. 1954. Laboratory and field tests of sounding leads. *Beach Erosion Board. Technical Memorandum No. 54.* Corps of Engineers. Fort Belvoir, VA 43 pp.

Whelan, T., J. T. Ishmail and G. B. Rainey. 1978. Gas-sediment interaction in Mississippi Delta sediments. *Proc. O.T.C. Vol. II. Paper 3166:*1029-1037.

Williams, D. J. A. and A. E. James. 1978. *Rheology of Brisbane and Rotterdam Mud*. Report Commissioned by Hydraulics Research Station, University College, Swansea. 8 pp.

Wood, A. B. 1941. *A Textbook of Sound*. McGraw-Hill, New York.

Yariv, S. and H. Cross. 1979. *Geochemistry of Colloid Systems*. Springer Verlag, New York. 450 pp.

RESUSPENSION POTENTIAL OF DEPOSITED COHESIVE SEDIMENT BEDS

A. J. Mehta, T. M. Parchure, J. G. Dixit and R. Ariathurai
Coastal and Oceanographic Engineering Department
University of Florida
Gainesville, Florida

Abstract: Surficial layers of fine, cohesive sediment beds in estuaries typically consist of partially consolidated sediments deposited from flow. Such layers tend to have a high water content and exhibit a non-uniform variation of the shear strength with depth. Predictive modeling of estuarial fine sediment transport requires an appropriate description of the erosive behavior of deposited beds. Most available descriptions however are based on laboratory tests using mechanically placed beds. Such beds possess relatively uniform properties and are representative of settled, consolidated estuarial beds. A laboratory test methodology for investigating the resuspension potential of deposited beds is described. Tests using beds of kaolinite show that the rate of erosion varies exponentially with the excess bed shear stress. The influence of bed consolidation time and water chemistry on the rate and the critical shear stress for erosion has been investigated. The derived expression for the rate of erosion can be suitably incorporated in the resuspension routines for estuarial fine sediment transport models.

Introduction

Fine sediments in estuaries consist mainly of particles in the clay and silt size ranges plus some organic matter. Such sediments characteristically are cohesive in nature due to the presence of usually attractive electro-chemical surface forces on the particles. Cohesion results in particle aggregation when inter-particle collisions occur, with each aggregate consisting of perhaps millions of individual clay particles. In the estuary, Brownian motion, local velocity gradients, and differences in the settling velocities of different-sized particles can, under suitable hydraulic conditions, result in collisions which promote further aggregation or cause disaggregation of large units. Consequently, a wide range of aggregates of different sizes, shear strengths and densities are found in suspension and in the deposit. An understanding of estuarial fine sediment dynamics requires a comprehension of the manner in which sediment aggregates undergo cycles of settling, deposition, consolidation, resuspension, and advective plus dispersive transport (Postma 1967; Gibbs 1967; Krone 1972).

This paper is concerned with the mechanism of resuspension of estuarial cohesive sediment beds. In mathematical models for the prediction of fine sediment transport, an appropriate routine for resuspension must be

ISBN 0-12-404070-5

included. A review of literature reveals that existing transport models utilize erosion rate expressions which are based on laboratory erosion tests in which mechanically placed beds, as opposed to beds deposited from flow, were investigated (Ariathurai et al. 1977; O'Connor and Tuxford 1980). Placed beds are representative of fully consolidated estuarial deposits. Since placed beds are characteristically different from flow deposited beds, it is inappropriate to use rate expressions derived using the former, particularly in the case of the surficial bed layers which are not fully consolidated. Also, insufficient information is available on the resuspension potential of deposited beds. These aspects are highlighted in the sequel by reviewing salient aspects of resuspension mechanics in estuaries and of laboratory tests designed to investigate the erodibility of cohesive sediment beds. The difference between placed and deposited beds is emphasized by reviewing available rate expressions for both types of beds. A new laboratory test methodology for investigating the resuspension potential of deposited beds is described. Test results are then presented and an expression for the erosion rate is developed.

Resuspension Phenomena

Resuspension in Estuary

A quasi-steady state description of fine sediment dynamics pertinent to resuspension in the estuarial mixing zone is schematized in Fig. 1. A binary fluid-sediment mixture can be considered to exist in four states: a suspension in horizontal transport; a stationary suspension which may possess a vertical velocity component; a consolidating deposit; or a settled, consolidated bed. A stationary suspension with practically no mechanical strength results from settling of aggregates in transport, particularly at times close to slack water. Under suitable conditions a bed deposit with a skeletal soil framework will

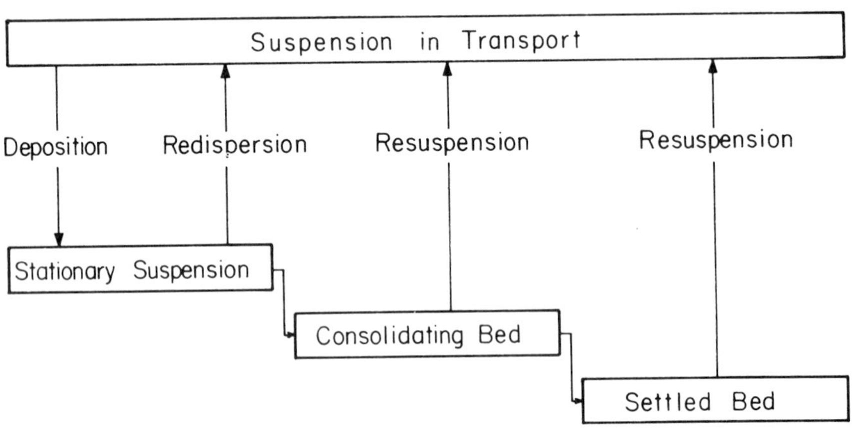

Figure 1. Schematic representation of the physical states of fine sediment in estuarial mixing zone.

develop. Such a deposit tends to exhibit a non-Newtonian behavior with a small but measurable shear strength. Consolidation of this deposit due to overburden and thixotropic rearrangement will result in a settled bed with a lower water content, a higher shear strength, and a more stable structural configuration.

Entrainment of a stationary suspension, which typically could occur shortly after current reversal following slack water, has been referred to as redispersion (Parker and Kirby 1977). Entrainment from a consolidating or a settled bed is referred to as resuspension, which is of interest here. Resuspension of even a few centimeters of the flow-deposited surficial layer of a bed can contribute measurably to the suspended load. For example, assuming a bed bulk density of 0.336 gm/cm^3, resuspension of an 8 cm thick bed in a 5 m deep estuarial water column will result in a suspended sediment concentration of 5.4 gm/liter, which is a significant amount.

Resuspension in Laboratory

Scientific interest in resuspension has been associated mainly with soil erosion-related beds. Investigations utilizing flow-deposited beds have been comparatively recent. When studying soil erosion, the practice is to place the soil in the apparatus manually or by pouring a pre-prepared slurry. The sediment is thus remolded and sometimes compacted. The erodibility of these placed beds, which often have a water content lower than the liquid limit, is found to be characteristically different from the erodibility of deposited beds. This is because among other factors the depth-variation of the aggregate shear strength is generally much more uniform in the case of a placed bed in comparison with a flow-deposited bed, whose water content is typically above the liquid limit. Both placed and deposited beds can exhibit two modes of failure. The first, known as surface erosion, involves aggregate-by-aggregate entrainment of the surficial sediment, and is of interest here. The second, known as mass erosion, results from a dynamic shear loading of the bed. In this case the plane of failure lies deep in the bed, and failure results in an almost instantaneous entrainment of the sediment above the plane.

Resistance to erosion is characterized by the properties of the bed. Therefore, the resuspension potential of a deposited bed, which is of main interest here, would be expected to be contingent upon the pre-erosion stress history, i.e. the mode of formation of the bed before its erosion. This becomes evident when it is recognized that the degree of aggregation of the settling sediment influences the initial structure of the deposit. Increasing aggregation implies the formation of aggregated units of decreasing density and rapidly decreasing shear strength. Thus, aggregates in a fresh deposit are often composed of weak and more "fluffy" units, in comparison with the settling aggregates (Krone 1972). As consolidation proceeds, the aggregates in layers below the surface are crushed, and comparatively stronger, more tightly packed units result. The bed is thus stratified, and the aggregate shear strength and density typically increase with depth monotonically, although not necessarily at a constant rate. When consolidation is complete after a comparatively long period of perhaps several days, the shear strength and the density distributions become invariant, and the bed acquires more uniform properties. Placed beds, by virtue of the manner in which they are prepared, are representative of fully consolidated beds.

Bed structures which can be represented by placed beds as well as by flow-deposited beds are found in the estuarine environment. With reference to modeling the erosive behavior of these beds, it is instructive to review some proposed expressions for the rate of erosion of placed and deposited beds.

Expressions for Erosion Rate

Placed Beds

A common feature of results from laboratory tests with placed beds appears to be that the rate of surface erosion becomes constant or nearly constant after a short initial period of relatively more rapid erosion. The observed constancy essentially precludes an exchange between the material in suspension and that on the bed. This is the case because exchange implies that the eroded sediment in suspension would redeposit, with the rate of deposition varying with the concentration of sediment in suspension. As this concentration would increase in time due to bed erosion, the rate of deposition would correspondingly increase with time. At some point in time the rate of deposition would become equal to the rate of erosion. The suspended sediment concentration would attain an equilibrium value, and the apparent erosion rate would be zero. In cohesionless sediment transport this equality between the rates of erosion and deposition results in a constant rate of horizontal sediment transport. The observed behavior in the case of erosion of placed beds can be explained if exchange is precluded. Essentially, settling aggregates are torn up and re-entrained near the bed where the highest shear stresses prevail, and a condition of no deposition, or complete reflection, exists. Furthermore, since the shear strength of the placed bed remains relatively invariant over depth, the rate of erosion under a constant applied shear stress remains constant over typical laboratory test time-scales. Partheniades (1977) has demonstrated the existence of such a condition of no exchange on the basis of stochastic considerations.

Expressions for the rate of erosion, ε, as a function of the time-mean bed shear stress, τ, are summarized in Table 1. In Eqs. 1 and 2, ω is a dummy variable, and in Eqs. 3, 4 and 5, τ_c is the critical shear stress for erosion. In Eq. 6, ε_{ch} is the erosion rate when $\tau = \tau_{ch}$, which is referred to as a characteristic shear stress. When $\tau > \tau_{ch}$, the bed erodes much more rapidly than when $\tau < \tau_{ch}$. Christensen and Das (1973) have also utilized τ_{ch} in place of τ_c in Eq. 4. The coefficients α and β in all the expressions are in general evaluated from experimental observations. Equation 3 has been extensively utilized in the resuspension routines of estuarial fine sediment transport models (Ariathurai et al. 1977; O'Connor and Tuxford 1980).

Table 1. Expressions for rate of erosion of placed beds.

Investigator(s)	*Expression*	*Equation Number*
Partheniades (1962)	$\varepsilon = \alpha_1 \left[1 - \frac{1}{\sqrt{2\pi}} \int_{-(\beta_1/\tau)-\beta_2}^{(\beta_1/\tau)-\beta_2} \exp(-\omega^2/2)d\omega\right]$	(1)
Christensen (1965)	$\varepsilon = \alpha_2\left[0.5 - \frac{1}{\sqrt{2\pi}} \int_{0}^{-6.10 + (6.18/\sqrt{\beta_3\tau})} \exp(-\omega^2/2)d\omega\right]$	(2)
Kandiah (1974); Arulanandan (1975)	$\varepsilon = \alpha_3 \left(\frac{\tau - \tau_c}{\tau_c}\right)$	(3)
Christensen and Das (1973); Raudkivi and Hutchison (1974); Gularte (1978)	$\varepsilon = \beta_4 \exp[\alpha_4(\tau - \tau_c)]$	(4)
Lambermont and Lebon (1977)	$\varepsilon = \alpha_5(\tau^{\beta_5} - \tau_c^{\beta_5})\tau^{1/2}$	(5)
Mehta (1981)	$\frac{\varepsilon - \varepsilon_{ch}}{\varepsilon_{ch}} = \alpha_6 \left(\frac{\tau - \tau_{ch}}{\tau_{ch}}\right)$	(6)

Deposited Beds

The rate of erosion of deposited beds typically decreases as erosion proceeds, but depending upon the test conditions, erosion may or may not be arrested. Some of the rate expressions are summarized in Table 2. The coefficients δ and λ are empirical constants. The parameter δ_3 and the critical shear stress τ_c in Eq. 9 are observed to vary with the depth z of erosion below the initial level of the fluid-bed interface. This is an important observation made by Thorn and Parsons (1980) following the work of Owen (1975). Unfortunately in all four reported investigations in Table 2, insufficient emphasis was given to defining the pre-erosion stress history of the bed. Conditions under which the bed is prepared are of considerable importance to the prediction of the resuspension potential of the deposited beds.

Table 2. Expressions for rate of erosion of deposited beds.

Investigator(s)	*Expression*	*Equation Number*
Krone (1962)	$\varepsilon = \delta_1 t^{-\lambda_1}$	(7)
Yeh (1979); Fukuda and Lick (1980)	$\varepsilon = \delta_2 \exp(-\lambda_2 t)$	(8)
Thorn and Parsons (1980)	$\varepsilon = \delta_3(z)[\tau - \tau_c(z)]$	(9)

Mehta and Partheniades (1979) presented experimental evidence which suggested that Eqs. 1 or 2 can be utilized in the case of a deposited bed in which the depth-variation of the cohesive aggregate strength is made nearly uniform by artificially mixing the deposit (in the vertical direction), and eliminating much of the stratification which otherwise would result from a preferential initial deposition of larger and stronger aggregates. Following this mixing the rate of erosion remained invariant, and the suspended sediment concentration did not attain an equilibrium value. As explained previously, this behavior implies that an exchange between the sediment in suspension and that on the bed did not occur. This observation appears to be corroborated by the laboratory tests of Krone (1962) and Yeh (1979) in which the suspended sediment concentration continued to increase, albeit at a decreasing rate, after several hundred hours. Under typical laboratory time-scales for erosion and deposition, exchange would result in an equilibrium concentration within a relatively short time period.

In the light of the above observations, the resuspension behavior of the deposited beds is explained when it is recognized that, by virtue of its mode of formation, the aggregate shear strength of the deposited bed increases measurably with depth. Under a constant applied bed shear stress, this increase in strength results in an "armoring" effect which in turn decreases the erosion rate as the bed is scoured. Deposited beds tend to be extremely weak. Utilization of mechanical means such as a penetrometer for measuring the bed strength is likely to produce results of doubtful quality.

Two aspects relative to the deposited bed become apparent. First, the relevant laboratory test methodology must give an explicit consideration to bed preparation. Second, the test procedure must be useful in giving an indication of the bed shear strength and its variation with depth. These aspects are elaborated in the sequel.

Methodology and Experiments

A convenient manner in which resuspension of deposited beds can be investigated in the laboratory is depicted schematically in Fig. 2, which shows the way in which the applied bed shear stress is varied in discrete time-steps. Such a variation results in three distinct phases of sediment transport. Phase I is a mixing stage in which a sediment of concentration C_o, previously equilibrated with the fluid, is mixed at a shear stress τ_m for a period T_m. The shear stress τ_m must be large enough to prevent any deposition of the suspended material.

In Phase II the flow is reduced to yield a comparatively lower shear stress τ_{d1}, which is maintained for a duration T_{d1}. Deposition will occur in this phase. Given sufficient T_{d1}, and τ_{d1} less than a certain τ_{min} which is found to be a characteristic value for the fluid-sediment mixture, the entire amount of sediment will deposit and the flow will be clarified. On the other hand, if $\tau_{d1} > \tau_{min}$, a certain fraction C^*_{eq} of C_o will be maintained in suspension indefinitely (Mehta and Partheniades 1975). In the latter case, a second lowering of the shear stress to τ_{d2} for a period T_{d2} will clarify the suspension. The last sequence in Phase II is a period T_{dc} with no flow. During this period the remaining small fraction of the sediment in suspension will deposit rapidly and the bed will consolidate. Phases I and II together define the pre-erosion stress history of the bed, which is characterized by C_o, τ_m, τ_{d1}, τ_{d2}, T_m, T_{d1}, T_{d2} and T_{dc}.

Resuspension will occur in Phase III in which a series of shear stresses τ_1, τ_2 etc., of increasing magnitudes are applied over corresponding durations T_1, T_2 and so on. In choosing the magnitudes of τ_1, τ_2 etc., it is convenient to select the normalized differential shear stress $\Delta\tilde{\tau}_i = (\tau_{i+1} - \tau_i)/\tau_i$, where i = 1, 2, etc. as an experimental parameter. The selection of $\Delta\tilde{\tau}_i$ and corresponding periods T_i is an important factor in experimental design (Parchure 1980).

The experiments were performed at the University of Florida in two flumes: a recirculating straight flume and a rotating annular flume. The steel recirculating flume was 18 m long, 0.6 m wide and 0.9 m deep, with an underflow-type control gate at the downstream end. The return pipe diameter was 0.2 m. One side of the flume was made of glass panels for visual observations. The two main components of the annular flume were: an an-

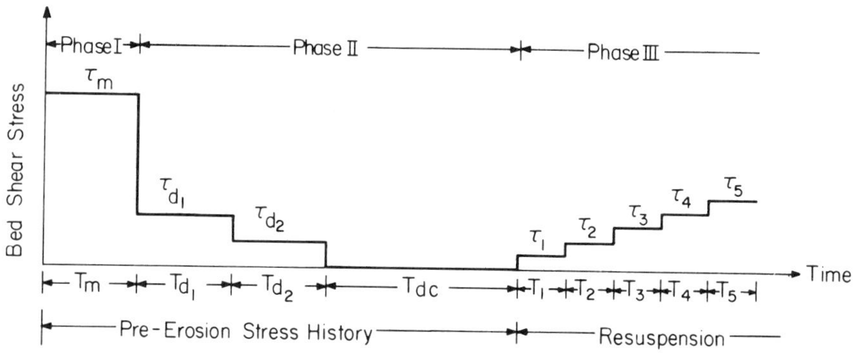

Figure 2. Schematic representation of the selected methodology for the variation of the applied bed shear stress during bed preparation and resuspension tests.

nular fibreglass channel (0.21 m wide, 0.46 m deep, and 1.5 m in mean diameter) containing the fluid-sediment mixture, and an annular ring of slightly smaller width positioned within the channel and in contact with the fluid surface. A simultaneous rotation of the two components in opposite directions generated a uniform turbulent shear field free from aggregate-disrupting elements such as pumps and diffusors in which very high shearing rates prevail. By a proper adjustment of the speeds of the two components the rotation-induced secondary currents were eliminated, and the distribution of the bed shear stress across the channel width was found to be uniform (Mehta and Partheniades 1975).

A commercially available kaolinite with a cation exchange capacity of approximately 11 milliequivalents per hundred grams was used in all tests. The fluids were a) tap water with a total salt concentration of 0.28 ppt (pH = 8.5, Sodium Adsorption Ratio = 0.012) in the recirculating flume, and b) tap water plus commercial grade sodium chloride at 35 ppt concentration (pH = 8.1; Sodium Adsorption Ratio = 12) in the annular flume. The sediment was equilibrated with the fluid for a minimum of two weeks prior to the tests.

The variation of the bed bulk density with depth at the end of Phase II, i.e. just prior to resuspension, was obtained for various periods, T_{dc}, with the help of a specially designed 2.5 cm diameter metal tube in which core samples were frozen *in situ*, using a mixture of alcohol and dry ice. These tests were carried out in a separate series in which conditions in Phases I and II were identical to those for the resuspension tests (Parchure 1980; Dixit 1982).

The parameters C_o, τ_m, T_m, τ_{d1}, T_{d1}, τ_{d2}, T_{d2} and T_i were generally held constant in the majority of the tests in each apparatus. Figure 3 gives typical parameters for tests in tap water and Fig. 4 for tests in salt water. The magnitudes of each parameter in the tests with the two fluids were generally of the same order. The fluid temperature varied somewhat from test to test due to ambient conditions. The mean temperature was 26° C in the tap water tests as well as in the salt water tests. Results are discussed in terms of the influence of T_{dc} and water chemistry on the critical shear stress and the rate of erosion.

Results

As an example, the concentration, C, versus time, t, relationship obtained during resuspension in the tap water tests with T_{dc} = 24 hr is shown in Fig. 3. In most cases, this relationship in a given time-step conformed to the expression

$$C(t) = C(T_{i-1}) + a_i \ln(1 + t - \sum_{n=1}^{n=i-1} T_n) \qquad (10)$$

where $C(T_{i-1})$ is the concentration at the end of time-step $i-1$ (or at the beginning of time-step i). Hence $dC/dt = a_i/(t+1)$, where a_i = the rate of

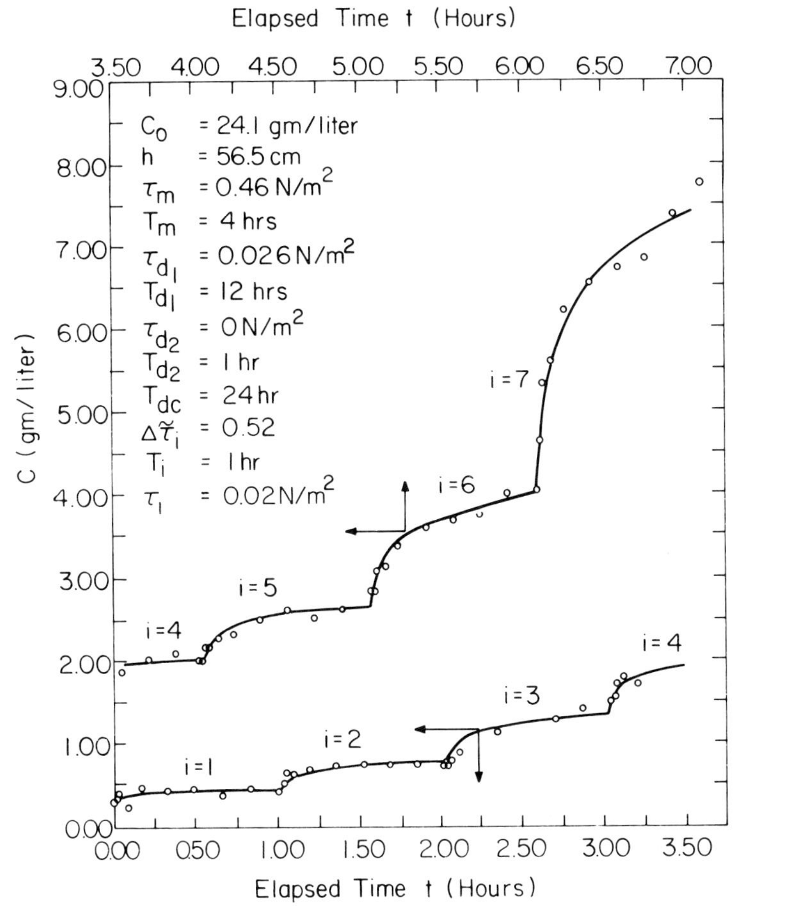

Figure 3. *Example (T_{dc} = 24 hrs) of experimentally observed variation of suspended sediment concentration with time during resuspension, using kaolinite in tap water.*

change of concentration at the beginning of the time-step. Since $\varepsilon = h(dC/dt)$, where h = the depth of flow, it follows that $a_i = \varepsilon_I/h$, where ε_I = the initial rate of erosion in the time-step. Inasmuch as T_i = 1 hr, the final rate in grams per centimeter squared per minute (gm/cm^2-min) would be $0.016\varepsilon_I$, which is small compared with ε_I.

An initial concentration (0.29 gm/liter in Fig. 3) at the beginning of step $i = 1$ was recorded in all tests in the recirculating flume because of the rather rapid resuspension of the deposited sediment in the return pipe and in other parts of the system. This rapid resuspension was due to the comparatively high shear stresses in the return flow and turbulent mixing elsewhere. This "background" concentration was subtracted from the measured instantaneous value in all tests.

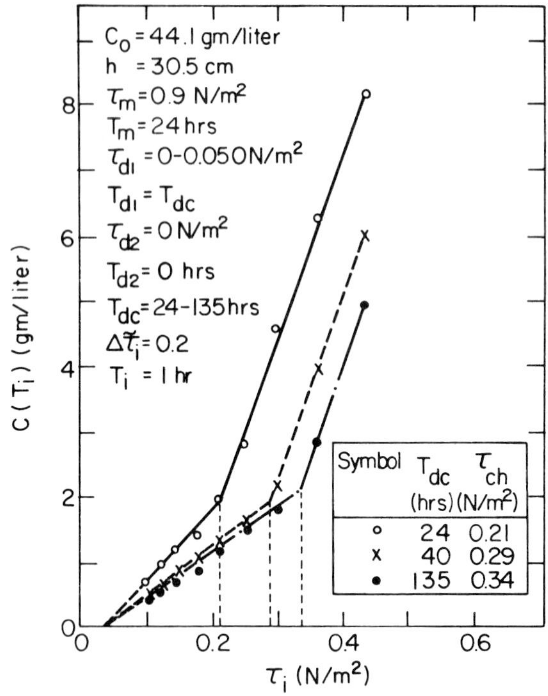

Figure 4. $C(T_i)$ versus τ_i for three values of T_{dc}, using kaolinite in salt water.

An important difference between the profiles in steps $i = 1$ through 5 and steps $i = 6, 7$ in Fig. 3 is that in the former (which will be referred to as Type I), the concentration appears to approach a constant magnitude, whereas in the latter (Type II), there appears to be a more significant increase in the concentration with time at the end of the step. This difference between Type I and Type II profiles is apparent in Fig. 5, in which $C(T_i)$, the concentration at the end of the step, i, is plotted against τ_i, for three tests using tap water with $T_{dc} = 11$, 24 and 144 hrs. When $\tau_i > \tau_{ch}$, $C(T_i)$ increased more rapidly with τ_i than in the case when $\tau_i < \tau_{ch}$. The value, τ_{ch} also increased with T_{dc}. The significance of τ_{ch} and the observed variation of $C(T_i)$ is understood when it is recognized that, a) the rate of erosion is controlled by the excess shear stress, $\tau_i - \tau_c$, and that b) the structure of the bed, as noted, is such that τ_c increases more significantly with depth above a certain depth level than below this level, at which $\tau_c = \tau_{ch}$. Type I profiles result when, in a given time-step of duration T_i, the bed erodes to a level at which the applied shear stress, τ_i, approaches the aggregate shear strength (which is equal to the critical shear stress). In other words, $\tau_i - \tau_c \rightarrow 0$, at which point further erosion is arrested, or nearly arrested. Given T_i, it is necessary to maintain the normalized differential shear stress $\Delta\tilde{\tau}$ below a certain value ($\Delta\tilde{\tau} = 0.52$ was maintained in the tap water tests and 0.20 in the salt water tests) in order to yield a suffi-

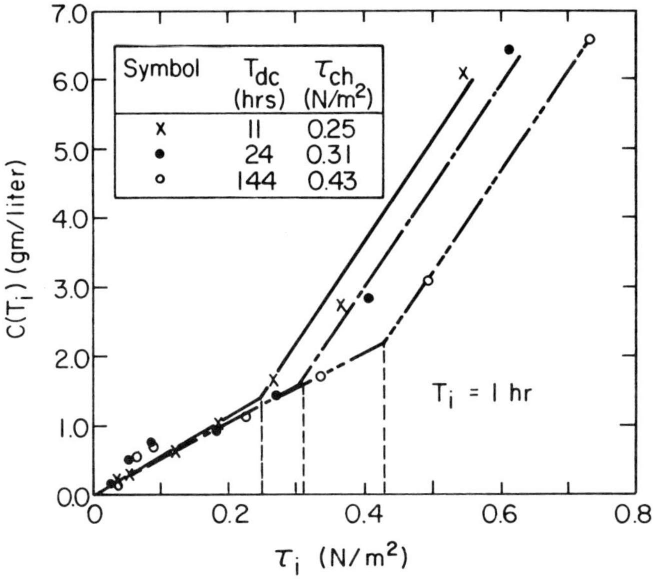

Figure 5. C(T_i) versus τ_i for three values of T_{dc}, using kaolinite in tap water.

cient number of Type I profiles, and, therefore, information on the variation of τ_c with depth. At depths below $\tau_c = \tau_{ch}$, bed properties are more uniform, and two cases can arise: 1) τ_c increases relatively slowly with depth, or 2) τ_c becomes nearly invariant. In case 1, the bed must be allowed to erode for a sufficiently long duration T_i in order for erosion to proceed to a depth at which τ_i equals the shear strength. The required magnitude of T_i can increase rapidly. Thus, referring to Fig. 3, computations yield estimated values for T_6 and T_7 to be at least 2 hrs and 6 hrs, respectively (Dixit 1982). In some cases, periods T_i of one or two orders of magnitude longer duration than these would have been required. Practical considerations however dictate that meaningful information of engineering accuracy must be obtained in the least possible time. This can be achieved through a proper interpretation of the test results such as those given in Fig. 3, as described later. In case 2, the bed will continue to erode as long as the applied shear stress remains constant, inasmuch as the condition $\tau_i - \tau_c \rightarrow 0$ will not be attained, for all practical purposes.

The increase in τ_{ch} with T_{dc} observed in Fig. 5 clearly indicates that the aggregate shear strength is increasing with consolidation. The same is evident in Fig. 4 which is analogous to Fig. 5, for tests using salt water. Figures 6a, b show normalized plots of the depth-variation of the bed bulk density for various periods T_{dc}, from tap water tests. After T_{dc} = 48 hrs, the variations seem to conform to the dimensionless relationship

$$\frac{\varrho}{\bar{\varrho}} = \zeta\left(\frac{z'}{H}\right)^{\xi} \tag{11}$$

where $z' = H - z$, H = bed thickness, $\bar{\varrho} = \varrho$ averaged over H, $\zeta = 0.794$ and $\xi = -0.288$. Equation 11 is in agreement with the reanalysed data of Owen (1975) and Thorn and Parsons (1980), for four natural muds which gave $\zeta = 0.660$ and $\xi = -0.347$ (Dixit 1982). It has been assumed in the following calculations that during the period of resuspension (Phase III), the density profile did not vary significantly with time.

Erosion Parameters

In order to yield a unified understanding of the erosive properties of the deposited bed, the experimental data were analyzed in three steps as follows: 1) assuming that at the end of time-step *i* of a Type I profile, the applied shear stress τ_i equals the critical shear stress τ_c at the corresponding depth of erosion, the dependence of τ_c on the depth z below the initial fluid-bed interface was obtained; 2) a corresponding expression for the rate of erosion for each time-step was empirically derived; and 3) utilizing the coefficients of the rate expression derived from the analysis of Type I profiles, Type II profiles were used to yield information on the dependence of τ_c on depth below the level at which $\tau_c = \tau_{ch}$, assuming the derived rate expression to be applicable throughout. Main results are illustrated by examples here (Parchure 1980; Dixit 1982).

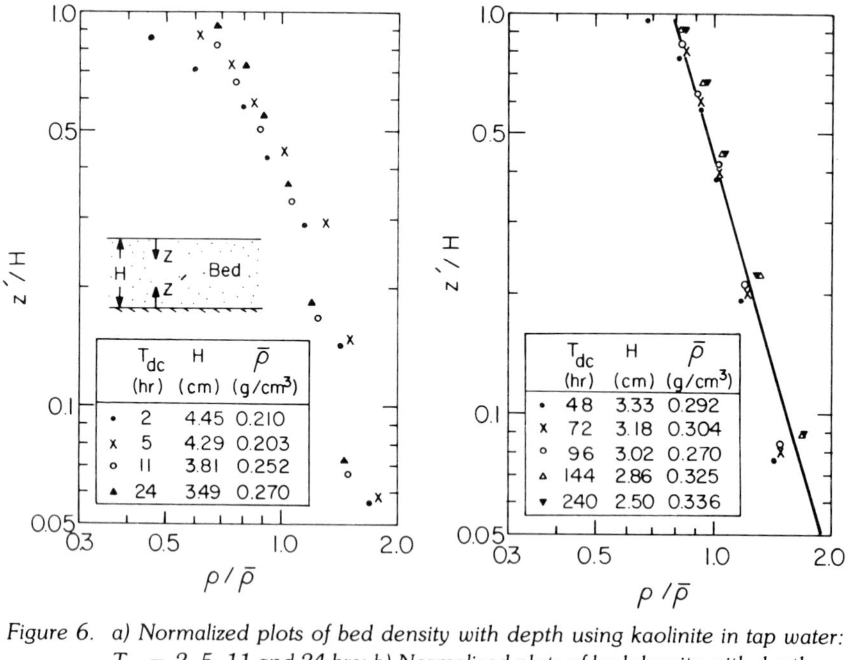

Figure 6. a) Normalized plots of bed density with depth using kaolinite in tap water: T_{dc} = 2, 5, 11 and 24 hrs; b) Normalized plots of bed density with depth using kaolinite in tap water: T_{dc} = 48, 72, 96, 144 and 240 hrs.

As shown in Fig. 7 (which corresponds to the test shown in Fig. 3), the rate of erosion shows an exponential dependence on the normalized excess shear stress, $(\tau_i - \tau_c)/\tau_c$, according to

$$\varepsilon_i = \varepsilon_{oi}\exp[\alpha_i \frac{\tau_i - \tau_c}{\tau_c}] \tag{12}$$

where α_i and ε_{oi} are empirical coefficients. Figure 8 shows a similar relationship for a test using salt water with T_{dc} = 40 hrs. Equation 12 is analogous to Eq. 4 which is based on a heuristic interpretation of the chemical reaction rates theory of Eyring (1936) and others. This theory involves the activation energy concept according to which a "threshold" energy barrier must be crossed for the conversion of reactants to products. As Paaswell (1973) has noted, Eq. 4 and hence Eq. 12 make possible the understanding of erosion as an internal energy/external energy system, where ε is a measure of the work done on the system (reflected in τ_i) and α_i and ε_{oi} represent measures of the internal energy of the deposit. Although α_i values in a given test varied, the

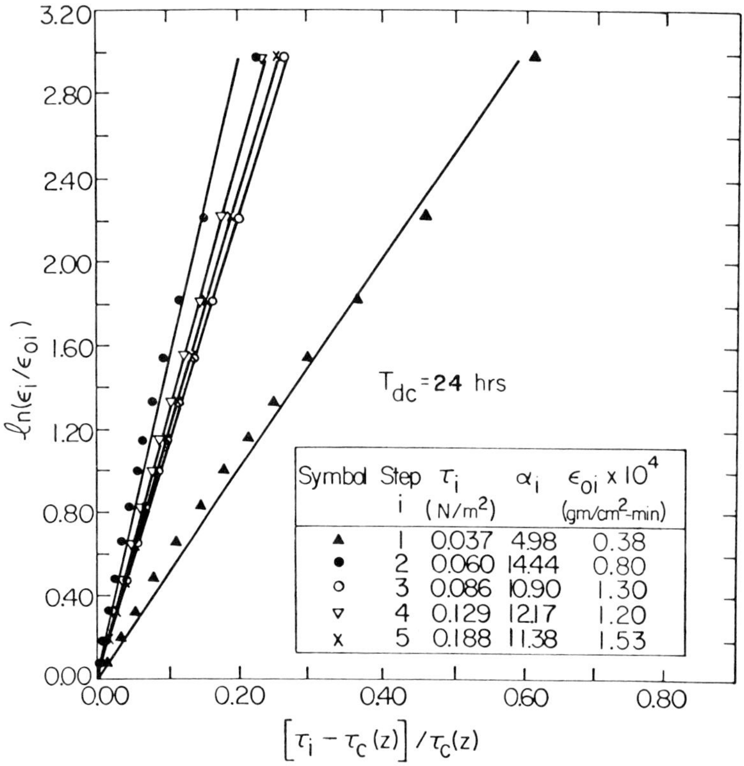

Symbol	Step i	τ_i (N/m^2)	α_i	$\epsilon_{oi} \times 10^4$ (gm/cm^2-min)
▲	1	0.037	4.98	0.38
●	2	0.060	14.44	0.80
○	3	0.086	10.90	1.30
▽	4	0.129	12.17	1.20
x	5	0.188	11.38	1.53

Figure 7. Normalized rate of erosion, $\varepsilon_i/\varepsilon_{oi}$, versus normalized excess shear stress, $[\tau_i - \tau_c(z)]/\tau_c(z)$, using kaolinite in tap water with T_{dc} = 24 hrs.

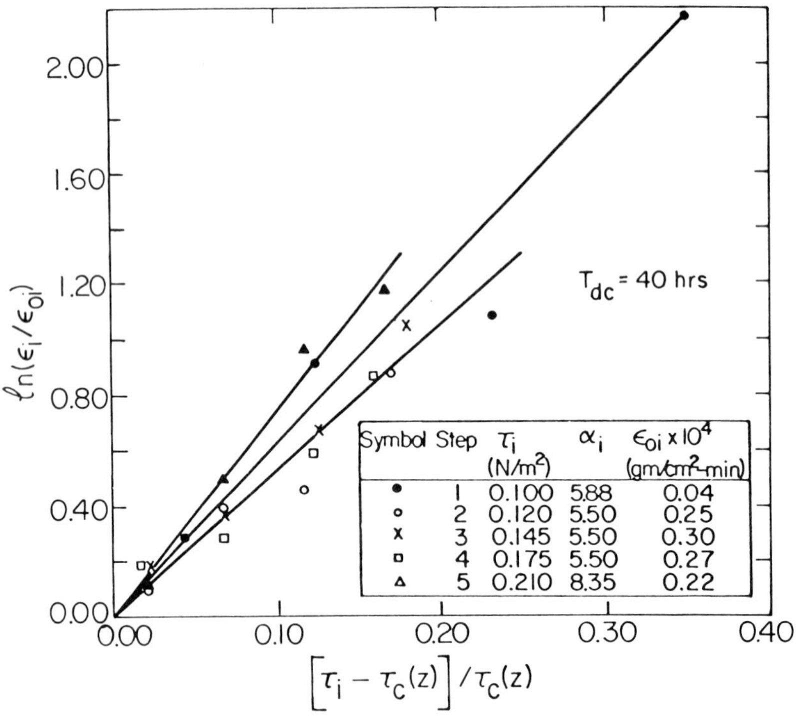

Figure 8. Normalized rate of erosion, $\varepsilon_i/\varepsilon_{oi}$, versus normalized excess shear stress, $[\tau_i - \tau_c(z)]/\tau_c(z)$, using kaolinite in salt water with $T_{dc} = 40$ hrs.

variations appeared to be random or nearly random about a mean value for tests with $T_{dc} \geq 48$ hrs. For tests with $T_{dc} < 48$ hrs, no clear trends were observed either, but additional experimental evidence is required for trend interpretation.

In the case of $T_{dc} > 48$ hrs, ε_{oi} appeared to increase somewhat with depth and approach a constant value. These trends, which are apparent from the stated values of α_i and ε_{oi} in Fig. 7 and also from Fig. 8 which shows results from salt water tests, can in general be construed to be indicative of the increasing uniformity and stabilization of the bed structure in the lower depths with increasing T_{dc}.

In Fig. 9, $\bar{\alpha}_i$ (the mean value of α_i from each test) is plotted against T_{dc}. The term $\hat{\bar{\alpha}}_i$ (the value of $\bar{\alpha}_i$ averaged over the duration T_{dc}) is larger in the tap water tests than in the salt water tests by a factor of 1.6. In Fig. 10, the corresponding $\bar{\varepsilon}_{oi}$ shows no variation with T_{dc}, and $\hat{\bar{\varepsilon}}_{oi}$ from the tap water tests is larger than that from the salt water tests by a factor of 2. These differences in the $\hat{\bar{\alpha}}_i$ and $\hat{\bar{\varepsilon}}_{oi}$ values imply that kaolinite is resuspended with greater facility in tap water than in salt water. This trend appears to be consistent with the flocculation characteristics and the corresponding observed depositional rates of kaolinite in the two fluids (Mehta and Partheniades 1975). The response of

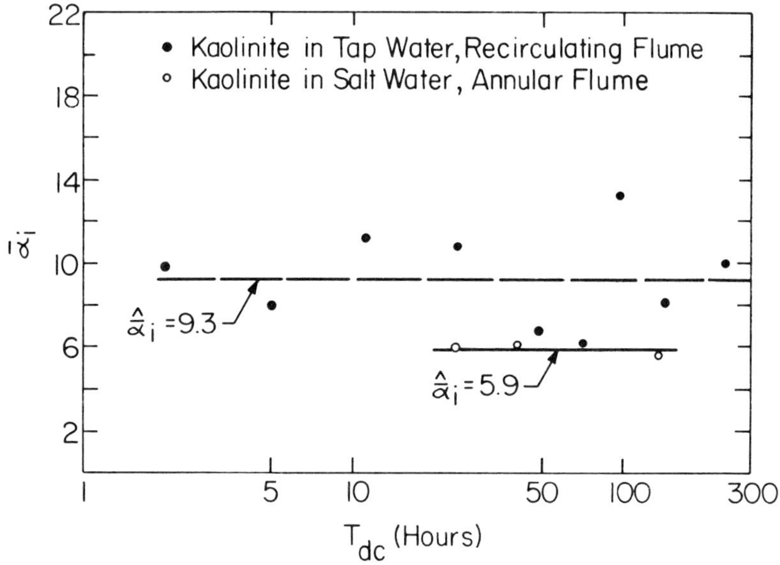

Figure 9. $\bar{\alpha}_i$ *versus* T_{dc} *from tests with kaolinite in tap water and with salt water.*

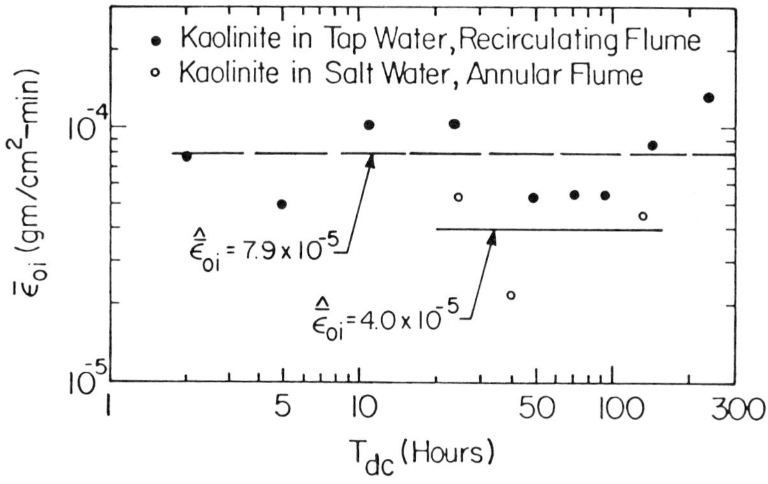

Figure 10. $\bar{\varepsilon}_{oi}$ *versus* T_{dc} *from tests with kaolinite in tap water and with salt water.*

estuarial beds with kaolinitic sediments in a fluid environment with variable salinity will be qualitatively similar to the observed trend in the laboratory. As for the observed quantitative difference in the rate of resuspension in the laboratory tests, the relatively higher flow shearing rates in the recirculating flume (in which a pump was present) in comparison with the annular flume is likely to have influenced the aggregation of the depositing sediment (and therefore the rate of resuspension) particularly in test results corresponding to $T_{dc}<48$ hrs. This is because, in the initial stages of bed formation, the degree of aggregation of the deposited sediment is an important factor which influences the bed structure (Krone 1972).

In Fig. 11, in which τ_c is plotted against z from three tests in salt water, the rate of change ($d\tau_c/dz$) decreased with depth as the bed structure approached greater uniformity. Furthermore, while the surficial value of τ_c was not altered by increasing T_{dc}, an observation which is consistent with a similar conclusion of Partheniades (1962), τ_c below the surface increased measurably with T_{dc}. The implication is that since α_i and ε_{oi} apparently do not vary with T_{dc}, the rate of erosion decreases significantly with increasing T_{dc}. This is because $(\tau_i - \tau_c)/\tau_c$ will decrease for a given τ_i, as τ_c increases with T_{dc}.

For $T_{dc}\geqslant 48$ hrs, the tap water test results yield the following approximate relationship between τ_c and $z'(= H-z)$:

$$\frac{\tau_c(z)}{\tau_i} = \frac{1}{1+ p_i\,[1 - q_i(z'^{\,\xi+1}_{i-1} - z'^{\,\xi+1})]} \tag{13}$$

where $z'_{i-1} = z'$ at the end of time-step $i-1$, $p_i = (1/\bar{\alpha}_i)\ell n(a_i h/\bar{\varepsilon}_{oi})$ and $q_i = \bar{\varrho}\zeta/[(\xi+1)H^{\xi} a_i\, h\ell n(a_i h/\bar{\varepsilon}_{oi})]$. In Fig. 12, Eq. 13 (with values of p_i and q_i as indicated) was compared with computed values derived from data from a test with $T_{dc}=$ 96 hrs (Dixit 1982).

Conclusions

Estuarial fine sediment beds, which are subject to resuspension under flows controlled by the tides, occur in various stages of consolidation. Partially consolidated beds, formed by deposition of suspended sediment at low flow velocities, are soft, with a high water content and low aggregate shear

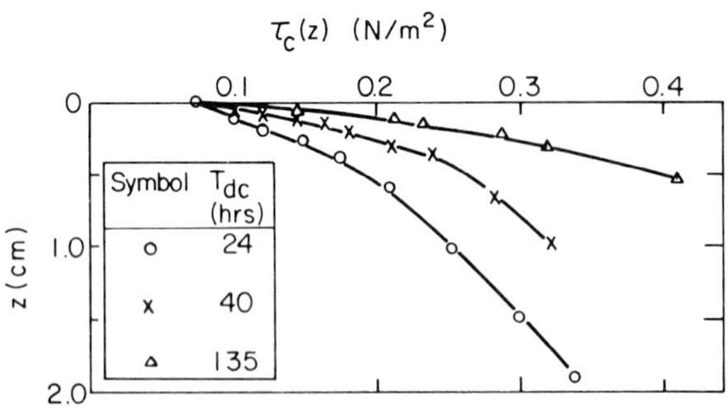

Figure 11. *$\tau_c(z)$ versus z from three tests using kaolinite in salt water corresponding to three values of T_{dc}.*

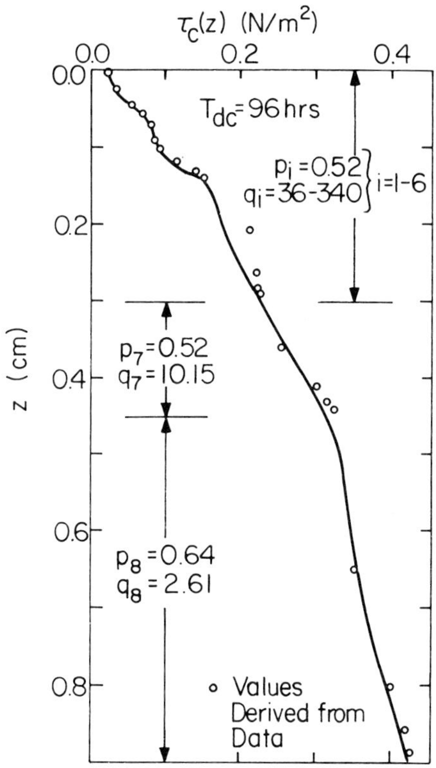

Figure 12. $\tau_c(z)$ versus z from test using kaolinite in tap water, with T_{dc} = 96 hrs; comparison between values derived from data and Eq. 13.

strength. The bed deposit tends to exhibit a marked variation in the bulk density and shear strength with depth. The bed material is resuspended when the applied shear stress equals the shear strength. The shear strength is thus equal to the critical shear stress. Results of laboratory experiments on kaolinitic beds presented here demonstrate that the critical shear stress increases rapidly with depth in the surficial layer. The thickness of this layer is of the order of one or two centimeters. Below this layer the rate of increase of the critical shear stress decreases. This is because the lower layers have properties which are more uniform in comparison with those of the top layer. During the first few hours the degree of aggregation of the depositing sediment and thixotropic rearrangement influence the properties of the bed structure. After approximately two days, consolidation due to overburden becomes the dominant factor in determining bed properties. The rate of erosion of such beds varies exponentially with the excess bed shear stress normalized by the critical shear stress. The critical shear stress and the rate of erosion are found to vary strongly with bed consolidation time and also with water chemistry.

In the resuspension routines for mathematical models of estuarial fine sediment transport, partially consolidated beds formed by deposition should be treated separately from settled, consolidated beds. The rate of erosion of deposited beds is given by Eq. 12. The rate of erosion of consolidated beds is appropriately obtained from laboratory experiments using placed beds. Equation 3, according to which the rate of erosion varies linearly with the normalized excess shear stress, is suitable for consolidated beds.

Acknowledgments

Guidance provided by Prof. E. Partheniades in the development of the bed density measuring procedure and assistance given by E. J. Hayter in the development of computer routines for data analysis are acknowledged. Support for this study was derived from the following sources: U. S. Environmental Protection Agency Grant R806684010; National Science Foundation Grant GK-31259; and State of Florida Special Appropriation to the Coastal and Oceanographic Engineering Department.

References Cited

Ariathurai, R., R. C. MacArthur and R. B. Krone. 1977. Mathematical model of estuarial sediment transport. Tech. Rept. D-77-12, U. S. Army Eng. Waterways Expt. Sta., Vicksburg, MS 162 pp.

Arulanandan, K. 1975. Fundamental aspects of erosion of cohesive soils. *J. Hydr. Div., ASCE.* 101(HY5):635-639.

Christensen, B. A. 1965. Discussion of erosion and deposition of cohesive soils (by E. Partheniades). *J. Hydr. Div., ASCE,* 91(HY5):301-308.

Christensen, R. W. and B. M. Das. 1973. Hydraulic erosion of remolded cohesive soils, pp. 8-19. *In:* National Research Council (ed.), *Soil erosion: causes and mechanisms, prevention and control,* Highway Res. Board Special Rept. 135, Washington, DC

Dixit, J. G. 1982. Resuspension potential of deposited kaolinite beds. M. S. Thesis, U. Florida, Gainesville, FL 167 pp.

Eyring, H. 1936. Viscosity, plasticity and diffusion as examples of absolute reaction rates. *J. Chem. Phys.* 4:283-291.

Fukuda, M. K. and W. Lick. 1980. The entrainment of cohesive sediments in fresh water. *J. Geophys. Res.* 85(C5):2813-2824.

Gibbs, R. J. 1967. The geochemistry of Amazon River system: Part I. The factors that control the salinity and the composition and concentration of the suspended solids. *Geol. Soc. Am. Bull.* 78:1203-1232.

Gularte, R. C. 1978. Erosion of cohesive marine sediment as a rate process. Ph. D. Thesis, U. Rhode Island, Kingston, RI 190 pp.

Kandiah, A. 1974. Fundamental aspects of surface erosion of cohesive soils. Ph. D. Thesis, U. California, Davis, CA 261 pp.

Krone, R. B. 1962. Flume studies of the transport of sediment in estuarial shoaling processes. Final Rept., Hydr. Engrg. Lab. and San. Engrg. Res. Lab., U. California, Berkeley, CA 118 pp.

Krone, R. B. 1972. A field study of flocculation as a factor in estuarial shoaling processes. Tech. Bull. 19, Comm. on Tidal Hydr., U. S. Army Corps of Engrs., Vicksburg, MS 125 pp.

Lambermont, J. and G. Lebon. 1978. Erosion of cohesive soils. *J. Hydr. Res.* 16(1):27-44.

Mehta, A. J. 1981. Review of erosion function for cohesive sediment beds. *Proc. First Indian Conf. on Ocean Engrg.,* Indian Inst. of Technology, Madras, India, I:122-130.

Mehta, A. J. and E. Partheniades. 1975. An investigation of the depositional properties of flocculated fine sediments. *J. Hydr. Res.* 13(4):361-381.

Mehta, A. J. and E. Partheniades. 1979. Kaolinite resuspension properties. *J. Hydr. Div., ASCE.* 105(HY4):409-416.

O'Connor, B. A. and C. Tuxford. 1980. Modeling siltation at dock entrances, pp. 359-371. *In:* H. S. Stephens (ed.), *Proc. Third Int. Symp. on Dredging Tech.*, BHRA, Paper F2, Bordeaux, France.

Owen, M. W. 1975. Erosion of Avonmouth mud. Rept. INT-150, Hydr. Res. Sta., Wallingford, UK 46 pp.

Paaswell, R. E. 1973. Causes and mechanisms of cohesive soil erosion: The state of the art, pp. 52-74. *In:* National Research Council (ed.), *Soil erosion: causes and mechanisms, prevention and control.* Highway Res. Board Special Rept. 135, Washington, DC

Parchure, T. M. 1980. Effect of bed shear stress on the erosional characteristics of kaolinite. M. S. Thesis, U. Florida, Gainesville, FL 186 pp.

Parker, W. R. and R. Kirby. 1977. Fine sediment studies relevant to dredging practice and control, pp. 15-26. *In:* H. S. Stephens (ed.), *Proc. Second Int. Symp. on Dredging Tech.*, BHRA, Paper B2, Texas A & M U., College Station, TX

Partheniades, E. 1962. A study of erosion and deposition of cohesive soils in salt water. Ph. D. Thesis, U. California, Berkeley, CA 192 pp.

Partheniades, E. 1977. Unified view of wash load and bed material load. *J. Hydr. Div., ASCE,* 103(HY9):1037-1057.

Postma, H. 1967. Sediment transport and sedimentation in the estuarine environment, pp. 158-179. *In:* G. H. Lauff (ed.), *Estuaries,* Am. Assoc. Adv. Sci., Washington, DC

Raudkivi, A. J. and D. L. Hutchison. 1974. Erosion of kaolinite by flowing water. *Proc. Roy. Soc. Lond.,* A.337:537-554.

Thorn, M. F. C. and J. G. Parsons. 1980. Erosion of cohesive sediments in estuaries: An engineering guide, pp. 349-358. *In:* H. S. Stephens (ed.), *Proc. Third Int. Symp. on Dredging Tech.*, BHRA, Paper F1, Bordeaux, France.

Yeh, H. Y. 1979. Resuspension properties of flow deposited cohesive sediment beds. M. S. Thesis, U. Florida, Gainesville, FL 118 pp.

SEDIMENTATION ASSOCIATED WITH TIDAL BARRIERS IN CHINA'S ESTUARIES AND MEASURES FOR ITS REDUCTION

Luo Zhaosen and Gu Peiyu
Nanjing Hydraulic Research Institute
Ministry of Water Conservancy and
Ministry of Communications
Nanjing, China

Abstract: We describe the severity of sedimentation in estuaries with tidal barriers in China and explain the causes in light of changes in hydraulic conditions associated with tidal barrier construction, i.e., decrease in river flow and tidal flow and deformation of tidal waves. We discuss siltation-reduction measures now in use, including hydraulic sluicing, agitation dredging, and tidal flushing. Each method has advantages and disadvantages that make it location specific.

Severity of Sedimentation

In the past thirty years, tidal barriers have been constructed in medium and small estuaries along the Chinese coast. These dams are important in saltwater intrusion prevention and for freshwater storage for industrial, agricultural, and civilian use; for facilitating abnormal tide and typhoon protection; and enhancing drainage of waterlogged lowlands. An analysis of the types of Chinese estuaries (Huang 1963) shows that most tidal barriers are constructed in estuaries where sediments are supplied from the sea or from both land and sea. Because of abundant sediment supplies, siltation has taken place either downstream or both upstream and downstream of the barriers after their construction. For example, there are 57 large- and medium-sized barriers with a drainage discharge of more than 100 $m^3\ sec^{-1}$ in the coastal regions in Jiangsu Province. Of these, four relatively large barriers, i.e., in Seyang estuary, Xinyang estuary, Dolung estuary, and Huangsha estuary, are responsible for controlling waterlogging over an area of 13,700 km^2 and for bringing 933,000 ha of cultivated land under irrigation by freshwater storage in the District of Leeshahe, Northern Jiangsu. From 32 to 34 million m^3 (39.6 million m^3 by 1980) of sediment were deposited in the 30 km of river channel downstream of the Barrier of Seyang after completion in 1956 (Table 1). Drainage capacity of the waterway decreased over 50%, as compared with that prior to barrier construction.

Originally there were adequate water depths in the Yong River estuary (Zhejiang Provence). In 1959 a tidal gate was built on the Yao River, a relatively large tributary of the estuary, and since construction about 20

ISBN 0-12-404070-5

Table 1. Statistics of amount of siltation in selected Chinese estuaries with tidal barriers or reservoirs.

Tidal Barrier	*Condition*	*Channel Volume (10^4 m^3)*	*Shoaling (10^4 m^3)*	*River Length (km)*	*Shoaling per km (10^4 m^3)*	*Reduction in Channel Volume (%)*	*Date*
Seyang Estuary	Prior to gate construction	6839		28.0			1954
	After gate construction	2876	3963	30.6	129.5	57.9	Oct. 1980
Xinyang Estuary	Prior to gate construction	3222.5		17.5			June 1957
	After gate construction	1122.4	2100.1	23.0	91.3	65.2	Sept. 1970
Xinyang Estuary	In initial stage of cutoff project	801.8		12.36			May 1972
	After project completion	602.7	199.1	12.36	16.1	24.8	May 1977
Yong River Estuary	Prior to gate construction	5388.7					June 1959
	After gate construction	3122.2	2266.5	22.1	102.6	42.1	June 1973
Fengho Horn	On project completion	74.07					Aug. 1964
	After shoaling	38.74	35.33	1.3	27.2	47.7	Apr. 1977
Yonglin River Estuary	Prior to reservoir construction	5800					June 1958
	After reservoir construction	4440	1360	44.2	30.8	23.5	July 1968

million m^3 of sediment have deposited over the entire 21.3 km of river from Ningbo to Zhenghai (Table 1), with the accretion thickness averaging roughly 2.3 m. As a result, passenger ships of 3,000 tons which once could enter or leave at any time before gate construction now have to do so at high water.

Most of the moderate- and small-sized estuaries in the provinces of Fujian and Guangdong have been brought under control by means of tidal barriers. However, they have not experienced heavy sedimentation because of abundant rainfall and terrigenous material-dominated deposition, although siltation has taken place wherever there is a supply of marine sediment. For example, sedimentation has taken place both above and below tidal barriers such as the Barrier of Fengho Horn (Guangdong Province) (Table 1) and the Gate of Nangxi River located on a tributary of the Jiulong River (Fujian Province).

Estuaries in the North of China are characterized by meager annual freshwater input and sediment supplies from both land and sea, resulting in heavy siltation above and below tidal barriers. For example, a tidal barrier was constructed at the mouth of the Hai River estuary (Hepei Province) in 1959. By October 1969, 14.07 million m^3 of sediment had deposited in the reaches about 75 km upstream of the gate with 9.78 million m^3 subsequently deposited over the lower reach from the gate to the bar area. Even the tidal barriers for harnessing the Hai River Basin and those constructed on the relief canals have given rise to heavy siltation, with the worst situation being riverbeds aggraded to high water level and river discharge capacity being reduced to about 10% of design capacity. The Gate of Shuangtaizhi River estuary (Liaoning Province), with a design flood discharge capacity of 3,000 m^3 sec^{-1} can now discharge only from 1,500 to 2,000 m^3 sec^{-1}. Industrial and agricultural water supplies have suffered from the shoaling above the gate.

In Shandong Province, tidal gates are located mainly in the vicinity of tidal limits; however, siltation has also taken place following gate construction. Construction of the Changtan Reservoir upstream of the tidal limit on the Yonglin River, a major tributary of Jiao River estuary (Zhejiang Province) has resulted in deposition over the entire 44-km tidal reach of the river (Table 1).

In short, tidal barrier construction has caused siltation in all estuaries with marine-derived sediments. When sediment derives from both land and sea, siltation has occurred above and below barriers (Table 1).

Changes in Hydraulic Conditions and Estuarine Sedimentation

1. Decrease in river runoff:

This comprises decreased total river flow and dry weather flow and changes in annual distribution of river flow. Because of different precipitation conditions, rivers in the South generally have experienced less alteration in flow volume compared with Northern rivers. For example, from 1950 to 1954, prior to gate construction, mean annual seaward flow of the Hai River estuary (Northern China) was 8,428 million m³. In 1960 to 1969, after gate construction, it was reduced to 3,165 million m³ and further reduced to 710 million m³ from 1970 to 1975. Also typical is the Yonglin River (Zhejiang Province) which had a mean annual flow of about 450 million m³ prior to construction of the Changtan Reservoir but no discharge in 1966, 1967, and 1970 after reservoir construction. Figure 1 shows the relations in Sheyang estuary between long term mean monthly volume of river discharge and quantity of deposition or riverbed erosion after barrier construction, and between the net quantity of scouring or silting and mean daily discharge at Shanghuanyang station before barrier construction (Dou 1963). There is a close relationship between river discharge and quantity of erosion sedimentation of a river channel. After barrier construction, river discharge decreased and the upstream balancing flow required to maintain a channel at the same cross section increased, leading to shoaling below the tidal barrier.

2. Decrease in tidal discharge:

(a) Tidal barrier construction on main rivers has resulted in increased high water level and decreased low water level, leading to a local increase in the tidal range. However, total tidal volume decreases due to the interception at the barrier site of the portion of tidal flow penetrating upstream to the tidal limit. Consequently, tidal flow below the barrier generally decreases. At the river mouth cross section, tidal flow might increase in the initial stage of barrier construction (Dou 1963), dependent on barrier location, but it decreases with siltation.

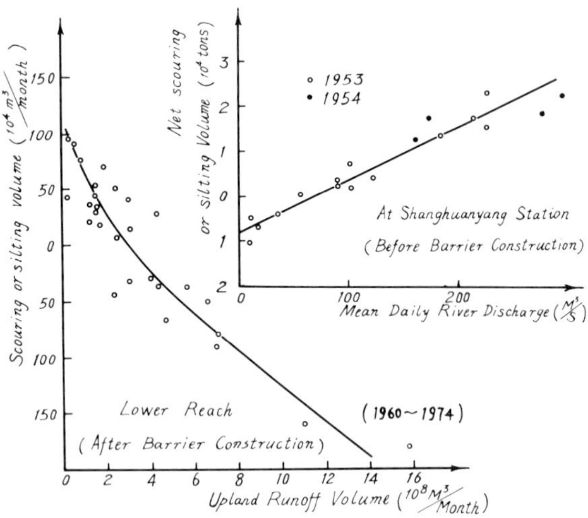

Figure 1. *Relation between volume of downstream river flow and quantity of scouring (values below 0) or silting (values above 0) in Sheyang estuary before and after tidal barrier construction.*

(b) Gate construction on one tributary has no influence on the distances tidal waves travel upstream in another tributary and causes little alteration to high and low tide levels. However, tidal flow in the tributary with the tidal gate is appreciably reduced.

The effect of decreased tidal discharge on the evolution of riverbeds can be explained using morphological relationships. The following are regime channel relations developed for plain tracts and tidal estuaries (Dou 1964):

$$B = 1.33 \left(\frac{gU_{os}SQ^5}{K\alpha^8U^8_{ob}} \right)^{1/9} \tag{1}$$

$$h = 0.807 \left(\frac{K\alpha^2U^2_{ob}Q}{gU_{os}S} \right)^{1/3} \tag{2}$$

$$A = 1.073 \left(\frac{K^2Q^8}{g^2U^2_{os}S^2\alpha^2U^2_{ob}} \right)^{1/9} \tag{3}$$

in which Q = average ebb discharge rate, including freshwater runoff; S = average ebb sediment concentration; U_{ob} = threshold velocity of bed material; U_{os} = critical velocity for deposition of suspended sediment; α = movability coefficient of river bottom relative to bank; K = sediment-transport coefficient; g = acceleration of gravity; B = channel width; h = channel depth; and A = cross-sectional area.

These relationships were applied to a study of effects of tidal discharge on changes in riverbeds before and after gate construction on the Yong River and the Seyang River estuaries; fair agreement was obtained with the field measurements.

For the purpose of simplifying calculations, it can be assumed that the cross-sectional area of a river channel has a relation mainly with the average ebb discharge rate for equal conditions of sediment supply from marine sources. An analysis of data obtained before and after barrier construction on the Seyang River, Yong River, Guan River, and Feiyun River estuaries and on Liu River (medium diameters of bed sediments are between 0.015 to 0.05 mm) yielded (Fig. 2):

$$A = 3.84Q^{8/9} \tag{4}$$

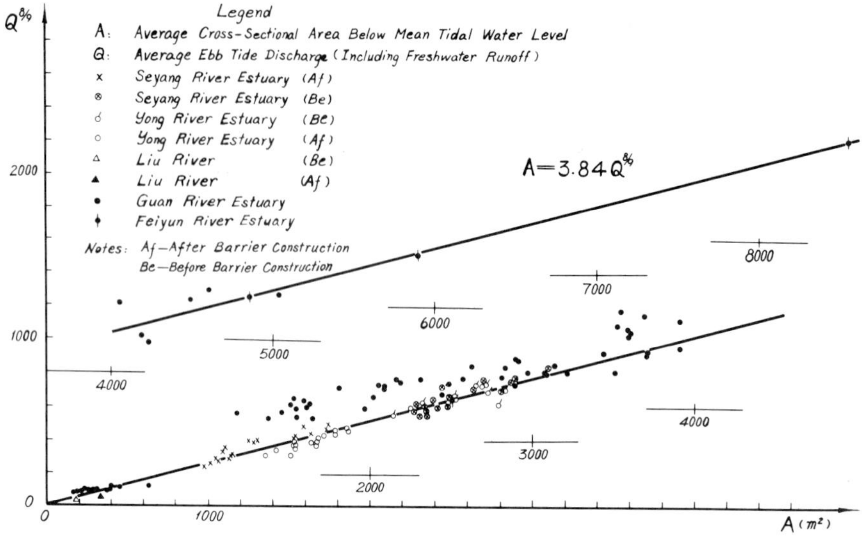

Figure 2. *Relation between cross-sectional area (A) and average ebb discharge rate ($Q^{8/9}$) with respect to river channel morphology of Chinese estuaries.*

which exhibits the same dependency on ebb discharge as Equation 3. Both equations show that a decrease in tidal discharge following gate construction must result in siltation below gates, thus modifying the dimensions of the cross section of river channels.

3. Deformation of tidal waves:

The nature of the tidal wave varies from place to place in the estuary, and its deformation following barrier construction also depends on how far the barrier is located upstream. Generally, when the barrier site is a short distance upriver, the tidal wave changes from a progressive wave to a standing wave or from a quasi-standing wave to a perfect standing wave subsequent to barrier construction. The progressive-wave and standing-wave properties of the tide below the Sheyang estuary barrier are shown in Fig. 3. Changes in wave properties due to barrier construction may be summarized as follows:

(1) $h\uparrow > h\downarrow$ changed into $h\uparrow' \approx h\downarrow'$
(2) $u\uparrow'/u\downarrow' > u\uparrow/u\downarrow$ or $u\uparrow < u\downarrow$ changed into $u\uparrow' > u\downarrow'$
(3) $T\uparrow \approx T\downarrow$ changed into $T\uparrow' < T\downarrow'$; and
(4) $+a\uparrow' > +a\uparrow$

in which h, u, and T represent average water depth, average velocity, and average duration, respectively; $+a$ represents acceleration; "↑" flood tide; "↓" ebb tide; and "′" means "subsequent to barrier construction". After barrier construction, u and du/dt (i.e., $+a'$) increased and h diminished due to the deformation of tidal waves during flood tide and vice versa during ebb tide (Fig. 3). In general, sediment-carrying capacity of a stream is directly proportional to u^3/ghw, and is also dependent upon acceleration in the case of unsteady flow. Therefore, for cases where tidal-current velocity is higher than that of incipient sediment motion, the sediment-carrying capacity of flood currents is increased over that of ebb currents when barriers are constructed, leading to sedimentation below the barrier.

The three factors discussed above interrelate with and interact on one another. Deposition due to a reduction in downstream river flow is followed by a further deformation of tidal waves and a reduction in the tidal prism, leading to decreased ebb discharge, including freshwater runoff, and a further diminution in the downstream transport capacity. Following sedimentation, the river channel readjusts to the new hydraulic conditions, reaching a new equilibrium. However, if no runoff discharge occurs, then the river channel will aggradate to high tide level under these conditions. In this case, without human action complete loss of drainage capability may occur.

Remedial Measures to Reduce Sedimentation

Hydraulic Sluicing

Hydraulic sluicing includes limiting or adjusting river flow or increasing

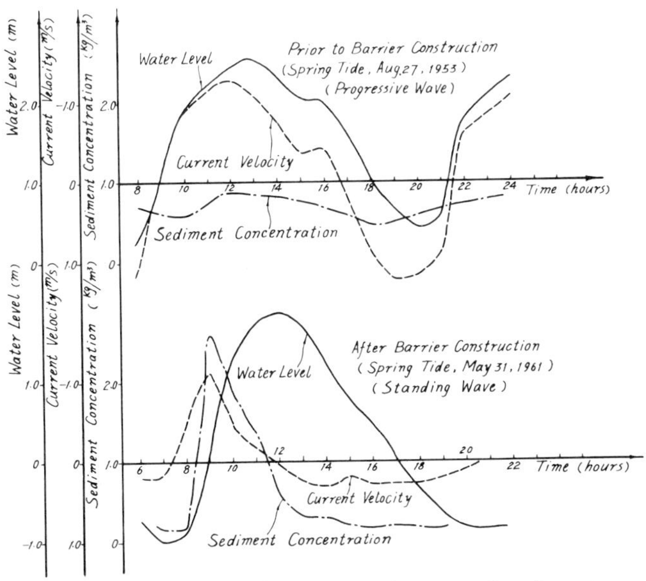

Figure 3. Hydrographs of tidal level, current velocity, and sediment concentration measured 5 km downstream from Sheyang Barrier, before and after barrier construction.

sediment transport in ebb tide and reducing it in flood tide, ultimately to obtain riverbed erosion.

1. Opening the gate at spring tide: If sedimentation is heavy during spring tide and light during neap tide, release of water at spring tide may remove deposited material and check upstream travel of turbid water. In Seyang River estuary, the measured amount of sediment accumulated during spring tide with the gate closed was four to five times larger than during neap tide. The quantity of silt removed by sluicing operations during spring tide was two to three times larger than during neap tide (Dou 1963). One cubic metre of water released during spring tide may be ten times more efficient than one released during neap tide.

2. Opening the gate by means of water head: This can increase transient discharge and transient scouring velocity, the effect of scouring in the reaches immediately below the gate being fairly good. In most cases, delaying opening a gate until a head of 1.0 m builds up results in no accretion within 1.0 km below the gate.

3. Opening the gate after lower high water: In some estuaries with appreciable diurnal tidal inequality, a tidal range difference exists between the two semidiurnal tides, producing different flushing effects. After higher high water, ebb tidal flow is relatively small, low water level is relatively high, and flow from upstream contributes little to velocity increase. After lower high water, ebb tidal flow is relatively great, the water level drops to its lowest level, and the flow from upstream produces a relatively large increase in ebb tidal-current velocity, thus leading to an appreciable scouring effect. Further, releasing water after lower high water also can serve to check a following flood tide with a relatively large tidal range, reducing upstream travel of turbid water. For example, when the Gate of Sayang River was opened after higher high water with an average discharge of 800 m^3 sec^{-1} (total volume of river water discharged = 1,440,000 m^3), and closed at the following falling tide, 18,000 tons of silt were accumulated during the two diurnal tides. Contrarily, when the Gate was opened after lower high water (average discharge was again 800 m^3 sec^{-1} and total volume of river water discharged was 1,440,000 m^3), and closed at the following falling tide, 47,000 tons of silt were scoured away during the two semidiurnal tides. Thus, for equal discharges the scouring effect depends on which tide is used when the gate is opened for discharge of sediments.

4. Diverted-water sluicing: If fresh water enters an estuary below a barrier, the gate can be opened briefly near high water to trap some of the fresh water from downstream to upstream. Opening the gate again when the tide is near low water level flushes the downstream deposited sediment. If there is salt water below the gate and some fresh water can be pumped from another channel to the upstream channel, the gate may be opened under certain water head to move away the deposited sediment.

Agitation Dredging

If there is a scanty water supply from upstream, or if the silt deposited below the gate is relatively firm, agitation dredging may be used to resuspend the deposited sediment which is then carried away by the ebb current. Mechanical agitation, if combined with water release through the gate, gives rise to good results. This method has been successively adopted at the various tidal gates located in Jiangsu and Hebei Provinces near the sea.

In field tests at Xinyang Estuary Gate, two boats of 80 hp each dragged a rake to agitate sediment when the gate was opened and water released. The scouring ability displayed by the same volume of water, with the use of mechanical agitation, was three times greater than without mechanical agitation. Flume experiments indicated that each kind of rake had an optimum dragging speed, that there was an optimum raking time for a given reach of river and a given type of tide, and that shape depended upon the soil type of the riverbed.

The observations in the estuary of New Ziya Floodway (Du et al. 1980) and others in Heipei Province showed that during a natural tidal period with no mechanical agitation, over 70% of the sediment transport by flood tide was deposited on the channel bed downstream of the tidal barrier. With adoption of mechanical agitation, sediment concentration on ebb tide became greater than that of flood tide. Dragging rakes removed deposited sediment in the channel downstream of the tidal barrier of Dou River estuary and the cross-section areas of channel were enlarged to about 80% of the designed area. The resulting cost was only about 10% of the cost of dredging.

Based on the principle that air released under water can produce intensive upward current and mixing, thus resuspending sediment, an aerated-type rake can be constructed from an ordinary rake with holes on its back, coupled with an air compressor. Laboratory experiments showed that such modified rakes increased agitation of bottom sediments and resulted in more vertically uniform sediment suspension, permitting sediment flushing by predominantly seaward surface currents.

In 1977, a field test with an aerated-type rake was carried out at the mouth of Yong River estuary by Zhenghai Harbour Construction Command of Ningbo City and Tianjin Waterways Transportation Engineering Research Institute of the Ministry of Communications. A towboat of 600 hp with a 4 m wide aerated rake was used. Air compressor output was 9 m^3 sec^{-1}. After raking ended, vertical suspended sediment concentration had increased 1.15 ~ 2.02 times compared with that caused by unmodified rakes.

In 1979, the Jiangsu Provincial Engineering Team for Silting Prevention conducted comparison tests with and without the use of aeration in the reach below the Gate of Xinyang estuary. Tests involved five tidal cycles, 58 vertical samples, and 280 measurements. At a mean water depth of 2.5 m, measured mean sediment concentration on the vertical was 7.1 kg m^{-3} for aerated-type rakes and 2.5 kg m^{-3} for nonaerated rakes, a 2.8 times increase.

Measured sediment concentration near the water surface was 2.5 kg m^{-3} for the aerated-type rake and 0.57 kg m^{-3} for the unmodified rake, a 4.4 times increase. For similar conditions, the cost for removing one cubic metre of silt with aerated-type rakes was only one-third of that with ordinary rakes.

Tidal Flushing by Enlarging Tidal Prism

If a tidal gate is located where river flow is low and water diversion from another basin is impossible, tidal flushing may be a solution to sedimentation. This concept and design were proposed by Dou Guoren in 1970 and two field experiments were performed at the Gate of Liangdu River (Jiangsu Province) in January 1973 and March 1978, respectively. As a result, a significant flushing effect occurred in the reach below the gate but shoaling took place above the gate. Laboratory model experiments were performed with a 34 m long tidal flume which was trapezoidal in cross-section (the model scales were: $\lambda_L = 250$, $\lambda_h = 40$). Corresponding to the prototype, the old gate is 4.13 km from the river mouth and the new one 8.5 km distant. With the downstream gate regulated, two series of experiments were undertaken: (1) removal of deposits by use of free discharge after tidal intrusion into the space between the gates; and (2) removal of deposits by use of the tidal fall associated with the closing of the old gate at high tide and reopening at low tide. Experimental results showed that it was possible to increase velocity of ebb current greatly and decrease its duration, to prolong duration of flood current, to increase depth of water during flood tide, and to decrease depth of water during ebb tide. These effects all contribute to strengthening the ability of the ebb current to transport sediment. Thus, this method of tidal flushing may be a solution to the problem of siltation below the gate. Its selection, however, is still dependent upon economic comparison.

Measurements at the Gate of Liangdu River show that it is better to introduce the tidal current at the turn of high water slack and ebb tide when there is a significant reduction in sediment concentration of tidal water below the gate. In order to reduce sediment influx between the gates, the structural design of the gate may be such that some surface fresh water will be admitted. However, some siltation in the intervening space is unavoidable. This silt can be removed by hydraulic sluicing using a small quantity of the water passing through the upper gate.

Comparison and Discussion of Various Measures

1. Hydraulic sluicing uses water energy effectively, is simple in operation, and is effective. Together with mechanical agitation it becomes more efficient. However, some conditions must be satisfied. Adequate water should be stored in network river channels for release at an opportune moment. This method is always applied to the gates along the coast and the banks of the Yangtze River estuary (Jiangsu Province). However, because there is scant

precipitation in northern China (e.g. Hepei Province), with rainfall mostly concentrated in a short rainy season, and because there are no network river channels for sufficient water storage, hydraulic sluicing can hardly be used.

2. Agitation dredging may be adoptable everywhere, the finer the deposited sediment the better. The magnitude of the threshold velocity of a sediment grain on the riverbed is inversely proportional to sediment grain size when its diameter $d_{50} < 0.2$ mm. At the same time, fine silt is more difficult to remove when it is compacted; however, fine sediment can be carried further at the same velocity once it is resuspended. In accordance with China's experience, this method is efficient when sediment grain $d_{50} < 0.1$ mm. Below tidal barriers in Baihai Bay (Hepei Province), deposited sediment suspended by mechanical agitation can be moved by ebb tide flow and run to the sea when $d_{50} = 0.005$ mm.

In deeper water (e.g. $h > 3$ m), more efficiency is achieved with aerated-type rakes. This tool can also be used for deepening the bar draft of an estuary easily and conveniently. It is sometimes cheaper than dredging. However, it is a complicated problem to grasp the opportune time to remove river sediment so as to save money and time.

3. Tidal flushing by enlarging the tidal prism requires construction of another gate, more expense, and complicated control. However, it uses tidal energy instead of mechanical energy. But, farmers do not want salt water to be diverted above a tidal barrier because it will occupy a segment of river channel for storage in place of fresh water for irrigation. Also, some siltation will occur between two gates, especially when upland runoff is low. These problems require further study.

Acknowledgments

Part of this paper had been presented in International Symposium on River Sedimentation held in Beijing on March 24-29, 1980 (Luo and Gu 1980). In this paper, some new data are discussed. We would like to thank Mr. Huang Sheng and Mr. Dou Guoren, Vice Directors of our Institute, for their concrete help and guidance while preparing our paper. Mr. Pan Yingfa, Mr. Ma Qinan, and Miss Haung Yimei joined in this work together. We also thank the concerned departments of water conservancy of provinces and cities along the Chinese coast, especially Jiangsu Province, for giving the information that helped us complete this research. We thank the Tianjin Waterway Transportation Engineering Research Institute for measurement data.

References Cited

Dou Guoren. 1963. Analyses of the sedimentation below tidal barrier of Sheyang River Estuary. Proceedings of Research Reports of the Nanjing Hydraulic Research Institute, River and Harbour Division (1958-1962), Vol. 3, 1963, Nanjing (Chinese).

Dou Guoren. 1964. Regime channel relations for plain tracts and tidal estuaries. *Journal of Chinese Society of Hydraulic Engineering,* No. 2, p. 1-13, 1964, Beijing (Chinese).

Du Guohan et al. 1980. Sedimentation caused by tidal barriers in muddy estuaries on the shore of Bohaibay. International Symposium on River Sedimentation. March 24-29, 1980, Beijing (Chinese, English abstract-in-depth).

Huang Sheng and Ge Zhizhen. 1963. A deliberation on the category of estuaries. Proceedings of Research Reports of the Nanjing Hydraulic Research Institute, River and Harbour Division, Vol. 2, 1963, Nanjing (Chinese).

Luo Zhaosen and Gu Peiyu. 1980. Process of sedimentation and measures of its reduction in estuaries below tidal barriers. International Symposium on River Sedimentation, March 24-29, 1980, Beijing (Chinese, English abstract-in-depth).

SEDIMENT PROBLEMS AND THEIR CONTROL IN U.S. NAVY PIER SLIPS IN SELECTED HARBORS OF THE UNITED STATES

John F. Hoffman
Oceanography Department
U.S. Naval Academy
Annapolis, Maryland

Abstract: United States Navy pier slips in the harbors of Hampton Roads, VA; Charleston, SC; Mayport, FL; and Alameda, CA experience sedimentation problems. This affects berthing of deep-draft vessels such as aircraft carriers where depths of water up to 45 ft (13.7 m) are required. In addition to problems involving grounding and clogging of cooling water systems, both of which render a vessel inoperative, increasing cost of dredging maintenance is encountered. Various means of sediment control in pier slips are discussed. These include the use of single row of water jets, water jet arrays, silt curtains, agitation dredging, and a sand crater-eductor method.

Introduction

Three characteristics desirable in a harbor for navy ships are: easy access to the open ocean, protection from waves and surges that occur in the open ocean, and deep water for berthing. Where water is shallow, channels and pier slips must be deepened by dredging to depths below those established by estuarine hydrodynamics and sediment movement. The initial process is referred to as new or capital dredging. Maintenance of these depths in channels and pier slips requires constant dredging. This repetitive process is called maintenance dredging.

The deepest draft vessels of the U.S. Navy range in draft from 37 ft (11.3 m) to 40 ft (12.2 m). Described below are sedimentation problems occurring in four harbors wherein U.S. Naval deep draft vessels are berthed, together with the associated dredging. These are: Alameda, CA; Charleston, SC; Mayport, FL; and Hampton Roads, VA. The information contained in this paper has been obtained to a large extent by interviews and site visits.

Sedimentation Problems

Alameda Naval Air Station

The ship channel into the Alameda Naval Air Station is roughly 4000 ft (1219 m) long by 1000 ft (305 m) wide, extending from deep water in San Francisco Bay to the eastern end of the Station breakwater (Fig. 1). Project depth of the ship channel, turning basin, and berthing area is 42 ft (12.8 m)

ISBN 0-12-404070-5

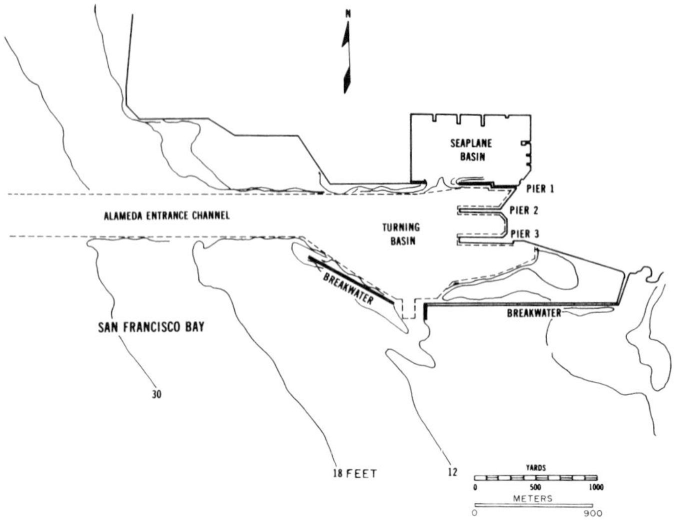

Figure 1. Details of the berthing and turning basin facilities of the Alameda Naval Air Station showing depth contours in feet and location of entrance channel.

below MLLW. Inside the breakwater, the channel widens along the southerly limits to provide a turning basin approximately 4000 ft (1219 m) long and 2500 ft (762 m) wide. Along the northerly edge of the turning basin and in line with the entrance to the berthing area there is a 650 ft (198 m) gap in the seawall which, in the past, permitted access to the seaplane base. Across from the turning basin and opposite the lagoon entrance is a 750 ft (229 m) gap in the breakwater that permits direct access for small vessels into the lagoon area.

The bottom sediments, typical Bay fine-grained silts and clays, are readily agitated, suspended, and transported by wind, wave, and tidal action, as well as by rotation of ships' propellers. Median grain diameters for two grab samples obtained and analyzed by the Corps of Engineers (San Francisco District) in 1971 were 3.8 and 4.6 μm.

Navy siltation records of the Alameda channel, basin, and berthing areas indicate that the rate of siltation from 1957 to 1959 was two to three times as great as from 1959 to 1961. Each period was preceded by dredging. The rate was approximately 5-6 ft (1.5-1.9 m) per year from 1957 to 1959, and 2 ft (0.6 m) or less per year from 1959-1961. Measurements were made in the channel, turning basin and adjacent to the south side of Pier No. 3. An evaluation made by the U.S. Army Corps of Engineers (Fig. 2) depicts areas to be dredged to bring channel, basin, and berthing areas to operational depth of 42 ft (12.8 m) plus 1 ft (0.3 m) overdredge.

Muds appeared to be migrating from the shallow seaplane basin into the main turning basin. There was a depth differential of about 25 ft (7.6 m) between the seaplane basin and the turning basin. A tongue of sediment ex-

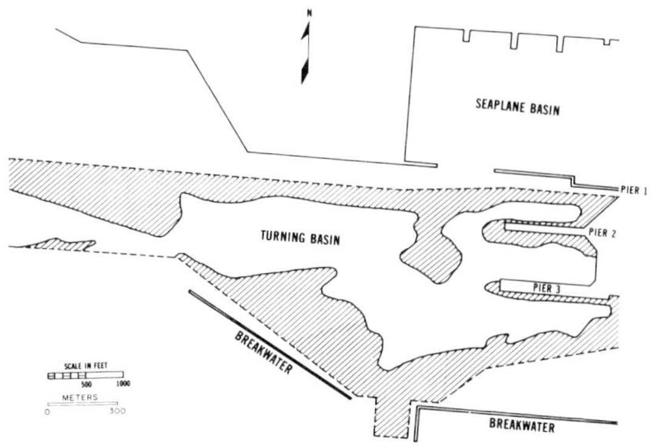

Figure 2. Siltation patterns in the waterway at Alameda Naval Air Station showing areas to be dredged in 1973 (cross-hatched).

tended outward some 1000 ft (305 m) into the turning basin from the entrance to the seaplane basin.

Annual maintenance dredging volumes in the period 1950-1977 ranged from <1 million yd^3 (0.8 million m^3) to >1.6 million yd^3 (1.2 million m^3). The average annual volume is 1 million yd^3 (0.8 million m^3). Maintenance dredging in the main ship channel and turning basin is by hydraulic and hopper dredges, and in the small boat dock and pier areas by bucket and scow. A dredge spoil disposal site in the vicinity of the entrance to Alameda Naval Air Station, approximately 58 ft (17.7 m) below MLLW, was utilized prior to 1971. In 1972, critically shoaled areas were excavated using a hydraulic dredge to pump the sediment to a land disposal site at the station. Currently, because of State and Federal environmental controls, dredge spoil is dumped in the vicinity of Alcatraz Island where the water depth and tidal action are believed to be adequate for rapid dispersion of the sediments.

Charleston Naval Complex

Charleston Harbor is formed by the Cooper, Ashley, and Wando Rivers. The Cooper River, the principal tributary of Charleston Harbor, is a tidal estuary formed by the confluence of its East and West Branches (locally termed "The Tee") at a point about 32 miles (51.5 km) upstream from the harbor entrance.

Several major sediment types have been deposited within the Charleston estuary. These include a longshore drift and continental shelf sand component being deposited over the major part of the estuary itself, and Holocene sand bars present within the landward rivers (Corps of Engineers 1976). The longshore drift shelf sand is concentrated both in the vicinity of the harbor mouth where it grades seaward into continental shelf sands as well as along the north half of the estuary to the vicinity of Mt. Pleasant. Bottom samples obtained by the Corps of Engineers (Charleston District) in the vicinity of the jetties and landward between Ft. Sumter and Ft. Moultrie contain >90% sand-size materials. Landward of these locations the sand fraction is intermixed with silt and clay increasing abruptly toward the west and more gradually toward the north (Corps of Engineers 1976).

Sand also occurs in bottom sediments in the Wando and Cooper Rivers. The accumulations here are related to Holocene and recent channel deposits. In the Ashley River, similar deposits occur at depth, but surficially are buried by recent silt and clay.

Between the two sand components the floor of the Charleston estuary is covered by dark-gray sludge composed of >75% silt and clay. Within the area of occurrence of the sludge there is no apparent relationship between physical size characteristics, water depth, and harbor currents. It is assumed that the silt-clay fractions are present in flocculated state such that their aggregate masses behave physically as much coarser particles.

Beginning in 1942, the rate of shoaling in Charleston Harbor increased greatly. Deposits of black muck material began to settle in the harbor. Large shoals began to form in the project channels. After removal by dredging they reformed quickly in the same locations so that frequent dredging became necessary to maintain project depths. Slips which were easily maintained before began to shoal rapidly to low tide level unless dredged. At some locations slips which had been dredged to full depth would refill in little over a year unless cleaned out (Corps of Engineers 1976).

Two physical changes affecting Cooper River had occurred just before the increased shoaling became evident. The Santee-Cooper hydroelectric project (Pinopolis hydroelectric station) began operation in early 1942 and the navigation project depth in Cooper River was increased from 30 to 35 ft (9.1 to 10.7 m) in 1941 and 1943. It was evident that the latter change could be of little effect for most of the project channel was naturally deeper than 35 ft (10.7 m) prior to 1940. Further model studies at the U.S. Army Waterways Experiment Station, Vicksburg, MS, showed that deepening the navigation channels would have increased shoaling by only 10% (all other factors remaining equal). Therefore, only the increase in the flow in the Cooper River (caused by diversion into the river as a result of the Santee-Cooper project) could account for the marked change. It was thought at first that the condition might be temporary pending a period of adjustment but this proved not to be the case.

The annual volume of maintenance dredging in the Charleston Harbor area increased from <0.5 million yd^3 (0.38 million m^3) prior to 1942 for the entire harbor to a volume estimated to be >10 million yd^3 (7.6 million m^3) at the present time.

Shoaling rates in Charleston Harbor vary from year to year. The most significant factor influencing these rates is the fresh water diverted into the harbor by the Santee-Cooper Hydroelectric Project. In comparing inflow rates and dredging rates it is readily apparent that dredging necessary to maintain navigability increases as the discharge into the Cooper River at the Pinopolis Hydroelectric Power-plant increases (Corps of Engineers 1976).

Dredging in the pier slips of the Charleston Naval Station and Naval Shipyard is accomplished by means of the Navy-owned and operated dredge "ORION", a hydraulic cutterhead dredge with a 20-in (50.8 cm) diameter suction line and a 16-in (40.6 cm) diameter discharge line. The depth maintained is essentially 35 ft (10.7 m) below MLW. The frequency of dredging and target depths are shown in Fig. 3. The average volume of spoil dredged annually is 2.7 million yd^3 (2.1 million m^3).

The navigation channel in that portion of the Cooper River abutting the Naval facilities is maintained to a depth of 35 ft (10.7 m) below MLW by the Corps of Engineers by means of another hydraulic cutterhead dredge. The dredge spoil amounts to about 1.3 million yd^3 (1 million m^3) annually. Spoil is pumped by pipeline to the two disposal areas (Clouter Creek and Yellow Creek).

Mayport Naval Station

Mayport Basin (Ribault Bay) is about 0.5 mi (0.8 km) long and about 700 yd (640 m) wide. It is located on the south side of St. Johns River, Florida, about 1.5 mi (2.4 km) west of the river entrance into the ocean

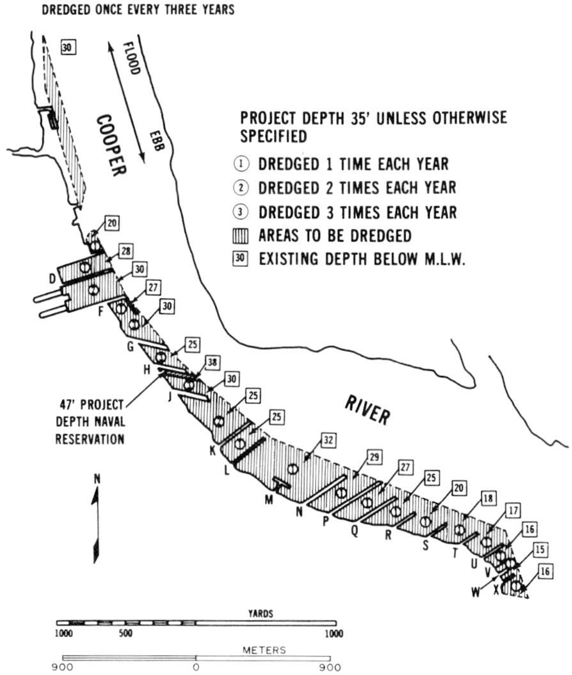

Figure 3. Areas dredged by the U.S. Navy in and around Navy piers and slips at Charleston, SC, Naval Facilities.

(Fig. 4). The river at the point of entry is kept open by means of jetties.

Based on test borings made for the Corps of Engineers (Jacksonville District) the bottom of Mayport turning basin is black soft sandy clayey silt about a foot (0.3 m) thick underlain by fine to medium quartz sand containing some silt and clay. Berth C-1 was reported by the Chief Engineer of a carrier in port to have a build-up of sediment from October 1978 to May 1979 from 42 ft (12.8 m) to 37 ft (11.3 m) below MLW. The shoaling rate was 5 ft (1.5 m) in seven months or at the average rate of about 0.7 ft (0.2 m) per month in berth C-1.

Piers C-1 and C-2 normally experience a rapid build-up of sediment. Scuba divers have observed an underkeel clearance of deep draft vessels of about 1.5 feet. With sediment as close as that to the sea chest and other water intakes, clogging of piping, heat exchangers, etc., becomes a problem.

The amount of spoil dredged annually ranged from 0.3 million yd^3 (0.2 million m^3) to 1.96 million yd^3 (1.5 million m^3) with the average dredging volume in the 20-year period 1959-1978 being 0.6 million yd^3 (0.5 million m^3). It is estimated that about 85% of the spoil comes from the turning basin with the remainder from the channel and from near the bulkheads.

The methods of dredging the slips and turning basin of Mayport Naval Station include hopper dredge, hydraulic pipeline-cutterhead dredge, and bucket and scow. The hopper dredge and scow transport the spoil to an offshore dumping site. The hydraulic dredge was used to pump the spoil onto waste land at the Naval Air Station. (This is no longer used as a disposal area.) Dredging the turning basin by hopper dredge removes only the accumulated sediments in the turning basin. The area between the circular configuration in which the hopper dredge moves and the rectangular shape of the waterway is not dredged. Consequently the sediment accumulates in the

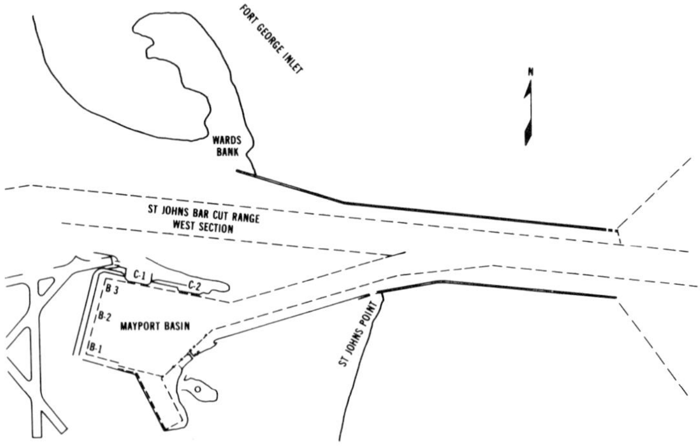

Figure 4. Mayport, Florida, channel approach and turning basin.

berthing areas for aircraft carriers (C-1 and C-2) and is not removed unless dredged by hydraulic pipeline-cutterhead dredge or by bucket and scow.

Norfolk Naval Station

Hampton Roads (or Norfolk harbor) is utilized by both military and commercial shipping. Four rivers flow into Norfolk harbor. These are the James, Elizabeth, Layfayette and Nansemond Rivers. At Hampton Roads, Entrance Reach separates into two channels – Newport News Channel and Norfolk Harbor Reach. Newport News Channel terminates at Newport News where the James River has a depth of from 46 to 64 feet (14 to 19.5 m). Norfolk Harbor Reach, the channel servicing the U.S. Naval Station at Sewell's Point, terminates at the Elizabeth River. All the aforementioned channels are maintained at a depth of 45 ft (13.7 m) below MLW.

Based on test borings made under the sponsorship of the engineering firm of Froehling & Robertson in 1978, sediments dredged from within the pier berths are silt to coarse clay probably resulting from the flocculation of suspended sediment and from bedload movement in the rivers flowing in to Hampton Roads. Available information as to the change of salinity of the harbor water with depth indicates that a tidal wedge exists in Hampton Roads (U.S. Naval Base, Norfolk, VA, 1979). Inflowing river water containing suspended sediment mixes with saline water in the harbor, causing a coalescence of the suspended particles which flocculate out in the salt wedge (Hoffman 1976). There is also a possibility that sediment is carried in from the lower part of Chesapeake Bay by the inflowing tidal current. On the other hand, Byrne (1972) in a study of harbor conditions relating to the effect of a proposed bridge-tunnel crossing of Hampton Roads near Craney Island indicates that "the tidal flow over Hampton Flats is such that flood currents dominate over the ebb currents. Winds from the northeast through the east to southwest have sufficient fetch to generate waves capable of stirring the bottom sediments of the Flats. Given the frequency and intensity of winds in the area coupled with tidal currents, the direction of the net bottom sediment-transport should be to the southwest from the Flats."

Shoaling rates for the individual pier slips at the U.S. Naval Station range from 1.5 to 2.0 ft/yr (0.5 to 0.6 m/yr) based on the volume of spoil dredged over a known time period (U.S. Naval Base, Norfolk, VA, 1979).

Some idea as to siltation in Pier 12 is shown in Fig. 5 (Brehmer et al. 1967). Depicted are areas where the berths shoaled from 1-2 ft (0.3 to 0.6 m) and from 2-3 ft (0.6 to 0.9 m) in an 18-month period from 1961-1963.

Dredging of the Naval Station pier slips is accomplished by means of a hydraulic cutterhead pipeline dredge. Volumes dredged annually from the slips at the Naval Station range from about 0.2-0.9 million yd^3 (0.15-0.7 million m^3).

Dredge spoil from Norfolk harbor channels and pier slips is pumped to Craney Island, a diked-disposal area located on the south side of Norfolk har-

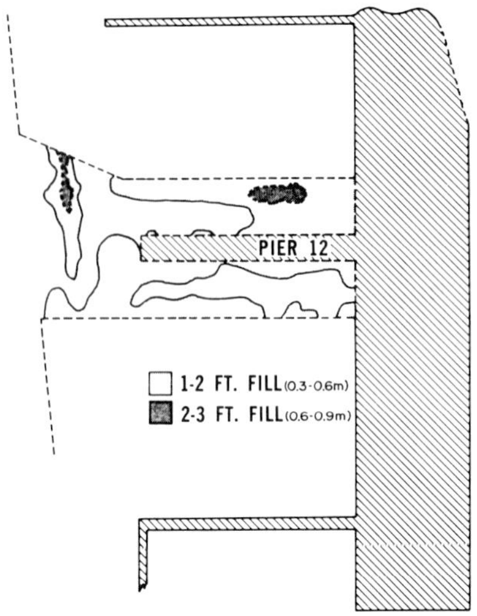

Figure 5. Distribution of siltation rates in Pier 12 basins during an 18-month period, 1961-1963 (after Brehmer et al. 1967). Cross-hatched area represents land and pier facilities.

bor. A permanently installed pipeline extending along the bottom of the harbor permits hydraulic cutterhead pipeline dredges to pump directly to the disposal area through a temporary connection without interfering with navigation.

Sediment Control

Sediments at any given location in an estuary are derived from four sources: the watershed, the tributary river channels, other portions of the estuary itself, and the open ocean. The complexity of the factors involved preclude control of the sediments at their source. Consequently, control of sediments in harbors, aside from dredging, involves diverting the sediments from critical areas, trapping or blocking the sediments before they reach critical areas, and/or resuspending the sediments and removing them from critical areas.

At present, control of shoaling in Navy pier slips is by dredging. However, increases in dredging costs and the problem of dredge spoil disposal have motivated the search for other methods of sediment control. The U.S. Navy has tested or is investigating four methods of controlling sediments in problem areas. These are:

- Water jets – both single line and arrays
- Silt curtains

- Eductors
- Agitation dredging

Water jets

Use of jets to flush a berth at the appropriate stage of tide represents one method of decreasing siltation in a pier slip. Briefly tested at Mare Island Ship Yard, Vallejo, CA, is such a system. One jet arrangement is a multi-branch array of hydraulic jets designed, fabricated, and emplaced on the pre-dredged mud bottom of a capital ship berth at Mare Island Naval Shipyard. Water is delivered at the rate of 1600 gpm from a 150 horsepower pump. This system has been designed by Scripps Institution of Oceanography under sponsorship of the Naval Facilities Engineering Command (R&D) NAVFAC.

Actuated twice daily for 35 min in concert with the ebb flow in Mare Island Strait, the sequentially-operated jets removed sediments deposited in the berth. An area in an adjacent berth having a similar size and exposure was used as a control. However, gradual settlement, side slumping, and disruption of the array elements by the dragging of anchors has led to a reconsideration of the initial design.

Silt Curtains

Presently under test in Mare Island Naval Shipyard, sponsored by NAVFAC R&D, is the effect of a silt curtain made of 5-ply Hypalon fabric. This curtain can be made to float vertically to prevent silt from entering the pier slip or can be collapsed to the bottom when a ship is scheduled to enter the berth (Jenkins et al. 1981).

Silt curtains are also being considered for use in the Port of Rotterdam, Holland (Hoffman 1978). The Office of Chief Engineer of the Port of Rotterdam has studied 26 different methods of the construction of silt screens across harbor entrances. The plan selected for use in one of Rotterdam's harbors involves a curtain that will be about 925 ft (282 m) long and will cost about $3,250,000 (1978) to construct. It is expected to reduce the dredging in the harbor by 30% which represents an annual savings of $1,000,000.

Eductors

Eductors for dredging pier slips are useful in non-cohesive material. The basic eductor works on the principle of the Venturi tube. When a jet of water is constricted in a tapered tube, a vacuum is created. Fluid from the surrounding environment moves towards the chamber. If the eductor rests on a sandy submarine bottom, both sand and water are sucked into the chamber and passed along with the flow. If the eductor is on a flexible hose, it sinks into the sand to form a crater. Fluidization of the sand by water jets increases the flow of the sand-water mixture to the eductor.

Such a system is used to pass littoral drift beneath two jetties with the twofold purpose of preventing shoaling of Rudee Inlet and preventing beach starvation at Virginia Beach, VA. Once deposited on the other side of the jetty the sand moves parallel to the shore in a northerly direction. The system was designed and installed by the Corps of Engineers Waterways Experiment Station and later sold to the Virginia Beach Erosion Commission at the latter's request (Hoffman 1977).

Agitation Dredging

Agitation dredging is the removal of sediment from pier slips and wharves located adjacent to shipping channels by suspending the sediment by agitation at the time of an ebbing tide. The suspended sediment is carried to the main channels by the outflowing water and thence down-channel. Agitation is accomplished by dragging an I-beam or similar device behind a tug or by means of the wash from boat propellers. At present the procedure is not employed by the U.S. Navy.

The dragged I-beam method of agitation dredging is best suited for dredging small accumulations of sediments such as one to three feet (Hussey et al. 1975). This process can dredge large areas in a very short time. Dragged I-beam dredging is advantageous in maintaining the project depth within very close limits thus avoiding the need to dredge overdepth for improved efficiency of the dredging process. Overdepth dredging in the Savannah River and the Cooper River has been shown to increase silting rates.

Agitation dredging of pier slips can also be accomplished by utilizing propeller wash. A craft (with a sizable propeller wash baffled downward) moving forward and backward during the ebbing tide can cause the deposited sediments to be resuspended and moved to the main channel.

The Corps of Engineers U.S. Army District, Portland, OR, sponsored a study of the agitation effect of a twin-propellered LCM vessel modified with a stern-mounted deflector plate to direct the propeller wash downward. The results of field studies made at four sites in Tillamook Bay, OR, where the water depth was about 10 ft (3 m), indicated that the initial rate of sediment movement was approximately 3300 yd^3/h (2440 m^3/h) tapering off in about 6 min to a steady-state rate of about 1700 yd^3/h (1300 m^3/h). The median grain diameter of the Tillamook sediment was 0.19 mm with a specific gravity of 2.68. At present this approach for Navy purposes has not been implemented, although some trials with sea-going tugs have been made.

References Cited

Brehmer, M. L., N. M. Nichols and D. R. Calder. 1967. Study and Control of Marine Fouling Organisms, Naval Base, Norfolk, VA. Contract Report NBY-46710, Virginia Institute of Marine Sciences, Gloucester Point, VA.

Byrne, R. J. 1972. Physical and Geological Studies of the Proposed Bridge-Tunnel Crossing of Hampton Roads Near Craney Island. Special Report No. 24. Applied Marine Science and Ocean Engineering. Virginia Institute of Marine Science, Gloucester Point, VA. pp. 225-234.

Corps of Engineers, 1976. Final Impact Statement – Maintenance Dredging of Charleston Harbor, Ashley River, and U.S. Navy Channels in Cooper River. March 1976.

Hoffman, J. F. 1976. Decrease in Harbor Maintenance Dredging Through the Use of Pile Dikes and Related Structures Together with an Analysis of Estuarine Sedimentation Problems. Report EPRD-29. U.S. Naval Academy, Annapolis, MD.

Hoffman, J. F. 1977. An Investigation of New Methods for the Maintenance Dredging of Pier Slips and an Investigation of Selected Dredging Problems in U.S. Navy-Connected Harbors. Report EPRD-37. U.S. Naval Academy, Annapolis, MD.

Hoffman, J. F. 1978. European Dredging – A Review of the State of the Art. Report R-12-78. Office of Naval Research (London).

Hussey, Gay, and Bell, Inc. 1975. Engineering Report on Agitation Dredging in Savannah Harbor, State of Georgia. Consulting Engineers, Savannah, GA.

Jenkins, S. A., D. L. Inman and W. G. Van Dorn. 1981. The Evaluation of Sediment Management Procedures, Final Report. Contract Report N00014076-C-0631 Scripps Institution of Oceanography, LaJolla, CA. Phase 4.

U.S. Naval Base, Norfolk, VA., 1979, Public Works Officer, Pers. Comm.

SEDIMENT CONTROL THROUGH DREDGING

Adolph W. Mohr
Corps of Engineers
30 Pryor Street
Atlanta, Georgia

Abstract: Sediment control is necessary, but its traditional benefit-cost comparison is no longer the sole criterion for its use. Control can be achieved in various forms of river training and dredging. Dredges are divided into hydraulic and mechanical dredges, their dominant types being the cutterhead, hopper, and endless chain bucket dredges. The operating principle of hydraulic dredges is more complex than that of mechanical dredges and is subject to several constraints. Yet, nearly all dredging in this country is hydraulic, with a more even distribution among the other types in the rest of the world. Dredged material density increases in importance due to increases in transport distances, environmental concerns, and fuel consumption. It also affects some sediment measurements which can be obtained in place or while the material is handled. Upon comparing dredging to river training, neither method is universally advantageous. Both methods will continue to exist side-by-side and be decided upon on a case-by-case basis.

Introduction

In this paper, the term "sediment control" is confined to methods aimed mainly to maintain navigable channels. This paper provides an overview of channeling for navigation, indicates several basic methods to achieve channeling, and outlines constraints. Obviously, no person is expert in all methods nor completely impartial. This conference is designed to discuss nondredging methods of sediment control. Yet, dredging must be included to complete the picture and serve as a measuring stick to evaluate nondredging methods. This is the main purpose of this paper.

Discussion

1. No Control Versus Control

Theoretically, restricting navigation to deep water or reducing vessel draft sufficiently eliminates the main need for sediment control. It is true that with increases in construction costs, pollution awareness, and energy scarcity, more thought ought to be given to the justification of some of our continuous efforts. Some moves in this direction are:

a. Construction of offshore unloading facilities for deep draft ocean-going vessels with further transport of goods by light draft vessels or pipeline.

b. Reduction in the use of man-made structures on eroding beaches and the number of beaches replenished with material.

ISBN 0-12-404070-5

c. Elimination of maintenance of authorized navigation projects which historically have a low benefit-to-cost ratio.

Nevertheless, aside from theory, in the real world the alternative of avoiding sediment control to aid navigation is unrealistic. In fact, there exist good reasons to expand existing channels in depth and width and to create new ones. Yet, where we consider it worth the effort to create channels and canals, experience shows that irrespective of our efforts, nature attempts to reverse them. This results in continual maintenance dredging.

2. Money and Other Costs

Until about 25 years ago, our efforts were nearly exclusively measured in terms of money. This simple system is now considered inadequate. Pollution control and energy conservation have been acknowledged as important parameters. These two constraints are not directly related to each other and generally increase cost. Efforts to quantify or measure them and relate them to cost have largely been unsuccessful. It is now recognized that fuel, regardless of its current market value, is an irreplaceable resource with a real value not yet defined. Still, available methods to achieve our goals are now evaluated as to their relative fuel consumption or fuel intensiveness.

Pollution concerns are even hazier. They lack meaningful measuring units and are not well defined. For instance, it is recognized that pollution may consist of turbidity, chemical, or biological pollution. Measuring units for these three types are vague and generally only limiting values are given. No attempt is made to combine the three basic pollution forms in one measuring system. Adding to these problems, such intangibles as aesthetics, public acceptance, and effects on fish and wildlife and recreation leave any evaluation to considerable judgment. This judgment is influenced by an individual's background, job, agency, and outlook on life, and is highly subjective.

It is generally accepted that water transport is least costly and least energy intensive. The Tennessee Tombigbee Waterways Authority (1) reports this energy comparison:

Mode of Transport	BTU per Ton-Mile
Water	500
Rail	750
Pipeline	1,850
Truck	2,400
Air	63,000

It appears to follow that the water transport mode requires the least BTU per Ton-Mile, and also pollutes the least. The foregoing is a strong justification for providing channels to permit water transportation.

3. Channeling Methods Defined

For the purposes of this discussion, the various methods of channeling can be shown and defined as outlined on Fig. 1. In addition to varying costs, each of these methods has its merits and demerits. For instance, river training does not pollute, but is considered to spoil aesthetics and interfere with water sports and wildlife. Dredging does not involve these demerits, is quicker and more predictable, but is considered to pollute, is repetitive, and is fuel intensive.

In regard to dredging depths and quantities, several terms are important which unfortunately are not uniformly defined and accepted. Some of them are (Fig. 2):

a. Allowable side slope: A slope equal to or slightly flatter than the natural angle of repose.

b. Required excavation yardage: The in-place volume of material to be

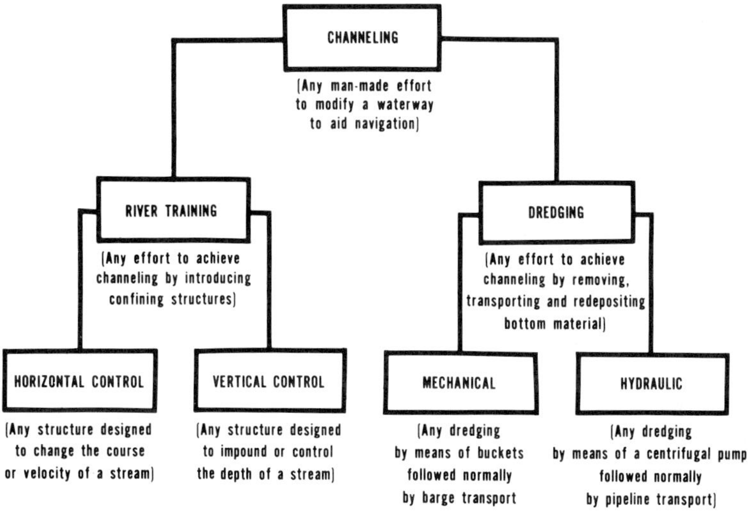

Figure 1. Methods of channeling.

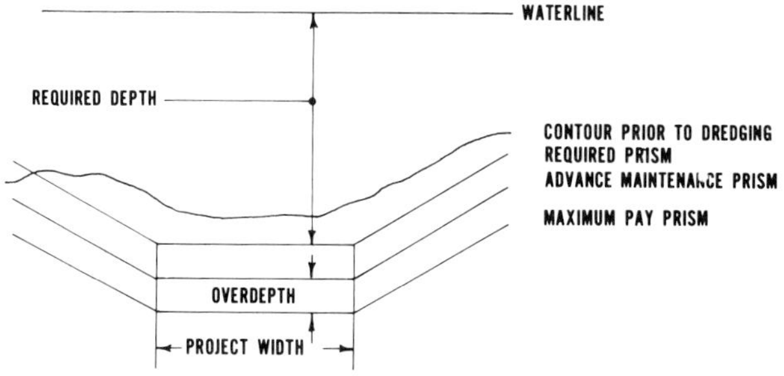

Figure 2. Dredging depths and quantities.

removed from within the required prism.

c. Advance maintenance yardage: The in-place volume of material to be removed from a prism a specified distance below the required prism (not always used).

d. Pay overdepth yardage: The in-place volume of material actually removed below the required prism (b) or advance maintenance prism (c) up to the maximum pay prism. This is thus a specified distance below the required prism.

Note: some agencies do not add side slopes to the advance maintenance and/or overdepth prisms. This appears unrealistic since as per (a) above the side slope material will slide to the bottom and will have to be dredged. Other agencies extend the required side slopes to the bottom of the advance maintenance depth. This appears undesirable since shoaling often concentrates at the sides and thus quickly decreases the required project depth.

4. Dredging in General

Dredging is an ancient profession which has added considerably to the development of civilization. Dredges have connected oceans and moved more material than any other means. A definition for a dredge would be: "An earth-moving machine specialized to remove subaqueous bottom material." Essentially, dredging is performed to gain bottom material or to increase water depth. Only in isolated instances are both goals desirable in a particular assignment.

The U.S. Army Corps of Engineers dredges approximately 350 million cubic yards annually for waterways, harbors and some beach nourishment, including some work for other federal agencies. Dredging for nonfederal accounts is not well documented but is believed to approximate the federal work load. At an average dredging cost of $1.50/cubic yard, this results in an annual nationwide dredging expenditure in excess of one billion dollars (exclusive of sand, gravel, and shell dredging).

In the context of this paper, we are essentially concerned with that form of dredging in which material removal is of prime importance and the material itself is an unwanted by-product. However, for either purpose, the basic function of a dredge is to remove material from the bottom of a body of water, lift it above the surface, and dispose of it in one of several ways. The machines performing this function can be divided into mechanical and hydraulic dredges. In terms of volume handled, these dredges account for nearly all dredging worldwide. In the United States, hydraulic dredging predominates, with a more even distribution of methods in the rest of the world. The earliest dredging efforts were all mechanical in nature. Hydraulic dredges have only been in existence since the middle of the last century.

5. Present Day Dredge Types

Modern dredge types can be classified as in Fig. 3. This figure shows three mechanical and three hydraulic dredge types with the bucket and pipe-

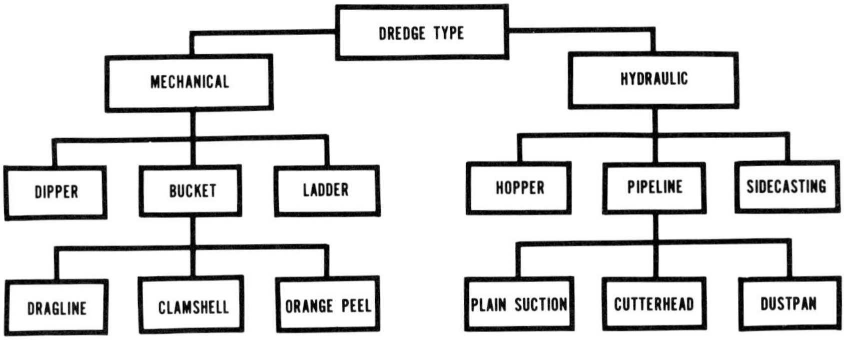

Figure 3. Present day dredge types.

line dredges again divided into three types each. The production for each dredge type varies considerably and depends greatly on dredge size and the material dredged. In terms of production, the ladder dredge predominates in the mechanical field and the cutterhead and hopper dredges in the hydraulic field. Only these dredge types are discussed here.

On a ladder dredge, an endless chain of buckets cuts material off the bottom, moves it to the top of the ladder, and dumps it on a chute to slide into a barge (Fig. 4). On a cutterhead dredge (attached to a pipeline) the cutterhead cuts material off the bottom adjacent to a suction inlet. The dredge pump moves the material up the ladder and discharges it through the pipeline (Fig. 5). On a hopper dredge a draghead picks material off the bottom (Fig.

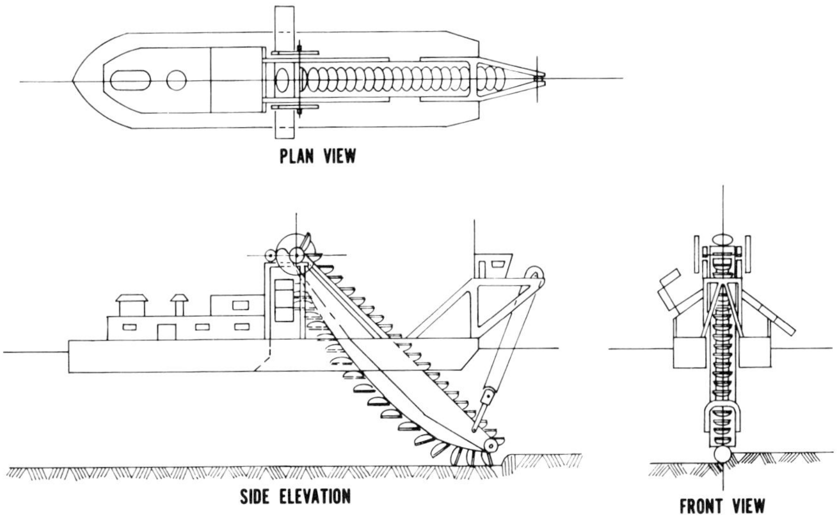

Figure 4. Bucket ladder dredge.

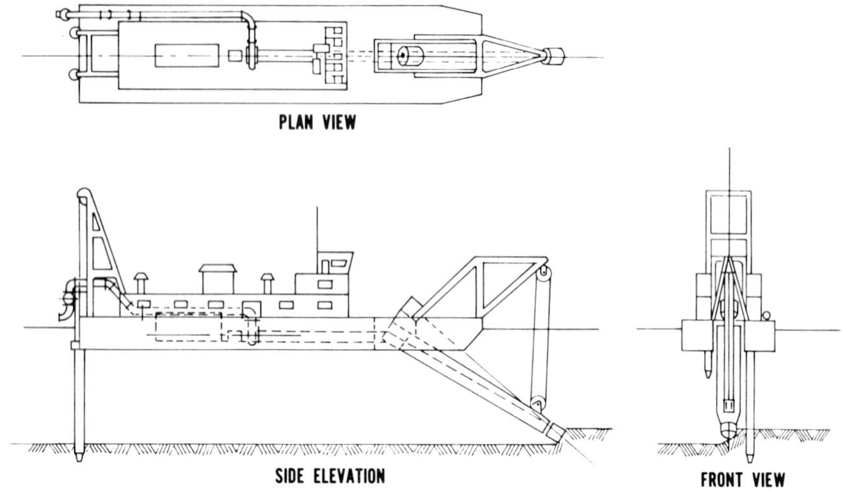

Figure 5. Pipeline dredge.

6). The pump delivers it into the hold of the vessel. The dredge then ceases dredging, proceeds to the disposal area, and dumps the material.

6. Hydraulic Operating Principle

While the mechanical dredging operation (ladder dredge) is simple and intuitively clear, the hydraulic dredge operation (Fig. 7) deserves a discussion. Atmospheric pressure is expressed as 30″ Hg, 34′ H_2O or 14.7 PSI. A

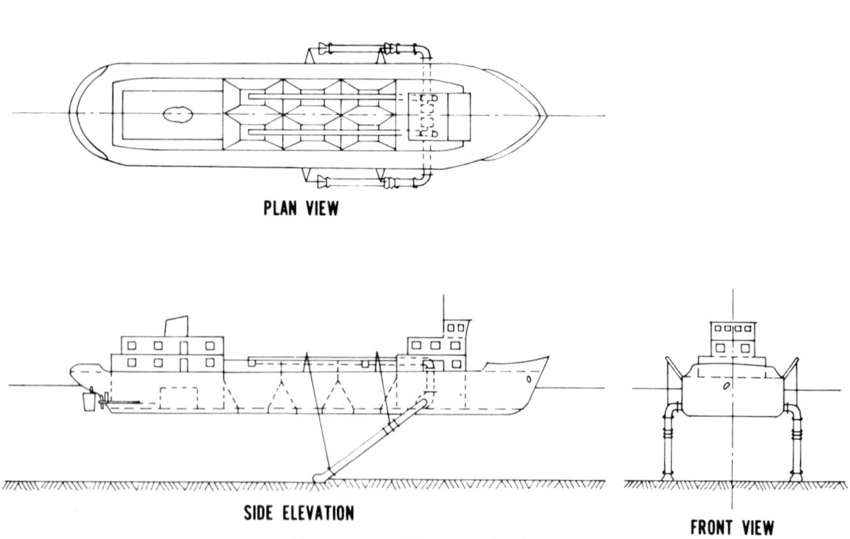

Figure 6. Hopper dredge.

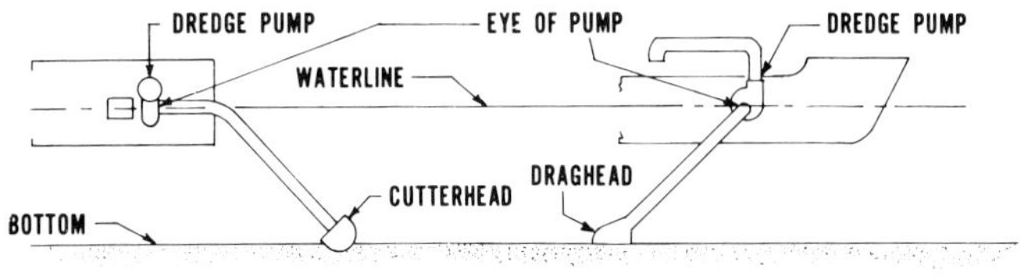

Figure 7. Hydraulic dredges.

dredge pump removes a part of the atmosphere at its eye, say 20″ Hg, leaving a remaining or absolute pressure of 10″ Hg. It is the atmospheric pressure on the outside which drives the dredged mixture to the pump. That is, it must form the dredge mixture, accelerate it, push it to the pump, and overcome friction on the way. This implies that the suction side of the pump is normally not limited by the available power but by the atmospheric pressure. Once in the pump, the pump can push it through a line length depending essentially on its power.

Hydraulic dredges are frequently limited by dredging depth, the material to be dredged, entrained gas, and debris. In addition, cutterhead dredges are limited by pump power, cutter power, or the speed of advance. Hopper dredges are limited by their maximum draft or their hopper volume.

7. Hydraulic Versus Mechanical

Past dredging developments have differed between the United States and Europe (2). Until the end of World War II, the bulk of dredging was performed by hydraulic dredges in the United States and by mechanical dredges in Europe. This difference in basic dredge types on both sides of the Atlantic for similar work is most likely the result of tradition, ingrained skills, and the relation of labor to fuel costs. These differences diminished after World War II. As a result, Europeans developed a modern hydraulic dredge fleet but maintained their mechanical ladder dredges. However, ladder dredges were not introduced in the United States.

In assignments where hydraulic and mechanical dredges can be used, the salient differences between these two basic dredge types are:

a. Mechanical dredges normally utilize barges for material transport. Hydraulic dredges normally transport the dredged material via a pipeline or in hoppers.

b. Mechanical dredges do not require dilution water and obtain the dredged material with a minimum of water. Hydraulic dredges require dilution water, which frequently amounts to 100% while dredging mud and 300% while dredging sand.

c. Mainly due to the absence of dilution water, mechanical dredges normally require appreciably less power than hydraulic dredges.

d. Mechanical dredges furnish the bottom material essentially as it is picked up, while the effluent from hydraulic dredges is frequently graded as it settles out.

e. Due to operating differences, mechanical dredges require slightly more personnel than hydraulic dredges.

8. Other Dredge Types and Methods

In addition to mechanical and hydraulic dredges, there are several other dredging methods. They are mentioned here essentially to complete the picture and because they may play a more important part in the future. Their total present-day production in cost or volume worldwide is estimated at about 1%.

a. Water jet method: As a dredging aid, a series of water jets in the entrance of a suction assemblage is directed toward the dredge pump. This provides kinetic energy to the flow stream and thus aids the pump. This method is often beneficial where the dredging process is limited by conditions on its suction side.

b. Air lift method: If finely dispersed air is applied to the entrance of a suction assemblage, the resulting air-water mixture will rise above the water level surrounding the assemblage. This principle has been used by divers to dredge relatively small quantities of material.

c. Pneuma method: Here, a tank is lowered to the bottom, emptied with compressed air, and then vented to the atmosphere. When a check valve in the tank bottom opens, water and scoured bottom material rushes in. The bottom check valve closes when compressed air is applied to push the contents of the tank into a discharge line. The tank is again vented to the atmosphere to repeat the cycle. Use of three tanks in sequence and an automatic valve-actuating mechanism results in a nearly continuous discharge. A modification of this principle applies suction after venting the tanks. As water depth increases, dredges using this principle are increasingly able to pick up material. The reverse is true for hydraulic dredges.

d. Agitation dredging: Here, material is dislodged from the bottom by agitation and is transported by the moving surrounding water. Agitation can occur mechanically, by water or air jets, and by normal or directed propeller wash. This is one of the oldest methods of dredging but is now frequently constrained due to associated turbulence. One large-scale operation remains in the Mississippi Passes where several million cubic yards of material are dredged annually.

e. Dredging with bars or rakes: This method is used to remove material from confined areas which are difficult to dredge, such as sites adjacent to piers or in boat slips. It normally requires that the material moved be rehandled by other means.

9. Density Considerations

Density of the material to be dredged is of utmost importance, since it affects dredging time, wear and tear, and volumetric measurements. The significant densities in dredging, and their ranges and vertical distribution, are shown in Fig. 8. Virgin material may be of any type and is normally more compacted and therefore denser than maintenance material. Maintenance materials occur most often as mud and sand. Figure 8 shows the average density ranges for the hydraulic dredged mixture, mud, sand, and solid rock. The dredged mixture density is essentially constant for all noncohesive materials (slightly less for broken rock). Since the particle density of all dredged material is that of rock (approx. 2650 gr/l), mud contains more interstitial water than sand. Sand bottoms are well defined. Light density mud bottoms are defined (1) by using depth soundings, (2) as the interface between a flowing medium over a stationary one, and (3) as interference with shipping. Mud bottom elevations determined by these three methods often disagree. Sand density changes little below the bottom, but mud density normally changes appreciably. Therefore, the mud density in the shoal must be averaged. As a rule, in the past the dredged material rate was maximized. Due to increases in transport distance, fuel consumption, and environmental concerns, the emphasis is shifting toward maximizing dredged mixture density.

10. Sediment Measurements

Sediment measurements determine the progress made in dredging and form the basis for payment. They can be performed in the dredging area (surveying the hole made) or in the disposal area (surveying the mound made). They can also be made while the material is in transport either flowing through piping or while stationary in a hopper. Practically, this results in four different types of measurements, none of which is universally superior to all others. That is, the best method must be selected for each application depending on its circumstances.

If dredging is of prime importance, surveying the dredging area is normally preferred. If dredging for fill, surveying the fill is preferred. If surveying in the dredging or disposal areas is not practical due to weather, waves, excessive shoaling, swelling of fine material, etc., the dredged material can be

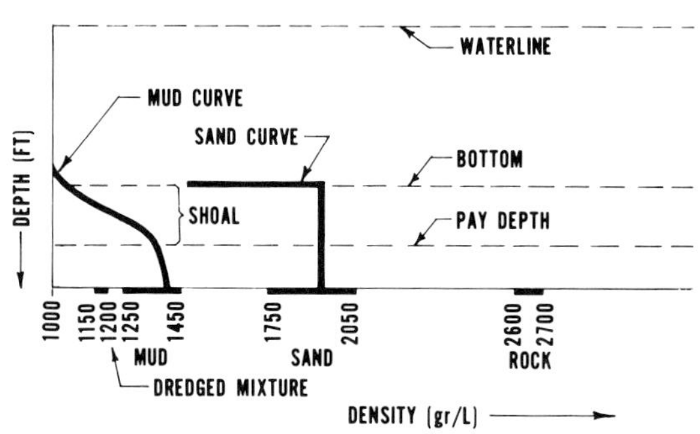

Figure 8. Density configurations.

measured while in transport. All hydraulic dredges can measure the material while it is flowing through their piping. Hopper dredges and barges can measure the material while it is stationary in their hoppers. The variables actually measured in a piping system depend on the type of instrument used (radiation, electric current, etc.) These instruments can approximate readouts in density or velocity in volumetric units since the relationship between these parameters and the variables actually measured is fairly well understood. The tools traditionally used to measure dredge production are within the comprehension of dredge personnel. Breakdowns are rare, equipment is easily calibrated, performance is reliable, and all work related to measurements is normally controlled by survey personnel. However, the use of these tools is necessarily slow and at times adversely affected by weather visibility and the applicability of tide data. Modern equipment alleviates the shortcomings of the traditional tools. However, it has high initial and maintenance costs and its technical function is no longer understood by dredge personnel. Also, equipment breakdown is frequent, a serviceman is required in case of breakdown or calibration, work related to measurements normally involves more than one department, and operation of the equipment is often erratic due to nearby vessels, oil tanks, or fish.

11. Typical Dredging Job

The following outlines a typical maintenance dredging job as currently administered by the U.S. Army Corps of Engineers.

a. Analysis of a periodic condition survey and/or groundings indicates the need for dredging. – MONTH 1

b. Funds are scheduled; bottom material is analyzed; required quantity to be dredged and overdepth quantity are computed; types of contract, material disposal, dredge type and approximate dredge size are decided upon; permits are obtained, and drawings and specifications are prepared. – MONTHS 2-6

c. Work is advertised, government estimate is prepared, and bids are opened. – MONTH 7

d. Work is awarded to low bidder, preconstruction conference is held, and notice to proceed is issued. – MONTH 8

e. Dredging equipment arrives at work site, predredging survey is performed, and work commences. – MONTH 9

f. Dredging operations are supervised for compliance with contract, and acceptance sections are surveyed when completed. – MONTHS 9-12

g. Dredging is completed, and post-dredging survey is performed. – MONTH 12

The U.S. Army Corps of Engineers is by far the largest customer of the United States dredging industry in terms of cost and quantity dredged. New work dredging is authorized by Congress. It follows all steps outlined above, except "a". The foregoing indicates the administrative work load, and the support and time required for a typical job without complications. It does not elaborate on procedures and additional steps and complications that frequently develop. The more frequent time delaying complications are: Insufficient funds, late permits, observing "environmental windows," protests of government estimate or low bidder, delays due to equipment limitations or natural elements, and encountering different site conditions or changed quantities.

12. Dredging Versus River Training

Within the context of this paper, dredging competes with river training as a means of sediment control essential to permit and sustain navigation. Both methods are compared in Table 1. Since neither of these methods is clearly advantageous in all instances, the optimum type of sediment control must be a compromise decided upon on a case-by-case basis. While dredging is admittedly largely repetitive, the history of many river training structures shows them to be far from permanent or even long range. Realizing that all views on this subject are influenced by an individual's background, the majority of the dredging community believes that river training alone will not assure safe navigation over a total river system. Examples of problems associated with river training can be cited. One result of the Long Range Spoil Disposal Study by the Philadelphia District of the Corps of Engineers (3) is that, based on model studies, even an extensive system of dikes did not show the desirable result of shifting the majority of the shoaling appreciably toward the mouth of the Delaware River. Various structures built and modified since 1938 to stabilize Barnegat Inlet have essentially failed, leaving the inlet more hazardous to navigation and ecologically less desirable than in 1938 (4).

The high failure rate of river training structures appears to indicate that their design lends itself little to precise engineering principles and is considerably the result of educated opinion. Since natural conditions are in near

Table 1. Comparison of advantages and disadvantages of river training and dredging.

River training		*Dredging*	
Advantages	*Disadvantages*	*Advantages*	*Disadvantages*
1. Considered by many to be permanent or long term solution.	1. High initial cost. 2. Long design and construction time. 3. Considered to spoil aesthetics and interfere with water sports and wildlife. 4. Often does not perform as predicted.	1. Accomplished in relatively short time. 2. Very fast in an emergency. 3. Predictable results.	1. Repetitive. 2. Considered to pollute in dredging and disposal areas. 3. Fuel intensive.

equilibrium, any structure, even if locally successful, will upset the natural balance in other places. Thus, dredging would appear to be the superior option in many instances.

Acknowledgments

Permission was granted by the Chief of Engineers, U.S. Army, to publish this paper which reflects the views of the writer.

References Cited

Tennessee-Tombigbee Waterway Authority pamphlet titled: Tennessee-Tombigbee Waterway, Facts for Media, No date. (1)

Mohr, A. W., 1975. Energy and Pollution Concerns in Dredging. Waterways Harbors and Coastal Engineering Div. (ASCE):405-417. (2)

U.S. Army Engineer District, Philadelphia, Corps of Engineers Pamphlet titled: Long Range Spoil Disposal Study, Part VI, Substudy 5, In-River Training Works, June 1969. (3)

Caccese, L. and Spies, R. paper and presentation titled: "Barnegat Inlet-Nature Prevails." Coastal Sediment Conference 1977, Charleston, SC, sponsored by ASCE. (4)

THE YANGTZE ESTUARY: ITS ORIGIN AND DEVELOPMENT

Convenors: *J. R. Schubel*
Marine Sciences Research Center
Stony Brook, New York

and

Chen Ji-yu
East China Normal University
Shanghai, China

THE CHANG JIANG (YANGTZE) ESTUARY: ESTABLISHING ITS PLACE IN THE COMMUNITY OF ESTUARIES

J. R. Schubel and D. J. Hirschberg
Marine Sciences Research Center
State University of New York
Stony Brook, New York

Abstract: Among the World's rivers, the Chang Jiang (Yangtze) ranks fourth in length; fifth in average water discharge at its mouth; and fourth in suspended sediment discharge. Its ranking among the World's community of estuaries is less secure. Like the World's other estuaries, the Chang Jiang estuary was formed by the most recent rise of sea level. Because of its large sediment discharge, the estuary has been largely filled. The sea is being expelled at a rate of nearly 40 m yr^{-1}. During the summer period of high river discharge, the estuary is displaced completely from its basin. At other times of the year, the head of the estuary extends inland only a few tens of kilometers. The estuary of the Chang Jiang not only is being driven seaward, but is being constricted at its mouth, and is migrating south. The remaining life span of the Chang Jiang estuary is short.

Introduction

The Chang Jiang River traverses Central China in an approximately westerly direction for nearly 6000 km before discharging through the Chang Jiang estuary into the East China Sea near Shanghai. The Chang Jiang is the longest river in China and its largest in terms of water discharge. It is a river of extremes. During the summer of 1981, the upper reaches of the Chang Jiang experienced severe flooding while the lower reaches of the river were experiencing the worst drought in more than 50 years.

In the four papers that follow, Chinese oceanographers present descriptions of (1) the geological development of the Chang Jiang estuary over the past 2000 years, (2) the morphology and sedimentary structure of its extensive river mouth bar system, (3) its tidal and non-tidal circulation patterns and the resulting patterns of sediment transport and accumulation, and (4) the use of LANDSAT imagery to study the dispersion of the Chang Jiang river plume in the East China Sea.

The purpose of this brief paper is to establish the place of the Chang Jiang River and estuary within the international community of rivers and estuaries, and in this way to provide a basis for estuarine comparisons, particularly on geomorphological grounds.

This is a difficult task. Criteria for comparing and ranking rivers are well established, readily available, and their significance apparent. They include

ISBN 0-12-404070-5

length, drainage area, water discharge, and sediment discharge. But, the important diagnostic comparative characteristics of estuaries, especially with regard to geological processes, are poorly established, even if basic data were available. For the Chang Jiang estuary the problems are compounded because of the lack of data. We have selected the length of the estuary, defined as the distance from the mouth to the landward limit of sea salt penetration, as the measure of the size of the estuary. Another useful geomorphological measure is the ratio of the volume of the estuarine basin to the volume of the average annual accumulation of sediment. Unfortunately, the data needed to calculate this ratio are readily available for only a very few estuaries and the Chang Jiang is not among them.

The River

Length of River

Among the World's rivers, the Chang Jiang ranks fourth in length (5980 km) behind the Nile (6650 km), the Amazon (6437 km), and the Mississippi (6020 km) (Table 1). The Chang Jiang is China's longest river, hence its name; Chang = long, Jiang = river, therefore Chang Jiang = Long River.

Water Discharge

Among the World's rivers, the Chang Jiang ranks fifth in water discharge at its mouth (22×10^3 m^3 s^{-1}) behind the Amazon (200×10^3 m^3 s^{-1}), the Congo (50×10^3 m^3 s^{-1}), the Orinoco (35×10^3 m^3 s^{-1}), and the Ganges-Brahmaputra (31×10^3 m^3 s^{-1}). Rankings of the World's top ten rivers by water discharge are summarized in Table 2.

Sediment Discharge

Among the World's rivers, the Chang Jiang ranks fourth in average suspended sediment discharge at its mouth (0.5×10^9 tons y^{-1}) behind the

Table 1. Rivers of the world ranked by length (from vander Leeden 1975).

Rank	*River*	*Country*	*Length (km)*
1	Nile	Egypt	6650
2	Amazon	Brazil	6437
3	Mississippi	USA	6020
4	Chang Jiang	China	5980
5	Yenisci	USSR	5540
6	Ob	USSR	5410
7	Huang Ho	China	4845
8	Congo	Congo	4700
9	Amur	USSR	4444
10	Lena	USSR	4400

Table 2. Rivers of the world ranked by water discharge ([1] from Milliman and Meade, in press; [2] from vander Leeden 1975; [3] from Chen Ji-yu, pers. comm. 1981).

Rank	*River*	*Country*	*Average Discharge at Mouth* ($\times 10^3\ m^3\ s^{-1}$)
1	Amazon	Brazil	200[1]
2	Congo	Zaire	50[2]
3	Orinoco	Venezuela	35[1]
4	Ganges-Brahmaputra	Bangladesh	31[1]
5	Chang Jiang	China	22[3]
6	Mississippi (including Atchafalaya)	USA	18[1]
7	Yenisei	USSR	17[2]
8	Lena	USSR	15[2]
9	Parana	Argentina	15[2]
10	Mekong	S. Vietnam	15[1]

Table 3. Rivers of the world ranked by suspended sediment discharge (from Milliman and Meade, in press).

Rank	*River*	*Country*	*Average Sediment Discharge at Mouth* ($\times 10^6$ *tons* yr^{-1})
1	Ganges-Brahmaputra	Zaire	1670
2	Huang Ho	China	1080
3	Amazon	Brazil	900
4	Chang Jiang	China	500
5	Irrawaddy	Burma	285
6	Magdalena	USA	210
7	Mississippi (including Atchafalaya)	Columbia	220
8	Orinoco	Venezuela	160
9	Mekong	S. Vietnam	160
10	Red River	USA	160

Ganges-Brahmaputra (1.7×10^9 tons yr^{-1}), the Huang Ho (1.1×10^9 tons yr^{-1}), and the Amazon (0.9×10^9 tons yr^{-1}) (Table 3). Two Chinese rivers – the Chang Jiang and her sister to the North, the Huang Ho (Yellow River) – discharge approximately 1.6 billion tons of sediment per year. This accounts for approximately 15 per cent of the total mass of sediment discharged by all the World's rivers to the Coastal Ocean.

Drainage Basin

Among the World's rivers and estuaries, the Chang Jiang ranks 9th in drainage area (1950 × 10^3 km^2). Ranking of the World's top ten rivers by drainage area is summarized in Table 4.

The Estuary

Length of Estuary

The length of an estuary can be defined as the distance from the mouth of the estuarine basin to the upstream limit of measurable sea salt in the lower layer. This length obviously varies from estuary to estuary. It also varies with time in any given estuary in response to changes in river discharge.

During the flood season in Southern China – July to October – riverflow is so strong that during much of this period the Chang Jiang estuary is expelled entirely from its basin; sea salt does not penetrate the mouth. Even during the dry season – November to May – the length of the estuary is short; sea salt rarely penetrates more than 40-60 km inland, approximately half way from the mouth of the estuarine basin to the mouth of the Huang Pu which enters just above the upstream end of Chongming Island.

The estuary has not always been so abbreviated. For the past 2000 years the estuary has been driven seaward at the rate of about 40-45 m yr^{-1}. Two thousand years ago the estuary was approximately 120 to 150 km long during the dry season, and even during the flood season there was an estuary, albeit a short one.

The lengths of the estuaries of the major rivers are small compared with the lengths of the estuaries of the rivers with lower water and sediment discharges. Presumably, this is because eustatic sea level has been stable long enough to now allow the major rivers to fill a larger proportion of their estuarine basins.

The lengths of the Chang Jiang estuary and of the estuaries of a number of the World's other major rivers, and the lengths of several of the World's major estuaries are summarized in Table 5.

Table 4. Rivers of the world ranked by drainage area (from vander Leeden 1975).

Rank	*River*	*Country*	*Drainage Area (× 10^3 km^2)*
1	Amazon	Brazil	5775
2	Congo	Zaire	4025
3	Mississippi	USA & Canada	3225
4	Nile	Egypt	2975
5	Yenisei	USSR	2600
6	Ob	USSR	2500
7	Lena	USSR	2425
8	Parana	Argentina	2300
9	Chang Jiang	China	1950
10	Amur	USSR	1850

Table 5. Lengths of selected estuaries during periods of high riverflow and low riverflow. (The length of the estuary is defined as the distance from the mouth of the estuarine basin inland to the last traces of measurable sea salt in the lower layer.)

Estuary	*High Riverflow Length (km)*	*Low Riverflow Length (km)*
Amazon	0	0
Chang Jiang	0	30-40
Chesapeake Bay	280	320
Congo	–	40
Huang Ho	0	0
Hudson	25	55
Long Island Sound	145	145
Mississippi	10	100
Rio de la Plata	150	300
San Francisco Bay	50	60
St. Lawrence	150	175

Population in Drainage Basin

More people live within the drainage basin of the Chang Jiang estuary than within the drainage basin of any other estuary in the World. The total population in the Chang Jiang's drainage basin exceeds that in the drainage basins of all United States estuaries combined; more than six times as many, 400 to 500 million people in all. The Chang Jiang is China's major navigational waterway and home of its major port and largest city (population >11 million) – Shanghai.

Other Unusual Features

Not only is the freshwater input to the Chang Jiang estuary very large, its tidal range is large and tidal currents are strong. The maximum tidal range of about 5m occurs in the upper reaches of the North Branch where maximum tidal currents average as much as 2.5 to 3.0 m s^{-1} (5 to 6 knots) (Shen Huan-ting, pers. comm. 1981).

Among the estuaries of the World's large rivers, the Chang Jiang is unusual in the strength of the tidal flow and, as a result, in the degree of mixing of salt and freshwaters. Over most of the year, the Chang Jiang estuary can be described best as a partially mixed (Type B) to vertically homogeneous (Type C) estuary. During the summer flood season, the lower reaches of the South Branch may exhibit characteristics of a salt wedge estuary (Shen Huan-ting, pers. comm. 1981).

Lifespan of Estuary

Like the World's other estuaries, the Chang Jiang estuary was formed by the most recent rise in sea level which began between 18,000 and 15,000 years ago. At that time sea level was approximately 100 m lower than today. Because the slope of the continental shelf off eastern China is so gentle (less than 2 minutes of arc), the rising sea advanced rapidly across the shelf reaching the present mouth of the estuary approximately 8000 to 10,000 years ago. Because of the large sediment discharge of the Chang Jiang its estuary has been driven seaward at the rate of 40 m yr^{-1}. The rate of infilling must have increased dramatically about 600 A.D. when widespread cultivation of the drainage basin of the middle and lower reaches of the river sent sediment yields soaring.

Large estuaries like the Chesapeake Bay, Delaware Bay, and Long Island Sound can expect geological lifetimes of an additional several tens of thousands of years if sea level remains near its present position, and if sediment inputs do not change dramatically. The remaining lifetime of the Chang Jiang estuary is far shorter: only about 1,000 years.

Acknowledgments

We thank Professors Chen Ji-Yu, Yun Cai-xing, Shen Huan-Ting and R. H. Meade for information used in this report. Contribution 317 of the Marine Sciences Research Center.

Reference Cited

Milliman, J.D. and R. H. Meade (in press). World-wide delivery of river sediment to the oceans. Bull. Geol. Soc. Amer.

vander Leeden, F. 1975. *Water Resources of the World.* Water Information Center Pub., Port Washington, NY 868 p.

THE MODEL OF DEVELOPMENT OF THE CHANG JIANG ESTUARY DURING THE LAST 2000 YEARS

Chen Ji-Yu, Yun Cai-xing and Xu Haigen
Institute of Estuarine and Coastal Research
Physics Department
East China Normal University
Shanghai, China

Abstract: This paper, based on field observations, hydrological surveys and historical literature, outlines the development of the Chang Jiang Estuary during the last 2000 years. The sequence of major events in its development may be summarized as follows: (1) gradual growth of shoals near the south bank; (2) attachment of shoals and islands to the north bank; (3) narrowing of the estuarine reach; (4) formation of normal channel, and (5) deepening of the channel. Such a developmental model is of significance in estuarine theory, as well as for the practice of estuarine management.

Introduction

The establishment of a model of estuarine development can have great significance both in theory and in practical application. Such a model reveals the basic pattern of estuarine development and consequently provides a theoretical basis for predicting the developmental tendency of an estuary and a reliable scientific base for effective management of that estuary. Since it is the outlet of the longest river in China, the Chang Jiang estuary has evolved in a relatively complicated manner (Figs. 1 & 2). Consequently, a summary of its evolutionary history for the establishment of its developmental model may have significant value for managing the estuary.

In recent years, while making a study of the historical development and modern evolution of the Chang Jiang estuary, the present writers have partitioned the model of development of the Chang Jiang estuary into the following five phases: (1) the growth of shoals at the south bank; (2) merging of sand bars into the north bank; (3) narrowing of the estuarine reach; (4) formation of the regular river channel; and (5) deepening of the channel. These five aspects will be dealt with separately in the following sections.

Growth of Shoals at The South Bank

It is a distinct phenomenon that the current at the Chang Jiang estuary takes different routes during ebb and flood periods. Over the past 2000 years, the Coriolis force on the ebb current has caused the ebb channel to con-

ISBN 0-12-404070-5

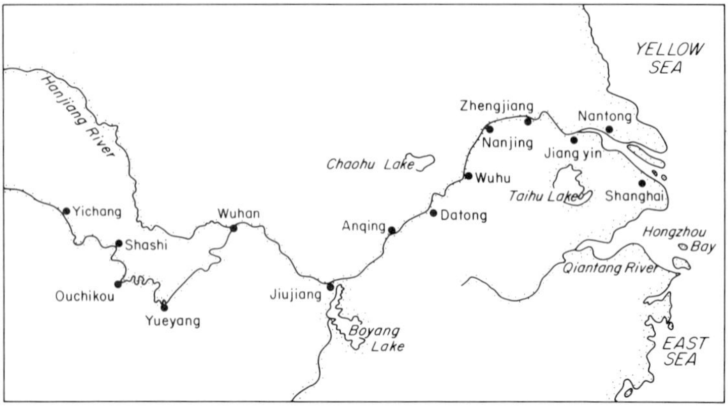

Figure 1. The lower reach of the Chang Jiang River.

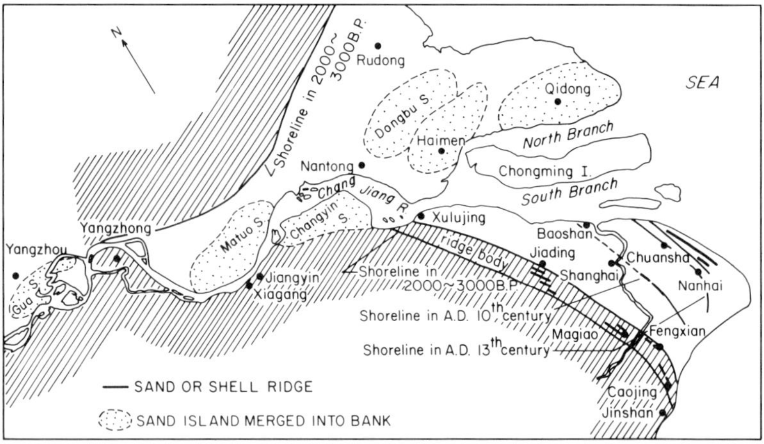

Figure 2. Historical changes of the Chang Jiang River Estuary.

tinuously shift southwards. The suspended load is transported to the sea with the ebb current, and it, too, has tended to go southwards in the process of dispersion. The remote sensing pictures taken by the Earth Resources Satellites have revealed this tendency clearly (Fig. 3). As a result, the south bank shoals of Chang Jiang River estuary became an important location for sedimentation. The gradual extension of the south shoals throughout the historical period caused the land to build out seaward and a succession of sea dykes were built one after another (Chen and Yun 1959a).

It is well known that in the shoreward part of the Tai Lake Plain, from Taichong at the north end to Caojing at the south, there are several parallel ridges of shells and sands which were called "ridge body" by local people (Tan 1973). As far back as 1000 years ago, the "ridge body" had already been

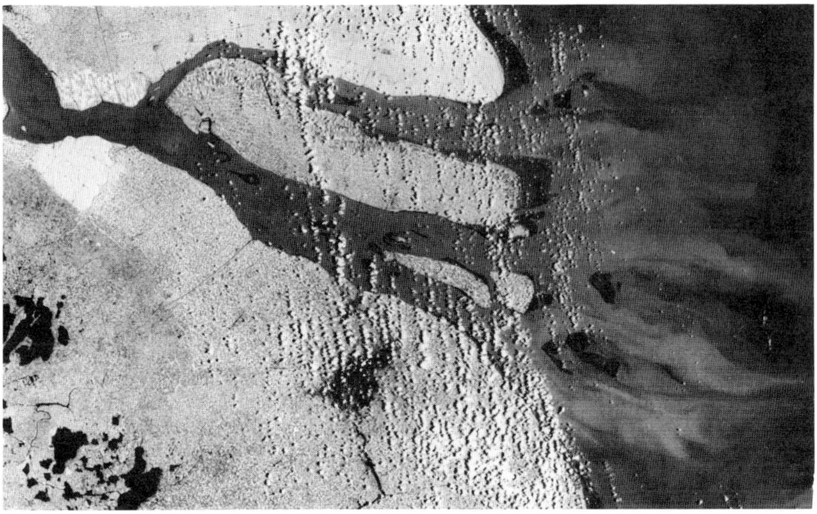

Figure 3. Earth resources satellite photograph of the mouth of the Chang Jiang Estuary.

mentioned in a book written by Jiatan. There are usually three ridges either of shells or of sand within the limits of the "ridge body." These include the Sand Ridge and a Bamboo Ridge in order of occurrence from Tai Lake Plain to the shore. There are also five ridges of shells at certain places in Jiading County.

The dates of formation of these ridges, which represent ancient shorelines, have not yet been determined. It has, however, been determined by the ^{14}C dating method that the top layer near Maqiao was formed 3995 ± 95 years ago and that the sediment of the ancient lagoon at the inner side of the sand ridge was formed 5785 ± 185 years ago. This enables us to determine that the "ridge body" began to take shape about 5000-6000 years ago (Chen et al. 1980).

Although the "ridge body" was formed during the postglacial period when inundation had nearly reached modern sea level, and remained so for a period, up to the present no cultural relics from before the Jin Dynasty have been discovered in the region to the east of the "ridge body." That is, during the 1st to 3rd Century B.C. the shoreline remained somewhere near the "ridge body" to its east. From the fact that the width of the "ridge body" is 8 km at Jiading and 4 km at Maqiao, we can readily infer that the seaward movement of the shoreline was very slow during 3000-4000 B.C. The average annual rate of seaward movement was only about 1-3 m. However, following the massive reclamation work in the mountain area of the Chang Jiang basin, suspended load in the flow increased, causing rapid progradation of the shoreline of the Chang Jiang Delta. From the historical literature we learn that the reclamation of the mountain areas of the Chang Jiang basin began mainly after the 3rd Century A.D. and flourished during the Sui and Tang Dynasties. At that time the method used was that of plowing and burning, resulting in rapid erosion of the soil. Consequently, the shoreline of Chang Jiang Delta moved rapidly seaward after the Sui and Tang Dynasties.

At the beginning of the 8th Century A.D., the shoreline was already located to the east of the Huangpu River, starting from 5 km south of Jinshan and turning north to pass Zhagang, Longhua, and Yanqiao and ending at the bank of the Huangpu River. There is a sand ridge between Xiasha and Zhoupu which indicates the probable location of the shoreline in the 10th Century. The distance between the shoreline and the "ridge body" had extended to about 30 km by the Song Dynasty. Recently the tombs of the 12th Century have been found successively in Nanhui and old Baoshan town. The fact that these relics were found outside the seawall of the Song Dynasty confirms that, at that time, there existed a new dyke outside the seawall. By the 14th Century there were three rows of sea dykes. To the west was the big dyke of the Song Dynasty, which also served as a reserve dyke. In the middle was the main protective dyke of the Yuan Dynasty. At the outside was the small dyke of the Yuan Dynasty. There were two sand ridges about 30 km in length and 50 km width beyond the small dyke. This indicates the shoreline of the 17th Century. The small dyke was still used for tide protection in the early Oing Dynasty. The famous Oingong dyke, built in 1733, simply utilized the Yuan Dynasty's small dyke as its base. Judging from this circumstance, we may presume that the south shoreline of Chang Jiang estuary moved outward very slowly during the 14th to 18th Centuries. At certain sections it even retreated landward. The reason is that the main stream of the Chang Jiang River flowed through the North Branch in that period (Chen 1957), which decreased sedimentation on the south shoals (Fig. 4). Later on, the main stream changed to the South Branch, making the south shoals again expand seaward. In the 1730's the local government built a small dyke and at the end of the 19th Century built an outer dyke. Up to the present, the shoal in Nanhui has continued to expand seaward at the rate of 1 km per 40 years.

Merging of Sand Bars into The North Bank

Due to the addition of sand to the shore, the north bank has been expanding continuously seaward over the last 2000 years. Historical data show that 2000 years ago the north shoreline of the Chang Jiang estuary was located near the line from Baipu to Xiaoyangkou (Chen 1957). The discharge of the Chang Jiang was substantial and the amount of sediment load carried also was large. There have long been sand bars and islands in the estuarine areas. Early in the Han Dynasty there were already documentary descriptions of the Dongbu Sands and Nanbu Sands in the estuarine reach. After formation, due to changes in dynamic condition, some bars collapsed and became part of the river while others merged into the bank.

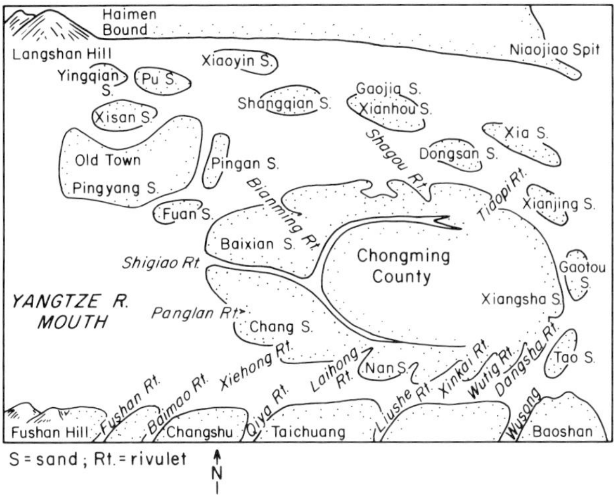

Figure 4. The Chongming Island's sketch during 1630-1640.

During the last 1000 years or more, there have been at least six important mergers of sands into the bank. About the 7th Century, the Dongbu sands merged into the bank, and so, in the 8th Century, did the Gua sands (Chen and Yun 1959b). In the 17th Century the Matuo sands merged into the bank, as did the sands at Haimen mainly in the 18th Century.

The sands in Qidong merged with the bank between the end of the 19th Century and the early part of the 20th Century (Fig. 5). In the 1920s, the Changyin sands were joined to the bank (Fig. 6). Of the above six mergers, only the last one was due to artificial factors. The others were all due to the forces of nature.

Why do the sand bars in the Chang Jiang estuary show a tendency to merge into the north bank? Through the analytical study of estuarine dynamics we may be able to answer this question. The tidal current flows by different routes during the ebb and flood periods and is a commonly observed phenomenon in most tidal estuaries. In an estuary like the Chang Jiang (Fig. 7), where ebb and flood discharges are both large, this phenomenon is more distinct. Although many factors are involved that could cause the ebb and flood current to take different courses, the Coriolis force plays a major role. The Nanjing Hydraulic Research Institute made some calculations of the Coriolis force intensity in the Chang Jiang estuary in 1964. At Qiyakou, for example, the Coriolis force will produce a pressure of 144 kg/m^2 upon the south bank when ebb velocity reaches 1.9 m/sec. Because of this factor the main current during the ebb period is close to the south bank and the main flood current is inclined to the north. This does not exclude an influence of the local river shape where a flood channel exists in the south and the main

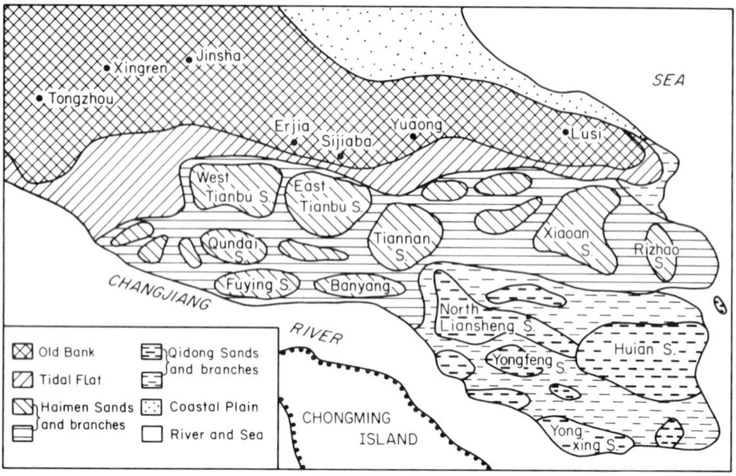

Figure 5. The formation and evolution of the Qidong-Haimen delta plain.

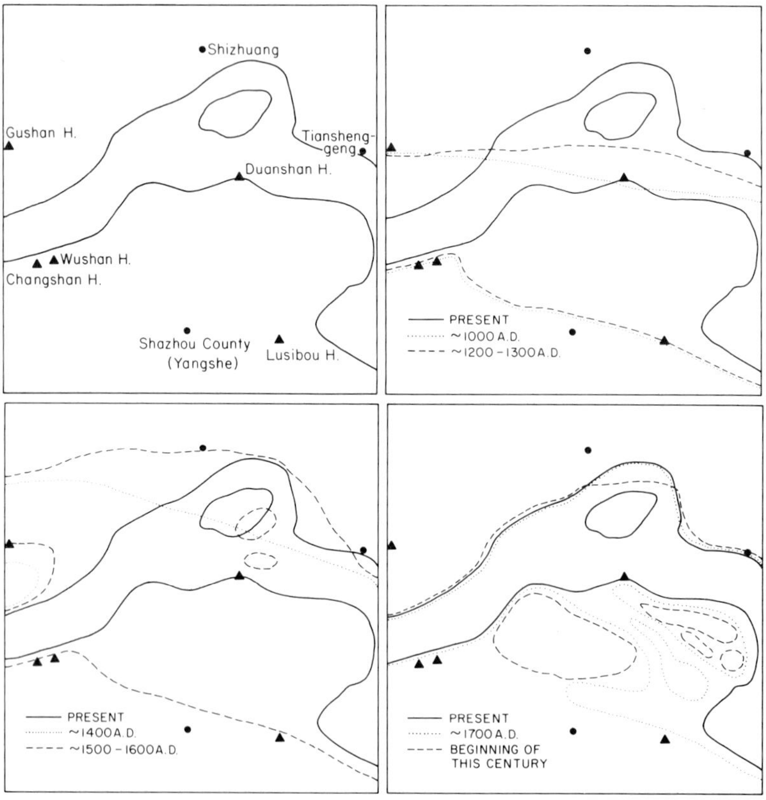

Figure 6. The historical changes of river bank near Duan Shan Hill.

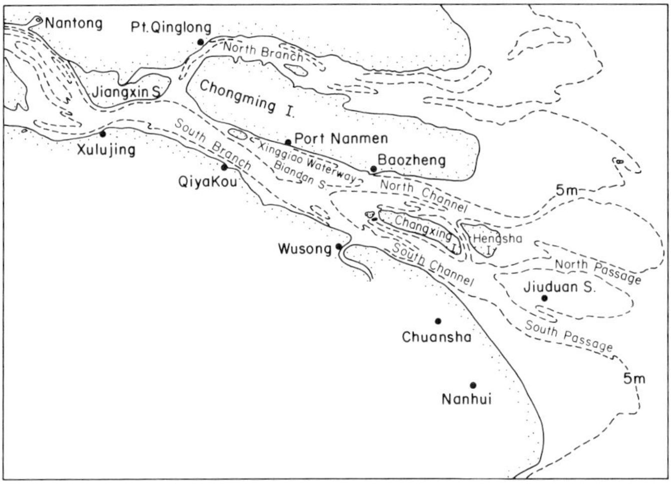

Figure 7. The present Chang Jiang River Estuary.

current is near the north bank. In our 1975 research work (Chen and Xu 1981), we have already described routes leading to the formation of sand bars in the Chang Jiang estuary. The slack flow area located between ebb and flood main currents will favor the deposition of sediment. Sand islands are developed gradually from submerged sand bars. Sometimes the channel at the north sand islands may become the main ebb route due to the swinging of the main current, but the Coriolis force action often forces the main ebb current to go south, thus making the channels at the north become flood channels. This now is the situation at the North Branch and also in the Xinqiao Waterway.

The flood tide of the North Branch is so powerful that it causes a tide wave to propagate up to the Yonglong Sands, where the wave crest breaks up to form a tidal bore. The flood current carries sediment to the South Branch, and during the dry season the flood current pours back into the South Branch through the North Branch. The data from the hydrologic survey made in January 1975 at the Xinqiao Waterway reveal that the flood current was the predominant current 44% of the time (Chen and Xu 1981).

In a channel with flood dominance the transport of sediments is net landward. Sediments carried upstream will not all flow back to sea with the ebb current, thus causing siltation of the channel. It must be pointed out that the flood-dominant channel will turn into an ebb-dominant each time the river reaches a flood peak, thus resulting in temporary erosion of the channel to maintain some flood channels. The 1975 survey data at the Xinquiao Waterway reflected this condition. However if the contour becomes unfavorable for the flood discharge to flow seaward, then the flood dominant channel finally will be silted, causing the merging of sand bars into the bank. At present, the

North Branch intersects the main stream at nearly a right angle near Xulujing, so only a very small amount of runoff volume goes through the North Branch. Therefore rapid siltation at the upper reach of the North Branch occurs and a tendency to merge Chongmin Island into the north bank is evident. In addition, owing to the existence of a phase difference of the tidal wave between the North and South Branches, the siltation in the North Branch near their confluence point is accelerated. This also contributes to the merger of sand islands into the bank.

Narrowing of The Estuarine Reach

The historical literature shows that the Chang Jiang Estuary was a funnel-shaped bay 2000 years ago. Its north point was located near Xiao Yangkou (which was called Liaojiaozui historically); its south point was located near Wangpan Mountain in the Hongzhou Bay, which in the past was connected with land (Chen and Yun 1959a). The distance between these two points was about 180 km. Now the distance between the north and south points of the Estuary is only about 90 km. The transect between Yangzhon and Zhenjiang historically was called “Hai Men,” which means sea gate. Songliao Mountain, a small mountain in the river, was called “mountain Haimen.” The plan view of the Chang Jiang estuary at that time was just like the narrow reach between Kan and Zhu Mountain prior to the opening of the north outlet of the Qiantang estuary. Although the river width between Yangzhou and Zhenjian was stated to be 20 km 2000 years ago, actually it was not so wide. According to historical records of the location of the river bank and measurements taken from the current map, the width was only about 12 km. Now, however, the width is only 2.3 km. This shows that the development of the Chang Jiang estuary has been a narrowing process over the past 2000 years. The record of the narrowing of each section is as follows:

Section	*River width before narrowing (km)*	*Present width (km)*	*Time of narrowing*
Zheng Jiang-Yangzhou	12	2.3	8th Century
Jiangyin	11	3.5	17th Century
Shiyiwei	18	7.5	20th Century
Xulujing	13	4.4	20th Century
Outfall of estuary	180	90	From about beginning of 1st Century to present

As a result of the growth of shoals, the merging of sand into the bank, and the narrowing of the estuary in the historical period, boundary conditions of the estuarine channel have undergone significant change, thus bringing about changes in the propagation of tidal waves, the dynamic condition of each section, and in man’s activities as well.

The range of tidal wave propagation in the historical period reached much farther upstream than it now does. "Ode to the River," written by Guo Pu, stated, "the end of propagation of the tidal wave reaches Chaisang." Chaisang is now called Jiujiang. It can be seen from this that the tidal limit was near Jiujiang between the 3rd and 4th Centuries and above Jiujiang 2000 years ago. It has now moved down to near Datong in Anhui province.

From the historical records the Chang Jiang estuary had a tidal bore. In the 2nd Century B.C., Mei Cheng in his article, "On Seven Enlightenments," said, "On the 15th of August by lunar calendar we are going to watch the tidal bore of the Qu River at Guangling (now Yangzhou)." In the 1st Century, Wang Chong, in his famous book "On Evaluation" also stated, "there are no bores at the Zhengjiang section of the great river, (main branch), but there are bores in the Qu River at Guangling." After the Tang Dynasty, the "Guangling Bore" disappeared due to the silting of the Qu River and merging of sands into the bank. After the Yuan Dynasty, someone threw doubt upon the existence of tidal bores.

As a matter of fact, the tidal bore exists not only at the Qiantang estuary, but also at many other tidal estuaries. As has been said above, the North Branch of the Chang Jiang Estuary also has a bore. In the 1930's and 1940's the tidal bore appeared in the flood channel located to the north of the Jiangxin Sands in Haimen county. As the width of the ancient funnel-shaped Chang Jiang estuary got narrower and narrower, upstream deformation of the tidal wave became stronger and stronger, thus providing conditions advantageous to the formation of a bore at Guangling. In addition, there was a sand island off the Yangzhou bank that served as a dividing line between the Great River to the south and the Qu River to the north. The Qu River was relatively shallow, so the tidal wave broke there to form the bore. According to historical records, the bore existed between the 2nd Century B.C. and the 8th Century A.D. (Chen 1957). Therefore, the description of the Quangling bore often appeared in the poems of the Six Dynasties and the Southern Qi Dynasty. Although in the 9th Century it was still recorded that there were tidal bores on the first and the 15th of each month by the lunar calendar in spring and autumn, after the merging of the Gua Sands into the bank there was no longer any tide in Yangzhou. So the Guangling bore had probably disappeared by the middle or later part of the 8th Century.

Before the merging of the Gua Sands into the bank, that section of the Zhengjiang-Yangzhou reach was wide, and a surge might propagate up to Nanjiang. It was recorded fourteen times that serious surges had occurred at Nanjiang between the 3rd and 6th Centuries. This is probably the earliest record of surges in the world.

It should be pointed out that the change of the river width had a distinct effect on human activities. In the lower part of the Chang Jiang before the Tang Dynasty, the ferry was mostly at Caishiji in Anhui province. After the Tang Dynasty, it was moved down to Yangzhou because of the narrowing of the river (Chen 1957).

Formation of Regular River Channel

Two thousand years ago, the Chang Jiang estuary took the shape of a funnel, and only above Yangzhou began to look like an ordinary river shape. Below Yangzhou a series of sand bars divided the river into many branches. However, as the sands merged into the bank and the river width became narrower and narrower, the reach consisting of the regular channel expanded downstream. In the 17th Century, the regular river shape began to form above Jiangyin, but below that point the river got very wide. Multiple branches existed simultaneously and the main stream changed frequently. In the 1950's, the reach above Xulujing gradually formed a regular river shape and only below that point did the river begin to branch to the sea.

What shape will the river channel now take in the lower reach of the Chang Jiang River? As is well known, river shapes may be classified into three types; meander type, central bar type, and braided type. Each has its own dynamic and boundary relationships with the incoming water and sediment. In the case of the Chang Jiang river, all the lower reach is of central bar type except the Jinjiang reach in Hupeh. The formation of this type of river shape is a concrete reflection of the dynamic and boundary conditions of the Chang Jiang River.

According to Fang (1964), when flood peak variation coefficient is less than 0.3 and ratio of sediment carrying capacity to sediment percentage is smaller than or equal to 1, good dynamic conditions will be provided for formation of the central bar type of river shape. The Chang Jiang River possessed just these conditions. Through the study of sands merging into the bank and the formation of a regular river channel below Zheng Jiang-Yangzhou, it is found that four sections of the central bar type and their intermediate sections have been formed over the past 1000 years. Two sections of the central bar type above Zheng Jiang-Yangzhou were also formed during this period. This indicates that the river shape will still be of a central bar type in future development of the Chang Jiang estuary as the formation of the regular river channel gradually reaches downstream. The dynamic conditions governing the formation of the regular river channel remain unchanged.

Below the regular river channel, the Chang Jiang estuary still keeps its branching form as it flows toward the sea. At present, the river course is divided into the North Branch and the South Branch near Baimaokou. Between the 14th and 18th Century the main current followed the North Branch to the sea. In the 18th Century, the main current shifted to the South Branch. Affected by different ebb and flood current routes, the South Branch was divided further into the South Channel and North Channel. In the 1940's, due to the splitting of the Jiuduan Sands, the South Channel was again divided into the South Passage and North Passage. Following this pattern of dynamic mechanism, the river channel flowing to the sea in the Chang Jiang estuary will continue to develop new sand bars near its mouth to form new branches of the waterway.

In the normal process of development of a regular river channel, although the channel is continually becoming more stable, the complete transformation to a steady state river channel usually takes a relatively long period. Especially after the merger of sand bars into a bank, the channel above the sands will need a certain period of adjustment to adapt to a new set of dynamic conditions. This often results in eroding and rearranging of the sands. For example, after the Matuo Sands merged into the bank, other sands above that point underwent significant changes. It was only after about 200 years that the channel reached steady state. After the merging of the Gua Sands into the bank, the upstream river section also took several centuries to reach a steady state condition.

In the reach downstream of regular river channels, the sands showed great changes over the historical period. One rose while others submerged. As the main stream shifts, a sand island may appear to silt up in the south and collapse in the north or vice versa. A well-known example is presented in the changing history of the Chongming Island. The great changes of the sand island and sand bars lead in turn to change of the river channel.

Deepening of River Channel

What has happened to the depth of the channel with the narrowing of the estuary and the formation of a regular channel during the last 2000 years? The characteristics of the river channel at different sections in the present Chang Jiang estuary may shed light on this question.

In the present Chang Jiang estuary, from the regular channel to the outlet to the sea, each section of channel possesses different characteristics. There is a big bar near the mouth. Here, the water surface is wide and the depth is shallow. On both sides are wide tidal flats. Although the depth at the top of the bar may vary somewhat, it generally is less than 6 m and is in a naturally-balanced state. The wide main channel appears to carry a predominant ebb current. At one side, or in some stretches on both sides, are the flood current channels. The plain view basically resembles the ebb tidal delta described by Hayes (1979). In the branched section below Xulujing, there is a distinct division into ebb and flood channel at each branch, forming a duplex channel with shoals in between. For example, the South Branch is divided into the main channel and the Xinqiao Waterway by the Biantan Sands. The former is an ebb channel while the latter is a flood channel. Even in the main channel of the South Branch, however, only the section from Qiyakou to Liuhe is single-track, namely a neutral channel for both ebb and flood current. The section below that still has a complex channel shape. The ebb channel is relatively deep. Local depths may reach 20-40 m, cutting into the silt layer of shallow sea facies. Taking the cross section as a whole, the average depth is less because of the broad river width and numerous sand bars. For instance, the width at the Liuhe cross section is 12.9 km and its average depth is only 6.9 m. In the branched reach, the channel is not steady due to diversified sands and shifting current lines. However, in the section where a regular channel has already been formed, the river width becomes narrower, water tends to flow in the channel, and the ebb and flood current do not have distinct routes. Here, there is basically a single-track channel form. For this reason, the average channel depth is relatively great. For example, in the Xiagang cross section of the Jiangyin reach, the river surface width is 1.4 km while the average depth is 13.4 m. We may conclude from this that the channel will become deeper and relatively steady with the formation of the regular river shape and narrowing of the river course.

From the above, we may summarize that the channels of the Chang Jiang estuary have been evolving into a relatively steady single-track channel as the formation of the regular river shape has extended downstream over the last 2000 years. Consequently, the tendency of development of the channel is favorable for navigation and flood control.

In conclusion, as has been explained above, we can see that the model of estuarine development may provide a reliable scientific basis for designing programs to harness the river. Up to the present, the Chang Jiang estuary has remained basically a natural estuary. In order to meet the requirements of industrial expansion, it is imperative to harness the estuary and to stabilize the river shape and deepen the channel. In recent years, we have applied the developmental model outlined above to construction design problems and obtained valuable results.

References Cited

Chen Ji-Yu. 1957. Notes on the development of the Yangtze Estuary. *Acta Geographica Sinica* 23:241-253.

Chen Ji-Yu and Xu Hai-Gen. 1981. Developmental processes of South Branch Channel of Yangtze Estuary. *Journal of East China Normal University, Natural Science Edition.* 2:93-108.

Chen Ji-Yu, Wang Baocan and Lu Cangzi. 1980. Coastal morphology of China. Chapter 11, *Physical Geography of China, (Geomorphology).* Science Press, pp. 313-348.

Chen Ji-Yu and Yun Caixing. 1959a. On the development of geography in the Chang Jiang Delta. *Acta Geographica Sinica* 25:201-220.

Chen Ji-Yu and Yun Caixing. 1959b. On the processes of evolution from Nanjing to Wusong. *Acta Geographica Sinica* 25:221-239.

Fang Zongdai. 1964. Analysis of river patterns and its application to river regulation. *Journal of Hydraulic Engineering* 1:1-12.

Hayes, M. O. 1979. Morphology of sand accumulation in estuary. An introduction to the symposium. *Estuarine Research* 2:3-22.

Tan Qixiang. 1973. On evolution of coastal line and development process in Shanghai region. *Archaeology* 1:2-10.

DESCRIPTIONS OF THE MORPHOLOGY AND SEDIMENTARY STRUCTURES OF THE RIVER MOUTH BAR IN THE CHANG JIANG ESTUARY

Chen Ji-Yu
Zhu Shu-Zheng
Lü Quan-Rong
Zhou Yueqin and
He Songlin
Institute of Estuarine and Coastal Research
East China Normal University
Shanghai, China

Abstract: The river mouth bar at the Chang Jiang Estuary has depths of less than 10 m over a broad area that extends across the entire estuary and out over the inner continental shelf. Flood and ebb channels are usually separate, with shallower areas between them. Over the past 100 years the 5 m contour has progressed seaward 5-12 km and the 10 m contour up to 14 km seaward (opposite the South Channel). The Heng Sha has migrated 5 km upstream and been brought under cultivation during that period, although there has been little change at the central core of the bar. Only about a 2 m change in channel depth (from 5 m to 7 m) has occurred. The bar is a part of and rests upon the Chang Jiang delta and cores bored into the bar reveal the longer-term history of the delta. Changing sediment characteristics and flux in the estuary have been recorded, as well as the changing form of the delta itself.

Introduction

There are large accumulated sediment bodies within or just outside many of the world's river mouths. These accumulated bodies are called river mouth bars. Due to differences in strength between runoff and flood tidal currents, the locations of river mouth bars vary. This problem was discussed in an earlier paper, "The sand bar of the Chientang Estuary" by Chen et al. (1964). The Chientang is a macrotidal estuary; its tidal currents are much stronger than the runoff. Therefore, the location of the bar is well within the estuary. However, the Chang Jiang River mouth bar is quite different from that of the Chientang Estuary. In the Chang Jiang, runoff and tidal currents are both stronger than in the Chientang and the bar is near the mouth. The basic characteristics and origin of the bar in the Chang Jiang Estuary are of interest to many researchers. The following discussion of the developmental history of this bar is based upon (1) observed topographic variations of the bar during the past one hundred years and (2) analysis of data from boring cores in Chang Jiang Estuary.

ISBN 0-12-404070-5

Topographic Characteristics of the Chang Jiang River Mouth Bar

The distribution of shallow water areas with depths less than 10 m around the mouth of Chang Jiang Estuary is shown in Fig. 1. The map shows that the 10 m depth contour in the channels of the Chang Jiang Estuary extends seaward near the mouth, and that there is a broad shallow water area with depths of less than 10 m that stretches from inside the mouth out onto the inner shelf. The shallow area reaches across the entire river mouth. This is the well-known bar system of the Chang Jiang Estuary. The bar extending out of the South Channel is about 60-80 km long and that extending out of the North Channel about 40-50 km long. On the inner face of the bar, the 10 m depth contour stretches upstream along the channels. The deepest places in the channels have depths of 40-50 m. Along its outer face, the bar slopes down toward the sea bottom and merges into the inner continental shelf at depths of about 20-50 m. Actually, the bar is part of the Chang Jiang delta. Because of local differences in dynamic flow conditions there are distinct morphological features on the bar. Flood and ebb currents follow different channels over the river mouth bar and there are often shoals between these channels. Figure 2 shows the distribution of shoals and channels on the bar of the Chang Jiang Estuary. Although the longitudinal profile of the bar is a convex upward curve (Fig. 3), the transverse section shows alternating channels and shoals (Fig. 4).

A comparison of successive bathymetric charts made over the past 100 years shows that the bar in Chang Jiang Estuary has been both complex and variable. The principal changes are shown in Fig. 5 and are described below:

(1). In general, the 5 m depth contour that marks the bar front has been extended toward the sea from 5 to 12 km. The 10 m depth contour has migrated seaward as well; the maximum progression is 14 km opposite the South Channel.

(2). Strong tidal currents have cut a new passage to the sea, the North

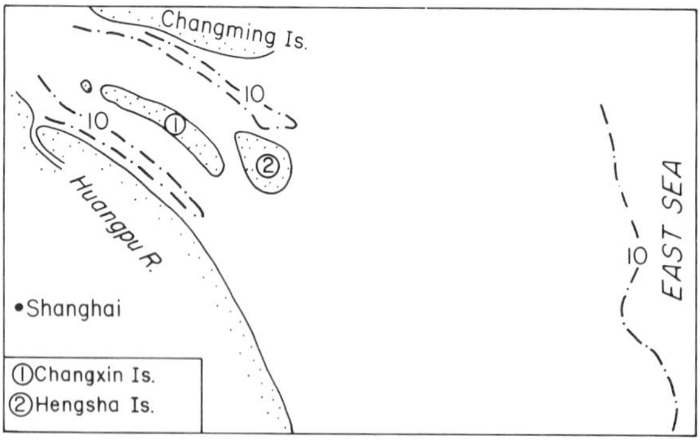

Figure 1. The distribution of the shallow areas (<10 m) in the Chang Jiang River mouth.

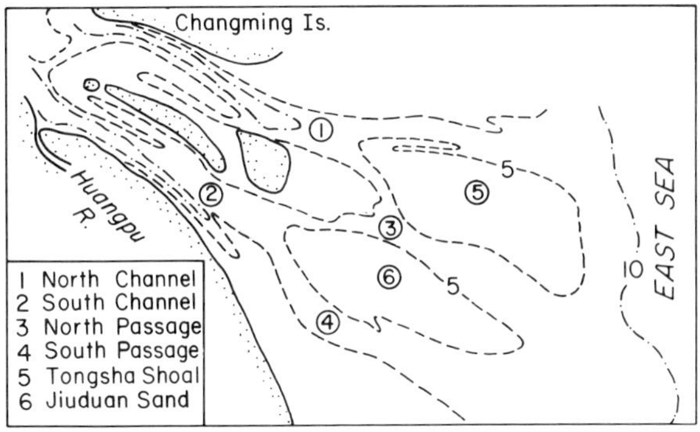

Figure 2. The distribution of shoals and channels on the Chang Jiang River mouth bar.

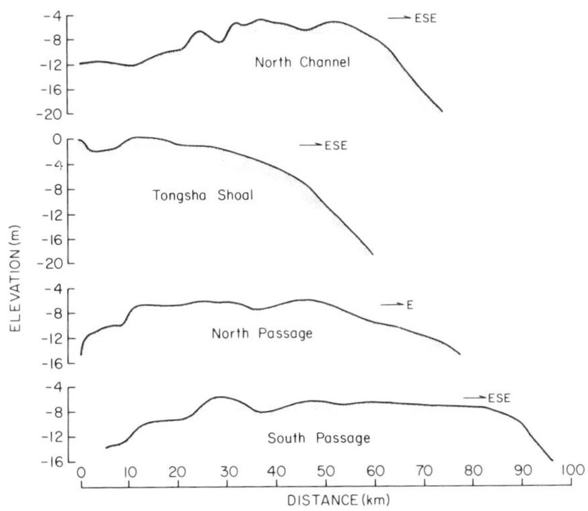

Figure 3. The longitudinal profile of the Chang Jiang River mouth bar.

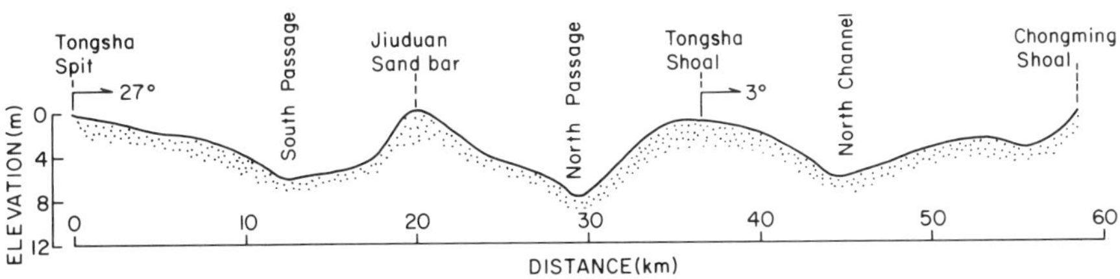

Figure 4. The cross-section of the Chang Jiang River mouth bar.

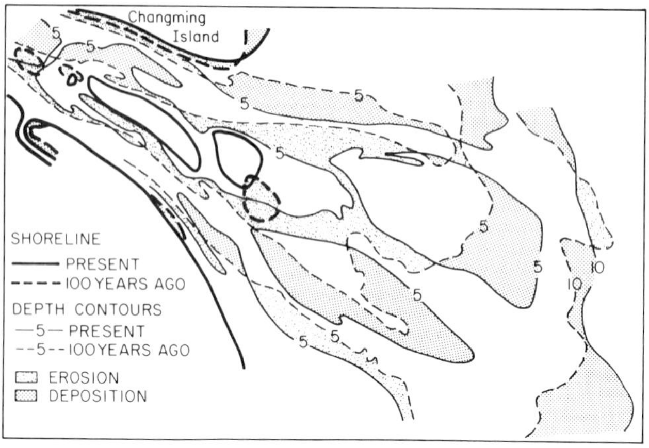

Figure 5. Changes in the topography in the Chang Jiang estuary over the last 100 years.

Passage, which formed the third-order branch in Chang Jiang Estuary. Between North Passage and South Passage a new shoal, Jiuduan Sha, was formed.

(3). The Heng Sha, located at the mouth of the Chang Jiang Estuary, migrated upstream 5 km, and was capable of being cultivated during that time. However, we must point out that the central core of the river mouth bar changed little over the same period.

There is always someone concerned with the locations of sand bars and the speed with which they grow, because the bars often provide land for agriculture and can affect navigation directly (Fig. 3). Since the minimum depth of the channels in the bar area of the Chang Jiang Estuary is only about 6 m below the theoretical base (sea) level the ingress and egress of large ships is restricted. Because of this, the shallow water area within the channel is called the river mouth bar by navigation departments. We refer to it as the "navigation channel bar" to distinguish it from the larger, more extensive, river mouth bar found in the Chang Jiang and other estuaries.

Although the form of the Chang Jiang River mouth bar has changed during the past 100 years, the change in its depth has been relatively small, only about 2 m (from a depth of 5 m to 7 m). This change is related to the alternation of the main discharge between South Channel and North Channel. When the main discharge of the Chang Jiang passes through the South Channel, the top of the bar at the South Channel is eroded slightly and the bar front builds up rapidly. When the main stream pours into the sea through the North Channel, the result there is similar. In recent years, the main discharge of the Chang Jiang into the sea has been through the North Channel, but the bar has not eroded yet because much of this flow is diverted toward the east shoal of Heng Sha Island (located between the North Channel and the North Passage). The depth at the bars in the Chang Jiang Estuarine Waterway is around 6 m and its variability is small. This depth is therefore suited to the basic hydrographic characteristics and sediment conditions in Chang Jiang Estuary and the bar has reached a natural equilibrium.

Sedimentary Structure of the Chang Jiang River Mouth Bar

In the evolutionary processes of in-filling and bar-building of the Chang Jiang Estuary, several elements were observed (Chen et al. 1982). These include narrowing of the estuarine reach, formation of the regular river channel, and deepening of the channel. Undoubtedly, the deep channel must have become elongated as the estuary prograded toward the sea. At the same time the top of the submerged delta was also extended seaward. Since the bar rests on the top of the delta, the sequences of boring cores obtained on the bar extend into the submerged delta and must reflect the developmental processes of the submerged delta as well (Fig. 6).

Figure 7 of a core segment from deep below the sediment surface

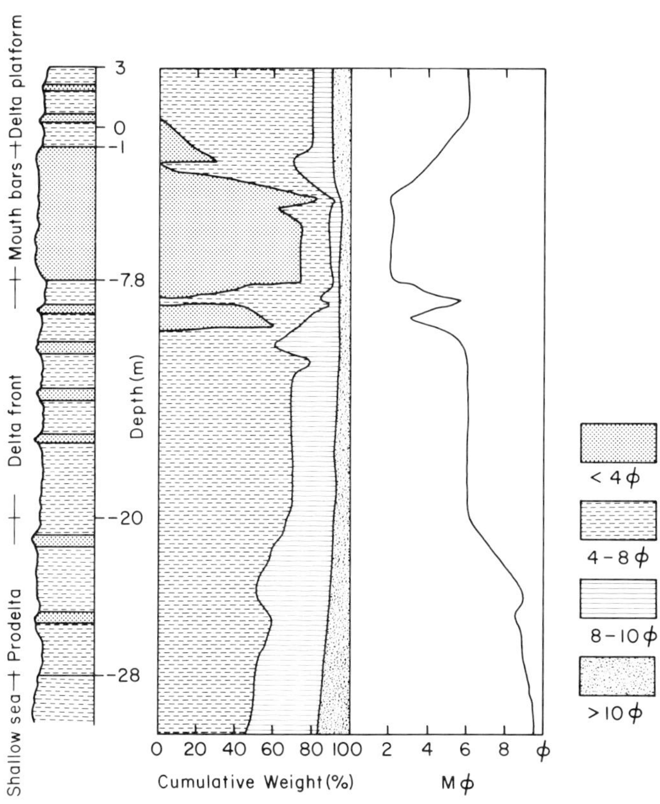

Figure 6. Typical boring core in the bar area of Chang Jiang estuary (mainly from south channel and Hengsha Island).

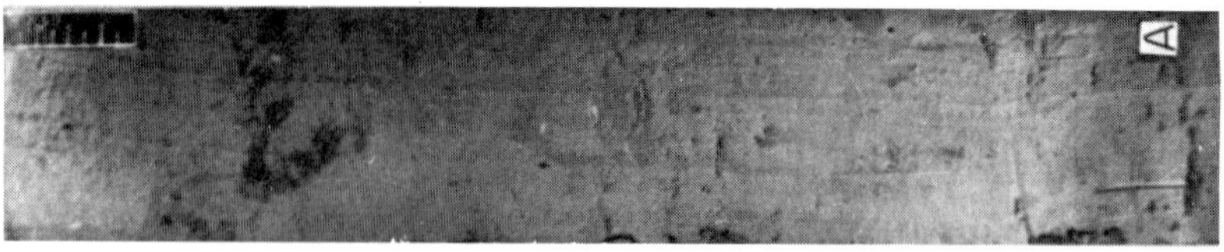

Figure 7. Shallow sea facies of the Chang Jiang Delta (Photo A). Top of core is to the left.

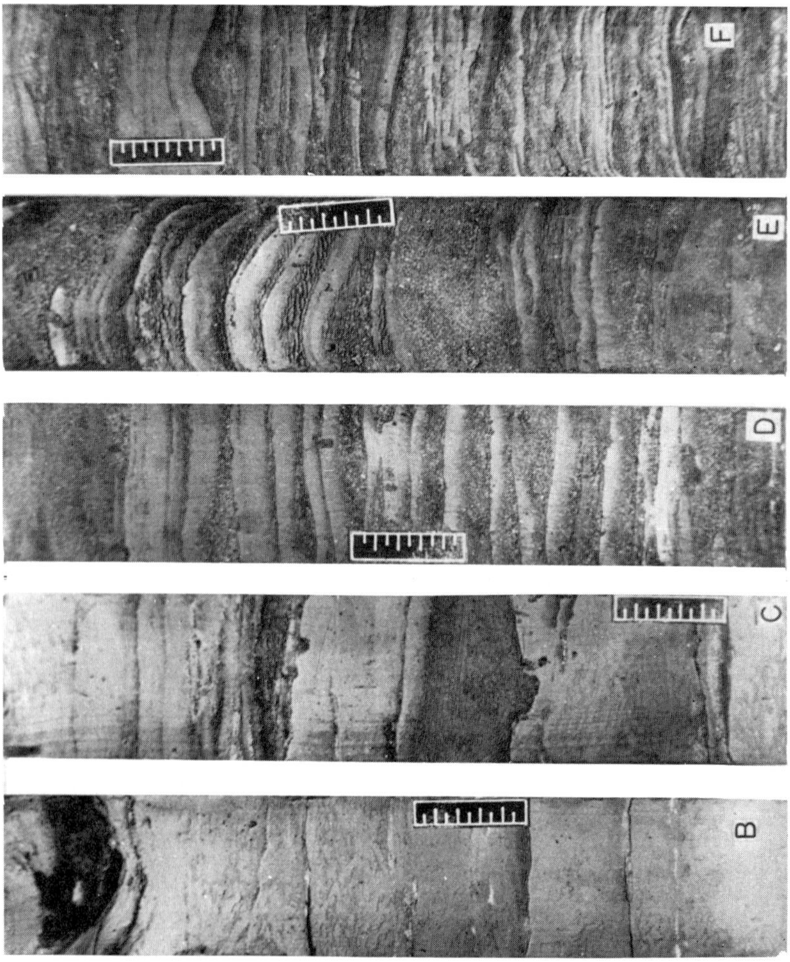

Figure 8. Delta front and pro-delta facies of the delta (Photos B-F). Top of core is to the left.

shows some profiles of shallow sea facies of the delta. The sediment is blue-grey silty clay. It consists of 45-65% clay, 55-35% silt and is structureless. It is found under the 28 m to 30 m sediment depth.

Figure 8 shows cores collected from the delta front sediment facies consisting of alternating layers of blue-grey silty clay and silt. The lamination is clear. The organic matter content is higher than in the shallow sea facies. The deeper layers are blue-grey silty clay with thin layers of silt between them. The mean grain-size is about 7.3 ϕ and there are horizontal or wavy laminae. Sometimes there are lenses of shelly sand. There are abundant remains of benthic organisms which have been identified as species found at depths greater than 20 m. The upper part of this facies is mainly blue-grey silty clay with many thin laminae of silt, or the two form intercalation beds. The grain-size of sediment is about 5.1 to 6.3 ϕ . There are lenses or inclusions of silt in some places with clear lamination. The oblique beddings and cross-beddings between them are major. Sometimes the cores contact erosional surfaces within the mud-sand beds.

Figure 9 shows the uppermost deposits of the bar. They consist of bluish-grey fine sand, silt (mostly coarse silt), and silt with some grey-brown mud and clay. The sand is pure, saturated, and flows easily. The fine sand and silt are also intercalated, with variations of the lithological characteristics as follows: lower part is sand to silt, and upper part is silt to fine sand.

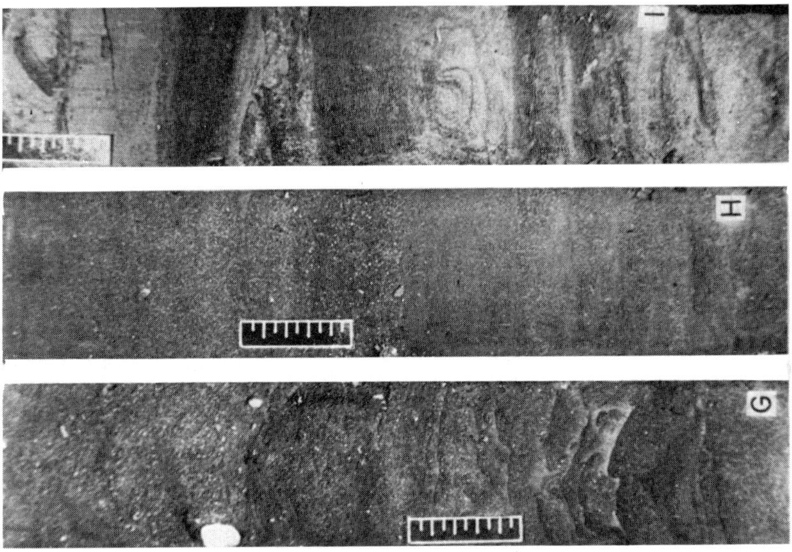

Figure 9. Deposits of the river mouth bar (Photos G-I). Top of core is to the left.

Figure 10 shows the topmost beds of the delta deposit. They consist of yellow, greyish-yellow, and yellowish-brown muddy silt, silty clay, and bluish-grey silt with horizontal laminae. They have plant root relicts, and often have rust spots.

The group of photographs described above shows the following depositional sequence (Fig. 6): shallow sea, pro-delta, delta front, bar top deposits, and delta plain facies. The differences among these facies are related to their age. When the materials of pro-delta facies were deposited, the present river mouth bar was within the old mouth of the Chang Jiang Estuary, far from its present position. The sediment from upstream of Chang Jiang River did not reach here, so the character of material accumulating in this area was fine-grained in size, with no clear bedding horizons. This was a relatively calm sedimentary environment. After progression of the delta into the sea, it produced depositional facies characteristic of a delta front. Its material became gradually coarser with interbedding of clay and silt. There was oblique bedding, which indicates that the flow at the bottom was strong enough during that time to resuspend and redistribute the sediments. Silt and sand lenses indicate motion of bottom loads. Sometimes sand waves occurred. The benthos diminished rapidly due to the rate of deposit, which increased gradually. There is an abundance of benthos only where the rate of deposit was reduced during a main stream swing out. When the main stream approached the mouth of the estuary, the top sediments characteristic of a river mouth bar were deposited and a bar formed there. This was due to the difference of dynamic conditions in the bar area. The alternation of sediment characteristics is also clear. The current pace through the main stream (or major distributary channels) is strong. The sediment becomes coarse, and oblique and cross bedding are so clearly apparent that there are obvious discontinuities between the deposited facies of the latter and former deposits. In the

Figure 10. Top-most beds of the delta (Photos J, K). Top of core is to the left.

shoal regions between the estuarine channels, the water flow is weakened and the sediment grain-size is fine with horizontal beddings and wavy beddings. When the shoal is gradually filled up to the lower tidal flat, fine sediments may be moved away because the wave action is sufficiently strong to resuspend the finer materials and winnow them out, leaving coarser materials behind. When the tidal flat reaches the middle tide level, the bar surface is built up because of marsh plant growth which traps sediments. The bar then develops gradually into an island. Sediments of the island resemble the sediments of the delta plain facies. They exhibit horizontal bedding that is sometimes indistinct. There are remnant plant roots within the profiles. After cultivation, the surface sediments are disturbed by human activity.

Acknowledgments

The authors would like to thank the following people for their help: Sun Jiemin and Tang Junbei.

References Cited

Chen Ji-Yu et al., 1964. The formation and its historical processes of the big sand bar inside the Chien Tang estuary. *Acta Geographica Sinica* 30(2):109-123.

Chen Ji-Yu, Yun Cai-xing and Xu Haigen. 1981. The model of development of the Chang Jiang estuary during the last 2000 years, pp. 655-666. *In:* V. S. Kennedy (ed.), *Estuarine Comparisons*. Academic Press, New York.

CIRCULATION OF THE CHANG JIANG ESTUARY AND ITS EFFECT ON THE TRANSPORT OF SUSPENDED SEDIMENT

Shen Huan-ting, Zhu Hui-fang and Mao Zhi-Chang
Institute of Estuarine and Coastal Research
East China Normal University
Shanghai, China

Abstract: Organizations responsible for the maintenance of navigation channels in the Chang Jiang estuary, and other organizations concerned with estuarine processes, have made hydrological, oceanographic, and sedimentological observations in the Chang Jiang estuary since the 1960s. These measurements are summarized in this paper and used to characterize mixing processes and salinity patterns from which circulation patterns and their effect on the transport of suspended sediment are inferred. It is shown that the Chang Jiang estuary is characterized by large fresh water runoff, abundant suspended sediments, and large intertidal volumes, all of which vary seasonally. As a result, stratification varies from place to place and time to time, ranging from well mixed to partially mixed. Fresh water runoff and tidal currents are shown to be the two most decisive factors in the formation of turbidity maxima, fluid mud layers, and the channel sand bars and their variations.

Introduction

Increasingly more attention is being paid to the causes and characteristics of the circulation patterns of the Chang Jiang estuary, and to the effects of circulation on siltation of navigation channels. To improve conditions of channels, organizations directly responsible for their maintenance and other organizations concerned with estuarine processes have made hydrological, oceanographic, and sedimentological observations in the Chang Jiang estuary since the 1960s.

The purposes of this paper are to use the existing field observations to characterize mixing processes and salinity patterns within the Chang Jiang estuary so as to infer the resulting circulation patterns and their effects on the transport of suspended sediment.

Estuarine Classifications

Because of variations in river flow, tidal flow, width, and depth, each estuary has distinct mixing patterns of fresh and salt water. Using these parameters as a criterion, Pritchard (1955) developed a classification scheme for estuaries and identified four distinct types: (1) Type A, a highly stratified or salt-wedge estuary, (2) Type B, a partially mixed estuary, (3) Type C, a well-

ISBN 0-12-404070-5

mixed, vertically homogeneous estuary, and (4) Type D, a thoroughly mixed, sectionally homogeneous estuary. Each of these types possesses a distinct density stratification and circulation pattern.

In general, an estuary moves from Type A toward Type D as the river flow decreases, the tidal flow increases, the width increases, or the depth decreases. Pritchard (1955) cited the mouth of the Mississippi River (U.S.) as an example of a Type A, salt wedge estuary. An example in China is the Zhu Jiang River estuary. Examples of Type B estuaries – partially mixed estuaries – are well known throughout the world and include Chesapeake Bay (U.S.A.) and the Hudson River estuary (U.S.A.) The Chang Jiang has the characteristics of a Type B estuary most of the year. Examples of Type C – vertically homogeneous estuaries – include Delaware Bay (U.S.A.) and the Qian Tang and Er Jiang estuaries in China. Type D estuaries may not exist in nature (Cameron and Pritchard 1963; Dyer 1973) but the definition is a useful one in that the assumption of vertical and/or lateral homogeneity in numerical and analytical models has produced realistic results.

It should be noted that these four types of estuaries are never well defined; they are merely four points on a more or less continuous spectrum or sequence. To complicate matters further, at any given time differing salinity structure and circulation patterns, i.e., types, may be present within the same estuary.

The degree of mixing in an estuary and hence its position in the estuarine sequence depends to a certain extent upon the relative importance of the freshwater runoff to the tidal flow. However, as noted earlier, geometry of the estuarine basin also plays a role. Because of these factors, an estuary may change from one type to another in response to natural changes in river and tidal flows or in response to changes in these parameters caused by man through freshwater diversion or dredging and filling.

Chong Ming Island divides the Chang Jiang estuary into a North Branch and a South Branch. The South Branch is, in turn, divided into a North Channel and a South Channel and the South Channel further subdivided into a North Passage and a South Passage (Fig. 1).

The North Branch, however, is slowly filling and at some time in the future, Chong Ming Is. will be joined with the mainland. This apparently results from the fact that suspended sediment carried seaward in the South Branch on the ebb is discharged into the North Branch by the succeeding flood tide since part of the tidal flow that enters the North Branch returns to the sea through the South Branch. In 1915, 25% of the river flow was discharged to the sea via the North Branch with the remainder through the South Branch. By 1959, only 1% of the river flow passed through the North Branch and it was closed to ship traffic. As may be seen from Fig. 1, the North Branch presently looks like a trumpet. As a result, its tides are amplified (25% higher than in the South Branch), mixing is intense, and the water column is well mixed. It is considered to be a Type C estuary. Prior to 1959, however,

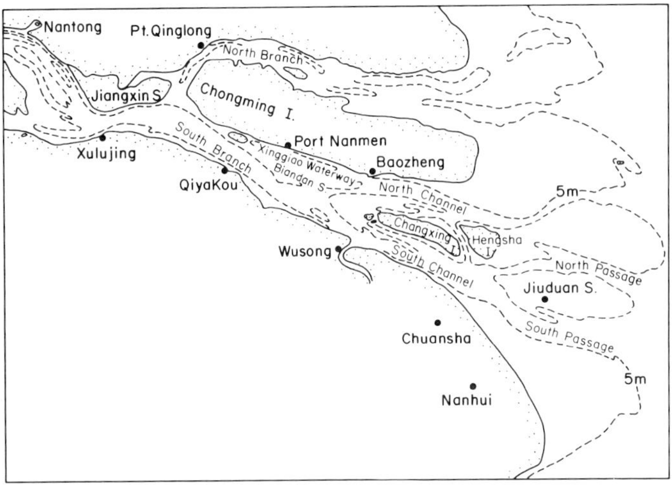

Figure 1. Chang Jiang estuary with station locations.

the river flow was much higher and the North Branch was partially mixed (Type B).

The three channels or passages of the South Branch, i.e., North Channel, South Channel (north passage), and South Channel (south passage), all differ from the North Branch in their freshwater discharge, tidal range, and geometry. Most of the time they are partially mixed but for short periods may be well mixed or poorly mixed.

The distribution of the fresh water discharge among the various branches/channels/passages of the Chang Jiang greatly affects the degree of mixing in those branches. For example, since 1958 the fresh water discharge component of the North Channel usually has been larger than that of the South Channel; since 1965, it has been consistently higher. In addition, for more than 10 years, the North Passage has carried more fresh water than the

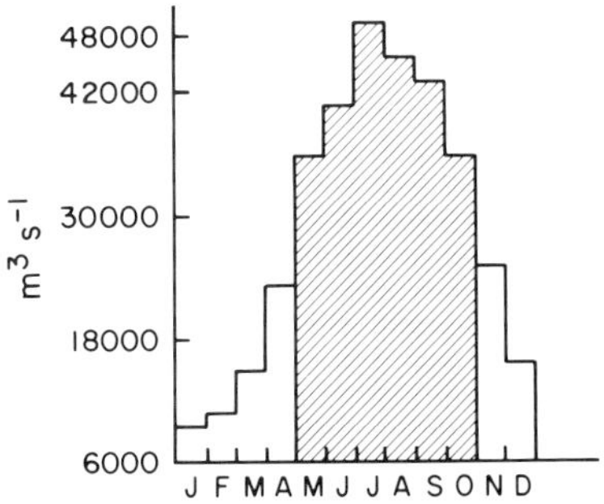

Figure 2. Variation of average monthly discharge at Datong Station in the Chang Jiang River.

South Passage. Given the same tidal ranges, the larger freshwater discharge should result in a more stratified North Channel than the South Channel and North Passage than South Passage. In general, however, all three channels are partially mixed (Type B).

Discharge of the Chang Jiang River also varies greatly between the flood season and the dry season (Fig. 2), and tidal currents vary greatly in speed between spring and neap tides. As a result of the variations in these driving mechanisms, mixing in the same branch, or channel, varies with time. Figure 3a shows the longitudinal salinity profile in the South Passage under spring and neap tides during the flood season; at this time the South Passage is partially mixed. Figure 3b shows that during the dry season the South Passage is partially mixed at times of mean tides and well mixed at times of spring tides. In summary, there are temporal changes in mixing within the Chang Jiang estuary. On an annual basis, stratification is stronger during the flood season than during the dry season. On shorter time scales, stratification is stronger during neap tides than during spring tides. The greatest mixing – the least stratification – is observed during the dry season during spring tides; at this time the South Passage is Type C.

The same general conclusions can be drawn from stratification – circulation diagrams (Hansen and Rattray 1966). These diagrams use two nondimensional parameters to determine mixing types: in the vertical, a stratification parameter defined as the ratio of the surface to bottom difference in salinity (δS) divided by the mean cross-sectional salinity ($\bar{S}_o$); and in the horizontal, a circulation parameter U_s/U_f defined as the ratio of the net surface velocity (U_s) to the sectionally averaged velocity (U_f). Because we did not have data across any cross-section, we used station vertical averages. For U_s we used the tidally averaged surface velocity (upper layer) and for U_f the tidally and vertically averaged velocity at the station. Using data from field observations taken in August 1978, we calculated the results shown in Fig. 4. This figure shows that "partially mixed" (Type B) was the predominant mode of mixing in South Passage, North Passage, and North Channel. North Channel was less well mixed than North Passage which was less well mixed than South Passage.

Estuarine Circulation Patterns

There are significant density gradients in the horizontal and the vertical directions at the lower reaches of the South Passage, the North Passage, and the North Channel, where Type B predominates. With such gradients, a density current develops which is unique to estuarine areas. These density currents are reflected in the vertical distribution of velocity (VVD).

Figure 5a shows the VVD over an entire tidal cycle at a station situated near an upper reach of salt water intrusion. Since the station is little affected by the salinity intrusion, the VVD shown in Figure 5a is similar to that of nontidal streams with a maximum near the surface and a decrease with depth.

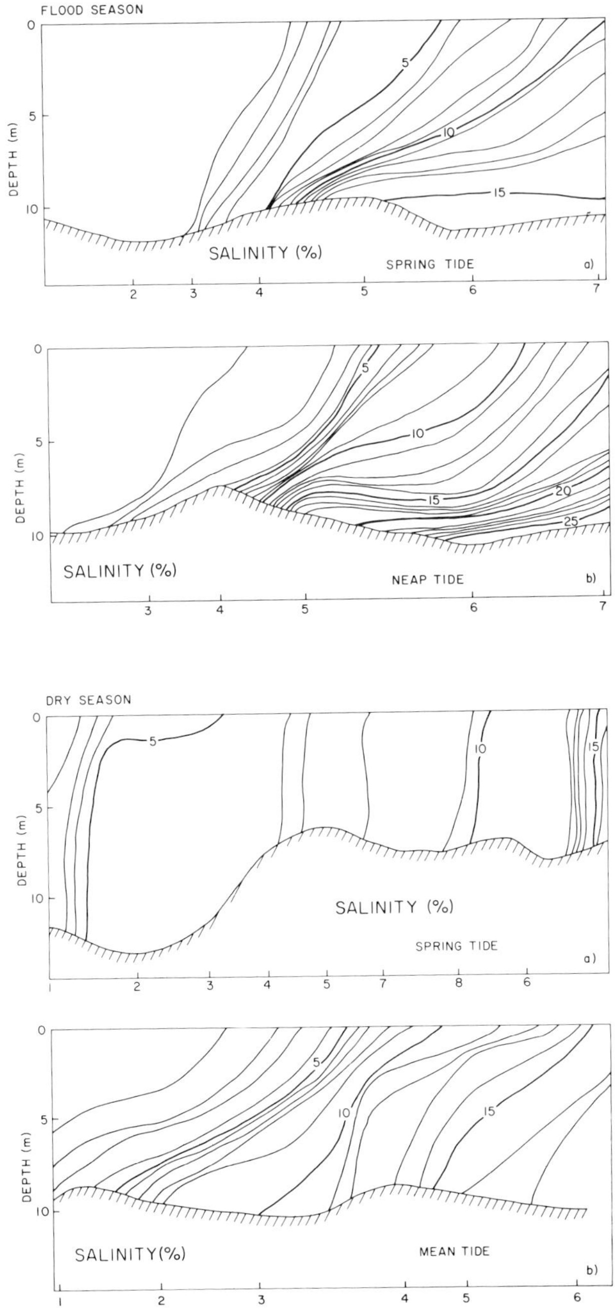

Figure 3. *(a-upper) Profile of the longitudinal salinity distribution in the South Passage under different tidal ranges during the flood season; (b-lower) As Fig. 3a, except for the dry season.*

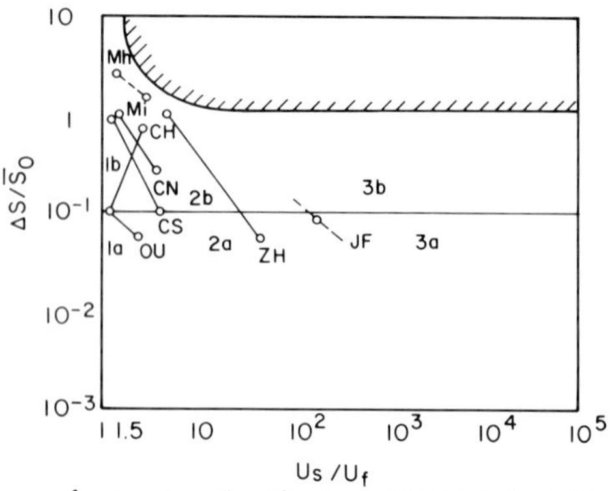

Figure 4. A diagram of estuarine classification. M-Mississippi River Mouth; JF-Straits of Juan de Fuca; ZH-Zhu Jiang; CS-South Passage; CN-North Passage; CH-North Channel; OU-Ou Jiang.

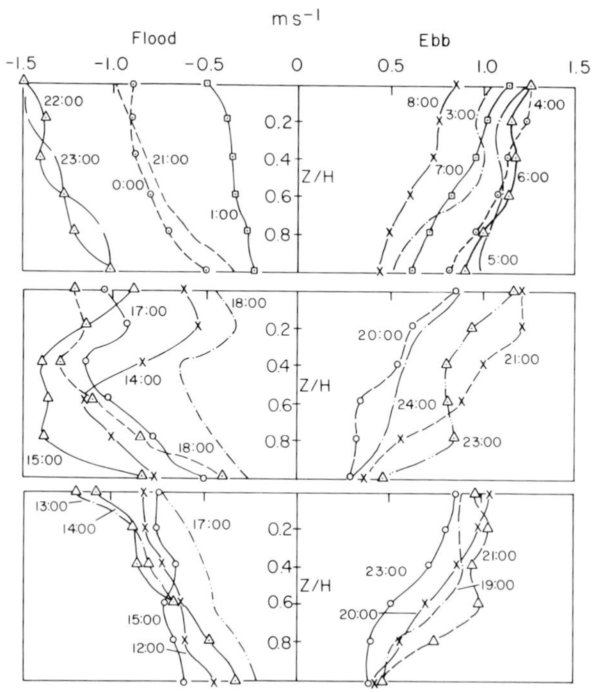

Figure 5. (a-upper) Vertical distribution of velocity landward of the null point (Z/H is the relative depth); (b-middle) as Fig. 5a, except near the null point; (c-lower) as Fig. 5a, except seaward of the null point.

Figure 5b shows the VVD over an entire tidal cycle at a station near the front of salt water intrusion with a distribution greatly different from that of Figure 5a – the flood velocity in the lower layer is considerably increased and the ebb velocity is decreased at depth. It is well known that during the flood tide, the density gradient in the lower layers agrees with the surface slope. This causes a maximum velocity to move down in the column to a position in the middle and lower layers. A density gradient opposing the surface slope decreases the ebb velocity in the lower layers during ebb periods.

Figure 5c shows the VVD over a complete tidal cycle at a site near the lower reach of the estuary away from the front of the sea water intrusion. Here the density gradient is so small that the current velocity distribution is not affected. Only during the ebb period are some slight distortions noted in the distribution.

In summary, there exists a net outflow at all depths from the surface to the bottom upstream of the salt water front with little influence of density gradient at any time. However, in those estuarine parts where the effects of a density gradient are felt, fresh water is discharged through the upper layer and there is a net inflow in the lower layer due to its longer flood excursion over the ebb excursion. While moving up the estuary, a part of the salt water in the inflow mixes up into the upper layer.

A longitudinal net non-tidal circulation diagram for the North Channel, the North Passage, and the South Passage has been calculated from simultaneously observed data taken in August, 1978 (Fig. 6). It represents an average over two tidal cycles (higher high water and lower high water). It is

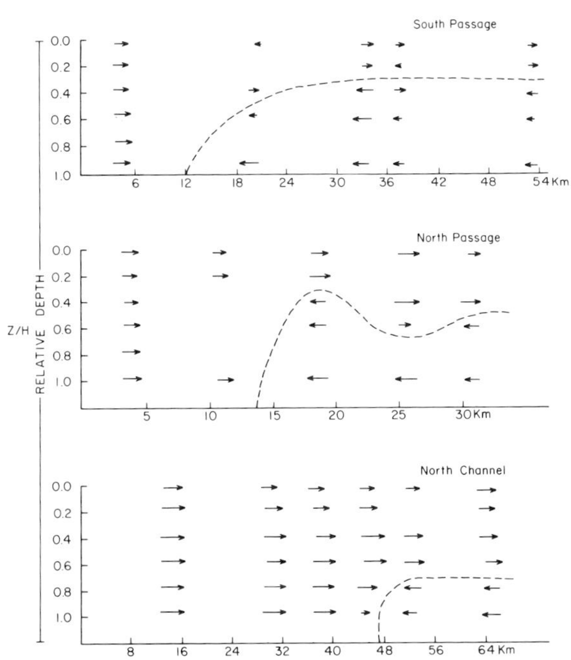

Figure 6. A diagram of longitudinal circulation in the Chang Jiang Estuary.

obvious that non-tidal circulations do exist in each of the three branches but their intensity and structure varies with the degree of mixing. This can be seen from the following:

(1) There is a null point with a zero net excursion where predominant outflow changes into predominant inflow. The position of this null point moves from well inside in the South Passage to well outside in the North Channel. This agrees with the freshwater discharge distribution between those three branches.

(2) A zero velocity plane with a zero net velocity is located at a section where outflow (downestuary) prevailing in the upper layer changes to inflow (upestuary) in the lower layer. The depth of this plane, deepest in the North Channel and shallowest in the South Passage with the North Passage in between, is in conformity with the tidal strengths in these branches. For the past several years we have noticed that the river channel in the South Passage has begun to change into a flood channel as a result of the gradual strengthening of the tidal actions as well as an enlarging of the residual current in the inflow.

(3) The flow structure can be extremely complex. For example, at Station 27 in South Passage, the current at the surface is out of the estuary; at a relative depth, Z/H, of 0.2 it is into the estuary; at 0.4 it is out; and at 0.6 it is into the estuary.

The predominant directions of flow in the surface and the bottom layers under different hydrographical conditions for the South Passage, the North Passage, and the North Channel are also shown in Figs. 7a, 7b, and 7c. The ordinates in these figures are the ratio of the flood excursion, i.e., the integral of the velocity over the flooding portion of the tidal cycle, to the sum of the flood and ebb excursions expressed as a percentage. For equal excursions it is 50%. Figures 7a, b, and c show that the surface currents are predominantly downstream (>50%) regardless of season or tidal condition. This confirms that the freshwater runoff is discharged to the sea through the upper layer. It can also be seen that the lower layer of all three branches is dominated by upestuary flow even though there are significant differences between the various branches. In general, the inflow excursion is larger during neap than spring tides.

The point where the flood and ebb excursions in the lower layer are equal, the null point, is shown on the figure as N. Figure 7c shows that this point is further up the estuary during the dry season than during the flood season. Since the position of this point responds to freshwater runoff and tide, a null zone is formed. This zone is located differently in the 3 branches. For the South Passage, it is usually located near the Tong Sa Shoals during the flood season and near the seaward end of the South Channel during the dry season. The null zone for the North Passage is close to the sand bar of the North Passage during flood season. The null zone of the North Channel occurs near the external shoals of its sand bar during flood season and near the internal shoals during the dry season.

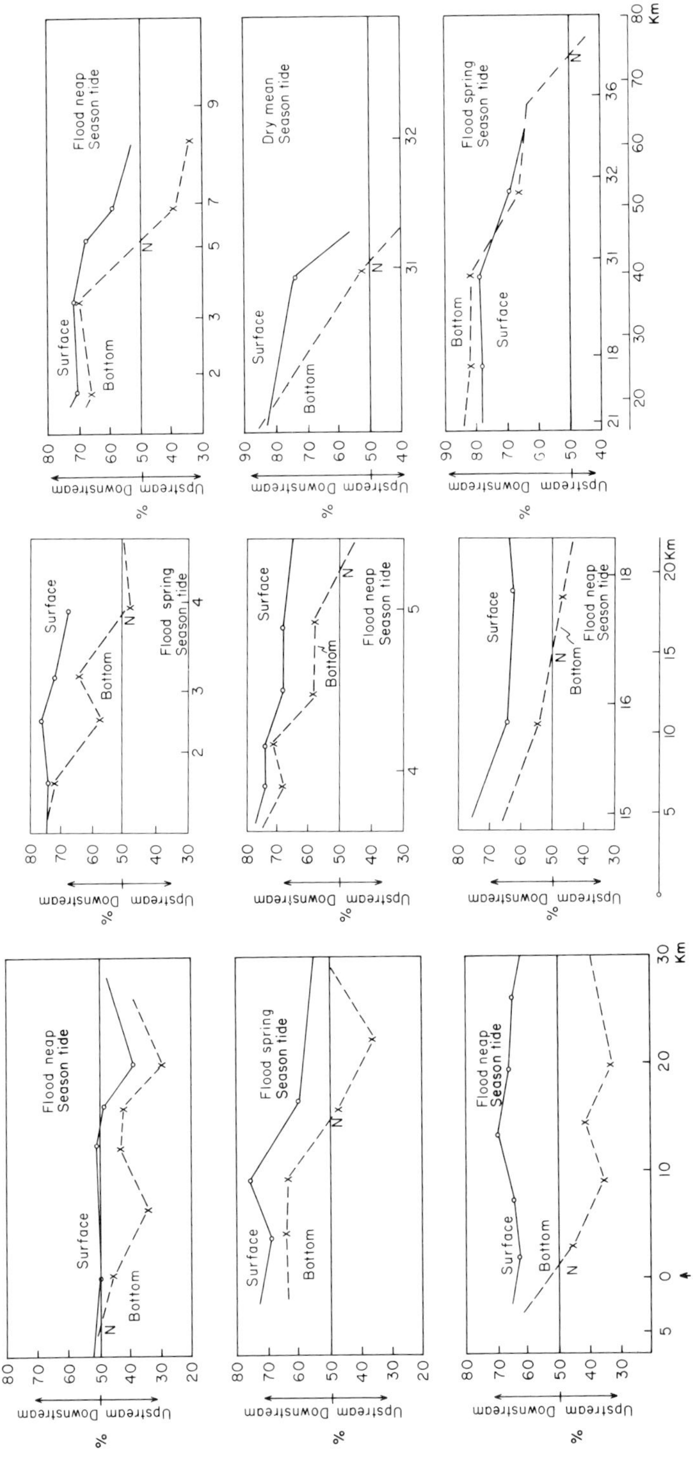

Figure 7. *(left) Longitudinal variation of the predominant flow direction in the South Passage; (center) North Passage; (right) North Channel.*

Although there are density currents in the Chang Jiang estuary, the freshwater runoff and tidal currents are very strong, several times larger than the density currents in volume flux. However, they are important near the null zone where runoff and tide are in balance.

Effects of Circulation on Shoaling in the Chang Jiang Estuary

The effects of circulation on the transportation and accumulation of sediments have been demonstrated in many estuaries throughout the world. In the Chang Jiang estuary, these effects are shown clearly in its turbidity maximum, fluid mud layer, and its complicated system of bars. The formation and variations of all of these features are related closely to the range of circulation patterns found in the Chang Jiang.

When suspended sediments are transported by a river into the tidal reaches of that river the vertical distribution patterns of suspended sediments are not altered by virtue of the fact that the flow is now oscillatory. However, when these suspended sediments are transported further downstream into the estuary proper where the water is no longer fresh, the vertical and horizontal distribution patterns do change. There are two fundamental reasons for these changes: flocculation and density currents.

Flocculation results from electrochemical actions on particle surfaces which cause small particles to clump together to form larger composite particles. This process leads to an increase in settling velocity causing a decrease in the concentration of suspended matter in the upper layer, an increase in the lower layer, and an increase in the sedimentation rate.

Since 1973 we have measured the relationship between salinity and the surface charges on suspended particles at seven stations along a longitudinal section in the North Passage. We used an automatic electropotential recorder to determine the solid-liquid boundary potential. In the following sections we describe, from the physical-chemical point of view, the depositing mechanisms of suspended sediment responsible for flocculation and its effects on the formation of the turbidity maximum.

Our analysis indicates that the main flocculating sections are located over the sand bars in the South Passage, the North Passage, and the North Channel (Shen et al. 1980) and that the variations in electropotential (§) of silt particles suspended in the water column are closely related to variations in salinity. When the salinity is $0^{0}/_{00}$, the value of § is very high. When the salinity increases to $5^{0}/_{00}$, the value of § drops rapidly. When the salinity is $13^{0}/_{00}$, the electropotential § is a minimum and when the salinity increases to $20^{0}/_{00}$, the value of § becomes high again. The lower the value of §, the thinner the electrostatic double layer on the particle surfaces and the greater the probability of collision between particles. Forces of attraction are greater than repellent forces, therefore the particles are able to flocculate, their settling velocities are increased, and sedimentation is increased. From a plot of the variation of § it may be predicted that in areas with salinities of $0^{0}/_{00}$ and $20^{0}/_{00}$ there is little, if any, flocculation. In areas with salinities of 5‰, 10‰, and 15‰ flocculation is more rapid, but still relatively slow. In an area with a salinity of 13‰, one would expect rapid flocculation.

The density-driven circulation is also important to sedimentation. Figure 8 shows the directions of sediment transport near the null zones in the South Passage, North Passage, and North Channel based on recent observations. The arrows represent net fluxes of suspended sediment through a unit cross section averaged over two tidal cycles.

Figure 8 shows that from the null point upstream, there is a net flux of suspended sediment seaward in both upper and lower layers whereas downstream from the null zone the flux is seaward in the upper layer and landward in the lower layer. Here the joint action of the density circulation together with flocculation produces sediment concentrations in the lower layer much higher than in the upper layer. In addition, since the net non-tidal current velocity is almost zero in the vicinity of the null point, the seaward-directed suspended sediments from the upper reach and the landward-directed material from the lower reach converge to form a turbidity maximum in the lower layer.

Near the top of the Tong Sa shoals between the North and South Passage, the net seaward water and sediment fluxes are different. The net

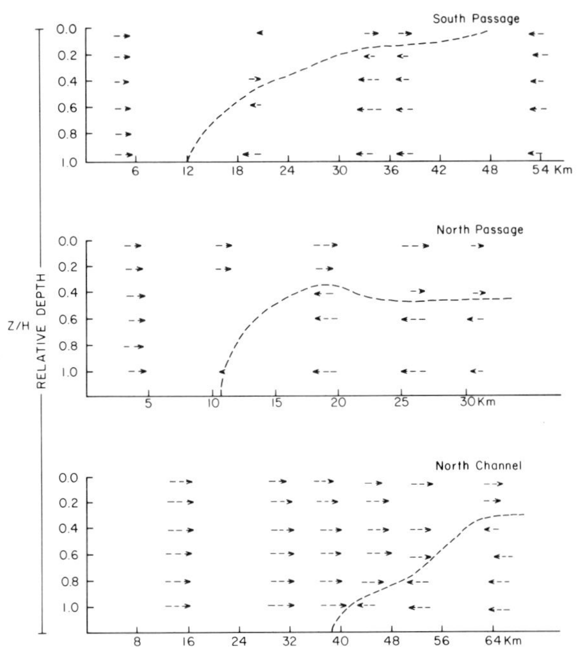

Figure 8. A diagram of longitudinal transportation of sediment in the Chang Jiang Estuary.

seaward sediment layer is limited to the surface, but the net water discharge may penetrate from the surface to the 0.4 m depth layer where the direction of the net flow is opposite to that of the net sediment flux. Always in the upper layer of a station seaward of the null point there is a net flow seaward and a net flux landward. Figures 9a and 9b are diagrams of the predominant longitudinal flows and sediment fluxes in the South Passage based on data obtained during July, 1963 which illustrates this effect. This effect is obviously related to the strong flood tide in this region. When the flood current velocity increases to a certain value the bottom sediments are stirred up, resuspended, and eventually carried up into the upper layer by vertical mixing. As a result, the sediment concentrations in the upper layer are higher during the flood tide than during the ebb resulting in the phenomenon of a net water outflow and a net sediment influx.

A turbidity maximum is found in all branches of the Chang Jiang estuary. It extends longitudinally about 25-46 km. Its concentrations vary from 0.1 kg/m^3 at the surface and from 1 to 8 kg/m^3 at the bottom. When the sediment concentration of the central part near the bottom becomes very high, a plastic, turbid fluid with high density and high viscosity (fluid mud) will occur. This is a serious threat to the maintainance of water depths in the navigational channels for even a well dredged channel will be shoaled by the fluid mud in only one or two days with a thickness up to 1.0-1.2 m.

The turbidity maximum varies with time and space in a complicated manner. However, these variations correspond to variations in the density

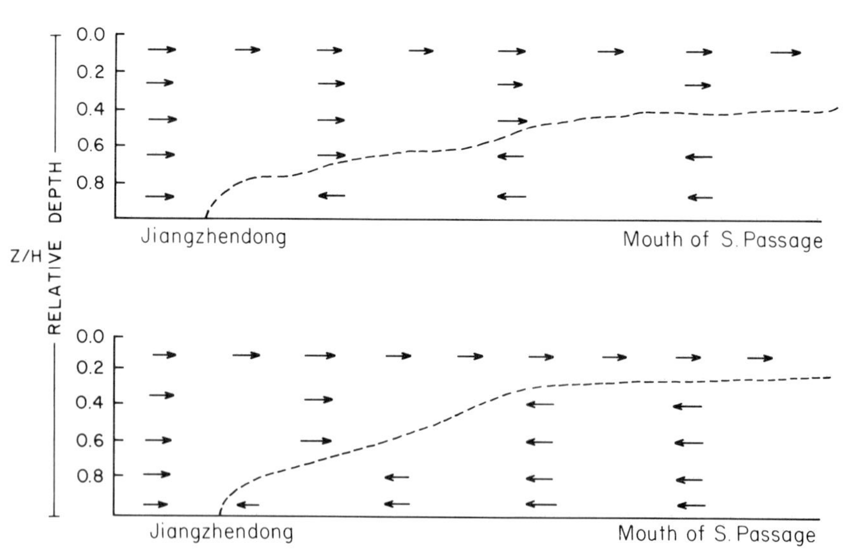

Figure 9. (a-upper) A diagram of predominant longitudinal flow in the South Passage; (b-lower) A diagram of predominant longitudinal sediment flux in the South Passage.

currents and its position coincides well with that of the salt water intrusion and of the null point (Fig. 10). This suggests that density currents are a major hydrodynamic factor for the formation of the turbidity maximum and fluid mud.

There are channel sand bars in the Chang Jiang estuarine branches. The depth from the surface to the tops of these sand bars has normally ranged from 5 to 7 m for many years. The bars fall into two groups according to their depositional positions: internal and external shoals. The external shoals are the main component of the sand bars in the channel and are rather stable in location. Their depths vary seasonally with deposition in the flood season and erosion in the dry season. In discussions of external shoals, some people have emphasized the action of density currents and others have stressed the effect of runoff and tidal currents. We hold that in addition to the bed-load transport they are mainly controlled by the deposition of suspended sediments. For example, the bed sediments of the Tong Sa shoals have silty mud as their essential material. This indicates that sedimentation of suspended materials is a controlling factor. Four factors determine the accumulation of fine sediments in this area: 1) the relative strengths of river runoff and tidal flow, 2) the difference in the locations of the ebb and flood channels, 3) the action of density currents, and 4) flocculation. Of the four, the first is probably the most important. The other three are controlled, in large part, by the first.

During the flood season, the position of the null zone is close to the location of the bars so sediment concentrations are high and deposition rapid.

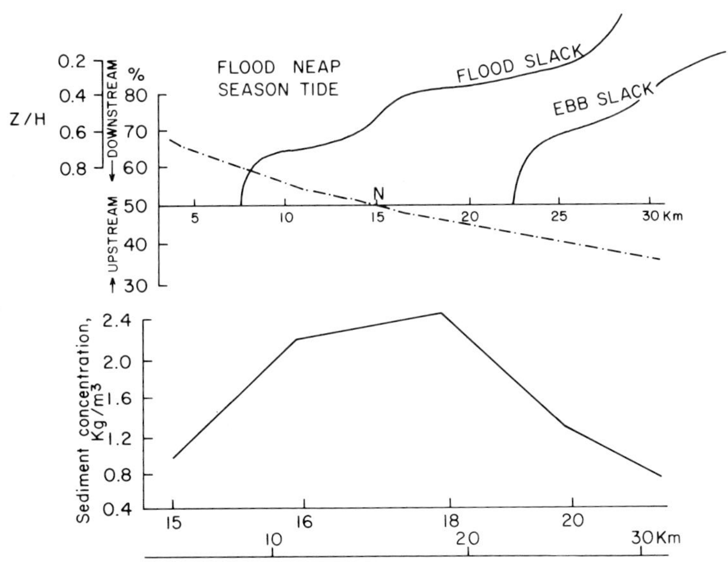

Figure 10. Longitudinal distribution of the front of salt-fresh water, predominant flow directions, and sediment concentration at the bottom in the North Passage.

During the dry season, the position of the null zone moves farther upstream as river flow decreases, and there is net erosion of the bars; the supply of sediment has been greatly diminished and flood tidal currents are more vigorous.

It can be seen from the above that the Chang Jiang estuary has very strong freshwater runoff and tidal currents. As a result, density currents are secondary in importance to the total circulation in the formation of sand bars and their variations, with the most important factor being the relative balance between runoff and tidal currents.

Conclusions

The Chang Jiang estuary is characterized by large fresh water runoff, abundant suspended sediments, and large intertidal volumes. The volume of runoff and the amount of suspended sediments vary with season. Tidal volumes change from spring to neap. The four branches (Fig. 1) differ from each other in geometry. As a result, mixing and circulation patterns are particularly complicated.

Estuarine classification types vary not only temporally from flood to dry seasons but also from spring to neap tides and from place to place. In the North Branch strong mixing predominates. In the three South branches, Type B prevails with Type C secondary. The degree of mixing in the North Passage is larger than that in the North Channel, but less than that in the South Passage.

Under the combined effects of freshwater runoff, tidal currents, and density currents, there exists in the three south branches density currents with a net outflow in the upper layer and a net inflow in the lower layer. The outflow layer in the North Passage is thinner than that in the North Channel but thicker than that in the South Passage. The null point moves up and down from flood to dry seasons and from spring to neap. Its position is most seaward in the North Channel and most landward in the South Passage. All of these are related to the present runoff distributions, i.e., North Channel, North Passage, and then South Passage. Freshwater runoff and tidal currents are the two most decisive factors in the formation of turbidity maxima, the fluid mud layers, and the sand bars in the channels and their variations.

Acknowledgments

The authors wish to express their sincere appreciation to Prof. Chen Ji-yu, Director of Institute of Estuarine and Coastal Research, for his helpful discussions and advice.

References Cited

Cameron, W. M. and D. W. Pritchard. 1963. Estuaries, pp. 306–324. *In:* M. N. Hill (ed.), *The Sea. Vol. 2,* John Wiley & Sons, New York.

Dyer, K. R. 1973. *Estuaries: A Physical Introduction.* John Wiley & Sons, London. 140 pp.

Han Nai-bin et al. 1973. Effects of density currents of salt wedge on channel with mouth bars in the Chang Jiang Estuary. Nanking Hydraulic Research Institute Technical Report.

Hansen, D. V. and M. Rattray, Jr. 1966. New dimensions in estuary classification. *Limnol. Oceanog.* 11:319-326.

Pritchard, D. W. 1955. Estuarine circulation patterns. *Proc. Amer. Civil Eng.* 81, No. 717.

Shen Huan-ting et al. 1980. A study on the variation of turbidity maximum in the Chang Jiang Estuary and the cause of its formation. *Proceedings of Estuaries and Coast Society, China,* 1980.

A STUDY OF DIFFUSION OF UPPER-LAYER SUSPENDED SEDIMENTS IN DISCHARGES FROM THE CHANG JIANG ESTUARY INTO THE SEA, BASED ON SATELLITE IMAGERY

Yun Cai-xing and Wan Jia-ruo
Institute of Estuarine and Coastal Research
Physics Department
East China Normal University
Shanghai, China

Abstract: The Chang Jiang River with its enormous discharge of water and huge quantity of suspended sediments influences the estuary as well as the natural environment of the adjacent offshore regions. This paper analyses the distribution, diffusion, and deposition of suspended sediments in the Chang Jiang estuary and in adjacent shelf areas as determined by satellite imagery. These data have important implications for construction and maintenance of navigation channels and harbors, and for utilizing and developing beach resources and aquatic resources. Good agreement of results of computer simulations based on satellite imagery with sea-truth data support the methodology used in this study.

Introduction

The Chang Jiang River, the longest river in China, with its enormous volume of water and huge quantity of sediments collected along its course, continually influences the hydrographic characteristics of its entire valley and reshapes its bed. Moreover, its discharges into the sea have contributed to the changes of the natural environment of its immediate estuarine area as well as the adjacent offshore regions. For many years, many experts have probed problems of the distribution of sediments discharged into the sea and the process of sedimentation, laying stress on such problems as the development of the Chang Jiang River delta and the dredging and training of the estuary (Chen 1957). In comprehensive oceanic investigations carried out in our country, it has been discovered repeatedly that the discharges of water and sediments from the Chang Jiang River have substantial influences on the East China Sea and the Yellow Sea regions (Mao et al. 1963). These influences involve the distribution of the nearshore current system off Mid-China and the problem of sources of deposits on the bottom of the continental shelf (Fan and Qin 1959). From the standpoint of production and construction, analysis of the discharges from the Chang Jiang River into the sea and the diffusion of suspended sediments might help in developing management strategies for developing deep water navigation channels within the Chang Jiang River estuary, for monitoring the trend of aquatic resources on the fishing grounds

ISBN 0-12-404070-5

of Chou San and Lu Si (south and north, respectively, of the estuary), and for appraising the quantity and quality of beach resources of Jiang-su and Zhejiang Provinces as well as Shanghai.

In this paper, the authors made use of various advantages that satellite images offer over other kinds of observations including: wide coverage, fine synchronization, and feasibility for trend observation. For our analysis we selected 20 sheets of satellite images of the Chang Jiang estuary taken by U.S. Landsat and Weather Satellite (NOAA-5) at different periods from 1976 to 1978. Through the classification of suspended sediment distribution, quantitative simulation tests, and checking with ground surveying data, an analysis was made of the pattern of distribution of the discharges from the river into the sea, the concentration of suspended sediments in the water bodies of the estuary and offshore; the modes of their diffusion; and the spheres of influence.

Classification of Sediment-Bearing Water Bodies

As turbid rivers flow into lakes, reservoirs, or seas, suspended sediments tend to disperse and deposit through a variety of mechanisms which include: advection, diffusion, settling, dilution, and flocculation. The results of these processes can be studied through analysis of remote sensing imagery. Research indicates that multispectral images MSS-5 (0.6-0.7 μm) and MSS-6 (0.7-0.8 μm) are more sensitive to reflectance of suspended sediments in water than other wave length bands, and their differences on a gray scale will match appropriate grades of sediment concentration (Shelley 1976). Therefore, computers can be used to determine the grades of sediment concentration through density classification of images. A trial was made on density classification and area estimates by means of the mini-computer type remote sensing image processing system (Fig. 1) prepared by our university. As an example, we used images of sediment diffusion in the discharge through the Chang Jiang estuary into the sea.

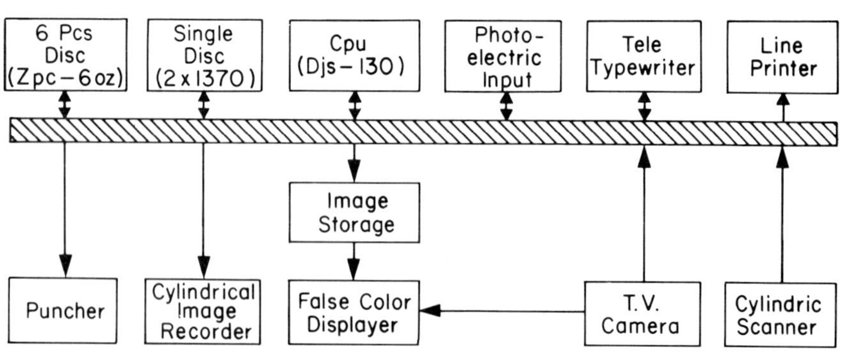

Figure 1. Block diagram of mini-computer image processing system.

Through computer processing, a well-graded and defined image of diffusion of suspended sediments in the discharges through the Chang Jiang estuary into the sea was obtained with the characteristics well matched with those images obtained through optical information and false color density slicing (Photos 1 and 2). According to the correlation of different sediment concentrations in water with image gray scale, the results of processes of diffusion of suspended sediments of the Chang Jiang River discharges into the sea produce different water types which may be classified into four grades: highly turbid tongues, a turbid water zone, plumes, and a diluted water zone. Highly turbid tongues constitute the main body of the Chang Jiang River discharge into the sea and have the highest sediment concentration. A turbid water zone is located between 122° 25′ - 122° 30′ E. It spreads like clouds due to the movement of rotary tidal currents and intermixing with sea water outside the estuary. Plumes represent more diffuse sediments with remarkably lower sediment concentration. The area of plumes generally ranges from 122° 30′ to 122° 35′ E. A diluted water zone involves peripheral front parts of the diffuse sediments, with its outside boundary running from 122° 40′ to 122° 45′ E.

Quantitative Analysis of the Diffusion of Suspended Sediments

The observation and determination of suspended sediment concentrations in estuarine and inshore areas is usually a complicated and cumbersome task. To determine sediment concentration of various turbid zones and the rates of deposition of sediments in the discharges from the Chang Jiang estuary, we made simulation experiments with associated correlation analysis based on the characteristics of spectral reflectance of sediment-laden water bodies, taking the sediment distribution pattern from a satellite image taken August 9th, 1978 as a standard. A preliminary expression of the relationship between sediment concentrations of the water bodies and density values of satellite images has been established, its coefficients being determined by the sediment concentrations observed at sea on the same date.

Simulation experiments were carried out through repeated photographing of specially prepared water samples of various sediment concentrations, employing filters matching wave bands of Landsat imagery. The result showed a good linear correlation between sediment concentrations and density values, or gray scale, of the negatives (Fig. 2). The mathematical formula can be expressed in the following form:

$$D = a_{(\lambda)} x + b_{(\lambda)}$$

where x represents the sediment concentration of a given water body; D, the density value of the negative; a and b are constants; and subscript λ denotes the wave band used.

After regression calculations and correlation analysis with the above formula, we found that the coefficient of correlation between sediment concen-

Photo 1. *Satellite image of the mouth of the Chang Jiang estuary taken on 9 August 1978.*

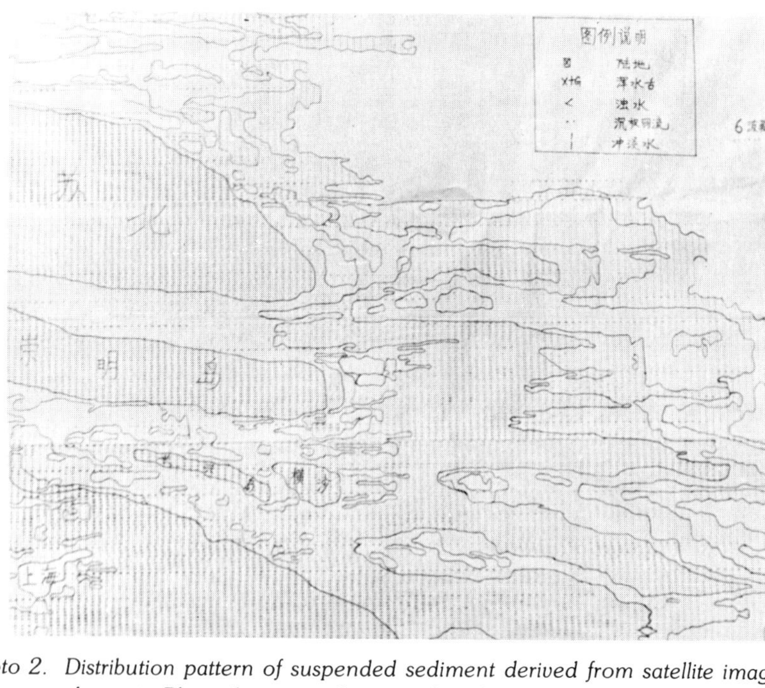

Photo 2. *Distribution pattern of suspended sediment derived from satellite imagery shown in Photo 1 using techniques described in this paper.*

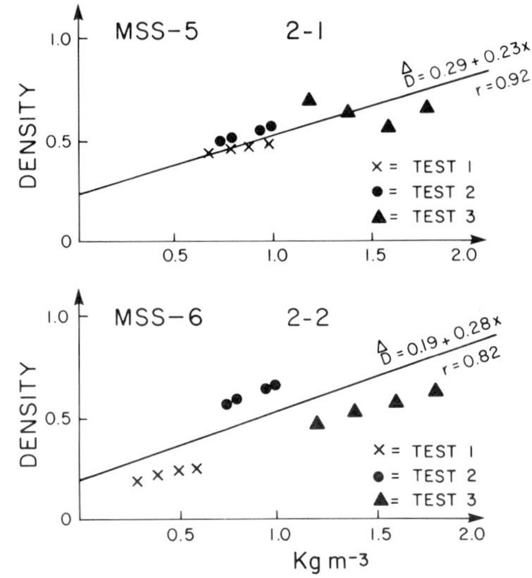

Figure 2. Correlation of suspended sediment concentration and density of negatives of Landsat images.

tration of a water body and density of the negative under test approaches 0.82-0.92. That of MSS-5 (0.6-0.7 μm) was the highest, with coefficients a and b equal to 0.23 and 0.29 respectively.

To obtain the sediment concentration corresponding to any gray scale area of satellite image, we took for computer processing the lowest grade – the lowest density – of the gray scale as approximating clear water. Then, a value of 1.23 kg/m^3 (the observed sediment concentration in the Chang Jiang estuary at the same time in the area of the highest gray scale) was substituted into the formula for calculation. As a result, we obtained average sediment concentration values representing different gray levels on satellite images of the Chang Jiang estuary (Table 1). After checking the calculated result in the table with the observed value in the 3rd gray scale area of satellite images, it was shown that the calculated sediment concentration value closely coincided with the observed value; its error was within 3.4%.

Based on the results of simulation tests, we obtained the different sediment concentrations, areas of distribution, and the rates of deposition for the four kinds of water bodies (Table 2).

From Table 2 we can see the trend that the total sediment quantity of the surface layer water decreases with gray scale levels (beginning from Level II) indicating that there exist two processes of diffusion of sediments in the discharges through the Chang Jiang estuary, i.e. the process of dilution and mixture with clear water outside the estuary, and the process of settling which leads to the deposition of fine-grained sediments. Taking the total quantity of sediments in the upper-layer water body of gray scale area II, we obtained a

Table 1. Comparison of calculated values of sediment concentration with the observed values at the Chang Jiang estuary.

	Turbid water level				
	I	*II*	*III*	*IV*	*V*
Gray scale grade	94.6	101.8	110.0	117.8	124.9
Calculated value of concentration (kg/m^3)	1.23	0.90	0.56	0.26	0
Observed value of concentration (kg/m^3)	1.23	<0.1	0.58	<0.1	<0.1

Table 2. Areas of distribution and total sediment content for four kinds of water bodies at the Chang Jiang estuary (Surface layer 1 m thick).

Gray scale area	*Kind of water body*	*Distribution Area (km^2)*	*Calculated sediment concentration (kg/m^3)*	*Total sediment content of 1 m thick surface layer (1000 tons)*
I	Highly turbid tongues	478	1.23	585
II	Turbid water	2260	0.90	2005
III	Plumes	1760	0.56	915
IV	Diluted water	1110	0.26	266

deposition rate of 54% at the transition from the 2nd gray scale to the 3rd gray scale area. This means that 54% of the suspended sediments in the surface layer water deposited, or settled to a depth of greater than 1m. A deposition rate of 71% was calculated from the 3rd to the 4th steps of the gray scale. For the transition from turbid water (II) to diluted water (IV), a total deposition of 87% from suspended sediments of upper-layer water was calculated. From this, it may be concluded that the area where plumes spread is the front fringe of sediment deposition off the Chang Jiang estuary.

Trend and Range of Influence of Upper-Layer Suspended Sediments

The Chang Jiang River with remote head waters and a long course, annually carries about 924 billion m^3 of water and 486 million tons of sediments into the sea through four branches; North Branch, North Channel, North Pass and South Pass. Due to the complicated dynamic factors and the interaction between run-off and tidal currents, the deposition in the estuary has become a serious problem obstructing the further development of navigable channels for a long time. The problem is related to such basic factors as the sources of sediments, the processes of diffusion and deposition in the estuary, and the directions of sediment transport. Landsat images, after density slicing, provide a basis for analysing the above-mentioned fundamental problems.

Some scientists maintain that the shift of the sediments along the north coast of Jiang-su Province contributes to the deposition of the Chang Jiang estuary. However, according to the classification of densities of Landsat images of the Chang Jiang estuary taken in flood season and the analysis of the distribution of sediments of the same region, it is apparent that the deposition of the estuary is caused directly by the sediments conveyed from its upper reaches.

Photo 3 shows that the highly turbid tongues represent the main body of these sediment-laden discharges with average sediment concentrations of up to 1.23 kg/m^3 being the highest observed. The turbid current takes the shape of belts due to the restriction of a trough present in the topography and the moment of inertia about the ebb axis. The width of the current generally ranges from 5 to 8 km, its seaward end having the shape of a tongue tip. The highly turbid tongues have different extensions into the sea, depending upon the discharges from the channels. These locations are consistent with the observed values of flows and sediments of these three branches at the same time. From this it is evident that, at present, North Channel ranks first in quantity of discharged sediments, followed by North Pass, then South Pass. The formation of highly turbid tongues at the estuary further demonstrates that in the vicinity of the estuary the discharges of sediments from the upper reaches into the sea usually take the form of axial current. This becomes the main outlet for the Chang Jiang River to drain its flood water and discharge its sediments, while the shoals and shallow waters located on both sides may be referred to as tidal shoals.

Plumes provide an important basis to indicate the directions of diffusion of the sediments discharged through the estuary into the sea. Four sheets of Landsat images representing typical cases were selected (Photos 3, 4, 5 and 6) for the purpose of explanation. The images taken on July 4 and August 9, 1978 reflect the situations of flood tide and ebb tide during the flood season respectively. The images taken on October 21 and November 26, 1976 reflect the situations of the final ebb and initial ebb respectively in the postflood season. The former two images show the northward deviation of the plumes, while the latter two show their southward deviation. Hence, it is ascertained that the suspended sediments in the discharges through the Chang Jiang estuary into the sea diffuse predominantly northward during the flood season, partly flowing back into the Chang Jiang River through its north branch and partly accompanying the northward current along the coast of the south part of North Jiang Su region, whereas they diffuse mainly southward during the post-flood and low water season.

It is noted that as the vast Chang Jiang River discharges its enormous volume of water and suspended sediments into the East China Sea, the sphere of influence of diffusion is exceedingly wide because of the dispersing and transporting effects of water masses of the outer sea and tidal currents of the shallow sea. Sixteen sheets of images taken by weather satellite from the

Photo 3. Satellite image of the mouth of the Chang Jiang estuary taken on 4 July 1978 during flood tide in the high runoff season.

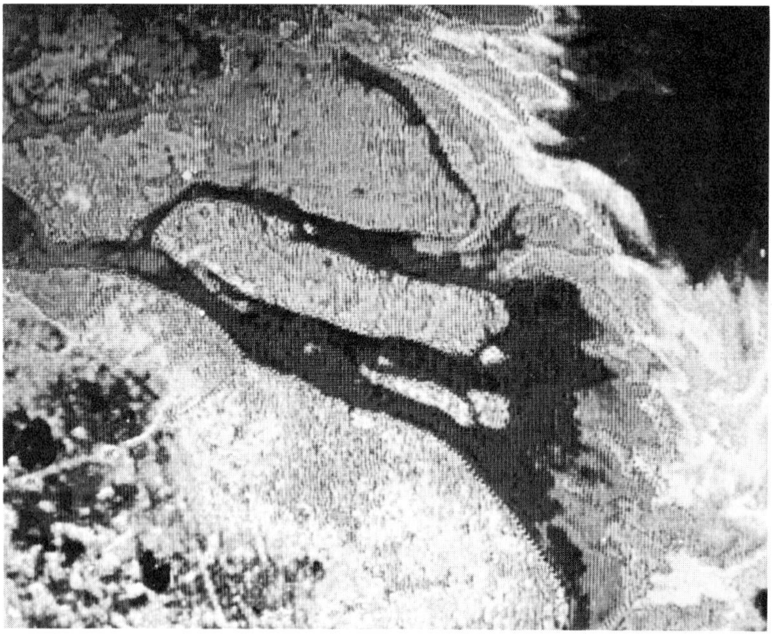

Photo 4. Satellite image of the mouth of the Chang Jiang estuary taken on 9 August 1978 during ebb tide in the high runoff season.

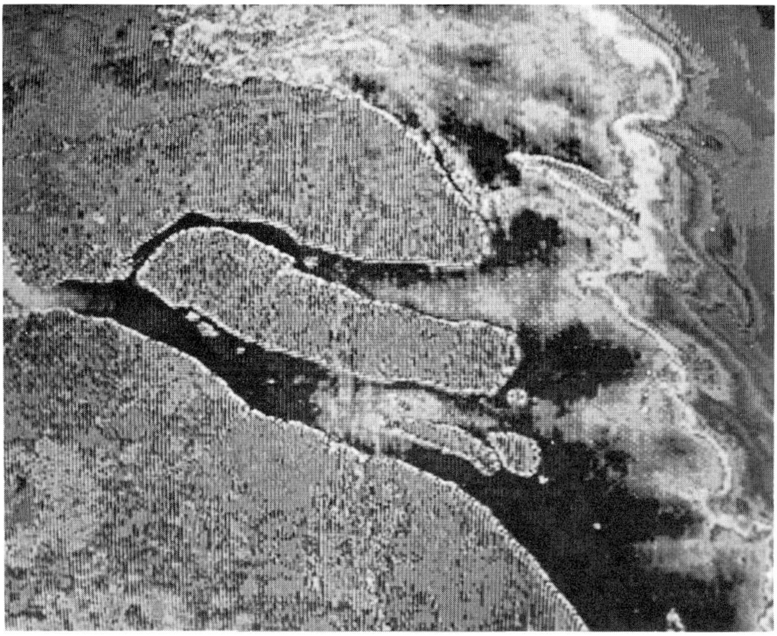

Photo 5. Satellite image of the mouth of the Chang Jiang estuary taken on 21 October 1976 near the end of ebb tide during a period of moderate runoff.

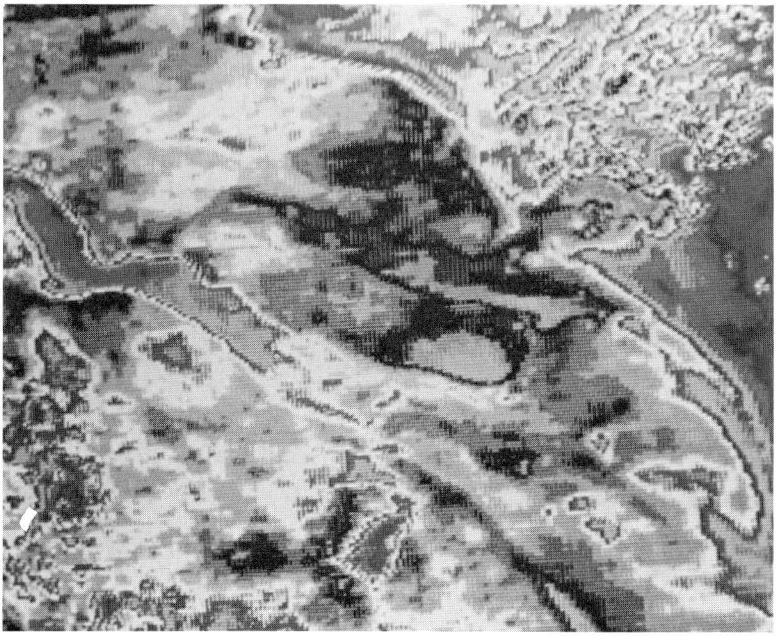

Photo 6. Satellite image of the mouth of the Chang Jiang estuary taken on 26 November 1976 near start of ebb tide during a period of low runoff.

winter, 1976, to the end of 1978 and available in the Shanghai Central Meteorological Observatory and Najing Meteorological Observatory show that the turbid water zones in the near-shore areas along the coasts of Jiang Su Province and Zhe Jiang Province are generaly several dozens of kilometers wide. The variation of width is dependent upon the following three factors: (1) run-off of the Chang Jiang River into the sea, (2) tides, and (3) the variation of direction and speed of wind. The last may cause considerable fluctuation in width. For example, although the image of Fig. 3-1 was taken at a time during spring tide of the flood season when the river flow and excursions of flood-tide and ebb-tide currents were greatest, the turbid water zone

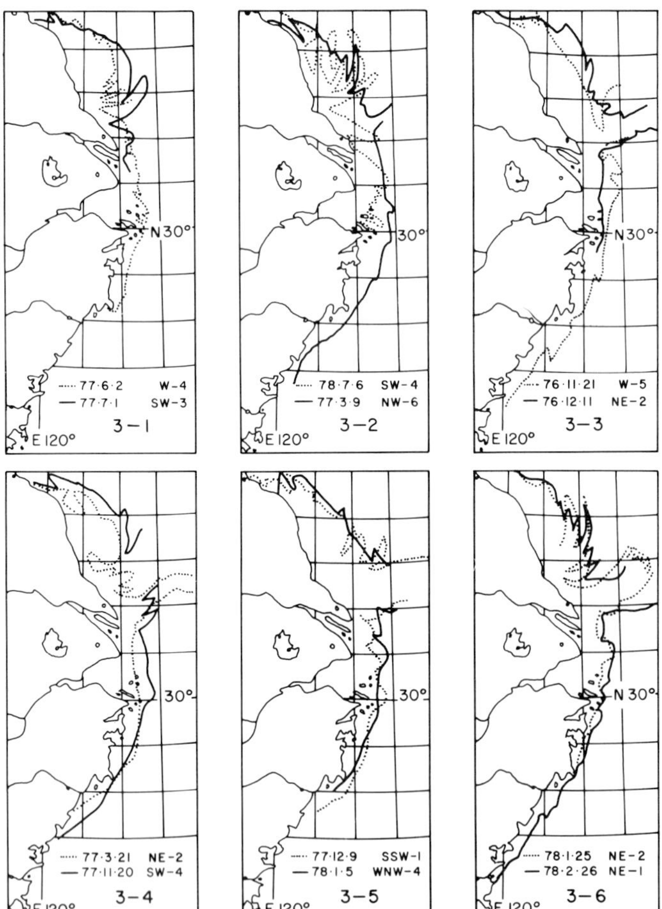

Figure 3. Distribution patterns of suspended sediment in the surface waters off the mouth of the Chang Jiang estuary based on satellite imagery. Contours mark the seaward limit of turbid waters on different dates and under different wind conditions.

on the image appeared relatively narrow. We attribute that to the fact that the direction of the prevailing wind was SW off the shore, its speed was low, the wind-run short, and the wave-generating conditions weak, so that suspended sediments and bottom deposits were not liable to be stirred. Figures 3-2 through 3-6 represent different current situations under different tide and wind conditions in winter and spring respectively. When the speed of north-westerly or northeasterly wind exceeded 7 m/sec, the width of turbid water zones outside the Chang Jiang estuary and along the coasts of Jiang Su Province and Zhe Jiang Province were greatly increased. From this, we conclude that the turbid water zones shown on the images may be ascribed partly to the resuspension of bottom deposits by wind waves.

The sphere where suspended sediments fluctuate *in situ* ranges from 20 to 30 km in width. The inshore region directly affected by the influence of diffusion of suspended sediments in the discharges from the Chang Jiang River into the sea ranges from 30 to 50 km. There are variations of southern and northern boundaries of this region with seasons. In the summer, under the influence of southeast monsoon and the branch of Kuroshio in the China Sea, the diffused sediments reach to Lu Si fishing ground to the north and to the mouth of the Ou Jiang River, Wen Chow, to the south (East China Normal University 1979). In the winter and spring, when northerlies or partially northerly winds prevail, the discharged sediments could reach the mouth of the Ming Jiang River in Fu Jian Province to the south.

Suspended sediments are good tracers of inshore surface currents. One characteristic appearing on the remote sensing imagery that warrants our attention is a belt of turbid water current 50 km wide, between 32°00′ and 33°00′ N spurting eastward. Presumably, this signifies the convergence of three different water bodies. This turbid water current adjoins in the north the extremely stable cold water mass of the Yellow Sea and in the south a branch of Kuroshio (Guan 1964), which flows constantly from south to north. The freshened inshore water from the Chang Jiang River under the influence of this branch of Kuroshio diverts toward the outer sea.

Conclusion

Through classification of grades of gray scale of satellite images and analysis of the patterns of diffusion of sediments, we have a better understanding of the sediments discharged through the Chang Jiang estuary: their main sources, their directions and modes of diffusion, their quantities of deposition in the outlet area, and their influence on neighbouring sea areas. Methodologically, by approach of simulation tests, we have inferred further the correlations between density values of image negatives and sediment concentrations of water bodies, and by means of computer image-processing system, we have conducted quantitative statistical calculations and classification. All these attempts are preliminary trials. But we are confident that the application of remote sensing technique not only will manifest its advantages in the sphere of research of movements of suspended sediments in estuarine and coastal regions, but also serve as efficient means for investigation of changes of coastal and delta regions, surveying sea beach resources, apprehension of inshore current systems and shallow water terrains, as well as management of sea coasts.

Acknowledgments

Special thanks are due to Messers Chan Kang-hong, Weng Muo-ying, Wang Chan-dao and Yee Chan-fang who are responsible for image processing of the present study and to Messers Hu Jia-ming, Cai Mong-yi, Zhou Cuo-ping and comrades from Shanghai Waterway Bureau who took part in test analysis.

References Cited

Chen Ji-yu. 1957. Notes on the development of the Yangtze Estuary. *Acta Geographica Sinica* 23(3):241-253.

East China Normal University. 1979. Estuarine and Coastal Research, National Sea Beach Resources Investigation Report of Wen Chou, 1979.

Fan Shi-qing and Qin Yun-shan. 1959. A preliminary study of bottom sediments in eastern China Sea and southern Yellow Sea. *Oceanologia et Limnologia Sinica* 2 (2).

Guan Bin-xian. 1964. A preliminary study of the distribution and variation of the velocity and volume transport of the Kuroshio and their relation to the topography. *Oceanologia et Limnologia Sinica* 6(3).

Mao Han-li, Gan Zi-jun and Lan Shu-fang. 1963. A preliminary study of the Yangtze diluted water and its mixing processes. *Oceanologia et Limnologia Sinica* 5(3).

Shelley, P. E. 1976. Sediment measurement in estuarine and coastal areas. Section IV, Indirect Measurement of Sediment, NASA CR-2769. pp. 70-92.

INDEX